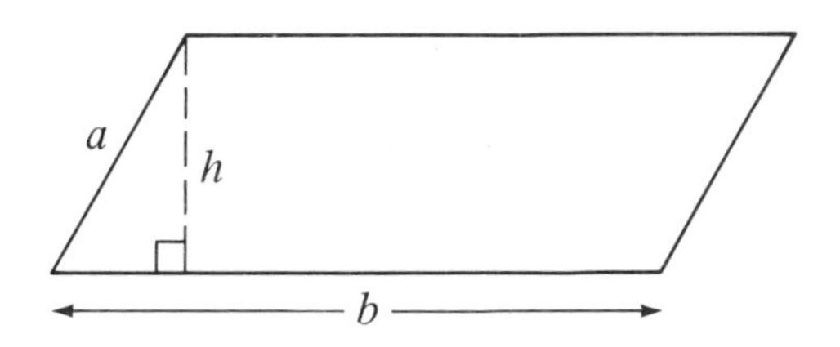

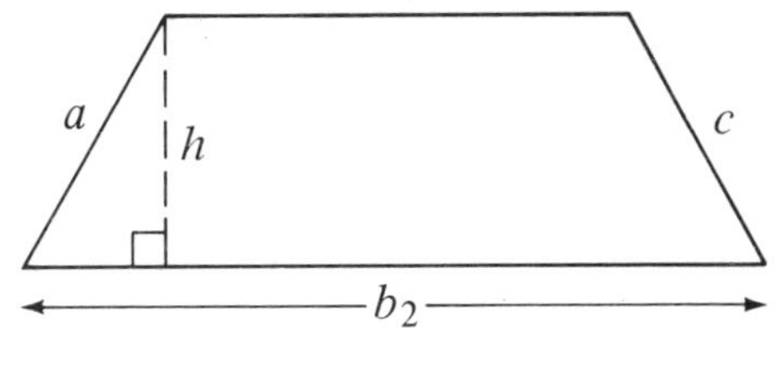

PARALLELOGRAM

Perimeter $= 2(a + b)$
Area $= bh$

TRAPEZOID

Perimeter $= a + b_1 + c + b_2$

Area $= \dfrac{(b_1 + b_2)h}{2}$

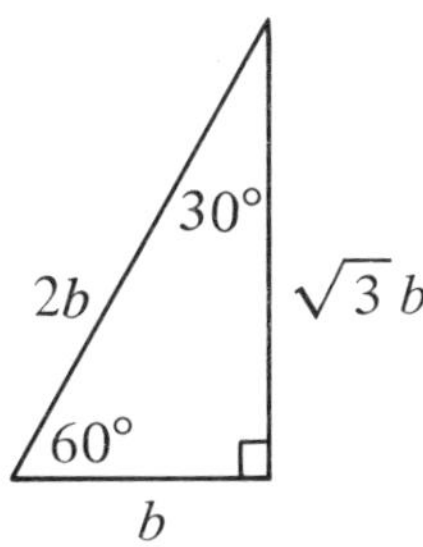

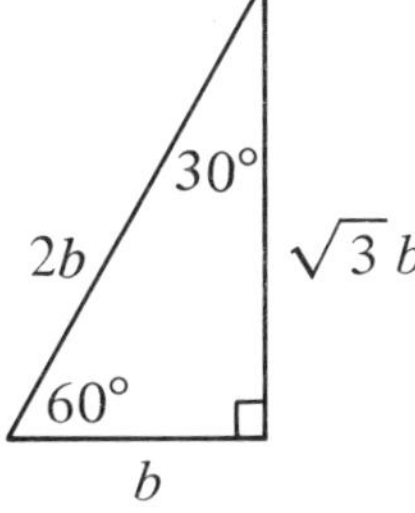

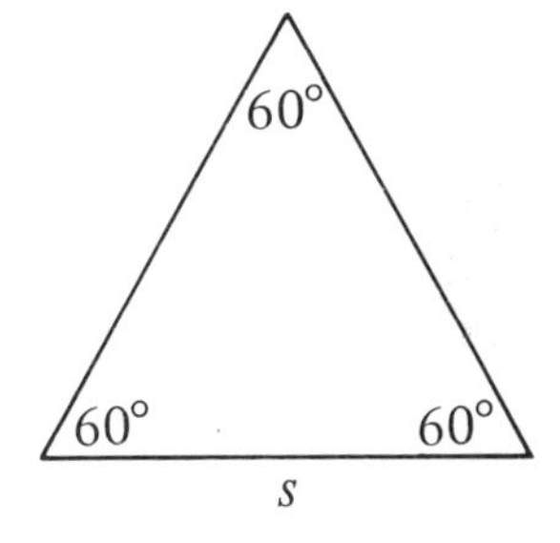

TRIANGLE

$\sin 30° = \cos 60° = \frac{1}{2}$

$\cos 30° = \sin 60° = \frac{\sqrt{3}}{2}$

$\tan 30° = \frac{\sqrt{3}}{3}$

$\tan 60° = \sqrt{3}$

EQUILATERAL TRIANGLE

Perimeter $= 3s$

Area $= \dfrac{s^2\sqrt{3}}{4}$

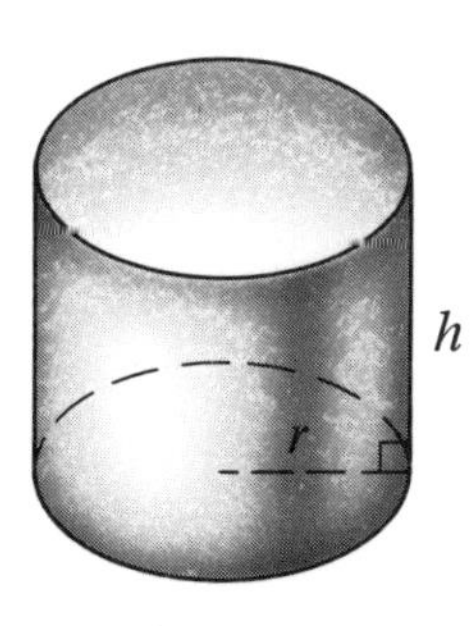

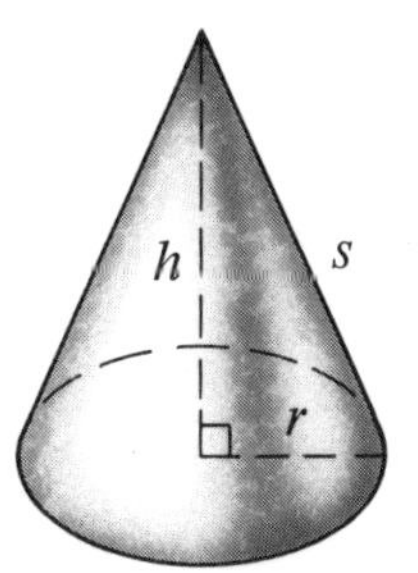

RIGHT CIRCULAR CYLINDER

Lateral Surface Area $= 2\pi rh$

Volume $= \pi r^2 h$

RIGHT CIRCULAR CONE

Lateral Surface Area $= \pi rs$

$s^2 = r^2 + h^2$

Volume $= \frac{1}{3}\pi r^2 h$

Algebra and Trigonometry

A Straightforward Approach

Other Titles by Martin M. Zuckerman

Sets and Transfinite Numbers
Intermediate Algebra: A Straightforward Approach
Elementary Algebra without Trumpets or Drums
Geometry: A Straightforward Approach
Arithmetic without Trumpets or Drums
Basic Mathematics

Algebra and Trigonometry

A Straightforward Approach

Martin M. Zuckerman
City College of the City University of New York

W • W • Norton & Company • Inc • New York

Published simultaneously in Canada by George J. McLeod
Limited, Toronto. Printed in the United States of America.

First Edition

Library of Congress Cataloging in Publication Data

Zuckerman, Martin M

Algebra and trigonometry.

Includes index.

1. Algebra. 2. Trigonometry. I. Title.

QA152.2.Z78 1980 512′.13 79-22909

ISBN 0-393-95020-4

1 2 3 4 5 6 7 8 9 0

Acknowledgments

The author gratefully acknowledges the many helpful comments of Ward D. Bouwsma, Southern Illinois University; George H. Bridgman, University of Minnesota, Duluth; Douglas S. Brown, Nassau Community College; David Cohen, University of California at Los Angeles; Stanley M. Lukawecki, Clemson University; and Cleon R. Yohe, Washington University, all of whom read through, and offered detailed critiques of, the entire manuscript. Many useful suggestions were also made by Clarence E. Bell, Mercer University; Thomas T. Bowman, University of Florida; Peter G. Casazza, The University of Alabama in Huntsville; Ronald C. Freiwald, Washington University; John H. Johnson, University of Wisconsin, Stevens Point; Daniel G. Lamet, Boise State University; Maurice L. Monahan, South Dakota State University; Richard F. Shephard, Publishing Consultant; and Robert F. Sutherland, Bridgewater State College. I also wish to express my appreciation to Joseph B. Janson II, my editor at Norton, as well as to Christopher P. Lang, Diane Nish, and Roy Tedoff, of Norton, and Lorraine Perrotta, who were involved in the production of this book, to Andrea Goodman, of Norton, who designed the book, and to Thomas Hayden, who prepared the art. For answer checking, thanks go to Florence Fong, Susan Hahn, Peter Leung, and Sharon Schwartz and for clerical work to Laura Jones, Ann Mulbry, and Maxine Murray.

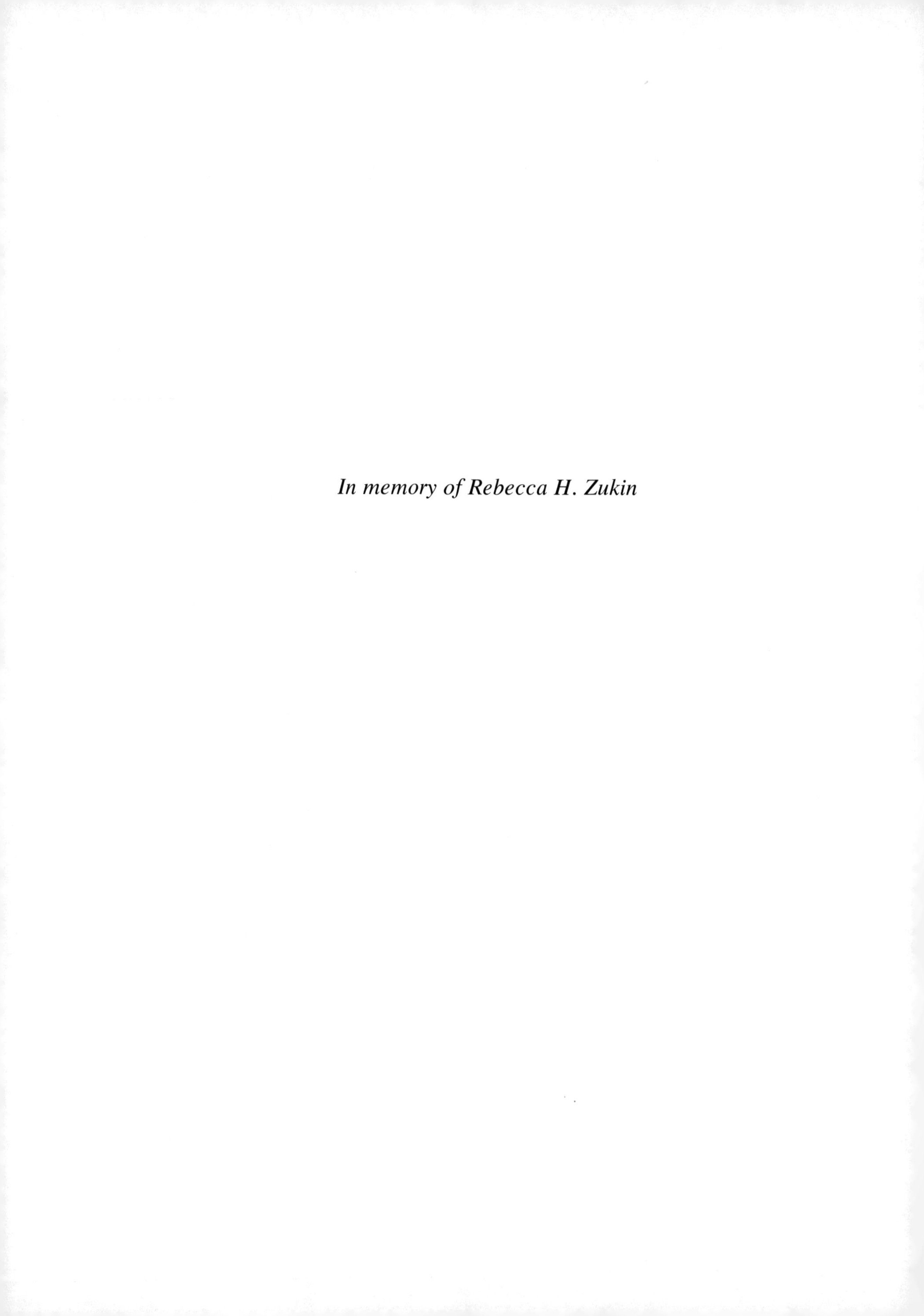

In memory of Rebecca H. Zukin

Contents

*Indicates optional topic

As the title implies, this text presents algebra and trigonometry to the student without unnecessary complications. The gradual, step-by-step approach taken is intuitive, with brief explanations given in simple, yet precise language. Common learning difficulties are recognized and dealt with as they occur. The order of topics is logical and cohesive.

- The book begins with a thorough review of elementary algebra for those who need it. There is also a complete development of exponents before the presentation of logarithm and higher-degree equations.
- Set-theoretic notation is used sparingly. Axiomatic development is completely avoided. Geometric arguments are often presented to explain algebraic concepts.
- Trigonometry is first introduced in terms of degree measurement and right triangles in order to provide motivation and immediate application. Only *after* the student has become familiar with these more basic notions are the unit circle and radian measure developed.
- Quadratic equations with complex roots are presented *before* discussing the arithmetic of complex numbers.
- Logarithmic and exponential functions are introduced only *after* a thorough treatment of the basic properties of exponents and logarithms.
- Conic sections are discussed with an emphasis on graphing; technical detail is minimized.
- Too often, students who have studied algebra have trouble recognizing algebraic concepts in calculus problems. Throughout this text many problems concern the algebraic and geometric techniques that are used in calculus.

- Strong emphasis is placed on practical, real-life problems. There are careful explanations of how to translate words into symbols.
- Over 600 figures illustrate important mathematical concepts. A second color is used functionally throughout the text and art.
- The textual material is interspersed with over 500 illustrative examples. Virtually every new idea is followed by an example demonstrating how various problems can be solved. An example is often furnished to illustrate a theorem *before* proving it; in general, a proof is given only when it contributes to student understanding.
- Student exercise sets occur at the end of each section, and there are review sets at the end of each chapter. Altogether, there are over 4400 exercises, varied in nature, ranging from drill exercises to those which challenge the superior student.
- The accompanying Workbook presents even more worked-out examples along with corresponding exercises.
- Sections 7.1–7.7, on logarithms, can be studied before covering Chapter 6, on quadratic equations and inequalities. Chapter 12, on systems of equations, can be studied after Chapter 6, if desired. For a more detailed breakdown of prerequisite sections as well as for suggestions on presenting material, see the "Chapter-by-Chapter Comments" in the Instructor's Manual.
- Answers to odd-numbered exercises and to all Review Exercises appear at the back of the book. Answers to the remaining even-numbered exercises appear in the accompanying Instructor's Manual.

1 Review of Elementary Algebra

1.1 Polynomials

real numbers

Algebra is primarily concerned with properties of **real numbers.** [See Figure 1.1 .] These numbers include:

Integers—An *integer* is a number such as 0, 1, 2, 17, -1, -2, -208. Among these, 1, 2, and 17 are **positive integers**, -1, -2, and -208 are **negative integers**, and 0 is neither positive nor negative.

Rational numbers—A *rational number* is a number, such as $\frac{2}{3}, \frac{8}{5}, \frac{-3}{10}, \frac{4}{1}$, that can be expressed as a *quotient of integers*, $\frac{M}{N}$, with the restriction $N \neq 0$. (N is *not* equal to 0.) Every integer M is a rational number because $M = \frac{M}{1}$.

Irrational numbers—An *irrational number* is a real number, such as $\sqrt{2}$, $\sqrt{3}$, and π, that *cannot* be expressed as a quotient of integers.

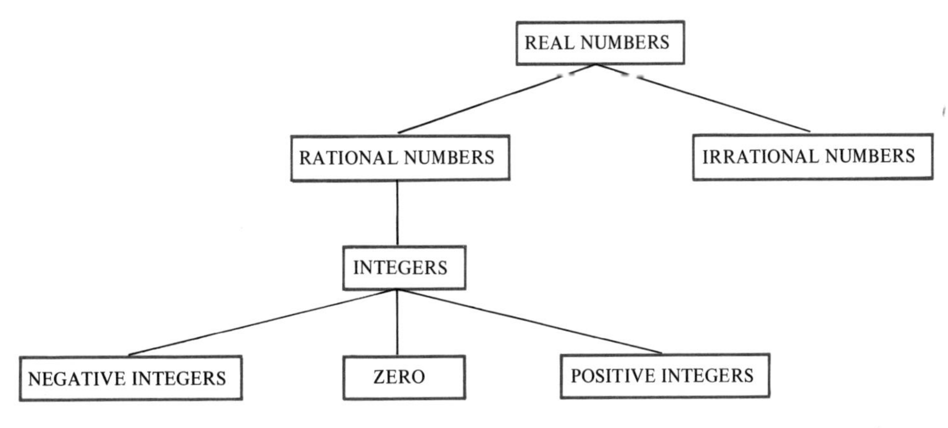

FIGURE 1.1

algebraic symbols

In algebra, symbols are used to represent numbers.

Definition. A **constant** is a symbol (or a combination of symbols) that designates a *specific* number.

The familiar symbols

$$0, \quad 1, \quad 2, \quad -1, \quad \frac{1}{2}, \quad \sqrt{2}, \quad \pi$$

are constants. Each of these symbols designates a *specific* number. Sometimes a letter, such as c or K, will also be used as a constant to designate a *specific* number.

Definition. A **variable** is a symbol that represents *any* one of a given collection of numbers.

The letters x, y, z, and t will be used as variables; other letters will also be used. Often a variable, such as x, will represent *any* real number. However, a variable can also be used to stand for only *some* of the real numbers. For example, if x represents the population of a country, then x would be a positive integer.

Sometimes it is convenient to attach **subscripts** to letters. For example, in a problem involving several positive and negative integers, you might distinguish between them by letting

p_1 read: p-sub-1

be the first *positive* integer considered, p_2 be the second such positive integer, p_3 be the third such positive integer, and so on, whereas n_1 might represent the first *negative* integer considered, n_2 could be the second such negative integer, and so on. Here p_1, p_2, and p_3 are symbols regarded as being *different* from one another, just as x, y, and z are different symbols.

Because constants and variables represent numbers, you can add, subtract, multiply, and divide them. For example, you can add two variables, add a variable and a constant, multiply two variables, or multiply a constant by a variable. You can also consider exponentiation (raising to a power) involving variables. You will consider such "algebraic expressions" as

$$x + 4, \quad x + y, \quad \frac{x-1}{x+2}, \quad x^3, \quad x^4 - 3x^2 + 4, \quad 2xy, \quad 2^x$$

EXAMPLE 1 Let x represent a real number. Express each of the following in algebraic symbols:

(a) two more than the number
(b) five less than the number
(c) half the number
(d) six less than twice the number
(e) half of the quantity that is six less than the number

Solution.

(a) $\underbrace{\text{two more than}}_{2\,+}\ \underbrace{\text{the number}}_{x}$ or, more precisely, $x + 2$

(b) five less than the number

$$\underbrace{-5+}_{\text{five less than}}\quad\underbrace{x}_{\text{the number}}\quad\text{or}\quad x-5$$

(c) half the number

$$\underbrace{\frac{x}{2}}_{\text{half the number}}$$

(d) six less than twice the number

$$\underbrace{-6+}_{\text{six less than}}\quad\underbrace{2x}_{\text{twice the number}}\quad\text{or}\quad 2x-6$$

Here first consider twice x; then subtract 6.

(e) Here first subtract 6 from x; then take half of this.

$$\frac{x-6}{2}$$

($x - 6$ ← six less than the number; half of → 2) □

terms The building blocks of algebraic expressions are *terms*.

Definition. A **term** is either

a constant, a variable, or
a *product* of constants and variables.

EXAMPLE 2 Each of the following is a term.

(a) -2 (This is a constant.)
(b) 0 (This is a constant.)
(c) x (This is a variable.)
(d) $-3y$ (This is the product of the constant -3 and the variable y.)
(e) xyz (This is the product of the variables x, y, and z.)
(f) y^6 $y^6 = y\cdot y\cdot y\cdot y\cdot y\cdot y$
(g) x^2y^3 $x^2y^3 = x\cdot x\cdot y\cdot y\cdot y$
(h) $(3x)(2y)$ (This equals the product of the constants 3 and 2 and the variables x and y.) □

EXAMPLE 3 (a) The expression

$$\frac{x}{4}$$

is a term because $\frac{x}{4} = \frac{1}{4}x$.

Thus $\frac{x}{4}$ is the *product* of the constant $\frac{1}{4}$ and the variable x.

(b) The expression

$$\frac{4}{x}$$

is *not* a term, because the definition of a term does *not* permit division *by a variable*. Note that $\frac{1}{x}$ is neither a constant nor a variable, and thus $\frac{4}{x}$ $\left(\text{which equals } 4 \cdot \frac{1}{x}\right)$ is not a term.

(c) The expression

$$x + 4$$

is *not* a term, because the definition of a term does *not* permit addition *of a variable and a constant*. □

Definition. The **(numerical) coefficient of a term** is the constant that occurs or the product of all constants that occur. If no constant is written, the coefficient is understood to be either 1 or -1.

EXAMPLE 4

(a) The coefficient of $4x^2$ is 4.

(b) The coefficient of $-3xy$ is -3.

(c) The coefficient of 10 is 10.

(d) The coefficient of $(3x)(2y)$ is 6.

(e) The coefficient of $(3 + 2)x$ is 5.

(f) The coefficient of x is understood to be 1.

(g) The coefficient of $-x^2$ is understood to be -1.

(h) The coefficient of $\frac{x}{2}$ is $\frac{1}{2}$.

(i) If a term has 0 as its coefficient, the term reduces to 0. For, when you multiply by 0, the product is always 0. Thus,

$$3 \cdot 0x^2 = 0 \text{ and } 0bc^3 = 0$$

□

polynomials

Definition. A **polynomial** is defined to be either

a term or a sum of terms.

EXAMPLE 5

(a) x^2, $3x$, xy, 4, and 0 are each terms; each of these is a polynomial. In particular, 4 and 0 are each **constant polynomials**. Furthermore, 0 is sometimes called the **zero polynomial**.

(b) $x^2 + 3x$ is a polynomial. It is the sum of the terms x^2 and $3x$.

(c) $x^2 + xy - \pi$ is a polynomial in 2 variables, x and y. It is the sum of the terms x^2, xy, and $-\pi$.

(d) $\frac{3x^2 + 1}{3}$, with the *constant* 3 as divisor, is a polynomial. For, as you will see,

$$\frac{3x^2 + 1}{3} = \frac{3}{3}x^2 + \frac{1}{3} = x^2 + \frac{1}{3}$$

□

EXAMPLE 6

$\frac{3x^2 + 1}{x + 3}$, with the *nonconstant polynomial* $x + 3$ as divisor, is *not* a polynomial. It is, rather, the quotient of polynomials, and is known as a *rational expression*. Rational expressions will be discussed beginning in Section 1.6 . □

evaluating polynomials Frequently, you evaluate a (nonconstant) polynomial for specific values of the variable (or variables). To do this, replace the variables by these numbers. If a variable occurs more than once, replace it by the same number each time it occurs.

EXAMPLE 7 Evaluate

$$x^3 - 5x^2 + 3x - 2$$

for the following values of x: (a) 1 (b) -1

Solution. (a) Replace each occurrence of x by 1 in the given polynomial

$$x^3 - 5x^2 + 3x - 2$$

to obtain:

$$1^3 - 5(1)^2 + 3 \cdot 1 - 2 = 1 - 5 + 3 - 2 = -3$$

(b) $x = -1$:

$$(-1)^3 - 5(-1)^2 + 3(-1) - 2 = -1 - 5 - 3 - 2 = -11$$ □

EXAMPLE 8 Evaluate $x^2 - 5xy + 8y^2$ when $x = 3$ and $y = 4$.

Solution. Replace each occurrence of x by 3 and each occurrence of y by **4:**

$$3^2 - 5(3)(4) + 8(4)^2 = 9 - 60 + 128 = 77$$ □

EXAMPLE 9 The area of a circle is given by πr^2, where r is the length of the radius. [See Figure 1.2 .] Determine the area when (a) r is 4 inches, (b) r is 9 centimeters.

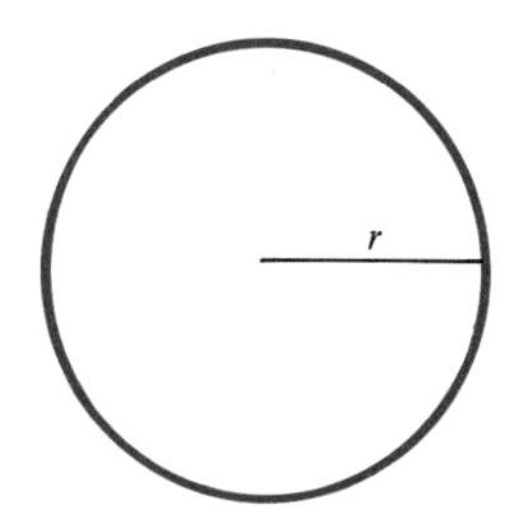

FIGURE 1.2

Solution.

(a) The area is $\pi \cdot 4^2$, or 16π, square inches.
(b) The area is $\pi \cdot 9^2$, or 81π, square centimeters. □

Two polynomials, P and Q, in the same variables, are said to be **equal** if they agree for *all* values of their variables. In this case, write

$$P = Q$$

For example,

$$x + 1 = 1 + x$$ [See page 8.]

For every value of x, both polynomials have the same value. Thus if $x = 0$, both polynomials equal 1; if $x = 10$, both polynomials equal 11; if $x = 100$, both polynomials equal 101.

A nonconstant polynomial can equal 0 for one or more values of its variables. For example,

$x - 5 = 0$	when	x is replaced by 5;
$x^2 - x = 0$	when	x is replaced by 0 and also
	when	x is replaced by 1.

But neither $x - 5$ nor $x^2 - x$ equals the zero polynomial.

Finally, when a polynomial P equals the constant polynomial c, write

$$P \equiv c$$

Thus if P is the zero polynomial, write

$$P \equiv 0.$$

EXERCISES

Assume all (italic) letters, possibly with subscripts, are used as variables.

1. Let x represent a real number. Express each of the following in terms of x.

(a) five more than the number **(b)** eight less than the number

(c) three times the number **(d)** one-fourth of the number

(e) the number increased by three **(f)** the number decreased by ten

2. Let x represent a real number. Express each of the following in terms of x.

(a) one more than half of the number **(b)** half of the quantity that is one more than the number

(c) two less than half of the number **(d)** half of the quantity that is two less than the number

Which expressions in Exercises 3–12 are terms?

3. $9x$ **4.** $-3x_1x_2y_1y_2$ **5.** -1 **6.** $x + 1$ **7.** a^2 **8.** $2x^2y^3$ **9.** $\frac{m}{x}$

10. $\frac{x + 1}{2}$ **11.** $\frac{x^2}{2}$ **12.** 0

In Exercises 13–20 determine the coefficient of each term.

13. $-19xy^2$ **14.** $\frac{z}{3}$ **15.** -7 **16.** $\frac{-x}{7}$ **17.** πxy **18.** $(5x)(8y)$

19. $\frac{2x_1x_2x_3}{3}$ **20.** $x^2(0 \cdot y)$

In Exercises 21–24 rewrite each term so that its coefficient appears first. Also, if necessary, simplify this coefficient.

21. $(5x)(2y)$ **22.** $\frac{x^2}{2}$ **23.** $x(-y)$ **24.** $3x^2 \cdot 5y \cdot 6z^2$

Which expressions in Exercises 25–32 are polynomials?

25. $2xy + x^2$ **26.** $\frac{-1}{x^2}$ **27.** $\frac{x^2}{5}$ **28.** $\frac{x^2}{5y}$ **29.** $x^4 + 3x^2 + 5x - 7 + \frac{1}{x}$ **30.** π

31. $\frac{x^2 + x}{3}$ **32.** $\frac{3}{x^2 + x}$

In Exercises 33–35 evaluate each polynomial for the specified value of the variable.

33. $x + 10$ when $x = 7$ **34.** $5y^2 - 7y + 1$ when $y = 2$ **35.** $x^8 - x^6 + 5x^3 - 1$ when $x = -1$

36. Evaluate $x^2 + 4$ when **(a)** $x = 0$, **(b)** $x = -1$, **(c)** $x = 4$, **(d)** $x = -4$.

37. Evaluate $t^3 + 10t^2 - t + 1$ when **(a)** $t = 10$, **(b)** $t = -10$, **(c)** $t = 100$, **(d)** $t = .1$.

38. Evaluate $x^7 - x^6 + x^2$ when **(a)** $x = 1$, **(b)** $x = -1$, **(c)** $x = 10$, **(d)** $x = .1$.

In Exercises 39–44 evaluate each polynomial for the specified values of the variables.

39. $x^2 - 2y$ when $x = 1, y = 2$

40. $2x + 3xy - 2y$ when $x = 2, y = 1$

41. $-y_1^2 + (-y_2)^2 + 1$ when $y_1 = 3, y_2 = -3$

42. $4t^2 + 2st - 3s + s^2$ when $s = 1, t = 2$

43. $3x^2y + 2xz - yz^2$ when $x = 4, y = -3, z = 1$

44. $2xyz - 3xy + xz - 10yz$ when $x = -2, y = -1, z = 4$

45. Evaluate $2xy + 3y^2$ when **(a)** $x = 2, y = -1$ **(b)** $x = 3, y = -2$.

46. Evalute $3abc + 2a^2 - b^2$ when **(a)** $a = 4, b = 2, c = -2$ **(b)** $a = 1, b = 2, c = -1$

In Exercises 47–52 refer to the diagrams on the inside front cover.

47. The area of a square is given by s^2, where s is the length of a side. Determine the area when s is 11 feet.

48. The volume of a cube is given by s^3, where s is the length of a side. Determine the volume when s is 10 meters.

49. The surface area of a box without a top is given by

$$lw + 2lh + 2wh,$$

where l is the length of the base, w is the width of the base, and h is the height of the box. Determine the surface area when l is 25 centimeters, w is 10 centimeters, and h is 6 centimeters.

50. The circumference of a circle is given by $2\pi r$, where r is the length of the radius. Determine the circumference when r is 5 meters.

51. The volume of a right circular cylinder is given by $\pi r^2 h$, where r is the length of the base radius and h is the length of the altitude. Determine the volume when r is 10 inches and h is 12 inches.

52. The lateral surface area of a right circular cylinder is given by $2\pi rh$, where r and h are as given in Exercise 51. Determine the lateral surface area when r is 10 centimeters and h is 12 centimeters.

53. The distance traveled by a car going at a constant rate is given by rt, where r is its rate and t is the time traveling. Determine the distance traveled if r is 60 kilometers per hour and t is 4 hours.

54. Suppose that an object in motion travels a distance given by

$$v_0(A + Bt),$$

where v_0 is its initial velocity and t is its time in motion. Determine the distance traveled when v_0 is 200 meters per second, A is 9 seconds, B is 13, and t is 7 seconds.

55. When a ball is thrown upward from the ground with an initial velocity of 64 feet per second, its height (in feet) after t seconds is given by

$$-16t^2 + 64t.$$

Determine its height (a) after 2 seconds, (b) after 3 seconds, (c) after 4 seconds.

56. It costs 45 cents per square foot to wallpaper the ceiling of a room and 25 cents per square foot to paper each of the four walls. How much does it cost to paper a room that is 20 feet long, 14 feet wide, and 9 feet high?

1.2 Addition and Subtraction of Polynomials

arithmetic and algebra

Some of the rules of arithmetic apply to the addition and subtraction of terms and of polynomials. Thus let a, b, and c be real numbers.

You can add numbers in either order and obtain the same sum.

$$a + b = b + a$$

This is known as the **Commutative Law of Addition.** And when you first add a and b, and then add c, you obtain the same sum as when you first add b and c, and then add this to a:

$$(a + b) + c = a + (b + c)$$

This is known as the **Associative Law of Addition.** Because of this equality, you may omit parentheses and write

$$a + b + c$$

for either sum.

EXAMPLE 1 (a) $6 + 4 = 4 + 6 = 10$ by the Commutative Law

(b) $(3 + 2) + 7 = 3 + (2 + 7) = 12$ by the Associative Law □

Subtraction *undoes* addition. Thus

$$8 - 5 = 3 \qquad \text{because} \qquad 5 + 3 = 8$$

In order to multiply a sum or a difference, parentheses must sometimes be used. *By convention,* if parentheses are omitted, multiply before adding or subtracting. Thus

$$\underbrace{5 \cdot 4}_{20} + 2 = 20 + 2 = 22 \qquad \text{and} \qquad \underbrace{5 \cdot 4}_{20} - 2 = 20 - 2 = 18,$$

whereas

$$5\underbrace{(4 + 2)}_{6} = 5 \cdot 6 = 30 \qquad \text{and} \qquad 5\underbrace{(4 - 2)}_{2} = 5 \cdot 2 = 10$$

Note that you can also multiply each number inside the parentheses by 5:

$$5(4 + 2) = 5 \cdot 4 + 5 \cdot 2 = 20 + 10 = 30$$
$$5(4 - 2) = 5 \cdot 4 - 5 \cdot 2 = 20 - 10 = 10$$

This method employs the **Distributive Laws** which state that

$$a(b + c) = a \cdot b + a \cdot c$$
$$(b + c)a = b \cdot a + c \cdot a$$

Similarly,

$$a(b - c) = a \cdot b - a \cdot c$$
$$(b - c)a = b \cdot a - c \cdot a$$

addition of terms

In order to define addition of terms, consider the following definitions.

Definition. Let x be a variable that occurs in a nonzero term. Then n is called the **exponent of** x if x^n is the power of x that occurs when the term has been simplified.

EXAMPLE 2 Consider the term $3x^2zyz^3$. Note that

$$3x^2zyz^3 = 3 \cdot x \cdot x \cdot z \cdot y \cdot z \cdot z \cdot z = 3x^2yz^4.$$

(a) The exponent of x is 2. (b) The exponent of y is 1.
(c) The exponent of z is 4. □

Definition. Two or more nonzero terms are said to be **similar** if

1. they contain precisely the same variables
   and
2. each variable has the same exponent in each term.

Thus *similar terms can differ only in their numerical coefficients or in the order of their variables.* For convenience, *0 is considered to be similar to every term.*

EXAMPLE 3 (a) $2xy$ and $-3xy$ are similar.
(b) $5x^2z$ and zx^2 are similar.
(c) $2zx^2$, $2x^2z$, and $2xzx$ are similar. (In fact, they are equal.)
(d) x^2y and xy are *dissimilar* (not similar) because the two terms differ in the exponent of x.
(e) Any two constant terms are similar because neither contains any variables. □

Similar terms can be added and subtracted by extending the Distributive Laws. For example, let a and b be constants and let x be a variable. Then

$$ax + bx = (a + b)x$$ **Addition of similar terms**

Also, $-(bx)$ is the **(additive) inverse of the term** bx and

$$-(bx) = (-1)(bx) = [(-1)b)]x = (-b)x$$

so that

$$\begin{aligned} ax - bx &= ax + [-(bx)] \\ &= ax + [(-b)x] \\ &= [a + (-b)]x \\ &= (a - b)x \end{aligned}$$ **Subtraction of similar terms**

EXAMPLE 4 (a) $10x + 3x = (10 + 3)x = 13x$
(b) $7x^2y - 5x^2y = (7 - 5)x^2y = 2x^2y$ □

Terms that are not similar cannot be combined in the preceding manner.

EXAMPLE 5 $5x + 3y$ cannot be further simplified. For, $5x$ and $3y$ are not similar. The Distributive Laws cannot be used as they were in Example 4. □

The Commutative and Associative Laws apply to the addition of terms. Thus when several terms are to be combined, arrange these so that similar terms are grouped together.

EXAMPLE 6

$$\begin{aligned} 3x + 7y + 5x &= 3x + (7y + 5x) \\ &= 3x + (5x + 7y) && \text{by the Commutative Law} \\ &= (3x + 5x) + 7y && \text{by the Associative Law} \\ &= 8x + 7y \end{aligned}$$

In practice, you would omit several of these steps. □

From now on, you will speak of a polynomial as a sum of terms, even when it contains just one term.

addition of polynomials

Polynomials are added in the same way as terms are. Thus to **add polynomials,** arrange their terms so that similar terms are grouped together. Write

$$P + Q$$

for the sum of the polymonials P and Q. For any polynomial P,

$$P + 0 = P$$

EXAMPLE 7 Let

$$\begin{aligned} P &= 4x + 3y - 5z, \\ Q &= 2x - 4y + 7z, \\ R &= -4x - 3y + 5z. \end{aligned}$$

Determine: (a) $P + Q$ (b) $P + R$

Solution. When several terms are involved, it is convenient to list similar terms in the same column.

(a)
$$\begin{array}{rl} P: & 4x + 3y - 5z \\ Q: & \underline{2x - 4y + 7z} \\ P + Q: & 6x - y + 2z \end{array}$$

(b)
$$\begin{array}{rl} P: & 4x + 3y - 5z \\ R: & \underline{-4x - 3y + 5z} \\ P + R: & 0 \end{array}$$

□

Example 7(b) suggests the following definition.

Definition. Let P be a polynomial. Then the **(additive) inverse of** P is the polynomial that must be added to P to obtain 0, just as in arithmetic, the **(additive) inverse of a number** a is the number that is added to a to obtain 0.

Write $-P$ for the (additive) inverse of P. Thus

$$P + (-P) \equiv 0$$

As you see from Example 7(b), *to obtain* $-P$, *replace the coefficient of each term of* P *by its (additive) inverse.*

EXAMPLE 8

$$-(-3 + x^2 + y - 2z) = 3 - x^2 - y + 2z$$

□

To **subtract polynomials,** define

$P - Q$ to be $P + (-Q)$.

EXAMPLE 9 Subtract $x^2 + 3xy - y^2$ *from* $2x^2 + 6xy + 5y^2$.

Solution. You want

$$2x^2 + 6xy + 5y^2 - (x^2 + 3xy - y^2) \qquad \text{or}$$

$$2x^2 + 6xy + 5y^2 + [-(x^2 + 3xy - y^2)].$$

The inverse of the polynomial $x^2 + 3xy - y^2$ is obtained by replacing each numerical coefficient by its inverse:

$$\begin{array}{r} 2x^2 + 6xy + 5y^2 \\ \begin{array}{r} - \quad\; - \quad\;\; + \\ \underline{+\; x^2 + 3xy - y^2} \end{array} \\ x^2 + 3xy + 6y^2 \end{array}$$

□

The Commutative and Associative Laws apply to the addition of polynomials:

Let P, Q, R be polynomials.

$$P + Q = Q + P \qquad \textbf{(Commutative Law)}$$
$$(P + Q) + R = P + (Q + R) \qquad \textbf{(Associative Law)}$$

Because of the Associative Law, you may omit parentheses, and write

$$P + Q + R.$$

EXAMPLE 10 Find the sum $P + Q + R$, where

$$P = 5x^2y^2 + 8x^2y + 3xy, \qquad Q = x^2y^2 - x^2y + 2xy^2,$$
$$R = 4x^2y + 6xy^2.$$

Solution. Rearrange the terms so that similar terms are in the same column.

$$\begin{array}{rl} P: & 5x^2y^2 + 8x^2y \qquad\qquad + 3xy \\ Q: & x^2y^2 - x^2y + 2xy^2 \\ R: & 4x^2y + 6xy^2 \\ \hline P + Q + R: & 6x^2y^2 + 11x^2y + 8xy^2 + 3xy \end{array}$$

□

EXAMPLE 11 Simplify: $5x - [3 - (4x - 5)]$

Solution. *Begin with the innermost parentheses and work outward.*

$$\begin{aligned} 5x - [3 - (4x - 5)] &= 5x - [3 - 4x + 5] \\ &= 5x - 3 + 4x - 5 \\ &= (5x + 4x) + (-3 - 5) \\ &= 9x - 8 \end{aligned}$$

□

As you develop facility in handling algebraic expressions, you may wish to skip elementary steps. For instance, in Example 11, you can combine steps, and write

$$5x - [3 - (4x - 5)] = 5x - 3 + 4x - 5 = 9x - 8.$$

Finally, a *nonzero* term is also called a **monomial.** The sum of two *dissimilar* monomials is called a **binomial.** The sum of three (mutually) *dissimilar* terms is called a **trinomial.** Thus $4x$ is a monomial, $2x + 5y$ is a binomial, and $x + y - z$ is a trinomial. Note that $x + 2x$ is a monomial because it is the sum of *similar* monomials.

EXERCISES

In Exercises 1–6 indicate which pairs of terms are similar.

1. $10z$ and $-5z$ **2.** $3xy$ and $4xz$ **3.** x^2y^3 and $-x^2y^3$ **4.** xzy and $4yxz$

5. x^3y^2 and x^2y^3 **6.** 19 and 35

In Exercises 7–12 group similar terms together.

7. $-xy, x, -x, 2xy, 3x$ *Answer:* $-xy, 2xy \quad | \quad x, -x, 3x$

8. $x^2y^2,\ -2x^2y,\ 3xy,\ -9x^2y^2,\ 5x^2y$

9. $14,\ x^2,\ x,\ -x^2,\ \frac{1}{2}x,\ \frac{1}{2}$

10. $2xy,\ -4xz,\ 3x^2z,\ 5zx,\ 6xz$

11. $-r^2st,\ rs^2t,\ 2r^2ts,\ 4rst^2,\ 3s^2rt$

12. $5xyz,\ 2xz,\ 3xzy,\ x^2zy,\ 5zy,\ -xyz,\ -2yz$

In Exercises 13–20 simplify wherever possible.

13. $5a + 4a$ **14.** $-7x - 2y$ **15.** $9b + 3b + 2b$

16. $3xy - 4xy - 6xy$ **17.** $a + b - a + b + 1$ **18.** $s^2 + t^2 - 2s^2 + t^2 + 2t^2$

19. $5r - (2s + 3s)$ **20.** $x^2 + 2y^2 - (3z^2 + 4x^2 - 5z^2)$

In Exercises 21–24 add the polynomials.

21. $\begin{array}{r} 2a + 3b \\ \underline{4a + b} \end{array}$ **22.** $\begin{array}{r} a - b + c \\ \underline{2a \qquad - c} \end{array}$ **23.** $\begin{array}{rrr} w & & -\ 3y \\ & x\ + & y + z \\ \underline{2w} & & \underline{+\ 2y + z} \end{array}$ **24.** $\begin{array}{r} x^2 - 2xy + y^2 + z \\ -\ 3xy - y^2 \\ 3x^2 - xy - 2y^2 - 3z \\ \underline{2x^2 \qquad + y^2 + z} \end{array}$

In Exercises 25–30 determine the additive inverse of each polynomial.

25. $2a$ **26.** $-5b$ **27.** $2a + 5b$ **28.** $3a^4 - 2a^2 + 4a$

29. $-abc + ab - ac + 4$ **30.** $x^3y - x^2y + xy - y + 3$

In Exercises 31–34 *subtract* the bottom polynomial from the top one.

31. $\begin{array}{r} 3x + 4y \\ \underline{2x + 2y} \end{array}$ **32.** $\begin{array}{r} 16a + 4b \\ \underline{12a + 4b} \end{array}$ **33.** $\begin{array}{r} 3x - 2y + z \\ \underline{2x - 2y - z} \end{array}$ **34.** $\begin{array}{r} a + b - c + 1 \\ \underline{-a - 2b + c - 3} \end{array}$

In Exercises 35–42 simplify each polynomial.

35. $a + 2b - [3a - (2a + b)]$

36. $x - (y + z) + (y - z) - (y - z)$

37. $m + (2n - r) - [(2n - 4) - m]$

38. $x^2 - (y^2 - z^2) + [x^2 - (y^2 - z^2)]$

39. $3a - [b - (2a + 3c) - (2a + b)]$

40. $5 + (2s - [3t + (2t - s)])$

41. $2a + 1 - [4b - (3 + 2a - 1)]$

42. $(2a - 3b) - (4a - 3b) - [(4a - 3b) - 2b]$

43. Add $2x + 3y - z$ to the sum of $3x - z$ and $2y + z$.

44. Subtract $2x - 3y$ from $3x - 2y$.

45. Subtract 0 from x^3.

46. Subtract x^2 from 0.

47. Subtract the sum of $3x$ and $5z$ from $2z$.

48. Evaluate $(x^2 + 5x + 1) - (x^2 + 2x - 2)$ when $x = 4$.

49. Evaluate $2x - (3x + 2y) + 8xy$ when $x = 2$ and $y = 3$.

50. Evaluate $x^3 - [x^2y + 2x - (3x + 5y)]$ when $x = -1$ and $y = -2$.

1.3 Multiplication of Polynomials

powers of the same variable

First note that

$$2^2 = 4 \quad \text{and} \quad 2^3 = 8.$$

Thus

$$2^2 \cdot 2^3 = 4 \cdot 8 = 32$$

Now observe that $32 = 2^5$. In other words,

$$2^2 \cdot 2^3 = 2^5 = 2^{2+3}$$

In Example 1, powers of the *same variable* are multiplied.

EXAMPLE 1 Multiply a^4 by a^2.

Solution. Observe that if you write this without exponents, you obtain

$$a^4 \cdot a^2 = (aaaa)(aa) = aaaaaa = a^6.$$

Thus

$$a^4 \cdot a^2 = a^{4+2} = a^6$$ □

Let m and n be positive integers. *To multiply two powers of the same variable (or of the same number),*

$$a^m \cdot a^n,$$

write down the base a and add the exponents, $m + n$*:*

$$\mathbf{a^m \cdot a^n = a^{m+n}}$$

This is because

$$a^m = \underbrace{a \cdot a \cdot a \ldots a}_{m \text{ times}} \quad \text{and} \quad a^n = \underbrace{a \cdot a \cdot a \ldots a}_{n \text{ times}}$$

Thus

$$a^m \cdot a^n = \underbrace{(a \cdot a \cdot a \ldots a)}_{m \text{ times}} \underbrace{(a \cdot a \cdot a \ldots a)}_{n \text{ times}} = \underbrace{a \cdot a \cdot a \ldots a}_{m + n \text{ times}} = a^{m+n}$$

The preceding rule can be extended to multiplying three or more powers of the same variable. For example, if r is also a positive integer,

$$\mathbf{a^m \cdot a^n \cdot a^r = a^{m+n+r}}$$

EXAMPLE 2 $y^5 \cdot y^3 \cdot y = y^{5+3+1} = y^9$ □

To multiply monomials, use the **Associative** and **Commutative Laws of Multiplication:**

$$(a \cdot b) \cdot c = a \cdot (b \cdot c) \qquad a \cdot b = b \cdot a$$

1. Group all numerical coefficients at the beginning.
2. Group powers of the same variable together.
3. Multiply the coefficients. Also multiply powers of the same variable according to the previous rules.

EXAMPLE 3

$$\begin{aligned}(2xy^3)(5x^4y^2) &= (2 \cdot 5)(x \cdot x^4)(y^3 \cdot y^2)\\ &= 10x^{1+4}y^{3+2}\\ &= 10x^5y^5\end{aligned}$$ □

multiplying polynomials To multiply two polynomials (not both monomials), first utilize the Distributive Laws.

EXAMPLE 4

$$\begin{aligned} 5a(a+b) &= (5a)a + (5a)b \\ &= 5(aa) + 5(ab) \\ &= 5a^2 + 5ab \end{aligned}$$

□

EXAMPLE 5 Multiply $2x + y$ by $3x + y^2$.

Solution. According to the Distributive Laws,

$$\begin{aligned} (2x+y)(3x+y^2) &= (2x+y)(3x) + (2x+y)(y^2) \\ &= (2x)(3x) + y(3x) + (2x)(y^2) + y(y^2) \\ &= 6x^2 + 3xy + 2xy^2 + y^3 \end{aligned}$$

□

Thus multiply each term of the first polynomial by each term of the second polynomial. The resulting polynomial is the sum of these products.

In practice, the following method is frequently preferred. Beneath the multiplication line, similar terms are placed in the same column, as in Example 6, which follows. These similar terms are then added in the last step.

EXAMPLE 6 Multiply $a^2 + 3ab - b^2$ by $a + 2b$.

Solution.

$$\begin{array}{rrrr} a^2 & + 3ab & - b^2 & \\ a & + 2b & & \\ \hline a^3 & + 3a^2b & - ab^2 & \\ & 2a^2b & + 6ab^2 & - 2b^3 \\ \hline a^3 & + 5a^2b & + 5ab^2 & - 2b^3 \end{array}$$

□

A product that you will frequently encounter is

$$(x+a)(x+b).$$

Of course, the letters may change. For example, you may see

$$(y+m)(y+n).$$

You will often see numbers in place of a and b, as in

$$(x+5)(x+3).$$

In general,

$$\begin{aligned} (x+a)(x+b) &= xx + ax + xb + ab \\ &= x^2 + ax + bx + ab \\ &= x^2 + (a+b)x + ab \end{aligned}$$

Therefore

$$\boldsymbol{(x+a)(x+b) = x^2 + (a+b)x + ab}$$

EXAMPLE 7

$$\begin{aligned} (x+5)(x+3) &= x^2 + \underbrace{(5+3)}x + \underbrace{5 \cdot 3} \\ &= x^2 + \quad 8 \quad x + 15 \end{aligned}$$

□

Let $b = a$. Then

$(x + a)(x + b) = (x + a)(x + a) = (x + a)^2$ and

$(x + a)^2 = x^2 + (a + a)x + aa = x^2 + 2ax + a^2$

Let $b = -a$. Then

$(x + a)(x + b) = (x + a)(x - a)$ and

$$(x + a)(x - a) = x^2 + \underbrace{(a - a)}_{0}x + a(-a) = x^2 - a^2$$

$$\begin{array}{r} x + a \\ x - a \\ \hline x^2 + ax \\ - ax - a^2 \\ \hline x^2 - a^2 \end{array}$$

Notice that the "x-terms", ax and $-ax$, "cancel out". Once again,

$$\mathbf{(x + a)^2 = x^2 + 2ax + a^2}$$
$$\mathbf{(x + a)(x - a) = x^2 - a^2}$$

△

EXAMPLE 8

$$(x + 4)^2 = x^2 + 2 \cdot 4x + 4^2 = x^2 + 8x + 16$$

□

EXAMPLE 9

$$(x + 5)(x - 5) = x^2 - 25$$

□

Finally, *by convention,* if parentheses are omitted, raise to a power before multiplying. Thus

$$3x^2 = 3 \cdot x \cdot x$$

whereas

$$(3x)^2 = 3x \cdot 3x = 9x^2$$

Also,

$$3(x + y)^2 = 3(x + y)(x + y)$$

whereas

$$[3(x + y)]^2 = 3(x + y) \cdot 3(x + y) = 9(x + y)^2$$

□

EXAMPLE 10

$$(2x + 1)^2 - 2(x + 1) = (2x + 1)(2x + 1) - (2x + 2) = 4x^2 + 4x + 1 - 2x - 2 = 4x^2 + 2x - 1$$

EXERCISES

In Exercises 1–36 multiply, as indicated.

1. $x^2 \cdot x$

2. $a^2 \cdot a^3 \cdot 6$

3. $b^8 \cdot b^4$

4. $m^{10} \cdot 2 \cdot m^{17} \cdot 7$

5. $z \cdot z^2 \cdot z \cdot z^3 \cdot z^2$

6. $(x^2y)y$

7. $(y^2z)(y^2z^2)$

8. $(2uv)(-5u^2v)$

9. $(7ab)(2a^2b)(3b^2)$

10. $(-2x^2y)(-yz)(xyz)$

11. $2(x - y)$

12. $-4(2x + 3y^2)$

13. $x(x + 3)$

14. $2x(x^2 + x^3)$

15. $-5x^2(x + xy)$

16. $2abc^2(a^2c - 4ab + 2ac^2)$
17. $-3r^2st^2(-rs + 6st - 2r^3st^{10})$
18. $(m + 8)(m + 1)$
19. $(m - 2)(m + 7)$
20. $-2(a - 9)(a + 5)$
21. $(x + 6)^2$
22. $(a^2 - 5)^2$
23. $(x + 9)(x - 9)$
24. $(x^2 + 3)(x^2 - 3)$
25. $(u^4 + 1)(u^4 - 1)$
26. $(c^3 - 1)(c^3 + 6)$
27. $(x^2 + x - 1)(x + 5)$
28. $(y^2 + 4y - 3)(y - 1)$
29. $(x^3 - 2y)(x^2 + y)$
30. $(2x + 3y)(x^2 - 4y)$
31. $(5a - 3b)(-2a + 6b)$
32. $(x^2 + 3x - 1)^2$
33. $(x + y)(x + y + z)$
34. $(x - y)(x - y + z)$
35. $(a + 3b)(2a + b + 5c)$
36. $(2x + y - 3)(x - y)^2$

In Exercises 37–56 simplify each polynomial.

37. $2a - 3(b + 2c)$
38. $5(6 - 2a) + 2(a + 1)$
39. $3x[x - 2(y - 2)]$
40. $x(x - y) + y(y - x)$
41. $(x - 2)(x + 1) + (y - 1)(y + 2)$
42. $x(x - [3 + (2 - a)])$
43. $(2a^2 + b^2) - 2a[3a - (2b + a)]$
44. $2y^2(y - 3) + 5[y^2 - y(2y + 1)]$
45. $(2x)^2 - 2x^2$
46. $2x^3 - (-x)^3$
47. $(x - 5)^3 - (2 - x)^2$
48. $3(x - 1)^2 - (3x - 1)^2$
49. $(a + 1)(a + 2)(a + 3)$
50. $(x^3 + x^2)(2x + 4) + (x^2 + 4x)(3x^2 + 2x)$
51. $(6x^2 + 4)(x^2 - 4x + 1) + (2x^3 + 4x)(2x - 4)$
52. $(3x^4 + 1)^2 - (2x^2 + 3)^2$
53. $(x^3 + 3x^2 - 2x + 1)(x^3 + x + 2)$
54. $(2y + 1)^2(y - 1)^2$
55. $(a + 2)^3 - 2(a - 1)^2 + (3a)^2$
56. $(2x^2 - x + 1)^2 - 3(x^2 + 2x - 3)^2 - x(x^2 - 1)$
57. Multiply $2x + 5$ by the sum of $3x + 1$ and $-2x$.
58. The square of $x - 1$ is multiplied by the cube of $2x - 3$. Determine the resulting polynomial.

1.4 Factoring, I

factoring integers

The process of writing integers as products of simpler integers is called *factoring*. It is helpful when combining rational numbers. There is also a factoring process that is useful when working with polynomials.

Definition. Let a and b be integers, with $b \neq 0$. Then b is said to be a **factor of** a (or a **divisor of** a) if

$$a = bc$$

for some integer c. In this case, a is called a **multiple of** b and of c.

For the present, only integral factors will be considered.

EXAMPLE 1

(a) 7 is a factor of 21 because $21 = 7 \cdot 3$. Clearly, 3 is also a factor of 21.

(b) 5 and -5 are each factors of 10 because $10 = 5 \cdot 2 = (-5)(-2)$. □

"Prime" numbers serve as "atoms" among the integers. By multiplying primes, you can obtain all other positive integers except 1.

Definition. An integer p ($\geqslant 2$) is said to be a **prime** if the only positive factors of p are 1 and p. An integer n ($\geqslant 2$) that is not a prime is called a **composite.**

Note that 1 is the only positive integer that is a factor of every integer. By convention, 1 is neither a prime nor a composite.

The first eight primes are 2, 3, 5, 7, 11, 13, 17, and 19. Observe that 4 is a composite because $4 = 2 \cdot 2$. And, by Example 1, both 10 and 21 are composites.

A **prime factorization** of an integer is an expression of the integer as a product of primes (possibly with only one factor). Thus $3 \cdot 5$ is a prime factorization of 15, $2 \cdot 2 \cdot 11$ is a prime factorization of 44, and 7 is a prime factorization of itself.

Fundamental Theorem of Arithmetic: Except for the order of the factors, *there is exactly one prime factorization of a composite.* The (only) prime factorization of a prime p is p itself.

It is convenient to use exponents in expressing prime factorization. Often, you can find the prime factors almost immediately. For example,

$$20 = 4 \cdot 5 = 2^2 \cdot 5$$

There is also a systematic way to determine prime factors.

EXAMPLE 2 Determine the prime factorizations of each:

(a) 60 (b) 72

Solution. Test powers of the primes 2, 3, 5, 7, and so on, in order of magnitude, to determine whether they are factors of the integer in question.

(a) 2, the smallest prime, is a factor of 60. Observe that 4 ($= 2^2$) is the highest power of 2 that is a factor. Next, divide $\frac{60}{4}$ (or 15) by 3, and obtain: $15 = 3 \cdot 5$

As you know, 3 and 5 are both primes. Therefore, the prime factorization is given by: $60 = 2^2 \cdot 3 \cdot 5$

(b) 2^3 (or 8) is a factor of 72, but 2^4 (or 16) is not.

$$\frac{72}{2^3} = \frac{72}{8} = 9 = 3^2$$

Thus

$$72 = 2^3 \cdot 3^2$$

□

You can also express a negative integer, $-m$, where $m > 2$, in terms of primes by considering the prime factorization of m. Thus

$$-72 = -(2^3 \cdot 3^2)$$

GCD The coefficients of (the terms of) a polynomial often have factors in common. To factor a polynomial, you must sometimes first find the "*greatest common divisor*" of its coefficients.

Definition. Let m and n be integers. Then the integer b is called a **common divisor of m and n** if b is a factor (divisor) of m as well as of n. The largest (positive) common divisor of m and n is called the **greatest common divisor of m and n**, written GCD (m, n).

The greatest common divisor of three or more integers can be defined similarly. Write GCD(25, 50, 90) for the greatest common divisor of these three integers.

If the integers are comparatively simple, their greatest common divisor can be determined at sight, as in parts (a) and (b) of Example 3. Otherwise, use prime factorization, as in part (c) of Example 3.

EXAMPLE 3 Determine: (a) GCD(6, 9) (b) GCD(25, 50, 90) (c) GCD(28, 98, 140)

Solution.

(a) The *positive* common divisors of 6 and 9 are 1 and 3. Thus GCD(6, 9) = 3

(b) GCD(25, 50, 90) = 5

(c) $28 = 2^2 \cdot 7$; $98 = 2 \cdot 7^2$; $140 = 2^2 \cdot 5 \cdot 7$

Determine the *smallest* power of each *common prime factor*. Then obtain the *product* of these smallest powers. Thus the common prime factors are 2 and 7. The smallest power of 2 is 2^1 (in the prime factorization of 98). If you were to use a higher power of 2, you would not obtain a factor of 98. Similarly, the smallest power of 7 is 7^1. The greatest common divisor is

$$2^1 \cdot 7^1 \quad \text{or} \quad 14$$ □

common factors Polynomials, like integers, can be *decomposed* into "prime factors". For example, the polynomial

$$5x^2 - 20$$

can be broken down as follows:

$$5x^2 - 20 = 5(x^2 - 4) = 5(x + 2)(x - 2)$$

Definition. Let P be an arbitrary *nonzero* polynomial with integral coefficients. The **greatest common factor of the terms of P**, or for short, **the common factor of P**, is defined to be the *product* of

1. the greatest common divisor of the coefficients of P
   and
2. the *smallest* power of every variable that occurs in *every* term of P.

Clause 2 is similar to the rule for determining the GCD of integers.

Definition. To **isolate the common factor Q of a polynomial P,** write

$$P = Q \cdot R,$$

where Q is the common factor of P, and R is another polynomial.

You isolate the common factor of a polynomial by applying the Distributive Laws.

EXAMPLE 4 (a) Let $P = 12a^2bc^3 - 24ac^4$. The coefficients of P are 12 and -24, and GCD$(12, -24) = 12$. (It is *not* necessary to factor the GCD into primes.) The smallest power of a is a^1. The smallest power of c is c^3. The variable b does not appear in *every* term of P, so it does not occur in the common factor of P. Thus the common factor Q of P is $12ac^3$. Isolate the common factor Q by writing

$$\underbrace{12a^2bc^3 - 24ac^4}_{P} = \underbrace{12ac^3}_{Q}\,\underbrace{(ab - 2c)}_{R}.$$

(b) Let $P = 20x^4 - 16x^3 + 40x^2 - 8x$. Then the common factor Q of P is $4x$. Isolate the common factor by writing

$$\underbrace{20x^4 - 16x^3 + 40x^2 - 8x}_{P} = \underbrace{4x}_{Q}\,\underbrace{(5x^3 - 4x^2 + 10x - 2)}_{R} \quad \square$$

Definition. Let P be an arbitrary polynomial with integral coefficients. To **factor** P, write

$$P = Q \cdot R_1 \cdot R_2 \ldots R_n$$

where Q is the common factor of P and where $R_1, R_2, \ldots, R_n$ are polynomials (with integral coefficients) that cannot be factored further. (If $Q = 1$, do not write it.)

Whenever possible, *first isolate the common factor* before applying the other factoring methods that will be explained. This usually reduces the amount of work involved.

difference of squares

Recall that

$$x^2 - a^2 = (x + a)(x - a).$$

This factoring method is known as the **difference of squares.**

EXAMPLE 5 Factor: $3x^3 - 48x$

Solution. First isolate the common factor, $3x$. Thus

$$\underbrace{3x^3 - 48x}_{P} = \underbrace{3x}_{Q}\,\underbrace{(x^2 - 16)}_{R}$$

$$\underbrace{x^2 - 16}_{R} = x^2 - 4^2 = \underbrace{(x + 4)}_{R_1}\,\underbrace{(x - 4)}_{R_2}$$

The polynomials R_1 and R_2 cannot be factored further. Thus the factorization is:

$$\underbrace{3x^3 - 48x}_{P} = \underbrace{3x}_{Q}\,\underbrace{(x + 4)}_{R_1}\,\underbrace{(x - 4)}_{R_2} \quad \square$$

Often you will factor polynomials of the form

$a^2x^2 - b^2y^2$.

By the Associative and Commutative Laws of multiplication,

$$a^2x^2 = (a \cdot a)(x \cdot x)$$
$$= (ax)(ax)$$
$$= (ax)^2$$

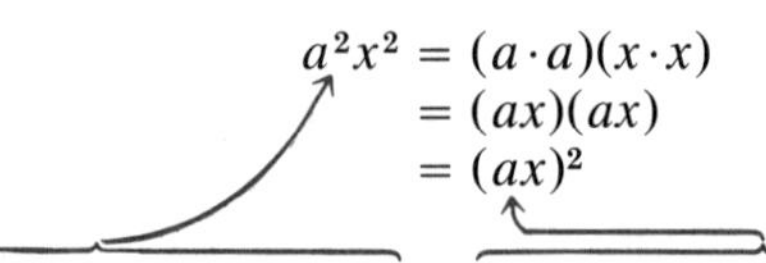

The product of the squares is the square of the product.

Similarly,

$$b^2y^2 = (by)^2$$

Thus

$$a^2x^2 - b^2y^2 = (ax)^2 - (by)^2 = (ax + by)(ax - by)$$

EXAMPLE 6 Factor: $9x^2 - 16y^2$

Solution. $9x^2 = (3x)^2$; $16y^2 = (4y)^2$. Thus

$$9x^2 - 16y^2 = (3x + 4y)(3x - 4y)$$ □

Even powers of a number are squares. Observe that

$$(a^2)^2 = a^2 \cdot a^2 = a^{2+2} = a^4.$$

Similarly,

$$(a^3)^2 = a^3 \cdot a^3 = a^{3+3} = a^6$$

Therefore

$$(a^2)^2 = a^{2 \cdot 2}, \qquad (a^3)^2 = a^{2 \cdot 3}$$

and

$$(a^n)^2 = a^{2n}$$

for every (positive) integer n. Thus

$$a^{10} = (a^5)^2, \qquad a^{34} = (a^{17})^2, \qquad 4a^8 = 2^2(a^4)^2 = (2a^4)^2$$

The difference of squares method can be extended to the case of the *difference of even powers.*

EXAMPLE 7 Factor: $a^6 - b^4$

Solution. $a^6 = (a^3)^2$; $b^4 = (b^2)^2$. Thus

$$a^6 - b^4 = (a^3 + b^2)(a^3 - b^2)$$ □

The difference of squares method can apply twice in the same example.

EXAMPLE 8

$$x^4 - y^4 = (x^2 + y^2)\underbrace{(x^2 - y^2)}_{\text{difference of squares}}$$
$$= (x^2 + y^2)(x + y)(x - y)$$

The factor $x^2 + y^2$, which is the *sum of squares,* cannot be factored (as a polynomial with integral coefficients). □

Finally, you may be given the difference between squares of polynomials.

EXAMPLE 9 Factor: $(3a + b)^2 - (x + y)^2$

Solution. Each of the polynomials $(3a + b)^2$ and $(x + y)^2$ is a square. Thus

$$(3a + b)^2 - (x + y)^2 = [(3a + b) + (x + y)][(3a + b) - (x + y)]$$

Remove the inner parentheses and write

$$(3a + b)^2 - (x + y)^2 = (3a + b + x + y)(3a + b - x - y) \quad □$$

EXERCISES

In Exercises 1–12 determine the prime factorization of each integer.

1. 8 **2.** 15 **3.** 40 **4.** 66 **5.** 84 **6.** 96 **7.** 100 **8.** 108

9. 125 **10.** 132 **11.** 160 **12.** 144 000 000

In Exercises 13–22 determine the greatest common divisor of the indicated integers.

13. 8, 24 **14.** 10, 15 **15.** −30, −40 **16.** 28, 48 **17.** 132, 144

18. 2, 8, 15 **19.** 4, 6, 10 **20.** 15, −20, 25 **21.** 100, 150, 175 **22.** 40, 60, 72, 120

23. Determine the smallest positive integer that is a product of three distinct primes.

24. Determine the smallest positive integer that is a product of four distinct primes.

In Exercises 25–40 isolate the common factor of each polynomial.

25. $5a - 15$ **26.** $18m^3 + 24$ **27.** $3x^2 + 5x$

28. $a^3 - a^2$ **29.** $2x^2 + 8x$ **30.** $12z^4 + 30z^2$

31. $108z^{10} - 96z^7$ **32.** $x^2y + xy^2$ **33.** $8a^2b^3 + 12a^3b^2$

34. $50m^5n^4 - 30mn^5$ **35.** $8x^2 + 4x + 12$ **36.** $3a^3 - 6a^2 + 3a$

37. $7b^5 + 14b^4 - 7b^2$ **38.** $5x^2y + 10xy - 15xy^2$ **39.** $7x^2y^3z + 10xyz^4 - 5xz^3$

40. $16a^3c^2b^4 + 20a^5b - 44a^3c^4 + 40$

In Exercises 41–64 factor each polynomial.

41. $x^2 - 1$ **42.** $y^2 - 49$ **43.** $1 - z^2$ **44.** $x^2 - y^2$

45. $x^2 - 9z^2$ **46.** $25x^2 - 4a^2$ **47.** $144r^2 - 121t^2$ **48.** $x^4 - 25$

49. $x^4 - 4y^6$ **50.** $100 - 49a^6$ **51.** $121a^8 - 100b^6$ **52.** $2x^2 - 2y^2$

53. $3a^2 - 12x^2$ **54.** $7a^4 - 28b^2$ **55.** $x^3 - x$ **56.** $x^5 - x^3$

57. $3ax^2 - 3ay^2$ **58.** $m^4 - 1$ **59.** $16y^{12} - 1$ **60.** $81z^4 - 16a^8$

61. $(2x + 3)^2 - y^2$ **62.** $z^2 - (a - 1)^2$ **63.** $(a + b)^2 - (a - b)^2$ **64.** $(9x^2 - 1)^3$

1.5 Factoring, II‡

factoring $x^2 + Mx + N$

EXAMPLE 1 Factor: $x^2 + 5x + 4$

Solution. To obtain $x^2 + 5x + 4$, try to multiply:

$$\begin{array}{l} x + \square \\ \underline{x + \square} \end{array}$$

Fill in the boxes with positive integers whose product is 4. You could try $2 \cdot 2$ or $4 \cdot 1$:

$$\begin{array}{lll} x + 2 & \qquad & x + 4 \\ \underline{x + 2} & & \underline{x + 1} \\ x^2 + 2x & & x^2 + 4x \\ \underline{\quad + 2x + 4} & & \underline{\quad + x + 4} \\ x^2 + 4x + 4 & & x^2 + 5x + 4 \\ \times & & \checkmark \end{array}$$

The correct coefficient of the middle term is 5. Thus

$$x^2 + 5x + 4 = (x + 4)(x + 1)$$

Observe that $4 \cdot 1 = 4$, and 4 is the constant term of $x^2 + 5x + 4$. Also, $4 + 1 = 5$, and 5 is the coefficient of $5x$ in the polynomial $x^2 + 5x + 4$. □

To factor a trinomial

$$x^2 + Mx + N,$$

where M and N are integers, try to find integers a and b such that

$$ab = N \qquad \text{and} \qquad a + b = M.$$

Then

$$\begin{array}{l} x + a \\ \underline{x + b} \\ x^2 + ax \\ \underline{\quad + bx \qquad + ab} \\ x^2 + \underbrace{(a + b)}_{M}x + \underbrace{ab}_{N} = x^2 + Mx + N \end{array}$$

In other words, you want factors, a and b, of N whose sum, $a + b$, is M.

‡For further examples of each type of factoring discussed here, see Section 1.5 of the accompanying Workbook.

One or both of the coefficients M or N may be negative. Note that N represents a product and M a sum.

The product of two positive numbers, or of two negative numbers, is positive. The product of a postive and negative number (in either order) is negative.

The sum of positive numbers is positive; the sum of negative numbers is negative. The sum of numbers unlike in sign can be positive, negative, or 0.

Table 1.1 is useful in determining the signs of a and b when factoring

$$x^2 + Mx + N, \qquad \text{that is,} \qquad x^2 + (a + b)x + ab.$$

TABLE 1.1

$M(= a + b)$	$N(= ab)$	a and b
positive	positive	both positive
negative	positive	both negative
positive	negative	one positive, one negative
negative	negative	one positive, one negative

Thus to factor $x^2 \underbrace{- 5x}_{M<0} \underbrace{+ 4}_{N>0}$, both factors of 4 must be *negative*.

$$x^2 - 5x + 4 = (x - 4)(x - 1)$$

EXAMPLE 2 Factor: $z^2 - z - 6$

Solution. $z^2 \underbrace{- 1z}_{M<0} \underbrace{- 6}_{N<0}$

One factor of -6 must be *positive;* the other factor must be *negative*. The possibilities are $6(-1)$, $(-6)\cdot 1$, $3(-2)$, $(-3)2$.

$z + 6$	$z - 6$	$z + 3$	$z - 3$
$z - 1$	$z + 1$	$z - 2$	$z + 2$
$z^2 + 6z$	$z^2 - 6z$	$z^2 + 3z$	$z^2 - 3z$
$- z - 6$	$z - 6$	$2z - 6$	$2z - 6$
$z^2 + 5z - 6$	$z^2 - 5z - 6$	$z^2 + z - 6$	$z^2 - z - 6$
×	×	×	✓

Thus the last pair, -3, 2, works.

$$z^2 - z - 6 = (z - 3)(z + 2)$$

□

factoring $Lx^2 + Mx + N$

Now consider trinomials

$$Lx^2 + Mx + N,$$

where L, M, and N are integers and where $L \neq 1$.

EXAMPLE 3 Factor: $2x^2 + 5x + 2$

Solution. First consider the factors of $2x^2$. Clearly

$$2x \cdot x = 2x^2$$

Thus consider:

$$\begin{array}{r} 2x + \square \\ \underline{x + \square} \\ 2x^2 + \end{array}$$

Try to fill in the boxes with the factors 1 and 2 of the constant term, 2. Observe that there are two possibilities:

$$\begin{array}{rl} 2x + 2 & \\ \underline{x + 1} & \\ 2x^2 + 2x & \\ \underline{ 2x + 2} & \\ 2x^2 + 4x + 2 & \\ \times & \end{array} \qquad \begin{array}{rl} 2x + 1 & \\ \underline{x + 2} & \\ 2x^2 + x & \\ \underline{ 4x + 2} & \\ 2x^2 + 5x + 2 & \\ \checkmark & \end{array}$$

Thus

$$2x^2 + 5x + 2 = (2x + 1)(x + 2)$$ □

To factor a polynomial in two variables, such as

$$6x^2 - 17xy + 5y^2,$$

it is helpful to replace y by 1 and first factor

$$6x^2 - 17x + 5$$

EXAMPLE 4 Factor: (a) $6x^2 - 17x + 5$ (b) $6x^2 - 17xy + 5y^2$

Solution.

(a) $6x^2 \underbrace{- 17x}_{M<0} \underbrace{+ 5}_{N>0}$

Both factors of 5 must be negative. Try all combinations for

$$\begin{array}{r} 6x - \square \\ \underline{x - \square} \end{array} \quad \text{and for} \quad \begin{array}{r} 3x - \square \\ \underline{2x - \square} \end{array}$$

$$\begin{array}{r} 6x - 5 \\ \underline{x - 1} \\ 6x^2 - 5x \\ \underline{- 6x + 5} \\ 6x^2 - 11x + 5 \\ \times \end{array} \qquad \begin{array}{r} 6x - 1 \\ \underline{x - 5} \\ 6x^2 - x \\ \underline{-30x + 5} \\ 6x^2 - 31x + 5 \\ \times \end{array}$$

$$\begin{array}{r} 3x - 5 \\ \underline{2x - 1} \\ 6x^2 - 10x \\ \underline{- 3x + 5} \\ 6x^2 - 13x + 5 \\ \times \end{array} \qquad \begin{array}{r} 3x - 1 \\ \underline{2x - 5} \\ 6x^2 - 2x \\ \underline{- 15x + 5} \\ 6x^2 - 17x + 5 \\ \checkmark \end{array}$$

Thus $6x^2 - 17x + 5 = (3x - 1)(2x - 5)$

(b) Modify the preceding factors:

$$\begin{array}{l} 3x \;-\; y \\ 2x \;-\; 5y \\ \hline 6x^2 - \;\; 2xy \\ \qquad - 15xy + 5y^2 \\ \hline 6x^2 - 17xy + 5y^2 \end{array}$$

□

factoring by grouping

Grouping of terms often enables you to factor polynomials. Rearrange the terms so as to *group together terms with a common factor*.

EXAMPLE 5 Factor: $ax + ay + bx + by$

Solution. The first two terms have the common factor a; the last two terms have the common factor b.

$$\begin{aligned} ax + ay + bx + by &= (ax + ay) + (bx + by) \\ &= a(x + y) + b(x + y) \end{aligned}$$

Note that $a(x + y)$ and $b(x + y)$ also have a common factor—namely, $x + y$.

$$a(x + y) + b(x + y) = (a + b)(x + y)$$

Thus

$$ax + ay + bx + by = (a + b)(x + y)$$

You could also have begun by grouping as indicated.

$$\begin{aligned} ax + ay + bx + by &= (ax + bx) + (ay + by) \\ &= (a + b)x + (a + b)y \\ &= (a + b)(x + y) \end{aligned}$$

Both groupings lead to the same factorization. □

EXAMPLE 6 Factor: $36 + a^2x^2 - 9a^2 - 4x^2$

Solution. First reorder the terms.

$$\begin{aligned} 36 + a^2x^2 - 9a^2 - 4x^2 &= (36 - 9a^2) + (-4x^2 + a^2x^2) \\ &= 9(4 - a^2) - x^2(4 - a^2) \\ &= (9 - x^2)(4 - a^2) \\ &= (3 + x)(3 - x)(2 + a)(2 - a) \end{aligned}$$

□

sum and difference of cubes

The factoring method

$$a^2 - b^2 = (a + b)(a - b)$$

is known as *the difference of squares*. However, *the sum of squares*,

$$a^2 + b^2,$$

cannot be factored (as a product of polynomials with integral coefficients).

In the case of cubes *you can factor both the sum and difference of cubes*. Observe that in the following products, the middle terms cancel:

$$\begin{array}{rrrr} a^2 & - ab & + b^2 & \\ a & + b & & \\ \hline a^3 & - a^2b & + ab^2 & \\ & + a^2b & - ab^2 & + b^3 \\ \hline a^3 & & & + b^3 \end{array} \qquad \begin{array}{rrrr} a^2 & + ab & + b^2 & \\ a & - b & & \\ \hline a^3 & + a^2b & + ab^2 & \\ & - a^2b & - ab^2 & - b^3 \\ \hline a^3 & & & - b^3 \end{array}$$

Thus

$$a^3 + b^3 = (a + b)(a^2 - ab + b^2)$$
$$a^3 - b^3 = (a - b)(a^2 + ab + b^2)$$

In both cases, the *binomial* factor has the *same signs* as the original; the *trinomial* factor has the *opposite sign* in the *middle* position.

EXAMPLE 7 Factor: $x^3 + 8$

Solution. $8 = 2^3$. Here $a = x$ and $b = 2$. Thus

$$x^3 + 8 = (x + 2)(x^2 - 2x + 4)$$ □

You will be factoring polynomials of the form

$$x^3 + (ab)^3, \qquad (xy)^3 - (ab)^3, \qquad \text{and so on.}$$

As was the case for squares,

the product of the cubes equals the cube of the product.

$$\begin{aligned} a^3b^3 &= (a \cdot a \cdot a)(b \cdot b \cdot b) \\ &= (ab)(ab)(ab) \\ &= (ab)^3 \end{aligned}$$

EXAMPLE 8

$$\begin{aligned} x^3 - 27a^3 &= x^3 - (3a)^3 \\ &= (x - 3a)[x^2 + 3ax + (3a)^2] \\ &= (x - 3a)(x^2 + 3ax + 9a^2) \end{aligned}$$ □

Note that

$$(a^2)^3 = a^2 \cdot a^2 \cdot a^2 = a^{2+2+2} = a^{3 \cdot 2}$$

Also,

$$(a^3)^3 = a^3 \cdot a^3 \cdot a^3 = a^{3+3+3} = a^{3 \cdot 3}$$

Similarly,

$$(a^n)^3 = a^{3n}$$

for every (positive) integer n. Thus a^{3n} is a cube. For example,

$$a^{15} = (a^5)^3, \qquad a^{27} = (a^9)^3, \qquad 8a^{12} = 2^3(a^4)^3 = (2a^4)^3$$

Thus, the present methods apply to binomials of the forms

$$a^{3m} + b^{3n} \; [\text{or } (a^m)^3 + (b^n)^3] \qquad \text{and}$$
$$a^{3m} - b^{3n} \qquad [\text{or } (a^m)^3 - (b^n)^3].$$

EXAMPLE 9 $$8a^3 + b^{24} = (2a)^3 + (b^8)^3$$
$$= (2a + b^8)[(2a)^2 - 2ab^8 + (b^8)^2]$$
$$= (2a + b^8)(4a^2 - 2ab^8 + b^{16})$$ □

EXERCISES

In Exercises 1–20 factor each trinomial.

1. $x^2 + 3x + 2$ **2.** $a^2 + 7a + 6$ **3.** $x^2 + 4x + 4$ **4.** $x^2 + 6x + 9$

5. $m^2 + m - 2$ **6.** $n^2 - n - 2$ **7.** $s^2 + s - 12$ **8.** $t^2 - 8t + 12$

9. $a^2 - 10a + 25$ **10.** $z^2 + 5z - 14$ **11.** $2x^2 + 3x + 1$ **12.** $2a^2 + 9a + 9$

13. $2m^2 + 5m - 3$ **14.** $2a^2 + 7a - 4$ **15.** $9x^2 - 6x + 1$ **16.** $4z^2 + 4z + 1$

17. $x^2 + 2xy + y^2$ **18.** $a^2 - 2ab + b^2$ **19.** $4u^2 + 4uv + v^2$ **20.** $r^2 + 3rs - 10s^2$

In Exercises 21–28 factor by first grouping together terms with a common factor.

21. $cu + cv + du + dv$ **22.** $ax - ay - bx + by$ **23.** $y^2 + by - ay - ab$

24. $cx - dx + 3cy - 3dy$ **25.** $4ax + 4ay - 6bx - 6by$ **26.** $5ay + 5az - 2by - 2bz$

27. $a^2 + 5a + ab + 5b$ **28.** $uv + u + v + v^2$

In Exercises 29–36 factor each binomial. (In Exercises 33 and 34 first use the difference of squares.)

29. $a^3 - 1$ **30.** $x^3 + 1000$ **31.** $x^3 - 8z^3$ **32.** $1000x^3 + 27y^3$

33. $y^6 - z^6$ **34.** $x^6 - 64$ **35.** $1\,000\,000 - z^9$ **36.** $x^{21}y^{24} + 1$

In Exercises 37–60 factor each polynomial, if possible. If the polynomial cannot be factored by any of the present methods, write "DOES NOT FACTOR".

37. $5a^2 + 20a + 20$ **38.** $6x^2 + 33x - 18$

39. $x^2 + x + 1$ **40.** $a^3 + a^2 + ab + b$

41. $45x^2 + 30x + 5$ **42.** $8ax^2 - 18ay^2 + 4bx^2 - 9by^2$

43. $-2y^2 - 24y - 70$ **44.** $a^3x^2 - 1000x^2$

45. $2b^2 + 3b + 2$ **46.** $144a^2y^2 - 36a^2z^2 - 64b^2y^2 + 16b^2z^2$

47. $a^2x^3 - b^2x^3 + a^2y^3 - b^2y^3$ **48.** $9x^4 + 27x^3 + 18x^2$

49. $8y^4 + 20y^3 - 100y^2$ **50.** $a^2 + 3ab - 4b^2$

51. $ax^2 + axy + ay^2 + bx^2 + bxy + by^2$ **52.** $2x^2 + 2x + 1$

53. $20x^6y^6 + 160$ **54.** $100x^5 - 700x^4 + 600x^3$

55. $a^2x^2 + 2a^2x + a^2 - b^2x^2 - 2b^2x - b^2$ **56.** $u^2 + 5uv - 14v^2$

57. $a^9b^9c^9d^2 - d^2$ **58.** $m^2 - 4m - 4$

59. $a^2 + 2ab + 2b^2$ **60.** $a^3 - (b + 1)^3$

1.6 Rational Expressions

Division *undoes* multiplication, just as subtraction *undoes* addition. Thus if $b \neq 0$,

$$\text{then } \frac{a}{b} = c, \qquad \text{if } a = bc,$$

$$\text{and if } \frac{a}{b} = c, \qquad \text{then } a = bc$$

division by 0

Division by 0 is undefined. To understand why not, first try to divide a *nonzero* number, such as 8, by 0. Then

$$\frac{8}{0} = c \qquad \text{would mean that} \qquad 8 = 0 \cdot c = 0.$$

This is clearly false. Next, try to divide 0 by 0. Observe that

$$0 = 0 \cdot 0, \qquad 0 = 0 \cdot 1, \qquad 0 = 0 \cdot 8$$

You would obtain

$$\frac{0}{0} = 0, \qquad \frac{0}{0} = 1, \qquad \frac{0}{0} = 8.$$

There would be no unique *quotient, c,* as there is when the *divisor, b,* is nonzero.

fractions and rational numbers

A **fraction** is an expression of the form $\frac{a}{b}$, where a and b are real numbers and $b \neq 0$. Here a is called the **numerator** and b the **denominator.** A fraction expresses division of real numbers. Recall that a *rational number* is a real number that can be expressed in the fractional form $\frac{M}{N}$, where M and N are *integers* and $N \neq 0$. A rational number expresses division of integers. Thus $\frac{1}{2}$ and $\frac{-3}{8}$ are rational fractions, whereas $\frac{\pi}{2}$ and $\frac{\sqrt{3}}{4}$ are irrational fractions.

rational expressions

Definition. Let P and Q be polynominals, with $Q \neq 0$. Then $\frac{P}{Q}$, which expresses division of polynomials, is called a **rational expression.** P is called the **numerator** and Q the **denominator** of the rational expression $\frac{P}{Q}$.

EXAMPLE 1 Each of the expressions (a)–(d) is a rational expression.

(a) $\dfrac{x + 7}{x - 2}$ (b) $\dfrac{x^2 + 5xy - 3y}{x^3y}$

(c) $\dfrac{7}{x^4 + 3x^2 - 1}$. Here the numerator is the constant polynomial 7.

(d) $\dfrac{x^4 + 3x^2 - 1}{7}$. Here the denominator is the nonzero constant polynomial 7. □

Every integer M can be regarded as the rational number $\dfrac{M}{1}$. Similarly, *every polynomial P can be thought of as the rational expression* $\dfrac{P}{1}$. Thus, the polynomial $x^2 - 5x + 2$ is a rational expression (with denominator 1). Note that *every rational number is a rational expression.* For example, the rational number $\dfrac{3}{5}$ has the (constant) polynomials 3 and 5 as numerator and denominator, respectively. For that matter, every real number a, rational or irrational, can be regarded as a constant polynomial, and hence as the rational expression $\dfrac{a}{1}$.

evaluating rational expressions

To evaluate a rational expression for given values of the variables, replace the variables by the given numbers. *Each time* a variable occurs, replace it by the *same number.* (This is the way polynomials are evaluated.)

EXAMPLE 2 Evaluate $\dfrac{x^2y^2 - 2x + y}{5xy + 5}$ when $x = 2$ and $y = 3$.

Solution. Substitute 2 for each occurrence of x and **3** for each occurrence of y:

$$\frac{2^2 \cdot \mathbf{3}^2 - 2 \cdot 2 + \mathbf{3}}{5 \cdot 2 \cdot \mathbf{3} + 5} = \frac{35}{35} = 1$$

For these values of the variables, the value of the rational expression is 1. □

Because division by 0 is undefined, a rational expression $\dfrac{P}{Q}$ is undefined for values of the variable(s) that make $Q = 0$. For example,

$$\frac{x - 1}{x} \text{ is not defined when } x = 0.$$

equivalence

In Example 2, $\dfrac{35}{35}$ "reduced to" 1.

Definition. Let M and N be integers with $N \neq 0$, and let k be a nonzero integer. Then $\dfrac{M}{N}$ and $\dfrac{Mk}{Nk}$ are said to be **equivalent forms of**

the same rational number. Write

$$\frac{M}{N} = \frac{Mk}{Nk}$$

Similarly, **two rational expressions are** said to be **equivalent** if one is of the form $\frac{P}{Q}$ and the other is of the form $\frac{P \cdot R}{Q \cdot R}$, where R is a nonzero polynomial. Write

$$\frac{P}{Q} = \frac{P \cdot R}{Q \cdot R}$$

Thus, an equivalent form of a rational number (or rational expression) is obtained if the numerator and denominator are each multiplied or divided by the same nonzero integer (or polynomial).‡

EXAMPLE 3 (a) $$\frac{2}{3} = \frac{2 \cdot 5}{3 \cdot 5} = \frac{10}{15}$$

Notice that the "cross-products", $\frac{2}{3} \times \frac{10}{15}$, are equal:

$$2 \cdot 15 = 10 \cdot 3$$

(b) $$\frac{1}{x+1} = \frac{1 \cdot 4}{(x+1) \cdot 4} = \frac{4}{4x+4}$$

Again the cross-products, $\frac{1}{x+1} \times \frac{4}{4x+4}$, are equal:

$$1(4x + 4) = 4(x + 1)$$ □

In general, let $\frac{A}{C}$ and $\frac{B}{D}$ represent rational numbers (or rational expressions).

If $\frac{A}{C} = \frac{B}{D}$, then $A \cdot D = B \cdot C$

And if $A \cdot D = B \cdot C$, then $\frac{A}{C} = \frac{B}{C}$

Similarly, an equivalent fraction is obtained if both the numerator and denominator are multiplied or divided by the same nonzero real number. Thus

$$\frac{\pi}{2} = \frac{2\pi}{4}$$

‡According to this definition,

$$\frac{1}{x^2+1} = \frac{x}{(x^2+1) \cdot x}$$

Observe that $\frac{1}{x^2+1}$ is defined for *all* values of x because its denominator, $x^2 + 1$, is never 0. However, $\frac{x}{(x^2+1) \cdot x}$ is not defined when $x = 0$ because then its denominator is 0. When the notion of a *function* is introduced (in Chapter 3) this distinction will be important.

Let a and b be *positive real numbers,* with $b \neq 0$. The following are *equivalent* ways of expressing the same negative fraction.

$$\frac{-a}{b} \quad (\text{“}-\text{” in the numerator})$$

$$\frac{a}{-b} \quad (\text{“}-\text{” in the denominator})$$

$$-\frac{a}{b} \quad (\text{“}-\text{” before the entire expression})$$

(The usual practice is to place “$-$” either in the numerator or else before the entire expression.) Observe that

$$\frac{-a}{b} = \frac{(-a)(-1)}{b(-1)} = \frac{a}{-b}.$$

In fact,

$$\frac{-a}{b} = \frac{a}{-b} = -\frac{a}{b}$$

dividing powers

To **simplify** (the form of) **a rational number (or rational expression),** write an equivalent expression in which numerator and denominator have no common factors other than 1 or -1. The following rules for dividing powers of a *nonzero* number or variable a apply:

Let m and n be positive integers. When you divide a^m by a^n, that is,

$$\frac{a^m}{a^n},$$

divide by as many a's as are common to both numerator and denominator. Thus consider:

$$\frac{\overbrace{a \cdot a \cdot a \ldots a}^{m \text{ times}}}{\underbrace{a \cdot a \cdot a \ldots a}_{n \text{ times}}}$$

1. If $m > n$, all of the factors a will divide out from the denominator. Each time you divide by a, replace it by 1:

$$\frac{\overbrace{\overbrace{\overset{1 \cdot 1 \cdot 1 \ldots 1}{\cancel{a} \cdot \cancel{a} \cdot \cancel{a} \ldots \cancel{a}}}^{n \text{ times}} \cdot \overbrace{a \cdot a \cdot a \ldots a}^{(m-n) \text{ times}}}^{m \text{ times}}}{\underbrace{\underset{1 \cdot 1 \cdot 1 \ldots 1}{\cancel{a} \cdot \cancel{a} \cdot \cancel{a} \ldots \cancel{a}}}_{n \text{ times}}}$$

The denominator is the product of 1's—hence it is 1. The numerator is 1 times the product of a's, with $(m - n)$ such factors. Hence the quotient is

$$\frac{a^{m-n}}{1} \quad \text{or} \quad a^{m-n}.$$

△

Thus if $m > n$, then $\dfrac{a^m}{a^n} = a^{m-n}$

2. If $m = n$, all factors a divide out from both numerator and denominator, each of which reduces to 1. The resulting quotient is therefore 1:

$$\frac{\overbrace{\overset{1}{\not a} \cdot \overset{1}{\not a} \cdot \overset{1}{\not a} \ldots \overset{1}{\not a}}^{m \text{ times}}}{\underbrace{\underset{1}{\not a} \cdot \underset{1}{\not a} \cdot \underset{1}{\not a} \ldots \underset{1}{\not a}}_{m(=n) \text{ times}}} \qquad \triangle$$

Thus, as you would expect,

$$\frac{a^m}{a^m} = 1$$

3. If $m < n$, then all factors a divide out from the numerator. But $(n - m)$ factors remain in the denominator:

$$\frac{\overbrace{\overset{1}{\not a} \cdot \overset{1}{\not a} \cdot \overset{1}{\not a} \ldots \overset{1}{\not a}}^{m \text{ times}}}{\underbrace{\underbrace{\underset{1}{\not a} \cdot \underset{1}{\not a} \cdot \underset{1}{\not a} \ldots \underset{1}{\not a}}_{m \text{ times}} \cdot \underbrace{a \cdot a \cdot a \ldots a}_{(n-m) \text{ times}}}_{n \text{ times}}} \qquad \triangle$$

Thus if $m < n$, then $\dfrac{a^m}{a^n} = \dfrac{1}{a^{n-m}}$

EXAMPLE 4 (a) $\dfrac{5^7}{5^4} = 5^{7-4} = 5^3$ (b) $\dfrac{3^6}{3^6} = 1$ (c) $\dfrac{11^5}{11^{10}} = \dfrac{1}{11^{10-5}} = \dfrac{1}{11^5}$ □

When both numerator and denominator are monomials, apply these rules in order to simplify this type of rational expression.

EXAMPLE 5

$$\frac{9x^4y^2}{6x^2y^3} = \frac{3x^{4-2}}{2y^{3-2}} = \frac{3x^2}{2y}$$

Note that you divided both the numerator and the denominator of $\dfrac{9x^4y^2}{6x^2y^3}$ by $3x^2y^2$. □

The rules for exponents can be applied to polynomial factors common to numerator and denominator.

EXAMPLE 6

$$\frac{8(a+b)^2(x-1)^5}{2(a+b)^4(x-1)^3} = \frac{4(x-1)^{5-3}}{(a+b)^{4-2}} = \frac{4(x-1)^2}{(a+b)^2} \qquad \square$$

factoring Factoring techniques can be employed to help simplify rational expressions.

EXAMPLE 7

$$\frac{3b^2c + 6b}{3bc} = \frac{3b(bc + 2)}{3bc} = \frac{bc + 2}{c} \qquad \square$$

EXAMPLE 8

$$\frac{x^2 - y^2}{(x + y)^2} = \frac{(x + y)(x - y)}{(x + y)^2} = \frac{x - y}{x + y} \qquad \square$$

EXAMPLE 9

$$\frac{x^2 - 5x + 6}{x^2 - 4} = \frac{(x - 3)(x - 2)}{(x + 2)(x - 2)} = \frac{x - 3}{x + 2} \qquad \square$$

In calculus you will frequently have to simplify complicated rational expressions. Example 10, which follows, illustrates this technique. (The problem arises in ''differentiating'' a quotient.)

EXAMPLE 10 Simplify: $\dfrac{(3x + 2)^2 \cdot 2 - (2x - 1) \cdot 2(3x + 2)}{(3x + 2)^4}$

Solution. First observe that $2(3x + 2)$ is a common factor of the numerator. Thus divide numerator and denominator by $3x + 2$.

$$\frac{(3x + 2)^2 \cdot 2 - (2x - 1) \cdot 2(3x + 2)}{(3x + 2)^4} = \frac{2(3x + 2)[(3x + 2) - (2x - 1)]}{(3x + 2)^4} = \frac{2[x + 3]}{(3x + 2)^3} \quad \text{or} \quad \frac{2x + 6}{(3x + 2)^3} \qquad \square$$

EXERCISES

Express positive fractions without minus signs; express negative fractions with the minus sign in the numerator.

In Exercises 1–12 simplify each fraction.

1. $\frac{2}{4}$ **2.** $\frac{-3}{-9}$ **3.** $\frac{8\pi}{-32}$ **4.** $-\frac{14}{20}$ **5.** $-\frac{-15}{-24}$ **6.** $\frac{-400}{-650}$ **7.** $\frac{2^6}{2^3}$

8. $-\frac{3^4}{-3^4}$ **9.** $\frac{5^5}{5^7}$ **10.** $\frac{5^2 \cdot 7}{5^3}$ **11.** $\frac{3^2 \cdot 5 \cdot \pi}{3 \cdot 5^2 \cdot \pi}$ **12.** $\frac{2^8 \cdot 5^2 \cdot 7^4}{2^7 \cdot 3 \cdot 5 \cdot 7^5}$

In Exercises 13–18 evaluate each rational expression for the specified value(s) of the variable(s).

13. $\frac{3}{x - 1}$ when $x = 2$ **14.** $\frac{x + 2}{x - 3}$ when $x = 1$ **15.** $\frac{u^4 - 3u^3 + 5u^2 - 2}{u^3}$ when $u = -1$

16. $\frac{1}{xy}$ when $x = 1, y = 2$ **17.** $\frac{x}{y + 1}$ when $x = 5, y = 7$ **18.** $\frac{4x^2 + 3x - 5}{y^2 + 2y + 1}$ when $x = 0, y = 10$

19. Evaluate $\frac{x^2 - 2}{x^2 + 2}$ when (a) $x = 0$, (b) $x = 1$, (c) $x = -1$.

20. Evaluate $\frac{1}{x^2 + 3y}$ when (a) $x = 2$, $y = -2$; (b) $x = -2$, $y = 2$.

21. The acceleration (or change in velocity) of a body is given by $\frac{v - u}{t}$, where u is its initial velocity (in meters per second), v is its final velocity (in meters per second), and t is the time in motion (in seconds). Determine the acceleration if $v = 88$, $u = 52$, and $t = 4$.

22. An object is attracted to the earth by a force of magnitude $\frac{500\,000\,000}{h^2}$ (kilograms), where h is the distance (in kilometers) of the object from the center of the earth. Determine the magnitude of the force when (a) $h = 5000$, (b) $h = 10\,000$.

23. The number of ties a salesman can sell is given by $\frac{4000}{x - a}$, where a is the cost and x the sales price (in dollars). How many ties can be sold if (a) $x = 6$, $a = 4$; (b) $x = 8$, $a = 4$?

24. An ice cream plant is able to produce x gallons of chocolate ice cream and y gallons of vanilla ice cream per day, where $y = \frac{800 - 2x}{10 + x}$. Determine the number of gallons of vanilla ice cream produced on a day when 40 gallons of chocolate ice cream are made.

25. **(a)** The number of gallons of paint required for a room that is l feet long, w feet wide, and h feet high is given by

$$\frac{lw + 2lh + 2wh}{g}$$

where 1 gallon covers g square feet of surface. How many gallons are needed to paint a room that is 18 feet long, 12 feet wide, and 8 feet high if 1 gallon covers 48 square feet of surface?

(b) If the paint costs \$12.50 per gallon, how much does it cost to paint this room?

26. The coefficient of linear expansion, E, of a metallic substance indicates the fractional change in the length of a solid rod of the substance per degree change in its (Celsius) temperature. E is given by

$$\frac{L - L_0}{L_0 t}$$

where L_0 is the length of the rod at the freezing point of water (0°C) and L is its length at any other temperature t. Find the coefficient of linear expansion of steel if a steel bridge that is 2000 meters long at 0°C expands to 2000.22 meters when the temperature is 10°C.

In Exercises 27–62 simplify each rational expression. Leave the answer in factored form and with the minus sign, if any remains, in the numerator.

27. $\frac{a^3bc^2}{bc}$

28. $\frac{9m^2n^3}{6mn}$

29. $\frac{-4a^2b^2c}{6a^2b^3c^2}$

30. $\frac{24r^9s^7t}{18r^8s^8t^8}$

31. $\frac{-9a^4x^2}{-18a^5x^2}$

32. $\frac{12a^2bc^3}{-12a^2bc^3}$

33. $\frac{(a + b)^2}{a + b}$

34. $\frac{x^4(y + z)}{x^2(y + z)^2}$

35. $\frac{(x + y)(m + n)}{(x - y)(m + n)^2}$

36. $\frac{16(m - n)^2(p + q)^4(r - s)^3(u + v)^2}{20(m - n)(p + q)(r - s)^{10}}$

37. $\frac{a^2b^2 + a^2b}{a}$

38. $\frac{a^2b^2 + a^2b}{ab}$

39. $\dfrac{z}{y^3z^5 + yz}$

40. $\dfrac{x^3y^2 + x^2y^3}{x^3}$

41. $\dfrac{9a^3b^2 + 6b^3}{-3b^2}$

42. $\dfrac{3a^2b^2 + 6ab^4}{9ab}$

43. $\dfrac{m^2}{5m^2n + m^2p - m^2}$

44. $\dfrac{14a^4b^3c^3 + 16a^2b^3c^2 - 28ab^2c^3}{4a^2b^2c^3}$

45. $\dfrac{6a - 6b}{6a + 6b}$

46. $\dfrac{a^2 - 4}{a - 2}$

47. $\dfrac{4m^2 - 4n^2}{2(m + n)^2}$

48. $\dfrac{s^2 + 9s + 8}{(s + 8)(s - 1)}$

49. $\dfrac{a^2 + 7a + 10}{a^2 + 4a + 4}$

50. $\dfrac{x - a}{(a - x)^2}$

51. $\dfrac{a^3 - z^3}{z^3 - a^3}$

52. $\dfrac{y^4 - 1}{y^6 - 1}$

53. $\dfrac{a^4 - 16}{3a^2 + 12}$

54. $\dfrac{ab + ac + b + c}{b^2 - c^2}$

55. $\dfrac{r^2 - s^2}{sx - rx - sy + ry}$

56. $\dfrac{x^8 - y^8}{(x^4 + y^4)(x - y)^3}$

57. $\dfrac{(2x - 5)\cdot 5 - (5x + 1)^2}{(2x - 5)^2}$

58. $\dfrac{(t^2 + 1)\cdot 2t - (t^2 - 2)\cdot 2t}{(t^2 + 1)^2}$

59. $\dfrac{(x^2 + x + 1) - (x - 3)(2x + 1)}{(x^2 + x + 1)^2}$

60. $\dfrac{(x + 1)^2\cdot 3x^2 - x^3\cdot 2(x + 1)}{(x + 1)^4}$

61. $\dfrac{(2x - 1)^2\cdot 2x - (x^2 + 4)\cdot 2(2x - 1)\cdot 2}{(2x - 1)^4}$

62. $\dfrac{(5x + 1)^3\cdot 3x^2 - (x^3 + 1)\cdot 3(5x + 1)^2\cdot 5}{(5x + 1)^6}$

1.7 Multiplication and Division of Rational Expressions

Arithmetic operations can be defined for rational expressions, just as they were for polynomials. First, multiplication and division of rational numbers will be defined. These definitions then carry over to multiplication and division of rational expressions.

multiplication of rational numbers

As you may know, the area of a rectangle equals its length times its width. Let

$$A = \text{area}, \qquad l = \text{length}, \qquad w = \text{width}.$$

Then $A = lw$. [See Figure 1.3 .]

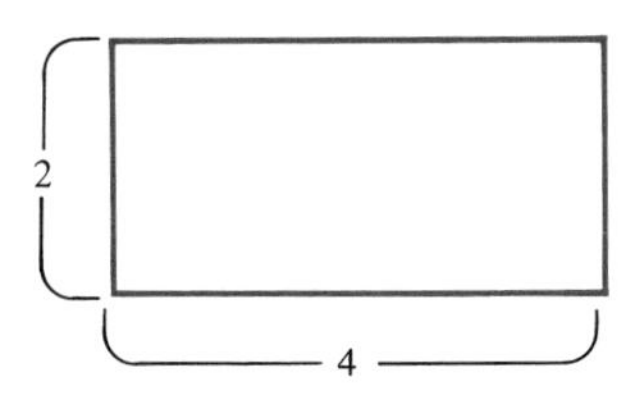

FIGURE 1.3 The area of the rectangle is 4·2, or 8, square units.

Now consider a square each of whose sides is of length 1 inch and whose area is therefore 1 square inch. Subdivide the square into 15 smaller rectangles, as in Figure 1.4 . The area of each small rectangle is $\frac{1}{15}$ square inch. The colored rectangle with dimensions $\frac{2}{3}$ inch $\times$ $\frac{2}{5}$ inch includes 4 small rectangles. It covers $\frac{4}{15}$ of the area

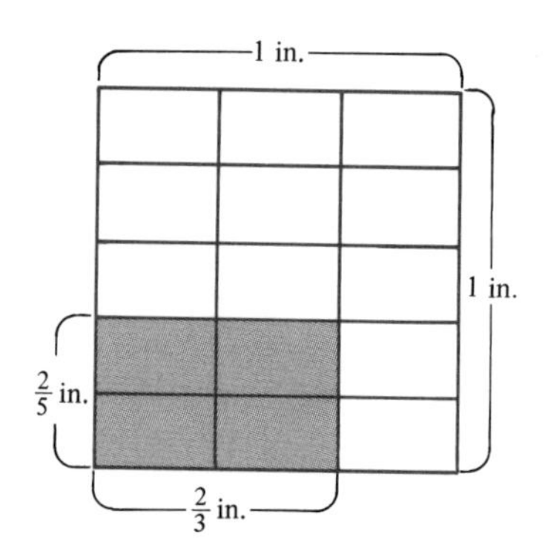

FIGURE 1.4

of the square. Thus its area should be $\frac{4}{15}$ square inch. Because

$$lw = A$$

it seems reasonable that

$$\frac{2}{3}\cdot\frac{2}{5} = \frac{4}{15}$$

In general, you can define **multiplication of rational numbers** by

$$\frac{a}{b}\cdot\frac{c}{d} = \frac{ac}{bd}$$

EXAMPLE 1

$$\frac{4}{9}\cdot\frac{3}{2} = \frac{4\cdot 3}{9\cdot 2} = \frac{12}{18} = \frac{2}{3}$$

In practice, *it is generally best to divide numerator and denominator by common factors before multiplying:*

$$\frac{\overset{2}{\cancel{4}}}{\underset{3}{\cancel{9}}}\cdot\frac{\overset{1}{\cancel{3}}}{\underset{1}{\cancel{2}}} = \frac{2}{3}$$

□

EXAMPLE 2

$$\frac{-96}{25}\cdot\frac{125}{-54} = \frac{-2^5\cdot 3}{5^2}\cdot\frac{5^3}{-2\cdot 3^3} = \frac{2^4\cdot 5}{3^2} = \frac{80}{9}$$

□

multiplication by the reciprocal

Definition. The **reciprocal of a nonzero number** a is defined to be $\frac{1}{a}$.

Because $a = \frac{a}{1}$, it follows that

$$a\cdot\frac{1}{a} = \frac{a}{1}\cdot\frac{1}{a} = 1$$

By the Commutative Law of Multiplication,

$$\frac{1}{a}\cdot a = 1$$

Thus *the reciprocal of a is the number multiplied by a to obtain 1.* (For this reason, the reciprocal of a is often called the "multiplicative inverse of a". Recall that $-a$, the (additive) inverse of a, is the number *added* to a to obtain 0.) Observe that for any number a and for $b \neq 0$,

$$\frac{a}{b} = \frac{a}{1}\cdot\frac{1}{b} = a\cdot\frac{1}{b}$$

Thus *division by b is the same as multiplication by* $\frac{1}{b}$.

Next, observe that

$$\frac{c}{d}\cdot\frac{d}{c} = \frac{cd}{dc} = 1$$

Thus by the definition of division of real numbers,

$$1 \div \frac{c}{d} = \frac{d}{c}$$

The reciprocal of $\frac{c}{d}$ can be written as $\frac{1}{\frac{c}{d}}$, which means $1 \div \frac{c}{d}$. Thus *the reciprocal of a rational number is obtained by inverting it, that is, by interchanging numerator and denominator.*

division of rational numbers

In order to discover the general rule for dividing by a rational number, fill in the blanks in Tables 1.2 and 1.3 by considering the pattern in the right-hand column of each table.

TABLE 1.2

$8 \div 8 =$	1
$8 \div 4 =$	2
$8 \div 2 =$	4
$8 \div 1 =$	8
$8 \div \frac{1}{2} =$	
$8 \div \frac{1}{4} =$	
$8 \div \frac{1}{8} =$	

TABLE 1.3

$54 \div 18 =$	3
$54 \div 6 =$	9
$54 \div 2 =$	27
$54 \div \frac{2}{3} =$	

Observe:

$$8 \div \frac{1}{2} = 8\cdot\frac{2}{1}, \quad 8 \div \frac{1}{4} = 8\cdot\frac{4}{1}, \quad 8 \div \frac{1}{8} = 8\cdot\frac{8}{1}, \quad \text{and}$$

$$54 \div \frac{2}{3} = 54\cdot\frac{3}{2}$$

Define **division of rational numbers** as follows:

$$\frac{a}{b} \div \frac{c}{d} = \frac{a}{b}\cdot\frac{d}{c} = \frac{ad}{bc}$$

Thus *to divide rational numbers, invert the divisor and multiply.*

EXAMPLE 3

$$\begin{aligned}\frac{12}{-25} \div \frac{42}{-5} &= \frac{12}{-25}\cdot\frac{-5}{42} \\ &= \frac{2^2\cdot 3}{-5^2}\cdot\frac{-5}{2\cdot 3\cdot 7} \\ &= \frac{2}{5\cdot 7} \\ &= \frac{2}{35}\end{aligned}$$

□

multiplication of rational expressions

The same concepts apply to rational expressions.

Definition. Let P, Q, R, S be polynomials with $Q \not\equiv 0$, $S \not\equiv 0$. Define

$$\frac{P}{Q}\cdot\frac{R}{S} = \frac{P\cdot R}{Q\cdot S}$$

Leave the resulting rational expression in factored form.

EXAMPLE 4

$$\frac{a^4}{2b^2}\cdot\frac{x+y}{x-y} = \frac{a^4(x+y)}{2b^2(x-y)}$$ □

In the next example simplify wherever possible. Divide by factors common to the numerator and denominator before multiplying.

EXAMPLE 5

$$\frac{x^2-y^2}{4ax}\cdot\frac{2x^3}{x^2+2xy+y^2} = \frac{(x+y)(x-y)}{4ax}\cdot\frac{2x^3}{(x+y)^2}$$
$$= \frac{(x-y)\cdot x^2}{2a(x+y)}$$ □

Products such as the following occur in calculus problems.

EXAMPLE 6 Simplify: $2\cdot\dfrac{2x-1}{x^2+1}\cdot\dfrac{(x^2+1)\cdot 2-(2x-1)\cdot 2x}{(x^2+1)^2}$

Solution.

$$2\cdot\frac{2x-1}{x^2+1}\cdot\frac{(x^2+1)\cdot 2-(2x-1)\cdot 2x}{(x^2+1)^2}$$
$$= \frac{2(2x-1)\cdot 2[x^2+1-(2x^2-x)]}{(x^2+1)^3}$$
$$= \frac{4(2x-1)[-x^2+x+1]}{(x^2+1)^3}$$

Leave the numerator in factored form. □

Note that for *nonzero* polynomials P and Q,

$$\frac{P}{Q}\cdot\frac{Q}{P} = 1$$

Therefore the **reciprocal of** $\dfrac{P}{Q}$, or $\dfrac{1}{\frac{P}{Q}}$, is $\dfrac{Q}{P}$. You obtain the reciprocal of $\dfrac{P}{Q}$ by inverting this rational expression. In particular, the polynomial P can be considered as the rational expression $\dfrac{P}{1}$. Thus the reciprocal of P is $\dfrac{1}{P}$. For example, the reciprocal of $\dfrac{5x+1}{2x-3}$ is $\dfrac{2x-3}{5x+1}$, and the reciprocal of $3x^2+5x+9$ is $\dfrac{1}{3x^2+5x+9}$.

division of rational expressions The rule for **division of rational expressions** is the same as for division of rational numbers: *Invert the divisor and multiply.*

$$\frac{P}{Q} \div \frac{R}{S} = \frac{P}{Q} \cdot \frac{S}{R} = \frac{P \cdot S}{Q \cdot R}$$

EXAMPLE 7

$$\frac{x^2y^3z}{abc} \div \frac{xyz^2}{a^2b} = \frac{x^2y^3z}{abc} \cdot \frac{a^2b}{xyz^2} = \frac{xy^2a}{cz}$$ □

EXAMPLE 8

$$\begin{aligned}\frac{27x^2 - 27a^2}{x^2 + a^2} \div \frac{3a - 3x}{x + a} &= \frac{27x^2 - 27a^2}{x^2 + a^2} \cdot \frac{x + a}{3a - 3x} \\ &= \frac{27(x + a)(x - a)}{x^2 + a^2} \cdot \frac{x + a}{-3(x - a)} \\ &= \frac{-9(x + a)^2}{x^2 + a^2}\end{aligned}$$ □

fractions The rules for multiplication and division of rational numbers also apply to fractions in general. Thus

$$\frac{\pi}{2} \cdot \frac{1}{2} = \frac{\pi}{4} \quad \text{and} \quad \frac{2\pi}{3} \div \frac{\pi}{2} = \frac{2\pi}{3} \cdot \frac{2}{\pi} = \frac{4}{3}$$

EXERCISES

In Exercises 1–8 multiply, as indicated. Simplify the resulting fractions.

1. $\frac{3}{5} \cdot \frac{1}{2}$
2. $\frac{-2}{3} \cdot \frac{-4}{9}$
3. $\frac{12}{25} \cdot \frac{-7}{24}$
4. $\frac{35}{36} \cdot \frac{9\pi}{-14}$
5. $\frac{48}{-7} \cdot \frac{-49}{18}$
6. $\frac{200}{3} \cdot \frac{63}{44}$
7. $\frac{144}{169} \cdot \frac{130}{81}$
8. $\frac{2^5 \cdot 7^2}{3} \cdot \frac{33}{16}$

In Exercises 9–24 multiply, as indicated. Simplify and leave the answers in factored form.

9. $\frac{x}{y} \cdot \frac{a}{b}$
10. $\left(\frac{x^2a^2}{4} \cdot \frac{2}{x}\right)^2$
11. $\frac{a^2b^3c}{3d} \cdot \left(\frac{bcd^2}{a^4}\right)^2$
12. $\frac{25a^2x^3}{(-7y^2z)^3} \cdot \frac{49yz}{(5ax)^2}$
13. $\frac{x + a}{x} \cdot \frac{4}{a} \cdot \frac{}{c}$
14. $\frac{a^2b}{x - 1} \cdot \frac{(x - 1)^2}{3b}$
15. $\frac{2(x + a)^2}{x - a} \cdot \frac{x + a}{6}$
16. $\frac{a^2(x - a)^3}{4b} \cdot \frac{12b^3}{a(x - a)^2}$
17. $\frac{a^2(x^2 - a^2)}{1 - x} \cdot \frac{x^2 - 1}{x + a}$
18. $\frac{x^2 + 3x + 2}{x^2 + 1} \cdot \frac{x + 1}{(x + 2)^2}$
19. $\frac{x^2 - 5x + 4}{5 - 5a^2} \cdot \frac{a^2 + 2a + 1}{x^2 - 1}$
20. $\frac{x^3a^3 - ax^3}{a^2 + 4a + 4} \cdot \frac{a^2 + 5a + 6}{a^2 - 8a + 7}$
21. $\frac{4x^2 - 12x + 9}{a^2x + a^2} \cdot \frac{x^2 - 1}{14x^2 - 21x}$
22. $\frac{5x^3 + 40}{2abc^3} \cdot \frac{ab^2 - a^2b}{(b^2 - a^2)(x + 2)^2}$
23. $\frac{4b^2}{b^2 - a^2} \cdot \frac{a^2 + 2ab + b^2}{12ab^3} \frac{(a - b)^3}{(a + b)^2}$
24. $\frac{x^2 + 8x + 15}{x^2 - 2x + 1} \frac{1 - x^2}{x^2 + 10x + 25} \cdot \frac{x^4 - 1}{1 + x^2}$

In Exercises 25–32 divide, as indicated. Simplify the resulting fractions.

25. $\frac{3}{2} \div \frac{1}{5}$
26. $-\frac{-20}{3} \div \frac{7}{12}$
27. $\frac{4}{9} \div \frac{3}{8}$
28. $\frac{4}{9} \div \frac{8}{3\pi}$

29. $\frac{-6}{25} \div \frac{36}{125}$ **30.** $\frac{24}{49} \div \frac{-12}{35}$ **31.** $\frac{98}{375} \div \frac{7}{75}$ **32.** $\frac{11 \cdot 13^4}{7^3\pi} \div \frac{7^2 \cdot 11^2 \cdot 13}{2\pi}$

In Exercises 33–45 divide, as indicated. Simplify and leave the answers in factored form.

33. $\frac{5a^2}{b} \div \frac{3}{c}$ **34.** $\frac{x^2y^2}{a} \div \frac{xa}{y}$ **35.** $\frac{4a^2bx}{9cy} \div \frac{12b^2x^2}{c^2y^2}$

36. $\frac{36m^2n^4}{25ax^3} \div \frac{-18m^3n^3}{75a^3x}$ **37.** $\frac{x-a}{x+a} \div \left(\frac{2}{x+a}\right)^2$ **38.** $\frac{3x+y}{1-x} \div \frac{3x+y}{x-1}$

39. $\frac{(x+1)^2}{x-3} \div \frac{x-3}{x+1}$ **40.** $\left(\frac{x^2+1}{x+1}\right)^2 \div \frac{x^2+1}{(x+1)^2}$ **41.** $\frac{a^2-4}{a^3} \div \frac{a^2+4a+4}{ab}$

42. $\frac{2ax-2a}{x^2+1} \div \frac{4a-4x}{(x^2+1)^2}$ **43.** $\frac{3y^4-27y^2}{4+4x^2} \div \frac{y^3+3y^2}{3x^2+3}$ **44.** $\frac{x^4-a^4}{a^2x^2+a^2} \div \frac{6x^2-6a^2}{3x^2+3}$

45. $\frac{1-y^2}{a^3-1} \div \frac{ay+a}{4a^2-8a+4}$ **46.** Divide $\frac{6x^3-18x^2}{6a^2}$ by $\frac{x^2-9}{4ax}$.

47. Divide the product of $\frac{(x+1)^2}{(x-1)^2}$ and $\frac{x^2-1}{x+1}$ by $\frac{x^2}{x-1}$.

48. The quotient of $\frac{3x+1}{x+3}$ divided by $\frac{9x^2+3x}{x^2-9}$ is multiplied by $\frac{x^2}{x-3}$. Determine the resulting product.

49. Determine the reciprocal of $\frac{x^2-2x}{(x+1)^3} \cdot \frac{x^2-1}{x^2+x-6}$.

50. What polynomial must be divided by x^2+4x+3 to obtain $\frac{1}{x+3}$?

51. What polynomial must be divided by x^2-25 to obtain $x+5$?

52. What rational expression must be divided by $\frac{x+3}{x+4}$ to obtain $\frac{(x+3)^2}{x+4}$?

In Exercises 53–56 you are given polynomials with *rational* coefficients. Factor each polynomial. (The factors will have rational coefficients.) For example, $x^2 - \frac{1}{4} = \left(x+\frac{1}{2}\right)\left(x-\frac{1}{2}\right)$

53. $a^2 - \frac{1}{16}$ **54.** $\frac{1}{4}x^2 - \frac{1}{9}$ **55.** $\frac{4y^2}{9} - \frac{1}{25}$ **56.** $\frac{u^4}{16} - \frac{1}{81}$

In Exercises 57–60 simplify each product. Leave the resulting expression in factored form.

57. $2 \cdot \frac{x+4}{x-3} \cdot \frac{(x-3)-(x+4)}{(x-3)^2}$ **58.** $2 \cdot \frac{x-1}{x^2+8} \cdot \frac{(x^2+8)-(x-1)\cdot 2x}{(x^2+8)^2}$

59. $2 \cdot \frac{x^3-x}{x^4+1} \cdot \frac{(x^4+1)(3x^2-1)-(x^3-x)4x^3}{(x^4+1)^2}$ **60.** $3 \cdot \left(\frac{2x+1}{x^2+4}\right)^2 \cdot \frac{(x^2+4)\cdot 2-(2x+1)\cdot 2x}{(x^2+4)^2}$

1.8 Addition and Subtraction of Rational Expressions

LCM of integers

In order to add and subtract rational numbers or rational expressions, find equivalent expressions having a common denominator. The key to doing this is the notion of "*least common multiple*".

Let a and b be integers with $b \neq 0$. Recall that a is called a multiple of b if $a = bc$ for some integer c. Thus *a is a multiple of b* means the same as *b is a factor of a*.

Definition. **The least common multiple (LCM)** of several integers is the smallest *positive* integer that is a multiple of each of them.

$$\text{LCM}(m_1, m_2, \ldots, m_n)$$

stands for the least common multiple of $m_1, m_2, \ldots, m_n$.

EXAMPLE 1

(a) The least common multiple of 2 and 3 is 6.

$$\text{LCM}\ (2, 3) = 6$$

For, $6 = 2 \cdot 3 = 3 \cdot 2$, and thus 6 is a multiple of 2 as well as of 3. No smaller positive integer is a multiple of *both* 2 and 3.

(b) $$\text{LCM}\ (2, 4) = 4$$

(c) $$\text{LCM}\ (-4, 6) = 12$$

In fact, 12 is the smallest *positive* integer that is a multiple of both -4 and 6. □

To find the LCM of several integers:

1. Determine their prime factorizations.
2. The LCM is the product of the *highest powers* of all primes that occur in these factorizations.

In Example 1(c), $4 = 2^2$, $6 = 2 \cdot 3$. The highest power of 2 that occurs is 2^2. The only other prime that occurs is 3; only 3^1 occurs. Thus the LCM is

$$2^2 \cdot 3 \qquad \text{or} \qquad 12$$

EXAMPLE 2

Determine: LCM (18, 45, 54)

Solution.

$$18 = 2 \cdot 3^2, \qquad 45 = 3^2 \cdot 5, \qquad 54 = 2 \cdot 3^3$$

$$\text{LCM}\ (18, 45, 54) = 2 \cdot 3^3 \cdot 5 = 270$$

□

LCM of polynomials

The same concepts apply to polynomials. Only polynomials with integral coefficients will be considered. *In the remainder of this section, "polynomial" will mean "polynomial with integral coefficients".*

Recall that when P and Q are polynomials with $Q \neq 0$, P is said to be a multiple of Q when $P = Q \cdot R$ for some polynomial R. Thus *P is a multiple of Q* means the same as *Q is a factor of P*. For example,

$$3x + 3 \text{ is a multiple of } x + 1 \text{ because } \underbrace{3x + 3}_{P} = \underbrace{(x + 1)}_{Q}\,\underbrace{3}_{R}.$$

Definition. Let $P_1, P_2, \ldots, P_n$ be polynomials. Express each polynomial in factored form. Then the product of the highest powers of all factors that occur in these factorizations is called the **least common multiple of** $P_1, P_2, \ldots, P_n$, and is written

$$\text{LCM}\,(P_1, P_2, \ldots, P_n).$$

EXAMPLE 3 Determine: LCM $(x^2 - 4, x^2 + 4x + 4)$

Solution.

$$x^2 - 4 = (x + 2)(x - 2), \qquad x^2 + 4x + 4 = (x + 2)^2$$
$$\text{LCM}\,(x^2 - 4, x^2 + 4x + 4) = (x + 2)^2(x - 2)$$ □

LCD *Definition.* The **least common denominator** (LCD) of several rational numbers or rational expressions is the LCM of the individual denominators.

EXAMPLE 4 Determine: LCD $\left(\frac{1}{4}, \frac{2}{3}, \frac{5}{18}\right)$

Solution. Determine the LCM of the *denominators:*

$$4 = 2^2, \qquad 3 = 3, \qquad 18 = 2 \cdot 3^2$$

Thus LCM $(4, 3, 18) = 2^2 \cdot 3^2 = 36$ and LCD $\left(\frac{1}{4}, \frac{2}{3}, \frac{5}{18}\right) = 36$ □

EXAMPLE 5 Determine: LCD $\left(\frac{1}{(x+1)^2}, \frac{-1}{x-3}, \frac{x}{x^2-1}\right)$

Solution. Again consider only the denominators:

$$(x + 1)^2, \qquad x - 3, \qquad x^2 - 1 = (x + 1)(x - 1)$$

Thus

$$\text{LCM}\,[(x + 1)^2, x - 3, x^2 - 1] = (x + 1)^2(x - 3)(x - 1)$$

and therefore

$$\text{LCD}\left(\frac{1}{(x+1)^2}, \frac{-1}{x-3}, \frac{x}{x^2-1}\right) = (x + 1)^2(x - 3)(x - 1)$$ □

In order to add and subtract rational numbers or rational expressions, you first rewrite them as equivalent expressions with the LCD as the denominator. For now, you will just practice writing these equivalent expressions. Afterward, you will perform the actual addition and subtraction.

EXAMPLE 6 (a) Determine: LCD $\left(\frac{3}{10}, \frac{1}{4}, \frac{-2}{5}\right)$

(b) Determine equivalent rational numbers with this LCD as the denominator.

Solution.

(a) First determine LCM (10, 4, 5).

$$10 = 2 \cdot 5, \qquad 4 = 2^2, \qquad 5 = 5$$

$$\text{LCM}(10, 4, 5) = 2^2 \cdot 5 = 20$$

$$\text{LCD}\left(\frac{3}{10}, \frac{1}{4}, \frac{-2}{5}\right) = 20$$

(b) To obtain each equivalent rational number, multiply the numerator and denominator by the appropriate number so that the equivalent form now has the LCD as its denominator.

$$\frac{3 \cdot 2}{10 \cdot 2} = \frac{6}{20}, \qquad \frac{1 \cdot 5}{4 \cdot 5} = \frac{5}{20}, \qquad \frac{-2 \cdot 4}{5 \cdot 4} = \frac{-8}{20}$$ □

In general, multiply out the numerator; this will often be necessary when you add and subtract expressions.

EXAMPLE 7 (a) Determine: $\text{LCD}\left(\frac{x}{x^2 - a^2}, \frac{a^2}{(x + a)^2}\right)$

(b) Determine equivalent rational expressions with this LCD as the denominator.

Solution.

(a) $x^2 - a^2 = (x + a)(x - a)$; $(x + a)^2$ is already factored.

$$\text{LCM}[x^2 - a^2, (x + a)^2] = (x + a)^2(x - a)$$

$$\text{LCD}\left(\frac{x}{x^2 - a^2}, \frac{a^2}{(x + a)^2}\right) = (x + a)^2(x - a)$$

(b)
$$\frac{x}{x^2 - a^2} = \frac{x}{(x + a)(x - a)} = \frac{x \cdot (x + a)}{(x + a)(x - a) \cdot (x + a)} = \frac{x^2 + ax}{(x + a)^2(x - a)}.$$

Also,

$$\frac{a^2}{(x + a)^2} = \frac{a^2 \cdot (x - a)}{(x + a)^2 \cdot (x - a)} = \frac{a^2x - a^3}{(x + a)^2(x - a)}$$ □

addition: same denominator

To add or subtract rational numbers or rational expressions with the *same denominator D:*

1. Add or subtract the numerators to obtain the numerator N.
2. Simplify the resulting expression $\frac{N}{D}$, if necessary.

EXAMPLE 8

$$\frac{3}{10} + \frac{1}{10} - \frac{9}{10} = \frac{3 + 1 - 9}{10} = \frac{-5}{10} = \frac{-1}{2}$$ □

EXAMPLE 9

$$\frac{1}{x^2} - \frac{2}{x^2} + \frac{a}{x^2} = \frac{1 - 2 + a}{x^2} = \frac{a - 1}{x^2}$$ □

addition: different denominators

To add or subtract rational numbers or rational expressions with *different denominators:*

1. Determine their LCD.
2. Determine equivalent rational numbers or rational expressions with this LCD as denominator.
3. Add or subtract the numerators.
4. Simplify the resulting expression, if necessary.

EXAMPLE 10 Determine: $\frac{1}{2} + \frac{2}{3} - 1$

Solution.

$$\text{LCD}\left(\frac{1}{2}, \frac{2}{3}, 1\right) = \text{LCM }(2, 3, 1) = 6$$

$$\frac{1}{2} = \frac{1 \cdot 3}{2 \cdot 3} = \frac{3}{6}, \qquad \frac{2}{3} = \frac{2 \cdot 2}{3 \cdot 2} = \frac{4}{6}, \qquad 1 = \frac{1}{1} = \frac{1 \cdot 6}{1 \cdot 6} = \frac{6}{6}$$

$$\frac{1}{2} + \frac{2}{3} - 1 = \frac{3}{6} + \frac{4}{6} - \frac{6}{6} = \frac{3 + 4 - 6}{6} = \frac{1}{6}$$ □

EXAMPLE 11 Determine: $\frac{a}{x^2yz^2} - \frac{b}{x^3y^2}$

Solution. Here the LCD is $x^3y^2z^2$.

$$\frac{a}{x^2yz^2} - \frac{b}{x^3y^2} = \frac{a \cdot xy}{x^2yz^2 \cdot xy} - \frac{b\ z^2}{x^3y^2 \cdot z^2} = \frac{axy - bz^2}{x^3y^2z^2}$$ □

EXAMPLE 12 Determine: $\frac{2}{x^2 - 6x + 5} - \frac{1}{x^2 - 1}$

Solution.

$$x^2 - 6x + 5 = (x - 5)(x - 1), \qquad x^2 - 1 = (x + 1)(x - 1)$$

The LCD is $(x - 5)(x - 1)(x + 1)$.

$$\frac{2}{x^2 - 6x + 5} - \frac{1}{x^2 - 1}$$
$$= \frac{2(x + 1)}{(x - 5)(x - 1)(x + 1)} - \frac{1(x - 5)}{(x + 1)(x - 1)(x - 5)}$$
$$= \frac{2(x + 1) - (x - 5)}{(x - 5)(x - 1)(x + 1)}$$
$$= \frac{2x + 2 - x + 5}{(x - 5)(x - 1)(x + 1)}$$
$$= \frac{x + 7}{(x - 5)(x - 1)(x + 1)}$$

The denominator, which involves three factors, is left in factored form. □

EXAMPLE 13 Determine: $\dfrac{1}{a^2 - ax} + \dfrac{1}{x^2 - ax}$

Solution. Note that $a - x = (-1)(x - a)$. Thus

$$a^2 - ax = a(a - x) = -a(x - a), \qquad x^2 - ax = x(x - a)$$

The LCD is $ax(x - a)$.

$$\frac{1}{a^2 - ax} + \frac{1}{x^2 - ax} = \frac{1\cdot(-x)}{-a(x-a)\cdot(-x)} + \frac{1\cdot a}{x(x-a)\cdot a}$$
$$= \frac{-x + a}{ax(x - a)}$$
$$= \frac{-(x - a)}{ax(x - a)}$$
$$= \frac{-1}{ax}$$

□

fractions The rules for addition and subtraction of rational numbers also apply to fractions in general. For example,

$$\frac{\pi}{2} + \frac{\pi}{4} = \frac{2\pi}{4} + \frac{\pi}{4} = \frac{3\pi}{4}$$

EXERCISES

In Exercises 1–8 determine the LCM of the indicated integers.

1. 4, 5 **2.** 5, 10 **3.** 12, 18 **4.** 72, 96 **5.** 2, 4, 5

6. 10, 12, 15 **7.** 50, 80, 125 **8.** 48, 72, 120

In Exercises 9–16 determine the LCM of the indicated polynomials.

9. x, y^3 **10.** x, x^2 **11.** x^2yz^2, y^2z

12. $x - a, (x - a)^2$ **13.** $x^2 - 9, ax + 3a$ **14.** $x^2 + 5x + 4, x^2 + 2x + 1$

15. $u^2 - a^2, (u + a)^3, a^2 - u^2$ **16.** $y^2 + 4y + 4, y^2 - 4, xy + y - 2x - 2$

In Exercises 17–23: (a) Determine the LCD of the indicated rational numbers. (b) Then determine equivalent rational numbers with this LCD as the denominator.

17. $\frac{1}{2}, \frac{1}{3}$ **18.** $\frac{3}{8}, \frac{-3}{4}$ **19.** $\frac{5}{8}, \frac{-1}{12}$ **20.** $\frac{1}{2}, \frac{2}{3}, \frac{-3}{4}$ **21.** $\frac{1}{4}, \frac{5}{6}, \frac{7}{12}$

22. $\frac{5}{16}, \frac{-1}{24}, \frac{11}{42}$ **23.** $\frac{8}{3\cdot 5^2}, \frac{-7}{2\cdot 3\cdot 5}, \frac{-3}{2^2\cdot 5^2}$

In Exercises 24–30: (a) Determine the LCD of the indicated rational expressions. (b) Then determine equivalent rational expressions with this LCD as the denominator.

24. $\frac{1}{a}, \frac{1}{x}$ **25.** $\frac{a}{x^2}, \frac{b}{x}$ **26.** $\frac{a}{x^2y}, \frac{b}{xy^2}$ **27.** $\frac{1}{x^2(x - a)}, \frac{-1}{x(x - a)^2}$

28. $\dfrac{1}{x^2 + 2ax + a^2}, \dfrac{1}{x^2 - a^2}$ **29.** $\dfrac{1}{9 - x^2}, \dfrac{x}{x + 3}, \dfrac{x^2}{x^2 + 6x + 9}$ **30.** $\dfrac{y + 2}{y^2 - 7y}, \dfrac{-1}{y - 7}, \dfrac{2}{7 - y}$

In Exercises 31–56 combine, as indicated.

31. $\dfrac{1}{3} + \dfrac{2}{3}$ **32.** $\dfrac{5}{12} - \dfrac{1}{12} + \dfrac{7}{12}$ **33.** $\dfrac{\pi}{3} + \dfrac{\pi}{4}$

34. $\dfrac{7}{12} + \dfrac{3}{4}$ **35.** $\dfrac{5}{16} + \dfrac{1}{24}$ **36.** $\dfrac{9}{40} + \dfrac{7}{30}$

37. $\dfrac{11}{12} - \dfrac{1}{4} + \dfrac{1}{3}$ **38.** $\dfrac{9}{10} + \dfrac{7}{20} - \dfrac{1}{40}$ **39.** $\dfrac{6}{25} + \dfrac{19}{75} + \dfrac{12}{125}$

40. $\dfrac{1}{2^3 \cdot 3^2 \cdot 5^4} + \dfrac{1}{2^5 \cdot 3 \cdot 5^3} - \dfrac{1}{2^4 \cdot 3^2 \cdot 5^2}$ **41.** $\dfrac{1}{m} + \dfrac{2}{m}$ **42.** $\dfrac{1}{m} + \dfrac{1}{n}$

43. $\left(\dfrac{5}{x^2} + \dfrac{1}{x}\right)^2$ **44.** $\left(\dfrac{5a}{b^2} - \dfrac{2b}{a^2}\right)^2$ **45.** $\dfrac{1}{a} + \dfrac{1}{b} - \dfrac{1}{c}$

46. $\dfrac{a^2}{x^2y^2} + \dfrac{b^2}{x^2z^2} + \dfrac{c^2}{y^2z^2}$ **47.** $\dfrac{1}{x - a} + \dfrac{2}{x - a}$ **48.** $\dfrac{1}{x^2 - 4} + \dfrac{2}{x - 2}$

49. $\dfrac{2}{x^2 + 5x + 4} + \dfrac{1}{x + 4}$ **50.** $\dfrac{1}{x + 3} - \dfrac{2}{x + 2}$ **51.** $\dfrac{2}{x^2 - 25} + \dfrac{1}{x + 5} - \dfrac{1}{x - 5}$

52. $\dfrac{1}{x + y} - \left(\dfrac{1}{x - y} - \dfrac{1}{x^2 - y^2}\right)$ **53.** $5x^3\left(x^2 - \dfrac{1}{x} + \dfrac{1}{x^2}\right)$ **54.** $\dfrac{a}{x^2y}\left(\dfrac{1}{a^2} + \dfrac{1}{a}\right)$

55. $\dfrac{x - 4}{x^2}\left(\dfrac{1}{x^2 - 16} - \dfrac{1}{x + 4}\right)$ **56.** $\dfrac{t^4}{t^2 - 1} \div \left(\dfrac{1}{t + 1} - \dfrac{1}{t - 1}\right)$

In Exercises 57–60 factor each polynomial. (The factors will have rational coefficients.)

57. $x^2 + x + \dfrac{1}{4}$ **58.** $x^2 + \dfrac{5}{6}x + \dfrac{1}{6}$ **59.** $x^2 - \dfrac{2}{3}x + \dfrac{1}{9}$ **60.** $x^2 + \dfrac{3}{10}x - \dfrac{1}{10}$

1.9 Complex Expressions

Definition. **Complex expressions** are rational numbers or rational expressions that contain other rational expressions either in their numerator or denominator (possibly in both).

Your task is to simplify these complex expressions. You can rewrite such expressions in terms of *division*.

EXAMPLE 1 Simplify: $\dfrac{\frac{1}{2}}{\frac{3}{4}}$

Solution 1. $\dfrac{\frac{1}{2}}{\frac{3}{4}}$ means $\dfrac{1}{2}$ *divided by* $\dfrac{3}{4}$. Thus

$$\frac{\frac{1}{2}}{\frac{3}{4}} = \frac{1}{2} \div \frac{3}{4} = \frac{1}{\not{2}_1} \cdot \frac{\not{4}^2}{3} = \frac{2}{3}$$

Solution 2. The LCD of $\frac{1}{2}$ and $\frac{3}{4}$ is 4. Multiply the numerator and denominator of the complex expression by this LCD, 4.

$$\frac{\frac{1}{2}}{\frac{3}{4}} = \frac{\frac{1}{\not{2}} \cdot \not{4}^2}{\frac{3}{\not{4}} \cdot \not{4}} = \frac{2}{3} \qquad \square$$

EXAMPLE 2

$$\frac{\frac{1}{3} + \frac{1}{6}}{9} = \left(\frac{1}{3} + \frac{1}{6}\right) \div 9 = \frac{2+1}{6} \cdot \frac{1}{9} = \frac{\not{3}^1}{\not{6}_2} \cdot \frac{1}{9} = \frac{1}{18} \qquad \square$$

EXAMPLE 3

$$\frac{1 + \frac{1}{x}}{x^2} = \left(1 + \frac{1}{x}\right) \div x^2 = \left(\frac{x}{x} + \frac{1}{x}\right) \div x^2$$

$$= \frac{x+1}{x} \cdot \frac{1}{x^2} = \frac{x+1}{x^3} \qquad \square$$

EXAMPLE 4

$$\frac{\frac{y}{y+1} + \frac{y-1}{y}}{\frac{y}{y+1} - \frac{y-1}{y}} = \left(\frac{y}{y+1} + \frac{y-1}{y}\right) \div \left(\frac{y}{y+1} - \frac{y-1}{y}\right)$$

$$= \frac{y \cdot y + (y-1)(y+1)}{y(y+1)} \div \frac{y \cdot y - (y-1)(y+1)}{y(y+1)}$$

$$= \frac{y^2 + y^2 - 1}{y(y+1)} \cdot \frac{y(y+1)}{y^2 - (y^2 - 1)}$$

$$= \frac{2y^2 - 1}{\not{y}(\not{y} + \not{1})} \cdot \frac{\not{y}(\not{y} + \not{1})}{1}$$

$$= 2y^2 - 1$$

A neater solution uses $y(y + 1)$, the LCD of the given rational expressions. Thus

$$\frac{\left(\frac{y}{y+1} + \frac{y-1}{y}\right) \cdot y(y+1)}{\left(\frac{y}{y+1} - \frac{y-1}{y}\right) \cdot y(y+1)} = \frac{y^2 + y^2 - 1}{y^2 - (y^2 - 1)}$$

$$= \frac{2y^2 - 1}{1}$$

$$= 2y^2 - 1 \qquad \square$$

EXERCISES

In Exercises 1–48 simplify.

1. $\dfrac{\frac{5}{8}}{2}$

2. $\dfrac{\frac{1}{3}}{12}$

3. $\dfrac{4}{\frac{2}{3}}$

4. $\dfrac{-8}{\frac{6}{7}}$

5. $\dfrac{\frac{\pi}{3}}{\frac{\pi}{6}}$

6. $\dfrac{\frac{3}{5}}{\frac{-9}{25}}$

7. $\dfrac{\frac{3}{4}}{\frac{15}{8}}$

8. $\dfrac{\frac{7}{10}}{\frac{21}{40}}$

9. $\dfrac{\frac{5}{42}}{\frac{15}{32}}$

10. $\dfrac{\frac{-27}{56}}{\frac{-9}{28}}$

11. $\dfrac{\frac{1}{2}+\frac{1}{4}}{\frac{1}{6}+\frac{5}{6}}$

12. $\dfrac{\frac{2}{3}-\frac{1}{3}}{\frac{3}{4}-\frac{1}{4}}$

13. $\dfrac{\frac{1}{2}+\frac{1}{8}}{\frac{5}{12}-\frac{1}{12}}$

14. $\dfrac{\frac{5}{9}+\frac{2}{9}}{\frac{11}{12}+\frac{1}{4}}$

15. $\dfrac{\frac{1}{16}+\frac{3}{4}}{2+\frac{1}{8}}$

16. $\dfrac{\pi-\frac{\pi}{4}}{\pi+\frac{\pi}{4}}$

17. $\dfrac{\frac{5}{27}-\frac{2}{9}}{\frac{1}{36}+\frac{5}{18}}$

18. $\dfrac{\frac{3}{100}+\frac{1}{20}}{\frac{7}{75}-\frac{1}{25}}$

19. $\dfrac{a}{\frac{b}{c}}$

20. $\dfrac{\frac{a}{b}}{c}$

21. $\dfrac{\frac{a}{b}}{\frac{c}{d}}$

22. $\dfrac{\frac{ax}{y}}{x}$

23. $\dfrac{\frac{ax^2}{yz^2}}{\frac{a^2x}{yz}}$

24. $\dfrac{\frac{m^2n}{xy^2}}{\frac{mn}{x^2y^2}}$

25. $\dfrac{\frac{abc^3}{xyz^3}}{\frac{c^2}{x^2y^2z^2}}$

26. $\dfrac{\frac{4ax}{yz}}{\left(\frac{12a^2}{xyz}\right)^2}$

27. $\dfrac{\left(\frac{2abc}{7xy}\right)^3}{\frac{4a^2b}{21}}$

28. $\dfrac{\frac{49a^2bc^2}{24xy^2}}{\frac{14abc}{9y}}$

29. $\dfrac{\frac{1}{x}+\frac{1}{y}}{1-\frac{1}{x}}$

30. $\dfrac{1+\frac{1}{a}}{1+\frac{1}{b}}$

31. $\dfrac{\frac{1}{a^2}+\frac{1}{a}}{a+\frac{1}{a^2}}$

32. $\dfrac{xy-\frac{1}{x}}{\frac{1}{x^2y}}$

33. $\dfrac{\frac{1}{a+1}-\frac{1}{a-1}}{\frac{1}{a^2-1}}$

34. $\dfrac{\frac{1}{a}+\frac{1}{a+1}}{\frac{1}{a}+\frac{1}{a^2}}$

35. $\dfrac{\frac{1}{x+3}-\frac{1}{x-3}}{\frac{1}{x^2-9}}$

36. $x-\dfrac{1}{\frac{1}{x}}$

37. $4y^2+\dfrac{\frac{1}{y}}{\frac{1}{y^3}}$

38. $\dfrac{\frac{1}{a^2-1}+\frac{1}{a}}{\frac{1}{a-1}+\frac{1}{a+1}}$

39. $\dfrac{\frac{x}{x^2-y^2}+\frac{1}{x+y}}{\frac{1}{x-y}+\frac{1}{x+y}}$

40. $\dfrac{\frac{1}{a^2-5a+6}+\frac{1}{a-2}}{\frac{a}{a-3}+\frac{1}{a-2}}$

41. $\dfrac{\frac{1}{x^2-1}-\frac{1}{x+1}}{\frac{1}{x-1}+\frac{1}{x^2-1}}$

42. $\dfrac{\frac{y-a}{y-b}-\frac{y-b}{y-a}}{\frac{1}{y-a}-\frac{1}{y-b}}$

43. $\dfrac{1-\frac{m^2}{n^2}}{1-\frac{2m}{-n}+\frac{m^2}{n^2}}$

44. $\dfrac{a+\frac{ab}{a-b}}{\frac{b^2}{b^2-a^2}-1}$

45. $\dfrac{\dfrac{1}{x} - \dfrac{1}{y}}{\dfrac{x - y}{x^2y^2}}$ **46.** $\dfrac{\dfrac{a}{b} - \dfrac{b}{a}}{\dfrac{1}{a^3b} - \dfrac{1}{ab^3}}$ **47.** $\dfrac{\left(\dfrac{2x - y}{x - 2y}\right) - y}{(x - 2y)^2}$ **48.** $\dfrac{\dfrac{a^2 + 2b^2}{a - b} + a + b}{(b^2 - a^2)^3}$

49. In calculus you will have to work with expressions such as

$$\frac{\dfrac{1}{2 + h} - \dfrac{1}{2}}{h}$$

Simplify this expression.

50. The modulus of rigidity, M, of an object measures the change in its shape while its volume is held constant. M is given by

$$\frac{\dfrac{F}{A}}{\dfrac{x}{h}},$$

where a force F acts (in opposite directions) on a surface area A. Also, x is the displacement of the top surface relative to the bottom surface, and h is the distance between the surfaces. Two clamps are fastened near the ends of a 5-inch long rectangular steel beam, whose rectangular cross section has an area of 5 square inches. A 1600-pound force exerted on each of these clamps, in opposite directions, causes a relative displacement of .000 136 inch. Find the modulus of rigidity of steel.

Review Exercises for Chapter 1

1. Rewrite each term so that its coefficient appears first. (If more than one constant appears, multiply these.)

(a) $(2x^2)(3y)$ **(b)** $4x \cdot 3y \cdot 2z$ **(c)** $x \cdot \dfrac{y}{4}$ **(d)** $(-2x)(-2y)(-z)$

2. Which of the following expressions are polynomials? **(a)** xy^2 **(b)** $x + y^2$ **(c)** $\dfrac{x}{y}$ **(d)** $\dfrac{x}{2} + \dfrac{y}{4}$

3. Evaluate $x^3 + 2x + 1$ when **(a)** $x = 1$ **(b)** $x = -1$ **(c)** $x = 2$ **(d)** $x = -2$.

4. The area of a rectangle is given by lw, where l is the length and w the width. Determine the area when l is 200 feet and w is 50 feet.

5. Add:

$$\begin{array}{rrrr} 2t + & x - & y + & z \\ & 3x + & y - & 2z \\ 4t & + & y - & z \\ & & 2y + & z \\ \hline \end{array}$$

6. Simplify: $2x - [4 - (3x + 2y) - (4x + 2y - 1)]$

7. Multiply: $x^2(xy^2 + xy - 3x^2)$

8. Multiply: $(2m - n)(m + n)$

9. Simplify: $(x - 2)(x + 1) - 3x(x^2 + x - 1)$

10. Determine the product of x^2 and the square of $x - 1$.

11. Determine the prime factorization of 900.

12. Determine GCD (75, 100, 225).

13. Factor each polynomial, if possible. Choose the appropriate technique in each case. *Warning:* Some of these polynomials cannot be factored by the present methods. Write "DOES NOT FACTOR" in this case.

(a) $5x^2 - 25x$ **(b)** $x^2 + 25$ **(c)** $y^2 - y - 20$

(d) $s^2t^2 - 16s^2$ **(e)** $a^3 + 8$ **(f)** $4x^2 + 8x + 3$

(g) $x^2 + 3ax + 2a^2$ **(h)** $ax + by - ay - bx$ **(i)** $2x^2 + 2x + 5$

14. Evaluate $\dfrac{2xy}{x+y}$ when $x = 3$ and $y = -2$.

15. Simplify each rational expression.

(a) $\dfrac{36x^2yz^2}{9xyz}$ (b) $\dfrac{20x^2y(x+y)(a-b)}{15xy^2(x-y)(a-b)}$ (c) $\dfrac{6x^2y - 9xy^2}{3xy}$ (d) $\dfrac{x^2+7x+6}{x^2-36}$ (e) $\dfrac{x^2-a^2}{a^3-x^3}$

16. Multiply or divide, as indicated. Simplify the result and leave it in factored form.

(a) $\dfrac{a^2-x^2}{x+2} \cdot \dfrac{x^2+4x+4}{x^2+2ax+a^2}$ (b) $\dfrac{a+b}{(a-b)^2} \cdot \dfrac{a^2-b^2}{b+1} \cdot \dfrac{1-b^2}{3a+3b}$ (c) $\dfrac{x^4-y^4}{x+y} \div \dfrac{3x^2-3y^2}{x^2+2xy+y^2}$

17. Add or subtract, as indicated. Simplify the result and leave it in factored form.

(a) $\dfrac{1}{x} + \dfrac{2}{x}$ (b) $\dfrac{1}{x^2yz^2} + \dfrac{y}{x^2z^2} - \dfrac{1}{x^2yz}$ (c) $\dfrac{1}{x^2-9} - \left(\dfrac{1}{x+3} - \dfrac{1}{3-x}\right)$

18. Combine, as indicated: $\dfrac{x}{x+1}\left(\dfrac{1}{x^2-1} - \dfrac{1}{x^2+x}\right)$

19. Simplify each expression:

(a) $\dfrac{\dfrac{ax^2y^2}{9bcz^2}}{\dfrac{a^2xy^2}{12b^2cz}}$ (b) $\dfrac{\dfrac{2}{x-2} - \dfrac{1}{x+2}}{\dfrac{1}{x^2-4}}$

Linear Equations and Inequalities

2

2.1 Roots of Equations

degree

Definition. The **degree** of a *nonzero* constant is defined to be 0. If a *nonzero* term contains a single variable, its degree is the exponent of that variable. If a *nonzero* term contains more than one variable, its degree is the *sum* of the exponents of these variables.

For technical reasons, the degree of the term 0 is undefined. Every nonconstant term has a positive degree. You speak of **first-degree terms, second-degree terms,** and so on.

EXAMPLE 1

(a) -7 is of degree 0.

(b) $3x^4$ is a fourth-degree term.

(c) $2x^2z$ is a third-degree term. The exponent of z is understood to be 1. Thus the sum of the exponents is $2 + 1$, or 3. □

Definition. Consider a *nonzero* polynomial in which similar terms have been combined. The **degree of this polynomial** is defined to be the highest degree of any of its terms.

EXAMPLE 2

(a) $3x^2 + 5x - 1$ is a *second-degree polynomial* (or a **quadratic polynomial**). The highest degree of any of its terms is 2.

(b) $x^3 + 2xy + xy^2$ is a *third-degree polynomial.* Two of its terms, x^3 and xy^2, are each of degree 3. The other term, $2xy$, is of degree 2.

(c) $x^2(x + 1)^2$ is a *fourth-degree polynomial* because

$$\begin{aligned} x^2(x + 1)^2 &= x^2(x^2 + 2x + 1) \\ &= x^4 + 2x^3 + x^2 \end{aligned}$$

(d) 14 is a polynomial of degree 0. □

equations

Definition. An **equation** is a statement of equality (or equivalence). An equation has the form:

$$\square = \square$$

Here $\square$ is known as the **left side of the equation;** $\square$ is the **right side.**

You will study equations involving constants and variables.

EXAMPLE 3 (a)

$$\frac{2}{6} = \frac{1}{3}$$

is an equation expressing the equivalence of rational numbers. $\frac{2}{6}$ is the left side and $\frac{1}{3}$ the right side. There are no variables present in this equation.

The following equations involve variables.

(b) $x + 1 = 6$

This is an example of a **linear** or **first-degree equation.** The highest-degree term present is a first-degree term. The left side of the equation is the first-degree polynomial $x + 1$; the right side is the constant 6.

(c) $3y - 7 = 8$

is also a first-degree equation.

(d) $x^2 + 4x + 1 = 0$

involves a second-degree polynomial and is known as a **quadratic** or **second-degree equation.** □

In the first three sections of this chapter, you will be primarily concerned with first-degree equations in a single variable. First-degree equations (in the variable x) can be written in the form

$$Ax + B = 0,$$

where A and B are constants and $A \neq 0$. For example, later you will see that

$x + 1 = 6$ can be rewritten as $x - 5 = 0$.

Thus $A = 1$ and $B = -5$

Equations involving only constants are either *true* or *false* statements. Thus

$2 + 2 = 4$ and $\frac{3}{9} = \frac{2}{6}$ are each *true*.

$2 \cdot 3 = 5$ and $\frac{1}{2} = \frac{3}{4}$ are each *false*.

roots

In Section 1.1 you evaluated a polynomial by substituting numbers for the variable(s) of the polynomial. You can also substitute a number for the variable of an equation. If the variable occurs more than once in an equation, substitute the same number for *each occurrence*.

Definition. A **root** or **solution** of an equation (in a single variable) is a number that when substituted in the equation yields a true statement. A root of an equation is said to **satisfy the equation.**

EXAMPLE 4 (a) 5 *is* a root of the equation

$$2x - 3 = 7$$

because when you substitute 5 for x,

$$2 \cdot 5 - 3 \quad 7$$

is a true statement. Thus 5 *satisfies* the given equation.

(b) -4 *is* a root of the equation

$$5x + 2 = -22 - x.$$

Here the variable x occurs twice. Substitute -4 for each occurrence of x, and obtain

$$5(-4) + 2 = -22 - (-4) \quad \text{or} \quad -18 = -18.$$

This is true.

(c) -3 *is not* a root of the equation

$$7 - 3x = 9 + x$$

For,

$$7 - 3(-3) \neq 9 + (-3) \quad \text{or} \quad 16 \neq 6$$

□

checking It is one thing to **solve an equation,** that is, to determine its roots. It is often a far simpler matter to **check** whether a given number is, indeed, a root of an equation. In parts (a) and (b) of Example 4, you *checked* that 5 is a root of the equation $2x - 3 = 7$ and that -4 is a root of the equation $5x + 2 = -22 - x$. In part (c) of Example 4 the *check* indicated that -3 is not a root of the equation $7 - 3x = 9 + x$. Hereafter, when checking a root, place a question mark over the equality sign until you determine the truth (or falsehood) of the statement in question. Then write a $\surd$ over the = if the statement (of equality) is true, and a $\times$ over the = if the statement is false.

EXAMPLE 5 Check that $\frac{1}{2}$ is a root of the equation $\frac{5}{y} = 9 + 2y$.

Solution. Substitute $\frac{1}{2}$ for each occurrence of y.

$$\frac{5}{\frac{1}{2}} \overset{?}{=} 9 + 2\left(\frac{1}{2}\right)$$

$$5 \cdot 2 \overset{?}{=} 9 + 1$$

$$10 \overset{\surd}{=} 10$$

□

solving equations You know how to *check* whether a given number is a root of an equation. It remains to determine how to find that root, that is, how to solve the equation in the first place!

Definition. **Equivalent equations** are equations with exactly the same roots.

EXAMPLE 6

$$5x - 52 = 2x + 8 \qquad \text{and} \qquad 3x = 60$$

are equivalent equations. Each has the root 20. For, replacing x by 20 in the first equation,

$$5 \cdot 20 - 52 \stackrel{?}{=} 2 \cdot 20 + 8$$
$$48 \stackrel{\checkmark}{=} 48$$

Also, replacing x by 20 in the second equation,

$$3 \cdot 20 \stackrel{\checkmark}{=} 60$$

Thus 20 is a root of each equation. Later you will see that 20 is the *only* root of each equation. Clearly the second equation, $3x = 60$, is the *simpler* one. □

An equation can be thought of as a balanced scale. *In order to preserve the balance, what is done to one side of the equation must be done to the other side.* You solve an equation by successively transforming it into simpler equivalent equations, until finally, you obtain the equation

$$x = c$$

with (the constant) root c. (Note that c is the *only root* of this last equation.)

Addition Principle

The **Addition Principle** *enables you to add the same quantity to both sides of an equation*:

Let R be any rational expression (possibly a polynomial or even a constant). Then

$$\square = \square$$

and

$$\square + R = \square + R$$

are equivalent equations.

Most frequently, the expression R to be added will be relatively simple. (See Exercise 55 for a restriction on the Addition Principle.) The symbol

$$R \,\Delta\!\!\!-\!\!\!-\!\!\!\Delta\, R$$

will indicate that the expression R is to be added to both sides of an equation.

Recall that $a - b = a + (-b)$; thus *you can also subtract by means of the Addition Principle.*

In order to simplify an equation, *bring expressions involving variables to one side, constants to the other side.* It does not matter which side of the simplified equation contains the variables. In part (a) of Example 7 the variable ends up on the left side and the constant on the right side. In part (b) of Example 7 the reverse is true.

EXAMPLE 7 (a) Solve $y - 5 = 0$

(b) Solve and check

$$2x - 9 = 3x - 7$$

Solution.

(a) Add 5 to both sides.

(b) Add $-2x + 7$ to both sides.

$$\begin{array}{rcl} y - 5 & = & 0 \\ +5 & \triangle\!\!\!\!\overline{\quad}\!\!\!\triangle & +5 \\ \hline y & = & 5 \end{array}$$

$$\begin{array}{rcl} 2x - 9 & = & 3x - 7 \\ -2x + 7 & \triangle\!\!\!\!\overline{\quad}\!\!\!\triangle & -2x + 7 \\ \hline -2 & = & x \end{array}$$

$$2(-2) - 9 \stackrel{?}{=} 3(-2) - 7$$
$$-13 \stackrel{\checkmark}{=} -13$$

□

(Usually, the check will *not* be given.)

Multiplication Principle

The second major tool in solving equations is the **Multiplication Principle**:

Let c be a *nonzero* constant. Then

$$\boxed{} = \boxed{}$$

and

$$\boxed{} \cdot c = \boxed{} \cdot c$$

are equivalent equations.

Recall that $\frac{a}{b} = a \cdot \frac{1}{b}$. Thus *the Multiplication Principle enables you to multiply or divide both sides of an equation by the same nonzero constant,* but *not* always by a variable. (See Exercise 57.)

EXAMPLE 8 Solve: (a) $\frac{x}{2} = 3$ (b) $5x = 28$

Solution.

(a) Multiply both sides by 2.

$$\frac{x}{2} \cdot 2 = 3 \cdot 2$$
$$x = 6$$

(b) Divide both sides by 5 (or equivalently, multiply both sides by $\frac{1}{5}$).

$$\frac{5x}{5} = \frac{28}{5}$$
$$x = \frac{28}{5}$$

□

both principles

The remaining examples use both the Addition and Multiplication Principles. Which principle do you apply first when both are used in solving an equation? Consider

$3x + 5 = 15,$ as opposed to $3(x + 5) = 15.$

To solve	To solve
$3x + 5 = 15,$	$3(x + 5) = 15,$
first add -5 to both sides:	first divide both sides by3:

$$\begin{array}{rcl} 3x + 5 & = & 15 \\ -5 & \triangle\!\!\!-\!\!\!\triangle & -5 \\ \hline 3x & = & 10 \\ \dfrac{3x}{3} & = & \dfrac{10}{3} \\ x & = & \dfrac{10}{3} \end{array}$$

$$\begin{array}{rcl} \dfrac{3(x + 5)}{3} & = & \dfrac{15}{3} \\ x + 5 & = & 5 \\ -5 & \triangle\!\!\!-\!\!\!\triangle & -5 \\ \hline x & = & 0 \end{array}$$

How else could you solve

$$3(x + 5) = 15?$$

You can now show that 20 is the one and only root of each of the equations of Example 6

$$5x - 52 = 2x + 8 \qquad \text{and} \qquad 3x = 60.$$

According to the Addition and Multiplication Principles, the following are equivalent equations:

$$\begin{array}{rcl} 5x - 52 & = & 2x + 8 \\ -2x + 52 & \triangle\!\!\!-\!\!\!\triangle & -2x + 52 \\ \hline 3x & = & 60 \\ \dfrac{3x}{3} & = & \dfrac{60}{3} \\ x & = & 20 \end{array}$$

Clearly, the last equation has the single root 20. Because all the equations are equivalent, each has 20 as its one and only root.

EXAMPLE 9 Solve the equation: $\frac{1}{2}[y - (3y - 4)] = 1$

Solution.

$$\frac{1}{2}[y - (3y - 4)] = 1$$

First, multiply both sides by 2.

$$y - (3y - 4) = 2$$

Next, remove parentheses.

$$\begin{array}{rcl} \underbrace{y - 3y} + 4 & = & 2 \\ -4 & \triangle\!\!\!-\!\!\!\triangle & -4 \\ \hline -2y & = & -2 \\ \dfrac{-2y}{-2} & = & \dfrac{-2}{-2} \\ y & = & 1 \end{array}$$

□

In the final example, an equation containing second-degree terms reduces to a first-degree equation.

EXAMPLE 10 Solve: $x^2 + 7 = (x - 1)^2$

Solution.

$$x^2 + 7 = (x - 1)^2$$

On the right side, square $x - 1$.

$$\begin{array}{rcl} x^2 + 7 & = & x^2 - 2x + 1 \\ -x^2 - 7 + 2x & \triangle\!\!\!-\!\!\!\triangle & -x^2 + 2x - 7 \\ \hline 2x & = & -6 \\ x & = & -3 \end{array}$$

□

EXERCISES

In Exercises 1–8 does the statement involve only constants? i. If so, is it true or is it false? ii. If not, is it a first-degree equation in a single variable?

1. $3x = 9$ **2.** $9 + 8 = 20 - 3$ **3.** $9 - 3 = 5 + 2$ **4.** $x = 2x$ **5.** $x = y$

6. $x^2 + 1 = 3x$ **7.** $\frac{y}{2} = \frac{3y}{4}$ **8.** $\frac{1}{x} + \frac{1}{2} = \frac{1}{3}$

In Exercises 9–22 check whether the number in color is a root of the given equation.

9. $x + 5 = 10,\quad 5$ **10.** $3y = 18,\quad 6$ **11.** $4x = -16,\quad 4$ **12.** $2x + 5 = x - 9,\quad 3$

13. $u - 7 = 6u + 2,\quad -1$ **14.** $10x + 1 = 0,\quad \frac{1}{10}$ **15.** $\frac{1}{x} = 6,\quad \frac{1}{6}$ **16.** $\frac{1}{x} = \frac{1}{6},\quad \frac{1}{6}$

17. $x = x,\quad -1$ **18.** $y = 2y,\quad 0$ **19.** $x = -x,\quad -1$ **20.** $x = -x,\quad 0$

21. $z = 2(z + 1),\quad 0$ **22.** $\frac{1}{x} + \frac{1}{2} = \frac{1}{3},\quad 1$

In Exercises 23–54 solve each equation. Check the ones so indicated.

23. $x - 2 = 3$ **24.** $z + 5 = -5$ **25.** $4x = 8$ **26.** $-3x = 9$ (Check.)

27. $\frac{x}{2} = 5$ **28.** $\frac{1}{3}x = -4$ **29.** $3x = \frac{1}{3}$ **30.** $.7x = .28$ (Check.)

31. $2x + 1 = 7$ **32.** $9 + \frac{1}{4}x = 10$ (Check.) **33.** $2x + 5 = 3x$ **34.** $5x - 2 = 7x + 2$

35. $5x - 9 = 3x + 9$ **36.** $10x + 62 = x - 1$ **37.** $y - (2 - 3y) = 18$ (Check.)

38. $3(z - 7) = -9$ **39.** $10\left(z - \frac{1}{4}\right) = \frac{3}{4}$ (Check.) **40.** $4(x - 7) = -5x - 1$

41. $-7(x + 5) = x - 3$ **42.** $3x - 1 = 5 - (x - 12)$

43. $x - [2 - (x - 3)] = 7 - x$ (Check.) **44.** $(3x - 2) - (6 + 5x) = 1$

45. $5x - 6 - (4x + 16) = 0$ **46.** $1 - (1 - [x - (1 - x)]) = -1$

47. $x^2 + 5x = x^2 - 10$ (Check.) **48.** $-5x + 2x^2 = 9x^2 + 35 - 7x^2$

49. $x^2 + 25 = (x - 5)^2$ **50.** $(x - 2)^2 + 3 = (x + 1)^2$ (Check.)

51. $(x + 5)(x - 3) = (x + 5)^2$ **52.** $(x + 1)(x - 2) = (x + 3)(x + 2)$

53. $2x - x^2 = 4x - (x + 2)^2$ **54.** $(x - 8)^2 + 6 = (x - 7)^2 + 1$ (Check.)

55. Are the equations

$$x + 1 = 1 \quad \text{and} \quad x + 1 + \frac{1}{x} = 1 + \frac{1}{x}$$

equivalent? Why or why not?

56. Are the equations

$$x + 1 = x + 2 \quad \text{and} \quad (x + 1)0 = (x + 2)0$$

equivalent? Why or why not?

57. Is 0 a root of either of the equations

$$x = 2 \quad \text{or} \quad x^2 = 2x$$

Are these equations equivalent? Why or why not?

2.2 Equations with Rational Expressions

constant denominators

When an equation involves fractions or rational expressions, to clear the equation of fractions, multiply both sides of the equation by the LCD of these expressions. Then solve the resulting equations as before. In the first two examples the denominators are constants.

EXAMPLE 1 Solve and check: $\frac{x-2}{3} = \frac{2x}{5}$

Solution.

$$\text{LCD}\left(\frac{x-2}{3}, \frac{2x}{5}\right) = 15$$

$$\frac{x-2}{3} = \frac{2x}{5}$$

First multiply both sides by 15 $\left(\text{or cross-multiply: } \frac{x-2}{3} \boxtimes \frac{2x}{5}\right)$

$$5(x-2) = 3 \cdot 2x$$

$$5x - 10 = 6x$$

$$-10 = x$$

Check.

$$\frac{-10-2}{3} \stackrel{?}{=} \frac{2(-10)}{5}$$

$$\frac{-12}{3} \stackrel{?}{=} \frac{-20}{5}$$

$$-4 \stackrel{\checkmark}{=} -4$$

□

EXAMPLE 2 Solve: $\frac{3y-1}{4} - \frac{y-3}{6} = 1$

Solution.

$$\frac{3y-1}{4} - \frac{y-3}{6} = 1$$

First multiply both sides by the LCD, 12.

$$3(3y-1) - 2(y-3) = 12$$

$$9y - 3 - 2y + 6 = 12$$

Now simplify the left side.

$$7y + 3 = 12$$

$$7y = 9$$

$$y = \frac{9}{7}$$

□

variable denominators

In Examples 3 and 4, the denominators contain variables.

EXAMPLE 3 Solve: $\frac{1}{x+3} = \frac{2}{x}$

Solution. Multiply both sides of the equation by the LCD, $x(x + 3)$ [or cross-multiply].

$$x = 2(x + 3)$$
$$x = 2x + 6$$
$$-6 = x$$

□

EXAMPLE 4 Solve: $\dfrac{1}{x^2 - 1} + \dfrac{1}{x - 1} = \dfrac{4}{x + 1}$

Solution.

$$\text{LCD}\left(\frac{1}{x^2 - 1}, \frac{1}{x - 1}, \frac{4}{x + 1}\right) = (x + 1)(x - 1)$$

$$\left[\frac{1}{x^2 - 1} + \frac{1}{x - 1}\right] \cdot (x + 1)(x - 1) = \frac{4}{x + 1} \cdot (x + 1)(x - 1)$$
$$1 + (x + 1) = 4(x - 1)$$
$$2 + x = 4x - 4$$
$$6 = 3x$$
$$2 = x$$

□

In several examples you multiplied both sides of an equation by an expression containing a *variable*. See Exercise 57 on page 57, Exercise 49 on page 60, and Section 6.7 to find out what problems this sometimes entails.

EXERCISES

Solve each equation. Check the ones so indicated.

1. $\dfrac{y}{9} = -2$ (Check.)
2. $\dfrac{-2x}{27} = \dfrac{2}{3}$
3. $\dfrac{x}{2} = \dfrac{-2}{3}$
4. $\dfrac{2y}{5} = \dfrac{1}{12}$ (Check.)

5. $\dfrac{x}{5} = \dfrac{x + 4}{6}$
6. $\dfrac{3x + 1}{2} = \dfrac{7x + 1}{3}$
7. $\dfrac{z - 1}{7} = \dfrac{2 - z}{5}$
8. $\dfrac{3x - 7}{14} = -x$

9. $\dfrac{x}{8} + \dfrac{x}{4} = 3$
10. $\dfrac{x}{3} + \dfrac{2x}{9} = 5$
11. $\dfrac{t}{4} - \dfrac{3t}{5} = -7$
12. $\dfrac{2u}{3} - \dfrac{5}{6} = \dfrac{3}{4} - \dfrac{u}{8}$

13. $\dfrac{5x}{6} + \dfrac{3x}{4} = \dfrac{5x}{3} - 1$ (Check.)
14. $\dfrac{5t}{3} - \dfrac{3t}{4} = \dfrac{2t}{3} + \dfrac{1}{2}$
15. $\dfrac{7z}{5} - \dfrac{3z}{2} = \dfrac{1}{10} - \dfrac{1}{4}$

16. $\dfrac{2x}{9} - \dfrac{x}{12} = \dfrac{5x}{6} + \dfrac{x}{8}$ (Check.)
17. $\dfrac{35}{x} = 5$ (Check.)
18. $\dfrac{9}{x + 6} = 3$ (Check.)

19. $\dfrac{-2}{x + 8} = -1$
20. $\dfrac{x}{x + 4} = \dfrac{1}{2}$
21. $\dfrac{x}{x + 1} = -1$ (Check.)

22. $\dfrac{5x}{x + 3} = 2$ (Check.)
23. $\dfrac{15}{y} = \dfrac{20}{y + 1}$
24. $\dfrac{18}{x + 5} = \dfrac{8}{x}$

25. $\dfrac{27}{x-2} = \dfrac{33}{x}$

26. $\dfrac{-12}{u-3} = \dfrac{3}{u-8}$ (Check.)

27. $\dfrac{x}{x+1} = \dfrac{5}{6}$

28. $\dfrac{z+5}{z} = -4$

29. $\dfrac{z+7}{z+2} = -4$ (Check.)

30. $\dfrac{y-10}{3y} = 0$

31. $\dfrac{t-2}{t+5} = \dfrac{2}{3}$

32. $\dfrac{1+t}{-2t} = -1$

33. $\dfrac{2}{x} + \dfrac{6}{x} = 2$ (Check.)

34. $\dfrac{5}{x} - \dfrac{2}{x} = 1$

35. $3x - \dfrac{1}{2} = \dfrac{1}{4}$

36. $\dfrac{1}{x} + \dfrac{1}{3} = \dfrac{1}{2}$

37. $\dfrac{3}{x} - \dfrac{1}{3} = \dfrac{2}{5}$

38. $\dfrac{5}{2x} + \dfrac{1}{2} = \dfrac{3}{2}$

39. $\dfrac{7}{x} - \dfrac{5}{2x} = \dfrac{3}{2}$

40. $\dfrac{1}{u} + \dfrac{5}{u-2} = \dfrac{4}{u-2}$

41. $\dfrac{5}{y+1} - \dfrac{18}{y+2} = \dfrac{-10}{y+1}$

42. $\dfrac{25}{t+1} - \dfrac{24}{t+4} = \dfrac{10}{t+1}$ (Check.)

43. $\dfrac{1}{u-2} + \dfrac{5}{u+4} = \dfrac{30}{(u-2)(u+4)}$

44. $\dfrac{18}{(z+2)(z-1)} + \dfrac{3}{z+2} = \dfrac{4}{z-1}$

45. $\dfrac{4}{x-1} + \dfrac{3}{x+1} = \dfrac{9}{x^2-1}$

46. $\dfrac{12}{y} - \dfrac{8}{y+2} = \dfrac{48}{y^2+2y}$ (Check.)

47. $\dfrac{4}{z+1} + \dfrac{3}{z+2} - \dfrac{21}{z^2+3z+2} = 0$

48. $\dfrac{8}{x+5} + \dfrac{4}{x^2+7x+10} = \dfrac{3}{x+2}$

49. Solve the equation $\dfrac{x}{x-2} = 5 + \dfrac{2}{x-2}$. Is 2 a root of this equation? (*Hint*: Check!)

2.3 Applications

Algebraic methods enable you to solve problems that arise in various situations. Often a problem is stated in words. Your first task is to translate the problem into mathematical symbols. Each problem to be considered can be formulated in terms of an equation. You solve the equation in order to solve the original problem.

distance

Suppose that an object travels at a *constant rate* of speed. Then *the distance that it travels equals its rate multiplied by the time in transit*. Let

$$d = \text{distance}, \qquad r = \text{rate}, \qquad t = \text{time}.$$

Thus

$$d = r \cdot t$$

For example, if an automobile travels at the constant rate of 60 miles per hour, in 5 hours it goes 300 miles.

You can use other forms of the distance formula. To determine time, use $t = \dfrac{d}{r}$; to determine the rate, use $r = \dfrac{d}{t}$.

Throughout this section all rates of speed are assumed to be constant.

EXAMPLE 1 Two cars leave a gas station at the same time, one traveling eastward, the other westward. The eastward-bound car travels at 50 miles per hour; the westward-bound car speeds at 65 miles per hour. How far apart are they after 3 hours?

Solution. Table 2.1 is useful.

TABLE 2.1

	r ·	t =	d
eastward	50	3	150
westward	65	3	195

The cars are traveling in *opposite* directions. Their distance apart d is the *sum* of the distances each has traveled in 3 hours:

$$d = 150 + 195 = 345$$

The distance apart is 345 miles. Note that each hour the cars move 65 + 50, or 115, miles apart, and $345 = 3 \cdot 115$. [See Figure 2.1(a).] □

In Example 1 the cars were traveling in *opposite* directions, and their distance apart was the *sum* of the distances each had traveled. Had they left the gas station at the same time but traveled in the *same* direction, their distance apart would have been the distance the faster car traveled *minus* the distance the slower car traveled. [See Figure 2.1(b).]

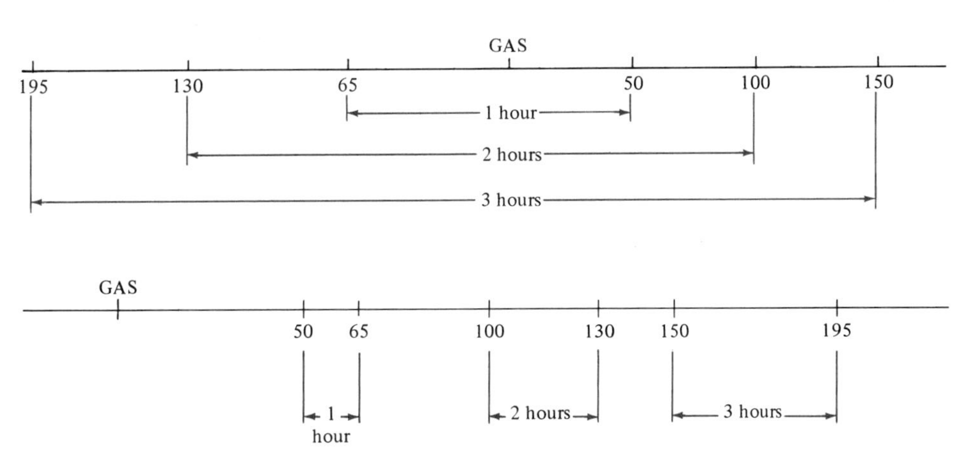

FIGURE 2.1 (a) The cars separate at the rate of 65 + 50 (or 115) miles per hour. After 3 hours they are 3· 115 (or 345) miles apart.
(b) The cars separate at the rate of 65 − 50 (or 15) miles per hour. After 3 hours they are 3· 15 (or 45) miles apart.

EXAMPLE 2 A bandit leaves Cheyenne, Wyoming in a stolen car at 4 a.m., traveling eastward at the constant rate of 60 miles per hour. At 6 a.m., a sheriff leaves Cheyenne and pursues him in a police car traveling at the constant rate of 75 miles per hour.

(a) At what time does the sheriff overtake the bandit?
(b) How far from Cheyenne is the scene of capture?

Solution.

(a) You want to find the time the bandit is overtaken. Thus let t be the time (in hours) the bandit travels before being captured. The sheriff, who leaves Cheyenne 2 hours later, travels for $(t - 2)$ hours. Here, the bandit and the sheriff, although traveling over *different periods of time, travel the same distance.*

TABLE 2.2

	r	$\cdot$ t	$=$ d
bandit	60	t	$60t$
sheriff	75	$t-2$	$75(t-2)$

The distance the bandit travels equals the distance the sheriff travels.

$$60t \stackrel{!}{=} 75(t-2)$$
$$60t = 75t - 150$$
$$150 = 15t$$
$$10 = t$$

The bandit, who starts out at 4 a.m., travels for 10 hours. The capture occurs at 2 p.m.

(b) $60t = 60 \cdot 10 = 600$

Thus the capture takes place 600 miles east of Cheyenne. □

interest

When money is invested, it earns interest over a period of time. To simplify matters, *the basic period of time will be taken as 1 year.* The **interest earned** *for 1 year* equals the yearly **interest rate** times the amount of money (or **principal**) invested. Let

I be the interest earned, R be the interest rate,
P be the principal.

Then

$$I = R \cdot P$$

Percent means hundredths. For example, 6% means the same as .06. Thus, \$100 invested for 1 year at 6% earns

.06(100) dollars, or \$6.

EXAMPLE 3 A man deposits \$800 in a bank whose annual (once a year) interest rate is 5%. How much interest does he earn if he leaves all the money, including the interest, in for 2 years?

TABLE 2.3

R	$\cdot$ P_1	$=$ I
.05	800	40

TABLE 2.4

R	$\cdot$ P_2	$=$ I
.05	840	42

Solution. To find the interest earned in the first year, let P_1 be his original principal (of \$800). The interest rate is 5%, or .05. [See Table 2.3.] After 1 year he earns \$40 interest. He leaves this in the bank and therefore has \$840 after 1 year. This is his principal, P_2, for the second year. The interest rate is the same. [See Table 2.4.] He earns \$42 interest the second year.

Thus in 2 years he earns \$82 interest. □

EXAMPLE 4 Smart Sam invests \$1000 more at 9% than in a safer investment at 6%. He earns the same as if the entire amount were invested at 8½%. How much money has Sam invested at 6%?

Solution. Let x be the amount of dollars invested at 6%. Then $x + 1000$ is the amount invested at 9%. The entire amount, which he could invest at $8\frac{1}{2}\%$, is $x + (x + 1000)$, or $2x + 1000$.

$$\text{Note that } \tfrac{1}{2}\% = \frac{.010}{2} = .005; \text{ hence } 8\tfrac{1}{2}\% = .08 + .005 = .085$$

TABLE 2.5

	R ·	P =	I
at 6%	.06	x	$.06x$
at 9%	.09	$x + 1000$	$.09(x + 1000)$
at $8\frac{1}{2}\%$	.085	$2x + 1000$	$.085(2x + 1000)$

$$\underbrace{.06x + .09(x + 1000)}_{\text{He earns}} \quad \underbrace{=}_{\text{the same as if}} \quad \underbrace{(2x + 1000)}_{\text{the entire amount}} \quad \underbrace{(.085)}_{\text{were at } 8\frac{1}{2}\%.}$$

Multiply both sides by 1000 to clear of decimals.

$$60x + 90(x + 1000) = (2x + 1000)85$$

$$60x + 90x + 90\,000 = 170x + 85\,000$$

$$5000 = 20x$$

$$250 = x$$

Sam has \$250 invested at 6%. □

mixtures

Several ingredients are combined to form a mixture. Your task is to determine the amount of a specific ingredient in the mixture. For example, if equal parts of alcohol and water are mixed to form 30 gallons of a solution, the amount of alcohol in the solution is $\frac{1}{2}$ (or 50%) of the solution, that is, 15 gallons.

In general, *the fraction of the substance times the amount of the mixture equals the amount of the substance in the mixture.*

Fraction of the Substance × Amount of Mixture = Amount of Substance in the Mixture

In the preceding paragraph, alcohol is the substance. Thus

$$\frac{1}{2} \text{ (fraction of the substance)} \times \begin{array}{c}\text{30 gallons of solution}\\ \text{(the mixture)}\end{array}$$
$$= \text{15 gallons of substance in the mixture}$$

EXAMPLE 5 How many gallons of a 12% salt solution should be combined with 10 gallons of an 18% salt solution to obtain a 16% solution?

Solution. Salt is the substance. Let x be the number of gallons of 12% solution. Then there are $(x + 10)$ gallons in the final mixture (the 16% solution).

TABLE 2.6

	Fraction of Salt ·	*Amount of Mixture* =	*Amount of Salt in Mixture*
12% solution	.12	x	$.12x$
18% solution	.18	10	1.8
16% solution	.16	$x + 10$	$.16(x + 10)$

$$\underbrace{\text{Amount of Salt in 12\% Solution}} + \underbrace{\text{Amount of Salt in 18\% Solution}} = \underbrace{\text{Amount of Salt in 16\% Solution}}$$

$$.12x + 1.8 = .16(x + 10)$$

Multiply both sides by 100.

$$12x + 180 = 16(x + 10)$$

$$12x + 180 = 16x + 160$$

$$20 = 4x$$

$$5 = x$$

Five gallons of 12% solution must be used. □

EXAMPLE 6 Alloy A is 1 part copper to 3 parts tin. Alloy B is 1 part copper to 4 parts tin. How much of alloy B should be added to 24 pounds of alloy A to obtain an alloy that is 2 parts copper to 7 parts tin?

Solution. Let *copper be the substance*. Let x be the number of pounds of alloy B.

TABLE 2.7

	Fraction of Copper ·	*Pounds of Alloy* =	*Pounds of Copper in Alloy*
A: 1 part copper, 3 parts tin	$\frac{1}{1+3}$ or $\frac{1}{4}$	24	6
B: 1 part copper, 4 parts tin	$\frac{1}{1+4}$ or $\frac{1}{5}$	x	$\frac{x}{5}$
C: 2 parts copper, 7 parts tin	$\frac{2}{2+7}$ or $\frac{2}{9}$	$24 + x$	$\frac{2}{9}(24 + x)$

$$\underbrace{\text{Pounds of Copper in alloy B}} + \underbrace{\text{Pounds of Copper in Alloy A}} = \underbrace{\text{Pounds of Copper in Alloy C}}$$

$$\frac{x}{5} + 6 = \frac{2}{9}(24 + x)$$

Multiply both sides by 45.

$$9x + 270 = 10(24 + x)$$

$$9x + 270 = 240 + 10x$$

$$30 = x$$

Thirty pounds of alloy B must be used. □

merchandising

There are merchandising problems closely related to the preceding two problems.

EXAMPLE 7 Peanuts at \$1.50 per pound are mixed with peanuts at \$1.00 per pound. How much of each must be used in order to have a 40-pound mixture that sells at \$1.20 per pound?

Solution. Let x be the number of pounds of the \$1.00 per pound peanuts. Then $(40 - x)$ pounds of the \$1.50 per pound peanuts are used.

TABLE 2.8

	Cents per Pound ·	*Number of Pounds* =	*Total Amount in Cents*
\$1.00 per pound peanuts	100	x	$100x$
\$1.50 per pound peanuts	150	$40 - x$	$150(40 - x)$
Mixture (\$1.20 per pound)	120	40	4800

$$\underbrace{\text{The value of the \$1.00 per pound peanuts}} + \underbrace{\text{The value of the \$1.50 per pound peanuts}} = \underbrace{\text{The value of the mixture}}$$

$$\begin{aligned} 100x + 150(40 - x) &= 4800 \\ 100x + 6000 - 150x &= 4800 \\ 1200 &= 50x \\ 24 &= x \\ 16 &= 40 - x \end{aligned}$$

Thus, 24 pounds of \$1.00 peanuts must be mixed with 16 pounds of \$1.50 peanuts. □

work

Suppose a job is done by several people, each working at a *constant rate*. You sometimes have to consider the fraction of work done by each person *working alone*. For example, if a man can paint a room in 4 hours, he paints $\frac{1}{4}$ of the room per hour. If he stops after 3 hours, he has done $\frac{3}{4}$ of the paint job. There may be a second worker involved. This will affect the time to complete the painting.

In work problems,

The fraction of work done in 1 time unit · Time = The fraction of work done

EXAMPLE 8 A gardener can mow a lawn in 9 hours. When his assistant aids him, it takes the two together only 6 hours. How long does it take the assistant alone to mow the lawn?

Solution. The gardener alone does $\frac{1}{9}$ of the work in 1 hour. Let x be the number of hours it takes the assistant alone to mow the lawn. The assistant does $\frac{1}{x}$ of the work in 1 hour.

TABLE 2.9

	Fraction of Work Done in 1 Hour ·	Hours (together) =	Fraction of Work Done (by each)
Gardener	$\frac{1}{9}$	6	$\frac{2}{3}$
Assistant	$\frac{1}{x}$	6	$\frac{6}{x}$

$$\underbrace{\text{Fraction of work done by gardener}} + \underbrace{\text{Fraction of work done by assistant}} = \underbrace{1 \text{ (total job)}}$$

$$\frac{2}{3} + \frac{6}{x} = 1 \qquad \text{Multiply both sides by } 3x.$$

$$2x + 18 = 3x$$

$$18 = x$$

The assistant alone mows the lawn in 18 hours. □

EXAMPLE 9 One pipe can *fill* a tank in 15 minutes and another pipe can *empty* the tank in 45 minutes. If the tank is half full when by mistake, both pipes are opened, how long will it take to fill the tank? [See Figure 2.2.]

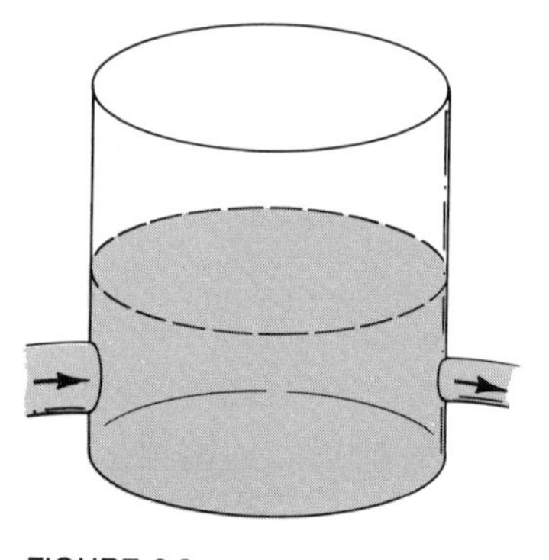

FIGURE 2.2

Solution. Here it is convenient to let x be the number of minutes necessary to fill the *empty* tank when both pipes are open. In 1 minute, the first pipe *fills* $\frac{1}{15}$ of the tank and second pipe *empties* $\frac{1}{45}$ of it. Because emptying a tank *undoes* filling it, you can say that the second pipe fills $-\frac{1}{45}$ of the tank.

The tank is half-filled to begin with. Only half the job of filling it remains. [See Table 2.10 on page 67.]

$$\underbrace{\text{Fraction of work done by first pipe}} + \underbrace{\text{Fraction of work done by second pipe}} = \underbrace{\frac{1}{2} \text{ (of total job)}}$$

$$\frac{x}{15} + \frac{-x}{45} = \frac{1}{2} \qquad \text{Multiply both sides by 90.}$$

$$6x - 2x = 45$$

$$4x = 45$$

$$x = \frac{45}{4}$$

TABLE 2.10

	Fraction of Work Done in 1 Minute	· *Minutes (together)* =	*Fraction of Work Done (by each)*
First Pipe	$\frac{1}{15}$	x	$\frac{x}{15}$
Second Pipe	$\frac{-1}{45}$	x	$\frac{-x}{45}$

It takes $\frac{45}{4}$ (or $11\frac{1}{4}$) minutes to fill the second half of the tank. □

EXERCISES

Unless otherwise specified, all rates are assumed to be constant.

1. How long will it take you to walk to class if you walk 3 miles per hour along a straight path, and if the classroom is a mile and a half from where you start?

2. An automobile headed westward travels for 2 hours at 70 miles per hour. It then slows down and continues westward at 60 miles per hour for the next hour and a quarter. How far does it travel?

3. Pensive Pete travels 60 miles per hour for 3 hours before realizing that he left his airplane tickets at his motel room. He returns at a rate of 80 miles per hour. How long does the return trip take?

4. Two cars leave the center of town at the same time traveling in opposite directions. One car is going 48 miles per hour; the other is going 62 miles per hour. How far apart are they after 6 hours?

5. Two trains approach one another along (straight) parallel tracks. They start out at the same time from stations 90 kilometers apart. One train travels at 100 kilometers per hour, the second train at 80 kilometers per hour. How far has the faster train gone when they pass each other?

6. Two bicyclists leave from the same place and at the same time traveling in the same direction along a straight road. One goes at 18 miles per hour, the other at 14 miles per hour. How far apart are they after an hour and a quarter?

7. A Mercedes and a Volkswagen leave a toll area at the same time traveling in the same direction along a straight road. The Volkswagen travels at two-thirds the rate of the Mercedes. At the end of 4 hours they are 100 miles apart. How fast is the Mercedes traveling?

8. A rowboat goes upstream at the rate of 4 kilometers per hour. It returns downstream at the rate of 8 kilometers per hour. If the round trip takes 3 hours, how far upstream did it go?

9. A hitchhiker begins to walk along a straight road at the rate of 3 miles per hour to the next town, which is 18 miles away. After 10 minutes a car picks him up, and in 15 more minutes he is in town. How fast was the car traveling?

10. Two cars leave the Hollywood Bowl at the same time and head for San Francisco. The first car takes 7 hours along the shorter route, which measures 420 miles. The other car, which travels 4 miles per hour faster, arrives a half hour later. How long is the second route?

11. A skier walks 3000 feet up a mountain road, rests for 10 minutes, and then skis down the road. She returns a half hour after she started. If the average rate at which she skis is four times as fast as the average rate at which she walks, what is the average rate at which she skis?

12. A cross-country runner can average 10 miles per hour over level ground and 6 miles per hour over hilly ground. Altogether it takes him an hour and forty minutes to cover 12 miles. How many of these miles are hilly?

13. How much interest is earned in 1 year on a principal of $560 if the annual interest rate is 6%?

14. What is the annual interest rate if $88 interest is paid in a year on a principal of $2200?

15. How much money must be invested for a year at 7% in order to earn $249.90 interest?

16. A sum of $2500 is left for 2 years in a bank that pays an annual interest rate of 5%. How much interest is earned?

17. A sum of $3600 is left for 3 years in a bank that pays an annual interest rate of 6%. How much interest is earned (to the nearest cent)?

18. If you have twice as much invested at 7% as at 5% and if your annual interest income from these two investments is $950, how much have you invested at each rate?

19. Part of a sum of $5400 is invested in 8% bonds and the remainder in 6½% tax-free bonds. The combined annual interest from these bonds is $375. How much money is invested in 8% bonds?

20. A woman invests $500 more at 8% than at 5.5%. She earns the same as if the entire sum were invested at 7%. How much money is invested at 8%?

21. A company invests $12 000 at 5% and $18 000 at 6%. At what rate should it invest its remaining $5000 in order to receive a combined interest on $2030?

22. How many gallons of a 20% salt solution should be combined with 18 gallons of a 28% salt solution to obtain a 22% solution?

23. How many gallons of a 15% salt solution should be combined with a 25% salt solution to obtain 40 gallons of a 21% solution?

24. How much pure acid should be added to 5 gallons of a 40% acid solution to obtain a 50% acid solution?

25. How many liters of a chemical that is 44% sulphuric acid must be combined with 12 liters of a chemical that is 56% sulphuric acid to obtain a 47% sulphuric acid mixture?

26. A martini mix contains 4 parts gin to 1 part vermouth. A drier mix contains 6 parts gin to 1 part vermouth. How much of the drier mix must be used for 15 ounces of a mixture that is 5 parts gin to 1 part vermouth?

27. How many ounces of an alloy containing 60% silver must be added to 50 pounds of an alloy containing 45% silver to obtain an alloy containing 51% silver?

28. How much cream that contains 33% butterfat must be blended with milk that contains 3% butterfat to obtain 10 gallons of half-and-half that contains 8% butterfat?

29. Ten gallons of chocolate ice cream that contains 18% butterfat is mixed with 8 gallons of vanilla ice cream that contains 12% butterfat to make chocolate ripple ice cream. What percent butterfat does the chocolate ripple contain?

30. An alloy is 2 parts zinc to 1 part tin. A second alloy is 3 parts zinc to 1 part tin. How much of each alloy should be used to make 200 tons of an alloy that is 7 parts zinc to 3 parts tin?

31. A grocer mixes 6 pounds of coffee that is 60% Colombian with one that is 80% Colombian. How much of the second type should be used to obtain a mixture that is 78% Colombian?

32. A candy store owner buys tootsie rolls at three for a dime and sells them at four for a quarter. After selling all of them, his profit is $4.20. How many tootsie rolls were there?

33. Coffee at $2.75 per pound is blended with coffee at $3.25 per pound. How much of each must be used to make 50 pounds of a mixture at $2.95 per pound?

34. Ten pounds of hazel nuts at $1.75 per pound and 12 pounds of pecans at $1.80 per pound are to be combined with walnuts at $1.90 per pound to obtain a mixture that will sell at $1.85 per pound. How many pounds of walnuts must be used?

35. Nine pounds of chocolate chip cookies at $4.10 per pound are mixed with 6 pounds of nut cookies at $3.80

per pound and with 10 pounds of ginger snaps at $3.53 per pound. How much should the assorted cookies cost per pound?

36. Ann can paint a barn in 10 hours. Joe can paint it in 12 hours. How long does it take them together to paint the barn?

37. Judd and Judy can clean their apartment together in 3 hours. They each work at the same speed. How long would it take Judy to clean it by herself?

38. Two privates sweep the mess hall together in 4 hours. One works twice as fast as the other. How long does it take the faster private to do the job alone?

39. A carpenter must saw several beams of wood. It takes him 2 days if he works with his assistant and 3 days if he works alone. How long does it take the assistant alone to saw these beams?

40. Three workers set out to plaster a ceiling. One can do the job alone in 2 hours, one in 3 hours, and the last one in 4 hours. How long does it take them together to plaster the ceiling?

41. A father and his two sons gather the autumn leaves. The father works twice as fast as one son and three times as fast as the other. If it takes the father 6 hours to gather the leaves by himself, how long does it take all three of them to do the job?

42. In Exercise 41, how long would it take the two boys to gather the leaves without aid from their father?

43. One pipe fills a tank in 50 minutes, another pipe in 75 minutes. How long does it take for both pipes together to fill the tank?

44. Two equal-sized pipes fill a tank in an hour. How long does it take for one of these pipes to fill the tank?

45. One pipe can fill a tank in 15 minutes. A second pipe can empty a full tank in 18 minutes. If, by error, both pipes are left open, how long does it take to fill the tank, when empty?

46. The hot water faucet fills a tub in 12 minutes, the cold water faucet in 8 minutes. When the stopper is lifted, a filled tub empties in 10 minutes. Both faucets are open and, by error, the stopper is lifted. How long does it take to fill two-thirds of the tub?

2.4 Literal Equations

Definition. A **literal equation** is one involving at least one letter other than the variable for which you are solving.

Sometimes these other letters all represent constants. The constants may be complicated numbers such as

$$-98\,936.243 \quad \text{or} \quad .000\,089\,140\,73$$

It is easier to designate them by literal constants, such as c or K. At other times several variables may be present. Thus in the equation

$$x + y = 20$$

there are two variables. You may sometimes want to solve this for x, and at other times for y. The methods already developed for solving equations apply here as well. Use the Addition and Multiplication Principles to bring terms containing *the variable for which you are solving* to one side. Treat all other letters as though they were constants. Thus bring all other terms to the other side.

EXAMPLE 1 Let a be a constant. Solve $2x - 1 = a$ for x.

Solution. The term containing x remains on the left side; the term -1 goes to the right side. Thus first add 1 to both sides.

$$2x - 1 = a$$

$$2x = a + 1$$

$$x = \frac{a+1}{2}$$

□

EXAMPLE 2 Let a, b, c be constants. Solve $3x + a = b - cx$ for x. Check your result.

Solution. Bring the terms containing x to the left side, the other terms to the right side.

$$3x + a = b - cx$$

$$\underline{+cx - a} \quad \triangle\!\!-\!\!\triangle \quad \underline{+cx - a}$$

$$3x + cx = b - a$$

Now isolate the common factor x on the left side.

$$(3 + c)x \quad b - a$$

Finally, divide both sides by $3 + c$.

$$x = \frac{b-a}{3+c}$$

Check. Substitute $\dfrac{b-a}{3+c}$ for x in the original equation.

$$3\left(\frac{b-a}{3+c}\right) + a \stackrel{?}{=} b - c\left(\frac{b-a}{3+c}\right)$$

Now multiply both sides by $3 + c$.

$$3(b - a) + a(3 + c) \stackrel{?}{=} b(3 + c) - c(b - a)$$

$$3b - \cancel{3a} + \cancel{3a} + ac \stackrel{?}{=} 3b + \cancel{bc} - \cancel{bc} + ac$$

$$3b + ac \stackrel{\checkmark}{=} 3b + ac$$

□

In Example 2, when you divide both sides of the equation by $3 + c$, you are assuming $c \neq -3$. Otherwise, $3 + c = 3 + (-3) = 0$. (Recall that division by 0 is undefined.) *In this section assume the constants and variables are such that division is defined.*

EXAMPLE 3 Suppose x, y, and z are variables. Solve

$$x = \frac{2y - 3z}{2 + z} \quad \text{(a) for } y, \quad \text{(b) for } z.$$

Solution.

(a) Leave the term in which y occurs on the right side; bring all other terms to the left side.

$$x = \frac{2y - 3z}{2 + z}$$

First multiply both sides by $2 + z$.

$$x(2 + z) = 2y - 3z$$

$$x(2 + z) + 3z = 2y$$

$$\frac{2x + xz + 3z}{2} = y$$

(b) Bring terms in which z occurs to the left side; bring all other terms to the right side.

$$x = \frac{2y - 3z}{2 + z}$$

$$(2 + z)x = 2y - 3z$$

$$2x + zx = 2y - 3z$$

$$zx + 3z = 2y - 2x$$

Now isolate the common factor z on the left side.

$$z(x + 3) = 2y - 2x$$

$$z = \frac{2y - 2x}{x + 3} \quad \text{or} \quad z = \frac{2(y - x)}{x + 3}$$ □

EXAMPLE 4 Solve $\frac{1}{x_1} + \frac{1}{x_2} = \frac{1}{x_3}$ for x_2.

Solution.

$$\text{LCD}\left(\frac{1}{x_1}, \frac{1}{x_2}, \frac{1}{x_3}\right) = x_1 x_2 x_3$$

$$\frac{1}{x_1} + \frac{1}{x_2} = \frac{1}{x_3}$$

First multiply both sides by $x_1 x_2 x_3$.

$$x_2 x_3 + x_1 x_3 = x_1 x_2$$

$$x_2 x_3 - x_1 x_2 = -x_1 x_3$$

Now isolate the common factor x_2 on the left side.

$$x_2(x_3 - x_1) = -x_1 x_3$$

$$x_2 = \frac{-x_1 x_3}{x_3 - x_1} \quad \text{or} \quad x_2 = \frac{x_1 x_3}{x_1 - x_3}$$ □

EXAMPLE 5 $$F = \frac{9}{5}C + 32$$

expresses degrees Fahrenheit, F, in terms of degrees Celsius, C.

(a) Solve for C. (b) Find the value of C when F = 98.6 .

Solution. [See Figure 2.3.]

(a)

$$F = \frac{9}{5}C + 32$$

$$F - 32 = \frac{9}{5}C$$

Now multiply both sides by $\frac{5}{9}$.

$$\frac{5}{9}(F - 32) = C$$

(b) Let F = 98.6 .

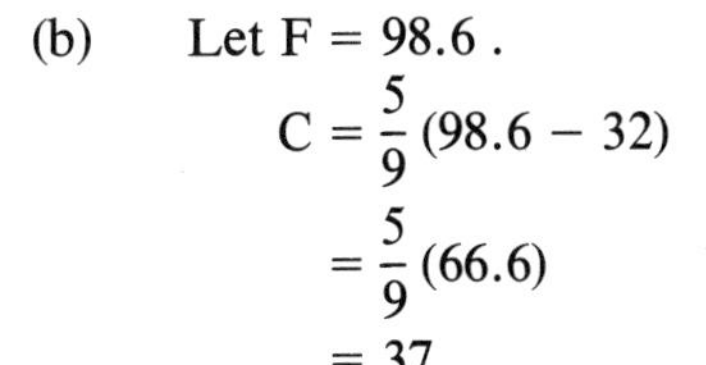

$$C = \frac{5}{9}(98.6 - 32)$$

$$= \frac{5}{9}(66.6)$$

$$= 37$$ □

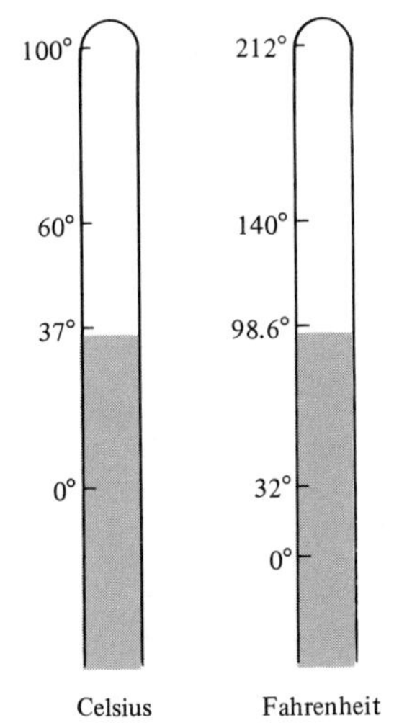

FIGURE 2.3 $F = \frac{9}{5}C + 32$

EXERCISES

Assume *a, b, c, and K* are constants. In Exercises 1–20 solve for x. Check the ones so indicated.

1. $5x = c$ (Check.)
2. $x - 7 = a$
3. $x + 13 = b - c$
4. $6x + c = b - x$
5. $ax + 7 = 3x - 2$ (Check.)
6. $2(x + a) - 1 = 5x + b$
7. $ax + 3 = (1 - a)x - b$
8. $ax - b(c + x) = 7x - a$
9. $\frac{3x - a}{b} = 9$
10. $\frac{a - 7bx}{2} = 1$
11. $\frac{4b - 3x}{6} = 5x - c$
12. $\frac{2 - 3b}{5x} = a - bc$
13. $\frac{4}{x} = a + b$
14. $\frac{1}{3 - x} = a - 2c$
15. $\frac{x}{x + a} = bc + 1$ (Check.)
16. $a^2x + a^2 = 1 - 2bc$
17. $a^2bx - 4 = 5 - 3a^2bx$
18. $\frac{c}{1 - x} = \frac{3}{4}$
19. $\frac{a}{4b - bx} = 7 - a$
20. $\frac{x}{2a - 3x} = -4$
21. Solve $a - 2t = 4 + b$ for t.
22. Solve $5x_1 + 7 = x_1x_2x_3 - 1$ for x_1.
23. Solve $\frac{3y_1 + 1}{y_2 + y_3} = \frac{y_1}{3}$ for y_1.
24. Solve $\frac{1}{K} = \frac{2}{z} - \frac{1}{5}$ for z.

In Exercises 25–36 assume *x, y, z, and t* are the variables. Solve for the indicated variables.

25. $2x + y = 1 + t$
(a) for y **(b)** for x

26. $3x - 2y = z + 1$
(a) for x **(b)** for y

27. $ax + by = cz$
(a) for x **(b)** for y

28. $3at - x = 1 + cz$
(a) for t **(b)** for z

29. $5ay - 3xz = 1$
(a) for y **(b)** for z

30. $\frac{1}{x} - \frac{1}{y} = 3z$
(a) for x **(b)** for y

31. $(x + y)(y + z) = 9$
(a) for x **(b)** for z

32. $xt - 3y + 3xy = 2t$
(a) for t **(b)** for y

33. $a(a - x) + 1 = b(t - b)$
(a) for x **(b)** for t

34. $\frac{4}{x} - \frac{1}{3t} = y$
(a) for x **(b)** for t

35. $\frac{K - 5t}{b} = \frac{3(x - 1)}{z}$
(a) for t **(b)** for z

36. $\frac{K}{t} + \frac{3}{xt} = -1$
(a) for x **(b)** for t

In Exercises 37–43 refer to the diagram on the inside front cover.

37. $C = 2\pi r$ gives the circumference C of a circle in terms of the radius length r. Solve for r.

38. $P = 2(l + w)$ gives the perimeter P of a rectangle in terms of the length l and width w. Solve for l.

39. $A = \frac{bh}{2}$ gives the area A of a triangle in terms of the base length b and the altitude length h. Solve for b.

40. $A = \frac{(b_1 + b_2)h}{2}$ gives the area A of a trapezoid in terms of the base lengths b_1 and b_2 and altitude length h.

(a) Solve for h. **(b)** Solve for b_1.

41. $V = lwh$ gives the volume V of a rectangular box in terms of its length l, width w, and height h. Solve for l.

42. $V = \pi r^2 h$ gives the volume V of a right circular cylinder in terms of the length, r, of the radius of the base and the length, h, of the altitude. Solve for h.

43. $V = \frac{1}{3}\pi r^2 h$ gives the volume V of a right circular cone in terms of the length, r, of the radius of the base and the length, h, of the altitude. Solve for h.

44. (a) Solve $S = 2\pi r(r + h)$ for h. (b) Determine h when $r = 5$ and $S = 100\pi$.

45. (a) Solve $h = \frac{v^2}{2g} + \frac{p}{c}$ for p. (b) Determine p when $h = 2$, $g = 32$, $v = 10$, $c = 5$.

46. (a) Solve $S = \frac{a - rl}{1 - r}$ for r. (b) Determine r when $S = 40$, $a = 10$, $l = 2$.

2.5 Inequalities and the Real Line

Equations are statements of *equality*. In order to study statements of *inequality*, you must reconsider real numbers from a geometric point of view. Various types of numbers—integers, rational numbers, positive numbers—will now be reexamined from this vantage point.

the real line

A horizontal line, L, extends indefinitely in both directions. Real numbers can be identified with points on L as follows. Choose an arbitrary point O on L and call this the **origin.** This point O corresponds to the number 0. Choose another point P on L to the *right* of O; the point P corresponds to the number 1. That part of the line L between O and P is called a **line segment,** and is denoted by $\overline{OP}$. The length of the line segment $\overline{OP}$ determines the basic **unit of distance.** [See Figure 2.4 .] The number 2 corresponds to the point 1 (distance) unit to the right of P; the number 3 is 1 unit farther to the right. This process can be continued indefinitely.

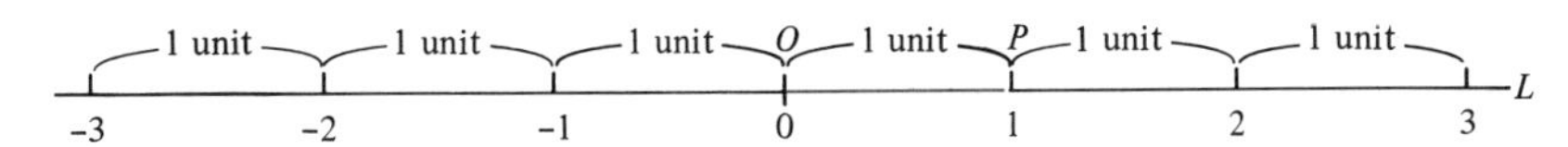

FIGURE 2.4 The line segment $\overline{OP}$ is in color. The length of $\overline{OP}$ is the unit of distance on L.

The point corresponding to the number -1 is located on L 1 unit to the *left* of O; the point corresponding to -2 is located 1 unit farther to the left, and so on.

Every integer can be represented by one of these processes.

EXAMPLE 1 On the line L:

(a) The integer 0 corresponds to the point O.

(b) The integer 63 corresponds to the point 63 units to the right of O.

(c) The integer -15 corresponds to the point 15 units to the left of O.

(d) Which integer corresponds to the point 31 units to the right of O?

(e) Which integer corresponds to the point 78 units to the left of O? □

Hereafter, when a real number a corresponds to a point on L, identify a with this point. Thus "the point 0" really means the point O, "the point 1" means the point P, and "the line segment from 0 to 1" is $\overline{OP}$.

rational numbers

Midway between 0 and 1 on L is the point corresponding to the *rational number* $\frac{1}{2}$. [See Figure 2.5 .] This point divides the line segment from 0 to 1 into 2 equal parts.

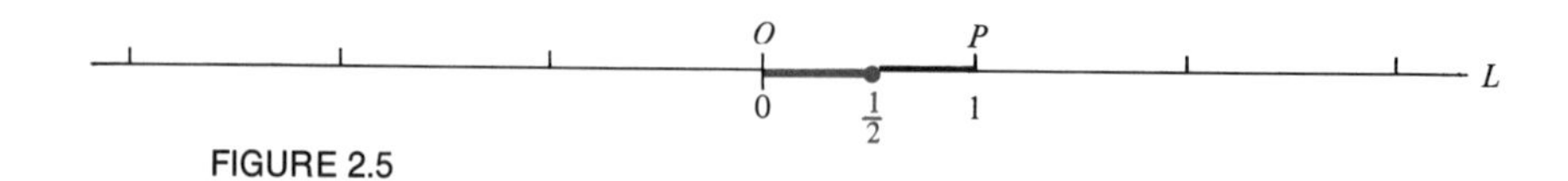

FIGURE 2.5

Recall that a rational number is a number, such as $\frac{1}{2}, \frac{3}{4}, \frac{-2}{3}$, that can be expressed as the quotient of two integers, $\frac{M}{N}$, $N \neq 0$. If N is *positive*, then in order to represent rational numbers $\frac{M}{N}$ on L, divide the line segments between integer points, as in Example 2.

EXAMPLE 2 (a) To represent the rational number $\frac{1}{5}$, divide the line segment between 0 and 1 into 5 equal parts. The first point of division to the *right* of 0 corresponds to $\frac{1}{5}$. [See Figure 2.6(a).]

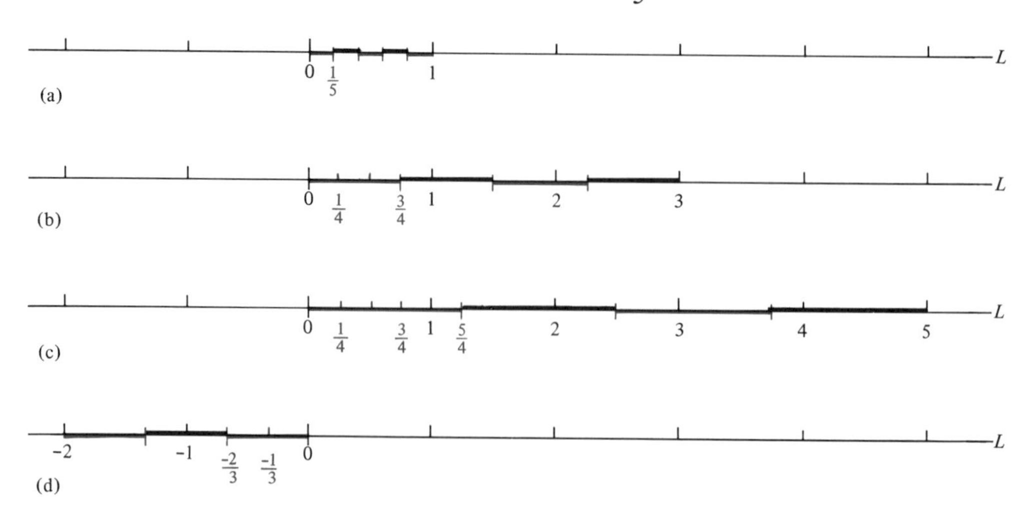

FIGURE 2.6

(b) To represent the rational number $\frac{3}{4}$, divide the line segment between 0 and 3 into 4 equal parts. The first point of division to the *right* of 0 corresponds to $\frac{3}{4}$. [See Figure 2.6(b).] Recall that $\frac{3}{4} = \frac{1}{4} \cdot 3$ and note that $\frac{3}{4}$ lies 3 times as far from the origin as $\frac{1}{4}$. Thus $\frac{3}{4}$ can also be obtained by dividing the line segment between 0 and 1 into 4 equal parts. Now the *third* point of division to the *right* of 0 corresponds to $\frac{3}{4}$.

(c) To represent the rational number $\frac{5}{4}$, or $1\frac{1}{4}$, divide the line segment between 0 and 5 into 4 equal parts. The first point of division to the *right* of 0 corresponds to $\frac{5}{4}$. [See Figure 2.6(c).] Note that $\frac{5}{4} = \frac{1}{4} \cdot 5$ and that $\frac{5}{4}$ lies 5 times as far from the origin as $\frac{1}{4}$.

(d) To represent $\frac{-2}{3}$, divide the line segment between -2 and 0 into 3 equal parts. The first point of division to the *left* of 0 corresponds to $\frac{-2}{3}$. [See Figure 2.6(d).] Note that $\frac{-2}{3} = \frac{1}{3} \cdot (-2)$ and that $\frac{-2}{3}$ lies twice as far from the origin as $\frac{-1}{3}$. □

irrational numbers

To see that the *irrational number* $\sqrt{2}$ represents the length of a line segment, observe in Figure 2.7 that a square whose side length is 1 has a diagonal of length $\sqrt{2}$. (See page 216.)

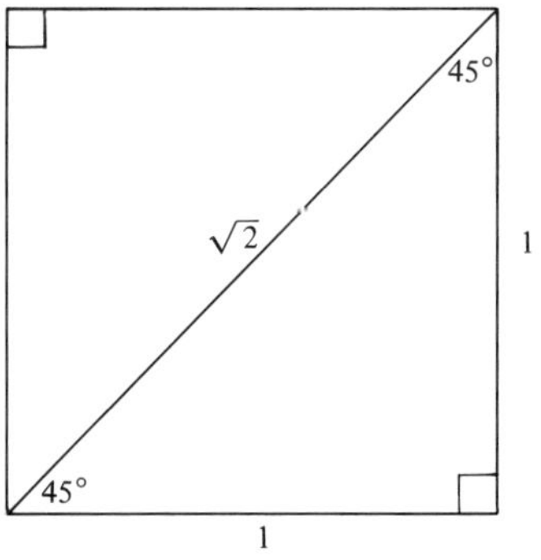

FIGURE 2.7

It can be shown that every rational number can be expressed as either an ordinary (terminating) decimal or as an **infinite repeating decimal**, such as

.6666. . ., .09 09 09. . .

The 3 dots are read, "and so on".

in which one or more of the **digits**

0, 1, 2, 3, 4, 5, 6, 7, 8, 9

repeat. For example, $\frac{3}{10} = .3$, and $\frac{3}{100} = .03$. Also,

$$\frac{2}{5} = .4 \qquad 5\overline{)2.0} = .4$$

$$\frac{2}{3} = .6666\ldots \qquad 3\overline{)2.0000\ldots} = .6666\ldots$$

$$\frac{1}{11} = .09\ 09\ 09\ldots \qquad 11\overline{)1.00\ 00\ 00\ldots} = .09\ 09\ 09\ldots$$

It is somewhat more difficult to show that every irrational number can be expressed as an **infinite nonrepeating decimal**, that is, a decimal in which the digits do *not* repeat in any pattern. For instance, the decimal representation of the irrational number $\sqrt{2}$ begins 1.414 21. . . . Thus $\sqrt{2}$ lies between the rational numbers 1.41 and 1.42 . And the decimal representation of the irrational number π begins 3.141 59 . . . . [See Figure 2.8 .]

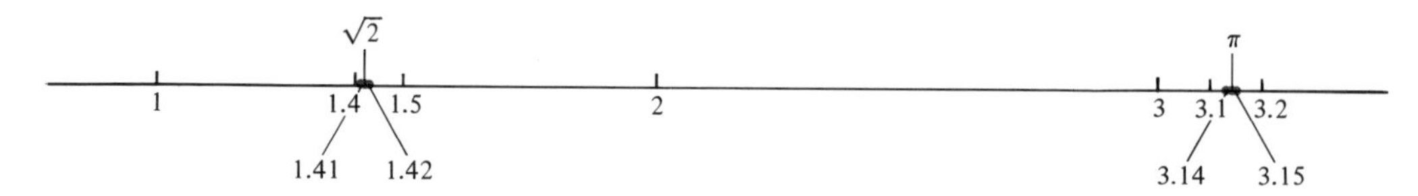

FIGURE 2.8 The irrational $\sqrt{2}$ lies between the rational numbers 1.41 and 1.42. The irrational number π lies between 3.14 and 3.15 .

Every point on L represents exactly one real number, and every real number corresponds to exactly one point on L. It is this correspondence between real numbers and points on the line L that enables you to picture numbers geometrically.

positive numbers
negative numbers

Definition. A real number is said to be **positive** if it lies to the *right* of 0 on L, and **negative** if it lies to the *left* of 0. (The number 0 is neither positive nor negative.) The **nonnegative numbers** consist of 0 together with the positive numbers.

Thus the numbers $5, \frac{3}{4}, \sqrt{2}$ are positive, whereas $-2, -\frac{14}{9}, -\pi$ are negative.

inequalities

Now you will consider various kinds of *inequalities*. Everyone knows that

3 *is less than* 7 .

But it is also true that

-7 *is less than* -3 .

Definition. Let a and b be real numbers. Say that "a **is less than** b" and write

$$a < b$$

if a lies to the *left* of b (*on* L). In this case you also say that "b **is greater than** a", and write

$$b > a$$

(Observe that in this notation, the symbols $<$, $>$ point to the *lesser* number.)

Let a, b, and c be real numbers.

If $a < b$ **and** $b < c$, **then** $a < c$ **(Transitivity of $<$)**

For, a lies to the left of b and b lies to the left of c; hence a lies to the left of c. [See Figure 2.9 .]

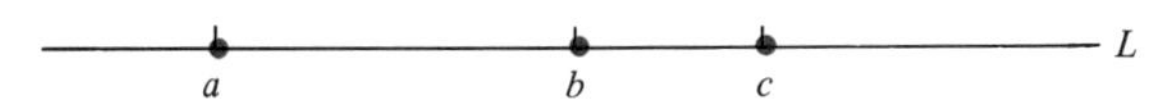

FIGURE 2.9 $a < b$ and $b < c$. a lies to the left of c. Thus $a < c$

EXAMPLE 3 Refer to Figure 2.10 .

(a) $2 < 6$ because 2 lies to the left of 6 on L. You can also write $6 > 2$.
(b) $-3 < 3$ because -3 lies to the left of 3.
(c) $-4 < -2$ because -4 lies to the left of -2.
(d) If $a < -4$ and $-4 < -2$, by Transitivity, $a < -2$. □

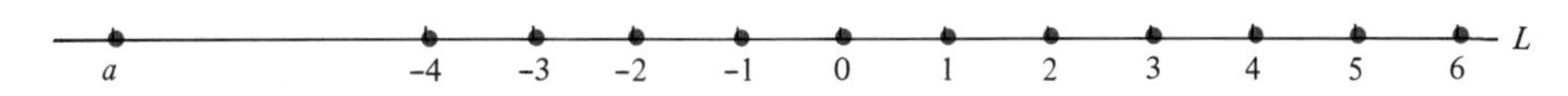

FIGURE 2.10 (a) $2 < 6$ (2 lies to the left of 6.) (b) $-3 < 3$ (-3 lies to the left of 3.) (c) $-4 < -2$ (-4 lies to the left of -2.) (d) If $a < -4$, then $a < -2$

Clearly, a number a is positive if $a > 0$ (or $0 < a$) and negative if $a < 0$. A negative number a is less than any positive number b. In fact,

$$a < 0 \quad \text{and} \quad 0 < b, \quad \text{and by Transitivity,} \quad a < b$$

For every two real numbers a and b exactly one of the following holds:

$$\mathbf{a = b; \quad a < b; \quad a > b} \quad \textbf{(The Law of Trichotomy)}$$

This is because on the real line L, a and b are the same point, or a lies to the left of b, or a lies to the right of b.

Definition. Let a and b be real numbers. Write

$$a \leqslant b$$

if either $a < b$ or $a = b$. Thus, $a \leqslant b$ indicates that a **is less than or equal to** b. Similarly, write

$$a \geqslant b$$

if either $a > b$ or $a = b$. Then $a \geqslant b$ indicates that a **is greater than or equal to** b.

Observe that for every number a,

$$a \leqslant a \qquad \text{and} \qquad a \geqslant a$$

are both true.

EXAMPLE 4 (a) $5 \leqslant 6$ because $5 < 6$
(b) $9 \geqslant 3$ because $9 > 3$
(c) $-7 \leqslant -4$ because $-7 < -4$
(d) $2 \leqslant 2$ because $2 = 2$; similarly, $2 \geqslant 2$ because $2 = 2$ □

Let a, b, and c be real numbers.

If $a \leqslant b$ and $b \leqslant c$, then $a \leqslant c$ (Transitivity of $\leqslant$)

The real line L extends indefinitely both to the right and to the left. Thus given any number a, no matter how far to the right it lies, there is always a number b to the right of a. Hence, b is greater than a. Similarly, no matter how far to the left a number c lies, there is always a number farther to the left, and thus less than c.

intervals

Instead of considering the *entire* real line, you may sometimes be interested only in the numbers "between a and b".

Let $a < x < b$ stand for both $a < x$ and $x < b$. Similarly, $a \leqslant x \leqslant b$ means $a \leqslant x$ and $x \leqslant b$.

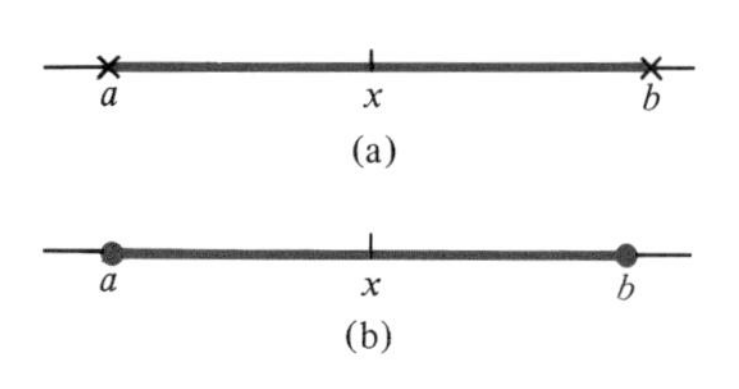

FIGURE 2.11 (a) The open interval (a, b) consists of all x that lie *both* to the right of a *and* to the left of b. The points a and b are excluded. (b) The closed interval $[a, b]$ consists of the numbers in (a, b) plus the end-points a and b.

Definition. Let a and b be real numbers and let $a < b$. The **open interval** (a, b) consists of all real numbers x such that $a < x < b$.The **closed interval** $[a, b]$ consists of all real numbers x such that $a \leqslant x \leqslant b$. Here a is called the **left end-point** and b the **right end-point** of each of these intervals.

Observe that *neither* end-point a nor b is in the *open* interval (a, b). *Both* end-points a and b are in the *closed* interval $[a, b]$.

Geometrically, the open interval (a, b) consists of all points that lie *both* to the right of a *and* to the left of b. [See Figure 2.11 .]

EXAMPLE 5

(a) The open interval (3, 5) consists of all real numbers x such that $3 < x < 5$. Thus 3.1, 4, and $\frac{9}{2}$ are in this open interval. For example, $3 < 3.1 < 5$. Neither 3 nor 5 is in this open interval. Also, such numbers as -4, 0, and 2 lie to the left of 3 and therefore are not in this interval. Similarly, 5.1, 6, and 10 lie to the right of 5, and are not in this interval.

(b) The closed interval [3, 5] consists of all the numbers of the open interval (3, 5) plus the two end-points 3 and 5. Thus 3, 3.7, 4.36, and 5 are in this closed interval; 2.9 and 5.01 are not.

(c) The open interval $(-7, -2)$ consists of all numbers x satisfying $-7 < x < -2$. Thus -6.6, -3, and -2.1 are in this open interval; -10, -7.5, -7, -2, -1.9, 0, and 4 are not. □

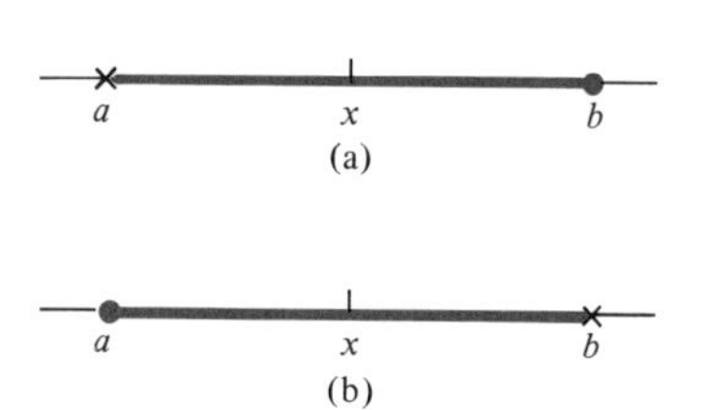

FIGURE 2.12 (a) The half-open interval $(a, b]$ consists of the numbers in (a, b) plus the right end-point b. (b) The half-open interval $[a, b)$ consists of the numbers in (a, b) plus the left endpoint a.

In trigonometry (Chapters 8 and 9) it is also necessary to consider **half-open** (or **half-closed**) **intervals** $(a, b]$ consisting of all real numbers x such that $a < x \leqslant b$, that is, such that $a < x$ *and* $x \leqslant b$ and $[a, b)$ consisting of all real numbers x such that $a \leqslant x < b$. [See Figure 2.12 .]

rays

Sometimes you will want to speak about all $x > 2$ or all $x \leqslant 5$.

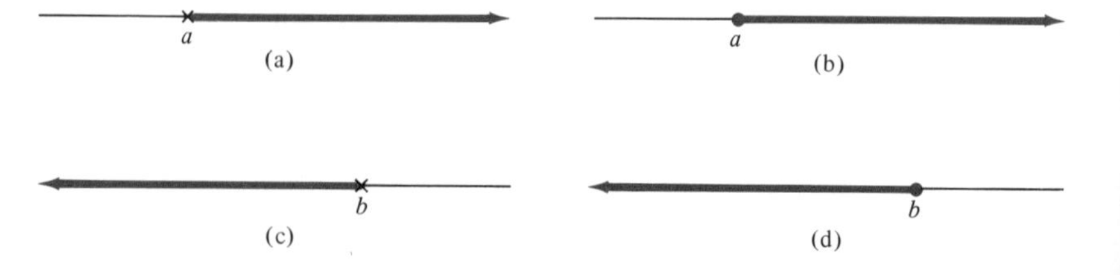

FIGURE 2.13
(a) The open right ray (a, ∞). The point a is excluded.
(b) The closed right ray $[a, \infty)$. The point a is included.

(continued on page 79)

FIGURE 2.13 *page 78 (cont.)*

(c) The open left ray $(-\infty, b)$. The point b is excluded.
(d) The closed left ray $(-\infty, b]$. The point b is included.

Definition. Let a and b be real numbers. The **open right ray** (a, ∞) consists of all real numbers x such that $a < x$. The **closed right ray** $[a, \infty)$ consists of all real numbers x such that $a \leqslant x$. The **open left ray** $(-\infty, b)$ consists of all real numbers x such that $x < b$. The **closed left ray** $(-\infty, b]$ consists of all real numbers x such that $x \leqslant b$.

Geometrically, the open *right* ray (a, ∞) consists of all x that lie to the *right* of a; the closed *right* ray $[a, \infty)$ includes the *left* end-point a. The open *left* ray $(-\infty, b)$ consists of all x that lie to the *left* of b; the closed *left* ray $(-\infty, b]$ includes the *right* end-point b. See Figure 2.13. [The symbol "∞" (*infinity*) is used to indicate that these lines extend indefinitely in one direction.]

EXAMPLE 6

(a) The open right ray $(3, \infty)$ consists of all real numbers x that lie to the right of 3. Thus 4 is in this ray, but 3 is not.
(b) The closed right ray $[-4, \infty)$ consists of -4 together with all numbers that lie to the right of -4. Among these are -3, 0, and all positive numbers.
(c) The open left ray $(-\infty, 0)$ consists of all numbers x that lie to the left of 0. Thus $(-\infty, 0)$ consists of the negative numbers.
(d) The closed left ray $(-\infty, 2]$ consists of 2 together with all numbers less than 2. □

EXERCISES

1. On a large sheet of paper draw a horizontal line L to represent the real numbers. Locate the following numbers on L:

(a) 0 (b) 1 (c) 2 (d) 3 (e) -1 (f) -2 (g) $\frac{1}{4}$
(h) $\frac{3}{4}$ (i) $1\frac{1}{2}$ (j) $\frac{7}{5}$ (k) π (l) 1.6 (m) -1.2 (n) 2.9

2. Each of the points P, Q, R, S, T, U in Figure 2.14 represents one of the following numbers. Indicate which number is represented by each point.

$.9, \quad -.9, \quad -1.05, \quad \frac{2}{5}, \quad \frac{1}{2}, \quad \frac{-1}{3}$

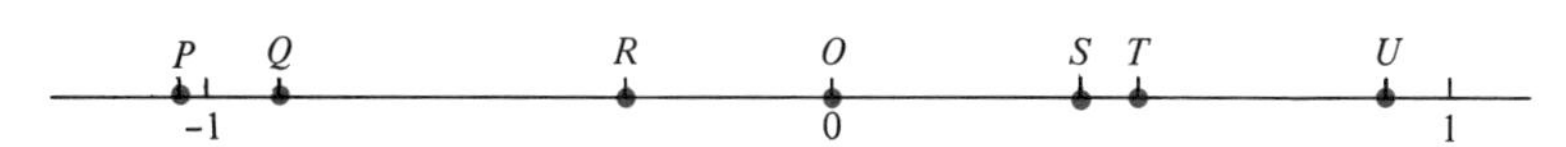

FIGURE 2.14

In Exercises 3–16 fill in "<" or ">".

3. 1 ____ 9 **4.** 13 ____ 12 **5.** 7 ____ 6.99 **6.** 8 ____ 0 **7.** -5 ____ 5
8. -1 ____ 10 **9.** -10 ____ 1 **10.** 0 ____ -3 **11.** -13 ____ -20 **12.** .09 ____ 1
13. $-.09$ ____ -1 **14.** $\frac{3}{4}$ ____ $\frac{1}{2}$ **15.** $\frac{-3}{5}$ ____ $\frac{2}{3}$ **16.** $\frac{9}{7}$ ____ $\frac{7}{9}$

In Exercises 17–26 fill in "left" or "right".

17. On L, 3 lies to the ____ of 8. **18.** On L, -3 lies to the ____ of 8.

19. On L, 3 lies to the ______ of -8. **20.** On L, -3 lies to the ______ of -8.

21. On L, $\frac{1}{3}$ lies to the ______ of $\frac{1}{4}$. **22.** On L, $-.8$ lies to the ______ of -1.

23. On L, π lies to the ______ of 3. **24.** On L, $\frac{-2}{3}$ lies to the ______ of $-.7$.

25. On L, .83 lies to the ______ of .829 . **26.** On L, $-$ 1.43 lies to the ______ of $-$ 1.427 .

In Exercises 27–30 rearrange the numbers so that you can write "$<$" between any two numbers.

27. $2, 8, -8, \frac{1}{2}, 0, -6$ **28.** 1.5, 1.439, 1.498, 1.501, 1.44, 1.51

29. $-2.7, -2.689, -2.69, -2.693, -2.701, -2.71$ **30.** $\frac{3}{8}, \frac{3}{4}, \frac{1}{4}, \frac{5}{8}, \frac{7}{8}, 2, \frac{1}{2}$

In Exercises 31–38 fill in "$\leq$" or "$\geq$". If both symbols apply, write "both".

31. 6 ______ 3 **32.** -6 ______ 3 **33.** -6 ______ -3 **34.** -3 ______ -6

35. -12 ______ 12 **36.** 9 ______ 9 **37.** .099 ______ 1 **38.** -3 ______ -3

In Exercises 39–46 describe each as either (a) an open interval (b) a closed interval (c) a half-open interval (d) an open right ray (e) a closed right ray (f) an open left ray or (g) a closed left ray.

39. $(4, 7)$ **40.** $[3, 6]$ **41.** $(-\infty, 6]$ **42.** $[0, \infty)$

43. $(-\infty, -4)$ **44.** $[100, 200]$ **45.** $(100, \infty)$ **46.** $(0, 100]$

In Exercises 47–56 answer "true" or "false".

47. 4 is in $[2, 7]$. **48.** -3 is in $(-4, 4)$. **49.** -2.1 is in $[-4, 1]$. **50.** 0 is in $(0, 6)$.

51. $\frac{1}{2}$ is in $[-1, 1]$. **52.** $\frac{-3}{2}$ is in $[-1, 0]$. **53.** 4 is in $[4, \infty)$. **54.** -2 is in $(-1, \infty)$.

55. 3 is in $(3, \infty)$. **56.** 3 is in $(-\infty, -3)$.

In Exercises 57–62 describe in one of the forms

$[a, b]$, (a, b), (a, ∞), $[a, \infty)$, $(-\infty, b)$, or $(-\infty, b]$

all numbers satisfying each of the following:

57. $2 \leq x \leq 9$ *Answer:* [2, 9]

58. $x > 6$ **59.** $x < -6$ **60.** $-100 < x < 100$ **61.** $x \geq 8$ **62.** $x \leq 8$

2.6 Arithmetic of Inequalities

There are several important arithmetic properties of inequalities.

Let a, b, c, d be real numbers.

addition and inequalities **(A) If $a < b$, then $a + c < b + c$ and $a - c < b - c$**

Property **(A)** asserts that *if you add or subtract the same number on both sides of an inequality, the sense (or direction) of the inequality is preserved.*

To illustrate Property **(A)**, let $a < b$. Then b lies to the right of a. Suppose, for example, $c > 0$. Then $a + c$ lies c units to the right of a, and $b + c$ lies *the same number of units* to the right of b. Thus $b + c$ lies to the right of $a + c$, and therefore

$a + c < b + c$ [See Figure 2.15 .] △

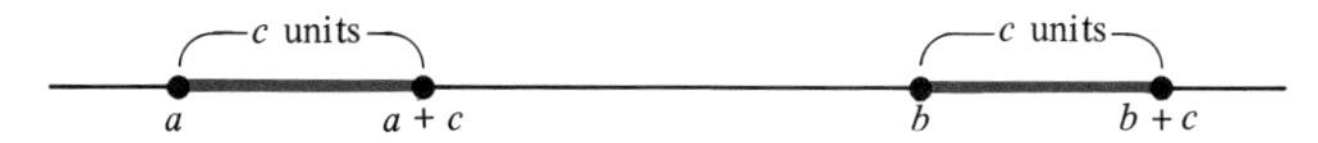

FIGURE 2.15 If $a < b$, then $a + c < b + c$. (The diagram is for the case of $c > 0$ and $a + c < b$.)

(B) If $\quad a < b \quad$ **and** $\quad c < d, \quad$ **then** $\quad a + c < b + d$

Thus inequalities may be added.

To verify Property **(B)**, suppose $a < b$. Then $a + c < b + c$ (by Property **A**). If $c < d$, then $b + c < b + d$ (by Property **A**). Thus

$$a + c < b + c \quad \text{and} \quad b + c < b + d$$

By Transitivity,

$a + c < b + d$ △

EXAMPLE 1 $\quad 2 < 4$

(a) By Property **(A)**,

$$\underbrace{2 + 5}_{7} < \underbrace{4 + 5}_{9}$$

(b) Also, by Property **(A)**,

$$\underbrace{2 - 5}_{-3} < \underbrace{4 - 5}_{-1}$$

□

EXAMPLE 2 (a) $3 < 6$ and $1 < 5$
By Property **(B)**,

$$\underbrace{3 + 1}_{4} < \underbrace{6 + 5}_{11}$$

(b) $-7 < -3$ and $-4 < -1$
Again by Property **(B)**,

$$\underbrace{(-7) + (-4)}_{-11} < \underbrace{(-3) + (-1)}_{-4}$$

□

multiplication and inequalities

Recall that c is *positive* if $c > 0$; c is *negative* if $c < 0$.

(C)$_1$ If $\quad a < b \quad$ **and** $\quad c > 0, \quad$ **then** $\quad ac < bc$

(C)$_2$ If $\quad a < b \quad$ **and** $\quad c < 0, \quad$ **then** $\quad ac > bc$

Thus *multiplication by a positive number preserves the sense of an inequality. Multiplication by a* **negative** *number* **reverses** *the sense of an inequality.* (The proofs are omitted.)

EXAMPLE 3 $\quad 2 < 3$

(a)

$$\underbrace{2(4)}_{8} < \underbrace{3(4)}_{12}$$

by Property **(C)$_1$**

(b)

$$\underbrace{2(-4)}_{-8} > \underbrace{3(-4)}_{-12}$$

by Property **(C)$_2$**. □

EXAMPLE 4 $-5 < -3$

(a) $\underbrace{(-5)(2)}_{-10} < \underbrace{-3)(2)}_{-6}$ by Property $(\mathbf{C})_1$ (b) $\underbrace{(-5)(-2)}_{10} > \underbrace{(-3)(-2)}_{6}$ by Property $(\mathbf{C})_2$. □

Division by a nonzero number c can be regarded as multiplication by $\frac{1}{c}$. Because $c > 0$ exactly when $\frac{1}{c} > 0$, the following hold.

$(\mathbf{C})_3$ If $\quad a < b \quad$ **and** $\quad c > 0, \quad$ **then** $\quad \frac{a}{c} < \frac{b}{c}$

$(\mathbf{C})_4$ If $\quad a < b \quad$ **and** $\quad c < 0, \quad$ **then** $\quad \frac{a}{c} > \frac{b}{c}$

These properties can also be stated in terms of $\leqslant$. For example:

(A′) If $a \leqslant b$, then $a + c \leqslant b + c$ and $a - c \leqslant b - c$

Property **(A)** implies that

if $a < b$, then $0 < b - a$.

For, if $a < b$, then $a - a < b - a$, that is, $0 < b - a$. Also, if $0 < b - a$, then by adding a to both sides, it follows that $a < b$. Thus $a < b$ is equivalent to $0 < b - a$.

EXERCISES

In Exercises 1–22 let a, c, x, and y be any numbers. Fill in: $<$ or $>$

1. $5 + c \;\square\; 10 + c$
2. $-2 + c \;\square\; -1 + c$
3. $-4 + c \;\square\; c$
4. If $b > 0$, $-4b \;\square\; -7b$
5. $a + 3 \;\square\; a + 2$
6. $a - 6 \;\square\; a - 5$
7. $a \;\square\; a - 2$
8. $a - 4 \;\square\; 2 + a$
9. $(-7) + (-2) \;\square\; (-4) + (-1)$
10. $(-8) + (-10) \;\square\; (-9) + (-7)$
11. $5(8) \;\square\; 3(8)$
12. $-2(8) \;\square\; -1(8)$
13. $6(-3) \;\square\; 15(-3)$
14. $(-6)(-2) \;\square\; (-4)(-2)$
15. $(-2)(-7) \;\square\; (-2)(-5)$
16. $(-3)(4)(7) \;\square\; (-3)(3)(6)$
17. If $x < y$ and $z < 0$, then $zx \;\square\; zy$
18. If $x < y$, then $y - x \;\square\; 0$
19. $\frac{1}{6} \;\square\; \frac{1}{5}$
20. If $r < 0$, then $\frac{-4}{r} \;\square\; \frac{-3}{r}$
21. $\frac{1}{9} + \frac{1}{4} \;\square\; \frac{1}{3} + \frac{1}{4}$
22. $\frac{2}{3}\left(\frac{-2}{5}\right) \;\square\; \left(\frac{-3}{4}\right)\frac{2}{3}$
23. Show that if $x < 4$, then $x + 2 < 6$.
24. Show that if $y \leqslant 9$, then $y - 4 \leqslant 5$.
25. Show that if $a \leqslant 4$, then $2a \leqslant 8$.
26. Show that if $b < -2$, then $\frac{b}{2} < -1$.

27. Show that if $x \leqslant 5$, then $-2x \geqslant -10$.

28. Show that if $x < 3$, then $3x < 10$.

29. Show that if $2x < 3$, then $x < 2$.

30. Show that if $x < 10$, then $\frac{-x}{10} > -1$.

31. One angle of a triangle is obtuse (that is, it measures more than 90° but less than 180°). Let x and y be the number of degrees of the other two angles. Find an inequality that expresses the restrictions on $x + y$.

32. A woman planning a party has at most $50 to spend on various appetizers. She wants to buy mixed nuts at $2 per pound, dried fruit at $3 per pound, and chocolate at $4 per pound. Let x, y, and z be, respectively, the number of pounds of nuts, dried fruit, and chocolate she can buy. Find an inequality that expresses her financial limitation.

2.7 Solving Inequalities

sets

In order to discuss the "solutions of an inequality," it is useful to introduce the notion of a *set*.

Definition. Any collection of (distinct) objects is known as a **set.**

Often the objects of a set are indicated by means of **braces:**

$$\{ \quad \}$$

The objects are listed between the braces. *The order in which the objects are listed does not matter.*

EXAMPLE 1

The set consisting of the first four positive integers is:

$$\{1, 2, 3, 4\}$$

Because the numbers can be listed in any order, this set can also be indicated by

$$\{2, 3, 4, 1\} \quad \text{or by} \quad \{4, 3, 2, 1\}, \quad \text{etc.}$$ □

A set is said to be **finite** if there is a specific number of objects in it: 4, or 8, or 50, or 1000.‡ Thus the set in EXAMPLE 1, which consists of four objects, is finite. In contrast, here are some **infinite** sets with which you are familiar.

EXAMPLE 2

Each of the following sets is infinite.

(a) the set of all real numbers

(b) the set of all positive integers

(c) $[0, 1]$

Among the numbers in this set are:

$$0, 1, \frac{1}{2}, \frac{1}{3}, \frac{1}{4}, \frac{1}{5}, \ldots$$ □

‡A finite set may even contain 0 objects, and thus be considered "*empty*".

solutions of inequalities

Recall that an equation is a statement of equality, and has the form:

$$\square = \square$$

Definition. An **inequality** is a statement of one of the forms:

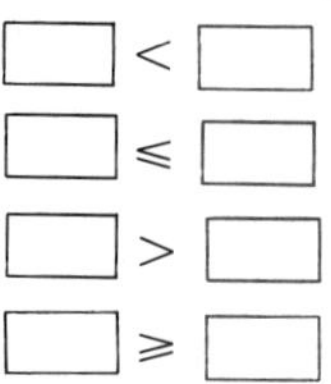

EXAMPLE 3 Each of the following is an inequality:

(a) $2 < 5$ (b) $x + 1 \leqslant 6$ (c) $2x - 3 \geqslant 4 - 3x$ □

A solution (or root) of an equation (in a single variable) is a number that when substituted in the equation yields a true statement. For most of the equations you have considered thus far, there was only one solution.

Definition. A **solution of an inequality** (in a single variable) is a number that when substituted in the inequality yields a true statement. The **solution set of an inequality** is the set of all solutions.

For example, the inequality $x < 0$ has as its solution set, the (infinite) set of negative numbers.

Equivalent equations are equations with exactly the same solutions (or roots).

Definition. **Equivalent inequalities** are inequalities with exactly the same solution sets.

You solve an inequality by successively transforming it into simpler equivalent inequalities. The arithmetic properties you studied in the preceding section enable you to simplify inequalities. Thus adding or subtracting the same number, or multiplying or dividing by the same *positive* number on both sides of an inequality *preserves* the sense of the inequality. On the other hand, multiplying or dividing both sides of an inequality by a *negative* number *reverses* the sense of the inequality. Here is how you use these properties to simplify an inequality.

Addition Principle (for inequalities)

The **Addition Principle** asserts that you can add the same quantity to both sides of an inequality:

Let R be any rational expression (possibly a polynomial or even a constant).

and

$$\square + R < \square + R$$

are equivalent inequalities.

EXAMPLE 4 Solve: $x + 5 < 10$

Solution. Use the Addition Principle as you would for the equation $x + 5 = 10$. Thus, subtract 5 from (or add -5 to) both sides of the given inequality:

$$\begin{array}{rcr} x + 5 & < & 10 \\ -5 & \triangle\!\!-\!\!\triangle & -5 \\ \hline x & < & 5 \end{array}$$

The solution set is the open left ray $(-\infty, 5)$. □

Multiplication Principle (for inequalities)

The **Multiplication Principle** asserts that *you can multiply both sides of an inequality by the same positive number.* But *if you multiply both sides by the same negative number,* the sense of inequality is reversed:

(a) Let $p > 0$. Then

$$\square < \square$$

and

$$\square \cdot p < \square \cdot p$$

are equivalent inequalities.

(b) Let $n < 0$. Then

$$\square < \square$$

and

$$\square \cdot n > \square \cdot n$$

are equivalent inequalities.

Because $\frac{a}{b} = a \cdot \frac{1}{b}$, $b \neq 0$, and b and $\frac{1}{b}$ are either both positive or both negative, the Multiplication Principle also enables you to divide both sides of an inequality by the same nonzero constant.

EXAMPLE 5 Solve: $4x > 24$

Solution. Divide both sides by the *positive* number 4 $\left(\text{or multiply by } \frac{1}{4}\right)$.

$$\frac{4x}{4} > \frac{24}{4}$$

$$x > 6$$

The solution set is the open right ray $(6, \infty)$. □

EXAMPLE 6 Solve: $\frac{-x}{2} < 4$

Solution. Multiply both sides by the *negative* number −2. This *reverses* the sense of inequality.

$$\frac{-x}{2} < 4$$

$$x > -8$$

The solution set is the open right ray $(-8, \infty)$. □

The Addition and Multiplication Principles also apply to inequalities involving $\leqslant$.

EXAMPLE 7 Solve: $3x + 6 \leqslant 9 - x$

Solution. Bring variables to one side, constants to the other.

$$\begin{array}{rcl} 3x + 6 & \leqslant & 9 - x \\ +x - 6 & \triangle\!\!-\!\!\triangle & -6 + x \\ \hline 4x & \leqslant & 3 \\ x & \leqslant & \dfrac{3}{4} \end{array}$$

The solution set is the closed left ray $\left(-\infty, \frac{3}{4}\right]$. □

EXAMPLE 8 Solve: $\dfrac{x}{2} + \dfrac{1}{3} \geqslant \dfrac{3x + 1}{4}$

Solution. The LCD is 12. Thus multiply both sides of the inequality by 12.

$$\begin{array}{rcl} 6x + 4 & \geqslant & 3(3x + 1) \\ 6x + 4 & \geqslant & 9x + 3 \\ -6x - 3 & \triangle\!\!-\!\!\triangle & -6x - 3 \\ \hline 1 & \geqslant & 3x \\ \dfrac{1}{3} & \geqslant & x \qquad \text{or } x \leqslant \dfrac{1}{3} \end{array}$$

The solution set is the closed left ray $\left(-\infty, \frac{1}{3}\right]$. □

simultaneous inequalities

Sometimes, you will be given two inequalities to solve simultaneously.

EXAMPLE 9 Solve: $3 < x + 2 < 9$

Solution. You are given two inequalities that hold for x:

$$3 < x + 2 \qquad \text{and} \qquad x + 2 < 9$$

You can work with both at the same time by subtracting 2 from each expression.

$$\begin{array}{ccccc} 3 & < & x+2 & < & 9 \\ -2 & & -2 & & -2 \\ \hline 1 & < & x & < & 7 \end{array}$$

The solution set is the open interval $(1, 7)$. □

EXAMPLE 10 Solve: $-10 \leqslant 3x - 5 \leqslant 10$

Solution. Add 5 to each expression.

$$-5 \leqslant 3x \leqslant 15$$

Divide each expression by (the positive number) 3.

$$\frac{-5}{3} \leqslant x \leqslant 5$$

The solution set is the closed interval $\left[\frac{-5}{3}, 5\right]$. □

EXAMPLE 11 Solve: $4 < -x < 10$

Solution. Multiply each expression by -1. This *reverses* the senses of both inequalities.

$$-4 > x > -10 \qquad \text{or} \qquad -10 < x < -4$$

The solution set is the open interval $(-10, -4)$. □

EXERCISES

In Exercises 1–56 solve the indicated inequalities.

1. $x + 2 < 6$ **2.** $x + 3 < 0$ **3.** $x - 8 < 5$ **4.** $x + 4 < -2$

5. $t + 6 \leqslant 10$ **6.** $2 - y \leqslant 8$ **7.** $y + 2 > 3$ **8.** $t - \frac{1}{2} \geqslant 1$

9. $2x < 6$ **10.** $3x \leqslant -6$ **11.** $\frac{t}{2} > 4$ **12.** $\frac{x}{4} \geqslant 0$

13. $-x < 1$ **14.** $-3s < 5$ **15.** $-2y > 2$ **16.** $-4z \geqslant -2$

17. $-\frac{x}{3} \leqslant \frac{1}{6}$ **18.** $\frac{-4x}{3} > 2$ **19.** $x + 2 < 5x - 1$ **20.** $2x - 7 \geqslant x - 9$

21. $15x + 2 \leqslant 12x - 7$ **22.** $\frac{t}{2} + 1 \leqslant 3t + 2$ **23.** $2(x - 1) + 3(x + 2) \geqslant x - 5$

24. $2(3x - 2) \leqslant 4(1 - x)$ **25.** $7(2x + 1) + 3 < 5\left(2x - \frac{1}{5}\right)$ **26.** $4(x + 1) - 3 \leqslant 2(1 - 2x) - 1$

27. $\frac{x}{4} + \frac{1}{2} \leqslant \frac{3}{8}$ **28.** $\frac{x}{5} > \frac{1}{3}$ **29.** $\frac{3x}{7} \geqslant \frac{-2}{9}$

30. $\frac{-2t}{5} > \frac{3}{10}$ **31.** $\frac{1 + y}{2} < \frac{3}{4}$ **32.** $\frac{2x - 1}{3} > \frac{3}{5}$

33. $\frac{1-2x}{10} \leq \frac{x}{5}$

34. $\frac{3-u}{3} < \frac{u}{4}$

35. $\frac{x}{3} - \frac{1}{6} \leq \frac{3x}{4}$

36. $\frac{5x-1}{2} + \frac{1}{5} > \frac{x+1}{10}$

37. $2 < x + 2 < 8$

38. $0 < x - 3 < 4$

39. $-1 \leq x + 8 \leq 5$

40. $\frac{1}{2} \leq x + \frac{1}{2} \leq 2$

41. $9 < 3x < 12$

42. $-10 \leq 2y \leq -6$

43. $-15 < 5z < 15$

44. $0 \leq \frac{x}{3} \leq 9$

45. $-1 < 2x + 1 < 8$

46. $-2 < 3x - 2 < 4$

47. $-9 < 5t - 1 < 9$

48. $-2 \leq \frac{2+x}{3} \leq 2$

49. $-1 \leq -x \leq 2$

50. $-4 < -x < 4$

51. $0 < -2x < 8$

52. $-3 < 1 - 2x < 9$

53. $-2 < 3 - 5x < 4$

54. $-34 \leq 1 - 7x < 22$

55. $\frac{1}{2} < \frac{3-2x}{4} < 1$

56. $\frac{1}{2} < \frac{3}{4} - 2x \leq 1$

57. A highway has a minimum speed limit of 40 miles per hour and a maximum speed limit of 55 miles per hour. A car travels on this highway for 3 hours within the legal range. What interval describes the distance traveled?

58. On an April morning the temperature in Boston ranges between 50° and 59° Fahrenheit. Celsius (C) can be determined from Fahrenheit (F) by the formula

$$C = \frac{5}{9}(F - 32)$$

On the Celsius scale what is the range of temperature in Boston that morning?

59. In London one December morning the temperature ranges between 0° and 5° Celsius. What is the range on the Fahrenheit scale? (See Exercise 58.)

60. During the preliminary phase of labor the expectant mother must take 6 to 9 deep breaths per minute. What is the range of breaths she should take during a 40-second period?

2.8 Absolute Value in Equations

distance from the origin

Definition. The **absolute value of a number** n is its distance from the origin.

Let $|n|$ denote the absolute value of n.

For every number n,

$$|n| = |-n|$$

because n and its inverse $-n$ are the same distance from the origin. Observe that (if $n \neq 0$) n and $-n$ are on opposite sides of the origin. Thus absolute value measures distance from the origin, but neglects direction.

EXAMPLE 1

(a) $|6| = 6$ because 6 is 6 units from the origin.

(b) $|-6| = 6$ because -6 is 6 units from the origin.

[See Figure 2.16 .] □

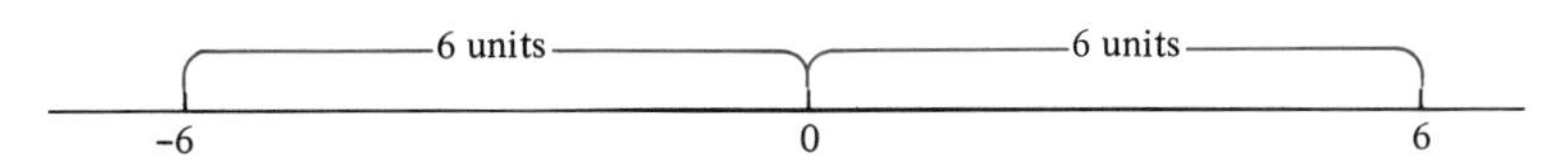

FIGURE 2.16 6 is 6 units to the right of the origin. −6 is 6 units to the left of the origin. Therefore $|6| = |-6| = 6$

EXAMPLE 2 A Volkswagen gets 20 miles to the gallon. It uses the same 3 gallons of gas to travel 60 miles west as it would to travel 60 miles east. [See Figure 2.17 .] Thus gas consumption relates to absolute value; you measure distance, but neglect direction. □

FIGURE 2.17

The absolute value of a number n can be described in the following way:

$$|n| = n, \qquad \text{if } n \text{ is positive or } 0.$$
$$|n| = -n, \qquad \text{if } n \text{ is negative.}$$

For, if n is negative, then $-n$ is positive; $-n$ is the distance between n and the origin. [See Figure 2.18(b).] Observe that

$$|n| \geqslant 0 \qquad \text{for every number } n.$$

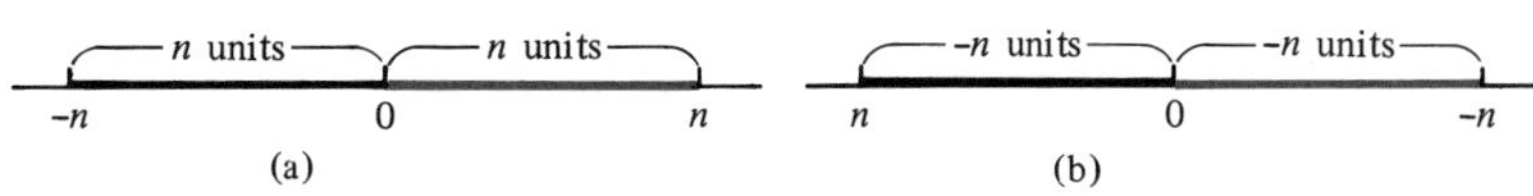

FIGURE 2.18 (a) If $n > 0$, then $|n| = n$. (b) If $n < 0$, then $-n > 0$ and $|n| = -n$

EXAMPLE 3 (a) $|8| = 8$ because 8 is positive.

(b) $|-8| = -(-8) = 8$ because -8 is negative.

(c) $|0| = 0$ □

distance between n and a

Clearly, $|n - 0| = |n| = |0 - n|$, whether n lies to the right or to the left of 0. Thus $|n - 0|$ is the distance between n and 0. In general, *$|n - a|$ is the distance between n and a*. [See Figure 2.19 .]

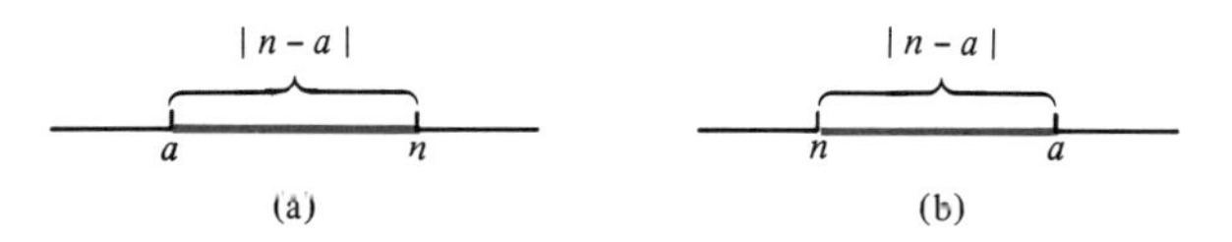

FIGURE 2.19 (a) $n > a$ (b) $n < a$

$|n - a| = n - a$ $\qquad$ $-(n - a) = a - n > 0$

$|n - a| = a - n$

In both cases, $|n - a|$ is the distance between n and a.

EXAMPLE 4 (a) $|8 - 3|$ is the distance between 8 and 3.

$$|8 - 3| = 5$$

(b) $|-2 - 3|$ is the distance between -2 and 3.

$$|-2 - 3| = |-5| = 5$$

[See Figure 2.20 on page 90.] □

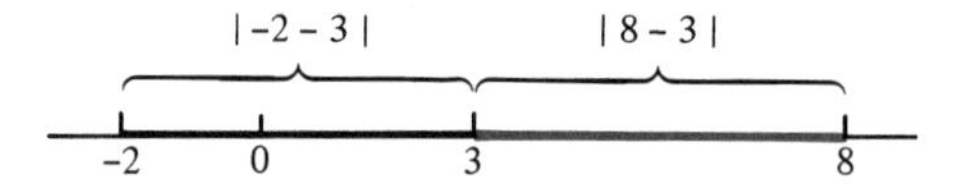

FIGURE 2.20 −2 and 8 are each 5 units from 3.

The absolute value of a product (or quotient) equals the product (or quotient) of the absolute values.

Thus let a and b be arbitrary real numbers.

$$|ab| = |a|\,|b|$$

and

$$\left|\frac{a}{b}\right| = \frac{|a|}{|b|}, \qquad \textbf{if } b \neq 0$$

EXAMPLE 5 (a) Let $a = 5$, $b = -3$.

$$|ab| = |5(-3)| = |-15| = 15$$
$$|a|\,|b| = |5|\,|-3| = 5 \cdot 3 = 15$$

(b) Let $a = -9$, $b = -2$.

$$|ab| = |(-9)(-2)| = |18| = 18$$
$$|a|\,|b| = |-9|\,|-2| = 9 \cdot 2 = 18$$

(c) Let $a = -3$, $b = 4$.

$$\left|\frac{a}{b}\right| = \left|\frac{-3}{4}\right| = \frac{3}{4}, \qquad \frac{|a|}{|b|} = \frac{|-3|}{|4|} = \frac{3}{4}$$

□

equations

There is a systematic way of solving equations involving absolute value. From the formula

$$|x| = \begin{cases} x, & \text{if } x \geq 0, \\ -x, & \text{if } x < 0, \end{cases}$$

it follows that the equation

$$|x| = c, \qquad \text{where } c > 0,$$

has two solutions, given by

$$x = c, \qquad x = -c.$$

Write these as $x = \pm c$.

For example, if $|x| = 9$, then $x = \pm 9$.
Also, the equation

$$|x| = 0$$

has 0 as its only solution. Similarly, the equation

$$|x - a| = c, \qquad c \geq 0,$$

is transformed into

$$x - a = \pm c$$

Thus $x = a \pm c$.

EXAMPLE 6 Solve: $|x - 3| = 5$

Solution.

$$x - 3 = \pm 5$$
$$x = 3 \pm 5$$

You obtain two equations:

$x = 3 + 5$	$x = 3 - 5$
$= 8$	$= -2$

The roots are 8 and -2.

Check:

for 8:	for -2:
$\lvert 8 - 3\rvert \stackrel{?}{=} 5$	$\lvert -2 - 3\rvert \stackrel{?}{=} 5$
$5 \stackrel{\checkmark}{=} 5$	$5 \stackrel{\checkmark}{=} 5$

□

EXAMPLE 7 Solve: $|2x + 1| = 6$

Solution.

$$2x + 1 = \pm 6$$
$$2x = -1 \pm 6$$
$$x = -\frac{1}{2} \pm 3$$

The roots are $\frac{5}{2}$ and $\frac{-7}{2}$. □

The equation

$$|x - 1| = -1$$

has no solution because the left side is the absolute value of a number, and is consequently nonnegative. However, the right side is negative.

$|a| = |b|$ Finally, if $|a| = |b|$, then by considering all combinations of signs for a and b, you will see that $a = \pm b$.

EXAMPLE 8 Solve $\left|\frac{2x - 3}{6x}\right| = 1$

Solution. Use $\left|\frac{a}{b}\right| = \frac{|a|}{|b|}$.

$$\frac{|2x - 3|}{|6x|} = 1$$
$$|2x - 3| = |6x|$$
$$2x - 3 = \pm 6x$$

$2x - 3 = 6x$	$2x - 3 = -6x$
$-3 = 4x$	$8x = 3$

$$\frac{-3}{4} = x \qquad\Bigg|\qquad x = \frac{3}{8}$$

Thus $\frac{-3}{4}$ and $\frac{3}{8}$ are the roots. □

EXERCISES

In Exercises 1–4 determine each absolute value.

1. $|15|$ **2.** $|-19|$ **3.** $|0|$ **4.** $\left|-\frac{2}{3}\right|$

5. Which numbers are 8 units from the origin?

6. Which numbers are 2 units from 5?

7. Which numbers are 5 units from 2?

8. Which numbers are 4 units from -1?

9. Which numbers are 3 units from -5?

10. Which numbers are $\frac{1}{2}$ unit from 2?

In Exercises 11–14 answer "true" or "false". If "false", find a *counterexample,* that is, find a numerical example that shows the statement to be false.

11. For all a, $|3a| = 3|a|$

12. For all a, $\left|\frac{a}{-10}\right| = \frac{|a|}{-10}$

13. For all a and b, $|-ab| = -ab$

14. For all a and b, $b \neq 0$, $\left|\frac{-a}{-b}\right| = \frac{|-a|}{|-b|}$

In Exercises 15–54 solve each equation. Check the ones so indicated.

15. $|x| = 7$ **16.** $|x| = 0$ **17.** $|x| = -2$ **18.** $|3x| = 6$

19. $|-2x| = 4$ **20.** $|-4x| = -4$ **21.** $|x - 1| = 6$ (Check.) **22.** $|x - 3| = 0$

23. $|x + 2| = 4$ **24.** $|x + 10| = 1$ **25.** $|2x - 1| = 9$ (Check.) **26.** $|3x + 1| = 4$

27. $\left|\frac{x}{2} - 5\right| = 3$ **28.** $\left|\frac{x - 5}{2}\right| = 3$ **29.** $|5x + 1| = 0$ **30.** $|4x - 3| = -3$

31. $|2x + 1| = 4x$ (Check.) **32.** $|3x - 2| = 2x - 3$ (Check.) **33.** $|x + 1| = 5 - 2x$ (Check.)

34. $|9 - x| = 2x + 1$ (Check.) **35.** $|x + 5| = 4x + 1$ (Check.) **36.** $|6x - 1| = 9 - 3x$ (Check.)

37. $|x + 1| = x$ (Check.) **38.** $|x + 1| = 2x$ (Check.) **39.** $|x| = x$

40. $\left|\frac{1 - 6x}{2}\right| = x + 2$ (Check.) **41.** $|2x| = |x + 1|$ **42.** $|3x - 1| = |1 + 5x|$ (Check.)

43. $|9 - x| = |6x|$ **44.** $|x + 4| = |8 - 2x|$ **45.** $|x + 3| = |x - 2|$ **46.** $|x + 4| = |1 - x|$

47. $\left|\frac{2x}{x + 2}\right| = 1$ **48.** $\left|\frac{x}{3x + 5}\right| = 2$ **49.** $\left|\frac{4x + 1}{x - 2}\right| = 3$ **50.** $\left|\frac{1 - x}{5x - 2}\right| = 2$

51. $\left|\frac{2x + 1}{5 - 2x}\right| = \frac{1}{2}$ **52.** $\left|\frac{3x - 7}{x + 2}\right| = \frac{3}{4}$ **53.** $\left|\frac{x + 2}{x - 1}\right| = 1$ **54.** $\left|\frac{4x - 3}{x + 5}\right| = 0$

*2.9 Absolute Value in Inequalities

$|x| < c$

As you know, $|x|$, or $|x - 0|$, is the distance between x and the origin. Thus $|x| < 4$ if its distance from the origin is less than 4. [See Figure 2.21 .] The distance between 3.5 and 0 is 3.5, which is less than 4. And the distance between -3 and 0 is 3, which is less than 4.

$$|-3 - 0| = |-3| = 3$$

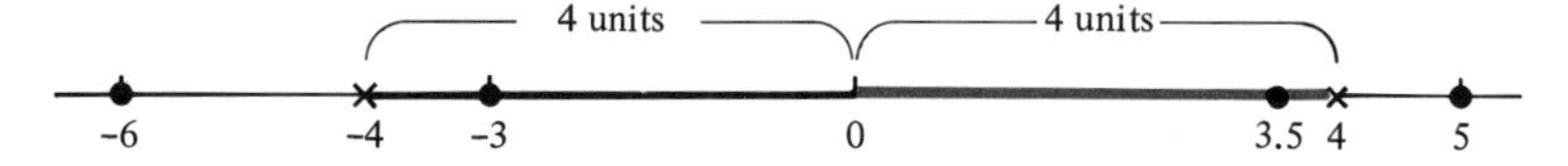

FIGURE 2.21. $|x| < 4$ means the same as $-4 < x < 4$.

On the other hand, the distance between 5 and 0 is 5, which is greater than 4. And the distance between -6 and 0 is 6, which is greater than 4.

$$|-6 - 0| = |-6| = 6$$

Thus, 3.5 and -3 are both solutions of $|x| < 4$. But 5 and -6 are not solutions of $|x| < 4$. In fact, the solutions of $|x| < 4$ are the numbers x with $-4 < x < 4$.

Let $c > 0$. In general,

$$|x| < c \qquad \text{means the same as} \qquad -c < x < c.$$

[See Figure 2.22 .]

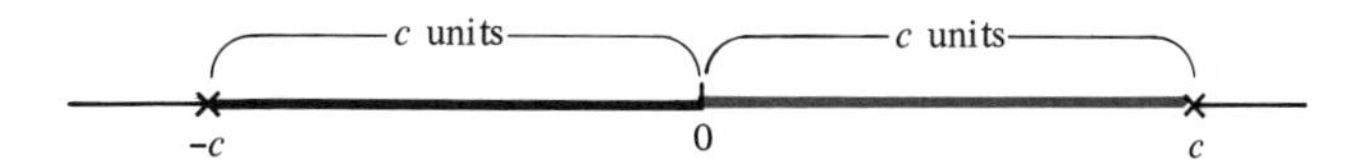

FIGURE 2.22. Let $c > 0$. Then $|x| < c$ means the same as $-c < x < c$.

EXAMPLE 1 Solve: $|x| < 3$

Solution.

$$|x| < 3 \qquad \text{means} \qquad -3 < x < 3.$$

The solution set is the *open* interval $(-3, 3)$. □

$|x - a| < c$

Next, observe that $|x - 2|$ is the distance between x and 2. Thus

$$|x - 2| < 1$$

means that x lies within 1 unit of 2. [See Figure 2.23 on page 94.] The distance between 2.8 and 2 is .8, which is less than 1. The distance between 1.5 and 2 is .5, which is less than 1.

$$|1.5 - 2| = |-.5| = .5$$

On the other hand, the distance between 3.1 and 2 is 1.1, which is

*Optional topic

greater than 1. And the distance between 0 and 2 is 2, which is greater than 1.

$$|0 - 2| = |-2| = 2$$

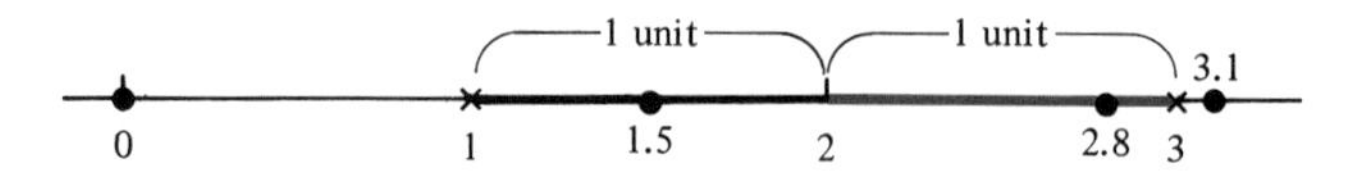

FIGURE 2.23. $|x - 2| < 1$ means $2 - 1 < x < 2 + 1$ or $1 < x < 3$.

Thus 2.8 and 1.5 are solutions of $|x - 2| < 1$, but 3.1 and 0 are not. The solutions of $|x - 2| < 1$ are the numbers x with

$$2 - 1 < x < 2 + 1 \qquad \text{or} \qquad 1 < x < 3.$$

For any number a, $|x - a|$ is the distance between x and a. Let $c > 0$. Then

$$|x - a| < c \qquad \text{means the same as} \qquad -c < x - a < c.$$

Add a to each expression and observe that this is equivalent to

$$a - c < x < a + c.$$

[See Figure 2.24 .]

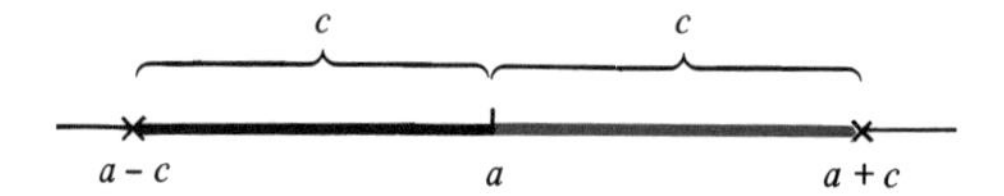

FIGURE 2.24. Let $c > 0$. Then $|x - a| < c$ means the same as $a - c < x < a + c$.

EXAMPLE 2 Solve: $|x - 6| < 4$

Solution.

$$-4 < x - 6 < 4 \qquad \text{Add 6 to all three expressions.}$$

$$2 < x < 10$$

The solution set is the *open* interval (2, 10). □

$|x - a| \leqslant c$

$$|x| \leqslant c \qquad \text{means the same as} \qquad -c \leqslant x \leqslant c,$$

and

$$|x - a| \leqslant c \qquad \text{is equivalent to} \qquad a - c \leqslant x \leqslant a + c.$$

EXAMPLE 3 Solve: $|5x - 2| \leqslant 18$

Solution.

$$-18 \leqslant 5x - 2 \leqslant 18$$

$$-16 \leqslant 5x \leqslant 20$$

$$\frac{-16}{5} \leqslant x \leqslant 4$$

The solution set is the *closed* interval $\left[\frac{-16}{5}, 4\right]$. □

Multiplication by a negative number reverses the sense of an inequality. When two inequalities are involved, both senses are reversed.

EXAMPLE 4 Solve: $|3 - 2x| < 5$

Solution.

$$-5 < 3 - 2x < 5$$
$$\underline{-3} \qquad \underline{-3} \qquad \underline{-3}$$
$$-8 < -2x < 2 \qquad \text{Multiply each expression by } \frac{-1}{2}.$$
$$4 > x > -1 \quad \text{or}$$
$$-1 < x < 4$$

The solution set is the *open* interval $(-1, 4)$. □

$|x| > c$ Next observe that $|x| > 5$ means that the distance between x and the origin is greater than 5. Also, if $c > 0$, then $|x| > c$ *means the distance between x and the origin is greater than c*. As you see in Figure 2.25,

$$\text{if } |x| > c, \qquad \text{then} \qquad x < -c \quad \text{or} \quad x > c.$$

The notation

$A \cup B$ (read: A **union** B)

indicates the set of objects that are in *at least one* of the sets A or B. Thus if $|x| > c$, then x lies in $(-\infty, -c) \cup (c, \infty)$.

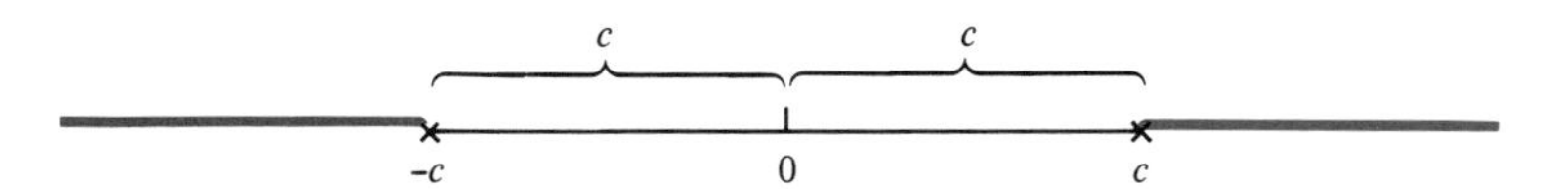

FIGURE 2.25. Let $c > 0$. Then $|x| > c$ means that x lies in $(-\infty, -c) \cup (c, \infty)$.

$|x - a| > c$ Similarly, $|x - a| > c$ *means that the distance between x and a is greater than c, and thus*

$$\text{either} \qquad x - a < -c \qquad \text{or} \qquad x - a > c$$

That is,

$$x < a - c \qquad \text{or} \qquad x > a + c$$

Thus if $|x - a| > c$, then x lies in $(-\infty, a - c) \cup (a + c, \infty)$. [See Figure 2.26 .]

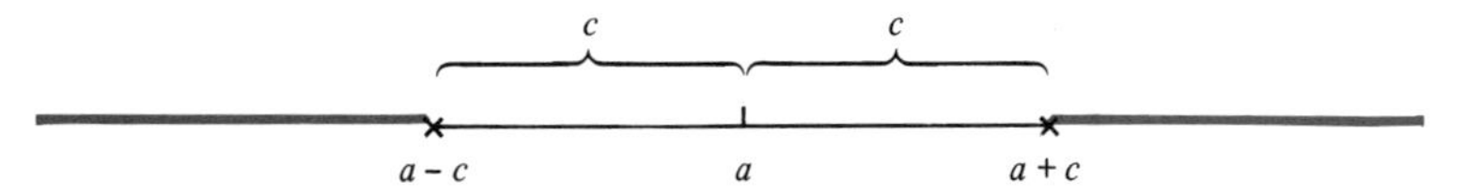

FIGURE 2.26. Let $c > 0$. Then $|x - a| > c$ means that x lies in $(-\infty, a - c) \cup (a + c, \infty)$.

EXAMPLE 5 Solve: $|x - 4| > 10$

Solution.

$$x - 4 < -10 \quad \text{or} \quad x - 4 > 10$$
$$x < -6 \quad \text{or} \quad x > 14$$

The solution set is $(-\infty, -6) \cup (14, \infty)$. □

$|x - a| \geqslant c$

$$|x| \geqslant c \quad \text{means that } x \text{ lies in} \quad (-\infty, -c] \cup [c, \infty)$$

and

$$|x - a| \geqslant c \quad \text{means that } x \text{ lies in} \quad (-\infty, a - c] \cup [a + c, \infty).$$

EXAMPLE 6 Solve: $|2x + 5| \geqslant 9$

Solution.

$$2x + 5 \leqslant -9 \quad \text{or} \quad 2x + 5 \geqslant 9$$
$$2x \leqslant -14 \quad \text{or} \quad 2x \geqslant 4$$
$$x \leqslant -7 \quad \text{or} \quad x \geqslant 2$$

The solution set is $(-\infty, -7] \cup [2, \infty)$. □

$x^2 < a^2$

Finally, observe that for $a > 0$, the inequality

$$x^2 < a^2 \quad \text{is equivalent to} \quad |x| < a,$$
$$\text{and hence to} \quad -a < x < a.$$

[See Figure 2.27.] For,

if $|x| < a$, then $x^2 = |x|^2 < a^2$, whereas

if $|x| \geqslant a$, then $x^2 = |x|^2 \geqslant a^2$. Similarly,

$$x^2 \leqslant a^2, \quad |x| \leqslant a, \quad \text{and} \quad -a \leqslant x \leqslant a$$

are all equivalent.

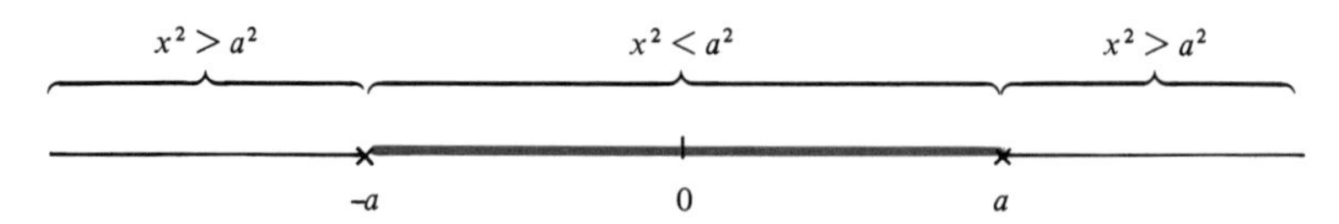

FIGURE 2.27. Let $a > 0$. Then $x^2 < a^2$ is equivalent to $|x| < a$ and hence to $-a < x < a$.

EXAMPLE 7 If $x^2 < 9$, then $-3 < x < 3$ □

$|a \pm b|$

The absolute value of a sum or difference is at most the sum of the absolute values:

$$|a + b| \leqslant |a| + |b|$$
$$|a - b| \leqslant |a| + |b|$$

These inequalities can be combined:

$$|a \pm b| \leqslant |a| + |b|$$

The signs of the numbers a and b determine which of the symbols

$$< \quad \text{or} \quad = \quad \text{holds in each case.}$$

EXAMPLE 8 (a) Let $a = 7$, $b = 5$.

$$|a + b| = \underbrace{|7 + 5|}_{12} = \underbrace{|7| + |5|}_{12} = |a| + |b|$$

Thus here, $|a + b| = |a| + |b|$

(b) Let $a = -1$, $b = 8$.

$$|a + b| = \underbrace{|-1 + 8|}_{7} < \underbrace{|-1| + |8|}_{9} = |a| + |b|$$

Thus here, $|a + b| < |a| + |b|$

(c) Let $a = 6$, $b = 3$.

$$|a - b| = \underbrace{|6 - 3|}_{3} < \underbrace{|6| + |3|}_{9} = |a| + |b|$$

Thus here, $|a - b| < |a| + |b|$

(d) Let $a = 6$, $b = -3$.

$$|a - b| = \underbrace{|6 - (-3)|}_{9} = \underbrace{|6| + |-3|}_{9} = |a| + |b|$$

Thus here, $|a - b| = |a| - |b|$ □

EXERCISES

In Exercises 1–36 solve each inequality.

1. $|x| < 10$ **2.** $|x| < \frac{1}{2}$ **3.** $|x| \leq 4$ **4.** $|x| \leq \frac{5}{3}$ **5.** $|x - 1| < 7$

6. $|x - 2| < 4$ **7.** $|x + 3| < 8$ **8.** $\left|x + \frac{1}{2}\right| < \frac{1}{2}$ **9.** $\left|x - \frac{1}{2}\right| < 4$ **10.** $\left|x + \frac{3}{4}\right| < \frac{1}{2}$

11. $|2x - 1| < 5$ **12.** $|3x - 2| < 7$ **13.** $|5x + 1| < 6$ **14.** $\left|\frac{x + 3}{2}\right| < 4$ **15.** $|2x + 1| < 0$

16. $|3x + 1| \leq 0$ **17.** $\left|\frac{3x - 1}{4}\right| < 2$ **18.** $\left|\frac{x}{2} - \frac{1}{3}\right| < \frac{1}{6}$ **19.** $|x| > 5$ **20.** $|x| > \frac{1}{3}$

21. $|x| \geq 40$ **22.** $|y| \geq \frac{3}{4}$ **23.** $1 + |y| \geq 2$ **24.** $|y| \geq -2$

25. $|z - 1| > 6$ **26.** $|t + 2| \geq 2$ **27.** $|3x + 1| > 4$ **28.** $\left|\frac{x}{2} + 1\right| \geq 11$

29. $|4x - 3| > 15$ **30.** $|2x + 5| > 3$ **31.** $\left|\frac{x}{2} - \frac{4}{3}\right| > \frac{1}{3}$ **32.** $\left|\frac{5x}{8} - \frac{3}{4}\right| \geq \frac{1}{2}$

33. $|5x - 3| \geq 0$ **34.** $|5x - 3| > 0$ **35.** $|5x - 3| > -1$ **36.** $|5x - 3| > 1$

In Exercises 37–52 determine c.

37. $-5 < x < 5$ means the same as $|x| < c$. **38.** $-2 \leq x \leq 2$ means the same as $|x| \leq c$.

39. x is in $(-\infty, -9) \cup (9, \infty)$ means the same as $|x| > c$.

40. Either $x \leq -1$ or $x \geq 1$ means the same as $|x| \geq c$.

41. $-4 < x < 6$ means the same as $|x - 1| < c$.

42. $-7 < x < 3$ means the same as $|x + 2| < c$.

43. $0 \leq x \leq 8$ means the same as $|x - c| \leq 4$.

44. $-9 < x < -5$ means the same as $|x + c| < 2$.

45. $-8 < x < -4$ means the same as $|x - c| < 2$.

46. $-2 < x < 3$ means the same as $|x - c| < \frac{5}{2}$.

47. $x^2 < 25$ means the same as $|x| < c$.

48. $x^2 \leq 100$ means the same as $|x| \leq c$.

49. $x^2 < \frac{1}{4}$ means the same as $-c < x < c$.

50. $x^2 \geq 1$ means the same as $|x| \geq c$.

51. $-6 < x < 6$ means the same as $x^2 < c$.

52. $-.1 \leq x \leq .1$ means the same as $x^2 \leq c$.

In Exercises 53–58 answer "true" or "false". If "false", find a *counterexample*; that is, find a numerical example that shows the statement to be false.

53. For all a, $|a + 6| = |a| + 6$

54. For all a, $|a - 4| = |a| - 4$

55. For $a > 0$ and $b > 0$, $|a + b| = |a| + |b|$

56. For $a > 0$ and $b > 0$, $|a - b| = |a| + |b|$

57. For $a > 0$ and $b < 0$, $|a - b| = |a| + |b|$

58. For $a < 0$ and $b < 0$, $|a + b| = |a| + |b|$

59. Suppose $|x - 2| < \frac{1}{10}$. Show that $|5x - 10| < \frac{1}{2}$.

60. Suppose $|x - 1| < 2$. Show that $|x + 1| < 4$.

Review Exercises for Chapter 2

1. Check whether the number in color is a root of the given equation.

(a) $x - 7 = 3$, 10 (b) $5x = -20$, 4 (c) $2x - 1 = 5 - 2x$, $\frac{3}{2}$ (d) $\frac{3}{x} = \frac{1}{5}$, 15

2. Solve each equation. Check the ones so indicated.

(a) $2u + 1 = 5 - 2u$ (Check.) (b) $\frac{x}{4} = \frac{3}{5}$ (c) $3(y - 2) = 1 - (4 - 2y)$

(d) $\frac{u + 3}{8} = \frac{u - 2}{3}$ (Check.) (e) $\frac{3}{y + 2} = \frac{6}{y + 7}$ (f) $\frac{2}{x + 2} + \frac{1}{x - 2} = \frac{10}{x^2 - 4}$

3. Two trains approach each other along (straight) parallel tracks. One train travels at 100 kilometers per hour and goes twice as fast as the other. If their distance apart is 225 kilometers, in how many hours will they pass each other?

4. If you have twice as much invested at 8% as at 5% and if your annual interest income from these two investments is \$315, how much is invested at each rate?

5. Alloy A is 1 part copper to 4 parts tin. Alloy B is 1 part copper to 7 parts tin. How much of alloy B should be added to 30 pounds of alloy A to obtain an alloy that is 1 part copper to 5 parts tin?

6. Bob can paint a room in 6 hours and Ed can paint it in 8 hours. How long does it take them to paint the room together?

7. (a) Solve $\frac{5u + a}{b} = 3$ for u. (b) Solve $\frac{4}{y} + \frac{1}{2x} = z$ for y. (c) Solve $\frac{A}{t} + \frac{1}{st} = -2$ i. for s, ii. for t.

8. Rearrange the following numbers so that you can write "<" between any two numbers.

$$\frac{1}{6}, \quad \frac{1}{3}, \quad \frac{2}{5}, \quad \frac{1}{2}, \quad \frac{1}{5}, \quad \frac{2}{3}$$

9. Determine the additive inverse of each number.

(a) 5 (b) -5 (c) 0 (d) $|-5|$

10. Determine each absolute value.

(a) $|12|$ (b) $|-4|$ (c) $|0|$ (d) $\left|\frac{-3}{4}\right|$

11. Answer "true" or "false".

(a) 2 is in $(2, 7)$. (b) -3 is in $[-5, -2]$. (c) 0 is in $(-\infty, 0)$. (d) $\frac{1}{2}$ is in $[-1, 1]$.

12. Describe in one of the forms

$$[a, b], (a, b), (a, \infty), [a, \infty), (-\infty, b), \text{ or } (-\infty, b]$$

the set of numbers satisfying each of the following:

(a) $5 \leqslant x \leqslant 7$ (b) $-2 < x < 3$ (c) $x \geqslant -1$ (d) $x < 3$

13. Fill in "<" or ">".

(a) $x + 3 \ \square \ x - 4$ (b) If $a < 0$, $-2a \ \square \ -4a$ (c) If $a > 0$, $\frac{a}{3} \ \square \ \frac{a}{4}$

14. Solve each inequality.

(a) $2(x + 1) + 3(x - 2) \geqslant 1$ (b) $\frac{1 + t}{3} < \frac{t}{2}$ (c) $-1 \leqslant 2t + 1 \leqslant 7$

15. (a) Which numbers are 6 units from 4? (b) Which numbers are 6 units from -4?

16. Solve each equation. Check the ones so indicated.

(a) $|x - 7| = 7$ (b) $|2 - x| = 3x$ (Check.) (c) $|x + 1| = |x - 3|$ (Check.)

17. Solve each inequality.

(a) $|x - 1| < 4$ (b) $|y + 3| \geqslant 3$ (c) $\left|\frac{x}{4} + \frac{1}{2}\right| \leqslant 0$

18. Find the value of c.

(a) $-3 \leqslant x \leqslant 3$ means the same as $|x| \leqslant c$. (b) $-1 < x < 7$ means the same as $|x - c| < 4$.

19. Determine all roots of the equation $\left|\frac{x}{x + 1}\right| = 1$.

3 Functions

3.1 What Is a Function?

definition of a function

You are now going to consider *correspondences between two sets.* In other words, you will *pair* the objects of one set with those of another set.

EXAMPLE 1 Consider the correspondence given by:

$1 \to 2$
$2 \to 4$
$3 \to 6$
$4 \to 8$
$5 \to 10$

To each object of the first set, $\{1, 2, 3, 4, 5\}$, there corresponds *exactly one* object of the second set, $\{2, 4, 6, 8, 10\}$, as indicated by the arrows. □

Definition. A **function** is a correspondence between two sets, called the **domain** and the **range** of the function. Furthermore, to each object of the domain there corresponds *exactly one* object of the range, and every object of the range corresponds to *at least one* object of the domain.

The correspondence of Example 1 represents a function with domain $\{1, 2, 3, 4, 5\}$ and range $\{2, 4, 6, 8, 10\}$. Each object of the domain is paired with exactly one object of the range. And *here,* each object of the range is paired with exactly one object of the domain. However, there are also functions in which several different objects of the domain are paired with the *same* object of the range.

EXAMPLE 2 (a) The correspondence indicated by

$5 \to 2$
$10 \to 3$
$15 \to 2$

represents a function with domain $\{5, 10, 15\}$ and range $\{2, 3\}$. Observe that *two different objects of the domain,* 5 and 15, *correspond to the same object, 2, of the range.* This is allowed in the definition of a function. For what is required is that

to each object of the domain there corresponds exactly one object of the range.

(b) The following is *also* a function, because *each object of the domain is paired with exactly one object of the range.*

$$5 \to 2$$
$$10 \to 2$$
$$15 \to 2$$

Note that there is only one object in the range. When this is the case, the function is called a **constant function.**

(c) The correspondence

$$2 \to 5$$
$$\searrow 15$$
$$3 \to 10$$

does *not* represent a function. For *there correspond two distinct objects,* 5 and 15, *of the "range" to the object* 2 *of the "domain".* □

The letters f, g, h, F, G, H will stand for functions.

Definition. Let f be a function with domain A and range B. The objects of A are called the **arguments of f**. The object b of B that corresponds to a particular argument a is called the **value of f at a**, or simply, a **function-value.**

Write $f(a) = b$ and read this as f of a equals b.

EXAMPLE 3 Suppose f is the function defined by the correspondence:

$$1 \to 10$$
$$2 \to 20$$
$$3 \to 30$$
$$4 \to 40$$

The domain is $\{1, 2, 3, 4\}$; the arguments of f are 1, 2, 3, and 4. The range is $\{10, 20, 30, 40\}$.

The value of f at 1 is 10; $f(1) = 10$. (f of 1 equals 10.)
The value of f at 2 is 20; $f(2) = 20$. (f of 2 equals 20.)
The value of f at 3 is 30; $f(3) = 30$. (f of 3 equals 30.)
The value of f at 4 is 40; $f(4) = 40$. (f of 4 equals 40.) □

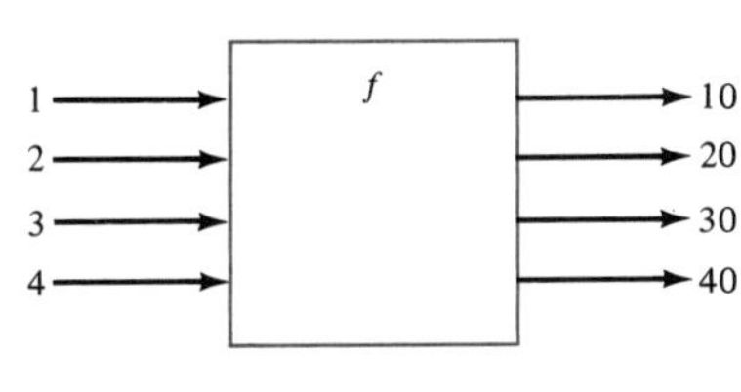

FIGURE 3.1

A function f can be thought of as a machine. The arguments of f are then the *inputs,* and the value of f at input a is then the corresponding *output.* The function of Example 3 can be pictured as in Figure 3.1 .

polynomial and rational functions

A function f is often defined by an equation. The equation specifies the value of f at an argument x. One side of the equation is $f(x)$; the other side is an algebraic expression, such as a polynomial or a rational expression. A function defined by a polynomial is called a **polynomial function,** and a function defined by a rational expression is called a **rational function.**

EXAMPLE 4 Let f be the polynomial function defined by

$$f(x) = 2x + 5$$

for every real number x. Determine the following values of f:

(a) $f(0)$, (b) $f(1)$, (c) $f(2)$, (d) $f(-1)$

Solution. To determine each value, substitute the specified argument for x in the polynomial $2x + 5$.

(a) $f(0) = 2 \cdot 0 + 5 = 5$

(b) $f(1) = 2 \cdot 1 + 5 = 7$

(c) $f(2) = 2 \cdot 2 + 5 = 9$

(d) $f(-1) = 2(-1) + 5 = 3$ □

If $f(x) = 6$ for every real number x, then f is a constant function. Write: $f(x) \equiv 6$. If $f(x) \equiv 0$, then f is called **the zero function.**

EXAMPLE 5 Let g be the rational function defined by

$$g(x) = \frac{x^3 - 1}{x^2 + 1}$$

for every real number x. Determine the following values of g:

(a) $g(1)$ (b) $g(10)$ (c) $g(.1)$ (d) $g(-1)$

Solution.

(a) $g(1) = \dfrac{1^3 - 1}{1^2 + 1} = \dfrac{0}{2} = 0$

(b) $g(10) = \dfrac{10^3 - 1}{10^2 + 1} = \dfrac{1000 - 1}{100 + 1} = \dfrac{999}{101}$

(c) $g(.1) = \dfrac{(.1)^3 - 1}{(.1)^2 + 1} = \dfrac{.001 - 1}{.01 + 1} = \dfrac{-.999}{1.01} = \dfrac{-999}{1010}$

(d) $g(-1) = \dfrac{(-1)^3 - 1}{(-1)^2 + 1} = \dfrac{-1 - 1}{1 + 1} = \dfrac{-2}{2} = -1$ □

A rational function is not defined when the *denominator* of the defining rational expression is zero. (Recall that division *by* 0 is not defined.) In Example 5 the denominator, $x^2 + 1$, is positive for all x. (Why?) The problem of a zero denominator never arose. Note that in part (a), the *numerator* is 0 and the denominator is nonzero; the quotient is 0. In Example 6, which follows, the *denominator* is 0 when $x = 0$. Thus, 0 is not in the domain of the function.

EXAMPLE 6 Let F be the rational function defined by

$$F(x) = \frac{x^2 - 3x + 5}{x}$$

for every nonzero real number x. Determine the following values of F:

(a) $F(2)$ (b) $F(5)$ (c) $F\left(\frac{1}{2}\right)$

Solution.

(a) $$F(2) = \frac{2^2 - 3(2) + 5}{2} = \frac{3}{2}$$

(b) $$F(5) = \frac{5^2 - 3(5) + 5}{5} = \frac{15}{5} = 3$$

(c) $$F\left(\frac{1}{2}\right) = \frac{\left(\frac{1}{2}\right)^2 - 3\left(\frac{1}{2}\right) + 5}{\frac{1}{2}} = \frac{\frac{1}{4} - \frac{3}{2} + 5}{\frac{1}{2}}$$

$$= \frac{1 - 3 \cdot 2 + 5 \cdot 4}{4} \div \frac{1}{2}$$

$$= \frac{15}{4} \cdot \frac{2}{1} = \frac{15}{2}$$

□

If nothing is said to the contrary, assume a function specified by an algebraic expression is defined for all possible (real) values for which the expression makes sense. Thus, a rational function is assumed to be defined except where the denominator is 0.

Frequently, however, the domain of a function is an interval. The defining equation applies *only to those arguments in this interval,* as in Example 7.

EXAMPLE 7 A ball is dropped from a height of 64 feet above ground level. Its height t seconds from the time it is dropped is given by

$$h(t) = 64 - 16t^2 \text{ (feet)}.$$

Thus

$$h(0) = 64 - 16 \cdot 0^2 = 64$$

$$h(1) = 64 - 16 \cdot 1^2 = 48$$

$$h\left(\frac{3}{2}\right) = 64 - 16 \cdot \left(\frac{3}{2}\right)^2 = 64 - 16 \cdot \frac{9}{4} = 28$$

$$h(2) = 64 - 16 \cdot 2^2 = 0$$

When $t = 2$, the ball hits the ground. Thus the function is defined only for $0 \leq t \leq 2$, that is, its domain is the closed interval [0, 2]. Observe that *h decreases* as *t increases*. The range of the function is [0, 64].

□

verbal descriptions Functions are often defined by verbal descriptions. Such functions arise in geometry and in applications to such fields as engineering, biology, and economics. You will frequently want to convert the verbal description into an algebraic description.

EXAMPLE 8 Part of a prairie is to be enclosed by a fence so as to form a rectangular region. The length of this rectangle must be three times the width. Determine the area of the enclosed region as a function of the width. What will be the area if the width of the fence is (a) 50 feet, (b) 200 feet? [See Figure 3.2 .]

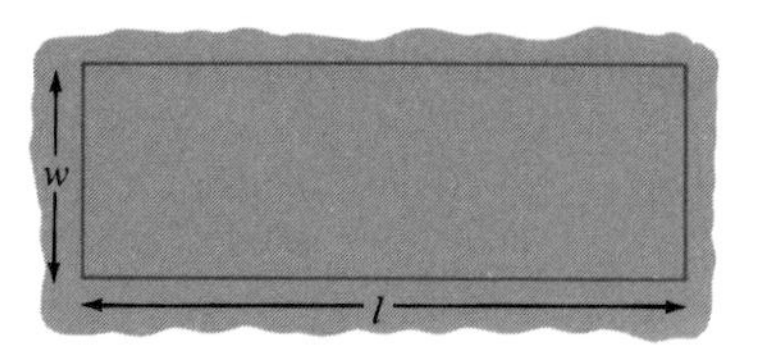

FIGURE 3.2 The length, l, of the fence is 3 times the width, w.

Solution. The area A of a rectangle is given by

$$A = l \cdot w$$

where l is the length and w the width. The length, l, of the rectangular fence is three times the width, w. Thus

$$l = 3w \quad \text{and} \quad A = 3w \cdot w = 3w^2$$

(a) Let $w = 50$. Then
$$A = 3 \cdot 50^2 = 7500$$
The enclosed area is 7500 square feet.

(b) Let $w = 200$. Then
$$A = 3 \cdot 200^2 = 120\,000$$
The enclosed area is 120 000 square feet. □

EXAMPLE 9 The profit derived from a novelty item is one-third of the cost, minus an initial expense of one hundred dollars. Determine the profit if the cost is (a) \$300, (b) \$3000, (c) \$21 000.

Solution. Let c be the cost (in dollars). The profit, $f(c)$, is given by

$$f(c) = \frac{c}{3} - 100.$$

(a) $$f(300) = \frac{300}{3} - 100 = 0$$

(b) $$f(3000) = \frac{3000}{3} - 100 = 900$$

(c) $$f(21\,000) = \frac{21\,000}{3} - 100 = 6900$$

Thus an entrepreneur would break even on a \$300 cost, and would derive profits of \$900 and \$6900 on costs of \$3000 and \$21 000, respectively. □

EXAMPLE 10 A wire of length 25 (centimeters) is cut into two pieces, of lengths x and $25 - x$, respectively. The first piece is bent into a circle, the second into a square. [See Figure 3.3 .] Express the *total area* as a function of x.

Solution. Let A_1 be the area of the circle. Because x represents the circumference of the circle, the radius, r, must be $\frac{x}{2\pi}$. Thus

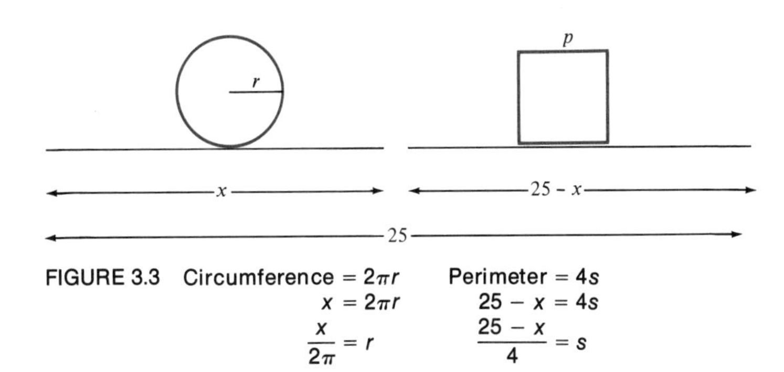

FIGURE 3.3 Circumference $= 2\pi r$ Perimeter $= 4s$

$$x = 2\pi r \qquad 25 - x = 4s$$

$$\frac{x}{2\pi} = r \qquad \frac{25 - x}{4} = s$$

$$\begin{aligned} A_1 &= \pi r^2 \\ &= \pi \left(\frac{x}{2\pi}\right)^2 \\ &= \pi \cdot \frac{x^2}{4\pi^2} \\ &= \frac{x^2}{4\pi} \quad \text{(square centimeters)} \end{aligned}$$

$$\left(\frac{a}{bc}\right)^2 = \frac{a^2}{b^2c^2}$$

Let A_2 be the area of the square. Because the perimeter of the square is $25 - x$, each side must have length $\frac{25 - x}{4}$. Thus

$$\begin{aligned} A_2 = s^2 &= \left(\frac{25 - x}{4}\right)^2 \\ &= \frac{625 - 50x + x^2}{16} \quad \text{(square centimeters)} \end{aligned}$$

The total area, A, is given by

$$\begin{aligned} A &= A_1 + A_2 \\ &= \frac{x^2}{4\pi} + \frac{625 - 50x + x^2}{16} \\ &= \frac{4x^2 + \pi(625 - 50x + x^2)}{16\pi} \\ &= \frac{(4 + \pi)x^2 - 50\pi x + 625\pi}{16\pi} \quad \text{(square centimeters)} \end{aligned}$$

□

EXERCISES

In Exercises 1 and 2 which of the following correspondences represent functions?

1.
$1 \to 3$
$2 \to 6$
$3 \to 9$
$4 \to 12$

2.
$1 \to 3$
$1 \to 6$
$1 \to 9$
$1 \to 12$

In Exercises 3 and 4 determine: (a) the domain and (b) the range of the indicated function.

3.
$-6 \to 0$
$-5 \to 0$
$-4 \to 0$
$-3 \to 1$
$-2 \to 1$
$-1 \to 1$

4.
$1 \to 1$
$2 \to 1$
$3 \to 1$
$4 \to 1$
$5 \to 1$

5. Suppose f is the function defined by the correspondence:

$$1 \to 0$$
$$2 \to 1$$
$$3 \to 0$$
$$4 \to 1$$

(a) What are the arguments of f?
(b) Determine the value of f at 1.
(c) Determine the value of f at 3.

6. Suppose g is the function defined by the correspondence:

$$\frac{1}{2} \to 1$$
$$\frac{2}{3} \to 2$$
$$\frac{3}{4} \to 3$$
$$\frac{4}{5} \to 4$$

(a) What are the arguments of g?
(b) Determine the value of g at $\frac{2}{3}$.
(c) Determine the value of g at $\frac{3}{4}$.

In Exercises 7–20 for each function determine the indicated function-values.

7. $f(x) = 2x$ **(a)** $f(1)$ **(b)** $f(2)$ **(c)** $f(3)$ **(d)** $f(4)$

8. $g(x) = x + 2$ **(a)** $g(2)$ **(b)** $g(5)$ **(c)** $g(10)$ **(d)** $g(18)$

9. $h(x) = x - 10$ **(a)** $h(0)$ **(b)** $h(5)$ **(c)** $h(10)$ **(d)** $h(-10)$

10. $F(x) = 2x + 3$ **(a)** $F(1)$ **(b)** $F(-1)$ **(c)** $F(3)$ **(d)** $F(-3)$

11. $G(x) = 1 - 3x$ **(a)** $G(1)$ **(b)** $G(0)$ **(c)** $G(-1)$ **(d)** $G\left(\frac{1}{9}\right)$

12. $H(x) = x^2 + 1$ **(a)** $H(0)$ **(b)** $H(1)$ **(c)** $H(2)$ **(d)** $H(3)$

13. $f(u) = u^4 - u^2 + 1$ **(a)** $f(0)$ **(b)** $f(1)$ **(c)** $f(-1)$ **(d)** $f(-2)$

14. $g(y) = y^5 - 2y^2 + 2y + 2$ **(a)** $g(0)$ **(b)** $g(-1)$ **(c)** $g(2)$ **(d)** $g\left(\frac{1}{2}\right)$

15. $h(t) = \frac{1}{t}$ **(a)** $h(1)$ **(b)** $h(4)$ **(c)** $h\left(\frac{1}{4}\right)$ **(d)** $h(.1)$

16. $F(x) = \frac{x + 1}{x - 1}$ **(a)** $F(2)$ **(b)** $F(10)$ **(c)** $F(99)$ **(d)** $F(-1)$

17. $G(x) = \frac{x^4 - 3x^3}{x^2 + 5}$ **(a)** $G(0)$ **(b)** $G(1)$ **(c)** $G(-1)$ **(d)** $G(-2)$

18. $H(x) \equiv 4$ **(a)** $H(1)$ **(b)** $H(2)$ **(c)** $H(4)$ **(d)** $H(-4)$

19. $f(t) = (t - 1)(t - 4)$ **(a)** $f(0)$ **(b)** $f(1)$ **(c)** $f(-1)$ **(d)** $f(4)$

20. $g(x) = \frac{(x^3 + 1)(x^2 - 5)}{x^2 + 3}$ **(a)** $g(0)$ **(b)** $g(2)$ **(c)** $g(-1)$ **(d)** $g(-2)$

In Exercises 21–26 determine the largest possible domain of a function defined by each equation. [*Hint:* First find the values for which the denominator is 0. Thus if $f(x) = \frac{1}{x - 1}$, the largest possible domain is the set of all real numbers other than 1.]

21. $f(x) = \frac{x + 3}{2x - 1}$ **22.** $g(x) = \frac{x + 2}{3x + 1}$ **23.** $h(x) = \frac{1}{x^2}$ **24.** $F(t) = \frac{1}{t^2 + 4}$

25. $G(t) = \frac{1}{t^2 - 4}$ **26.** $H(t) \equiv 0$

In Exercises 27–34 determine: (a) the domain and (b) the range of each function

27. $f(x) = x$ for $0 < x < 4$

28. $g(x) = x + 1$ for $-2 \leqslant x \leqslant 2$

29. $h(x) = x + 1$ for $-2 < x < 2$

30. $f(t) = 3t$ for $1 \leqslant t \leqslant 4$

31. $g(t) = 2t + 1$ for $t \geqslant 1$

32. $h(t) = t + 2$ for $t \leqslant -2$

33. $F(x) = 10 - x$ for $4 \leqslant x \leqslant 10$

34. $G(x) = x^2$ for $1 \leqslant x \leqslant 5$

In Exercises 35–50, when necessary, refer to the geometric figures on the inside front cover.

35. **(a)** Express the volume of a sphere as a function of r, the length of the radius.

(b) Determine the volume if $r = 1$. **(c)** Determine the volume if $r = 5$.

36. Suppose the length, l, of a rectangle is twice the width, w. Express the area of the rectangle as a function of

(a) l, **(b)** w. **(c)** Determine the area when $l = 16$. **(d)** Determine the area when $w = 16$.

37. Suppose the base length, b, of a triangle is two-thirds of h, the length of the corresponding altitude. Express the area of the triangle as a function of

(a) h, **(b)** b. **(c)** Determine the area when $h = 18$. **(d)** Determine the area when $b = 18$.

38. Water is pouring into a reservoir at the rate of 100 cubic feet per minute.

(a) Express the volume of the water as a function of t, the number of minutes after the water begins pouring. **(b)** What is the volume of the water after 10 minutes? **(c)** What is the volume of the water after an hour?

39. Harry is 5 years older than his sister, Harriet. **(a)** Express Harry's age as a function of Harriet's age, h. **(b)** Express Harriet's age as a function of Harry's age, H.

40. The height, in feet, of an airplane after t seconds is given by $h(t) = \frac{t^2}{5}$.

(a) How high is the plane after 5 seconds? **(b)** How high is the plane after half a minute?

(c) How long does it take for the plane to rise 500 feet?

41. The number of cars an automobile salesman has sold equals two more than four times the number of weeks since he began working. Let $t = 1, 2, \ldots$ be the number of weeks he has been working.

(a) Express the number of cars he has sold as a function of t.

(b) How many cars has he sold after 52 weeks?

42. An insurance saleswoman earns an annual salary of \$12 000 plus a 10% commission on total annual premiums up to \$300 000 and a 15% commission on total annual premiums above \$300 000. Express her annual earnings as a function of her total annual premiums.

43. A ball is thrown directly upward. Its height, $h(t)$, in feet, after t seconds is given by $h(t) = 128t - 16t^2$. Determine the height after **(a)** 2 seconds, **(b)** 4 seconds, **(c)** 6 seconds, **(d)** 8 seconds.

(e) What should be the domain of this function?

44. A piece of wire 60 centimeters long is cut into two pieces. One piece is bent to form a circle, and the other piece is bent to form a square. Express the length, s, of a side of the square as a function of r, the length of the radius of the circle.

45. A poster is printed on a rectangular piece of cardboard of length x inches and area 240 square inches. The margins at top and bottom are each 3 inches and at each side are 2 inches. Express the printed area as a function of x. [See Figure 3.4 .]

46. A window has the shape of a square, of side length s (inches), topped by a semicircle. [See Figure 3.5 .] Express the total area of the window as a function of s.

47. A rectangular pasture is to be fenced off along a river bank. No fencing is required alongside the river bank. The fencing along the parallel side costs 10 dollars a meter, whereas the fencing for the other two sides costs 5 dollars a meter. If 1000 dollars is appropriated for the fencing, express the area of the pasture as a function of l, the length of the side parallel to the river bank.

48. An open rectangular storage bin is to be constructed. Its base is to be a square of side length s (meters), and is to cost 5 dollars per square meter. Its height is $\frac{2}{3}s$. The sides will each cost 3 dollars per square meter. Express the cost of construction as a function of s.

49. Boxes without tops are made from square sheets of cardboard of side length 50 centimeters by cutting out equal squares from each of the four corners, and then turning up the sides. [See Figure 3.6 .]

(a) Express the volume of a box as a function of x, the side length (in centimeters) of each square corner.

(b) What is the domain of the function?

50. Express the area of each colored region in Figure 3.7 as a function of x.

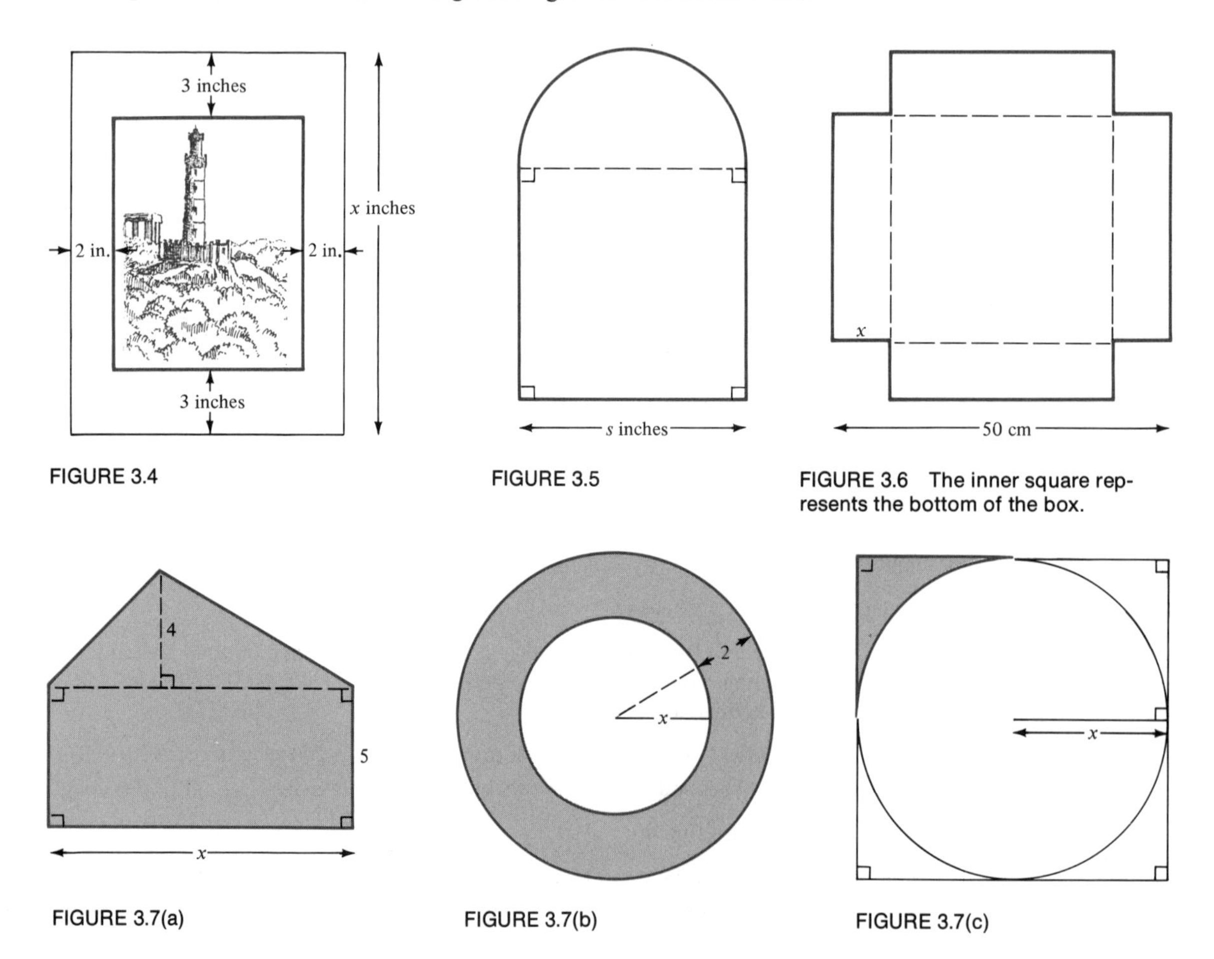

FIGURE 3.4

FIGURE 3.5

FIGURE 3.6 The inner square represents the bottom of the box.

FIGURE 3.7(a)

FIGURE 3.7(b)

FIGURE 3.7(c)

3.2 Graphs

rectangular coordinates

In order to picture functions geometrically, it is convenient to introduce a **rectangular** (or **Cartesian**) **coordinate system** on the plane.

In Chapter 2 you saw how real numbers correspond to points on a horizontal number line L. Through the origin on L draw another line perpendicular to L. The second line is thus vertical. [See Figure 3.8 .] From now on, call the horizontal line the ***x*-axis** and the vertical line the ***y*-axis.** Together the two lines are called the **coordinate axes.** The **origin,** which is the intersection of the coordinate axes, will also correspond to 0 on the y-axis.

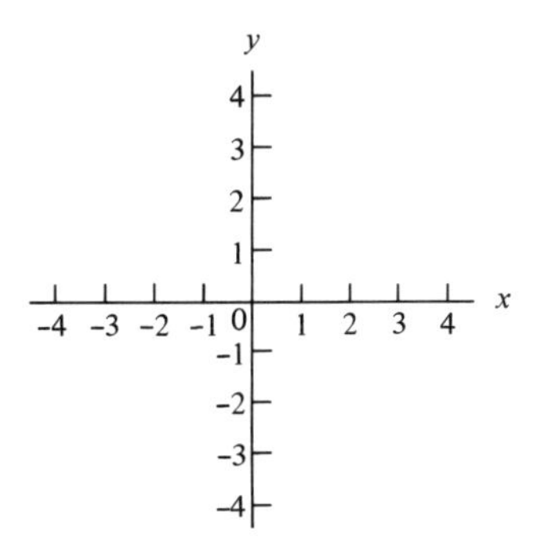

FIGURE 3.8

There is already a distance unit on the x-axis in terms of which positive numbers are marked off to the right of 0 and negative numbers are marked off to the left of 0. For convenience, choose the same distance unit on the y-axis. Mark off positive numbers *upward from* 0 and negative numbers *downward,* as in Figure 3.8 .

Every point on the x-axis corresponds to a real number. So too, every point on the y-axis corresponds to a real number. Moreover, every real number corresponds to exactly one point on *each* of these coordinate axes. Just as you identified points on the x-axis with the numbers to which they correspond, so too, will you identify points on the y-axis with the numbers to which they correspond.

Points on the plane will correspond to *pairs* of real numbers, *in a definite order*. Let P be a point on the plane. Draw a perpendicular from P to each of the coordinate axes (as in Figure 3.9). Let the vertical line from P intersect the x-axis where $x = a$, and let the horizontal line from P intersect the y-axis where $y = b$. The point P corresponds to the numbers a and b, *in this order.* To indicate the ordering of a and b, write

$$(a, b)$$

Call (a, b) an **ordered pair.** Thus an ordered pair (a, b) indicates two numbers a and b, *in the order written*. The number a is called the **first coordinate** or ***x*-coordinate** of P; b is called the **second coordinate** or ***y*-coordinate** of P. You say that the coordinates of P are (a, b) and write

$$P = (a, b)$$

[See Figure 3.9 .]

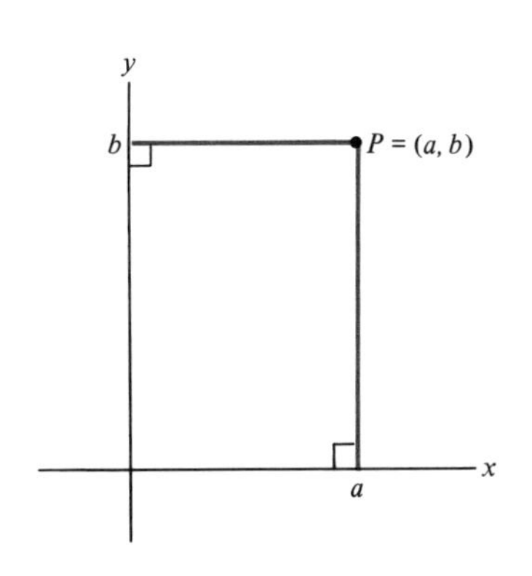

FIGURE 3.9

EXAMPLE 1 Several points on the plane are represented in Figure 3.10 . Their coordinates are indicated below Figure 3.10 on page 110. □

Starting with an ordered pair (c, d) of numbers, you can also find the corresponding point Q on the plane. Locate c on the x-axis; locate d on the y-axis. Draw perpendiculars through each of these points. The intersection of these perpendiculars is the point Q corresponding to the ordered pair (c, d).

EXAMPLE 2 Plot the following points on a rectangular coordinate system on the plane:

(a) $P = (4, 6)$ (b) $Q = \left(\frac{1}{2}, -3\right)$ (c) $R = \left(0, \frac{-1}{2}\right)$

(d) $S = (-3.5, -2.5)$ (e) $T = (-5, 0)$

Solution. See Figure 3.11 . □

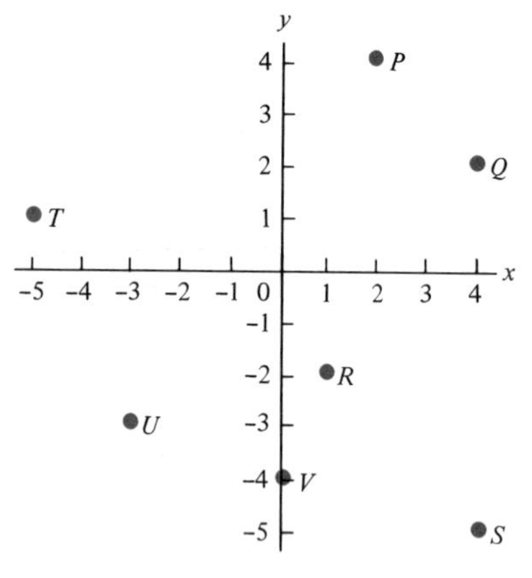

FIGURE 3.10 $P = (2, 4)$, $Q = (4, 2)$ $R = (1, -2)$. $S = (4, -5)$, $T = (-5, 1)$, $U = (-3, -3)$, $V = (0, -4)$

FIGURE 3.11

It is important to observe that *two ordered pairs are equal only when they agree in both coordinates.* Agreement in a single coordinate is not enough. For example,

$$(4, 5) \neq (4, 6)$$

These ordered pairs correspond to different points on the plane. [See Figure 3.12(a).] Also

$$(3, 1) \neq (4, 1);$$

these ordered pairs correspond to different points. [See Figure 3.12(b).] Finally,

$$(2, 3) \neq (3, 2);$$

the coordinates are reversed. The first coordinates do not agree $(2 \neq 3)$; nor do the second coordinates agree. [See Figure 3.12(c).]

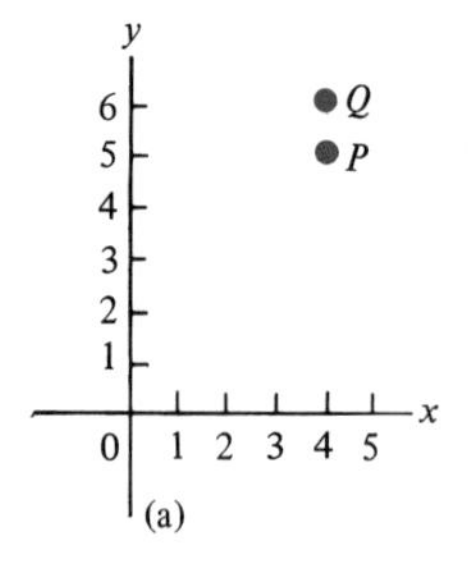

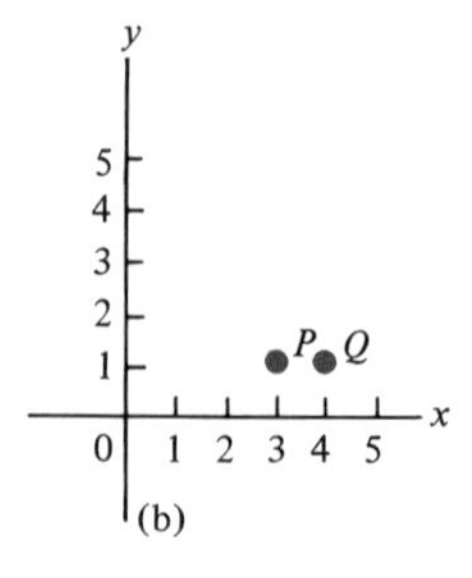

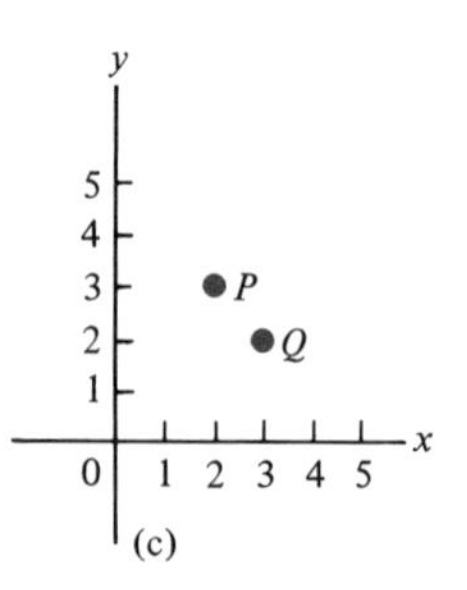

FIGURE 3.12 (a) $P = (4, 5) \neq (4, 6) = Q$ (b) $P = (3, 1) \neq (4, 1) = Q$ (c) $P = (2, 3) \neq (3, 2) = Q$

quadrants

The coordinate axes divide the plane into 4 different regions, called **quadrants.** These quadrants are numbered in a counterclockwise direction, as indicated in Figure 3.13 . *Every point that is not on a coordinate axis* (that is, for which *neither* coordinate is zero) *lies in exactly one quadrant.* Points on a coordinate axis are not considered to be in any quadrant. The signs of the coordinates determine the quadrants, as indicated in Table 3.1 .

You will see references to "the first quadrant" (meaning Quadrant I), "the second quadrant" (Quadrant II), and so on.

EXAMPLE 3 [Refer to Figure 3.13 .]

(a) (4, 1) lies in the first quadrant. Both coordinates are positive.
(b) $(-4, 1)$ lies in the second quadrant. Here the x-coordinate is negative and the y-coordinate is positive.

TABLE 3.1

Quadrant	x	y
I	+	+
II	−	+
III	−	−
IV	+	−

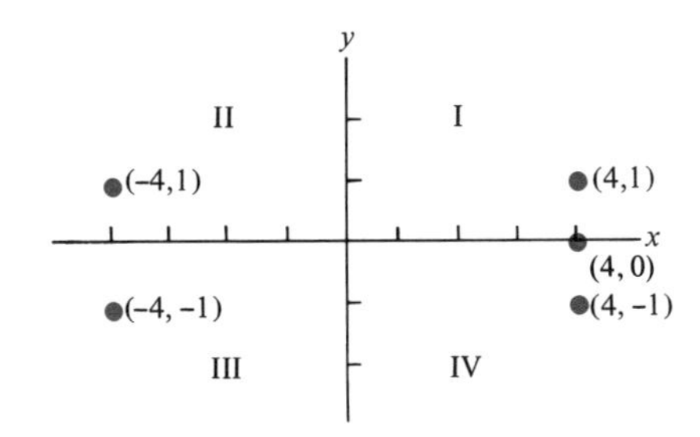

FIGURE 3.13

(c) $(-4, -1)$ lies in the third quadrant. Both coordinates are negative.
(d) $(4, -1)$ lies in the fourth quadrant. The x-coordinate is positive and the y-coordinate is negative.
(e) (4, 0) does not lie in any quadrant, for one of its coordinates is zero. This point is on the x-axis. □

graphing

Let f be a function and suppose that $f(a) = b$. This relationship between a and b can be represented by the ordered pair (a, b) and therefore by the point (a, b) of the plane. A function can be considered as a set of ordered pairs. The set of first coordinates of the ordered pairs is the domain of the function, and the set of second coordinates is the range.

Thus, the function of part (a) of Example 2, page 100, which was formerly indicated by

$$5 \rightarrow 2$$
$$10 \rightarrow 3$$
$$15 \rightarrow 2$$

can now be written as

$$\{(5, 2), (10, 3), (15, 2)\}$$

The domain is the set of first coordinates, $\{5, 10, 15\}$, and the range is the set of second coordinates, $\{2, 3\}$.

You can represent a function geometrically by locating the points corresponding to the ordered pairs of the function.

Definition. The **graph of a function** f is the pictorial representation of the function on the plane. The graph consists of all points (x, y) such that $f(x) = y$.

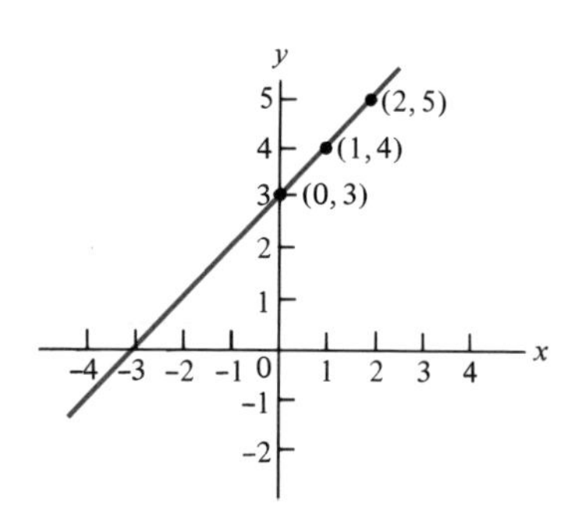

FIGURE 3.14
The graph of $y = x + 3$

Figure 3.14 depicts the graph of the function defined by the equation

$$y = \underbrace{x + 3}_{f(x)}.$$

The graph is a line. Observe that the points (0, 3), (1, 4), and (2, 5) lie on the line and that $f(0) = 3$, $f(1) = 4$, and $f(2) = 5$.

When a function f is defined by an equation of the form

$$y = f(x),$$

you will often refer to the arguments of f as the ***x*-values**; you will also refer to the corresponding ***y*-values**. If nothing further is stated, assume the domain is the set of all possible real numbers. You can often graph the function as follows:

1. Assign several numerical values to x.
2. Compute the corresponding y-values.
3. Plot the points (x, y).
4. Join these points by a "smooth" curve. (This curve may, in fact, be a line, as in Figure 3.14 .)

This procedure will yield the graph of the function provided the function is "relatively simple" and the numerical values in Step 1 are "well chosen". Examples 4–8 illustrate the usual procedure.

lines

The graphs in Examples 4–6 will be lines. The graph of any function of the form

$$y = \underbrace{mx + b}_{f(x)} \tag{3.1}$$

is a line. Here m and b can be any real numbers. Thus, the graph of each of the functions

$y = x + 3$	$(m = 1,\ \ b = 3$, as in Figure 3.14)
$y = -2x$	$(m = -2,\ b = 0$, as in Figure 3.15)
$y = 3x - 2$	$(m = 3,\ \ b = -2$, as in Figure 3.16)
$y \equiv 4$	$(m = 0,\ \ b = 4$, as in Figure 3.17)

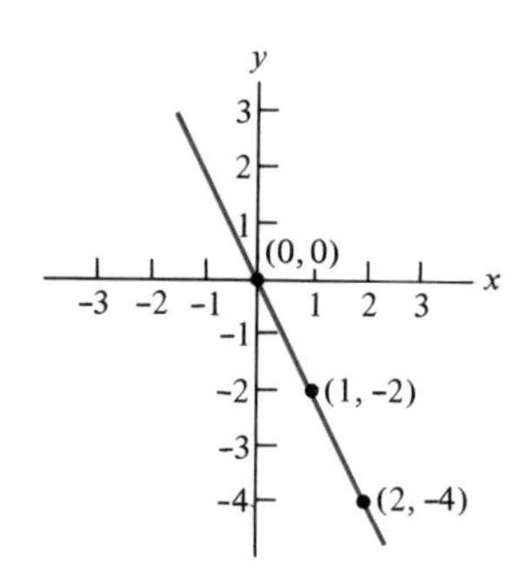

FIGURE 3.15
The graph of $y = -2x$

is a line. In Chapter 4 you will learn that m expresses the "slope" (or steepness) of the line and that the line intersects the y-axis at $(0, b)$.

A line is determined by two points. Thus when a function of the form **(3.1)** is given, you will recognize that its graph is a line. You need only plot two points and draw the line that passes through them. (You may also wish to plot a third point as a check. If your plotting is accurate, a (straight) line passes through the three points.)

EXAMPLE 4 Graph the function defined by

$$g(x) = -2x.$$

TABLE 3.2

x	$y = g(x) = -2x$
0	0
1	−2
2	−4

Solution. Corresponding values of x and y are given in Table 3.2 .

The graph of the function is a line through the origin. [See Figure 3.15 .] □

EXAMPLE 5 Graph the function defined by

$$h(x) = 3x - 2.$$

TABLE 3.3

x	$y = h(x) = 3x - 2$
0	−2
1	1
2	4

Solution.

[See Figure 3.16 .]

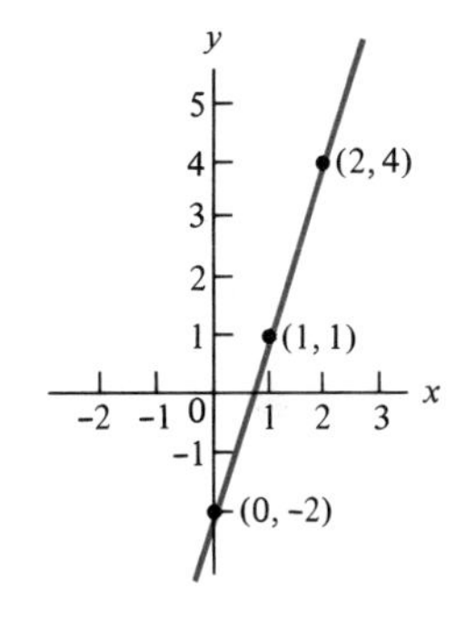

FIGURE 3.16
The graph of $y = 3x - 2$ □

EXAMPLE 6 Graph the function defined by

$$F(x) \equiv 4.$$

Solution.

The graph of this function is a horizontal line that crosses the y-axis at (0, 4). Note that the y-coordinate of every point on the line is 4. [See Figure 3.17 .] □

TABLE 3.4

x	$y = F(x) \equiv 4$
0	4
1	4
2	4

FIGURE 3.17
The graph of $y \equiv 4$

In Examples 4 and 5, the domain and range of each function were the set of all real numbers. In Example 6, the domain was the set of all real numbers, but the range was $\{4\}$.

squaring function; parabolas

The function defined by

$$f(x) = x^2$$

is known as the **squaring function.** Because

$$x^2 \geqslant 0 \qquad \text{for all } x,$$

the range of the squaring function is the set of all *nonnegative (real) numbers,* that is, the set consisting of all positive numbers and 0.

EXAMPLE 7 Graph the function defined by

$$f(x) = x^2.$$

TABLE 3.5

x	$y = x^2$
0	0
$\frac{1}{4}$	$\frac{1}{16}$
$-\frac{1}{4}$	$\frac{1}{16}$
$\frac{1}{2}$	$\frac{1}{4}$
$-\frac{1}{2}$	$\frac{1}{4}$
$\frac{3}{4}$	$\frac{9}{16}$
$-\frac{3}{4}$	$\frac{9}{16}$
1	1
−1	1
2	4
−2	4

Solution. Let $y = f(x)$. The defining equation is *not* of the form $y = mx + b$. You must plot several points to determine the graph. It will be important to consider fractional values of x as well as integral values. [See Table 3.5 .] Note that

$$\left(\frac{1}{4}\right)^2 = \frac{1}{4}\cdot\frac{1}{4} = \frac{1}{16}, \qquad \left(-\frac{1}{4}\right)^2 = \left(-\frac{1}{4}\right)\cdot\left(-\frac{1}{4}\right) = \frac{1}{16},$$

$$\left(\frac{1}{2}\right)^2 = \frac{1}{2}\cdot\frac{1}{2} = \frac{1}{4}, \qquad \left(\frac{3}{4}\right)^2 = \frac{3}{4}\cdot\frac{3}{4} = \frac{9}{16}.$$

These points are plotted in Figure 3.18(a); a smooth curve is drawn connecting them in Figure 3.18(b).

As you see from Table 3.5, $f(1) = f(-1)$ and $f(2) = f(-2)$. Thus, the same y-value can correspond to different x-values. In general,

$$f(-x) = (-x)^2 = (-x)(-x) = x^2 = f(x)$$

Thus whenever (x, y) is on the graph, so is $(-x, y)$, its *mirror image with respect to the y-axis.* The graph is *symmetric with respect to the y-axis.* This curve is known as a **parabola.** The "turning point" on a parabola, here (0, 0), is called the **vertex** of the parabola. □

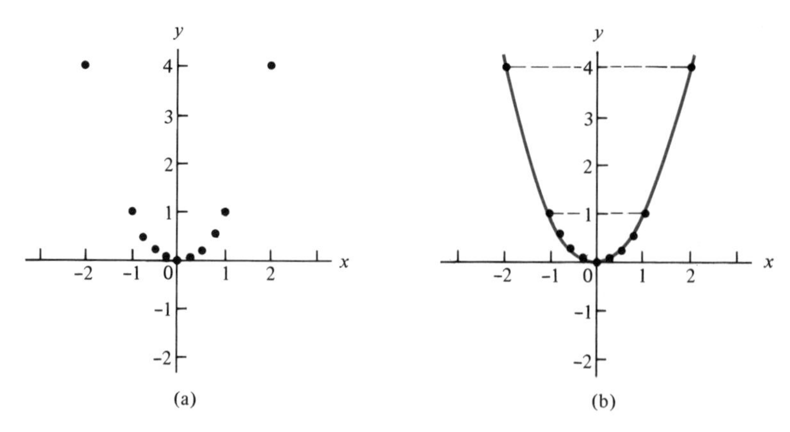

FIGURE 3.18
The parabola given by $y = x^2$ with vertex (0, 0)

EXAMPLE 8 Graph the functions defined by

(a) $F(x) = x^2 - 1$ (b) $G(x) = 2x^2$.

TABLE 3.6(a)

x	x^2	$F(x) = x^2 - 1$
-2	4	3
-1	1	0
$-\frac{1}{2}$	$\frac{1}{4}$	$-\frac{3}{4}$
0	0	-1
$\frac{1}{2}$	$\frac{1}{4}$	$-\frac{3}{4}$
1	1	0
2	4	3

Solution.

(a) *First* square x. *Then* subtract 1.

The graph of F is given in Figure 3.19(a). The vertex of this parabola is $(0, -1)$.

(b) *First* square x. *Then* multiply this by 2.

The graph of G is given in Figure 3.19(b). The vertex of this parabola is the origin.

In part (a), the range is the closed ray $[1, \infty)$. In part (b), the range is $[0, \infty)$, the set of all nonnegative numbers. □

TABLE 3.6(b)

x	x^2	$G(x) = 2x^2$
-2	4	8
-1	1	2
$-\frac{1}{2}$	$\frac{1}{4}$	$\frac{1}{2}$
0	0	0
$\frac{1}{2}$	$\frac{1}{4}$	$\frac{1}{2}$
1	1	2
2	4	8

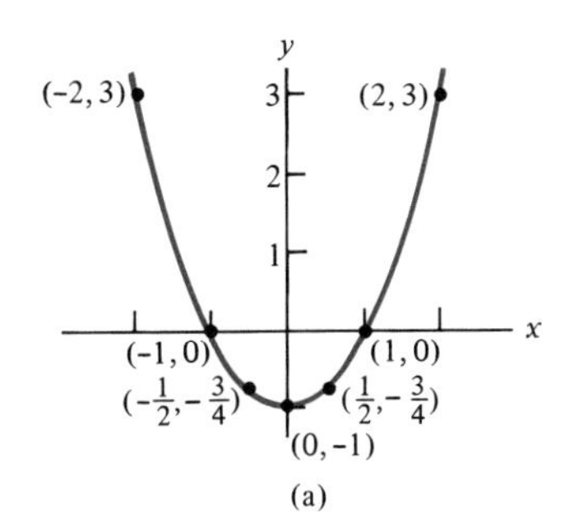

FIGURE 3.19(a)
The graph of $F(x) = x^2 - 1$

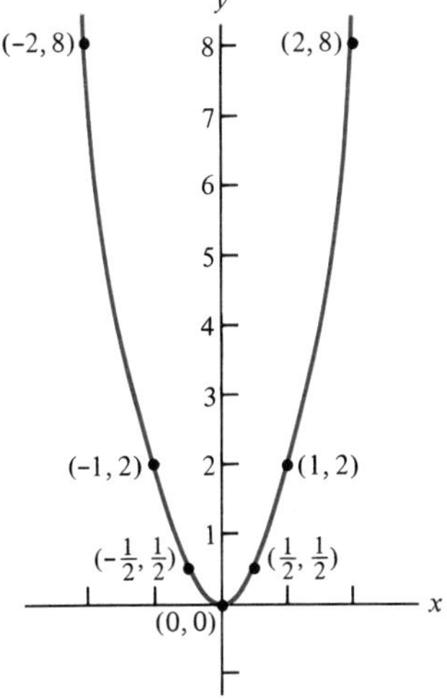

FIGURE 3.19(b)
The graph of $G(x) = 2x^2$ (b)

absolute-value function

Consider the **absolute-value function,** defined by $f(x) = |x|$ for all real numbers x. Thus f can be specified by two rules:

$$f(x) = \begin{cases} x, & \text{if } x \geq 0 \\ -x, & \text{if } x < 0 \end{cases}$$

Here the graphing procedure must be modified. The graph consists of two rays:

$$y = x, \qquad \text{for} \qquad x \geq 0$$

and

$$y = -x, \qquad \text{for} \qquad x < 0$$

TABLE 3.7

x	$f(x) = \|x\|$
-4	4
-3	3
-2	2
-1	1
0	0
1	1
2	2
3	3
4	4

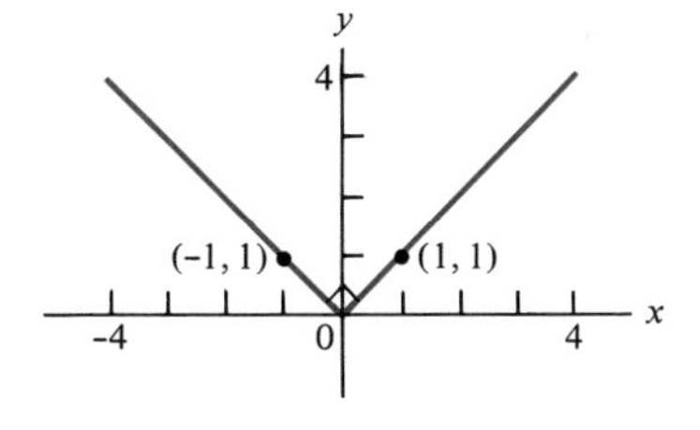

FIGURE 3.20 The graph of $f(x) = |x|$

The two rays meet at the origin at a right angle. [See Figure 3.20 .] Observe that the same y-value can correspond to different x-values. For example,

$$f(1) = f(-1) = 1$$

The graph is symmetric with respect to the y-axis.

EXAMPLE 9 Graph the functions defined by

(a) $g(x) = |x + 2|$ (b) $h(x) = |x| + 2$

Solution. Note that

$|x + 2|$ means *first* add 2 to x; *then* take the absolute value.

[See Table 3.8(a) and Figure 3.21(a).]

TABLE 3.8 (a)

x	$x + 2$	$g(x) = \|x + 2\|$
−5	−3	3
−4	−2	2
−3	−1	1
−2	0	0
−1	1	1
0	2	2
1	3	3
2	4	4
3	5	5

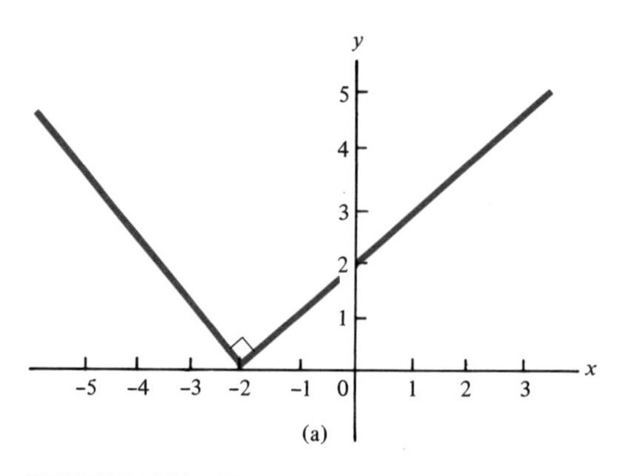

FIGURE 3.21(a) The graph of $g(x) = |x + 2|$

$|x| + 2$ means *first* take the absolute value; *then* add 2 to this.

[See Table 3.8(b) and Figure 3.21(b).]

TABLE 3.8 (b)

x	$\|x\|$	$h(x) = \|x\| + 2$
−4	4	6
−3	3	5
−2	2	4
−1	1	3
0	2	2
1	1	3
2	2	4
3	3	5
4	4	6

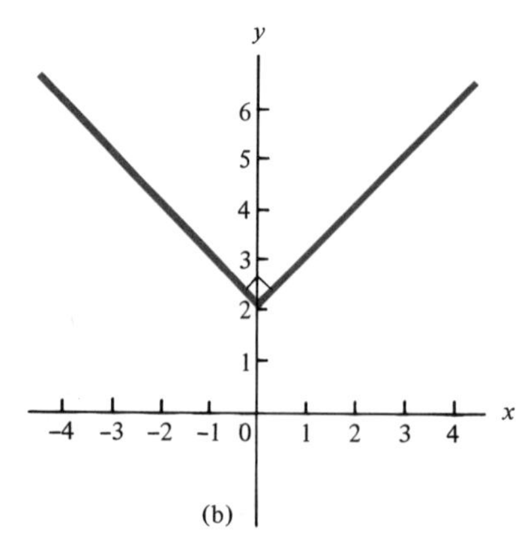

FIGURE 3.21(b) The graph of $h(x) = |x| + 2$

greatest-integer function

A function f can be specified by any rule(s) or verbal description that indicates how $f(x)$ corresponds to x for each x in the domain.

EXAMPLE 10 Let $f(x)$ be the *greatest integer that is less than or equal to* x. Graph this function.

Solution. What is the greatest integer ≤ 0? Observe that for each *negative* integer n, clearly, $n < 0$ and therefore $n \leq 0$. Thus, $-100 \leq 0$, $-10 \leq 0$, $-5 \leq 0$, $-1 \leq 0$. Also, $0 = 0$ and therefore $0 \leq 0$. But for each *positive* integer p, $p > 0$, and thus p is *not* less than or equal to 0. Among the integers that are less than or equal to 0, 0 *itself* is the greatest. (Every other integer that qualifies is less than 0.) Thus $f(0) = 0$.

Similarly, for each *integer* n, it follows that $n \leq n$. Every other *integer* that is less than or equal to n is actually *less than* n. Thus, n *itself* is the greatest integer less than or equal to n. For example, $f(1) = 1$, $f(2) = 2$, $f(-1) = -1$, $f(-2) = -2$, etc.

Next, what is the greatest *integer* $\leq \frac{1}{2}$? Observe that $0 \leq \frac{1}{2}$, but $1 > \frac{1}{2}$. Every other *integer* is either less than 0, or else greater than 1 and hence greater than $\frac{1}{2}$. Thus the greatest *integer* less than or equal to $\frac{1}{2}$ is 0, that is,

$$f\left(\frac{1}{2}\right) = 0.$$

Similarly, for each x in the half-open interval $[0, 1)$, $f(x) = 0$.

For x in the half-open interval $[1, 2)$, the greatest *integer* less than or equal to x is 1, that is, $f(x) = 1$. Thus

$$f(1) = 1, \quad f\left(\frac{3}{2}\right) = 1, \quad f(1.99) = 1$$

For x in the half-open interval $[2, 3)$, $f(x) = 2$. Thus

For the *negative* number -1.5, observe that $-2 \leq -1.5$, but $-1 > -1.5$.

The greatest *integer* less than or equal to -1.5 is -2, that is, $f(-1.5) = -2$. Similarly, for each x in the half-open interval $[-2, -1)$, $f(x) = -2$.

For x in the half-open interval $[-1, 0)$, $f(x) = -1$.

The graph of f is drawn in Figure 3.22 . Because of the appearance of the graph, f is an example of a **step function.** Observe that the y-values "jump" *before* every integer x-value. The domain of f is the set of all real numbers; the range is the set of integers.

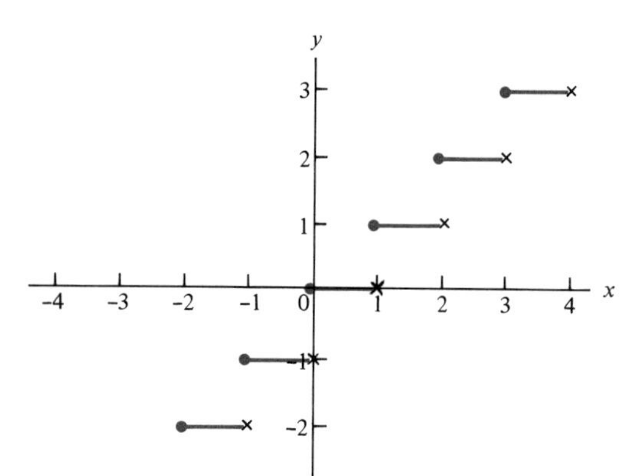

FIGURE 3.22 The graph of the "greatest integer function". The points (0, 0), (1, 1), (2, 2), etc., are included. The points (1, 0), (2, 1), (3, 2), etc., are excluded.

The greatest integer less than or equal to x is frequently denoted by $[\![x]\!]$. Thus

$$[\![1]\!] = 1, \qquad [\![2.8]\!] = 2, \qquad [\![-7]\!] = -7, \qquad [\![-3.4]\!] = -4 \qquad \square$$

EXERCISES

1. In Figure 3.23 determine the coordinates of each point.

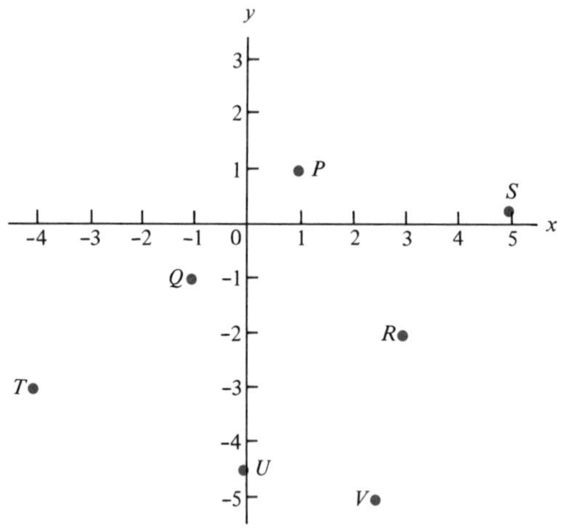

FIGURE 3.23

On a coordinate system plot the points specified in Exercises 2–8.

2. $(5, 4)$ **3.** $(-4, 2)$ **4.** $(0, 4)$

5. $(-4, -3)$ **6.** $(-3.5, 0)$ **7.** $(0, 0)$

8. $(1.5, -2.5)$

In Exercises 9–14 determine the quadrant of each point, or indicate that the point lies on a coordinate axis.

9. $(1, -1)$ **10.** $(-2, 3)$ **11.** $(-3, -7)$

12. $(6, 0)$ **13.** $\left(\frac{1}{2}, \frac{2}{5}\right)$ **14.** $(0, 0)$

In Exercises 15–44 graph each function.

15. $y = x + 1$ **16.** $y = x - 2$ **17.** $f(x) = 2x$ **18.** $g(x) = -x$ **19.** $h(x) = -3x$

20. $y = 2x - 1$ **21.** $y = 3x + 4$ **22.** $y = 5 - 2x$ **23.** $y = 1 - 4x$ **24.** $y \equiv 6$

25. $y \equiv -2$ **26.** $y = x^2 + 1$ **27.** $F(x) = x^2 + 5$ **28.** $G(x) = 2 - x^2$ **29.** $H(x) = -2x^2$

30. $y = \dfrac{x^2}{2}$ **31.** $y = |x + 1|$ **32.** $y = |x - 2|$ **33.** $f(x) = |x| - 1$ **34.** $g(x) = |x| + 3$

35. $h(x) = \left|\dfrac{x}{2}\right|$ **36.** $y = -2\,|x|$

37. $y = \begin{cases} x + 1, & \text{if } x < 0 \\ 1 - x, & \text{if } x \geqslant 0 \end{cases}$ **38.** $y = \begin{cases} -x, & \text{if } x < 0 \\ x, & \text{if } x \geqslant 0 \end{cases}$ **39.** $y = \begin{cases} 2x, & \text{if } x \leqslant 0 \\ -2x, & \text{if } x > 0 \end{cases}$

40. $F(x) = \begin{cases} x + 4, & \text{if } x \leq 0 \\ 4, & \text{if } x > 0 \end{cases}$ **41.** $G(x) = \begin{cases} 1, & \text{if } x \leq 0 \\ 2, & \text{if } x > 0 \end{cases}$ **42.** $H(x) = \begin{cases} 4, & \text{if } x \leq 2 \\ 2x, & \text{if } x > 2 \end{cases}$

43. $y = \begin{cases} x, & \text{if } x < 0 \\ 2x, & \text{if } 0 \leq x \leq 2 \\ 4, & \text{if } x > 2 \end{cases}$ **44.** $y = \begin{cases} 1, & \text{if } x < -1 \\ -x, & \text{if } -1 \leq x \leq 1 \\ -1, & \text{if } x > 1 \end{cases}$

In Exercises 45–50: (a) Graph the indicated function. (b) What is the domain of the function? (c) What is the range of the function?

45. $y = x, \quad 0 \leq x \leq 4$ **46.** $y = 2 - x, \quad -1 \leq x \leq 3$ **47.** $y = x + 4, \quad 0 \leq x \leq 5$

48. $y = 2x, \quad -2 \leq x \leq 2$ **49.** $y = -2x, \quad -2 \leq x \leq 2$ **50.** $y = x^2, \quad -2 \leq x \leq 2$

51. Let $F(x)$ be the greatest integer *less than* x. Determine: **(a)** $F(0)$ **(b)** $F(1)$ **(c)** $F(-1)$ **(d)** $F\left(\frac{1}{2}\right)$ **(e)** $F\left(-\frac{1}{2}\right)$ **(f)** Graph this function.

52. Let $G(x)$ be the greatest integer less than or equal to $x + 1$. Determine: **(a)** $G(0)$ **(b)** $G(1)$ **(c)** $G(-1)$ **(d)** $G\left(\frac{1}{2}\right)$ **(e)** $G\left(-\frac{1}{2}\right)$ **(f)** Graph this function.

53. Let $H(x)$ be one more than the greatest integer less than or equal to x. Determine: **(a)** $H(0)$ **(b)** $H(1)$ **(c)** $H(-1)$ **(d)** $H\left(\frac{1}{2}\right)$ **(e)** $H\left(-\frac{1}{2}\right)$ **(f)** Graph this function.

54. In 1978 the U.S. domestic postal rate was raised to 15 cents for each ounce, or part of an ounce (up to 4 pounds). Let x represent the number of ounces and let $f(x)$ represent the postal charge (in cents) for x ounces. Determine: **(a)** $f\left(\frac{1}{2}\right)$ **(b)** $f(1)$ **(c)** $f(2.2)$ **(d)** $f(3)$
(e) Graph f for $0 \leq x \leq 4$. (Use a smaller scale on the y-axis.)

55. A mechanics union assesses dues according to the following salary schedule.

Salary	*Dues*
salary < \$ 5 000	\$ 50
\$ 5 000 ≤ salary < \$10 000	\$100
\$10 000 ≤ salary < \$15 000	\$150
\$15 000 ≤ salary	\$200

Let each unit on the x-axis represent \$5000 in salary; let each unit on the y-axis represent \$50 in union dues. Graph the indicated function.

3.3 More on Functions and Graphs

evaluating functions

It is important to realize that the rule(s) or description that defines a function f must be adhered to *mechanically*, for each argument in the domain of f, however that argument is given. For example, let

$$f(x) = x + 5$$

If $x + 1$ is in the domain of f, then

$$f(x + 1) = (x + 1) + 5 = x + 6.$$

And if $x + h$ is in the domain of f, then

$$f(x + h) = (x + h) + 5 = x + h + 5.$$

You simply add 5 to each argument $x + h$ to find the corresponding function-value.

EXAMPLE 1 Let $f(x) = x^2 + 2$ for all x. Find

(a) $f(2x)$, (b) $f(-x)$, (c) $f\left(\frac{1}{x}\right)$, for $x \neq 0$, (d) $f(x + 1)$,

(e) $f(x + h)$, (f) $\frac{f(x + h) - f(x)}{h}$, for $h \neq 0$.

Solution. $f(x) = x^2 + 2$

(a) $$f(2x) = (2x)^2 + 2 = 2x \cdot 2x + 2 = 4x^2 + 2$$

(b) $$f(-x) = (-x)^2 + 2 = (-x)(-x) + 2 = x^2 + 2$$

Note that here

$$f(-x) = f(x).$$

(c) Assume $x \neq 0$.

$$f\left(\frac{1}{x}\right) = \left(\frac{1}{x}\right)^2 + 2 = \frac{1}{x^2} + 2 = \frac{1 + 2x^2}{x^2}$$

(d) $$\begin{aligned} f(x + 1) &= (x + 1)^2 + 2 \\ &= (x^2 + 2x + 1) + 2 \\ &= x^2 + 2x + 3 \end{aligned}$$

(e) $$\begin{aligned} f(x + h) &= (x + h)^2 + 2 \\ &= (x^2 + 2xh + h^2) + 2 \\ &= x^2 + 2xh + h^2 + 2 \end{aligned}$$

(f) $$\begin{aligned} f(x + h) - f(x) &= \overbrace{(x^2 + 2xh + h^2 + 2)}^{f(x + h)} - \overbrace{(x^2 + 2)}^{f(x)} \\ &= \cancel{x^2} + 2xh + h^2 + \cancel{2} - \cancel{x^2} - \cancel{2} \\ &= 2xh + h^2 \\ &= h(2x + h) \end{aligned}$$

Thus for $h \neq 0$,

$$\begin{aligned} \frac{f(x + h) - f(x)}{h} &= \frac{\cancel{h}(2x + h)}{\cancel{h}} \\ &= 2x + h \end{aligned}$$

difference quotient For a function f, the expression

$$\frac{f(x + h) - f(x)}{h}, \qquad h \neq 0,$$

is known as the **difference quotient of** f.

Thus in Example 1(f), you found the difference quotient for the function defined by $f(x) = x^2 + 2$.

The difference quotient is used to define the *derivative of a function*, one of the fundamental concepts of calculus.

Here is a physical interpretation of the difference quotient.

Suppose that an object travels at a *varying rate*. Its distance at time t can be specified as a function of t. Let $s(t)$ be the distance (in meters) traveled in t seconds. For example, if

$$s(t) = t^2,$$

the object has traveled 9^2, or 81, meters in 9 seconds and 10^2, or 100, meters in 10 seconds. Consider the difference quotient, now written as

$$\frac{s(t + h) - s(t)}{h}, \qquad h \neq 0.$$

Take $t = 9$, $h = 1$, so that $t + h = 10$. Then

$$\frac{s(10) - s(9)}{1} = \frac{100 - 81}{1} = 19 \text{ (meters per second)}$$

Observe that the dividend, $s(10) - s(9)$, represents the distance traveled during the *tenth* second (that is, from $t = 9$ to $t = 10$). The divisor represents the time interval, 1 second. The difference quotient, which involves distance divided by time, represents the *average velocity* of the object during the tenth second.

equality of functions

Two functions, f and g, are defined to be **equal**, written $f = g$, if

1. they have the *same domain* and
2. $f(a) = g(a)$ for each a in their (common) domain.

According to this definition, the functions

$$f(x) = x + 2 \qquad \text{and} \qquad g(x) = x + 2, \quad 0 \leqslant x \leqslant 3$$

are *not* equal because the domain of f is the set of all real numbers, whereas the domain of g is $[0, 3]$. [See Figure 3.24 (a) and (b).] Here is another such example.

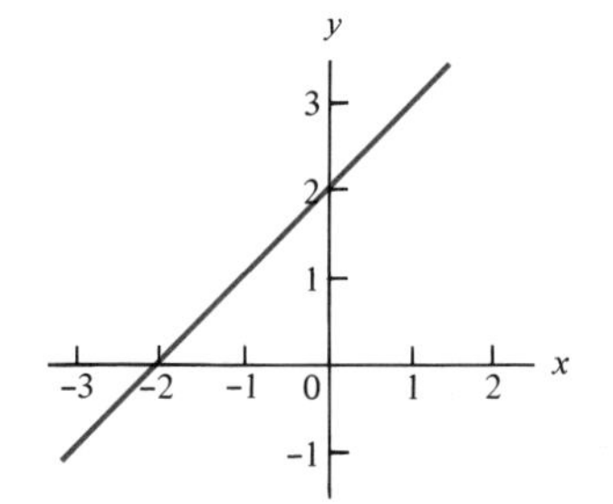

FIGURE 3.24(a) The graph of $f(x) = x + 2$

EXAMPLE 2 Let $f(x) = x + 2$ and $h(x) = \dfrac{x^2 - 4}{x - 2}$. Does $f = h$?

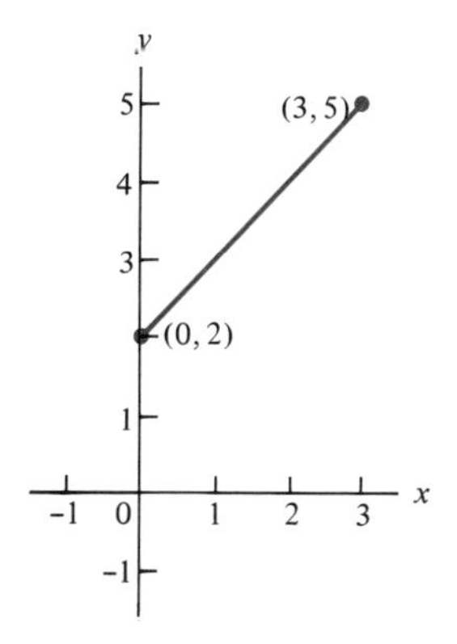

FIGURE 3.24(b) The graph of $g(x) = x + 2$, for $0 \leqslant x \leqslant 3$

Solution. Note that

$$h(x) = \frac{x^2 - 4}{x - 2} = \frac{(x + 2)(x - 2)}{x - 2} \tag{3.2}$$

At first, it might appear that you could continue:

$$\begin{aligned} h(x) &= \frac{(x + 2)\cancel{(x - 2)}}{\cancel{x - 2}} = x + 2 \\ &= f(x) \end{aligned} \tag{3.3}$$

However, there is a subtle point here. When you set $x = 2$ in the expression **(3.2)** for $h(x)$, you obtain

$$h(2) = \frac{2^2 - 4}{2 - 2} = \frac{0}{0}.$$

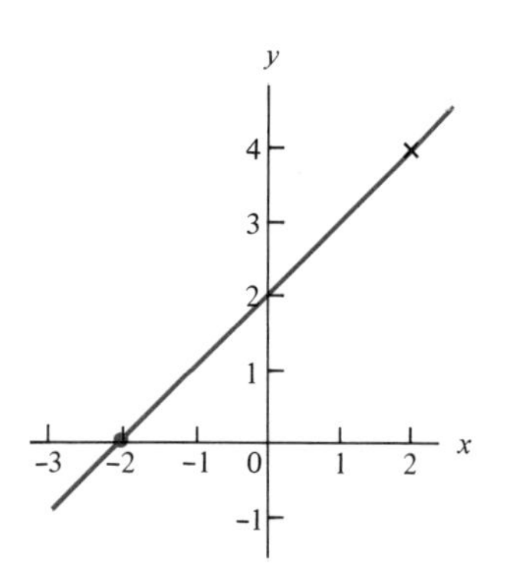

FIGURE 3.24(c) The graph of $h(x) = \dfrac{x^2 - 4}{x - 2} = x + 2$, for $x \neq 2$
The point (2, 4) is excluded.

But $\dfrac{0}{0}$ is undefined! Thus $h(x)$ is undefined at $x = 2$, and you can only write

$$h(x) = \frac{x^2 - 4}{x - 2} \qquad \text{for } x \neq 2.$$

Thus, statement **(3.3)** should really say

$$h(x) = x + 2 \qquad \textit{only for } x \neq 2,$$

whereas

$$f(x) = x + 2 \qquad \text{for } \textit{all } x.$$

Thus, the domains of the two functions differ. The graph of f is the line drawn in Figures 3.24(a). The graph of h is the *punctured line* (that is, the line with a point missing) of Figure 3.24(c). □

graph of a function

The definition of a function requires that

to each argument there corresponds exactly one value.

A single argument a cannot have two different values corresponding to it; you cannot have both

$$a \to b$$
$$a \to c$$

that is, you cannot have both

$$(a, b) \qquad \text{and} \qquad (a, c)$$

on the graph.

Thus, two points with the same x-coordinate cannot lie on the graph of a function. Because two points with the same x-coordinate determine a vertical line, it follows that *a vertical line cannot intersect the graph of a function more than once.*

You can immediately tell whether a curve on the plane represents the graph of *some* function.

1. If *every* vertical line intersects the curve *at most once,* the curve is the graph of a function.
2. If there is *at least one* vertical line that intersects the curve *more than once,* the curve is *not* the graph of a function.

EXAMPLE 3 In Figure 3.25, curves (a), (b), and (c) are graphs of functions; curves (d), (e), and (f) are not. In each of the graphs (a), (b), and (c), every vertical line intersects the curve at most once. In each of the graphs (d), (e), and (f), there is at least one vertical line that intersects the curve more than once. Notice that in (e) the curve is itself a vertical line, V. Thus a *vertical line is not the graph of a function.* However, a *horizontal line* [Figure 3.25(b)] is *the graph of a function.* □

intercepts

An ***x*-intercept** of a graph is the x-coordinate of an intersection with the x-axis, and the ***y*-intercept** is the y-coordinate of an intersection

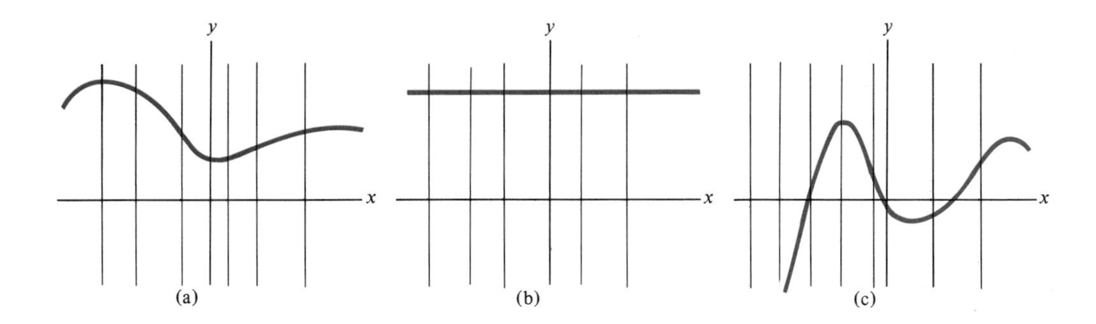

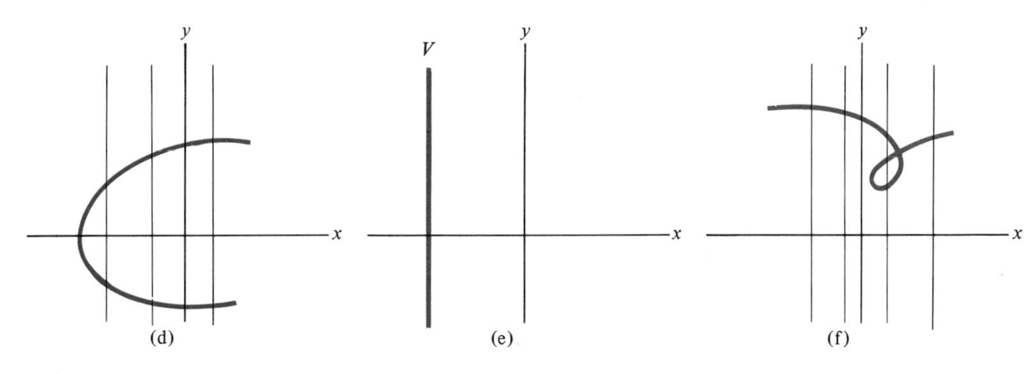

FIGURE 3.25

with the y-axis. *The graph of a function has at most one y-intercept* because a vertical line intersects such a graph at most once. For the function

$$y = x^2 - 4$$

depicted in Figure 3.26, the x-intercepts are -2 and 2, and the y-intercept is -4.

You can also find the intercepts algebraically. To this end, observe that the x-axis has the equation

$$y = 0$$

and the y-axis has the equation

$$x = 0. \quad \text{[See Figure 3.27 .]}$$

Thus to find the x-intercepts, set $y = 0$ in the given equation(s) of the function; to find the y-intercepts, set $x = 0$. For the function defined by

$$y = x^2 - 4,$$

in order to find the x-intercepts algebraically, set $y = 0$.

$$0 = x^2 - 4$$
$$4 = x^2$$

Note that

$$2^2 = 4 \qquad \text{and} \qquad (-2)^2 = 4.$$

Thus, the x-intercepts are 2 and -2.

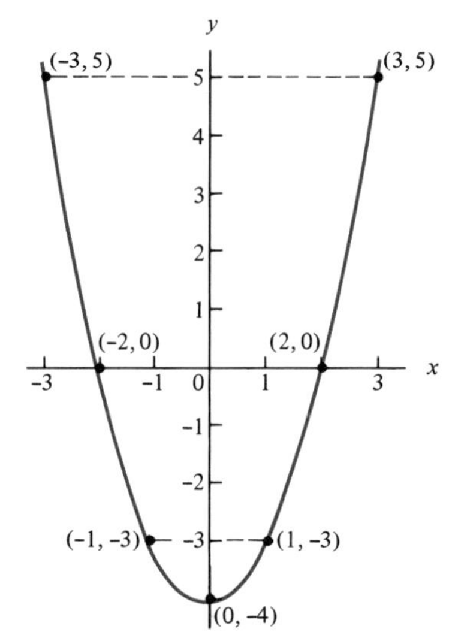

FIGURE 3.26 For $y = x^2 - 4$, the x-intercepts are -2 and 2, and the y-intercept is -4.

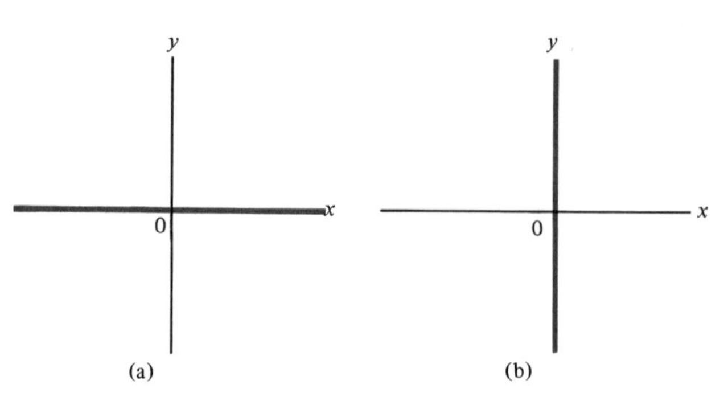

FIGURE 3.27
(a) The x-axis has the equation $y = 0$.
(b) The y-axis has the equation $x = 0$.

To find the y-intercept, set $x = 0$:

$$y = 0^2 - 4 = -4$$

symmetry

The graph of the function defined by

$$f(x) = x^2 - 4$$

is *symmetric with respect to the y-axis*. [See Figure 3.26 .] Note that

$$f(1) = f(-1) = -3$$
$$f(2) = f(-2) = 0$$
$$f(3) = f(-3) = 5$$

The points

$(1, -3)$ and $(-1, -3)$
$(2, 0)$ and $(-2, 0)$
$(3, 5)$ and $(-3, 5)$

are mirror images of each other with respect to the y-axis. In fact, for each x,

$$f(-x) = (-x)^2 - 4 = x^2 - 4 = f(x)$$

In general, the graph of a function F is **symmetric with respect to the y-axis** if for each x in its domain, $-x$ is also in its domain and

$$F(-x) = F(x)$$

Thus

if (x, y) is on the graph, so is $(-x, y)$.

In this case, the function F is called an **even function.** (Note that the exponent, 2, of x^2 is even.)

Now consider the graph of the function defined by

$$g(x) = x^3.$$

$g(1) = 1^3 = 1$ and $g(-1) = (-1)^3 = -1$
$g(2) = 2^3 = 8$ and $g(-2) = (-2)^3 = -8$
$g(3) = 3^3 = 27$ and $g(-3) = (-3)^3 = -27$

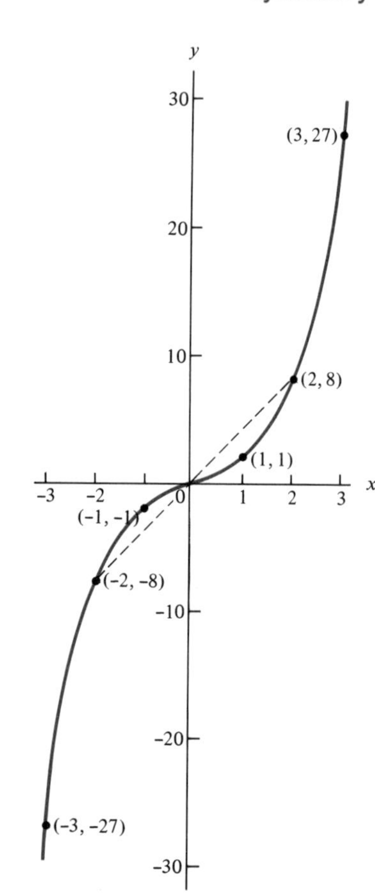

FIGURE 3.28 The graph of $g(x) = x^3$. Different scales have been used on the x- and y-axes.

Note that the points

$(1, 1)$ and $(-1, -1)$

$$\begin{array}{lcl} (2,\ 8) & \text{and} & (-2,\ -8) \\ (3, 27) & \text{and} & (-3, -27) \end{array}$$

are "reflections" of each other through the origin. [See Figure 3.28 . See also page 217.] For each x,

$$g(-x) = (-x)^3 = (-x)(-x)(-x) = -\underbrace{x^3}_{g(x)} = -g(x)$$

In general, the graph of a function F is **symmetric with respect to the origin** if for each x in its domain, $-x$ is also in its domain and

$$F(-x) = -F(x)$$

Thus

if (x, y) is on the graph, so is $(-x, -y)$.

In this case, the function F is called an **odd function.** (Note that the exponent, 3, of x^3 is odd.)

EXAMPLE 4 Determine whether the function defined by each equation is even, odd, or neither of these.

(a) $f(x) = -2x$ (b) $g(x) = x^4$
(c) $h(x) = x^4 - 2x$

Solution.

(a)
$$\begin{aligned} f(-x) &= -2(-x) \\ &= -\underbrace{(-2x)}_{f(x)} \\ &= -f(x) \end{aligned}$$

Thus f is odd.

(b)
$$\begin{aligned} g(-x) &= (-x)^4 \\ &= (-x)(-x)(-x)(-x) \\ &= x^4 \\ &= g(x) \end{aligned}$$

Thus g is even.

(c)
$$\begin{aligned} h(-x) &= (-x)^4 - 2(-x) \\ &= x^4 + 2x \end{aligned}$$

Because

$$\begin{aligned} h(x) &= x^4 - 2x, \\ h(-x) &\neq h(x) \end{aligned}$$

Because

$$\begin{aligned} -h(x) &= -x^4 + 2x, \\ h(-x) &\neq -h(x) \end{aligned}$$

Thus h is neither even nor odd. □

EXAMPLE 5 Determine whether the absolute value function, defined by

$$f(x) = |x|$$

is even, odd, or neither of these.

Solution. For each x,

$$f(x) = |x| = \begin{cases} x, & \text{if } x \geq 0 \\ -x, & \text{if } x < 0 \end{cases}$$

If x is positive, $-x$ is negative,

$$|x| = x \quad \text{and} \quad |-x| = -(-x) = x$$

If $x = 0$, then $-x = 0$ and

$$|x| = |-x| = 0$$

If x is negative, $-x$ is positive,

$$|x| = -x \quad \text{and} \quad |-x| = -x$$

For example, if $x = -5$, then $-(-5) = 5$, so that

$$|-5| = -(-5) = 5 \quad \text{and} \quad |-(-5)| = |5| = 5$$

In all cases,

$$f(x) = |x| = |-x| = f(-x)$$

Thus f is even. [See Figure 3.20 on page 115.] □

The zero function is both even and odd. In fact, if f denotes the zero function, then for every x, both

$$f(x) = 0 \quad \text{and} \quad f(-x) = 0.$$

Thus

$$f(-x) = f(x) = -f(x).$$

EXERCISES

1. Let $f(x) = 5x$. Determine: **(a)** $f(10)$ **(b)** $f(-10)$ **(c)** $f(\pi)$ **(d)** $f(x + 2)$ **(e)** $f(x - 3)$

2. Let $g(x) = x + 10$. Determine:

(a) $g(-10)$ **(b)** $g(x + 2)$ **(c)** $g(x - 2)$ **(d)** $g(x + h)$ **(e)** $\dfrac{g(x + h) - g(x)}{h}$, $h \neq 0$

3. Let $F(x) = 10x$. Determine:

(a) $F(x + 1)$ **(b)** $F(x + h)$ **(c)** $\dfrac{F(x + h) - F(x)}{h}$, $h \neq 0$ **(d)** $F(t)$ **(e)** $F(t^2)$

4. Let $G(u) = 2u + 3$. Determine:

(a) $G(u + 2)$ **(b)** $G(u - 3)$ **(c)** $G(2u)$ **(d)** $G(u + h)$ **(e)** $\dfrac{G(u + h) - G(u)}{h}$, $h \neq 0$

5. Let $f(t) = t^2$. Determine:

(a) $f(t + 2)$ **(b)** $f(t + h)$ **(c)** $\dfrac{f(t + h) - f(t)}{h}$, $h \neq 0$ **(d)** $f(-t)$ **(e)** $f(t^2)$

6. Let $g(x) = x^2 + x$. Determine:

(a) $g(3)$ **(b)** $g(-3)$ **(c)** $g(x + 2)$ **(d)** $g(x + h)$ **(e)** $\dfrac{g(x + h) - g(x)}{h}$, $h \neq 0$

7. Let $F(x) = \frac{1}{x}$, $x \neq 0$. Determine:

(a) $F(-1)$ (b) $F\left(\frac{1}{4}\right)$ (c) $F\left(\frac{1}{x}\right)$ (d) $F(x + h)$‡ (e) $\frac{F(x+h) - F(x)}{h}$, $h \neq 0$

8. Let $G(x) = \frac{1}{x^2}$, $x \neq 0$. Determine:

(a) $G(-1)$ (b) $G\left(\frac{1}{4}\right)$ (c) $G(-x)$ (d) $G\left(\frac{1}{x}\right)$ (e) $\frac{G(x+h) - G(x)}{h}$, $h \neq 0$

9. Let $f(x) = \frac{1}{x+1}$, $x \neq -1$. Determine:

(a) $f(0)$ (b) $f\left(\frac{1}{2}\right)$ (c) $f\left(-\frac{1}{2}\right)$ (d) $f\left(\frac{1}{x}\right)$, $x \neq 0$ (e) $\frac{f(x+h) - f(x)}{h}$, $h \neq 0$

10. Let $g(x) = \frac{2}{x^2 + 1}$. Determine:

(a) $g\left(\frac{1}{2}\right)$ (b) $g(-x)$ (c) $g(x^2)$ (d) $g\left(\frac{1}{x}\right)$, $x \neq 0$ (e) $\frac{g(x+h) - g(x)}{h}$, $h \neq 0$

11. Let $F(x) = x^2 + 3x - 4$. Determine: (a) $F(10)$ (b) $F(x + 1)$ (c) $F(x^2)$

(d) $F\left(\frac{1}{x}\right)$, $x \neq 0$ (e) $F(-x)$ (f) $\frac{F(x+h) - F(x)}{h}$, $h \neq 0$

12. Let $G(x) = \frac{x+1}{x-1}$. (a) For which x is G undefined? (b) Determine $G(2x)$.

(c) For which x is $G(2x)$ undefined? (d) Determine $G(x + 1)$.

(e) For which x is $G(x + 1)$ undefined? (f) Determine $\frac{G(x+h) - G(x)}{h}$, $h \neq 0$.

13. Suppose $f(x) = x + 6$ for all x, and $g(x) = x + 6$ for $x \geqslant 0$. Does $f = g$?

14. Suppose $f(x) = x^2$ for $1 \leqslant x \leqslant 3$, and $g(x) = x^2$ for $1 < x < 3$. Does $f = g$?

15. Suppose $f(x) = x^2 - 1$ for all x, and $g(x) = \begin{cases} x^2 - 1, & \text{for } x \neq 0 \\ -1, & \text{for } x = 0 \end{cases}$ Does $f = g$?

16. Suppose $f(x) = x + 3$ for all x, and $g(x) = \frac{x^2 - 9}{x - 3}$. Does $f = g$?

17. Suppose $f(x) = x - 5$ for all x, and $g(x) = \begin{cases} \frac{x^2 - 25}{x + 5}, & \text{for } x \neq -5 \\ -10, & \text{for } x = -5 \end{cases}$ Does $f = g$?

18. Suppose $f(x) = x + 1$ for all x, and $g(x) = \begin{cases} \frac{x^2 - 1}{x - 1}, & \text{for } x \neq 1 \\ 1, & \text{for } x = 1 \end{cases}$ Does $f = g$?

‡In Exercise 7, $F(x + h)$ is not defined when the denominator equals 0. Thus, $F(x + h)$ is not defined if $x = -h$. Similarly, in Exercises 8, 9, and 12, care must be taken in defining the functions at $x + h$.

19. Suppose $f(x) = x^2 + 2x + 4$ and the domain of f is $[0, 2]$. Assume $f = g$. Find

(a) $g(0)$ (b) $g(1)$ (c) $g(2)$ (d) the domain of g.

20. Suppose $f(0) = 0$ and $f(x) = \dfrac{1}{x}$ for $x \neq 0$. Which of the following are true?

(a) $f\left(-\dfrac{1}{2}\right) = -2$; (b) domain f is the set of all real numbers; (c) f is an odd function;

(d) f is an even function; (e) range f is the set of all real numbers.

In Exercises 21–32 [Figure 3.29] determine which diagrams represent graphs of functions.

21.

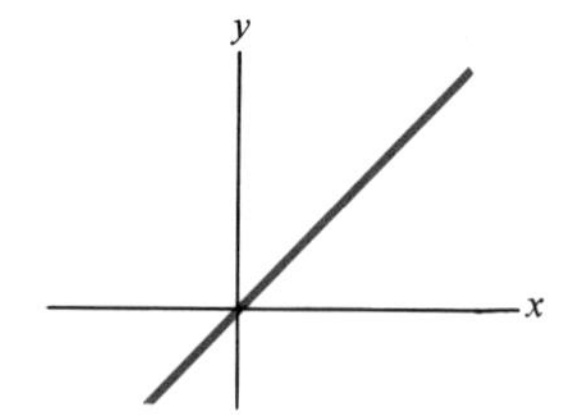

22.

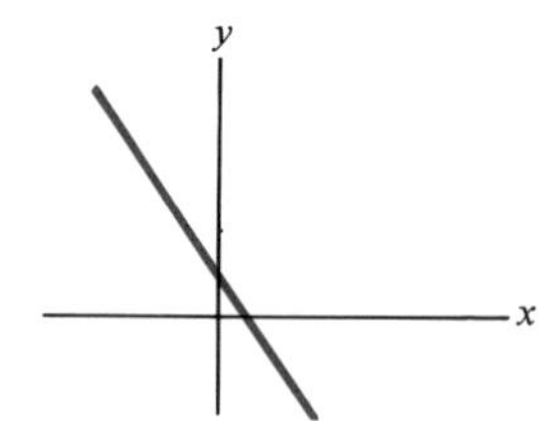

23.

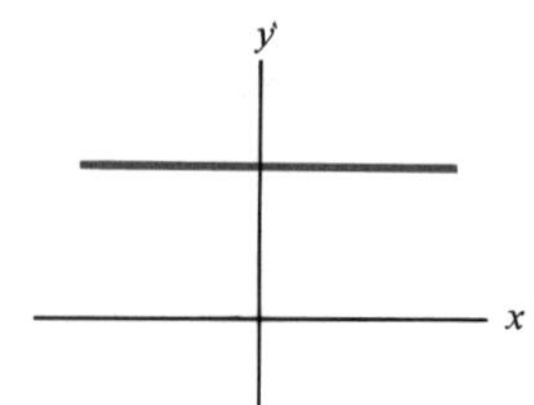

24.

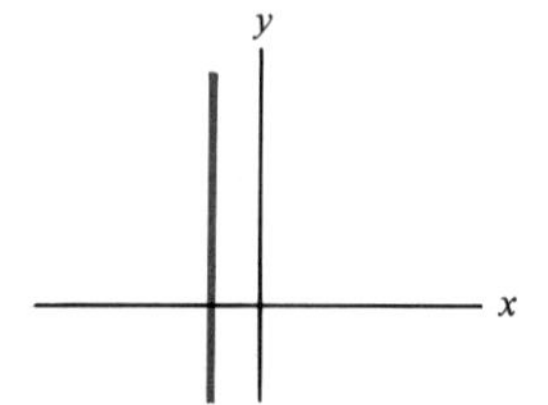

25.

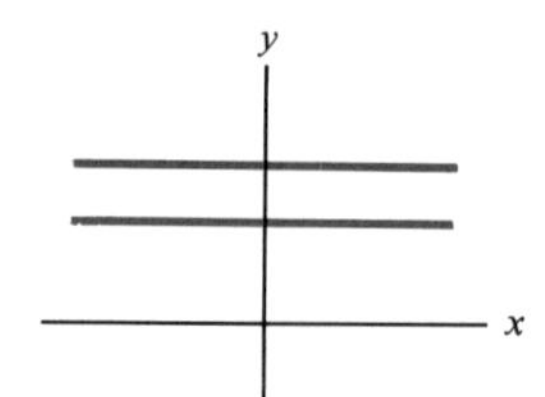

26.

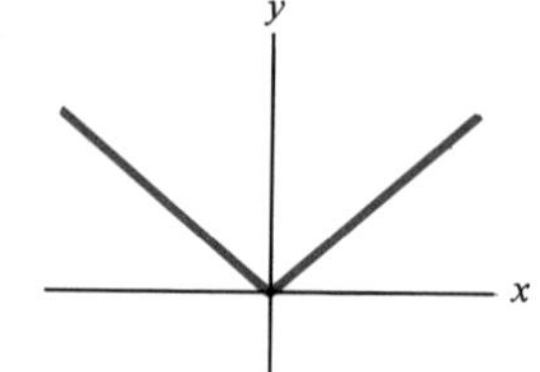

27.

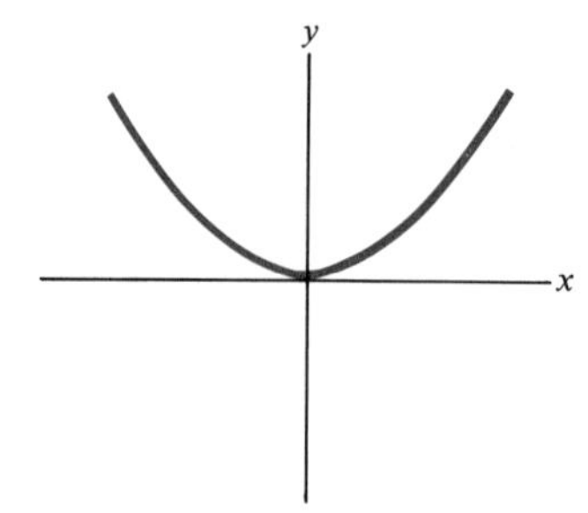

28.

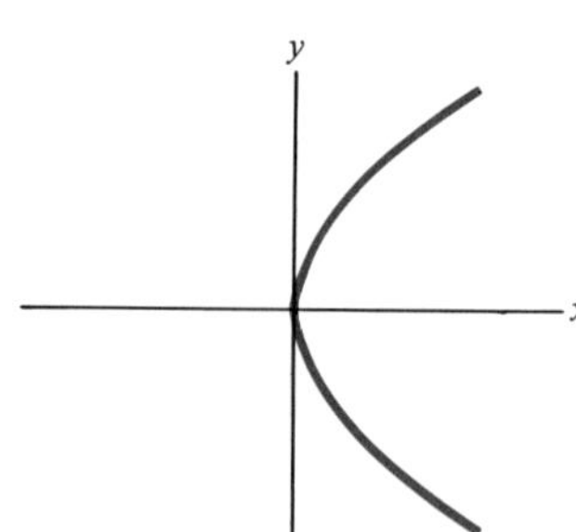

29.

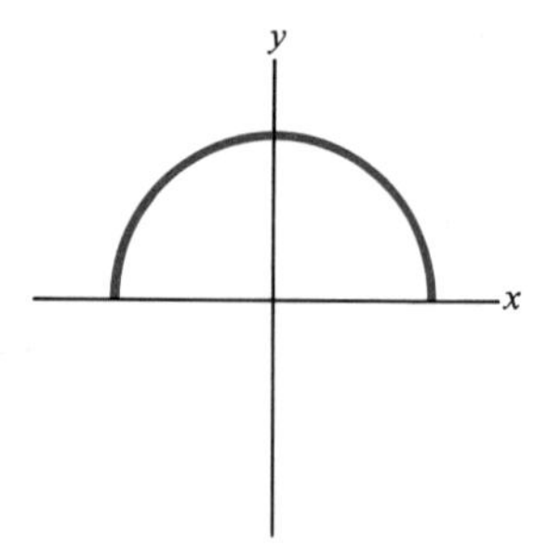

30.

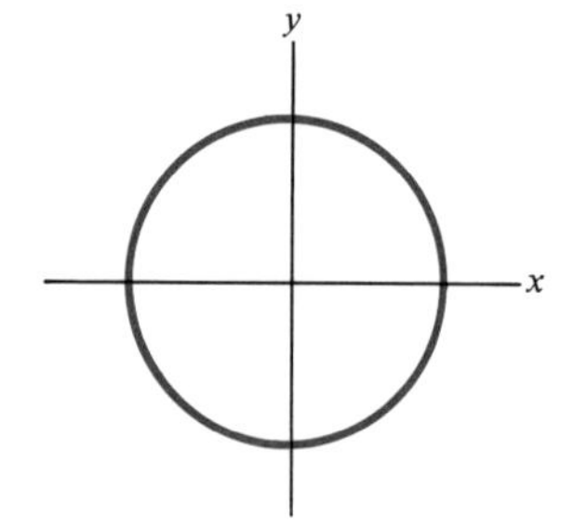

31.

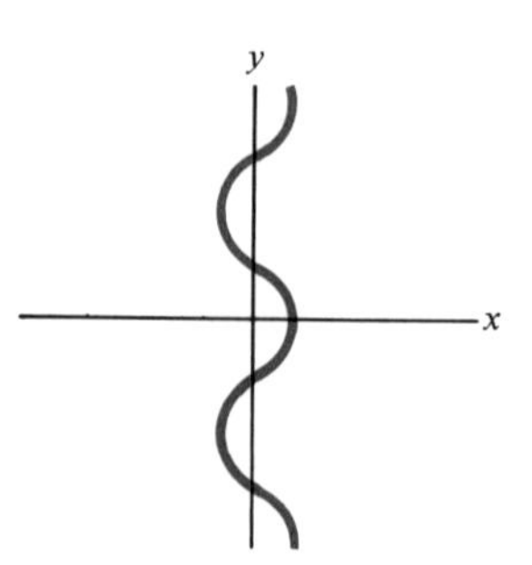

32.

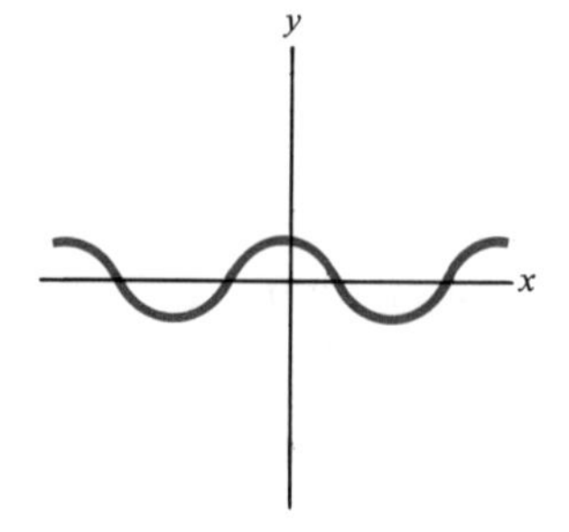

FIGURE 3.29

In Exercises 33–40 [Figure 3.30]:
(a) Find all x-intercepts, if any. (b) Find the y-intercept, if there is one. (c) Is there symmetry with respect to the y-axis? (d) Is there symmetry with respect to the origin?

33.

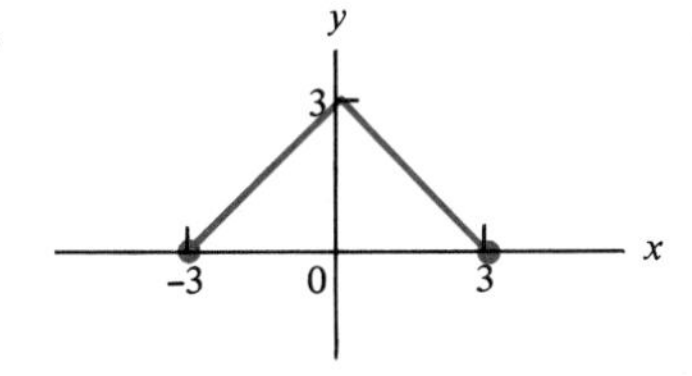

34.

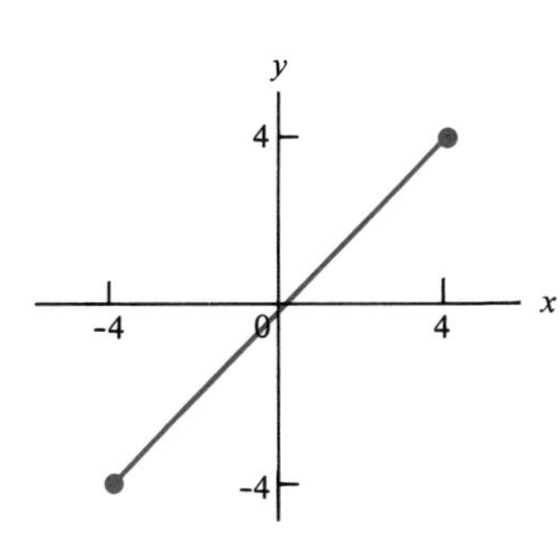

35.

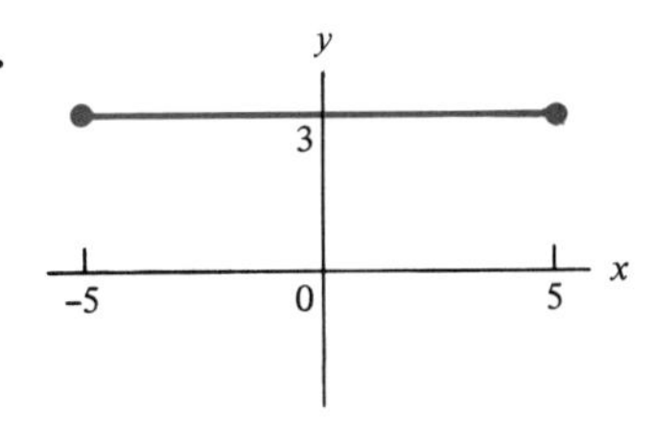

36.

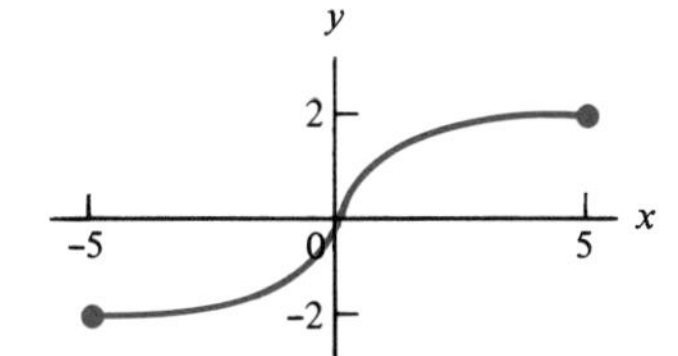

37.

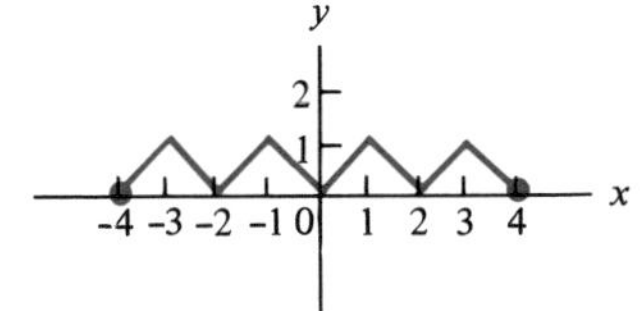

38.

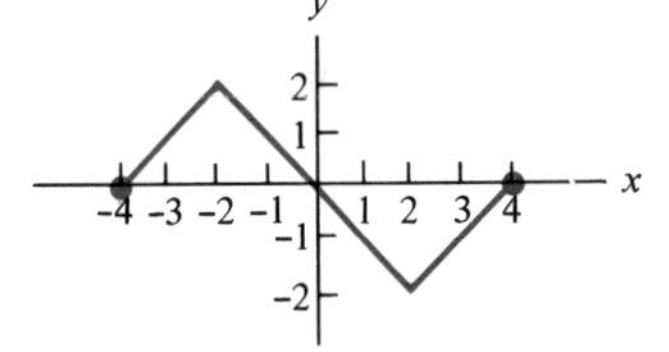

39.

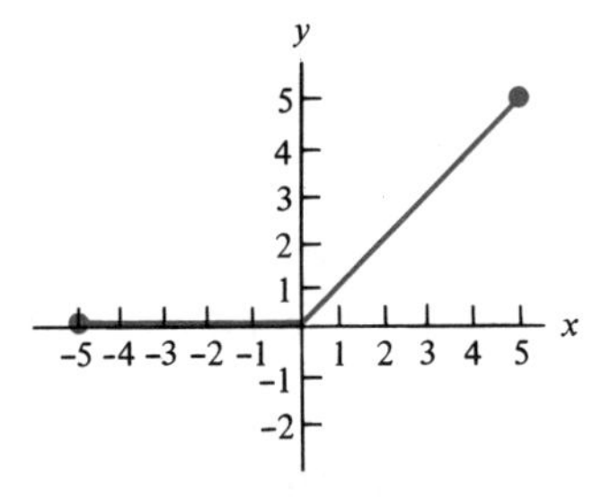

40.

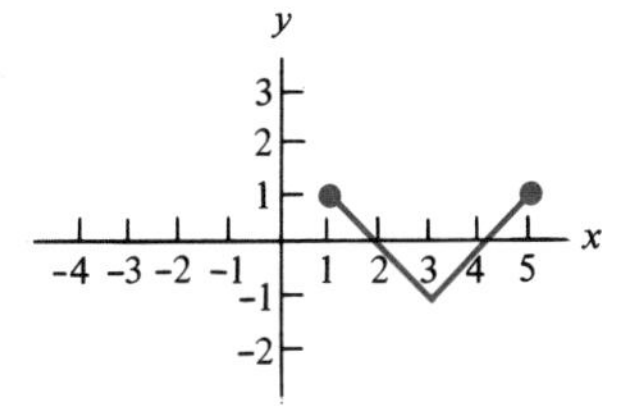

FIGURE 3.30

In Exercises 41–48:
(a) Find all x-intercepts, if any. (b) Find the y-intercept, if there is one. (c) Does the equation define an even function? (d) Does the equation define an odd function?

41. $f(x) = 4x$ **42.** $g(x) = x + 4$ **43.** $h(x) = 2x - 4$ **44.** $F(x) = 3 - x$

45. $G(x) = x^2 - 9$ **46.** $H(x) = x^2 - 36$ **47.** $f(x) = |x| - 1$ **48.** $g(x) = |x - 1|$

In Exercises 49–51 show that each equation defines an even function.

49. $f(x) = x^4$ **50.** $g(x) = 2x^4 + x^2 - 3$ **51.** $h(x) = \dfrac{x^2 + 1}{x^4 - 2}$

In Exercises 52–54 show that each equation defines an odd function.

52. $f(x) = x^5$ **53.** $g(x) = x^5 - 2x^3 + 4x$ **54.** $h(x) = \dfrac{x^3 + 2x}{x^2 + 3}$

55. Suppose f is an even function whose domain is the set of all real numbers. Suppose, further, that $f(1) = 5$, $f(-2) = 3$, $f(4) = 0$, and $f(-5) = -\pi$. Determine: **(a)** $f(-1)$ **(b)** $f(2)$ **(c)** $f(-4)$ **(d)** $f(5)$

56. Suppose g is an odd function whose domain is the set of all real numbers. Suppose, further, that $g(1) = 2$, $g(-2) = -4$, $g(3) = 0$, $g(4) = -4$. Determine: **(a)** $g(-1)$ **(b)** $g(2)$ **(c)** $g(-3)$ **(d)** $g(-4)$ **(e)** $g(0)$

57. A helicopter travels $s(t)$ meters in t seconds, where $s(t) = 2t^2$. Find the average velocity of the airplaine during the sixth second.

58. An airplane travels $s(t)$ kilometers in t minutes, where
$$s(t) = t^2 + 3t.$$
Find the average velocity during the first half hour.

59. Suppose $C(x)$ is the cost (in dollars) of manufacturing x tractors. Then the difference quotient
$$\frac{C(x + h) - C(x)}{h}$$
represents the average cost of manufacturing each tractor as production changes from x tractors to $(x + h)$ tractors. Let
$$C(x) = 4000x + 2500 - \frac{1000}{x}.$$
Find the average cost of manufacturing a tractor as production increases from 100 to 200 tractors.

3.4 Combining Functions

Functions are often combined in various ways to make new functions.

arithmetic of functions

Let f and g be functions with the same domain, D. Let c be any real number. Define

Addition of functions: $(f + g)(x) = f(x) + g(x)$ for all x in D

Subtraction of functions: $(f - g)(x) = f(x) - g(x)$ for all x in D

Multiplication of a function by a constant: $(cf)(x) = c \cdot f(x)$ for all x in D

Multiplication of functions: $(f \cdot g)(x) = f(x) \cdot g(x)$ for all x in D

Division of functions: $\left(\frac{f}{g}\right)(x) = \frac{f(x)}{g(x)}$ for all x in D for which $g(x) \neq 0$. (The quotient is not defined if g is the zero function.)

EXAMPLE 1 Suppose $f(1) = 2$, $g(1) = 3$, $f(2) = 10$, $g(2) = 5$. Then

$$(f + g)(1) = f(1) + g(1) = 2 + 3 = 5$$
$$(f + g)(2) = f(2) + g(2) = 10 + 5 = 15$$
$$(f - g)(1) = f(1) - g(1) = 2 - 3 = -1$$
$$(f - g)(2) = f(2) - g(2) = 10 - 5 = 5$$

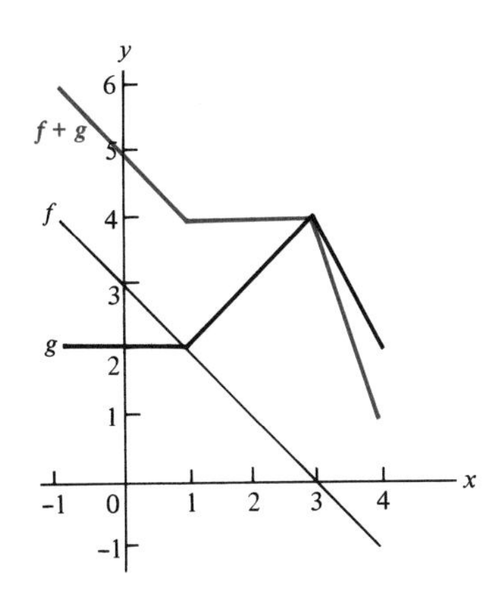

$$(4f)\,(1) = 4 \cdot f(1) = 4 \cdot 2 = 8$$
$$(4f)\,(2) = 4 \cdot f(2) = 4 \cdot 10 = 40$$
$$(f \cdot g)\,(1) = f(1) \cdot g(1) = 2 \cdot 3 = 6$$
$$(f \cdot g)\,(2) = f(2) \cdot g(2) = 10 \cdot 5 = 50$$
$$\left(\frac{f}{g}\right)(1) = \frac{f(1)}{g(1)} = \frac{2}{3}$$
$$\left(\frac{f}{g}\right)(2) = \frac{f(2)}{g(2)} = \frac{10}{5} = 2$$ □

These new functions are defined *point-wise*. In general, you can construct the graphs of $f + g$, $f - g$, etc. from the graphs of f and g by combining the values of $f(x)$ and $g(x)$ for each x. [See Figure 3.31 .]

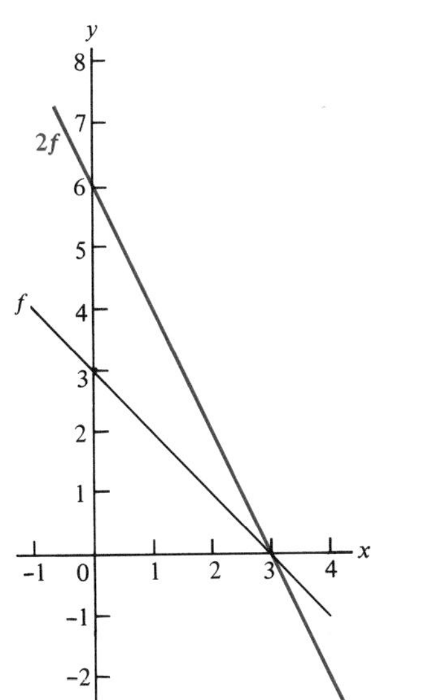

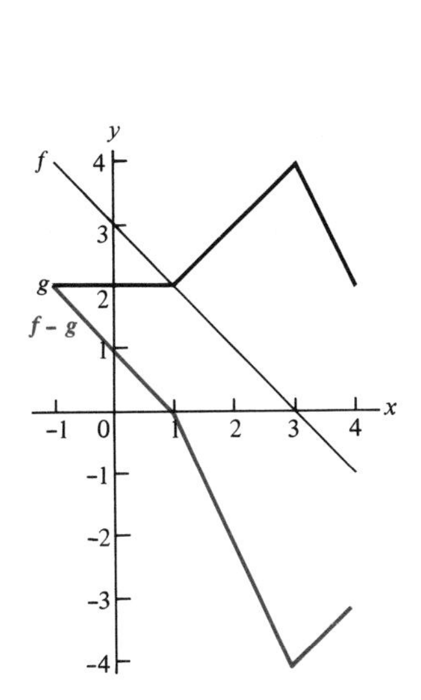

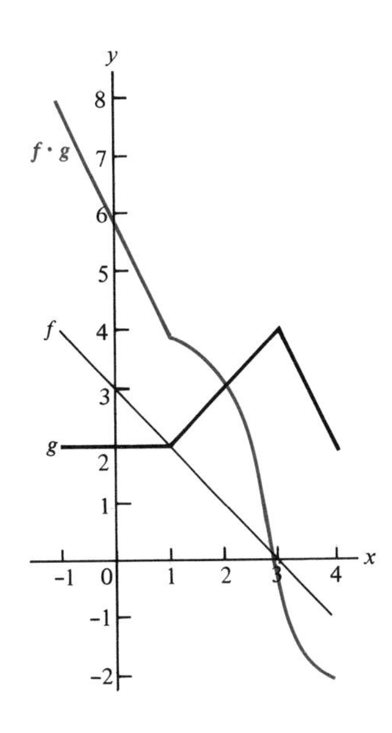

FIGURE 3.31

FXAMPLE 2 Suppose

$$f(x) = x^2, \quad \text{for all } x;$$
$$g(x) = 2x + 1, \quad \text{for all } x.$$

Then for all x,

$$(f + g)\,(x) = x^2 + 2x + 1$$
$$(f - g)\,(x) = x^2 - (2x + 1) = x^2 - 2x - 1$$
$$(g - f)\,(x) = 2x + 1 - x^2$$
$$(3f)\,(x) = 3x^2$$
$$(-3g)\,(x) = -3(2x + 1) = -6x - 3$$
$$(f \cdot g)\,(x) = x^2(2x + 1) = 2x^3 + x^2$$

Also,

$$\left(\frac{f}{g}\right)(x) = \frac{x^2}{2x+1}, \quad \text{if } 2x + 1 \neq 0, \quad \text{that is, if } x \neq \frac{-1}{2}$$

And

$$\left(\frac{g}{f}\right)(x) = \frac{2x+1}{x^2}, \quad \text{if } x \neq 0$$

□

You can also combine three or more functions. For example, if $f(x) = x^2$, $g(x) = 5x$, $h(x) = 3$, then

$$(f + g + h)(x) = x^2 + 5x + 3$$

Polynomial functions (in x) are constructed in this way. And *rational functions* are quotients of polynomial functions. Thus, if $f(x) = x^3 + x - 2$ and $g(x) = x - 3$, then

$$\left(\frac{f}{g}\right)(x) = \frac{x^3 + x - 2}{x - 3}, \qquad x \neq 3,$$

and $\frac{f}{g}$ is a rational function.

The sum or difference of odd functions is odd; the product or quotient of odd functions is even. The sum, difference, product, or quotient of even (nonzero) functions is even. The product or quotient of an odd and an even function is odd; the sum or difference of an odd and an even function (neither of which is the zero function) is neither odd nor even. A polynomial function in which *all* the exponents are odd (respectively, even) is odd (respectively, even). Here the constant term is regarded as involving the "0th power of x". (See Section 5.1 .)

EXAMPLE 3 For all x, let

$$f(x) = x^5, \qquad g(x) = 4x^3, \qquad F(x) = -2x^4, \qquad G(x) = 5x^2.$$

Then f and g are odd; F and G are even. Observe that

$$(f + g)(x) = x^5 + 4x^3,$$

so that

$$\begin{aligned}(f + g)(-x) &= (-x)^5 + 4(-x)^3 \\ &= -x^5 - 4x^3 \\ &= -(x^5 + 4x^3) \\ &= -[(f + g)(x)].\end{aligned}$$

This shows that $f + g$ is odd. Also,

$$(F + G)(x) = -2x^4 + 5x^2$$

so that

$$\begin{aligned}(F + G)(-x) &= -2(-x)^4 + 5(-x)^2 \\ &= -2x^4 + 5x^2 \\ &= (F + G)(x).\end{aligned}$$

This shows that $F + G$ is even. Furthermore, using the rules for combining odd and even functions,

$$(f - g)(x) = x^5 - 4x^3 \quad \text{(odd)}$$

$$(F - G)(x) = -2x^4 - 5x^2 \quad \text{(even)}$$

$$(f \cdot g)(x) = 4x^8 \quad \text{(even)} \qquad (F \cdot G)(x) = -10x^6 \quad \text{(even)}$$

$$\left(\frac{f}{g}\right)(x) = \frac{1}{4}x^2 \quad \text{(even)} \qquad \left(\frac{G}{F}\right)(x) = \frac{-5}{2x^2} \quad \text{(even)}$$

$$(f \cdot F)(x) = -2x^9 \quad \text{(odd)} \qquad \left(\frac{g}{G}\right)(x) = \frac{4}{5}x \quad \text{(odd)}$$

$$\left(\frac{f+g}{F+G}\right)(x) = \frac{x^5 + 4x^3}{-2x^4 + 5x^2} \left(\frac{\text{odd function}}{\text{even function}} = \text{odd function}\right)$$

$$(f + F)(x) = x^5 - 2x^4 \qquad \text{(neither odd nor even)}$$

□

composition of functions

The value v of a truckload of topsoil is a function of the weight w of the topsoil, which, in turn, is a function of the time t a laborer spent digging and shoveling it. This situation can be described by

$$v = f(w),$$
$$w = g(t).$$

The two functions can be *composed* as follows.

$$v = f(\underbrace{g(t)}_{w})$$

Here, *composition of functions* expresses the value of the topsoil *directly* in terms of the laborer's time.

In general, **composition of functions** is defined as follows. Let f and g be functions. For each x in the domain of g, for which $f(g(x))$ is defined, let

$$(f \circ g)(x) \qquad \text{stand for} \qquad f(g(x))$$

And for each x in the domain of f, for which $g(f(x))$ is defined, let

$$(g \circ f)(x) \qquad \text{stand for} \qquad g(f(x))$$

EXAMPLE 4 Let $f(x) = x^2$, for all x, $g(x) = x + 2$, for all x. Then for all x,

$$\begin{aligned}(f \circ g)(x) &= f(x + 2)\\ &= (x + 2)^2\\ &= x^2 + 4x + 4\end{aligned}$$

Thus

$$(f \circ g)(0) = 0^2 + 4 \cdot 0 + 4 = 4$$
$$(f \circ g)(1) = 1^2 + 4 \cdot 1 + 4 = 9$$
$$(f \circ g)(2) = 2^2 + 4 \cdot 2 + 4 = 16$$

And

$$(g \circ f)(x) = g(x^2) = x^2 + 2$$

In particular,

$$(g \circ f)(0) = 0^2 + 2 = 2$$
$$(g \circ f)(1) = 1^2 + 2 = 3$$
$$(g \circ f)(2) = 2^2 + 2 = 6$$

[See Figure 3.32.] □

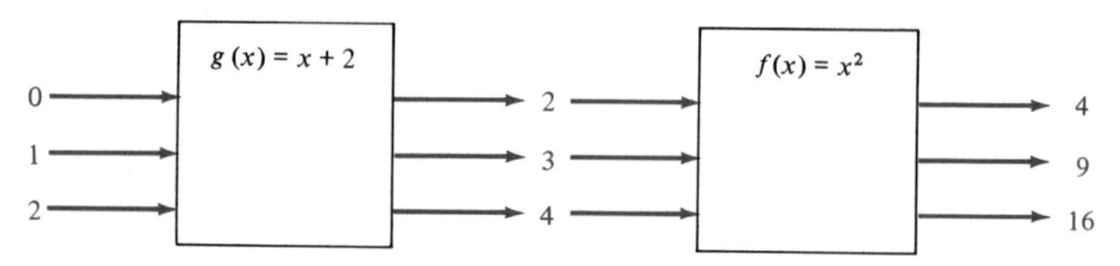

FIGURE 3.32(a) The $f \circ g$ machine

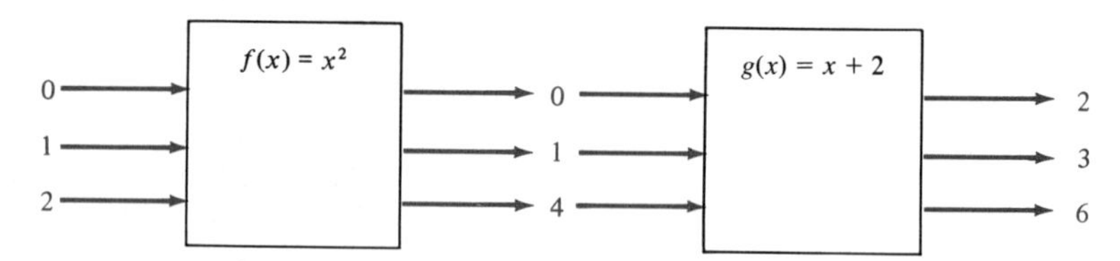

FIGURE 3.32(b) The $g \circ f$ machine

In the case of $g \circ f$, the function f is applied first; for $f \circ g$, the function g is applied first.

Observe that in general,

$$(f \circ g)(x) \neq g \cdot f(x)$$

Also,

$f \circ g$ denotes *composition* of functions,

whereas

$f \cdot g$ indicates *multiplication* of functions.

EXAMPLE 5 Let $f(x) = x + 1$, for all x, $g(x) = \frac{1}{x}$, for $x \neq 0$.

Then for x $\neq 0$,

$$(f \circ g)(x) = f\left(\frac{1}{x}\right) = \frac{1}{x} + 1 = \frac{1 + x}{x}$$

For instance,

$$(f \circ g)(1) = \frac{1+1}{1} = 2$$
$$(f \circ g)(2) = \frac{1+2}{2} = \frac{3}{2}$$
$$(f \circ g)(-1) = \frac{1+(-1)}{-1} = 0$$

And

$$(g \circ f)(x) = g(x+1) = \frac{1}{x+1}, \quad x \neq -1$$

Thus

$$(g \circ f)(1) = \frac{1}{2}, \qquad (g \circ f)(2) = \frac{1}{3},$$

$$(g \circ f)(-2) = \frac{1}{-2+1} = -1$$

□

EXERCISES

Assume the largest possible domain for each function.

In Exercises 1–8 find: **(a)** $(f + g)(x)$ **(b)** $(f + g)(2)$ **(c)** $(f - g)(x)$ **(d)** $(f - g)(0)$ **(e)** $(g - f)(x)$ **(f)** $(g - f)(-1)$

1. $f(x) = x + 2, \quad g(x) = x + 1$

2. $f(x) = x + 4; \quad g(x) = x - 3$

3. $f(x) = 2x, \quad g(x) = x + 3$

4. $f(x) = 10x, \quad g(x) \equiv 5$

5. $f(x) = 4x, \quad g(x) = \dfrac{1}{x^2 + 1}$

6. $f(x) = 3x + 5, \quad g(x) = 2x -$

7. $f(x) = x^2 + 2, \quad g(x) = x^2 + x - 1$

8. $f(x) = x^3 + 2x^2 - x + 1, \quad g(x) = x^3 - x + 4$

In Exercises 9–16 find: **(a)** $(2f)(x)$ **(b)** $(2f)(4)$ **(c)** $(-3g)(x)$ **(d)** $(-3g)(1)$ **(e)** $(f \cdot g)(x)$ **(f)** $(f \cdot g)(0)$

9. $f(x) = 3x, \quad g(x) = x + 2$

10. $f(x) = -2x, \quad g(x) = x - 1$

11. $f(x) \equiv 4, \quad g(x) = 4 - x$

12. $f(x) = x + 3, \quad g(x) \equiv 0$

13. $f(x) = 2x + 1, \quad g(x) = 3x - 1$

14. $f(x) = 5 - x, \quad g(x) = 5 + x$

15. $f(x) = x^2 + 2, \quad g(x) = 1 - x^2$

16. $f(x) = x^2 + x + 2, \quad g(x) = 2x^2 + 3x - 1$

In Exercises 17–20 find:

(a) $\left(\dfrac{f}{g}\right)(x)$ **(b)** $\left(\dfrac{f}{g}\right)(2)$ **(c)** $\left(\dfrac{g}{f}\right)(x)$ **(d)** $\left(\dfrac{g}{f}\right)(-2)$ **(e)** domain $\left(\dfrac{f}{g}\right)$ **(f)** domain $\left(\dfrac{g}{f}\right)$

17. $f(x) = x - 4, \quad g(x) = x + 2$

18. $f(x) = 2x + 6, \quad g(x) \equiv 2$

19. $f(x) = x^2 + 1, \quad g(x) = x^2 - 1$

20. $f(x) = x + 4, \quad g(x) = x^2 - 16$

In Exercises 21–32 determine whether each of the indicated functions is: (a) odd (b) even (c) neither of these

21. $f(x) = x^4 + x^2$

22. $g(x) = x^7 - x^3$

23. $h(x) = x^5 + 7x^3 - x$

24. $F(x) = x^8 - x^4 + 5x^2 - 2$

25. $G(x) = \dfrac{x^2 + 5}{x}$

26. $H(x) = \dfrac{x - x^3}{2x - x^5}$

27. $f(x) = x^4 + 2x^3$

28. $g(x) = \dfrac{x^{12} + x^6 - x^2}{3x^7 - 2x^5}$

29. $h(x) = 5x^{16} + 3x^{12} - x^{10} + x^4 - 1$

30. $F(x) = \dfrac{x^8 - x^6 - 3x^4 + 2x^2}{x^4 + 5x^2 - 3}$

31. $G(x) = \dfrac{1}{x^4 - x^3}$

32. $H(x) = x^4 + x^3 - x^2 + x - 7$

In Exercises 33–42 find: **(a)** $(f \circ g)(x)$ **(b)** $(f \circ g)(1)$ **(c)** $(g \circ f)(x)$ **(d)** $(g \circ f)(1)$ **(e)** domain $(f \circ g)$ **(f)** domain $(g \circ f)$

33. $f(x) = x + 4, \quad g(x) = x - 2$

34. $f(x) = 2x, \quad g(x) = x - 5$

35. $f(x) = -2x, \quad g(x) = 4x + 2$

36. $f(x) = 4 - 2x, \quad g(x) = 5x + 1$

37. $f(x) = |x|, \quad g(x) = x + 2$

38. $f(x) = x^2, \quad g(x) = x + 3$

39. $f(x) = x^2, \quad g(x) = \frac{1}{x}$, for $x \neq 0$

40. $f(x) = 2x + 5, \quad g(x) = x^2 + 1$

41. $f(x) \equiv 0, \quad g(x) = x^2 + 4$

42. $f(x) = \begin{cases} 2x + 1 & \text{for } x < 0 \\ x + 1 & \text{for } x \geq 0 \end{cases} \qquad g(x) = \begin{cases} -x & \text{for } x < 0 \\ x^2 + x & \text{for } x \geq 0 \end{cases}$

43. Suppose $f(x) = \frac{-1}{x}$, for $x \neq 0$, and $g(x) = x^2$. Find:

(a) $(f \circ g)(x)$ (b) $(f \circ g)(2)$ (c) $(f \circ g)(x + h)$ (d) Is $(f \circ g)(0)$ defined?

44. Let $f(x) = \frac{x^2}{x - 1}$, for $x \neq 1$, and let $g(x) = x^2$. Find:

(a) $(f \circ g)(2)$ (b) $(g \circ f)(2)$ (c) $(g \circ f)(x^2)$ (d) Is $(f \circ g)(1)$ defined?

45. If $g(x) = x + 1$ and $(g \circ f)(x) = 2x$, what is $f(x)$?

46. If $g(x) = \frac{x + 1}{x - 2}$ and $(g \circ f)(x) = 2x$, what is $f(x)$?

47. If $(f \circ g)(x) = (g \circ f)(x)$, $g(x) = x + 2$, and $f(0) = 10$, find: (a) $f(2)$ (b) $f(4)$ (c) $f(6)$ (d) $f(20)$ (e) What can you say about $f(-2)$? (f) What can you say about $f(1)$?

48. A circle is inscribed in a square. Express the area of the square as a function of C, the circumference of the circle.

49. A rectangular field is 25 feet wide. Express its area as a function of its perimeter.

50. Assume a train travels at the constant rate of 90 miles per hour. A ticket costs 10 cents per mile traveled. Express the cost of a ticket as a function of the number of minutes in transit.

51. Rain is pouring into a circular puddle. At the end of t seconds the radius is $\frac{t^2}{t + 1}$ centimeters. Express the area of the puddle as a function of time.

52. Helium is being pumped into a spherical balloon. At the end of t seconds the radius is $(t + 2)$ centimeters. Express the volume of the balloon as a function of time.

3.5 Inverse of a Function

When a function is represented by a set of ordered pairs, distinct ordered pairs have distinct *first* coordinates. Every *vertical* line in the plane intersects the graph of a function *at most once*.

one-to-one functions

Definition. A function is said to be **one-to-one** if its distinct ordered pairs have distinct *second* coordinates (as well as distinct first coordinates).

Thus a function defined by $y = f(x)$ is one-to-one if distinct y-values always correspond to distinct x-values.

Geometrically, every *horizontal* line in the plane (as well as every vertical line) intersects the graph of a one-to-one function at most once.

Figure 3.33(a) depicts a one-to-one function. Distinct points

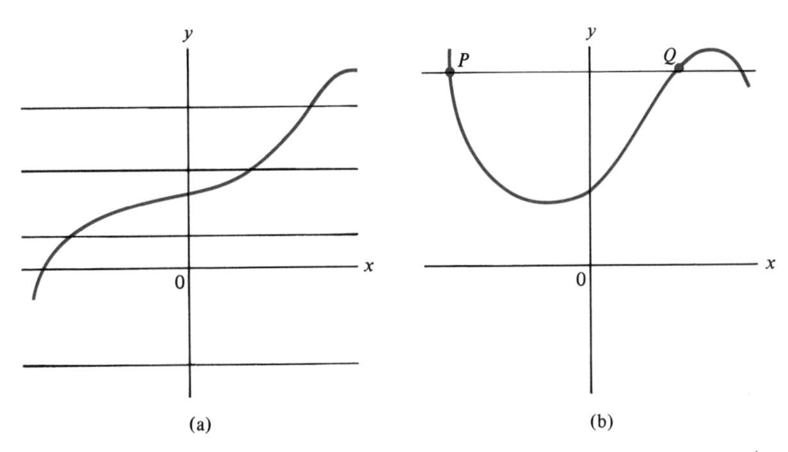

FIGURE 3.33(a) depicts a one-to-one function. Distinct points have distinct second coordinates. Every horizontal line cuts the graph at most once.
FIGURE 3.33(b) represents a function that is *not* one-to-one. There are 2 points P and Q with the same second coordinate. There is at least one horizontal line that cuts the graph more than once.

have distinct second coordinates. Every horizontal line cuts the graph at most once. Figure 3.33(b) represents a function that is *not* one-to-one. There are 2 points indicated, P and Q, with the same second coordinate. There is at least one horizontal line that cuts the graph more than once.

EXAMPLE 1 Let $f(x) = 2x$. Show that f is one-to-one.

Solution. Let x_1 and x_2 be *distinct* x-values. Can $(x_1, 2x_1)$ and $(x_2, 2x_2)$ have the same y-value, that is, can

$$2x_1 = 2x_2?$$

Divide both sides by 2.

But then

$$x_1 = x_2$$

However, x_1 and x_2 are *distinct*. $[x_1 \neq x_2]$ This is a contradition! Therefore

$$2x_1 \neq 2x_2,$$

and distinct ordered pairs,

$$(x_1, 2x_1) \qquad \text{and} \qquad (x_2, 2x_2),$$

have distinct second coordinates. This shows that the given function f is one-to-one. [See Figure 3.34 .] □

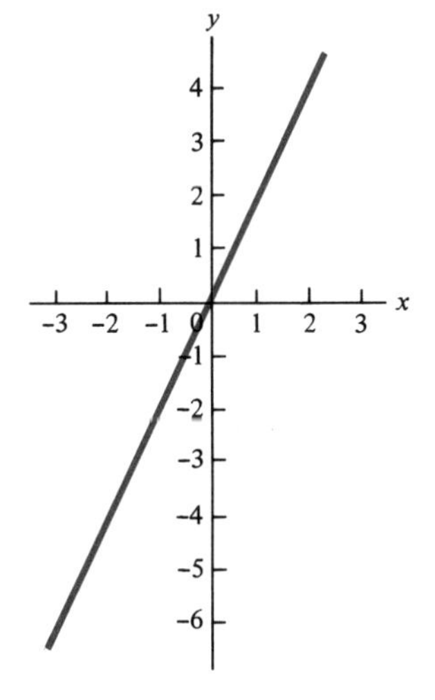

FIGURE 3.34 The graph of $f(x) = 2x$. Every horizontal line intersects this graph exactly once.

EXAMPLE 2 Let $g(x) = x^2$, for all x. Show that g is *not* one-to-one.

Solution. In order to show that g is *not* one-to-one, you must find two different x-values that have the same function-value.

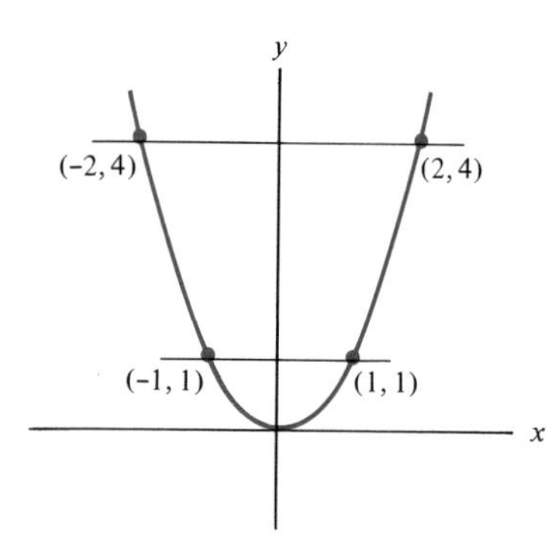

FIGURE 3.35 The graph of $y = x^2$. There are horizontal lines that intersect this graph more than once.

$$g(1) = 1^2 = 1 \qquad \text{and} \qquad g(-1) = (-1)^2 = 1$$

If g were one-to-one, $g(1)$ would have been different from $g(-1)$. Thus g is *not* one-to-one. [See Figure 3.35 .] □

increasing functions

The one-to-one functions with which you will be concerned in this course are either "increasing" or "decreasing" functions.

Definition. Let f be a function. Then f is said to be **increasing** if for *all* x_1 and x_2 in its domain,

$$\text{if} \quad x_1 < x_2, \qquad \text{then} \quad f(x_1) < f(x_2).$$

And f is said to be **decreasing** if for *all* x_1 and x_2 in its domain,

$$\text{if} \quad x_1 < x_2, \qquad \text{then} \quad f(x_1) > f(x_2).$$

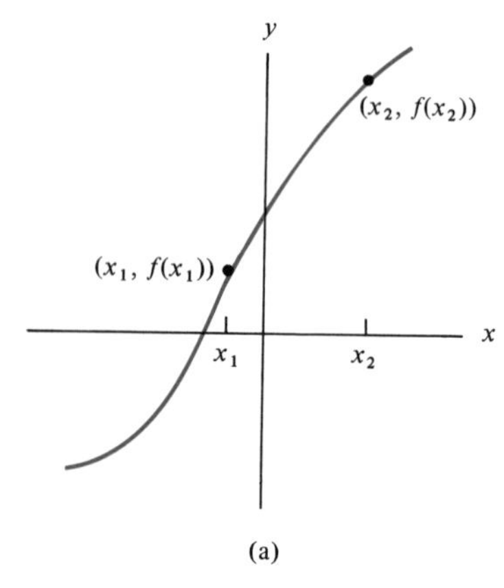

FIGURE 3.36(a) An increasing function. If $x_1 < x_2$, then $f(x_1) < f(x_2)$

A function is *increasing* if as you move to the *right on the x-axis,* the corresponding *y-values become greater.* [See Figure 3.36(a) .] A function is *decreasing* if as you move to the *right on the x-axis,* the corresponding *y-values become less.* [See Figure 3.36(b) .]

An increasing function f is one-to-one. In fact, suppose x_1 and x_2 are *distinct* x-values. One of these is less than the other. Arbitrarily assume that $x_1 < x_2$. Then because f is increasing,

$$f(x_1) < f(x_2)$$

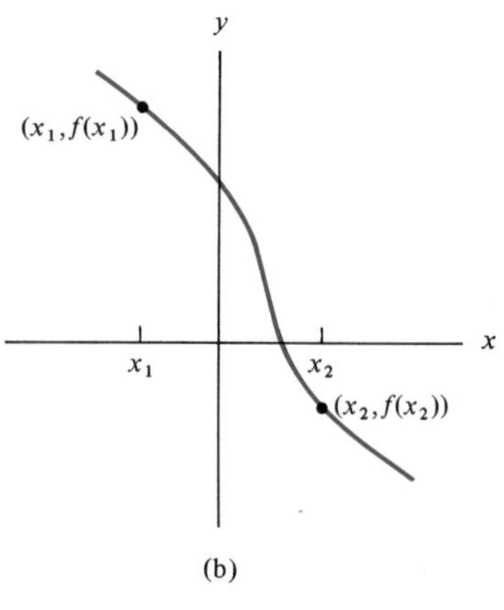

FIGURE 3.36(b) A decreasing function. If $x_1 < x_2$, then $f(x_1) > f(x_2)$

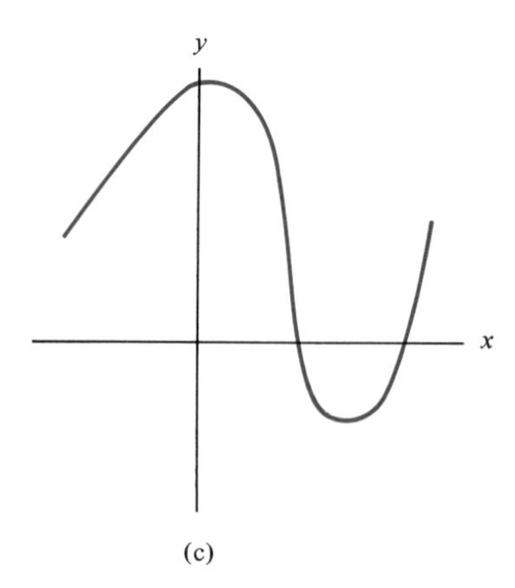

FIGURE 3.36(c) This function is neither an increasing nor a decreasing function.

Therefore

$$f(x_1) \neq f(x_2),$$

and f is one-to-one.

Similarly, it can be shown that a decreasing function is one-to-one.

The function $f(x) = 2x$ of Example 1 is an increasing function. [See Figure 3.34 on page 137.] The function $g(x) = -2x$ is a decreasing function. [See Figure 3.15 on page 112.]

inverse of a function

The notion of a one-to-one function is closely related to the concept of the "inverse of a function".

Suppose $f(x) = 2x$, as in Example 1. Then

$$f(1) = 2$$
$$f(2) = 4$$
$$f(3) = 6$$

and the points (1, 2), (2, 4), (3, 6) are on the graph of f. Interchange the coordinates.

Instead of:	Write:
(1, 2)	(2, 1)
(2, 4)	(4, 2)
(3, 6)	(6, 3)

A function f^{-1} will be defined that inverts the action of f. [See Figure 3.37 .]

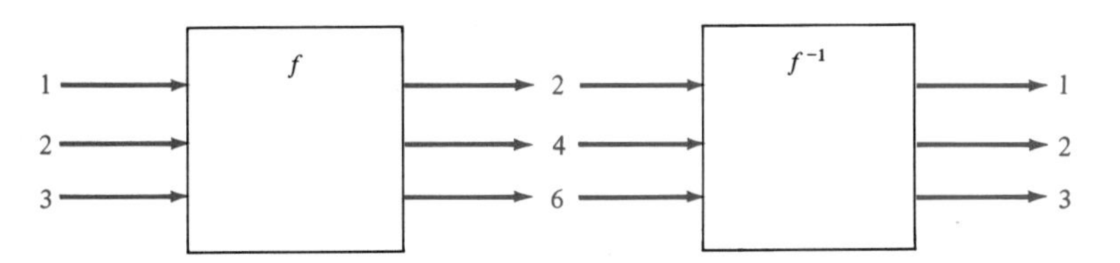

FIGURE 3.37

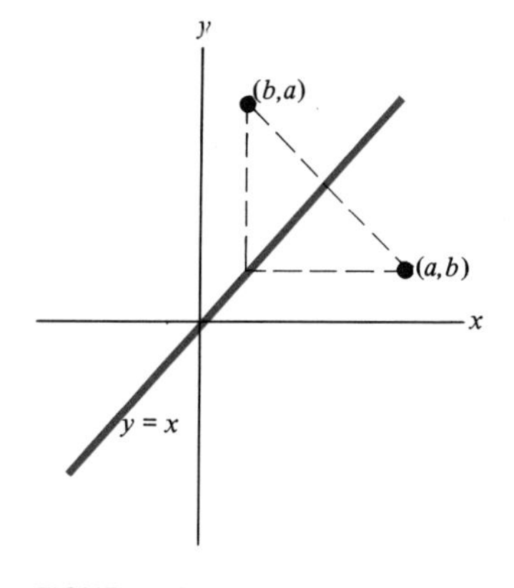

FIGURE 3.38

Definition. Let f be a one-to-one function. Interchange the coordinates of each ordered pair of f. The resulting correspondence is called the **inverse of** f.

Write f^{-1} for "the inverse of f".

If (a, b) is an ordered pair of f, then (b, a) is an ordered pair of f^{-1}. Thus if $f(a) = b$, then $f^{-1}(b) = a$.

Geometrically, *the point* (b, a) *is the reflection in the line* $y = x$ *of the point* (a, b) *on the graph of* f. [See Figure 3.38 .]

If a one-to-one function is given by an equation in x and y, to find its inverse:

1. Interchange the roles of x and y in the defining equation.
2. Solve for y in terms of x.

EXAMPLE 3 Let f be given by:

$$f(x) = 3x + 2$$

Determine f^{-1}.

Solution. (The verification that f is one-to-one is omitted.) Let $y = f(x)$. Then

$$y = 3x + 2$$

First interchange x and y in the defining equation:

$$x = 3y + 2$$

Now solve for y in terms of x:

$$x - 2 = 3y$$

$$\frac{x - 2}{3} = y$$

Thus f^{-1} is given by

$$f^{-1}(x) = \frac{x - 2}{3}.$$

□

For every one-to-one function f,

Domain f = range f^{-1}; range f = domain f^{-1}

It can be shown that *the inverse of a one-to-one function is itself a one-to-one function.* Moreover, if you interchange the ordered pairs of the function twice, you get back the original ordered pairs:

$$(a, b) \rightarrow (b, a) \rightarrow (a, b)$$

Thus, *the inverse of the inverse of a function is the function itself:*

$$(f^{-1})^{-1} = f$$

inverse and composition

What happens when you compose a one-to-one function with its inverse? Let x be in the domain of f.

If $y = f(x)$, then $f^{-1}(y) = x$

Thus for x in the domain of f,

$$(f^{-1} \circ f)(x) = f^{-1}(\underbrace{f(x)}_{y}) = f^{-1}(y) = x$$

Similarly, for y in the domain of f^{-1},

$$(f \circ f^{-1})(y) = f(\underbrace{f^{-1}(y)}_{x}) = f(x) = y$$

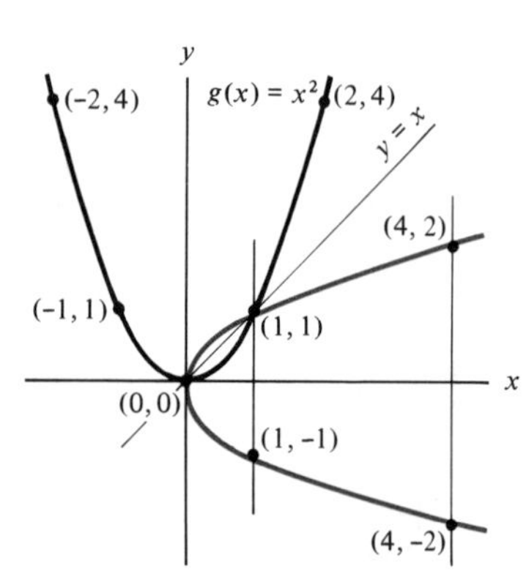

FIGURE 3.39(a) The inverted graph (in color) does *not* represent a function. There are vertical lines that intersect the graph twice.

To summarize:

For each x in the domain of a one-to-one function f,

$$(f^{-1} \circ f)(x) = x$$

And for each y in the domain of f^{-1},

$$(f \circ f^{-1})(y) = y$$

restricting the domain If a function given by $y = g(x)$ is not one-to-one, when you interchange its x- and y-values, you do *not* obtain a function.

EXAMPLE 4 Let $g(x) = x^2$.

As was shown in Example 2, g is *not* one-to-one. Interchange the coordinates of the ordered pairs of g.

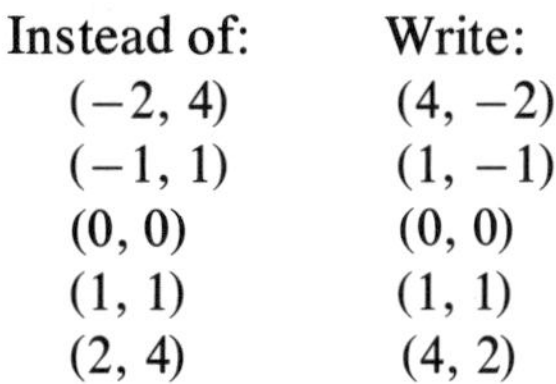

Instead of:	Write:
$(-2, 4)$	$(4, -2)$
$(-1, 1)$	$(1, -1)$
$(0, 0)$	$(0, 0)$
$(1, 1)$	$(1, 1)$
$(2, 4)$	$(4, 2)$

In the second column, both $(1, -1)$ and $(1, 1)$ have the same first coordinate, as do $(4, -2)$ and $(4, 2)$. The graph of the inverted correspondence is given in Figure 3.39(a). Observe that this does *not* represent a function because there are vertical lines that intersect the graph twice. However, if you *restrict the domain of g to nonnegative values, you obtain a function.* Define

$$G(x) = x^2 \quad \text{for} \quad x \geq 0.$$

This *restricted* function, G, is increasing [Figure 3.39(b)]—hence one-to-one. It follows that G has an inverse, G^{-1} [Figure 3.39(c)]. G^{-1} is known as the **(principal) square root function,** and will be considered further in Chapter 5. □

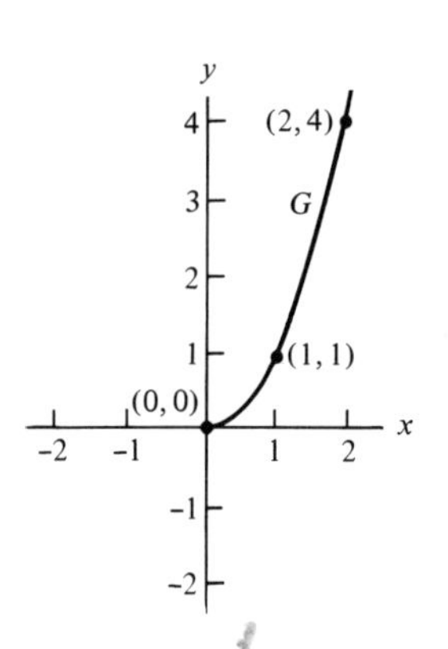

FIGURE 3.39(b) $G(x) = x^2$, for $x \geq 0$

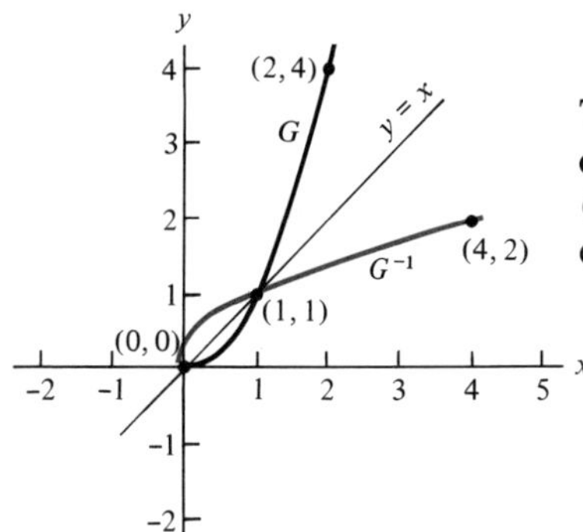

FIGURE 3.39(c)

EXERCISES

In Exercises 1–8 (Figure 3.40) which diagrams represent one-to-one functions?

1.

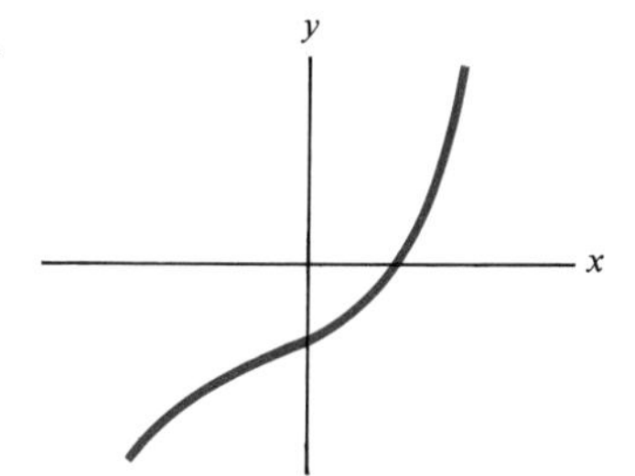

2.

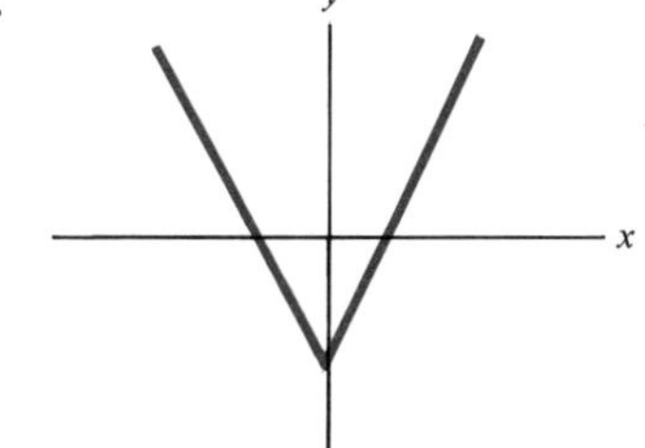

3.

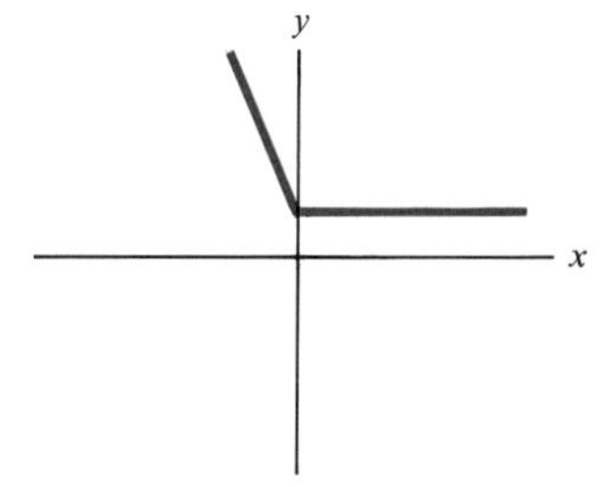

FIGURE 3.40

4.

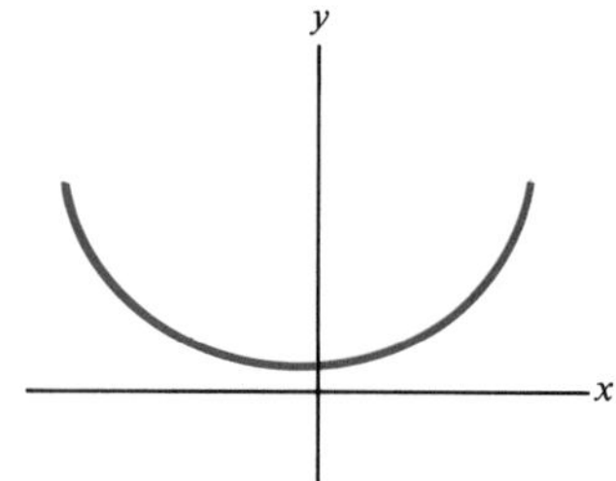

5.

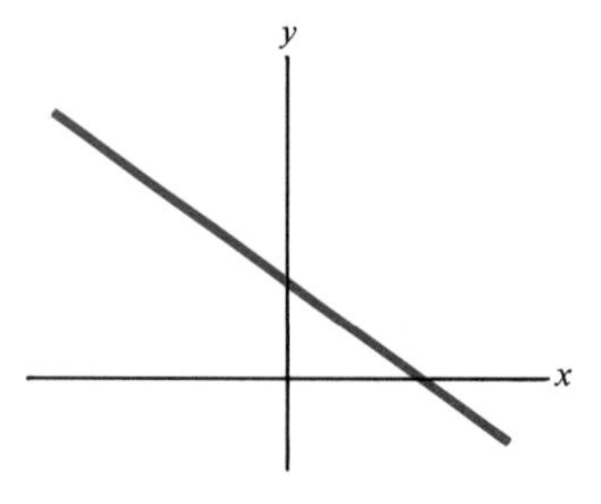

6.

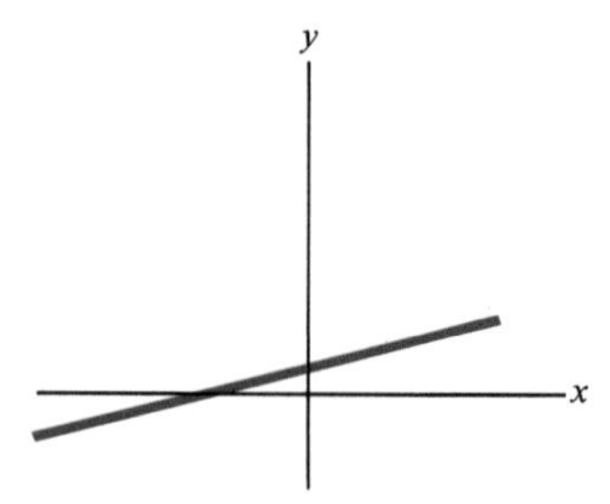

7.

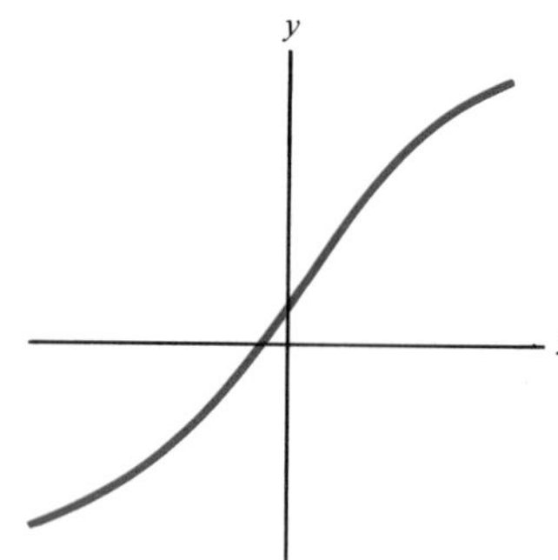

8.

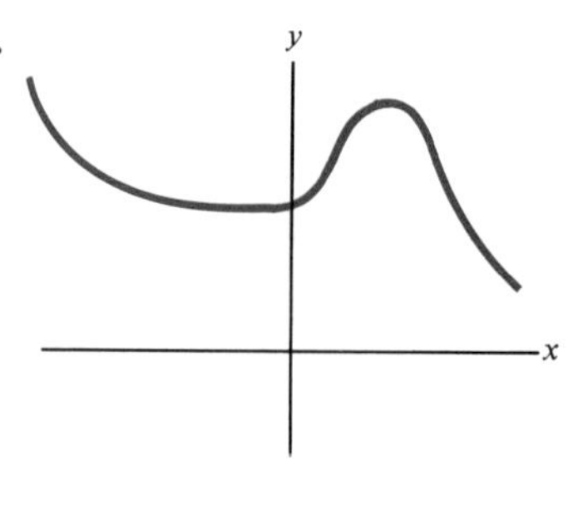

FIGURE 3.40 (*Continued from page 141*)

9. Suppose f is one-to-one, $f(1) = 2$, $f(2) = 3$, $f(3) = 6$, and $f(4) = 10$. Find: **(a)** $f^{-1}(2)$ **(b)** $f^{-1}(3)$ **(c)** $f^{-1}(6)$ **(d)** $f^{-1}(10)$

10. Suppose g is one-to-one, $g(0) = 1$, and $g(1) = 4$. For each of the following points: (0, 1), (1, 0), (1, 1), (1, 4), (4, 1), (4, 5), (5, 4), indicate that **(a)** it is on the graph of g^{-1}; **(b)** it is *not* on the graph of g^{-1}; or **(c)** there is insufficient information to determine **(a)** or **(b)**.

In Exercises 11–17 show that the function defined by each equation is *not* one-to-one. The domain of each function is the set of all real numbers. [*Hint:* For each function, find two different x-values that have the same function-value.]

11. $f(x) \equiv 4$ **12.** $f(x) = x^2 - 1$ **13.** $f(x) = -2x^2$ **14.** $f(x) = 3x^2 + 1$

15. $f(x) = \begin{cases} -x, & \text{if } x < 0 \\ x, & \text{if } x \geq 0 \end{cases}$ **16.** $f(x) = \begin{cases} x + 1, & \text{if } x < 0 \\ x^2, & \text{if } x \geq 0 \end{cases}$ **17.** $f(x) = \begin{cases} -1, & \text{if } x < 0 \\ 1, & \text{if } x \geq 0 \end{cases}$

18. In Figure 3.41 match the graphs of functions that are inverses of one another.

In Exercises 19–30 determine the inverse of the indicated one-to-one function.

19. $y = 2x$ **20.** $y = -3x$ **21.** $y = x + 8$ **22.** $y = x - 1$ **23.** $y = 9 - x$

24. $y = \frac{1}{2}x + \frac{1}{4}$ **25.** $y = 3x + 7$ **26.** $y = -2x + 5$ **27.** $y = 7 - 2x$ **28.** $y = \frac{1}{3}(x - 3)$

29. $y = \frac{1}{4}x + \frac{1}{2}$ **30.** $y = \frac{2x - 3}{4}$

In Exercises 31–34 determine: **(a)** f^{-1} **(b)** domain f^{-1} **(c)** range f^{-1}

31. $f(x) = x + 2,\quad 0 \leq x \leq 1$ **32.** $f(x) = 2x,\quad 1 \leq x \leq 4$

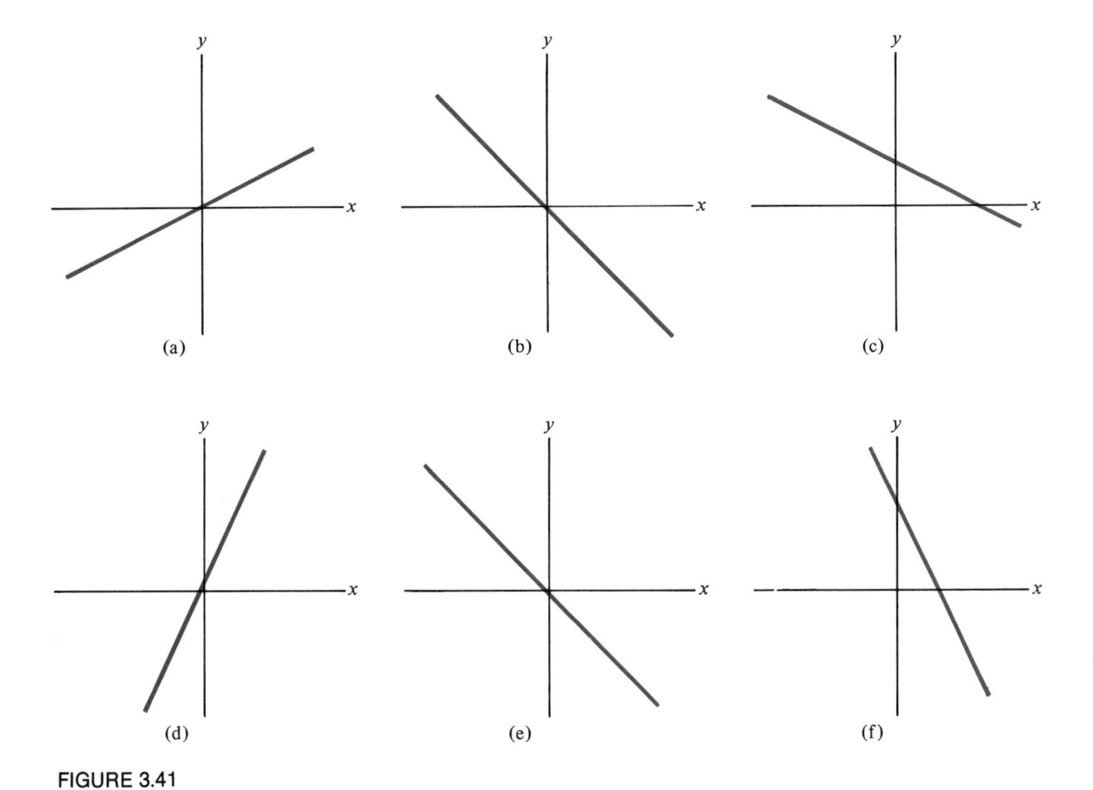

FIGURE 3.41

33. $f(x) = 3x + 1, \quad -1 \leqslant x \leqslant 2$ **34.** $f(x) = \dfrac{x+4}{2}, \quad 6 \leqslant x \leqslant 10$

Review Exercises for Chapter 3

1. Suppose $f(x) = 2x + 5$. Determine the following function-values.

(a) $f(2)$ **(b)** $f(10)$ **(c)** $f\left(\frac{1}{2}\right)$ **(d)** $f\left(\frac{-1}{2}\right)$

2. Suppose $g(x) = \dfrac{x-1}{x+1}$. Determine the following function-values.

(a) $g(1)$ **(b)** $g(-2)$ **(c)** $g\left(\frac{1}{4}\right)$ **(d)** $g(.1)$

3. Determine the largest possible domain of the function defined by $f(x) = \dfrac{x+5}{x-4}$.

4. **(a)** Determine the volume of a cube as a function of the length of a side, s.

(b) Determine the volume if the side is of length 5 inches.

(c) Determine the volume if the side is of length .2 inch.

5. The rental charge of a computer is a flat fee of \$300 plus \$200 per hour. (a) Find the rental charge as a function of the number of hours of computer time. (b) How much does it cost to rent the computer for 5 hours? (c) If an agency can spend \$2500, how much computer time can it purchase?

6. In Figure 3.42 determine the coordinates of each point.

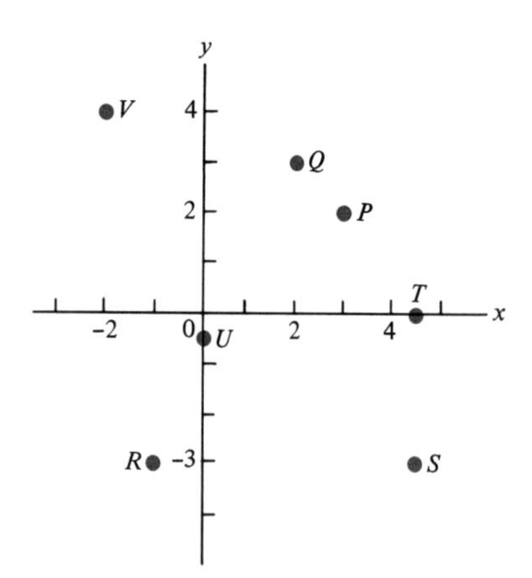

FIGURE 3.42

7. On a coordinate system plot the following ordered pairs:

(a) $(1, 5)$ (b) $(3, -1)$ (c) $(0, 4)$ (d) $\left(\frac{1}{2}, \frac{-1}{2}\right)$ (e) $(-3, -4)$ (f) $(-2, -2.5)$

8. Determine the quandrant of each of the following points, or indicate that the point lies on a coordinate axis.

(a) $(4, 3)$ (b) $(-4, 3)$ (c) $(-2, -5)$ (d) $(0, 6)$ (e) $(1, -4)$ (f) $\left(\frac{1}{2}, \frac{3}{2}\right)$

9. Which of the graphs in Figure 3.43 represents a function?

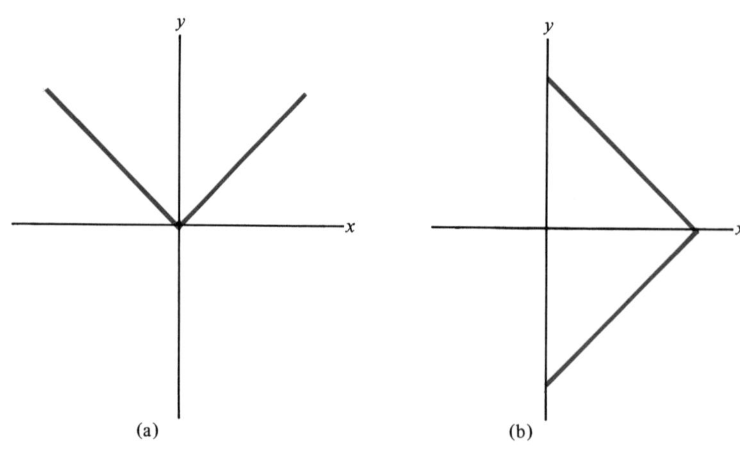

FIGURE 3.43

10. Graph the indicated function.

(a) $y = 3 - 2x$ (b) $y = x^2 - 2$ (c) $y = |x - 3|$ (d) $y = \begin{cases} 2, & \text{if } x \leqslant 0 \\ x + 2, & \text{if } x > 0 \end{cases}$

11. Let $f(x) = 3x + 1$. Determine:

(a) $f(-1)$ **(b)** $f(x + 1)$ **(c)** $f(2x)$ **(d)** $f(x + h)$ **(e)** $\dfrac{f(x + h) - f(x)}{h}$, $h \neq 0$

12. Let $f(x) = x - 4$ for all x and let

$$g(x) = \begin{cases} \dfrac{x^2 - 16}{x + 4}, & \text{if } x \neq -4 \\ -8\ , & \text{if } x = -4 \end{cases}.$$ Does $f = g$?

13. Let $f(x) = \dfrac{1}{x^2}$, for $x \neq 0$. Which of the following are true?

(a) f is an odd function. **(b)** f is an even function.

(c) The graph of f is symmetric with respect to the y-axis.

(d) The graph of f is symmetric with respect to the x-axis.

(e) The graph of f has neither an x-intercept nor a y-intercept.

14. Assume f and g are each defined for all real numbers. Find: **(a)** $(f + g)(4)$ **(b)** $(f - g)(0)$ **(c)** $(-2f)(-2)$ **(d)** $\left(\dfrac{f}{g}\right)(2)$

i. $f(x) = x + 2$, $g(x) = x - 4$ ii. $f(x) = x^2$, $g(x) = x - 1$
iii. $f(x) \equiv 0$, $g(x) = 3 - x$

15. Let $f(x) = x + 5$; $g(x) = \dfrac{1}{x}$, $x \neq 0$. Find: **(a)** $(f \circ g)(x)$ **(b)** $(f \circ g)(-1)$ **(c)** $(g \circ f)(x)$

(d) $(g \circ f)(-1)$ **(e)** domain $(f \circ g)$ **(f)** domain $(g \circ f)$

16. Which of the graphs in Figure 3.44 represents a one-to-one function?

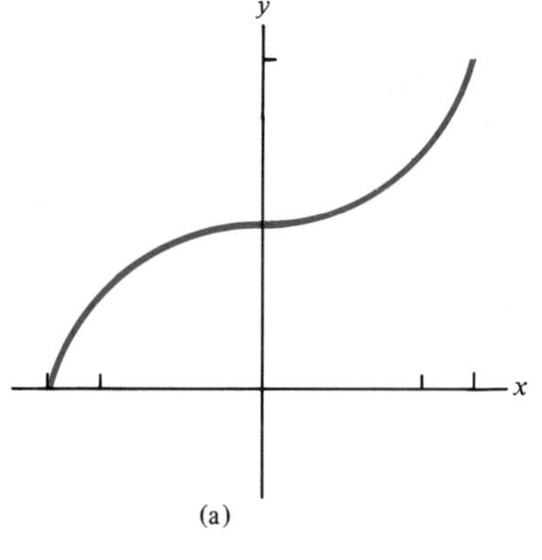

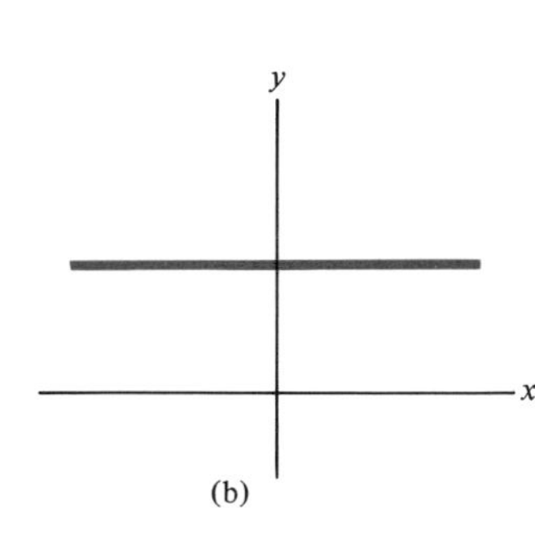

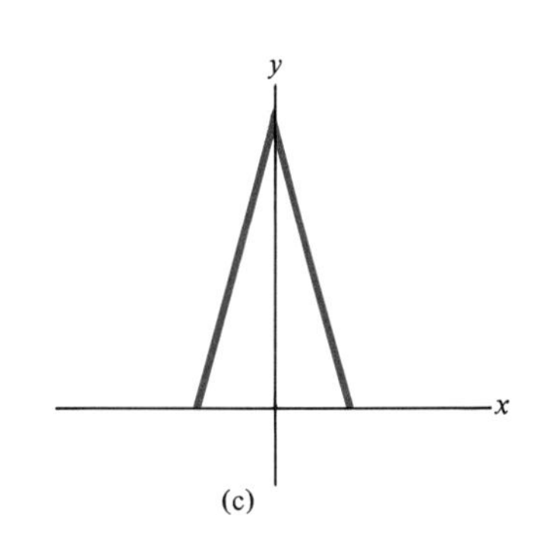

FIGURE 3.44

17. Determine the inverse of the function defined by the equation $y = 5x - 4$.

18. Show that the function defined by $f(x) = x^4 - 1$, for all x, is *not* one-to-one.

4 Lines

4.1 Proportions and Similar Triangles

proportions

Definition. A **proportion** is a statement that two or more fractions are equivalent.

$$\frac{3}{5} = \frac{6}{10}, \qquad \frac{-1}{-2} = \frac{4}{8}, \qquad \text{and} \qquad \frac{-5}{2} = \frac{10}{-4} = \frac{-20}{8}$$

are each proportions. The numerators and denominators of the equivalent fractions are said to be **proportional.**

EXAMPLE 1

Find the value of d in the proportion

$$\frac{-3}{7} = \frac{12}{d}.$$

Solution. Multiply both sides of the equation by $7d$ (or cross-multiply):

$$-3d = 12 \cdot 7$$

Thus

$$d = -4 \cdot 7 = -28$$

□

Many problems are solved by setting up proportions.

EXAMPLE 2

A psychologist needs 9 pounds of grain to feed 100 rats. After some time the rat population increases to 150. How much grain is needed?

Solution. Let x be the number of pounds of grain needed to feed 150 rats. Set up the proportion:

$$\frac{9}{100} = \frac{x}{150}$$

Multiply both sides by 300, the LCD.

$$\frac{9}{\cancel{100}_1} \cdot \cancel{300}^3 = \frac{x}{\cancel{150}_1} \cdot \cancel{300}^2$$

$$\frac{27}{2} = x$$

Thus $13\frac{1}{2}$ pounds are needed. □

similar triangles

Proportions are used to characterize *similar triangles,* which in turn, will be used to obtain an important property of lines. Intuitively, two triangles are similar if they have the same *shape,* though not necessarily the same *size.* [See Figure 4.1 .] Here is a more precise definition.

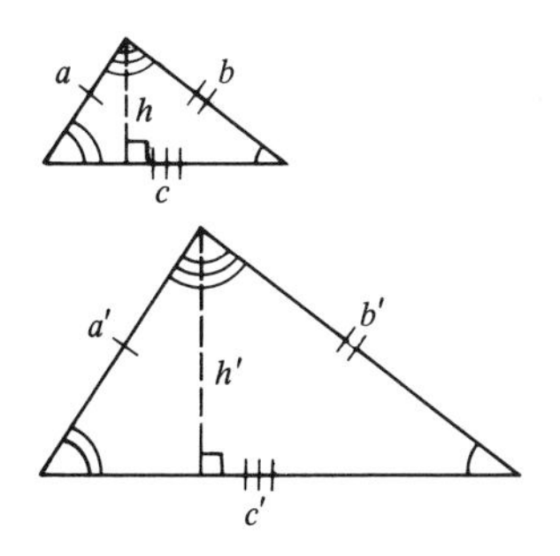

FIGURE 4.1

(a) Similar triangles: $\frac{a}{a'} = \frac{b}{b'} = \frac{c}{c'} = \frac{h}{h'}$

Definition. **Similar triangles** are triangles in which *the lengths of corresponding sides (and of corresponding altitudes) are proportional.* Alternatively, similar triangles are those in which *corresponding angles are equal (in measure).*

If two triangles satisfy either condition, they also satisfy the other. Also, if the lengths of corresponding sides are *equal,* the triangles are said to be **congruent.**

In Figure 4.1(a), the triangles are similar. Corresponding sides are marked with the same number of bars. The sides of lengths a, b, c, and an altitude of length h of one triangle correspond, respectively, to the sides of lengths a', b', c', and an altitude of length h' of the other. Thus any two of the following fractions are equivalent.

$$\frac{a}{a'} = \frac{b}{b'} = \frac{c}{c'} = \frac{h}{h'}$$

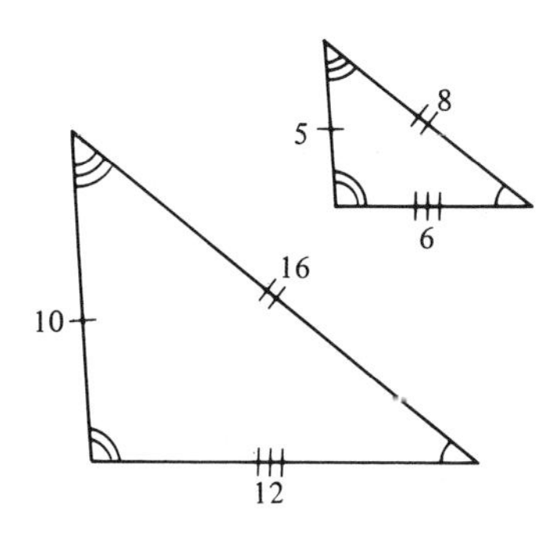

(b) Similar triangles: $\frac{5}{10} = \frac{8}{16} = \frac{6}{12}$

Corresponding angles are equal. The triangles of Figure 4.1(b) are also similar.

Note that the equation $\frac{a}{a'} = \frac{b}{b'}$ is equivalent to the equation $\frac{a}{b} = \frac{a'}{b'}$. In fact, multiply both sides of the first equation by $\frac{a'}{b}$ to obtain the second equation. When the lengths of corresponding sides of a triangle are proportional, both equations hold.

EXAMPLE 3 Assume the triangles in Figure 4.2 (page 148) are similar. Determine b and c.

Solution.

$$\frac{4}{16} = \frac{10}{b} = \frac{12}{c}$$

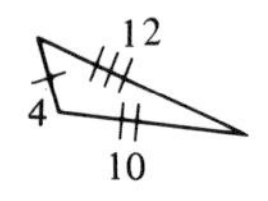

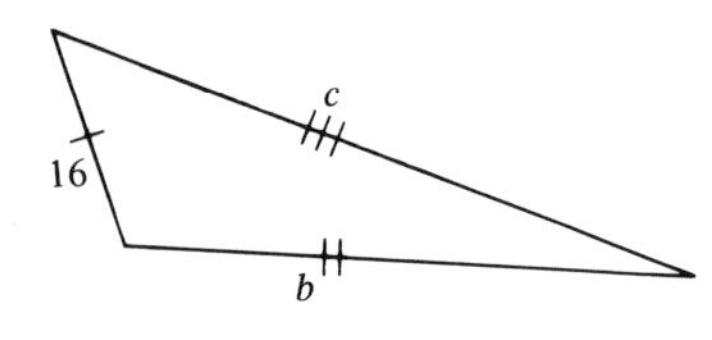

FIGURE 4.2

First solve for b:

$$\frac{4}{16} = \frac{10}{b}$$

$$4b = 160$$

$$b = 40$$

Now solve for c:

$$\frac{4}{16} = \frac{12}{c}$$

$$c = 48$$

□

Similarity of triangles is used in the solution of many problems that arise in calculus. Here is an example that illustrates this essential usage.

EXAMPLE 4 A water trough has a vertical cross section in the form of a triangle with base length 2 meters and altitude length 4 meters. If the trough is 8 meters long, express the volume of water it contains as a function of h, the water level (in meters).

Solution. In Figure 4.3, $\triangle ABC$ is similar to $\triangle ADE$. Thus

$$\frac{\overline{BC}}{2} = \frac{h}{4}$$

$$\overline{BC} = \frac{h}{2}$$

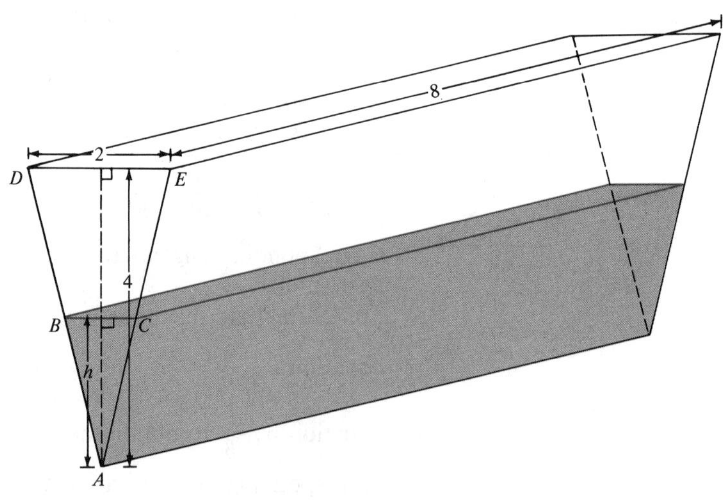

FIGURE 4.3

The volume V of water is given by

$$V = \text{length} \cdot \text{area } \triangle ABC.$$

Because the length of the trough is 8 meters,

$$V = 8 \cdot \left(\frac{1}{2} \cdot \frac{h}{2} \cdot h\right) = 2h^2$$

□

EXERCISES

In Exercises 1–10 find all possible values of a, b, c, or d in each proportion.

1. $\frac{a}{5}=\frac{4}{2}$ **2.** $\frac{-1}{-2}=\frac{c}{12}$ **3.** $\frac{9}{6}=\frac{3}{d}$ **4.** $\frac{-7}{2}=\frac{c}{-4}$ **5.** $\frac{\frac{1}{2}}{2}=\frac{c}{4}$ **6.** $\frac{\frac{1}{3}}{\frac{2}{3}}=\frac{\frac{3}{7}}{d}$

7. $\frac{5}{6}=\frac{7}{d}$ **8.** $\frac{-7}{3}=\frac{1}{d}$ **9.** $\frac{a}{8}=\frac{2}{a}$ **10.** $\frac{.01}{.1}=\frac{c}{2}$

11. A dozen eggs cost 90 cents. How much do 14 eggs cost?

12. Bel Paese sells for \$2.40 per pound. How much does a 6-ounce slice cost?

13. If 20 pounds of fertilizer costs \$4, how much do 90 pounds cost?

14. If 4 pounds of potatoes cost 54 cents, how much do 6 pounds cost?

15. Oranges sell for a dollar a dozen. How much do 15 oranges cost?

16. If 24 hungry freshmen can eat 96 hamburgers, how many hamburgers should be prepared for 43 hungry freshmen?

17. Tom earns \$12.56 for 4 hours of work. At this rate how much will he earn for 10 hours of work?

18. A pint of cider costs 45¢. At this rate how much would you pay for 2 quarts of cider?

19. On a highway map, $\frac{1}{3}$ of an inch represents 10 miles. How many miles are represented by an inch and a half?

20. Bob owned a $\frac{3}{4}$ interest in some property. When it was sold, Bob received \$27 000 for his share. For how much was the property sold?

21. A 27-meter rope is cut into 2 pieces that are in the ratio of $\frac{7}{2}$. How long is each piece?

22. A batter has 2 hits for every 7 times at bat. If he keeps up this rate, how many hits will he have after 350 at-bats?

23. Suppose Vida Blue wins 3 games for every game he loses. How many games does he win in a season in which he loses 8 games?

In each of Exercises 24–28 [Figures 4.4–4.8] the triangles are similar. Corresponding sides are marked with the same number of bars. Determine the lengths of the indicated sides.

24.

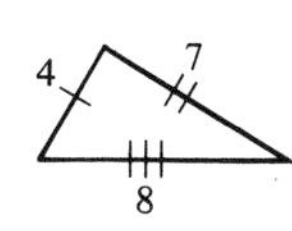

8 b c

FIGURE 4.4

25.

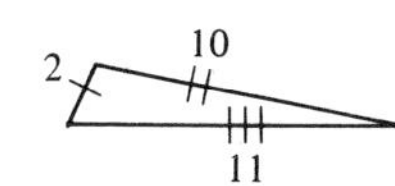

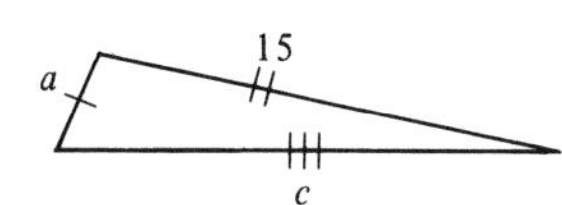

FIGURE 4.5

26.

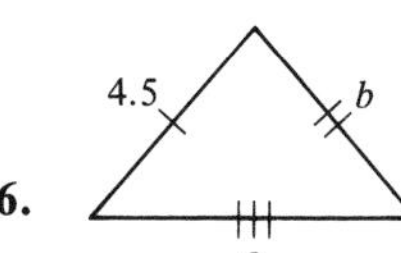

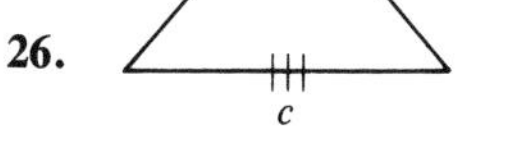

6 6 8

FIGURE 4.6

27.

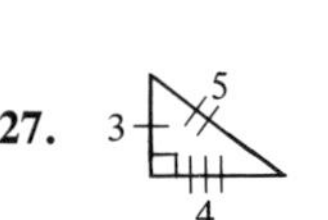

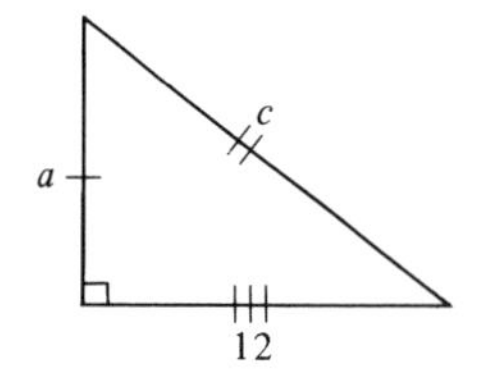

FIGURE 4.7

28.

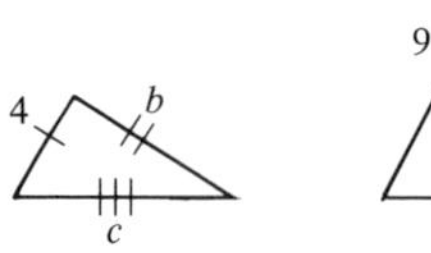

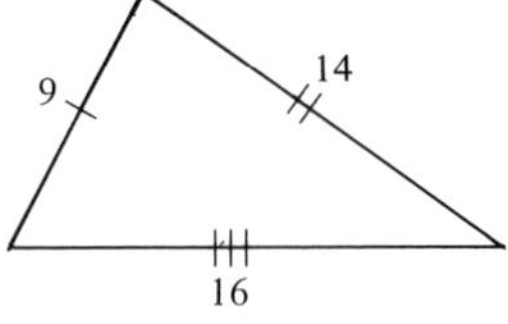

FIGURE 4.8

FIGURE 4.9

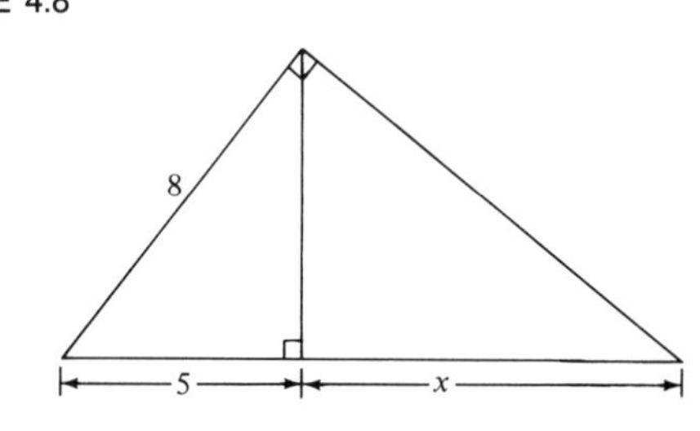

FIGURE 4.10

29. In Figure 4.9, find x.

30. In Figure 4.10, find x.

31. The width and length of a 5×7 (rectangular) photograph are enlarged proportionately. Express the area A of the enlargement as a function of its length l.

32. A 5-foot woman casts a 4-foot shadow. Express the length l of her son's shadow as a function of h, the son's height.

33. A man 6 feet tall walks at the rate of 5 feet per second directly away from a street lamp 15 feet high. Express the length l of his shadow as a function of t, the number of seconds he has walked.

34. A coffee-maker in the shape of an inverted (right-circular) cone has an altitude of length 16 centimeters and a base radius of length 18 centimeters. [See Figure 4.11 .] Express the volume V of coffee as a function of h, its level (in centimeters).

35. Express the volume V of the right-circular cylinder in Figure 4.12 as a function of r, the length of its base radius.

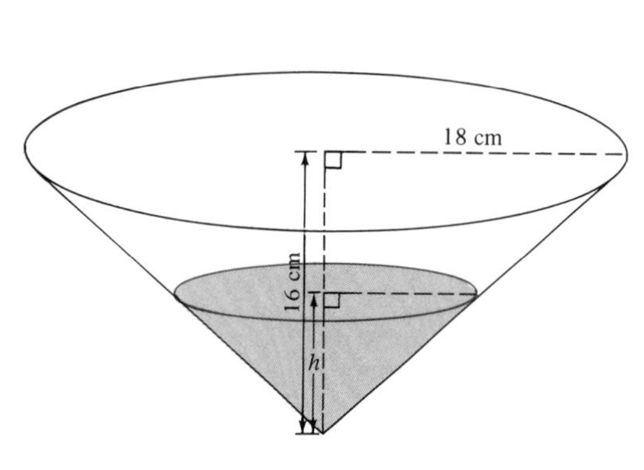

FIGURE 4.11

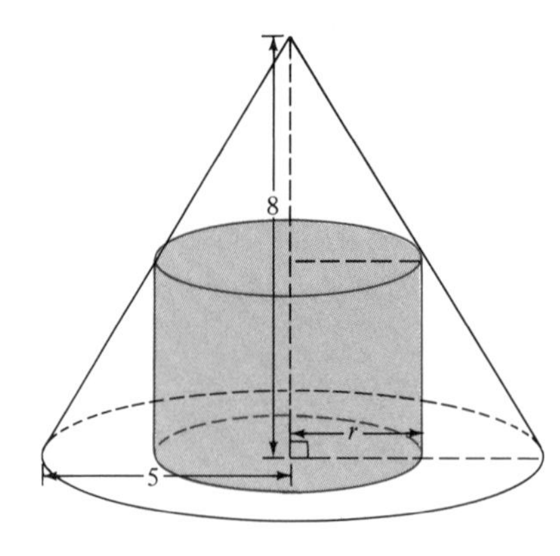

FIGURE 4.12

4.2 Point-Slope Equation of a Line

Consider a linear equation in two variables, such as

$$5x - y = 9 - x.$$

An ordered pair (x_1, y_1) **satisfies** such an equation, or is a **solution** of the equation, if a *true* statement results when x_1 replaces x and y_1 replaces y in the equation. Thus (2, 3) satisfies the preceding equation because

$$5 \cdot 2 - 3 = 9 - 2 \qquad \text{or} \qquad 7 = 7.$$

As you have seen in Chapter 3, a line can be described by a certain type of equation in the two variables x and y. The ordered pairs (x_1, y_1) corresponding to points on the line (and only these ordered pairs) satisfy the equation of the line. When you are given such an equation, you can graph the line by plotting two points and drawing the line through them.

It is one thing to draw a line; it is another matter to determine its equation. If you have sufficient information about a line, you can determine its equation.

slope

The "slope" of a line expresses its slant or inclination. How fast do the y-values change along the line as the x-values change? Do the y-values increase or decrease as the x-values increase?

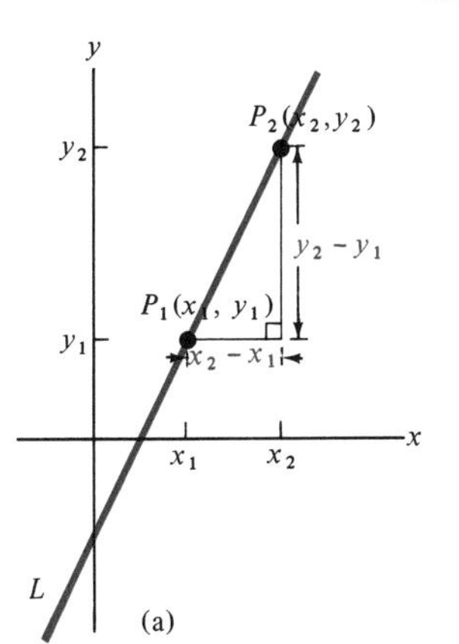

FIGURE 4.13
(a) The rise, $y_2 - y_1$, and run, $x_2 - x_1$, are both positive. L slopes upward to the right.

Let P_1 and P_2 be two different points on a line L. Suppose $P_1 = (x_1, y_1)$ and $P_2 = (x_2, y_2)$. The **rise** or the **change in y** (along L from P_1 to P_2) is defined to be

$$y_2 - y_1.$$

The **run** or the **change in x** (along L from P_1 to P_2) is defined to be

$$x_2 - x_1.$$

In Figure 4.13(a), the line slants *upward to the right*. The rise, $y_2 - y_1$, and the run, $x_2 - x_1$, are *both positive*. In Figure 4.13(b), the line slants *downward to the right*. The rise, $y_2 - y_1$, is *negative* although the run, $x_2 - x_1$, is still *positive*.

In the following, assume that the lines L discussed are not vertical. Then distinct points on each line L will have distinct x-values. Thus, if (x_1, y_1) and (x_2, y_2) are on L, the run, $x_2 - x_1$, is nonzero. Division by $x_2 - x_1$ is therefore defined.

Definition. Let L be a nonvertical line and let P_1 and P_2 be *any* two distinct points on L. Suppose $P_1 = (x_1, y_1)$ and $P_2 = (x_2, y_2)$. The slope of L is defined to be

$$\frac{y_2 - y_1}{x_2 - x_1}.$$

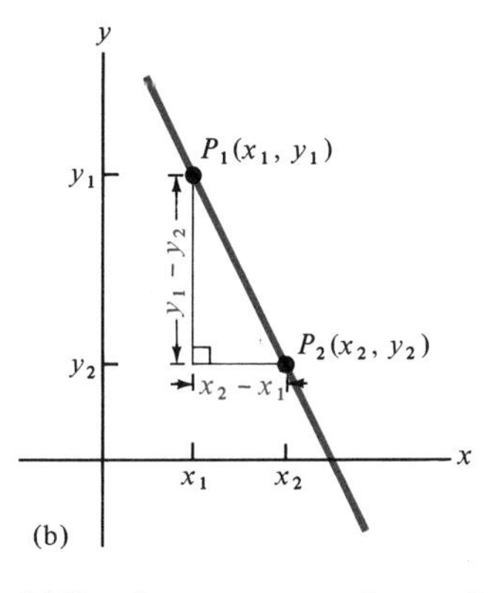

(b) The rise, $y_2 - y_1$ or $-(y_1 - y_2)$, is negative. The run, $x_2 - x_1$, is positive. The line slopes downward to the right.

It is important to recognize that *no matter which two distinct points are chosen on L, this quotient is always the same*. In Figure 4.14 consider the rise and run from P_1 to P_2, as well as the rise and run from P_3 to P_4. Note that right triangles P_1P_2Q and P_3P_4R are similar because corresponding angles are equal. The lengths of corresponding sides of similar triangles are proportional. Thus

$$\frac{y_2 - y_1}{x_2 - x_1} = \frac{y_4 - y_3}{x_4 - x_3}$$

Either of these quotients defines the slope of the line L.

In determining the slope of L, it does not matter whether you go from P_1 to P_2 or from P_2 to P_1 because

$$\frac{y_2 - y_1}{x_2 - x_1} = \frac{(-1)(y_1 - y_2)}{(-1)(x_1 - x_2)} = \frac{y_1 - y_2}{x_1 - x_2}.$$

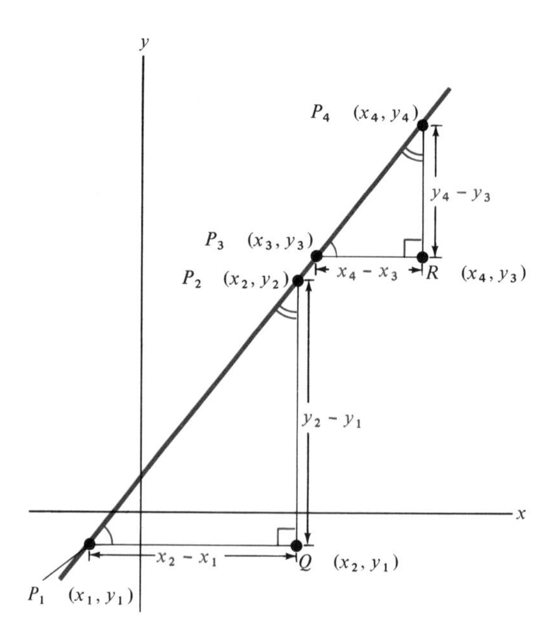

FIGURE 4.14 Triangles P_1P_2Q and P_3P_4R are similar.

$$\frac{y_2 - y_1}{x_2 - x_1} = \frac{y_4 - y_3}{x_4 - x_3}$$

because the lengths of corresponding sides of similar triangles are proportional.

EXAMPLE 1 Determine the slope of the line through the points (1, 2) and (2, 5).

Solution. Let $(x_1, y_1) = (1, 2)$ and $(x_2, y_2) = (2, 5)$.

$$\text{Slope} = \frac{\text{rise}}{\text{run}} = \frac{y_2 - y_1}{x_2 - x_1} = \frac{5 - 2}{2 - 1} = 3$$

The slope is positive; the line slopes upward to the right. [See Figure 4.15(a).] Note that

$$\frac{y_1 - y_2}{x_1 - x_2} = \frac{2 - 5}{1 - 2} = \frac{-3}{-1} = 3.$$

□

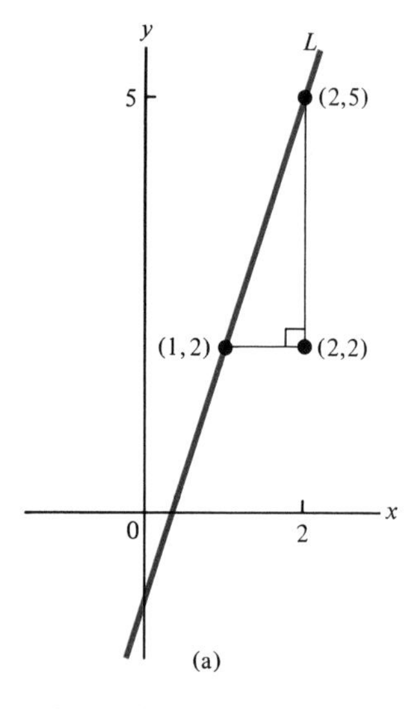

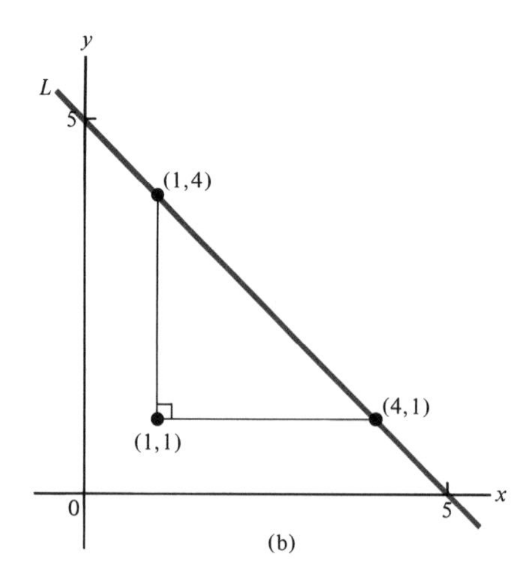

FIGURE 4.15

(a) slope $L = \frac{\text{rise}}{\text{run}} = \frac{5 - 2}{2 - 1} = 3$

(b) slope $L = \frac{\text{rise}}{\text{run}} = \frac{1 - 4}{4 - 1} = -1$

EXAMPLE 2 Determine the slope of the line through the points (1, 4) and (4, 1).

Solution. Let $(x_1, y_1) = (1, 4)$ and $(x_2, y_2) = (4, 1)$.

$$\text{Slope} = \frac{\text{rise}}{\text{run}} = \frac{y_2 - y_1}{x_2 - x_1} = \frac{1-4}{4-1} = \frac{-3}{3} = -1$$

The slope is negative; the line slopes downward to the right. [See Figure 4.15(b).] □

EXAMPLE 3 Consider the line L through $(-1, 2)$ and $(1, 4)$. Determine the y-coordinate of the point on L with x-coordinate 2.

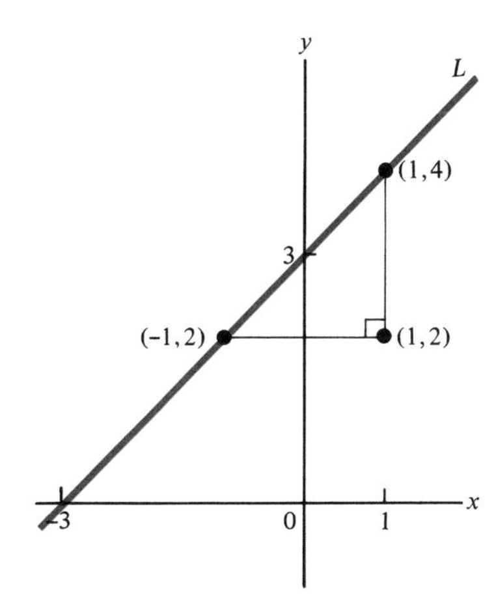

FIGURE 4.16
$$\text{slope } L = \frac{\text{rise}}{\text{run}} = \frac{4-2}{1-(-1)} = 1$$

Solution. Let $(x_1, y_1) = (-1, 2)$ and $(x_2, y_2) = (1, 4)$. [See Figure 4.16 .]

$$\text{Slope } L = \frac{\text{rise}}{\text{run}} = \frac{y_2 - y_1}{x_2 - x_1} = \frac{4-2}{1-(-1)} = \frac{2}{2} = 1$$

Let (x_3, y_3) be the point on L with x-coordinate 2. Thus $x_3 = 2$, and you must determine y_3. The slope of L can be expressed in terms of the point (x_3, y_3) and one other point—say (x_2, y_2). You already know that the slope of L is 1. Thus

$$\frac{y_3 - y_2}{x_3 - x_2} = \frac{y_3 - 4}{2-1} = \text{slope } L = 1$$
$$y_3 - 4 = 1$$
$$y_3 = 5$$

□

For an application of the slope to graphs of functions, see the footnote.‡

‡Here is a geometric interpretation of the *difference quotient of a function f* that will be useful in calculus. Let $(x, f(x))$ and $(x + h, f(x + h))$ be two points on the graph of f. The line connecting these two points is called a **secant.** [See Figure 4.17 .] The numerator

$$f(x+h) - f(x)$$

of the difference quotient

$$\frac{f(x+h) - f(x)}{h}, \quad h \neq 0,$$

represents the change in the function-values. The *denominator,*

$$h, \text{ or } (x+h) - x,$$

represents the change in the x-values. Thus

$$\frac{f(x+h) - f(x)}{h}$$

represents the *slope of the secant line*. The denominator h can be positive or negative, but *not* 0.

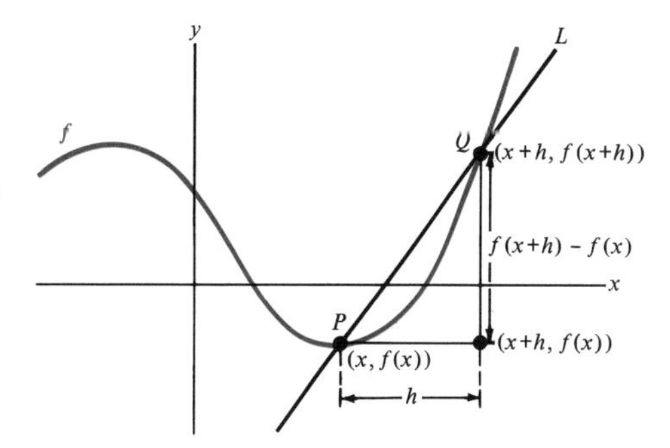

FIGURE 4.17 L is the secant connecting P and Q.

$$\text{slope } L = \frac{f(x+h) - f(x)}{h}$$

[In the diagram both $f(x + h) - f(x)$ and h are positive.]

point-slope form The fact that the slope of a line L can be obtained from *any* two distinct points on L suggests how to find the equation of L. Let (x_1, y_1) be a fixed point on L and let (x, y) be an arbitrary point on L other than (x_1, y_1). Let m be the slope of L. Then

$$\frac{y - y_1}{x - x_1} = m \qquad \text{or} \qquad y - y_1 = m(x - x_1)$$

The second equation is known as the **point-slope form of the equation of a line.** If you know a point (x_1, y_1) on L as well as the slope m of L, you can determine the equation of L.

As you will see, there are various forms in which the equation of a line can be written. Any two equations of a line L are equivalent because ordered pairs corresponding to the points of L (and only these ordered pairs) satisfy each of these equations. Thus you can speak of *the* equation of a line.

EXAMPLE 4 Determine the equation of the line L through (1, 3) with slope $\frac{1}{2}$.

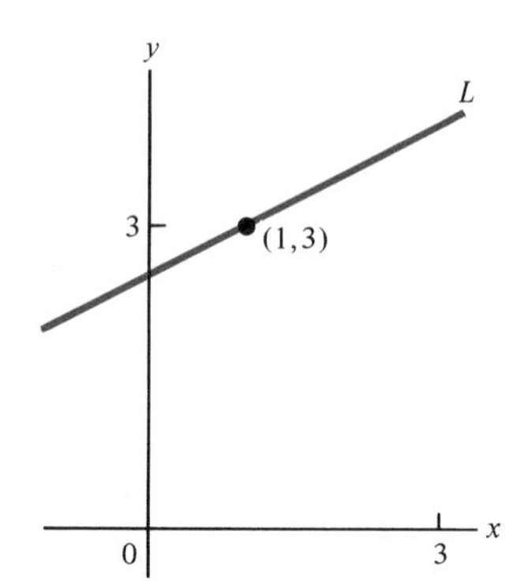

FIGURE 4.18 The line L given by $2(y - 3) = x - 1$

Solution. Here $(x_1, y_1) = (1, 3)$; $m = \frac{1}{2}$. Thus, substituting the indicated numbers in the equation

$$y - y_1 = m(x - x_1),$$

you obtain

$$y - 3 = \frac{1}{2}(x - 1).$$

This is the equation of L. Alternatively, multiply both sides by 2 to eliminate fractions, and obtain:

$$2(y - 3) = x - 1$$
$$2y = x + 5$$

[See Figure 4.18.] □

EXAMPLE 5 (a) Determine the equation of the line L through (0, 4) with slope -2.

(b) Determine the x-coordinate of the point on L with y-coordinate -2.

(c) Graph L.

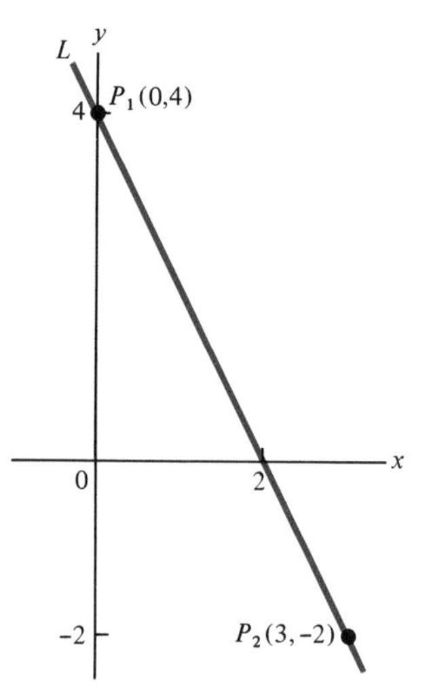

FIGURE 4.19 The line L given by $y - 4 = -2x$.

Solution.

(a) Here $(x_1, y_1) = (0, 4)$, $m = -2$. Thus

$$y - y_1 = m(x - x_1)$$
$$y - 4 = -2(x - 0)$$
$$y - 4 = -2x$$

(b) Replace y by -2 in the preceding equation:

$$-2 - 4 = -2x$$
$$3 = x$$

(c) Two distinct points determine the graph of L. Use the points P_1 with coordinates (0, 4), and P_2 with coordinates (3, −2). [See Figure 4.19.] □

horizontal lines

On a *horizontal* line L, all of the y-values are the same. For any two distinct points (x_1, y_1) and (x_2, y_2) on L,

$$y_1 = y_2$$

Thus

$$\text{rise } L = y_2 - y_1 = 0$$

Therefore

$$m = \text{slope } L = \frac{\text{rise}}{\text{run}} = \frac{y_2 - y_1}{x_2 - x_1} = \frac{0}{x_2 - x_1} = 0$$

In other words, *a horizontal line has slope* 0. The equation of a horizontal line is

$$y - y_1 = m(x_2 - x_1) = 0 \qquad \text{or} \qquad y = y_1.$$

Thus the equation of a *horizontal* line is determined by a *single y-value,* y_1. [See Figure 4.20 .]

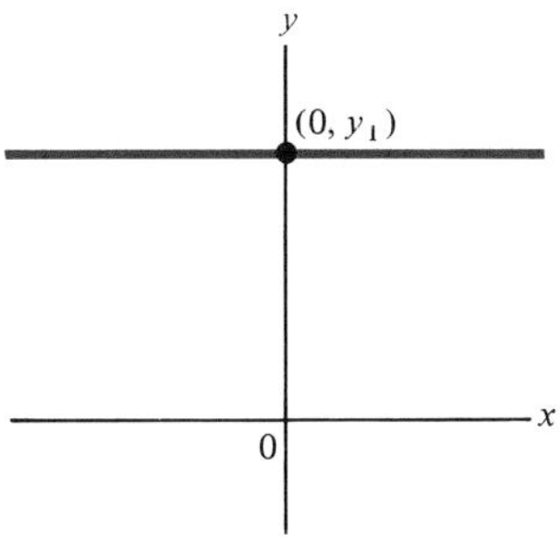

FIGURE 4.20 A horizontal line has the equation $y = y_1$.

EXAMPLE 6 Determine the equation of the horizontal line L through (1, 3).

Solution. Here $(x_1, y_1) = (1, 3)$. The equation of L is

$$y = 3.$$

[See Figure 4.21 .] □

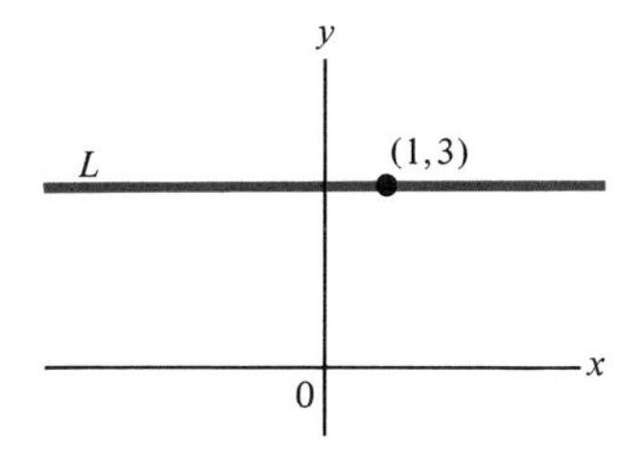

FIGURE 4.21 The line L given by $y = 3$

vertical lines

What about a *vertical* line? So far, vertical lines have been excluded from the discussion because *all x-values are the same.* Recall that

$$\text{slope} = \frac{\text{rise}}{\text{run}} = \frac{y_2 - y_1}{x_2 - x_1}.$$

For a vertical line, x_2 always equals x_1, so that $x_2 - x_1 = 0$. Because division by 0 is undefined, *the slope of a vertical line is undefined.* However, the fact that all x-values are the same immediately yields the *equation* of a vertical line. Determine one x-value, x_1, and you know all x-values. The equation of a vertical line is then

$$x = x_1.$$

[See Figure 4.22 .]

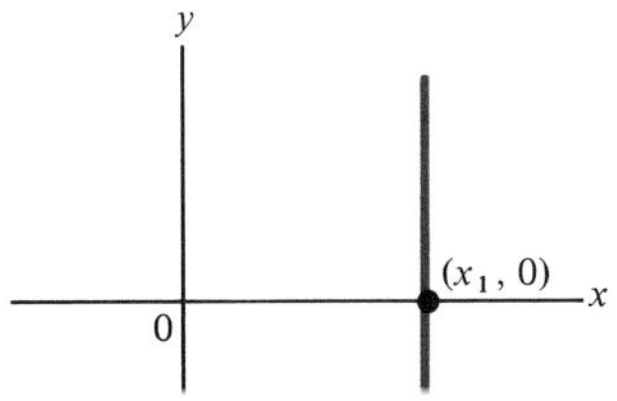

FIGURE 4.22 A vertical line has the equation $x = x_1$.

EXAMPLE 7 Determine the equation of the vertical line through (– 4, 2).

Solution. Here $(x_1, y_1) = (-4, 2)$. The equation of L is

$$x = -4.$$

[See Figure 4.23 .] □

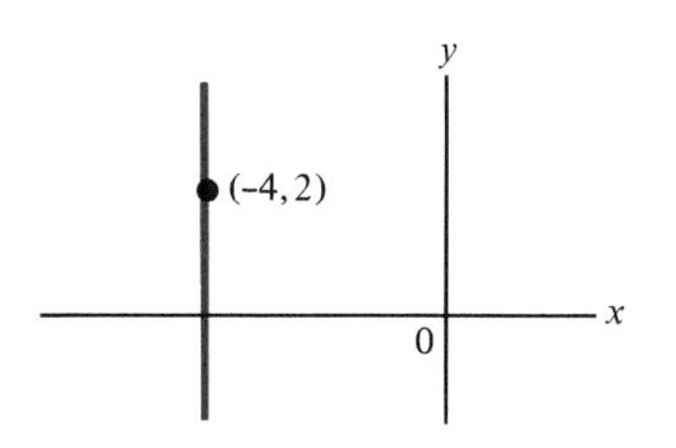

FIGURE 4.23 The line given by $x = -4$

EXERCISES

In Exercises 1–10 determine the slope of the line through the given points.

1. $(1, 1)$ and $(3, 3)$
2. $(1, 6)$ and $(-1, -6)$
3. $(6, 3)$ and $(0, 4)$
4. $(2, 8)$ and $(10, 2)$
5. $(0, 8)$ and $(1, 1)$
6. $(1, 3)$ and $(5, -3)$
7. $(3, 2)$ and $(-7, -4)$
8. $\left(\frac{3}{4}, \frac{-2}{3}\right)$ and $\left(0, \frac{-1}{3}\right)$
9. $\left(-3, \frac{1}{2}\right)$ and $\left(0, \frac{1}{4}\right)$
10. $(.6, 1)$ and $(.2, -.2)$
11. A line L has slope 2 and passes through $(2, 4)$. Determine the x-coordinate of the point on L with y-coordinate 1.
12. A line L has slope -1 and passes through $(0, 4)$. Determine the x-coordinate of the point on L with y-coordinate -2.
13. A line L has slope 5 and passes through $(-1, 3)$. Determine the y-coordinate of the point on L with x-coordinate 1.
14. A line L has slope $\frac{1}{3}$ and passes through the origin. Determine the y-coordinate of the point on L with x-coordinate 9.
15. A line L passes through $(1, -3)$ and the origin. Determine the y-coordinate of the point on L with x-coordinate 5.
16. A line L passes through $(-2, -4)$ and $(-1, -1)$. Determine the y-coordinate of the point on L with x-coordinate 1.

In Exercises 17–26 determine the equation of the line that has slope m and that passes through the point P.

17. $m = 3$, $P = (0, 4)$
18. $m = -1$, $P = (2, 2)$
19. $m = 2$, $P = (-1, 3)$
20. $m = 4$, $P = (0, 0)$
21. $m = \frac{1}{2}$, $P = (1, 2)$
22. $m = \frac{-1}{2}$, $P = (4, 0)$
23. $m = -3$, $P = (-1, -1)$
24. $m = \frac{-1}{4}$, $P = (1, 4)$
25. $m = 0$, $P = \left(\frac{1}{2}, \frac{3}{4}\right)$
26. $m = \frac{-3}{4}$, $P = (-1, 5)$

In Exercises 27–32: (a) Determine the equation of the line that has slope m and that passes through the point P. (b) Graph this line.

27. $m = 2$, $P = (3, 3)$
28. $m = -4$, $P = (1, 0)$
29. $m = 4$, $P = (1, 5)$
30. $m = 0$, $P = (1, 4)$
31. $m = \frac{-1}{2}$, $P = (4, 2)$
32. $m = \frac{1}{3}$, $P = (3, 6)$
33. Which of the following lines are horizontal? Which are vertical? Which are neither horizontal nor vertical?

    (a) $y = x - 4$ **(b)** $y = 2x + 1$ **(c)** $y = 1$ **(d)** $x = 3$ **(e)** $y = 0$ **(f)** $x = -2$
34. Determine the equation of the vertical line through $(5, 5)$.
35. Determine the equation of the horizontal line through $(5, 5)$.
36. A line passes through $(2, 8)$ and $(-2, 8)$. Determine its equation.
37. A line passes through $(1, -4)$ and $(1, 0)$. Determine its equation.
38. Determine the equation of the vertical line through $(0, -2)$.

39. A nonvertical line L has slope m and passes through the origin. It intersects the line with equation $y = 4$ when $x = x_0$. Express x_0 as a function of m.

40. A nonvertical line passes through $(-2, 3)$. Express its slope m as a function of its y-intercept b.

4.3 Alternate Forms of Linear Equations

slope-intercept form

The point-slope form of the equation of L requires knowing one point on L in addition to the slope, m. Let this point be $(0, b)$. Thus b is the y-intercept. Then the equation of L is

$$y - b = m(x - 0) \qquad \text{or} \qquad y = mx + b.$$

This second equation is in **slope-intercept form.** Thus, the slope and y-intercept of a (nonvertical) line L determine the slope-intercept form of the equation of L. [See Figure 4.24 .]

EXAMPLE 1

Let L have slope -2 and y-intercept 4. Determine the slope-intercept form of the equation of L. [See Figure 4.25 .]

Solution. Here $m = -2$ and $b = 4$. The slope-intercept form of the equation of L is

$$y = -2x + 4.$$

□

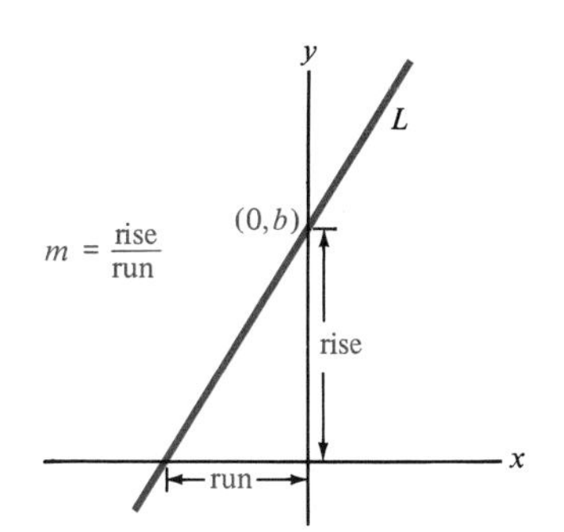

FIGURE 4.24 The slope-intercept form of the equation of L: $y = mx + b$

linear functions

A **linear function** is a function f given by

$$f(x) = mx + b,$$

where m and b can be any real numbers. Every linear function determines a line with equation

$$y = mx + b.$$

The function defined by

$$f(x) = -2x + 4$$

is a linear function. For each x, the ordered pair $(x, f(x))$ lies on the line L of Figure 4.25 .

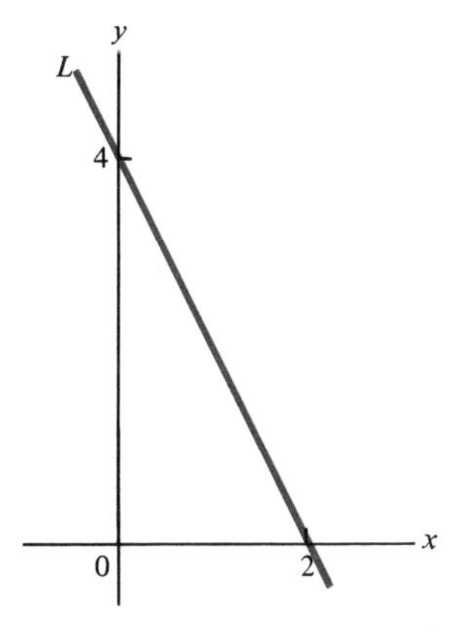

FIGURE 4.25 The line L given by $y = -2x + 4$

general form There is one more important form of the equation of a line, known as the **general form:** The equation of a line can be written as

$$Ax + By + C = 0, \tag{4.1}$$

where A and B are not *both* 0.

EXAMPLE 2 Determine the general form of the line given by

$$y - 7 = -3(x + 5).$$

Solution. Add $3(x + 5)$ to both sides, simplify, and arrange the terms, as in Equation (**4.1**).

$$y - 7 = -3(x + 5)$$
$$y - 7 + 3(x + 5) = 0$$
$$y - 7 + 3x + 15 = 0$$
$$3x + y + 8 = 0$$

This is the general form of the equation. Here $A = 3$, $B = 1$, $C = 8$ □

EXAMPLE 3 (a) Determine the slope-intercept form of the equation

$$2x - 3y + 1 = 0$$

of a line L.

(b) Determine the x-coordinate of the point on L with y-coordinate -6.

(c) Graph L.

Solution.

(a) To find the slope-intercept form, solve the equation for y. Thus,

$$2x + 1 = 3y$$

Change sides and divide by 3:

$$y = \frac{1}{3}(2x + 1)$$
$$y = \frac{2}{3}x + \frac{1}{3}$$

This is the slope-intercept form. The slope is $\frac{2}{3}$ and the y-intercept is $\frac{1}{3}$.

(b) Let $y = -6$ in the original equation:

$$2x - 3(-6) + 1 = 0$$
$$2x + 18 + 1 = 0$$
$$2x = -19$$
$$x = \frac{-19}{2}$$

(c) You now know two points on the line:

$$\left(0, \frac{1}{3}\right) \quad \text{and} \quad \left(\frac{-19}{2}, -6\right)$$

Draw the line through these two points, as in Figure 4.26 . □

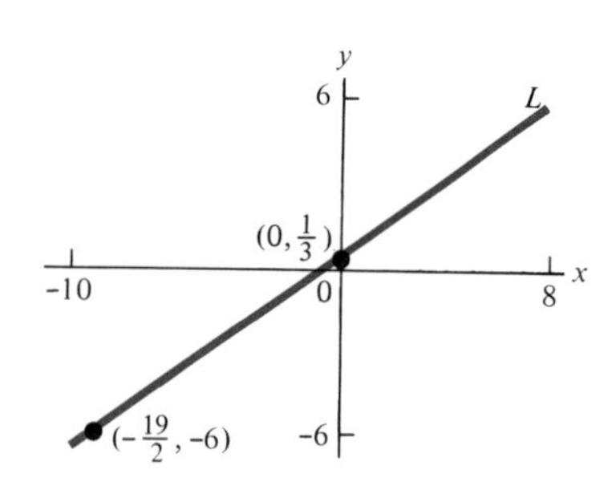

FIGURE 4.26

EXERCISES

In Exercises 1–6 determine: (a) the x-intercept and (b) the y-intercept of each of the indicated lines

1. $y = 2(x + 7)$ **2.** $y + 1 = -4(x - 3)$ **3.** $y = 2x - 5$ **4.** $y = -3x + 1$

5. $2x + 3y + 5 = 0$ **6.** $5x - y = 0$

In Exercises 7–12 determine the slope-intercept form of the equation of the indicated line L.

7. $L: y - 5 = 2(x - 1)$ **8.** $L: y + \frac{1}{2} = x - \frac{3}{4}$

9. L has slope 3 and y-intercept 1. **10.** L has slope 2 and x-intercept 2.

11. L passes through (8, 1) and (3, 4). **12.** L has x-intercept 4 and y-intercept -3.

In Exercises 13–18 determine the general form of the equation of the indicated line L.

13. $L: y - 2 = x + 8$ **14.** $L: y - 7 = 2(x - 5)$

15. L passes through (1, 8) and the origin. **16.** L is the vertical line through (2, 5).

17. L is the horizontal line through (2, 5). **18.** L has slope 4 and y-intercept -5.

In Exercises 19–24: (a) determine the slope of the indicated line L; (b) graph L.

19. $L: y + 1 = 5(x - 4)$ **20.** $L: 3x + 4y - 7 = 0$

21. $L: 2x - y - 2 = 0$ **22.** L passes through $(-1, 5)$ and $(-2, -1)$.

23. L has x-intercept 5 and y-intercept 1. **24.** L has x-intercept -2 and y-intercept -4.

In Exercises 25–32 determine the equation of the line L in any form.

25. L passes through $(-1, -1)$ and (4, 2).

26. L passes through (2, 4) and has x-intercept 5.

27. L passes through $(2, -3)$ and has y-intercept 5.

28. L passes through $(3, -1)$ and has x-intercept 3.

29. L passes through $(7, -2)$ and has y-intercept -2.

30. L passes through the origin and has slope -3.

31. L passes through (1, 5) and has no x-intercept.

32. L passes through $(-2, 5)$ and has no y-intercept.

In Exercises 33–40 indicate which functions are linear.

33. $g(x) = -3x$ **34.** $h(x) = 6x + 7$ **35.** $F(x) = 4 - x$ **36.** $G(x) = \frac{1}{x}$

37. $H(x) = \frac{1}{2x - 4}$ **38.** $f(x) \equiv 4$ **39.** $g(x) \equiv 0$ **40.** $h(x) = 2x^2 + 1$

In Exercises 41–44 suppose f is a linear function.

41. If $f(1) = 4$ and $f(2) = 7$, find $f(3)$. **42.** If $f(-2) = 8$ and $f(2) = -2$, find $f(0)$.

43. If $f(3) = 7$ and $f(6) = 1$, find $f(4)$. **44.** If $f(3) = -4$ and $f(8) = -4$, find $f(-1)$.

45. Let $f(x) = mx + b$, where $m \neq 0$. Show that f^{-1} is a linear function.

46. Suppose f and g are each linear functions. Show that $g \circ f$ is a linear function.

47. Let F = degrees Fahrenheit and C = degrees Celsius. Show that F is a linear function of C. Then show that C is a linear function of F.

48. **(a)** Suppose a line is neither vertical nor horizontal and it does not pass through the origin. Show that its equation can be expressed in the form

$$\frac{x}{a} + \frac{y}{b} = 1$$

(b) What do a and b represent here?

49. Consider a line given by

$$y = mx + b, \quad \text{where } m \neq 0.$$

Find the x-intercept.

50. Consider a line given by

$$Ax + By + C = 0, \quad \text{where } A \neq 0 \text{ and } B \neq 0.$$

(a) Find the slope. **(b)** Find the y-intercept. **(c)** Find the x-intercept.

4.4 Parallel and Perpendicular Lines

parallel lines

Definition. **Parallel lines** on the plane are lines that do not intersect.

If L_1 and L_2 are parallel lines, write $L_1 \parallel L_2$. In this case L_1 is also said to be **parallel to** L_2 (and L_2 **parallel to** L_1).

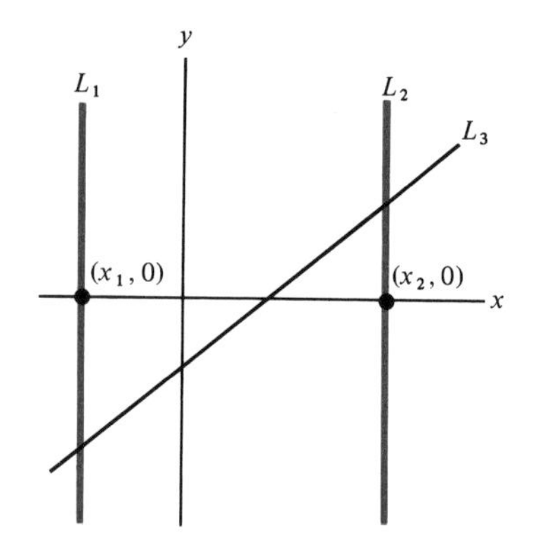

FIGURE 4.27 $L_1 \parallel L_2$
Each of the vertical lines L_1 and L_2 intersects the nonvertical line L_3. Thus neither is parallel to L_3.

Two vertical lines

$$x = x_1 \quad \text{and} \quad x = x_2,$$

where $x_1 \neq x_2$, *are parallel.*

A vertical line intersects a nonvertical line; hence these lines are not parallel. [See Figure 4.27 .]

Let L_1 and L_2 be distinct nonvertical lines. Then it can be shown that *if they are parallel, their slopes are equal, and if their slopes are equal, the lines are parallel.* (A proof will be given in Chapter 8, p. 319.) Thus let

$$m_1 = \text{slope } L_1 \quad \text{and} \quad m_2 = \text{slope } L_2.$$

If $L_1 \parallel L_2$, then $m_1 = m_2$. And if $m_1 = m_2$, then $L_1 \parallel L_2$. [See Figure 4.28 .]

EXAMPLE 1 (a) The lines given by

$$y = 4x + 3 \quad \text{and} \quad y = 4x - 5$$

are different because the first has y-intercept 3 and the second has y-intercept -5. These lines are parallel because each has slope 4. [See Figure 4.29 .]

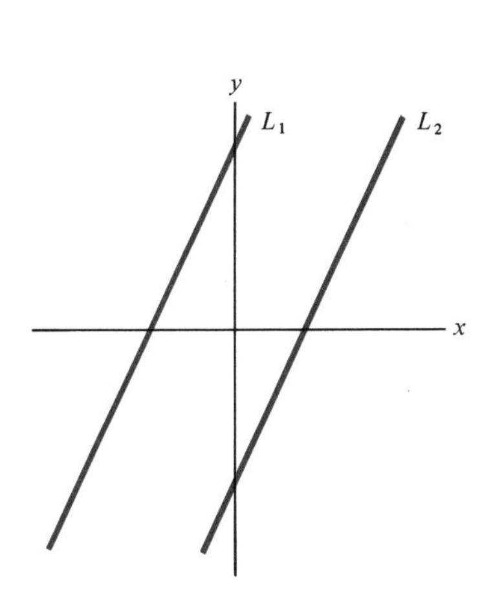

FIGURE 4.28 $L_1 \parallel L_2$
m_1 = slope L_1 = slope L_2 = m_2

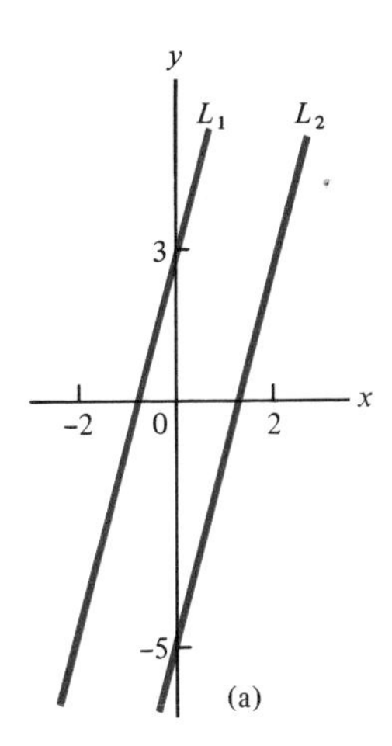

FIGURE 4.29 L_1 is given by $y = 4x + 3$ and L_2 by $y = 4x - 5$. slope $L_1 = 4$ = slope L_2. Thus L_1 and L_2 are parallel.

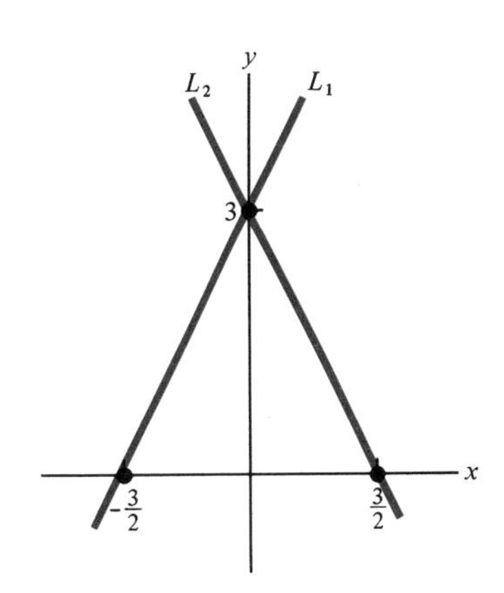

FIGURE 4.30 L_1 is given by $y = 2x + 3$ and L_2 by $y = -2x + 3$. slope $L_1 = 2$, slope $L_2 = -2$. Thus L_1 and L_2 are not parallel.

(b) The lines given by

$$L_1: y = 2x + 3 \qquad \text{and} \qquad L_2: y = -2x + 3$$

are not parallel because L_1 has slope 2, whereas L_2 has slope -2. [See Figure 4.30 .] □

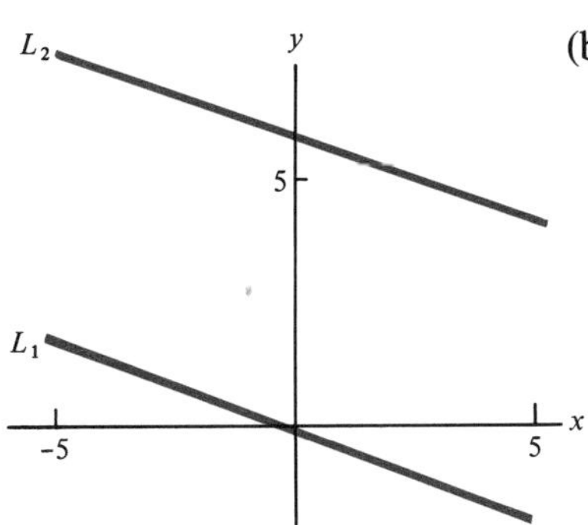

FIGURE 4.31

EXAMPLE 2 The lines given by

$$L_1: 2x + 6y + 1 = 0 \qquad \text{and} \qquad L_2: 3(y - 4) = 5 - x$$

are parallel. Here

$$m_1 = \text{slope } L_1 = \frac{-2}{6} = \frac{-1}{3}, \qquad m_2 = \text{slope } L_2 = \frac{-1}{3}$$

[Check these computations! Also check that L_1 and L_2 are *different* lines. See Figure 4.31 .] □

Let L be a given line and let P be a point that is *not* on L. Then *there is exactly one line that passes through P and is parallel to L.* [See Figure 4.32 .]

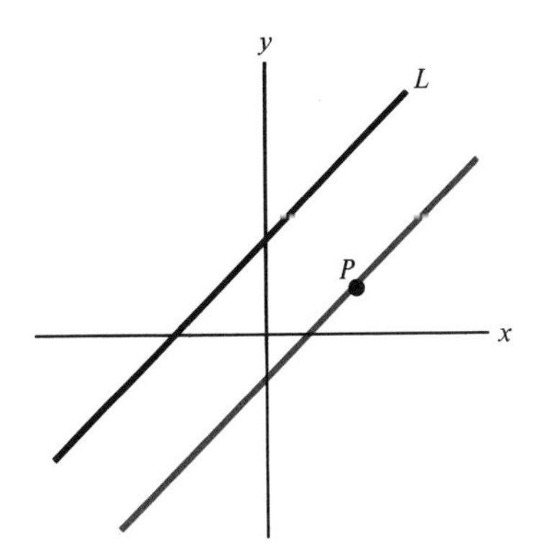

FIGURE 4.32 There is exactly one line that passes through P and is parallel to L.

EXAMPLE 3 Determine the equation of the line through (1, 4) that is parallel to the line given by

$$y = 2x - 5.$$

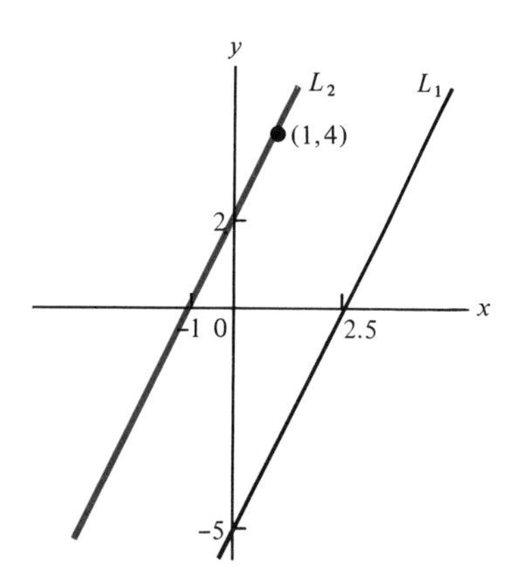

FIGURE 4.33 L_1 is given by $y = 2x - 5$. $L_2 \parallel L_1$ and L_2 passes through (1, 4). Thus L_2 is given by $y - 4 = 2(x - 1)$.

Solution. The line given by

$$y = 2x - 5$$

has slope 2. Therefore, write the equation of the line that has slope 2 and that passes through (1, 4). The point-slope form of the equation of this line is thus

$$y - 4 = 2(x - 1).$$

[See Figure 4.33 .] □

perpendicular lines

Definition. Two lines are **perpendicular** if they intersect at right angles.

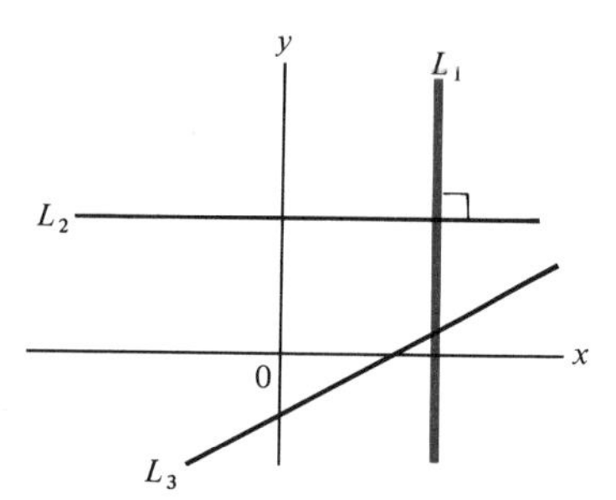

FIGURE 4.34 L_1 is vertical; L_2 is horizontal; $L_1 \perp L_2$, L_3 is not horizontal. L_1 is not perpendicular to L_3.

If lines L_1 and L_2 are perpendicular, write

$$L_1 \perp L_2$$

A vertical line and a horizontal line are perpendicular. A vertical line is not perpendicular to any nonhorizontal line. [See Figure 4.34 .]

Draw a line through (0, 0) and (1, 3). Now draw another line through (0, 0), but perpendicular to the given line. Observe that the second line appears to pass through (3, −1). [See Figure 4.35 .] Compute the slopes of both lines, and observe that these slopes, 3 and $\frac{-1}{3}$, are "negative reciprocals".

FIGURE 4.35 Slope $L_1 = \frac{3}{1} = 3$, slope $L_2 = \frac{-1}{3}$

Let L_1 and L_2 be distinct nonvertical lines. Then it can be shown that *if L_1 and L_2 are perpendicular, their slopes are negative reciprocals, and if their slopes are negative reciprocals, the lines are perpendicular*. (A proof will be given in Chapter 8, p. 320.) Thus let

$$m_1 = \text{slope } L_1 \qquad \text{and} \qquad m_2 = \text{slope } L_2.$$

If $L_1 \perp L_2$, then $m_1 = \frac{-1}{m_2}$. Also, if $m_1 = \frac{-1}{m_2}$, then $L_1 \perp L_2$. Note that

$$\text{if} \quad m_1 = \frac{-1}{m_2}, \quad \text{then}$$

$$m_1 \neq 0, \qquad m_2 = \frac{-1}{m_1}, \qquad \text{and} \qquad m_1 m_2 = -1.$$

EXAMPLE 4 (a) The lines given by

$$L_1\colon y = 2x + 3 \qquad \text{and} \qquad L_2\colon y = \frac{-1}{2}x$$

are perpendicular. Here

$$m_1 = \text{slope } L_1 = 2 \qquad \text{and} \qquad m_2 = \text{slope } L_2 = \frac{-1}{2}$$

[See Figure 4.36 .]

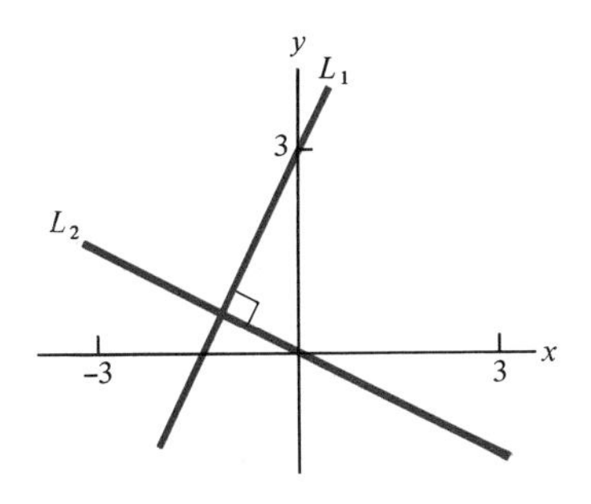

FIGURE 4.36 L_1 is given by $y = 2x + 3$ and L_2 by $y = \frac{-1}{2}x$. Here $m_1 = 2$, $m_2 = \frac{-1}{2}$. Thus $L_1 \perp L_2$

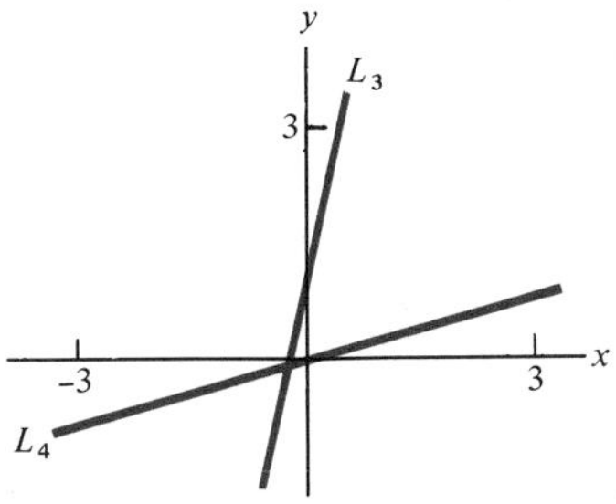

FIGURE 4.37 L_3 is given by $y = 4x + 1$ and L_4 is given by $y = \frac{x}{4}$. Here $m_3 = 4$, $m_4 = \frac{1}{4}$. Thus L_3 and L_4 are not perpendicular. $(m_3 \cdot m_4 = 1 \neq -1)$

(b) The lines given by

$$L_3\colon y = 4x + 1 \qquad \text{and} \qquad L_4\colon y = \frac{x}{4}$$

are *not* perpendicular. Here

$$m_3 = \text{slope } L_3 = 4 \qquad \text{and} \qquad m_4 = \text{slope } L_4 = \frac{1}{4}$$

But the *negative reciprocal* of 4 is $\frac{-1}{4}$. [See Figure 4.37 .] □

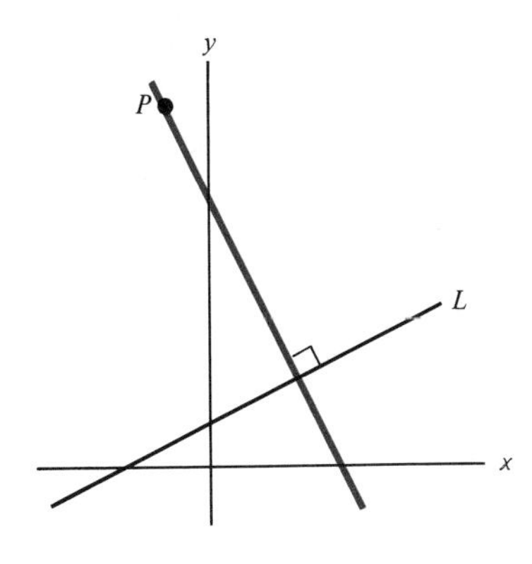

FIGURE 4.38

Let L be a given line and let P be a point. Then *there is exactly one line that passes through P and is perpendicular to L.* [See Figure 4.38 .]

EXAMPLE 5 Determine the equation of the line through $(-2, -2)$ that is perpendicular to the line given by

$$2y = x + 1.$$

Solution. The equation of the given line can be written in the form

$$y = \frac{x}{2} + \frac{1}{2}.$$

This line has slope $\frac{1}{2}$. Now find the line that has slope -2 $\left(\text{that is, } \frac{-1}{\frac{1}{2}}\right)$ and that passes through $(-2, -2)$. The point-slope form of the equation of this line is then

$$y - (-2) = -2(x - (-2)) \qquad \text{or} \qquad y + 2 = -2(x + 2).$$

[See Figure 4.39 .] □

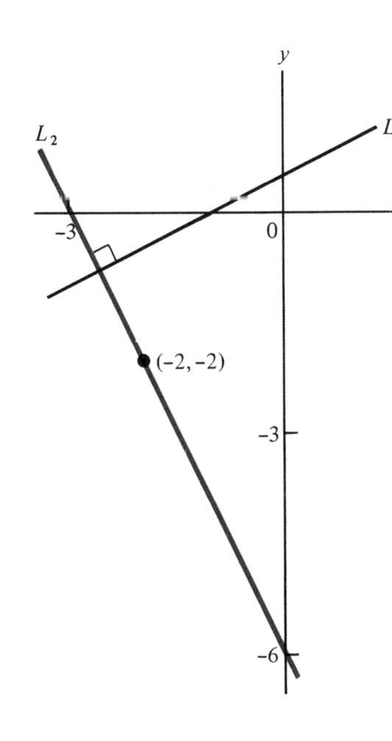

FIGURE 4.39 L_1 is given by $2y = x + 1$ and L_2 by $y + 2 = -2(x + 2)$. Here $m_1 = \frac{1}{2}$, $m_2 = -2$. Thus $L_1 \perp L_2$

EXAMPLE 6 Consider the lines:

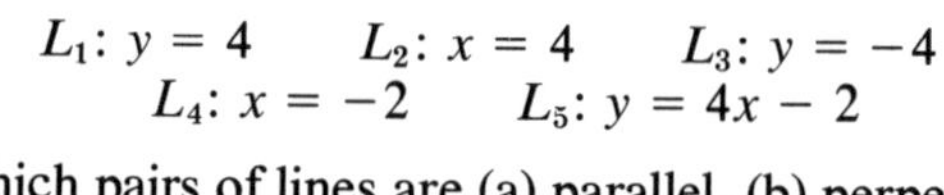

$$L_1: y = 4 \qquad L_2: x = 4 \qquad L_3: y = -4$$
$$L_4: x = -2 \qquad L_5: y = 4x - 2$$

Which pairs of lines are (a) parallel, (b) perpendicular?

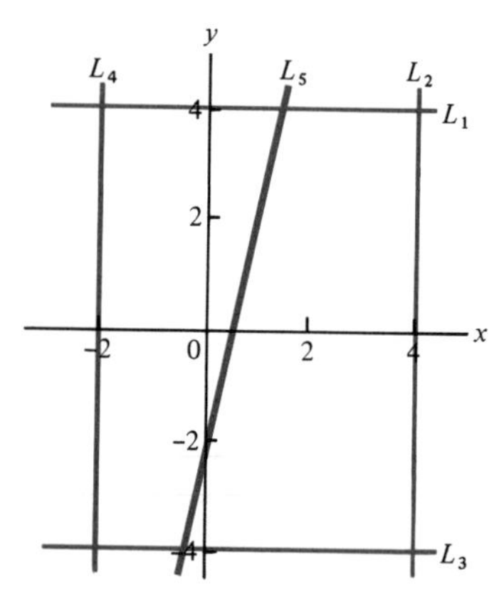

FIGURE 4.40

Solution. L_1 and L_3 are horizontal; L_2 and L_4 are vertical. L_5 is neither horizontal nor vertical.

(a) Two horizontal lines are parallel. (Each has slope 0.) Two vertical lines are parallel. Thus,

$$L_1 \parallel L_3; \qquad L_2 \parallel L_4$$

(b) A vertical line is perpendicular to a horizontal line. Thus,

$$L_1 \perp L_2, \qquad L_1 \perp L_4, \qquad L_3 \perp L_2, \qquad L_3 \perp L_4$$

[See Figure 4.40 .] □

EXAMPLE 7 Show that the lines L_1 through (6, 2) and (2, 6) and L_2 through the origin and (5, 5) are perpendicular.

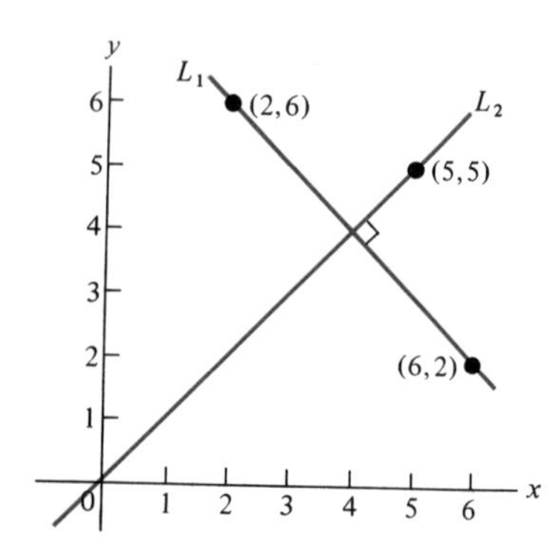

FIGURE 4.41 $m_1 = \text{slope } L_1 = -1$, $m_2 = \text{slope } L_2 = 1$, $m_1 = \dfrac{-1}{m_2}$. Thus $L_1 \perp L_2$

Solution. Consider the slopes of these lines.

$$m_1 = \text{slope } L_1 = \frac{6-2}{2-6} = -1$$

$$m_2 = \text{slope } L_2 = \frac{5-0}{5-0} = 1$$

$$m_1 = \frac{-1}{m_2} \qquad \text{and thus} \qquad L_1 \perp L_2$$

[See Figure 4.41 .] □

EXERCISES

In Exercises 1–18 determine whether the given lines are (a) parallel, (b) perpendicular, or (c) neither parallel nor perpendicular.

1. $y = 2x + 1$ and $y = 2x + 5$

2. $y = -x + 5$ and $y = -x - 3$

3. $y = 3x + 1$ and $y = \dfrac{-1}{3}x + 4$

4. $y + 1 = x - 1$ and $y + 4 = 1 - x$

5. $4x + 7y + 3 = 0$ and $7x + 4y + 3 = 0$

6. $x + y + 1 = 0$ and $5x - 5y + 2 = 0$

7. $y - 4 = \frac{1}{2}(x - 2)$ and $3x - 6y = 4$

8. $y = 2$ and $x = 2$

9. $y - 3 = 0$ and $y = 4$

10. $x + 7 = 0$ and $x - 7 = 0$

11. $2x + 8y = 3$ and $y = 4(3 - x)$

12. $y = 6x$ and $y = \frac{x}{6}$

13. $y = 3x$ and $x = \frac{-1}{3}y$

14. $y = 2x$ and $x = -2y$

15. $y = 3$ and $y = 3x$

16. $y = 0$ and $y = x$

17. $x + y + 1 = 0$ and $y = x$

18. $3x - 3y = 9$ and $y = x - 3$

In Exercises 19–24 determine whether the lines through P and Q and through R and S are (a) parallel, (b) perpendicular, or (c) neither parallel nor perpendicular.

19. $P = (3, 3),\ Q = (-4, -4);\ R = (2, -2),\ S = (-1, 1)$

20. $P = (3, 2),\ Q = (5, 3);\ R = (1, 4),\ S = (5, 6)$

21. $P = (3, 9),\ Q = (2, 8);\ R = (5, 3),\ S = (-2, 1)$

22. $P = (-1, -2),\ Q = (3, 0);\ R = (5, 6),\ S = (1, 4)$

23. $P = (3, 5),\ Q = (3, -2);\ R = (4, 3),\ S = (4, 5)$

24. $P = (1, 7),\ Q = (-2, 7);\ R = (5, 9),\ S = (5, 0)$

In Exercises 25–34 determine the equations of the lines that pass through P and are (a) parallel to, (b) perpendicular to, the given line L.

25. $P = (1, 1), \quad L: y = 2x + 8$

26. $P = (4, 0), \quad L: y = 5 - x$

27. $P = (0, -3), \quad L: 2x + y + 1 = 0$

28. $P = (0, 0), \quad L: 4(y - 3) = 3(x + 2)$

29. $P = (2, 5), \quad L: x + y = 2$

30. $P = (-3, 1), \quad L: x - 2y = 4$

31. $P = (2, 8), \quad L: y = -2$

32. $P = (3, 4), \quad L: x + 3 = 0$

33. $P = (0, 0), \quad L: x = 9$

34. $P = (2, 0), \quad L: y = 2$

In Exercises 35–38 determine all pairs of (a) parallel lines, (b) perpendicular lines:

35.
$L_1: y = 3x$
$L_2: y = -3x$
$L_3: y = \frac{x}{3}$
$L_4: y = 3(x - 2)$
$L_5: 3y = 2 - x$

36.
$L_1: 2x + 3y = 5$
$L_2: 9y = 5 - 6x$
$L_3: 3x - 2y = 7$
$L_4: 3x + 2y = 1$
$L_5: y = \frac{3x}{2}$

37.
$L_1: x + y = 4$
$L_2: x - y = 7$
$L_3: y = 3 - x$
$L_4: y = x$
$L_5: y = -x$

38.
$L_1: x = 5$
$L_2: y = -5$
$L_3: x = \frac{-1}{5}$
L_4: the x-axis
L_5: the y-axis

In Exercises 39–41 let L_1, L_2, and L_3 be lines in the plane.

39. Suppose $L_1 \parallel L_2$ and $L_2 \parallel L_3$. Is $L_1 \parallel L_3$?

40. (a) Let L_1, L_2, and L_3 be *distinct* lines. Suppose $L_1 \perp L_2$ and $L_2 \perp L_3$. What can be said about L_1 and L_3?
(b) In (a) if the word "distinct" is omitted, what can be said about L_1 and L_3?

41. Suppose $L_1 \perp L_2$ and $L_2 \parallel L_3$. What can be said about L_1 and L_3?

4.5 Graphing Inequalities

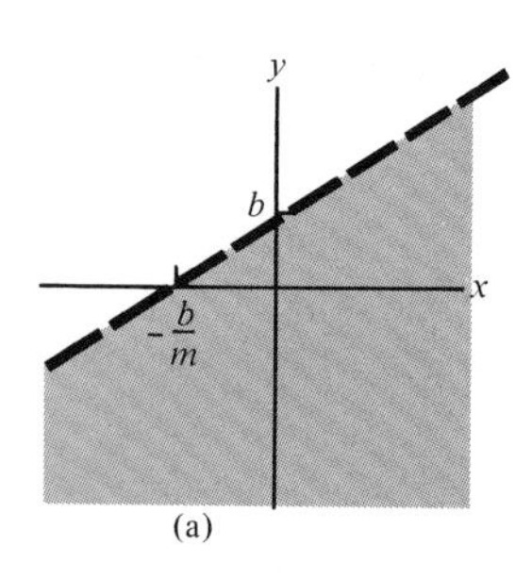

FIGURE 4.42
(a) The graph of the inequality $y < mx + b$. The line is exluded. (In the diagram $m > 0$, $b > 0$)

The graph of the *equation*

$$y = mx + b$$

is the line with slope m and y-intercept b. Suppose (x_1, y_1) lies on this line. Then

$$y_1 = mx_1 + b.$$

If $\quad y_2 < mx_1 + b,$

then the point (x_1, y_2) lies *below* (x_1, y_1) and hence *below* the line with equation $y = mx + b$. [See Figure 4.42(a).] Also, if

$$y_3 > mx_1 + b,$$

then (x_1, y_3) lies *above* (x_1, y_1) and hence *above* the line

$$y = mx + b.$$

[See Figure 4.42(b).] Thus this line divides the plane into two regions. Points lying *below* the line satisfy the *inequality*

Points lying *above* the line satisfy the inequality

$$y > mx + b.$$

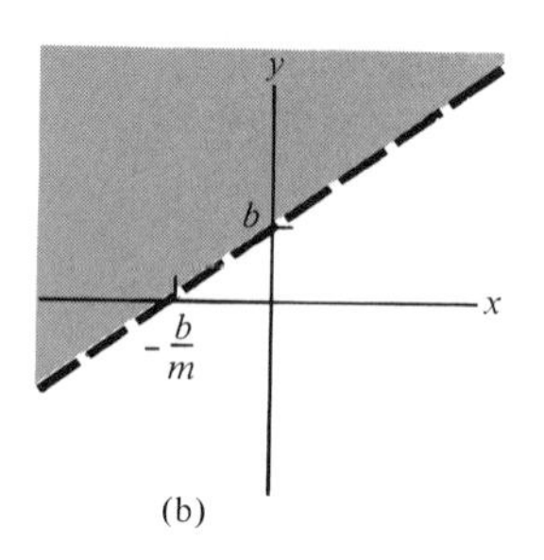

(b) The graph of the inequality $y > mx + b$. The line is excluded. (In the diagram $m > 0$, $b > 0$)

EXAMPLE 1 Consider the line given by the equation

$$y = x.$$

[See Figure 4.43 .]

(a) The point (2, 1) lies *below* this line. Observe that the y-value is 1, the x-value is 2, and

$$1 < 2.$$

Thus (2, 1) satisfies the inequality

$$y < x.$$

(b) The point (1, 2) lies *above* the line and

$$2 > 1.$$

Thus (1, 2) satisfies the inequality

$$y > x.$$

□

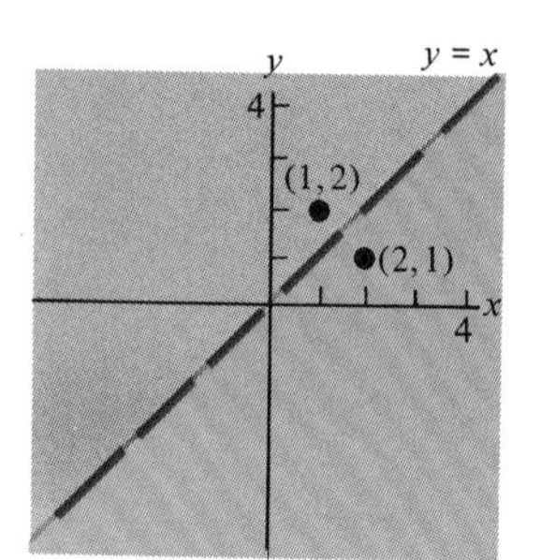

FIGURE 4.43 The graphs of $y < x$ (color) and $y > x$ (gray).

EXAMPLE 2 Graph the inequality:

$$y < 2 - x$$

Solution. First draw the line given by

$$y = 2 - x,$$

as in Figure 4.44 . The graph of the inequality

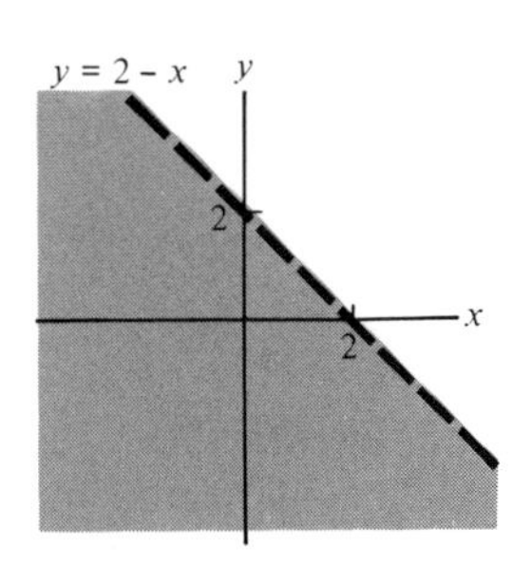

FIGURE 4.44

$$y < 2 - x$$

consists of all points lying *below* this line. □

The graph of an inequality

$$y \leqslant mx + b$$

consists of all points lying *on or below* the line given by the equation

$$y = mx + b.$$

[See Figure 4.45(a).] The graph of an inequality

$$y \geqslant mx + b$$

consists of all points lying *on or above* the line given by

$$y = mx + b.$$

[See Figure 4.45(b).]

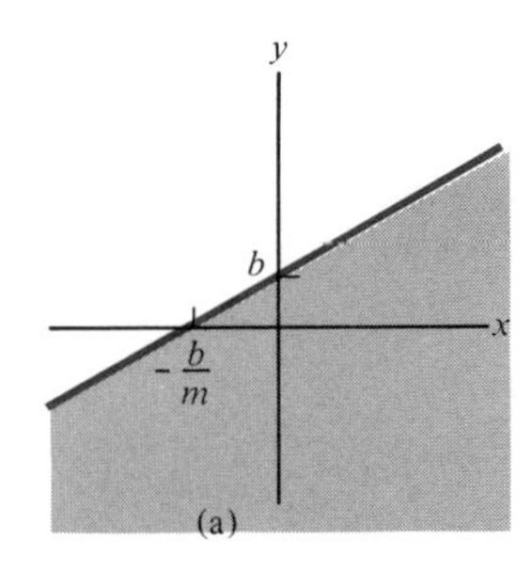

FIGURE 4.45
(a) The graph of the inequality $y \leqslant mx + b$. The line is included. (In the diagram $m > 0$, $b > 0$)

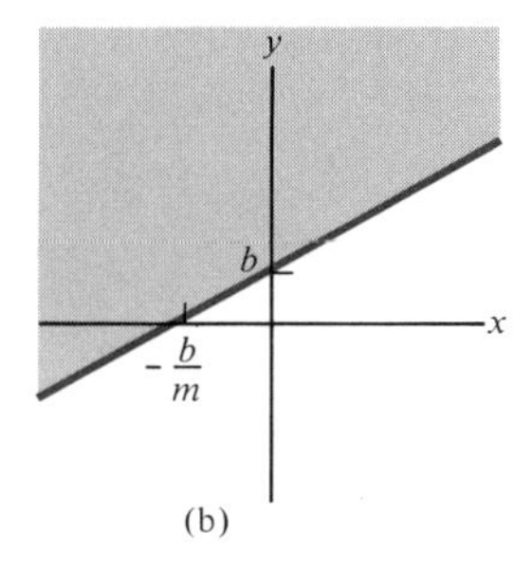

(b) The graph of the inequality $y \geqslant mx + b$. The line is included. (In the diagram $m > 0$, $b > 0$)

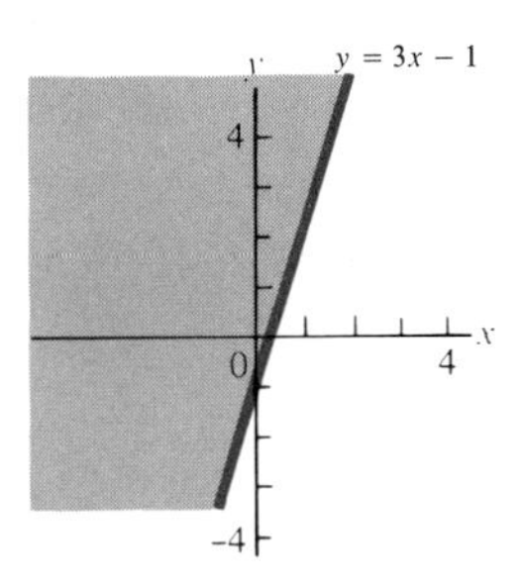

FIGURE 4.46 The graph of $y \geqslant 3x - 1$

EXAMPLE 3 Graph the inequality:

$$y \geqslant 3x - 1$$

Solution. First draw the line given by

$$y = 3x - 1,$$

as in Figure 4.46 . The graph of the inequality

$$y \geqslant 3x - 1$$

consists of all points lying *on or above* this line. □

EXAMPLE 4 Graph the inequality:

$$2x - y \leqslant 4$$

Solution. Rewrite the inequality in the form

$$2x - 4 \leqslant y \qquad \text{or} \qquad y \geqslant 2x - 4.$$

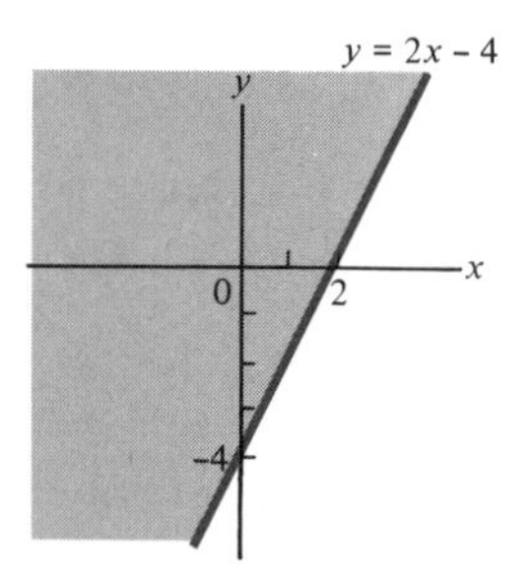

FIGURE 4.47 The graph of $2x - y \leq 4$

Now draw the line with equation

$$y = 2x - 4$$

as in Figure 4.47 . The graph of the given inequality consists of all points lying *on or above* this line. □

The graph of an inequality does *not* depict a function. For, the graph of a function is such that every vertical line intersects the graph at most once. In fact, every vertical line intersects the graph of each inequality previously described infinitely many times. [See Figure 4.48 .]

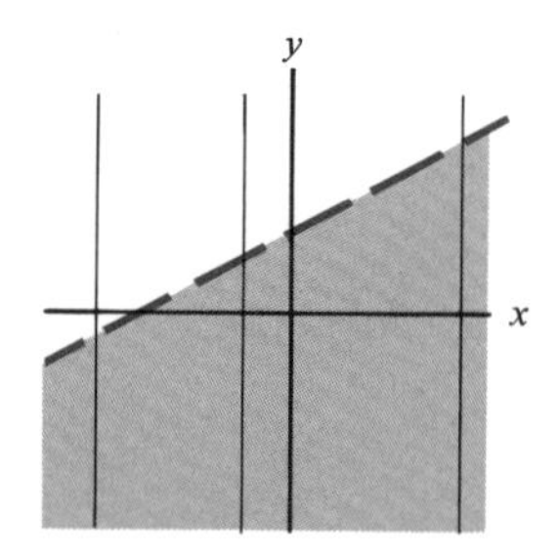

FIGURE 4.48 Every vertical line intersects the graph of this inequality infinitely many times.

A vertical line is given by an equation of the form

$$x = c,$$

where c is a real number. Thus the point (a, b) satisfies the inequality

$$x < c$$

if $a < c$. In this case, the point lies to the *left of* the vertical line. For example, (2, 5) satisfies the inequality

$$x < 4$$

But (5, 2) does not. (Again, a function is not defined.) [See Figure 4.49 .]

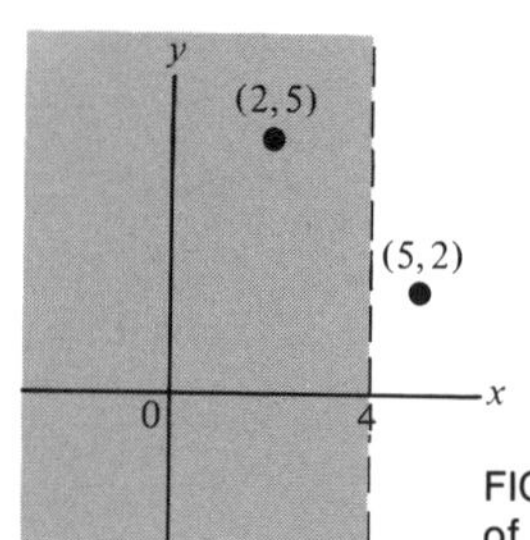

FIGURE 4.49 The graph of $x < 4$

EXERCISES

In Exercises 1–20 indicate which points satisfy the given inequality.

		(a)	(b)	(c)	(d)
1.	$y < 2x$	(1, 1)	(1, 2)	(1, 3)	(−1, −1)
2.	$y < 5x$	(2, 0)	(2, 3)	(3, 12)	(0, 0)
3.	$y < 1 - 3x$	(0, 0)	(1, 1)	(1, −1)	(−1, 3)
4.	$x < 2 + y$	(0, 2)	(−1, 0)	(−2, −2)	(2, 0)
5.	$y > x$	(3, 5)	(5, 3)	(−3, −5)	(−5, −3)
6.	$x > 1 - y$	(2, 0)	(0, 2)	(−1, 2)	(−1, 3)
7.	$y > 2x + 3$	(0, 4)	(1, 6)	(−2, 0)	(3, 9)

8. $y > 5 - 3x$ **(a)** $(1, 2)$ **(b)** $\left(\frac{1}{3}, 3\right)$ **(c)** $\left(\frac{1}{2}, 4\right)$ **(d)** $(2, 0)$

9. $y \leqslant 4x$ **(a)** $(1, 4)$ **(b)** $(0, 0)$ **(c)** $(-1, -4)$ **(d)** $(1, -4)$

10. $y \leqslant 2x + 1$ **(a)** $(1, 1)$ **(b)** $(2, 5)$ **(c)** $(-1, -1)$ **(d)** $(0, 1)$

11. $y \leqslant \frac{x + 1}{2}$ **(a)** $\left(0, \frac{1}{2}\right)$ **(b)** $(1, 2)$ **(c)** $(1, 1)$ **(d)** $(2, 2)$

12. $y \leqslant x + \frac{1}{2}$ **(a)** $(1, 1)$ **(b)** $(1, 2)$ **(c)** $\left(1, \frac{3}{2}\right)$ **(d)** $(-1, 0)$

13. $y \geqslant x + 1$ **(a)** $(5, 6)$ **(b)** $\left(0, \frac{1}{2}\right)$ **(c)** $(3, 3.9)$ **(d)** $(-4, -3)$

14. $y \geqslant 2$ **(a)** $(2, 0)$ **(b)** $(3, 1)$ **(c)** $(2, 2)$ **(d)** $(2, 3)$

15. $2y \geqslant x$ **(a)** $(1, 1)$ **(b)** $(1, 2)$ **(c)** $(2, 1)$ **(d)** $(2, 2)$

16. $y + 1 \geqslant 2x$ **(a)** $(1, 1)$ **(b)** $(1, 2)$ **(c)** $(2, 1)$ **(d)** $(2, 3)$

17. $x - 2y < 5$ **(a)** $(0, 0)$ **(b)** $(2, 1)$ **(c)** $(3, -1)$ **(d)** $(5, 1)$

18. $2x + 3y \leqslant 1$ **(a)** $(1, 1)$ **(b)** $\left(0, \frac{1}{3}\right)$ **(c)** $\left(\frac{1}{2}, 0\right)$ **(d)** $(-1, 1)$

19. $3x - 7y + 1 \leqslant 0$ **(a)** $(0, 0)$ **(b)** $(3, 7)$ **(c)** $(7, 3)$ **(d)** $(2, 1)$

20. $2x - 5 > 3y$ **(a)** $(1, 1)$ **(b)** $(0, 0)$ **(c)** $(1, 2)$ **(d)** $(4, 1)$

In Exercises 21–36 graph each inequality.

21. $y < 2x$ **22.** $y < x + 1$ **23.** $y < 2x - 1$ **24.** $y < 1 - 2x$ **25.** $y > x$

26. $y > x - 2$ **27.** $y > \frac{x}{2}$ **28.** $2x < 1$ **29.** $y \leqslant 3$ **30.** $y \leqslant 2x + 1$

31. $y \leqslant 4 - 2x$ **32.** $y \geqslant -3$ **33.** $y \geqslant x - 4$ **34.** $x \geqslant 10$ **35.** $2x + y < 1$

36. $x - 4y - 1 \geqslant 0$

4.6 Direct and Inverse Variation

There are many situations in science and everyday life in which two variables, x and y, change, yet the quotient $\frac{y}{x}$ (or product xy) remains constant.

direct variation

If a rectangular piece of material is cut from a roll of silk of fixed width w, the area A of the piece depends on its length l. Thus in Figure 4.50 on page 170,

$$A_1 = l_1 w \quad \text{or} \quad \frac{A_1}{l_1} = w$$

and

$$A_2 = l_2 w \quad \text{or} \quad \frac{A_2}{l_2} = w$$

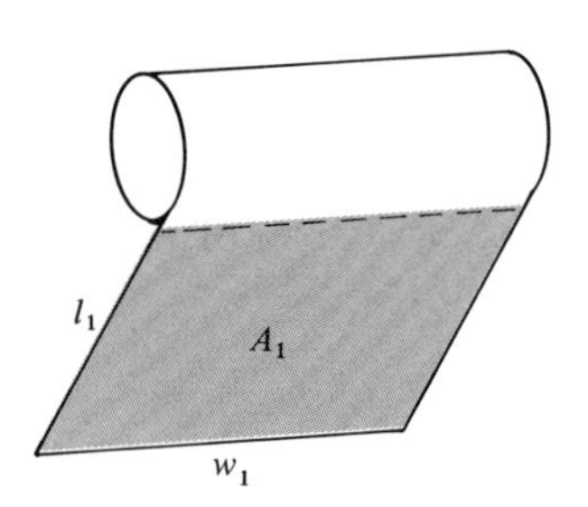

Therefore

$$\frac{A_1}{l_1} = \frac{A_2}{l_2} = w$$

No matter where the material is cut,

$$\frac{\text{area}}{\text{length}} \text{ equals (the constant) width.}$$

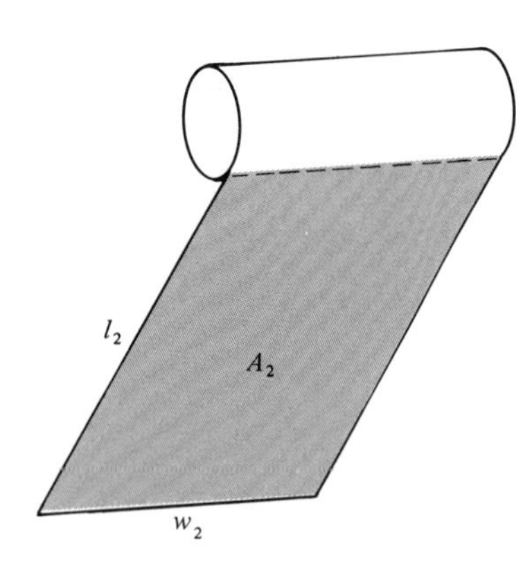

FIGURE 4.50

Definition. Let x and y be variables. Then y **varies directly as** x if

$$y = kx$$

for some constant k ($\neq 0$). In variational formulas, k is called the **constant of variation.**

Note that if y varies directly as x, then x also varies directly as y. In fact,

$$x = \frac{1}{k}y,$$

and $\frac{1}{k}$ ($\neq 0$) is the constant of variation. Also,

$$\frac{y}{x} = k,$$

when $x \neq 0$. Thus *if y varies directly as x, the quotient $\frac{y}{x}$ remains constant.* Therefore, if x_1 and x_2 are (nonzero) values of x and if y_1 and y_2 are the corresponding values of y, then

$$\frac{y_1}{x_1} = \frac{y_2}{x_2} = k.$$

Corresponding values of y and x are proportional.

EXAMPLE 1 Suppose $y = 2x$. Then y varies directly as x. The constant of variation is 2. As x varies over nonzero numbers, the corresponding values of y and x are proportional and the quotient $\frac{y}{x}$ remains 2. A few of these corresponding values are given in Table 4.1 . Fill in the blanks.

TABLE 4.1

x	$y = 2x$
1	2
2	4
$\frac{5}{2}$	5
4.1	
	14

Solution. When $x = 4.1$,

$$y = 2(4.1) = 8.2$$

When $y = 14$,

$$14 = 2x \qquad \text{and thus} \qquad x = 7$$

□

When y varies directly as x, a linear function

$$y = kx$$

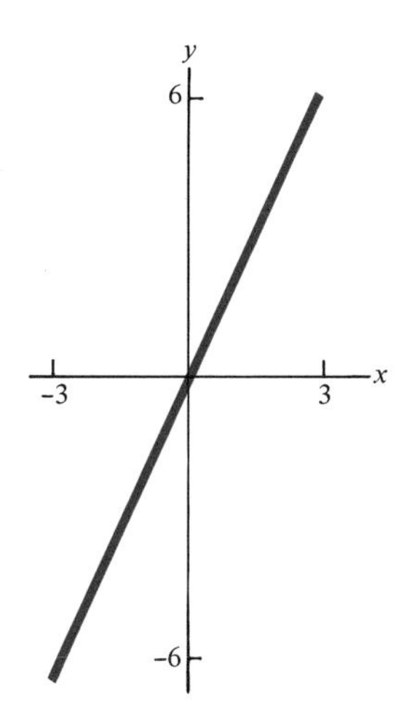

FIGURE 4.51 The graph of the function $y = 2x$.

is defined. The graph of this function is the line through the origin with slope k. In Figure 4.51 the function described in Example 1 (with $k = 2$) is graphed.

Sometimes y varies directly, not as x, but as a power of x.

Definition. Let x and y be variables and let n be a positive integer. Then y **varies directly as the nth power of** x if

$$y = kx^n$$

for some nonzero constant k.

EXAMPLE 2 Suppose y varies directly as x^2. Suppose, further, that $y = 12$ when $x = 2$. Find y when $x = 3$.

Solution.

$$y = kx^2$$

Substitute **12** for y and 2 for x.

$$\mathbf{12} = k(2^2)$$

$$3 = k$$

The constant of variation is 3, and

$$y = 3x^2.$$

Let $x = 3$.

$$y = 3(3^2) = 3 \cdot 9 = 27$$

□

EXAMPLE 3 The cost of building an expansion bridge varies as the cube of its length. If a 100-foot expansion bridge costs \$40 000 to build, how much does a 500-foot bridge cost?

Solution. Let c be the cost (in dollars) and let l be the length of the bridge (in feet). Then

$$c = kl^3$$

When $l = 100$, $c = \mathbf{40\ 000}$.

$$\mathbf{40\ 000} = k(100^3)$$

$$40\ 000 = 1\ 000\ 000k$$

$$k = \frac{4}{100} = \frac{1}{25}$$

Now let $l = 500$ in the formula

$$c = \frac{l^3}{25}.$$

Then

$$c = \frac{500^3}{25} = \frac{125\,000\,000}{25} = 5\,000\,000$$

The cost of a 500-foot bridge is \$5 000 000. □

inverse variation

Definition. Let x and y be variables. Then y **varies inversely as** x if

$$y = \frac{k}{x}$$

for some nonzero constant k.

Note that if y varies inversely as x, then, by cross-multiplication,

$$xy = k,$$

so that *the product of x and y remains constant.* Furthermore,

$$x = \frac{k}{y},$$

so that *x varies inversely as y* (with the same constant of variation k).

EXAMPLE 4 Suppose $y = \dfrac{12}{x}$.

(a) Determine y when $x = 2$ and when $x = 6$.
(b) Determine x when $y = 4$ and when $y = 9$.

Solution. y varies inversely as x. The constant of variation is 12.

(a) When $x = 2$,

$$y = \frac{12}{2} = 6$$

When $x = 6$,

$$y = \frac{12}{6} = 2$$

(b) When $y = 4$,

$$4 = \frac{12}{x}$$

$$x = \frac{12}{4} = 3$$

When $y = 9$,

$$9 = \frac{12}{x}$$

$$x = \frac{12}{9} = \frac{4}{3}$$ □

The graph of

$$y = \frac{12}{x} \quad \text{or} \quad xy = 12$$

is called a **hyperbola** and consists of two curves, known as **branches.** On each branch, as a point moves further away from the origin, it gets closer and closer to one of the coordinate axes. [See Figure 4.52 .]

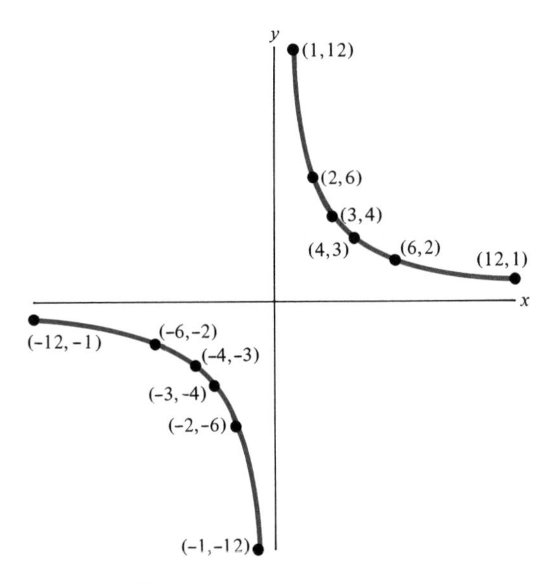

FIGURE 4.52

In general, when y varies inversely as x, a function is defined by

$$y = \frac{k}{x}.$$

The graph of this function is a hyperbola.

EXAMPLE 5 Boyle's Law states that at constant temperature, the pressure p of a compressed gas varies inversely as the volume v of gas. [See Figure 4.53 .] Suppose the pressure is 50 pounds per square inch when the volume is 200 cubic inches. Determine the pressure when the gas is compressed to 125 cubic inches.

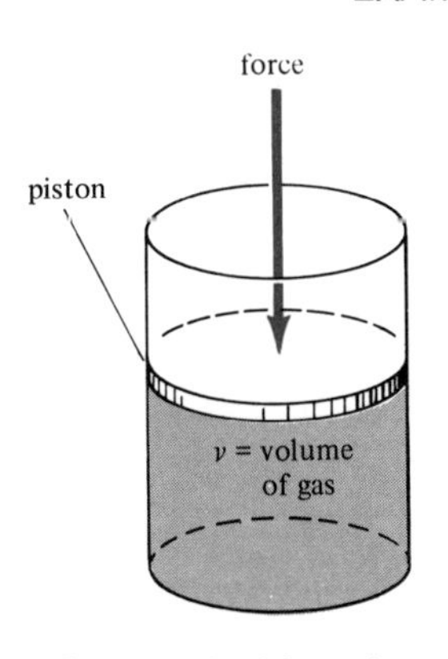

FIGURE 4.53 A force is exerted on a piston, producing a pressure p, where $p = \frac{k}{v}$, $k\ (\neq 0)$ constant.

Solution. p varies inversely as v. Thus

$$p = \frac{k}{v}$$

When $p = 50$, $v = 200$:

$$50 = \frac{k}{200}$$

$$10\,000 = k$$

Therefore

$$p = \frac{10\,000}{v}$$

Let $v = 125$.

$$p = \frac{10\,000}{125} = 80$$

When the volume of gas is 125 cubic inches, the pressure is 80 pounds per square inch. □

Just as y can vary *directly* as a *power of* x, so too, y can vary *inversely* as a *power of* x.

Definition. Let x and y be variables and let n be a positive integer. Then y **varies inversely as the nth power of** x if

$$y = \frac{k}{x^n}$$

for some constant $k \neq 0$.

EXAMPLE 6 Suppose $y = \frac{72}{x^2}$, $x \neq 0$. Determine y when:

(a) $x = 2$ (b) $x = -3$

Solution. Here y varies inversely as x^2. The constant of variation is 72.

(a) Let $x = 2$.

$$y = \frac{72}{2^2} = \frac{72}{4} = 18$$

(b) Let $x = -3$.

$$y = \frac{72}{(-3)^2} = \frac{72}{9} = 8$$ □

Inverse square variation, as in Example 6, occurs widely in physical applications, such as in laws of gravitational, electrical, and magnetic attraction, and of light and sound intensity.

EXERCISES

In Exercises 1–4 assume y varies directly as x. Determine the constant of variation in each case.

1. $y = 6$ when $x = 2$ **2.** $y = -6$ when $x = 3$ **3.** $y = -1$ when $x = -3$ **4.** $y = 9$ when $x = 4$

In Exercises 5–8 assume y varies directly as x.

5. Suppose $y = 8$ when $x = 4$. Find y when $x = 6$.

6. Suppose $y = 5$ when $x = 2$. Find y when $x = \frac{1}{10}$.

7. Suppose $y = -2$ when $x = 1$. Find x when $y = 1$.

8. Suppose $y = 5$ when $x = 3$. Find x when $y = 3$.

In Exercises 9–12 assume y varies directly as x^2. Determine the constant of variation in each case.

9. $y = 4$ when $x = 2$ **10.** $y = 18$ when $x = 3$ **11.** $y = 24$ when $x = -2$ **12.** $y = 12$ when $x = \frac{3}{2}$

In Exercises 13–16 assume y varies directly as x^2.

13. Suppose $y = 12$ when $x = 2$. Find y when $x = 4$.

14. Suppose $y = 45$ when $x = 3$. Find y when $x = 1$.

15. Suppose $y = 2$ when $x = 4$. Find y when $x = 1$.

16. Suppose $y = 1$ when $x = -1$. Find y when $x = 2$.

17. Suppose y varies directly as x^3 and suppose $y = 54$ when $x = 3$. What is the constant of variation?

18. Suppose y varies directly as x^4. If $y = 8$ when $x = 2$, find y when $x = 1$.

19. Suppose y varies directly as x, and $y = 6$ when x is 2.
(a) What function is defined? **(b)** Graph this function.

20. Suppose y varies directly as x, and $y = -5$ when $x = 5$.
(a) What function is defined? **(b)** Graph this function.

21. Let y vary directly as x. Find y when $x = 0$.

22. Fill in:
(a) The ____________________ of a circle varies directly as the radius length.
(b) The ____________________ of a circle varies directly as the radius length squared.

23. The amount of bread consumed in a town varies directly as the population. If 80 000 pounds per week are consumed in a town of 20 000 inhabitants, how many pounds of bread per week are consumed in a town of 15 000 inhabitants?

24. At a constant rate the distance traveled varies directly as time. A car travels 110 miles in 2 hours. At this rate, how far does it travel in 3½ hours?

25. A typist finds that the (approximate) number of errors she makes varies as the square of her typing speed. If she makes 2 errors per page at 40 words per minute, how many errors per page does she make at 80 words per minute?

26. According to Hooke's Law, the distance a vertical spring stretches varies directly with the weight placed on it. If a 50-kilogram weight stretches a vertical spring 4 centimeters, how large a weight must be used to stretch the spring 14 centimeters?

27. The distance a ball rolls down an inclined plane varies directly as the square of its time in motion. If the ball rolls 3 meters during the first 2 seconds, how much *further* will it roll during the *next* 2 seconds?

28. The base of a statue is uniformly dense and is cubic in shape. Suppose its side length is doubled.
(a) By what factor does its weight increase?
(b) Assuming the bottom is not painted, by what factor does the cost of painting the base increase?

In Exercises 29–32 assume that y varies inversely as x. Determine the constant of variation in each case.

29. $y = 4$ when $x = 2$ **30.** $y = 9$ when $x = -4$ **31.** $y = \frac{1}{2}$ when $x = \frac{1}{3}$ **32.** $y = \frac{1}{20}$ when $x = 10$

In Exercises 33–36 assume that y varies inversely as x.

33. Suppose $y = 3$ when $x = 2$. Find y when $x = 3$.

34. Suppose $y = -20$ when $x = 2$. Find y when $x = -4$.

35. Suppose $y = 9$ when $x = 4$. Find x when $y = 12$.

36. Suppose $y = 10$ when $x = 7$. Find x when $y = \frac{1}{2}$.

In Exercises 37–40 assume that y varies inversely as x^2. Determine the constant of variation in each case.

37. $y = 5$ when $x = 2$ **38.** $y = 4$ when $x = 3$ **39.** $y = -2$ when $x = 5$ **40.** $y = \frac{1}{5}$ when $x = 10$

In Exercises 41–44 assume that y varies inversely as x^2.

41. Suppose $y = 4$ when $x = 4$. Find y when $x = 8$.

42. Suppose $y = 10$ when $x = 2$. Find y when $x = 1$.

43. Suppose $y = 4$ when $x = 2$. Find y when $x = 8$.

44. Suppose $y = 1$ when $x = 1$. Find y when $x = 3$.

45. Suppose y varies inversely as x, and $y = 3$ when $x = 2$.
(a) What function is defined? **(b)** Graph this function. **(c)** What figure is represented?

46. Suppose y varies inversely as x, and $y = \frac{1}{2}$ when $x = 2$.
(a) What function is defined? **(b)** Graph this function.

47. Let y vary inversely as x. Can $y = 0$?

In Exercises 48–51, fill in "directly" or "inversely".

48. When length is held constant, the area of a rectangle varies ________ as the width.

49. When width is held constant, the length of a rectangle varies ________ as the area.

50. When area is held constant, the length of a rectangle varies ________ as the width.

51. If all seats are $2.50, the gross receipts of a movie theatre vary ________ as the number of customers.

52. Suppose the demand for an art book varies inversely as the price per book. If 6000 volumes are sold at $24 per book, how many would have been sold at $18 per book?

53. The beetle population of a field varies inversely as the amount of pesticide used. If 1000 beetles are in the field when 40 pounds of pesticide are used, how many pounds of pesticide are needed to bring the bettle population down to 100?

54. The weight of a body varies inversely as its distance from the center of the earth. Assume the length of the earth's radius is 4000 miles. If a man weighs 150 pounds on earth, how much does he weigh 1000 miles above the surface of the earth?

55. The resistance of an electrical wire of fixed length varies inversely as the square of its thickness. When the wire is .005 inch thick the resistance is 20 ohms. What is the resistance of a wire that is .01 inch thick?

*4.7 Joint Variation

Definition. Let x, y, z, and t be variables. Then t **varies jointly as** x **and** y if

$$t = kxy$$

*Optional topic

for some constant k ($\neq 0$). Similarly, t **varies jointly as** x, y, **and** z if

$$t = kxyz$$

for some constant k ($\neq 0$).

EXAMPLE 1 Assume t varies jointly as x and y. Suppose $t = 500$ when $x = 5$ and $y = 20$. Find t when $x = 10$ and $y = 15$.

Solution.

$$t = kxy$$

Let $t = 500$, $x = 5$, $y = 20$. Then

$$500 = k \cdot 5 \cdot 20$$
$$5 = k$$

Thus

$$t = 5xy$$

When $x = 10$ and $y = 15$,

$$t = 5 \cdot 10 \cdot 15 = 750$$

□

Note that t can vary jointly as various combinations of *powers* of x, y, and z. For example, t **varies jointly as** x **and the square of** y if

$$t = kxy^2$$

for some constant k ($\neq 0$). Similarly, t **varies jointly as the square of** y **and the cube of** z if

$$t = ky^2z^3$$

for some $k \neq 0$.

EXAMPLE 2 Assume t varies jointly as x, y, and z^2. If $t = 450$ when $x = 2$, $y = 3$, and $z = 5$, find t when $x = 4$, $y = 1$, and $z = 10$.

Solution.

$$t = kxyz^2$$

Let $t = 450$, $x = 2$, $y = 3$, and $z = 5$. Then

$$450 = k \cdot 2 \cdot 3 \cdot 5^2$$
$$3 = k$$

Therefore

$$t = 3xyz^2$$

When $x = 4$, $y = 1$, and $z = 10$, then

$$t = 3 \cdot 4 \cdot 1 \cdot 10^2 = 1200$$

□

Definition. Let x, y, and t be variables. Then t **varies inversely as** x **and** y if

$$t = \frac{k}{xy}$$

for some constant k ($\neq 0$).

(The word "jointly" is not usually used in the case of inverse variation.)

EXAMPLE 3 Assume t varies inversely as x and y. If $t = 6$ when $x = 2$ and $y = 5$, find x when $t = 15$ and $y = 2$.

Solution.

$$t = \frac{k}{xy}$$

Let $t = 6$, $x = 2$, and $y = 5$. Then

$$6 = \frac{k}{2 \cdot 5}$$

$$60 = k$$

$$t = \frac{60}{xy}$$

Let $t = 15$ and $y = 2$. Then

$$15 = \frac{60}{2x}$$

$$x = \frac{60}{2 \cdot 15} = 2$$

□

Other possibilities for inverse variation can occur. For example, t **varies inversely as** x, y, **and the cube of** z, if

$$t = \frac{k}{xyz^3}$$

for some constant k ($\neq 0$).

Finally combinations of direct and inverse variations can occur. For example, t **varies directly as** x **and inversely as** y if

$$t = \frac{kx}{y}$$

for some constant k ($\neq 0$). Also, t **varies jointly as** x **and the square of** y **and inversely as** z if

$$t = \frac{kxy^2}{z}$$

for some constant k ($\neq 0$).

EXAMPLE 4 The resistance of an electrical wire varies directly as the length of the wire and inversely as the square of its thickness. The resistance is 40 ohms when the wire is 2000 feet long and .1 inch thick. Determine the resistance when the wire is 5000 feet long and .2 inch thick.

Solution. Let R = resistance, l = length of the wire (in inches), and t = thickness (in inches). Then

$$R = \frac{kl}{t^2}$$

Also, $R = 40$ when $l = 2000 \times 12 = 24\,000$ (inches) and $t = .1$. Therefore

$$40 = \frac{k \cdot 24\,000}{(.1)^2}$$

$$40(.01) = 24\,000k$$

$$k = \frac{.4}{24\,000} = \frac{1}{60\,000}$$

Thus

$$R = \frac{l}{60\,000t^2}$$

When $l = 5000 \times 12 = 60\,000$ and $t = .2$,

$$R = \frac{60\,000}{60\,000\,(.2)^2} = \frac{1}{.04} = \frac{100}{4} = 25$$

The resistance is 25 ohms. □

EXERCISES

In Exercises 1–4 assume t varies jointly as x and y. Determine the constant of variation in each case.

1. $t = 24$ when $x = 2$ and $y = 3$

2. $t = -100$ when $x = -2$ and $y = -20$

3. $t = 1$ when $x = 4$ and $y = \frac{1}{2}$

4. $t = 3$ when $x = 2$ and $y = 5$

In Exercises 5–8 assume t varies jointly as x and y.

5. Suppose $t = 18$ when $x = 2$ and $y = 3$. Find t when $x = 3$ and $y = 4$.

6. Suppose $t = 30$ when $x = 5$ and $y = 2$. Find y when $t = 45$ and $x = 3$.

7. Suppose $t = 12$ when $x = 8$ and $y = 9$. Find y when $t = 2$ and $x = 3$.

8. Suppose $t = .1$ when $x = .02$ and $y = .5$. Find t when $x = .4$ and $y = .05$.

9. Suppose t varies jointly as x, y, and z. If $t = 16$ when $x = 2$, $y = 1$, and $z = 2$, find t when $x = 4$, $y = 3$, and $z = 3$.

10. Suppose t varies jointly as u, v, and w. If $t = 20$ when $u = \frac{1}{2}$, $v = 4$, and $w = 5$, find u when $t = 40$, $v = 10$, and $w = 4$.

11. Suppose t varies jointly as x and y^2. If $t = 96$ when $x = 3$ and $y = 4$, find t when $x = 5$ and $y = -3$.

12. Suppose t varies jointly as x and y^3. If $t = 270$ when $x = 5$ and $y = 3$, find t when $x = 3$ and $y = 2$.

13. Suppose t varies jointly as x, y, and z^2. If $t = 400$ when $x = 5$, $y = 2$, and $z = 4$, find t when $x = -1$, $y = -4$, and $z = 5$.

14. Suppose t varies jointly as x^2, y^3, and z^4. If $t = 96$ when $x = \frac{1}{2}$, $y = 2$, and $z = -2$, find t when $x = \frac{1}{3}$, $y = 3$, and $z = -3$.

In Exercises 15–18 assume t varies inversely as x and y. Determine the constant of variation in each case.

15. $t = 5, x = 2, y = 2$ **16.** $t = 4, x = 1, y = 3$ **17.** $t = 6, x = -2, y = \frac{1}{2}$ **18.** $t = \frac{2}{3}, x = \frac{1}{4}, y = 9$

In Exercises 19–22 assume t varies inversely as x and y.

19. If $t = 12$ when $x = 1$ and $y = 3$, find t when $x = 2$ and $y = 6$.

20. If $t = -2$ when $x = -5$ and $y = -3$, find t when $x = 6$ and $y = \frac{1}{2}$.

21. If $t = \frac{1}{2}$ when $x = \frac{1}{2}$ and $y = \frac{1}{2}$, find x when $t = 1$ and $y = \frac{1}{4}$.

22. If $t = 50$ when $x = \frac{1}{2}$ and $y = \frac{1}{5}$, find y when $t = 2$ and $x = 5$.

23. Suppose t varies inversely as x, y, and z. If $t = 4$ when $x = 2$, $y = 6$, and $z = 3$, find t when $x = 9$, $y = 2$, and $z = \frac{4}{3}$.

24. Suppose t varies inversely as x, y, and z^3. If $t = 6$ when $x = 2$, $y = -1$, and $z = 1$, find t when $x = 4$, $y = 2$, and $z = -2$.

25. Suppose t varies directly as x and inversely as y. If $t = 40$ when $x = 4$ and $y = 3$, find t when $x = 6$ and $y = 2$.

26. Suppose t varies directly as x and inversely as y. If $t = 100$ when $x = 5$ and $y = 2$, find x when $t = 20$ and $y = 4$.

27. Suppose t varies jointly as x and y and inversely as z. If $t = 10$ when $x = 5$, $y = 4$, and $z = 8$, find t when $x = 20$, $y = 2$, and $z = 5$.

28. Suppose t varies jointly as x and y and inversely as z^2. If $t = 100$ when $x = 5$, $y = 10$, and $z = 3$, find t when $x = 9$, $y = 4$, and $z = 6$.

29. Suppose t varies jointly as x^2 and y and inversely as z. If $t = 10$ when $x = 2$, $y = 5$, and $z = 8$, find t when $x = 4$, $y = 2$, and $z = 16$.

30. Suppose t varies directly as x^3 and inversely as y and z^2. If $t = 12$ when $x = 2$, $y = 4$, and $z = 3$, find t when $x = 3$, $y = 18$, and $z = 3$.

31. The weight of a (right-circular-) cylindrical can of peas varies jointly as the height and the square of the length of the base radius. The weight is 250 grams when the height is 20 centimeters and the base radius measures 5 centimeters. Find the height when the weight is 960 grams and the base radius measures 8 centimeters.

32. Several particles rotate about an axis, each with constant velocity. The kinetic energy of each particle varies jointly as its mass and the square of its velocity. One particle has mass 5 kilograms, velocity 12 meters per second, and kinetic energy 30 joules. Find the kinetic energy of a second particle with mass 30 kilograms and velocity 10 meters per second.

33. The wind force on a vertical sail varies jointly as the area of the sail and the square of the wind velocity. At noon the wind is blowing at 12 kilometers per hour on a small craft whose triangular sail has base length 3 meters and altitude length 5 meters. Later in the day, the wind is blowing at 18 kilometers per hour on a larger craft whose trapezoidal sail has base lengths 6 meters and 14 meters and altitude length 9 meters. By what factor is the wind force greater on the larger craft?

34. The stiffness of a homogeneous rectangular beam varies jointly as the square of its width and the cube of its depth. Two such beams are constructed from the same material. The second beam is twice as deep, but only half as wide, as the first beam. How many times as stiff as the first beam is the second?

35. The safe-load of a homogeneous rectangular beam varies jointly as its width and the square of its depth, and inversely as the length between supports. For a beam of width 8 inches, of depth 5 inches, and of length 20 feet between supports, the safe-load is 3000 pounds. What is the safe-load of a beam that is of width 10 inches, of depth 10 inches, and of length 30 feet between supports?

36. Let d be the distance between two particles of masses m and M. Gravitational attraction varies jointly as m and M and inversely as d^2. A 10-pound mass and a 5-pound mass that are 100 feet apart attract each other with a force F. Determine the gravitational attraction for a 40-pound mass and a 15-pound mass that are 200 feet apart. (Your answer should be a multiple of F.)

37. The maximum load of a (right-circular-) cylindrical pillar varies directly as the fourth power of the length of its diameter and inversely as the square of its height. If a pillar 2 meters in diameter and 12 meters high can support 15 000 kilograms, what weight can be supported by a pillar (of the same material) 4 meters in diameter and 8 meters high?

38. The illumination from a source of light varies directly as the intensity of the light and inversely as the square of the distance from the light. A woman is standing 2 meters from a street lamp. On the other side of town a man is standing 3 meters from a street lamp that is twice as intense. Who has the better illumination, and by what factor?

39. According to Coulomb's Law, the force between two small electrical charges varies directly as the product of the charges and inversely as the square of the distance between them. The distance between the charges is doubled and one of the charges is tripled. By what factor must the other charge be increased in order to maintain the original force between them?

Review Exercises for Chapter 4

1. Find the values of a, b, and c in the following proportions:

(a) $\frac{a}{3} = \frac{3}{9}$ **(b)** $\frac{5}{b} = \frac{-10}{14}$ **(c)** $\frac{\frac{1}{3}}{\frac{2}{9}} = \frac{c}{4}$

2. Lemons sell for 93 cents a dozen. How much do 8 lemons cost?

3. The triangles in Figure 4.54 are similar. Corresponding sides are marked with the same number of bars. Determine the lengths of sides a and b.

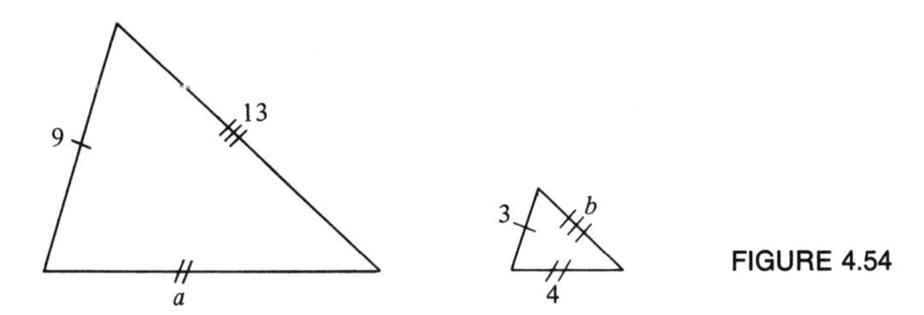

FIGURE 4.54

4. Determine the slope of the line through $(-2, 1)$ and $(2, -2)$.

5. **(a)** Determine the equation of the line that has slope 2 and that passes through the point $(3, -1)$.

(b) Graph this line.

6. Which of the following lines are horizontal? Which are vertical?

(a) $y = -1$ **(b)** $x = 2$ **(c)** $y = x$

7. Determine (a) the x-intercept and (b) the y-intercept of the line with equation $y + 2 = 2(x + 2)$.

8. Determine the slope-intercept form of the equation of the line that passes through (2, 5) and (5, 2).

9. Determine the equation of the line that has x-intercept 2 and y-intercept -4.

10. Determine the equations of the lines that pass through (2, 5) and are (a) parallel to, (b) perpendicular to the line with equation $y = 2x - 1$.

11. Determine all pairs of (a) parallel lines, (b) perpendicular lines:

$$L_1\colon y = 2x \qquad L_2\colon y = \frac{-x}{2} \qquad L_3\colon y = 4 - 2x \qquad L_4\colon y = 1 - \frac{x}{2} \qquad L_5\colon y = 1 + \frac{x}{2}$$

12. Indicate which points satisfy the inequality $2x - y \geq 1$.

(a) (1, 1) (b) (2, 2) (c) (0, 0) (d) (4, 6)

13. Graph the inequality $x - y \leq 1$.

14. Suppose y varies directly as x, and $y = 12$ when $x = -3$. Find y when $x = 6$.

15. Suppose y varies directly as x^2, and $y = 20$ when $x = 2$. Find y when $x = 3$.

16. Suppose y varies inversely as x, and $y = 2$ when $x = 3$. Find y when $x = -6$.

17. Suppose y varies inversely as x^2, and $y = 2$ when $x = 4$. Find x when $y = 8$.

18. Suppose t varies jointly as x and y and inversely as z. If $t = 30$ when $x = 10$, $y = 3$, and $z = 2$, find t when $x = 2$, $y = 5$, and $z = 1$.

19. A manufacturer constructs rectangular boxes of the same height. The cost of construction varies jointly as the length and width. If the cost is 12 cents when the length is 8 inches and the width is 3 inches, what is the cost when the length is 1 foot and the width is 10 inches?

Exponents, Roots, and Radicals

5

5.1 Zero and Negative Exponents

You are already familiar with *positive integral exponents*. Thus in the expression

a^7 (the 7th *power* of a),

a is the *base* and 7 the *exponent*. It will be useful to define *zero and negative integral exponents,* as well.

Recall that

$$2^4 = 2 \times 2 \times 2 \times 2 = 16,$$
$$2^5 = 2 \times 2 \times 2 \times 2 \times 2 = 32.$$

TABLE 5.1

$2^5 =$	32
$2^4 =$	16
$2^3 =$	8
$2^2 =$	4
$2^1 =$	2
$2^0 =$	
2^{-1} –	
$2^{-2} =$	
$2^{-3} =$	
$2^{-4} =$	
$2^{-5} =$	

In order to see how both zero and negative exponents should be defined, fill in the blank spaces in the right-hand columns of Tables 5.1 and 5.2 by observing the patterns:

In Table 5.1 observe that every time the exponent is decreased by 1, the number in the right-hand column is *divided by* 2. And in Table 5.2, every time the exponent is decreased by 1, the number in the right-hand column is *divided by* 10. It seems natural to complete these tables as follows:

$$2^1 = 2 \qquad 10^1 = 10$$
$$2^0 = 1 \qquad 10^0 = 1$$
$$2^{-1} = \frac{1}{2} = \frac{1}{2^1} \qquad 10^{-1} = \frac{1}{10} = \frac{1}{10^1}$$
$$2^{-2} = \frac{1}{4} = \frac{1}{2^2} \qquad 10^{-2} = \frac{1}{100} = \frac{1}{10^2}$$
$$2^{-3} = \frac{1}{8} = \frac{1}{2^3} \qquad 10^{-3} = \frac{1}{1000} = \frac{1}{10^3}$$
$$2^{-4} = \frac{1}{16} = \frac{1}{2^4} \qquad 10^{-4} = \frac{1}{10\,000} = \frac{1}{10^4}$$
$$2^{-5} = \frac{1}{32} = \frac{1}{2^5} \qquad 10^{-5} = \frac{1}{100\,000} = \frac{1}{10^5}$$

These patterns suggest the following definition.

TABLE 5.2

10^5 =	100 000
10^4 =	10 000
10^3 =	1000
10^2 =	100
10^1 =	10
10^0 =	
10^{-1} =	
10^{-2} =	
10^{-3} =	
10^{-4} =	
10^{-5} =	

Definition. Let $a \neq 0$ and let m be a positive integer. (Then $-m$ is a negative integer.) Define

$$a^0 = 1 \qquad \text{and} \qquad a^{-m} = \frac{1}{a^m}.$$

Thus any *nonzero* number to the 0th power equals 1. And any *nonzero* number to a *negative* (integral) power equals the *reciprocal* of the corresponding *positive* (integral) power.

As you learned in Section 1.6, for $a \neq 0$,

$$\frac{a^5}{a^2} = \frac{\not{a} \cdot \not{a} \cdot a \cdot a \cdot a}{\not{a} \cdot \not{a}} = a^3,$$

$$\frac{a^2}{a^2} = \frac{\not{a} \cdot \not{a}}{\not{a} \cdot \not{a}} = 1,$$

and

$$\frac{a^2}{a^5} = \frac{\not{a} \cdot \not{a}}{\not{a} \cdot \not{a} \cdot a \cdot a \cdot a} = \frac{1}{a^3}$$

You also learned that *when m and n are positive integers and $m > n$, then*

$$\frac{a^m}{a^n} = a^{m-n}.$$

Thus

$$\frac{a^5}{a^2} = a^3 = a^{5-2}$$

Now observe that with this definition of zero and negative exponents,

$$\frac{a^m}{a^n} = a^{m-n} \qquad \text{for } a \neq 0 \text{ and } \textit{for } \underline{\textit{all}} \textit{ integers } m \textit{ and } n$$

(positive, negative, or zero). For example,

$$\frac{a^2}{a^2} = 1 = a^0 = a^{2-2} \qquad \text{and} \qquad \frac{a^2}{a^5} = \frac{1}{a^3} = a^{-3} = a^{2-5}$$

Also,

$$\frac{a^{-3}}{a^2} = \frac{\frac{1}{a^3}}{a^2} = \frac{1}{a^3} \cdot \frac{1}{a^2} = \frac{1}{a^5} = a^{-5}$$

Thus

$$\frac{a^{-3}}{a^2} = a^{-3-2}$$

Similarly,

$$\frac{a^{-3}}{a^{-2}} = \frac{\frac{1}{a^3}}{\frac{1}{a^2}} = \frac{1}{a^3} \cdot \frac{a^2}{1} = \frac{1}{a} = a^{-1}$$

Thus

$$\frac{a^{-3}}{a^{-2}} = a^{-3-(-2)}$$

Observe that 0^m is defined only for positive m. To see why negative m are excluded, let $m = -2$. The definition would yield $0^{-2} = \frac{1}{0^2}$, which is undefined. A similar argument shows why 0^m is not defined for any negative integer m. For $m = 0$, recall that the definition of a^0 was motivated by considering quotients such as $\frac{a^2}{a^2}$; but if $a = 0$, this quotient would have the form $\frac{0}{0}$, which is not defined. Thus 0^0 is also not defined.

Here are some examples illustrating these definitions.

EXAMPLE 1 (a) $3^0 = 1$ (b) $3^{-1} = \frac{1}{3}$ (c) $3^{-2} = \frac{1}{3^2} = \frac{1}{9}$

(d) $3^{-3} = \frac{1}{3^3} = \frac{1}{27}$ (e) $3^{-4} = \frac{1}{3^4} = \frac{1}{81}$ □

The base a can be a fraction.

EXAMPLE 2 (a) $\left(\frac{2}{7}\right)^{-1} = \frac{1}{\frac{2}{7}} = \frac{7}{2}$ (b) $\left(\frac{2}{7}\right)^{-2} = \frac{1}{\left(\frac{2}{7}\right)^2} = \frac{1}{\frac{4}{49}} = \frac{49}{4}$ □

Now, consider the law:

$$\frac{a^m}{a^n} = a^{m-n}, \qquad \text{for } a \neq 0, \qquad \text{for all integers } m \text{ and } n$$

EXAMPLE 3 (a) $\frac{5^2}{5^2} = 5^{2-2} = 5^0 = 1$ (b) $\frac{5^1}{5^3} = 5^{1-3} = 5^{-2} = \frac{1}{25}$ □

EXAMPLE 4 Let $a \neq 0$, $b \neq 0$.

(a) $\frac{a^4}{a^7} = a^{4-7} = a^{-3}$ (b) $\frac{a^0}{a^9} = a^{0-9} = a^{-9}$

(c) $\frac{a^3 \cdot b^5}{a^6 \cdot b^6} = \frac{a^3}{a^6} \cdot \frac{b^5}{b^6} = a^{3-6}b^{5-6} = a^{-3}b^{-1}$ □

Finally, note that

$$a^{-n} = \frac{1}{a^n}, \qquad \text{for } a \neq 0,$$

for every integer n—positive, negative, or zero. For example, if $n = -2$, then

$$a^{-(-2)} = a^2 \qquad \text{and} \qquad \frac{1}{a^{-2}} = \frac{1}{\frac{1}{a^2}} = 1 \cdot \frac{a^2}{1} = a^2$$

Therefore

$$a^{-(-2)} = \frac{1}{a^{-2}}$$

Also, for $a \neq 0$ and for all integers m and n,

$$\frac{1}{a^{-n}} = \frac{1}{\frac{1}{a^n}} = a^n$$

and

$$\frac{a^{-m}}{a^{-n}} = \frac{\frac{1}{a^m}}{\frac{1}{a^n}} = \frac{1}{a^m} \cdot \frac{a^n}{1} = \frac{a^n}{a^m}$$

Thus

$$\frac{a^{-m}}{a^{-n}} = \frac{a^n}{a^m}$$

Accordingly, if $a \neq 0$,

$$\frac{1}{a^{-4}} = a^4 \quad \text{and} \quad \frac{a^{-6}}{a^{-9}} = \frac{a^9}{a^6} = a^3$$

EXERCISES

In Exercises 1–48 assume $a \neq 0$, $b \neq 0$, $x \neq 0$, and $y \neq 0$.‡
In Exercises 1–26 evaluate each expression.

1. 10^{-1} **2.** 3^{-2} **3.** 6^0 **4.** 2^{-3} **5.** 2^{-5} **6.** 7^{-2} **7.** π^0

8. 8^{-2} **9.** 12^{-2} **10.** 3^{-3} **11.** 10^{-4} **12.** 10^{-6} **13.** $(-5)^{-1}$ **14.** $(-9)^0$

15. $(-2)^{-3}$ **16.** $(-5)^{-2}$ **17.** $\left(\frac{1}{2\pi}\right)^{-1}$ **18.** $\left(\frac{2}{3}\right)^{-2}$ **19.** $\left(\frac{2}{5}\right)^{-1}$ **20.** $\left(\frac{5}{2}\right)^{-2}$ **21.** $\left(\frac{4}{7}\right)^{0}$

22. $\frac{2^{-2}}{3^{-3}}$ **23.** $\frac{2^{-2}}{3}$ **24.** $\frac{(2\cdot 5)^{-1}}{3^{-2}}$ **25.** $(2^{-2})^{-3}$ **26.** $\frac{(2^4)^{-1}}{2^{-2}}$

In Exercises 27–38 determine the exponent m.

27. $\frac{1}{b^1} = b^m$ **28.** $\frac{1}{a^2} = a^m$ **29.** $\frac{1}{\pi} = \pi^m$ **30.** $\frac{1}{8} = 2^m$ **31.** $1 = 2^m$ **32.** $2 = \left(\frac{1}{2}\right)^m$

33. $\frac{1}{25} = 5^m$ **34.** $\frac{1}{27} = 3^m$ **35.** $\frac{3}{2} = \left(\frac{2}{3}\right)^m$ **36.** $\frac{4}{9} = \left(\frac{3}{2}\right)^m$ **37.** $\frac{1}{32x^5} = (2x)^m$ **38.** $\frac{2}{x^5} = 2x^m$

‡Strictly speaking, it should be stipulated that each denominator is also nonzero. The same remark applies to other Exercise Sets. Thus here, in Exercise 45, $x \neq -y$, and in Exercise 46, $x \neq y$

In Exercises 39–48 simplify and express your answer without writing a fraction. Use negative exponents, if necessary.

39. $\dfrac{a^4b^2}{a^6b}$ *Answer:* $a^{-2}b$

40. $\dfrac{x^5}{x^8}$ **41.** $\dfrac{a^7}{a^7}$ **42.** $\dfrac{a^3b^3}{a^6b^0}$ **43.** $\dfrac{a^5b^8}{a^7b^9}$ **44.** $\dfrac{a^2x^3y^4}{a^4x^3y^2}$

45. $\dfrac{(x+y)^4}{(x+y)^5}$ **46.** $\dfrac{a^3(x-y)^2}{a^6(x-y)^6}$ **47.** $\dfrac{1}{x^{-10}}$ **48.** $\dfrac{y^{-7}}{y^{-2}}$

49. Let $f(x) = x^{-1}$. Find: **(a)** $f(1)$ **(b)** $f(2)$ **(c)** $f\left(\frac{1}{2}\right)$ **(d)** $f(-10)$ **(e)** What is the largest possible domain of this function? **(f)** Is f even, odd, or neither? **(g)** Graph this function.

50. Let $g(x) = -2x^{-2}$. Find: **(a)** $g(1)$ **(b)** $g(4)$ **(c)** $g(-4)$ **(d)** $g\left(\frac{1}{4}\right)$ **(e)** What is the largest possible domain of this function? **(f)** Is g even, odd, or neither? **(g)** Graph this function.

51. Let $F(x) = 4x^0$. Find: **(a)** $F(1)$ **(b)** $F(-1)$ **(c)** $F(3)$ **(d)** $F\left(\frac{2}{3}\right)$ **(e)** What is the largest possible domain of this function? **(f)** Is F even, odd, or neither? **(g)** Graph this function.

52. Let $G(x) = (x+3)^{-1} - 3x^{-3}$. Find: **(a)** $G(1)$ **(b)** $G(-1)$ **(c)** $G(2)$ **(d)** $G(-2)$ **(e)** What is the largest possible domain of this function? **(f)** Is G even, odd, or neither?

5.2 The Rules for Integral Exponents

powers of the same base

Clearly,

$$4^2 \cdot 4^3 = (4 \cdot 4)(4 \cdot 4 \cdot 4) = 4 \cdot 4 \cdot 4 \cdot 4 \cdot 4 = 4^5$$

Thus

$$4^2 \cdot 4^3 = 4^{2+3}$$

Now let a be any number, and let m and n be *positive* integers. Then recall that

$$a^m \cdot a^n = \underbrace{(a \cdot a \cdot a \ldots a)}_{m \text{ times}} \underbrace{(a \cdot a \cdot a \ldots a)}_{n \text{ times}} = \underbrace{a \cdot a \cdot a \ldots a}_{m+n \text{ times}} = a^{m+n}.$$

Therefore

(A) $\boldsymbol{a^m \cdot a^n = a^{m+n}}$

Thus, *to multiply powers of the same base, add the exponents.* Observe that (for $a \neq 0$)

$$a^m \cdot a^0 = a^m \cdot 1 = a^m = a^{m+0}$$

Thus

$$a^m \cdot a^0 = a^{m+0}$$

(Here m can be any integer—positive, negative, or zero.)

In Section 5.1, you learned that

(B) $$\frac{a^m}{a^n} = a^{m-n}$$

for $a \neq 0$ and for *all* integers m and n.

Let $a \neq 0$. **Rule (B) can be used to show that Rule (A),**

$$a^m \cdot a^n = a^{m+n},$$

holds for *all* integers m and n. For example, let $m = 6$ and $n = -9$. Then

$$a^6 \cdot a^{-9} = a^6 \cdot \frac{1}{a^9} = \frac{a^6}{a^9} = a^{6-9} = a^{6+(-9)}$$

EXAMPLE 1 (a) $a^8 \cdot a^5 = a^{8+5} = a^{13}$, for any a

(b) $7^4 \cdot 7^2 = 7^{4+2} = 7^6$

(c) $b^6 \cdot b^{-10} = b^{6+(-10)} = b^{-4} = \frac{1}{b^4}$, $b \neq 0$

(d) $8^{-3} \cdot 8^{-2} = 8^{-3+(-2)} = 8^{-5} = \frac{1}{8^5}$ □

Similarly, for $a \neq 0$ and for all integers m, n, and r,

$$a^m \cdot a^n \cdot a^r = a^{m+n+r}$$

Thus

$$a^5 \cdot a^3 \cdot a^2 = a^{5+3+2} = a^{10}$$

and for $b \neq 0$,

$$b^{-7} \cdot b^2 \cdot b^{-3} = b^{-7+2+(-3)} = b^{-8} = \frac{1}{b^8}$$

power of a power

Next, what is the cube of the square of a number a?

$$(a^2)^3 = a^2 \cdot a^2 \cdot a^2 = a^{2+2+2} = a^6 = a^{2 \cdot 3}$$

Thus

$$(a^2)^3 = a^{2 \cdot 3}$$

Now, what is the nth power of the mth power of a?

Suppose m and n are positive integers. Then for any number a,

$$(a^m)^n = \underbrace{a^m \cdot a^m \cdot a^m \cdots a^m}_{n \text{ times}}$$

$$= \underbrace{\underbrace{a \cdot a \cdot a \cdots a}_{m \text{ times}} \cdot \underbrace{a \cdot a \cdot a \cdots a}_{m \text{ times}} \cdots \underbrace{a \cdot a \cdot a \cdots a}_{m \text{ times}}}_{n \text{ of these}}$$

$$= \underbrace{a \cdot a \cdot a \cdots a}_{mn \text{ times}}$$

$$= a^{mn}$$

Does this result hold when n is a negative integer? Let $a \neq 0$ and consider:

$$(a^2)^{-3} = \frac{1}{(a^2)^3} = \frac{1}{a^{2\cdot 3}} = \frac{1}{a^6} = a^{-6} = a^{2(-3)}$$

Thus

$$(a^2)^{-3} = a^{2(-3)}$$

In fact, for $a \neq 0$ and for all integers m and n,

(C) $(a^m)^n = a^{mn}$

(To find a power of a power, write down the base and multiply the corresponding exponents.)

EXAMPLE 2 (a) $(5^3)^4 = 5^{3\cdot 4} = 5^{12}$, whereas $5^3 \cdot 5^4 = 5^{3+4} = 5^7$

(b) $(9^2)^{-5} = 9^{2(-5)} = 9^{-10} = \dfrac{1}{9^{10}}$

(c) $(7^{-2})^{-4} = 7^{(-2)(-4)} = 7^8$ □

EXAMPLE 3 Let a be any number.

(a) $(a^4)^2 = a^{4\cdot 2} = a^8$ (b) $a^{(4^2)} = a^{16}$

Thus

$$(a^4)^2 \neq a^{(4^2)}$$

Observe that

$$(a^4)^2 = a^4 \cdot a^4 = a^{4+4},$$

whereas

$$a^{(4^2)} = a^{4\cdot 4}.$$ □

powers of products and quotients

How do you find the cube of a product? By the Commutative and Associative Laws, for all numbers a and b,

$$(ab)^3 = ab \cdot ab \cdot ab = (aaa)(bbb) = a^3b^3$$

In general, for any nonzero numbers a and b and for any integer m,

(D) $(ab)^m = a^m b^m$

(The m*th power of a product is the product of the* m*th powers.)*

EXAMPLE 4 Let a, x, y, and z be any numbers.

(a) $(xy)^7 = x^7y^7$ (b) $(2a)^5 = 2^5a^5 = 32a^5$

(c) $(2\cdot 5)^2 = 2^2 \cdot 5^2 = 4 \cdot 25 = 100$

Note further that

$$(2\cdot 5)^2 = 10^2 = 100.$$

(d) $(3xyz)^3 = 3^3x^3y^3z^3 = 27x^3y^3z^3$ □

How do you find the cube of a quotient? Assume a is any number and $b \neq 0$.

$$\left(\frac{a}{b}\right)^3 = \frac{a}{b}\cdot\frac{a}{b}\cdot\frac{a}{b} = \frac{a\cdot a\cdot a}{b\cdot b\cdot b} = \frac{a^3}{b^3}$$

In general, for any nonzero numbers a and b and for any integer m,

(E) $$\left(\frac{a}{b}\right)^m = \frac{a^m}{b^m}$$

(The mth power of a quotient is the quotient of the mth powers.)

EXAMPLE 5 Assume $a \neq 0$, $x \neq 0$, $y \neq 0$.

(a) $\left(\frac{x}{y}\right)^4 = \frac{x^4}{y^4}$ (b) $\left(\frac{3}{4}\right)^2 = \frac{3^2}{4^2} = \frac{9}{16}$

(c) $\left(\frac{10}{3}\right)^3 = \frac{10^3}{3^3} = \frac{1000}{27}$ (d) $\left(\frac{3a^{-2}b^2}{5a}\right)^3 = \left(\frac{3b^2}{5a^3}\right)^3 = \frac{27b^6}{125a^9}$

(e) $\left(\frac{x}{y}\right)^{-2} = \frac{x^{-2}}{y^{-2}} = \frac{\frac{1}{x^2}}{\frac{1}{y^2}} = \frac{1}{x^2}\cdot\frac{y^2}{1} = \frac{y^2}{x^2}$ □

EXERCISES

Assume $a \neq 0$, $b \neq 0$, $x \neq 0$, $y \neq 0$, and $z \neq 0$.

In Exercises 1–42 simplify and write your answer using only positive exponents. If *necessary,* write as a fraction.

1. $a^2\cdot a$ **2.** $a^6\cdot a^2$ **3.** $a^4\cdot a^5$ **4.** $a^4\cdot a^{-3}$ **5.** $a^4\cdot a^{-5}$

6. $\frac{a^8}{a^4}$ **7.** $\frac{a^{-3}}{a^7}$ **8.** $\frac{a^{-3}}{a^{-7}}$ **9.** $(a^2)^5$ **10.** $(a^{-3})^2$

11. $(a^2x)^2$ **12.** $(a^3xy)^{-1}$ **13.** $\frac{10^{-2}}{a^{-3}}$ **14.** $\frac{a^0}{a^{-7}b}$ **15.** $\frac{(a^2)^3(b^3)^2}{a^6b^4}$

16. $\frac{3^2(xy^2)^3}{(9x)^2}$ **17.** $\left(\frac{2x^2y^3}{xy^2}\right)^2$ **18.** $\frac{(3x^4y)^{-2}}{(9xy^2)^{-1}}$ **19.** $\frac{(2a^2)^3\pi}{4a^5\pi^{-1}}$ **20.** $\left(\frac{10^2xy}{10xy^0}\right)^{-3}$

21. $\frac{(2a^{-3}b)^{-2}}{(8ab^{-4})^{-1}}$ **22.** $\left(\frac{25ax^2y^3}{5a^4x^{-2}y^0}\right)^{-2}$ **23.** $\frac{a^4\cdot a^3\cdot a^9}{a^{-2}}$ **24.** $a^2\cdot(a^5)^2\cdot a^{-3}$ **25.** $(x+y)^{-1}$

26. $\frac{(a+b)^{-2}}{(a+b)^{-4}}$ **27.** $3^{-1}+3$ **28.** $2^{-1}+2^{-2}$ **29.** $\pi+\pi^{-1}$ **30.** $x^2+5x^{-2}+10x^{-4}$

31. $y^{-1}-2y^{-2}+3y^{-3}$ **32.** $xy+x^{-1}y-xy^{-2}+x^{-3}y^{-3}$ **33.** $\frac{a^{-1}}{b^{-1}}-\frac{b}{a}$ **34.** $\frac{a^{-1}}{b^{-1}}+\frac{b^{-1}}{a^{-1}}$

35. $\frac{(x^2)^{-1}}{y^2}+\frac{2}{3x^2y^2}$ **36.** $(a^{-1}+b^{-1})^{-1}$ **37.** $(a^{-1}-b^{-1})ab$ **38.** $\frac{x^{-1}-y^{-1}}{x^{-2}-y^{-2}}$

39. $\left(\frac{a}{b^2}\right)^{-3}-\frac{b^{-2}}{a^{-3}}$ **40.** $\frac{x^{-2}+y^{-1}}{xy^{-1}-y}$ **41.** $\frac{a^3b^{-1}+ba^{-2}}{ab^2-a^{-1}b^{-1}}$ **42.** $\frac{x^{-1}yz^{-2}-x^{-1}y+1}{xz^{-2}-x^{-1}y^{-2}-1}$

43. **(a)** Determine the square of the cube of 2.

(b) Determine the cube of the square of 2.

44. What power of 2 is $\frac{1}{64}$?

5.3 Scientific Notation

Scientific notation is a convenient way of expressing large numbers, such as 676 000 000 000 000, and small positive numbers, such as .000 000 004. It is commonly used in fields such as chemistry, biology, engineering, physics, and astronomy, where such numbers frequently occur. You will use scientific notation in your study of logarithms (Chapter 7).

The powers of 10 play a key role in scientific notation. Observe the pattern:

$$\begin{array}{lllll}
10^1 = 10 & & 10^0 = 1 & & 10^{-1} = \frac{1}{10} = .1 \\
10^2 = 100 & & & & 10^{-2} = \frac{1}{100} = .01 \\
10^3 = 1000 & & & & 10^{-3} = \frac{1}{1000} = .001 \\
10^4 = 10\,000 & & & & 10^{-4} = \frac{1}{10\,000} = .0001 \\
10^5 = 100\,000 & & & & 10^{-5} = \frac{1}{100\,000} = .000\,01 \\
10^6 = 1\,000\,000 & & & & 10^{-6} = \frac{1}{1\,000\,000} = .000\,001
\end{array}$$

converting to scientific notation

Every *positive* number N can be expressed as the product of

(a) a number between 1 and 10

and

(b) an integral power of 10.

More precisely,

$$N = M \cdot 10^m,$$

where $1 \leq M < 10$ and m is an integer. This form of writing a number is called **scientific notation.**

EXAMPLE 1 To express 4730 in scientific notation, note that

$$4730 = \frac{4730}{1000} \times 1000 = 4.730 \times 10^3$$

Observe that *division by* 1000 (or 10^3) is accomplished by *moving the decimal point* 3 *places to the left* (to obtain 4.730). Balance this by multiplying by 10^3. □

EXAMPLE 2 To express .000 56 in scientific notation, note that

$$.000\ 56 = (.000\ 56 \times 10\ 000) \cdot \frac{1}{10\ 000} = 5.6 \times 10^{-4}$$

Here, *multiplication by* 10 000 (or 10^4) is accomplished by *moving the decimal point* 4 *places to the right* (to obtain 5.6). Balance this by multiplying by 10^{-4}. □

Let N be a positive number, expressed as a decimal. To obtain the scientific notation

$$M \cdot 10^m$$

for N:

1. Place the decimal point after the first *nonzero* digit to obtain M.

2a. If you moved the decimal point k places to the *left*, you *divided by* 10^k. To balance this, multiply by 10^k. Thus let $m = k$.

2b. If you moved the decimal point k places to the *right*, you *multiplied by* 10^k. To balance this, multiply by 10^{-k}. Thus let $m = -k$.

2c. If $1 \leqslant N < 10$, then $m = 0$

EXAMPLE 3 Express in scientific notation:

(a) 763 (b) .008 406 (c) 9.058 (d) 34.81

Solution.

(a) $N = 763$
Place the decimal point after the 7. Thus

$$M = 7.63$$

You moved the decimal point 2 places to the *left:*

763. to 7.63

Therefore $m = 2$ and

$$763 = 7.63 \times 10^2$$

(b) $N = .008\ 406$
The first *nonzero* digit is 8. Thus

$$M = 8.406$$

(The zeros to the left of 8 can be omitted.) You moved the decimal point 3 places to the *right:*

.008 406 to ~~00~~8.406

Therefore $m = -3$ and

$$.008\ 406 = 8.406 \times 10^{-3}$$

(c) $1 \leqslant 9.058 < 10$
Thus $m = 0$ and

$$9.058 = 9.058 \times 10^0$$

If N lies between 1 and 10, as is the case here, the factor $10^0(= 1)$ is usually omitted. Thus the scientific notation of 9.058 is 9.058 itself.

(d) $34.81 = 3.481 \times 10^1 = 3.481 \times 10$ Write "10^1" as "10". □

EXAMPLE 4 Express in scientific notation:

(a) 6 380 000 (b) 000 000 53 (c) .4 (d) $\frac{1}{4}$

Solution.

(a) $6\,380\,000 = 6.38 \times 10^6$

(b) $.000\,000\,53 = 5.3 \times 10^{-7}$

(c) $.4 = 4 \times 10^{-1}$

(d) $\frac{1}{4} = .25 = 2.5 \times 10^{-1}$ □

converting to decimal notation

To convert back from scientific notation

$$M \cdot 10^m, \qquad \text{where } m \neq 0,$$

to the usual decimal notation:

1. Move the decimal point m places to the *right* if $m > 0$. (Here you multiply by 10^m, which is *greater than* 1.)
2. Move the decimal point m places to the *left* if $m < 0$. (Here you multiply by 10^m, which is *less than* 1.

Note that this *reverses* the former process.

EXAMPLE 5 (a) $1.48 \times 10^4 = 14\,800$

Here $M = 1.48$ and $m = 4 > 0$. The decimal point is moved to the *right* 4 places.

(b) $6.04 \times 10^{-3} = .006\,04$

Here $M = 6.04$ and $m = -3 < 0$. The decimal point is moved to the *left* 3 places. □

uses of scientific notation

Here are several important examples of the use of this notation in the sciences.

The earth's mass is approximately 6×10^{27} grams.

All gases occupying equal volumes at the same temperature and pressure contain the same number of molecules *(Avogadro's Number)*, which has been determined to be 6.023×10^{23} molecules per gram molecular weight.

The speed of light and of every other electromagnetic wave (in a vacuum) is 3×10^{10} centimeters per second.

An electron has a charge of 4.8×10^{-10} electrostatic units.

Computations are often simplified by expressing numbers in terms of powers of 10. Instead of using scientific notation in the computation, it is usually best to express a number as the product of an *integer* and a power of 10. The final result is then written in scientific notation.

EXAMPLE 6 Find the value of:

$$\frac{25\,000 \times 160\,000}{800\,000 \times .0125}$$

Express your answer in scientific notation.

Solution.

$$\frac{25\,000 \times 160\,000}{800\,000 \times .0125} = \frac{\not{25} \times 10^3 \times \overset{2}{\not{16}} \times 10^4}{\not{8} \times 10^5 \times \underset{5}{\not{125}} \times 10^{-4}}$$

$$= \frac{2}{5} \times 10^{3+4-(5-4)}$$

$$= .4 \times 10^6$$

$$= 4 \times 10^5$$

□

EXERCISES

In Exercises 1–18 express each number in scientific notation.

1. 14.9 **2.** 362 **3.** 5084 **4.** 7.82 **5.** .695

6. .008 02 **7.** .9 **8.** 8036.2 **9.** 248 000 **10.** 3 062 000 000

11. .000 820 001 **12.** 843×10^4 **13.** 27.9×10^{-2} **14.** 9.36×10^{-5} **15.** $\dfrac{3 \times 4 \times 5}{100}$

16. .01 **17.** $5 \times 80 \times 10^6$ **18.** .000 03

In Exercises 19–32 convert scientific notation back to the usual decimal notation.

19. 8.34×10^1 **20.** 6.06×10^4 **21.** 4.44×10^8 **22.** 3.09×10^{-2} **23.** 5.15×10^0

24. 6.092×10^{-4} **25.** 8.181×10^{-1} **26.** 6×10^7 **27.** 6.2×10^{-2} **28.** 3.87×10^{-5}

29. 1.11×10^1 **30.** 3.93×10^{-7} **31.** $3.896\,24 \times 10^2$ **32.** 3.796×10^{-8}

In Exercises 33–40 find the value of each number. Express your answer in scientific notation.

33. $10^8 \times 10^4 \times 10^{-2}$ **34.** $\dfrac{10^7 \times 10^9}{10^3 \times 10^8}$ **35.** $60\,000 \times 5000 \times 20\,000$

36. $9000 \times .0004$ **37.** $25\,000 \times .002 \times 500 \times .000\,04$ **38.** $\dfrac{270 \times 8000}{.002 \times 300\,000}$

39. $\dfrac{200 \times .0004 \times 64\,000}{4000 \times .032 \times .08}$ **40.** $\dfrac{144\,000 \times .000\,36}{.0072 \times 960\,000}$

41. The human eye is normally capable of responding to light waves whose lengths are in the range of 3.8×10^{-5} to 7.6×10^{-5} centimeters. Which of the following centimeter readings lie within this range?

(a) .000 45 **(b)** .000 045 **(c)** 60×10^{-6} **(d)** $.039 \times 10^{-3}$

42. The coefficient of linear expansion indicates the fractional change in the length of a rod per degree change of temperature. For tungsten the coefficient is 4.4×10^{-6} per degree Celsius. What is the fractional change in the length of a tungsten rod when the temperature rises from 10° to 30° Celsius?

43. A light year is the distance light can travel in a year's time. One light year is approximately 6×10^{12} miles. If a star is 250 light years away from earth, how far away is it in miles?

44. One electron volt is the kinetic energy an electron acquires when it is accelerated in an electric field produced by a difference of potential of 1 volt.

1 electron volt = 1.60×10^{-19} joule

How many joules equal five million electron volts?

5.4 Rational Exponents

square roots

Suppose you are asked: What *positive* number times itself is 9? In other words, you are asked to find a *positive* number b such that $b^2 = 9$. Clearly, 3 satisfies this condition. 3 will be called the "principal square root of 9". Observe, further, that

$$(-3)^2 = (-3)(-3) = 9.$$

-3 will be called the "negative square root of 9".

Definition. Let a be positive. Then b is called the **principal square root of** a, or, for short, the **square root of** a if b is positive and if

$$b^2 = a.$$

In this case, $-b$ is called the **negative square root of** a. Also, the **(principal) square root of** 0 is 0.

If $a^{\frac{1}{2}}$ denotes the *principal* square root of a for $a \geqslant 0$, then

$$a^{\frac{1}{2}} \cdot a^{\frac{1}{2}} = a = a^1 = a^{\frac{1}{2}+\frac{1}{2}} \quad \text{and}$$

$$(a^{\frac{1}{2}})^2 = a^{\frac{1}{2}} \cdot a^{\frac{1}{2}} = a = a^1 = a^{\frac{1}{2}\cdot 2},$$

so that the properties

$$a^m \cdot a^n = a^{m+n} \quad \left(\text{where } m = n = \frac{1}{2}\right) \quad \text{and}$$

$$(a^m)^n = a^{mn} \quad \left(\text{where } m = \frac{1}{2}, n = 2\right)$$

are preserved. Also, because it is difficult to read in print,

$a^{\frac{1}{2}}$ will be written as $a^{1/2}$

Thus,

$$9^{1/2} = 3,$$

The *principal* square root of 9 is 3.

whereas

$$-9^{1/2} \quad \text{means} \quad -(9^{1/2}).$$

Thus

$$-9^{1/2} = -3$$

The *negative* square root of 9 is -3.

(The notation $\sqrt{a}$ is also used for the *principal* square root of a, and will be discussed in Section 5.6 .)

EXAMPLE 1 (a) $4^{1/2} = 2$ because $2^2 = 4$ and $2 > 0$ (b) $-4^{1/2} = -2$
(c) $100^{1/2} = 10$ (d) $0^{1/2} = 0$ □

A *positive* number has two square roots that are real numbers. But these square roots *may not be rational.* For example, $2^{1/2}$ and $3^{1/2}$ are irrational numbers. Recall that an irrational number can be expressed as an *infinite nonrepeating decimal.* You can *approximate* $2^{1/2}$ to 3 places by 1.414 and $3^{1/2}$ by 1.732. However, *these are not the exact values.* Except in applications, it is best to leave these square roots in the form $2^{1/2}$, $3^{1/2}$, and so on.

The square of a real number b is at least 0.

$b^2 \geqslant 0$

Thus a negative number, such as -9, cannot be the square of a real number. In fact,

$3 \cdot 3 = 9$ and $(-3)(-3) = 9$

There is no real number that when multiplied by itself yields -9. Therefore,

$(-9)^{1/2}$, "the square root of -9",

is not defined *in the real number system*. [See Section 6.3 .]

Recall that the function $g(x)$ defined by

$g(x) = x^2$ for *all* real numbers x

is *not* one-to-one. For example,

$g(1) = g(-1) = 1$ and $g(2) = g(-2) = 4$

However, the function $G(x)$ defined by

$G(x) = x^2$ for $x \geqslant 0$

is one-to-one and therefore has an inverse. The inverse function is given by

$G^{-1}(x) = x^{1/2}$ for $x \geqslant 0$.

[See Figure 5.1 .]

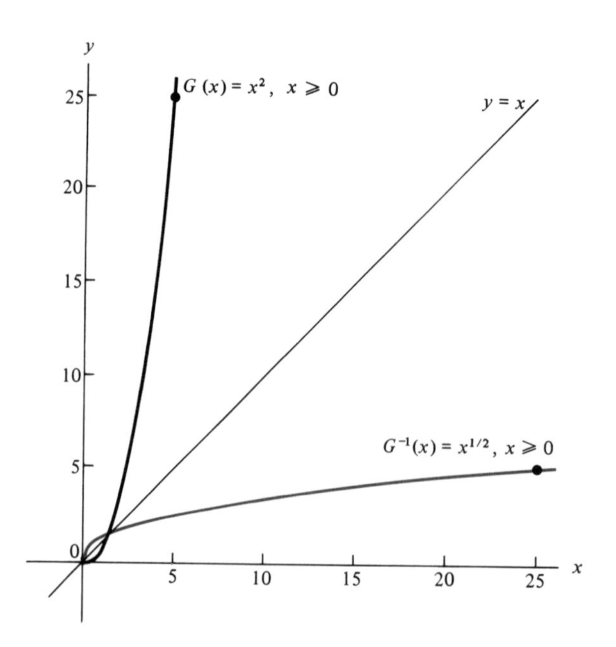

FIGURE 5.1 Note that $G(5) = 25$ $(5^2 = 25)$, whereas $G^{-1}(25) = 5$ $(\sqrt{25} = 5)$.

cube roots ***Definition.*** Let a be any real number. The real number b is called the **principal cube root of** a if $b^3 = a$.

Write $b = a^{1/3}$ if b is the principal cube root of a.

EXAMPLE 2 (a) $8^{1/3} = 2$ because $2^3 = 8$ (b) $(-8)^{1/3} = -2$ because $(-2)^3 = -8$

(c) $27^{1/3} = 3$ (d) $0^{1/3} = 0$ □

Every real number—positive, negative, or zero—has exactly one cube root that is a real number‡. Thus, the function $h(x)$ defined by

$$h(x) = x^3 \quad \text{for all real numbers } x$$

is one-to-one, and has an inverse given by

$$h^{-1}(x) = x^{1/3} \quad \text{for all real numbers } x.$$

[See Figure 5.2 .]

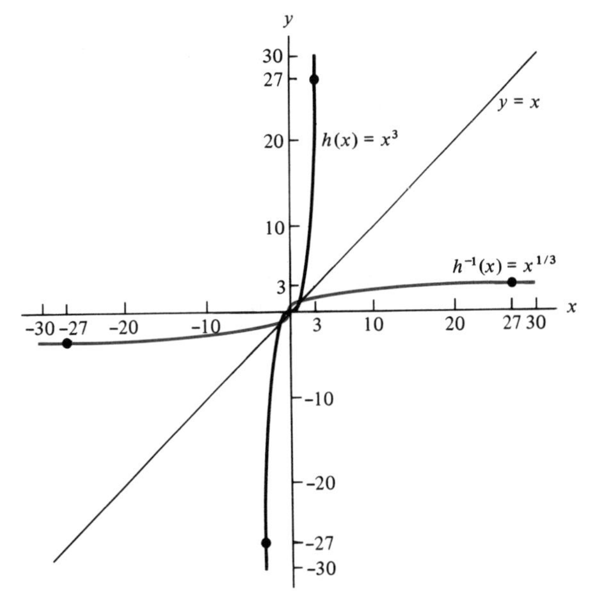

FIGURE 5.2 Note that $h(3) = 27$ $(3^3 = 27)$, whereas $h^{-1}(27) = 3$ $(27^{1/3} = 3)$.

*n*th roots In general, you can consider the nth roots of numbers, for $n = 1, 2, 3, 4, 5, \ldots$

Definition. Let n be an *even* positive integer: 2, 4, 6, . . . , and let $a > 0$. Then b is called the **principal *n*th root of** a if b is *positive* and if

$$b^n = a.$$

In this case, $-b$ is called the **negative *n*th root of** a. Also, the **(principal) *n*th root of** 0 is 0.

‡In Section 10.4 you will learn how to find the two *nonreal* cube roots of a nonzero real number.

Let n be an *odd* positive integer: 1, 3, 5, . . . , and let a be *any* real number. Then (the real number) b is called the **principal nth root of** a if $b^n = a$.

Write $b = a^{1/n}$ if b is the *principal nth root* of a.

Only nonnegative numbers have real nth roots for $n = 2, 4, 6, \ldots$. *But every real number—positive, negative, or zero—has a real nth root for* $n = 1, 3, 5, 7, \ldots$

EXAMPLE 3

(a) $81^{1/4} = 3$ because $3^4 = 81$ and $3 \geqslant 0$

(b) $32^{1/5} = 2$ because $2^5 = 32$

(c) $(-32)^{1/5} = -2$ because $(-2)^5 = -32$

(d) -5 has no real 6th root because negative numbers have real nth roots only for odd n.

(e) $1^{1/n} = 1$ for every positive integer n because $1^n = 1$

(f) $(-1)^{1/n} = -1$ for every *odd* positive integer n because *odd* powers of -1 equal -1. □

rational powers

For *rational numbers* $\frac{m}{n}$, rational powers $a^{m/n}$ can now be defined so as to extend the law

$$(a^m)^n = a^{mn}.$$

Definition. Let m be a nonzero integer, let n be a positive integer, and let a be a real number (with $a \geqslant 0$ if n is even, and $a \neq 0$ if $m < 0$). Define

$$a^{m/n} = (a^{1/n})^m.$$

In other words, *the* $\left(\frac{m}{n}\right)$ *th power of a is the mth power of the principal nth root of a.* Thus

$$8^{2/3} = (8^{1/3})^2 = 2^2 = 4$$

It can be shown that (with the previous restrictions)

$$(a^{1/n})^m = (a^m)^{1/n}.$$

In fact,

$$8^{2/3} = (8^2)^{1/3} = 64^{1/3} = 4$$

Thus $a^{m/n}$ is *also the principal nth root of the mth power of a.* However, in practice, it is often easier to first take roots, then integral powers, for you will usually be working with smaller and more easily recognized numbers. For example, consider $25^{3/2}$. First,

$$25^{3/2} = (25^{1/2})^3 = 5^3 = 125$$

Next,

$$25^{3/2} = (25^3)^{1/2} = 15\,625^{1/2} = 125$$

Note the difficulty in the second method.

EXAMPLE 4

(a) $16^{3/4} = (16^{1/4})^3 = 2^3 = 8$

(b) $4^{3/2} = (4^{1/2})^3 = 2^3 = 8$

(c) $(-27)^{2/3} = [(-27)^{1/3}]^2 = (-3)^2 = 9$

(d) Consider $25^{-1/2}$. Here the exponent is $\frac{-1}{2}$.

$$25^{-1/2} = (25^{1/2})^{-1} = 5^{-1} = \frac{1}{5}$$ □

If $a < 0$ and m and n are both even integers, then $(a^m)^{1/n}$ is defined, but $a^{1/n}$ and, hence, $(a^{1/n})^m$ are not. For example, suppose $a = -2$, $m = 2$, and $n = 2$. Then

$$(a^m)^{1/n} = [(-2)^2]^{1/2} = 4^{1/2} = 2,$$

but

$$(a^{1/n})^m \qquad \text{or} \qquad [(-2)^{1/2}]^2$$

would involve "the square root of a negative number". Note further that $a^{m/n} = (-2)^{2/2} = (-2)^1 = -2$, so that $a^{m/n} \neq (a^{1/n})^m$ in this case.

If $a < 0$, m is an odd integer, and n is an even integer, then neither $(a^m)^{1/n}$ nor $(a^{1/n})^m$ is defined. For example, suppose $a = -2$, $m = 1$, and $n = 2$. Then

$$(a^m)^{1/n} = [(-2)^1]^{1/2} = (-2)^{1/2} \quad (\textit{undefined}) \qquad \text{and}$$

$$(a^{1/n})^m = [(-2)^{1/2}]^1 = (-2)^{1/2} \quad (\textit{undefined})$$

Recall that

$$\frac{-1}{2} = -\frac{1}{2} = \frac{1}{-2}.$$

So far, only $\frac{-1}{2}$ has been defined as an exponent. Thus

$$25^{-1/2} = \frac{1}{5}$$

You will write

$$25^{-(1/2)}$$

when the exponent is $-\frac{1}{2}$, according to the following definition.

Definition. Let $a \neq 0$ and let m and n be positive integers (with a positive if n is even). Define:

$$a^{-(m/n)} \text{ to be } (a^{1/n})^{-m} \text{ and } a^{m/-n} \text{ to be } (a^{1/n})^{-m}$$

Thus, by definition,

$$a^{-(m/n)} = a^{m/-n} = a^{-m/n} = (a^{1/n})^{-m}$$

EXAMPLE 5 (a) $8^{-(1/3)} = (8^{1/3})^{-1} = 2^{-1} = \frac{1}{2}$

(b) $8^{2/-3} = (8^{1/3})^{-2} = 2^{-2} = \frac{1}{2^2} = \frac{1}{4}$

(c) $(-32)^{-3/5} = [(-32)^{1/5}]^{-3} = (-2)^{-3} = \frac{1}{(-2)^3} = \frac{1}{-8} = \frac{-1}{8}$ □

EXERCISES

In Exercises 1–30 simplify.

1. $36^{1/2}$ **2.** $1^{1/2}$ **3.** $64^{1/2}$ **4.** $121^{1/2}$ **5.** $144^{1/2}$ **6.** $64^{1/3}$

7. $125^{1/3}$ **8.** $(-125)^{1/3}$ **9.** $16^{1/4}$ **10.** $81^{1/4}$ **11.** $64^{2/3}$ **12.** $(-8)^{2/3}$

13. $(-1)^{4/3}$ **14.** $16^{5/4}$ **15.** $32^{2/5}$ **16.** $9^{-1/2}$ **17.** $16^{-1/2}$ **18.** $81^{-1/2}$

19. $100^{-1/2}$ **20.** $1000^{-1/3}$ **21.** $16^{-1/4}$ **22.** $16^{-3/4}$ **23.** $81^{-1/4}$ **24.** $81^{-3/4}$

25. $9^{-3/2}$ **26.** $125^{-2/3}$ **27.** $10\,000^{-3/4}$ **28.** $100^{-5/2}$ **29.** $36^{-(1/2)}$ **30.** $49^{1/-2}$

In Exercises 31 and 32 find the length of the hypotenuse of a right triangle, given the lengths of the other sides.

31. 5 centimeters and 12 centimeters **32.** 1 inch and 2 inches

33. The base of a right triangle is of length 24 centimeters and the hypotenuse is of length 25 centimeters. Find the length of the altitude.

34. The altitude of a right triangle is of length 3 inches and the hypotenuse is of length 4 inches. Find the length of the base.

35. The lengths of the sides of a triangle are 5 centimeters, 6 centimeters, and 8 centimeters, respectively. Show that this is *not* a right triangle.

36. The base of a right triangle is of length 10. Express the length y of the altitude as a function of x, the length of the hypotenuse.

37. The area of a square tablecloth is 3600 square inches. Find the length of a side.

38. The volume of a cubic box is 64 cubic meters. Find the length of a side.

39. The area of a circle is 144π square centimeters. Find the length of the radius.

40. The volume of a sphere is $\frac{4000\pi}{3}$ cubic inches. Find the length of the radius.

41. Express the length of the radius of a circle as a function of the area.

42. Express the length of the radius of a sphere as a function of the volume.

43. Determine the square root of the square of -1. **44.** Determine the cube root of the cube of -1.

45. Determine the square of the cube root of -1. **46.** Determine the cube root of the square of -1.

In Exercises 47–56 find: (a) the largest possible domain and (b) the corresponding range of each function.

47. $F(x) = x^2$ **48.** $f(x) = x^{1/2}$ **49.** $G(x) = x^3$ **50.** $g(x) = x^{1/3}$

51. $F(x) = x^2 - 4$ **52.** $f(x) = x^{1/2} - 4$ **53.** $H(x) = (x - 4)^2$ **54.** $h(x) = (x - 4)^{1/2}$

55. $G(x) = \frac{1}{(x-4)^2}$ **56.** $g(x) = \frac{1}{(x-4)^{1/2}}$

57. A rectangle is inscribed in a circle of radius length 10. Express the area of the rectangle as a function of its length l. [See Figure 5.3 .]

58. A right-circular cylinder is inscribed in a sphere of radius length 8. Express h, the altitude length of the cylinder as a function of r, the length of the base radius. [See Figure 5.4 .]

59. A boy flies a kite at a level height of 100 meters (above his hand). The kite moves in a horizontal line away from the boy at the rate of 7 meters per second. Express the length of the kite string as a function of t, the number of seconds after the kite is directly above him.

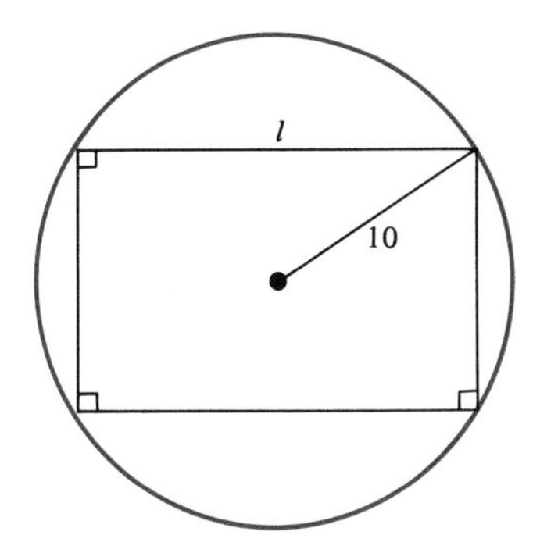

FIGURE 5.3

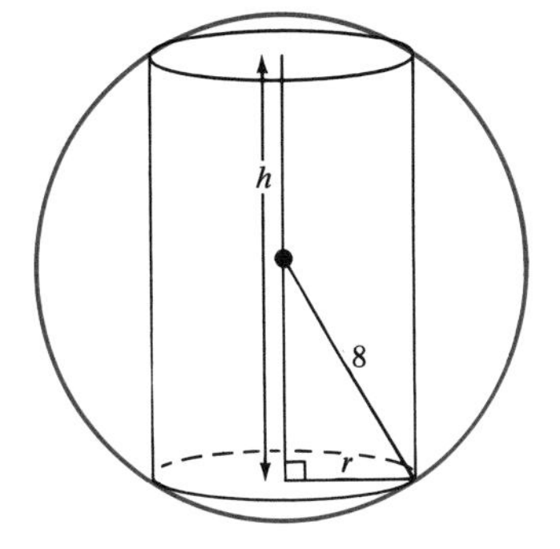

FIGURE 5.4

5.5 The Rules for Rational Exponents

The rules for integral exponents apply also to *rational* exponents. Thus, let a and b be nonzero numbers, let r and s be *rational numbers,* and suppose that a^r and a^s are both defined (as real numbers).

(A) $a^r \cdot a^s = a^{r+s}$

(B) $\dfrac{a^r}{a^s} = a^{r-s}$

(C) $(a^r)^s = a^{rs}$

Also, suppose b^r is defined.

(D) $(ab)^r = a^r b^r$

(E) $\left(\dfrac{a}{b}\right)^r = \dfrac{a^r}{b^r}$

The requirement that a^r and a^s are both defined is crucial. For example, in **(C)**, suppose that $a = -2$, $r = 2$, and $s = \dfrac{1}{2}$. Then

$$(a^r)^s = [(-2)^2]^{1/2} = 4^{1/2} = 2$$

But

$$a^{rs} = (-2)^{2 \cdot 1/2} = (-2)^1 = -2$$

Thus here,

$$(a^r)^s \neq a^{rs}$$

because although a^r is defined $[a^r = (-2)^2 = 4]$, a^s, which would equal $(-2)^{1/2}$, is not defined.

Rules (A), (B), and (C)

To apply Rules **(A)**, **(B)**, and **(C)**, combine the fractions r and s.

EXAMPLE 1 Let $x > 0$.

(a) $x^{1/4} \cdot x^{3/4} = x^{1/4+3/4} = x^1 = x$

(b) $x^{1/2} \cdot x^{1/3} = x^{1/2+1/3} = x^{3/6+2/6} = x^{(3+2)/6} = x^{5/6}$

(c) $x^{5/6} \cdot x^{-1/4} = x^{5/6-1/4} = x^{10/12-3/12} = x^{(10-3)/12} = x^{7/12}$ □

EXAMPLE 2 Let $a > 0$.

(a) $\frac{a^{2/5}}{a^{1/5}} = a^{2/5-1/5} = a^{1/5}$ (b) $\frac{a^{1/2}}{a^{1/4}} = a^{1/2-1/4} = a^{(2-1)/4} = a^{1/4}$

(c) $\frac{a^{2/3}}{a^{1/5}} = a^{2/3-1/5} = a^{10/15-3/15} = a^{(10-3)/15} = a^{7/15}$ □

EXAMPLE 3 Let $b > 0$.

(a) $(b^{1/2})^{1/3} = b^{(1/2)(1/3)} = b^{1/6}$ (b) $(b^{2/5})^{3/2} = b^{(2/5)(3/2)} = b^{3/5}$

(c) $(b^{-3/4})^{-2/9} = b^{(-3/4)(-2/9)} = b^{1/6}$ □

Rules (D) and (E) Now consider Rules **(D)** and **(E)**.

EXAMPLE 4 Let $x > 0$ and $y > 0$.

(a) $(xy)^{3/4} = x^{3/4}y^{3/4}$ (b) $(32x)^{-2/5} = 32^{-2/5}x^{-2/5} = \frac{1}{4x^{2/5}}$

(c) $\left(\frac{8}{27}\right)^{2/3} = \frac{8^{2/3}}{27^{2/3}} = \frac{4}{9}$ □

EXAMPLE 5 Let $a > 0$, $b \neq 0$, and $c > 0$.

(a) $(a^{1/2}b^{2/3})^{12} = a^{(1/2)(12)}b^{(2/3)(12)} = a^6b^8$

(b) $\frac{(a^2bc^{1/4})^4}{(a^{1/2}b^3c^2)^2} = \frac{a^8b^4c}{ab^6c^4} = \frac{a^7}{b^2c^3}$

Note that in part (b) each variable appears only once and with only *positive* exponents in the *simplified* form $\frac{a^7}{b^2c^3}$. □

To *simplify* the numerical expression $8^{1/2}$, separate all square factors of 8, and use Rule **(D)**. The simplified form will involve the rational power of a smaller integer.

EXAMPLE 6 (a)

$$\begin{aligned} 8^{1/2} &= (4\cdot 2)^{1/2} \\ &= 4^{1/2}\cdot 2^{1/2} && \text{by Rule (D)} \\ &= 2\cdot 2^{1/2} \end{aligned}$$

(b)

$$\begin{aligned} \left(\frac{1}{32}\right)^{1/4} &= \frac{1^{1/4}}{32^{1/4}} && \text{by Rule (E)} \\ &= \frac{1}{(16\cdot 2)^{1/4}} \\ &= \frac{1}{16^{1/4}\cdot 2^{1/4}} && \text{by Rule (D)} \\ &= \frac{1}{2}\cdot\frac{1}{2^{1/4}} \end{aligned}$$

□

factoring

EXAMPLE 7 Assume $x \neq 0$. Factor: $x^{2/3} - 2x^{-1/3} + x^{-4/3}$

Solution. The *lowest* power of x that occurs is $x^{-4/3}$. Note that each exponent equals $\frac{-4}{3}$ + *an integer:*

$$\frac{2}{3} = \frac{-4}{3} + 2, \qquad \frac{-1}{3} = \frac{-4}{3} + 1$$

Thus, first isolate the common factor $x^{-4/3}$.

$$\begin{aligned} x^{2/3} - 2x^{-1/3} + x^{-4/3} &= x^{-4/3}(x^2 - 2x + 1) \\ &= x^{-4/3}(x - 1)^2 \end{aligned}$$

□

EXERCISES

Assume all letters represent positive numbers.

In Exercises 1–40 simplify and express your answer using only positive exponents.

1. $x^{1/2} \cdot x^{1/2}$ **2.** $x \cdot x^{1/2}$ **3.** $x^2 \cdot x^{1/4}$ **4.** $a^{3/4} \cdot a^{5/4}$ **5.** $\frac{y^{3/4}}{y^{1/4}}$

6. $\frac{b^2}{b^{1/2}}$ **7.** $(a^{2/3})^6$ **8.** $(x^{1/2})^{1/4}$ **9.** $(a^{2/3})^{3/2}$ **10.** $\frac{(b^{5/6})^{3/2}}{b^{1/4}}$

11. $x^{1/4} \cdot x^{1/2}$ **12.** $a^{2/3} \cdot a^{1/6}$ **13.** $\frac{y^{3/4}}{y^{1/2}}$ **14.** $\frac{z^{3/5}}{z^{1/10}}$ **15.** $x^{1/2} \cdot x^{1/3}$

16. $b^{3/8} \cdot b^{1/6}$ **17.** $x^{7/10} \cdot x^{-1/5}$ **18.** $y^{2/3} \cdot y^{-1/9}$ **19.** $\frac{x^{3/4}}{x^{1/3}}$ **20.** $\frac{a^{3/5}}{a^{1/2}}$

21. $\frac{b^{1/6}}{b^{-1/3}}$ **22.** $\frac{y^{4/5}}{y^{-3/10}}$ **23.** $a^{3/4} \cdot a^{1/4} \cdot a^{1/2}$ **24.** $x^{1/2} \cdot x^{5/8} \cdot x^{-1/4}$ **25.** $\frac{y^{2/3}y^{1/4}}{y^{1/2}}$

26. $\frac{z \cdot z^{3/4}}{z^{1/2} \cdot z^0}$ **27.** $\frac{(x^{1/4}y^{2/3})^2}{x^{1/2}\, y^{1/3}}$ **28.** $\frac{(a^4 b^6)^{-1/2}}{a^{1/2}\, b^{-1}}$ **29.** $\frac{(ab^2c^3)^{1/6}}{(a^{-1}b^{-2})^{-1/2}}$ **30.** $\frac{(27xyz)^{1/3}}{(25x^2y^{1/2})^{1/2}}$

31. $(x^{1/2}y^{1/3})^6 \cdot \left(\frac{x}{y^2}\right)^{1/3}$ **32.** $\left[\left(\frac{xy^{1/2}}{z}\right)^{3/2}\right]^4$ **33.** $\left(\frac{x^2y^{1/3}}{z^{1/9}}\right)^3 \left(\frac{x^{-1/2}z^{1/4}}{y^{1/2}}\right)^2$ **34.** $\frac{(9a^4b^{4/3}c^{2/3})^{-1/2}}{(125a^{3/5}b^9)^{-1/3}}$ **35.** $27^{1/2}$

36. $\left(\frac{1}{8}\right)^{1/2}$ **37.** $(32a^4)^{1/2}$ **38.** $(-8a^6b^9)^{2/3}$ **39.** $(75a^2b^5)^{-1/2}$ **40.** $\frac{(63a^4)^{3/2}}{(28b^6)^{1/2}}$

In Exercises 41–50 simplify.

41. $(16x)^{1/2}$ **42.** $\left(\frac{27}{8}\right)^{1/3}$ **43.** $\left(\frac{4}{9}\right)^{3/2}$ **44.** $\left(\frac{100}{49}\right)^{-1/2}$ **45.** $\left(\frac{-1}{125}\right)^{-2/3}$

46. $\left(\frac{16}{81}\right)^{-3/4}$ **47.** $3^{1/3} \cdot 3^{2/3}$ **48.** $9^{1/2} \cdot 16^{3/4}$ **49.** $\frac{25^{3/2}}{5}$ **50.** $27^{-1/3} \cdot 9^{3/2}$

In Exercises 51–56 factor by first isolating the lowest powers.

51. $x^{3/2} + 4x^{1/2} + 4x^{-1/2}$ **52.** $y^{2/5} - y^{-3/5} - 2y^{-8/5}$ **53.** $a^{1/3} - a^{-5/3}$

54. $(y + 1)^{1/2} - (y + 1)^{-1/2}$ **55.** $(x^2 + 4)^{1/6} - (x^2 + 4)^{-5/6}$ **56.** $4y^2(2y^2 - 1)^{-1/3} - 8y^4(2y^2 - 1)^{-4/3}$

In Exercises 57–60 simplify by first factoring the numerator (or denominator).

57. $\frac{x^{1/2} - 2x^{-1/2}}{x - 2}$ **58.** $\frac{a^{3/2} + 5a^{1/2} + 6a^{-1/2}}{a^2 - 4}$ **59.** $\frac{(y + 4)^{2/3} - (y + 4)^{-4/3}}{(y + 4)^{2/3}}$

60. $\dfrac{(y^5 + 2y^4)^{1/2} - (16y + 32)^{1/2}}{y - 2}$

61. The base of a right triangle is of length 5 centimeters and the hypotenuse is of length 10 centimeters. Find the length of the altitude.

62. At noon ship A is 84 kilometers due east of ship B. Ship A sails due west at 12 kilometers per hour and ship B sails due north at 18 kilometers per hour. Express the distance between the ships as a function of t, the number of hours after noon (a) but before 7 P.M. that day, (b) at 7 P.M. (c) after 7 P.M.

5.6 Radical Notation

rational exponents and radicals

Definition. Let n be a positive integer, and let a be a real number (with $a \geqslant 0$ if n is even). Define:

$$\sqrt[n]{a} = a^{1/n}$$

For the case of $n = 2$, write

$$\sqrt{a} \qquad \text{in place of} \qquad \sqrt[2]{a}$$

Thus $\sqrt{a}$ indicates the *principal* square root of a; $\sqrt[n]{a}$ indicates the *principal* nth root of a.

EXAMPLE 1

(a) $\sqrt{9} = 9^{1/2} = 3$ (b) $\sqrt[3]{8} = 8^{1/3} = 2$

(c) $\sqrt[3]{-8} = (-8)^{1/3} = -2$ (d) $\sqrt[4]{16} = 16^{1/4} = 2$

(e) $\sqrt[4]{-16}$ is not defined (as a real number) because $-16 < 0$ and 4 is *even.* □

$\sqrt[n]{a}$ is called **radical notation;** the symbol $\sqrt{\ }$ is the **radical sign.** Radical notation is widely used. You will now reconsider the material on rational exponents in terms of radicals.

Recall that if m is a nonzero integer, n is a positive integer, and a is any real number,

$$a^{m/n} \qquad \text{was defined as} \qquad (a^{1/n})^m,$$

provided that $a \geqslant 0$ if n is even, and $a \neq 0$ if $m < 0$. You also learned that with these restrictions,

$$(a^{1/n})^m = (a^m)^{1/n}.$$

In radical notation,

$$(a^{1/n})^m = (\sqrt[n]{a})^m \qquad \text{and} \qquad (a^m)^{1/n} = \sqrt[n]{a^m}$$

Thus

$$a^{m/n} = (\sqrt[n]{a})^m = \sqrt[n]{a^m}$$

For example,

$$8^{2/3} = \left(\sqrt[3]{8}\right)^2 = 2^2 = 4 \qquad \text{and} \qquad 8^{2/3} = \sqrt[3]{8^2} = \sqrt[3]{64} = 4$$

EXAMPLE 2 Let $a \geq 0$ and $x \neq 0$. Express in radical notation:

(a) $a^{3/4}$ (b) $2^{2/3}$ (c) $\dfrac{1}{x^{3/5}}$

Solution. (There are several possible forms for the final answer.)

(a) $a^{3/4} = \left(\sqrt[4]{a}\right)^3 = \sqrt[4]{a^3}$ (b) $2^{2/3} = \left(\sqrt[3]{2}\right)^2 = \sqrt[3]{2^2} = \sqrt[3]{4}$

(c) $\dfrac{1}{x^{3/5}} = \dfrac{1}{\left(\sqrt[5]{x}\right)^3} = \dfrac{1}{\sqrt[5]{x^3}}$ □

EXAMPLE 3 Let a be any number. Express in terms of rational exponents:

(a) $\sqrt[5]{17}$, (b) $\sqrt[3]{a^2}$, (c) $\left(\sqrt[7]{a}\right)^4$

Solution.

(a) $\sqrt[5]{17} = 17^{1/5}$ (b) $\sqrt[3]{a^2} = a^{2/3}$ (c) $\sqrt[7]{a^4} = a^{4/7}$ □

Rules (**D**) and (**E**) for rational exponents $\dfrac{1}{n}$ can now be restated in terms of radical notation. Let a and b be nonzero numbers, let m and n be rational numbers, and suppose that $\sqrt[n]{a}$ and $\sqrt[n]{b}$ are defined (as real numbers).

(**D**) $\sqrt[n]{ab} = \sqrt[n]{a} \cdot \sqrt[n]{b}$

(**E**) $\sqrt[n]{\dfrac{a}{b}} = \dfrac{\sqrt[n]{a}}{\sqrt[n]{b}}$

EXAMPLE 4 (a) $\sqrt{9\,000\,000} = \sqrt{9 \times 10^6} = \sqrt{9}\sqrt{10^6} = 3(10^3) = 3000$

(b) $\sqrt[4]{\dfrac{16}{81}} = \dfrac{\sqrt[4]{16}}{\sqrt[4]{81}} = \dfrac{2}{3}$

(c) For any numbers a, b, and c,

$$\sqrt[3]{\frac{a^6 b^9 c}{27}} = \frac{\sqrt[3]{a^6}\,\sqrt[3]{b^9}\,\sqrt[3]{c}}{\sqrt[3]{27}} = \frac{a^2 b^3 \sqrt[3]{c}}{3}$$ □

When applying Rules (**A**), (**B**), and (**C**), first convert to rational exponents.

EXAMPLE 5 Let $x > 0$. Express in terms of rational exponents, and simplify.

(a) $\sqrt{x} \cdot \sqrt[4]{x}$ (b) $\dfrac{x}{\sqrt{x}}$ (c) $\sqrt[3]{x^9}$

Solution.

(a) $\sqrt{x} \cdot \sqrt[4]{x} = x^{1/2} \cdot x^{1/4} = x^{3/4}$ (b) $\dfrac{x}{\sqrt{x}} = \dfrac{x}{x^{1/2}} = x^{1-1/2} = x^{1/2}$

(c) $\sqrt[3]{x^9} = (x^9)^{1/3} = x^3$ □

arithmetic of radicals

Expressions with radicals (or rational exponents) can be added, subtracted, multiplied, factored, or divided. The *Distributive Laws,*

$$a(b + c) = ab + ac \quad \text{and} \quad (b + c)a = ba + ca,$$

play an important role in these combinations.

In the remainder of this section, if nothing is said to the contrary, assume all letters represent positive numbers.

EXAMPLE 6 Combine, as indicated.

(a) $7(10^{1/3}) - 4(10^{1/3}) + 10^{1/3}$

(b) $\sqrt{25x^3} + \sqrt{64x^3} - \sqrt{x^3}$

(c) $\dfrac{\sqrt{4x^3y^2}}{3} + \sqrt{\dfrac{x^3y^2}{4}}$

Solution.

(a) $7(10^{1/3}) - 4(10^{1/3}) + 10^{1/3} = (7 - 4 + 1)(10^{1/3}) = 4(10^{1/3})$

(b) First note that

$$\sqrt{25x^3} = \sqrt{25}\,\sqrt{x^2}\,\sqrt{x} = 5x\sqrt{x},$$
$$\sqrt{64x^3} = \sqrt{64}\,\sqrt{x^2}\,\sqrt{x} = 8x\sqrt{x},$$

and

$$\sqrt{x^3} = \sqrt{x^2}\,\sqrt{x} = x\sqrt{x}.$$

Thus

$$\sqrt{25x^3} + \sqrt{64x^3} - \sqrt{x^3}$$
$$= 5x\sqrt{x} + 8x\sqrt{x} - x\sqrt{x} = 12x\sqrt{x}$$

(c) $$\frac{\sqrt{4x^3y^2}}{3} = \frac{\sqrt{4}\,\sqrt{x^2}\,\sqrt{x}\,\sqrt{y^2}}{3} = \frac{2xy\sqrt{x}}{3}$$

Here 3 is not under the radical sign. However, in the following, 4 is under the radical sign:

$$\sqrt{\frac{x^3y^2}{4}} = \frac{\sqrt{x^2}\,\sqrt{x}\,\sqrt{y^2}}{\sqrt{4}} = \frac{xy\sqrt{x}}{2}$$

Thus

$$\frac{\sqrt{4x^3y^2}}{3} + \sqrt{\frac{x^3y^2}{4}} = \frac{2xy\sqrt{x}}{3} + \frac{xy\sqrt{x}}{2}$$
$$= \frac{4xy\sqrt{x} + 3xy\sqrt{x}}{6}$$
$$= \frac{7xy\sqrt{x}}{6}$$ □

In general,

$$\sqrt{a} + \sqrt{b} \neq \sqrt{a+b}$$

For example, $\sqrt{4} + \sqrt{9} = 2 + 3 = 5$, but $\sqrt{4+9} = \sqrt{13}$
Thus

$$\sqrt{4} + \sqrt{9} \neq \sqrt{4+9}$$

EXAMPLE 7 (a) $\sqrt{2}(3 - \sqrt{3}) = \sqrt{2}\cdot 3 - \sqrt{2}\sqrt{3} = 3\sqrt{2} - \sqrt{6}$

(b) $\sqrt{2}\,(\sqrt{2} + 1) = \sqrt{2}\,\sqrt{2} + \sqrt{2}\cdot 1 = 2 + \sqrt{2}$

(c) $(ab^{1/2} + 1)(a^{1/2}b^{1/2} - 2) = ab^{1/2}\cdot a^{1/2}b^{1/2} + 1\cdot a^{1/2}b^{1/2} - 2ab^{1/2} - 2$
$= a^{3/2}b + a^{1/2}b^{1/2} - 2ab^{1/2} - 2$ □

factoring — Factoring expressions with radicals may lead to further simplification.

EXAMPLE 8 (a) Factor: $2 + \sqrt{12}$ (b) Simplify: $\dfrac{2 + \sqrt{12}}{4}$

Solution.

(a)
$$\begin{aligned} 2 + \sqrt{12} &= 2 + \sqrt{4\cdot 3} \\ &= 2 + \sqrt{4}\,\sqrt{3} \\ &= 2 + 2\sqrt{3} \\ &= 2(1 + \sqrt{3}) \end{aligned}$$

(b)
$$\frac{2 + \sqrt{12}}{4} = \frac{2(1 + \sqrt{3})}{4} = \frac{1 + \sqrt{3}}{2}$$
□

EXAMPLE 9 Simplify:

(a) $5\sqrt{18} - 2\sqrt{8} + \sqrt{50}$ (b) $\sqrt{x^5y^4z} - 2\sqrt{xy^2z^3}$

Solution.

(a) Although the radicals are all different, they have a common factor, $\sqrt{2}$. Thus

$$\sqrt{18} = \sqrt{9}\,\sqrt{2} = 3\sqrt{2}$$
$$\sqrt{8} = \sqrt{4}\,\sqrt{2} = 2\,\sqrt{2}$$
$$\sqrt{50} = \sqrt{25}\,\sqrt{2} = 5\sqrt{2}$$
$$\begin{aligned} 5\sqrt{18} - 2\sqrt{8} + \sqrt{50} &= 5\cdot 3\sqrt{2} - 2\cdot 2\sqrt{2} + 5\sqrt{2} \\ &= 15\sqrt{2} - 4\sqrt{2} + 5\sqrt{2} \\ &= 16\sqrt{2} \end{aligned}$$

(b)
$$\begin{aligned} \sqrt{x^5y^4z} - 2\sqrt{xy^2z^3} &= x^2y^2\sqrt{xz} - 2yz\sqrt{xz} \\ &= (x^2y - 2z)y\sqrt{xz} \end{aligned}$$
□

absolute value — Finally, observe that for every real number x,

$$\sqrt{x^2} = \begin{cases} x, & \text{if } x \geqslant 0 \\ -x, & \text{if } x < 0 \end{cases}$$

For example,

$$\sqrt{5^2} = 5 \quad \text{and} \quad \sqrt{(-5)^2} = 5 = -(-5)$$

Observe that

$$|5| = 5 \quad \text{and} \quad |-5| = 5 = -(-5).$$

In fact, recall that for all x,

$$|x| = \begin{cases} x, & \text{if } x \geq 0 \\ -x, & \text{if } x < 0 \end{cases}$$

Therefore for every real number x,

$$\sqrt{x^2} = |x|$$

EXERCISES

In Exercises 1–58 assume all letters represent positive numbers.
In Exercises 1–6 simplify.

1. $\sqrt{25}$ **2.** $\sqrt[3]{-27}$ **3.** $\sqrt[5]{32}$ **4.** $\sqrt[4]{10\,000}$ **5.** $\sqrt{(-1)^2}$ **6.** $\left(\sqrt[4]{16}\right)^3$

In Exercises 7–12 express in radical notation.

7. $b^{2/3}$ **8.** $c^{-3/4}$ **9.** $a^{1/2}b^{1/3}$ **10.** $3(ab)^{2/5} + 3ab^{2/5}$ **11.** $(a + b)^{1/2}$ **12.** $a + b^{1/2}$

In Exercises 13–20 express in terms of rational exponents.

13. $\sqrt{x}\sqrt[4]{y}$ **14.** $\sqrt[3]{xy}$ **15.** $\sqrt[3]{x} \cdot y$ **16.** $\left(\sqrt[3]{5x^2y}\right)^2$ **17.** $\sqrt[3]{(xyz)^2}$ **18.** $\sqrt{2 + x}$

19. $2 + \sqrt{x}$ **20.** $\dfrac{\sqrt{x}}{\sqrt[3]{yz}}$

In Exercises 21–30 simplify, if possible.

21. $\sqrt{25a^2}$ **22.** $\sqrt{\dfrac{a^6b^2}{c^2}}$ **23.** $\sqrt[3]{\dfrac{8a^9b^3}{c^{12}}}$ **24.** $\sqrt{a^2} - \sqrt{b^2} + \sqrt{a^2 - b^2}$, $a \geq b$

25. $\sqrt{.01} + \sqrt{.0004}$ **26.** $\sqrt{12a^4c^7}$ **27.** $\left(\sqrt{\dfrac{1}{25a^4}}\right)^3$ **28.** $\sqrt{ab^2}\sqrt[3]{a^9b^6}$

29. $\sqrt[4]{32a^7b^8}$ **30.** $\sqrt[5]{-32a^6bc^5}$

In Exercises 31–44 combine as indicated.

31. $3\sqrt[4]{5} + 2\sqrt[4]{5} - \sqrt[4]{5}$ **32.** $\sqrt{9a} + \sqrt{4a}$ **33.** $\sqrt{\dfrac{13}{4}} + \dfrac{\sqrt{13}}{2}$ **34.** $\dfrac{\sqrt{11}}{3} + \dfrac{\sqrt{11}}{2}$

35. $\dfrac{\sqrt{12}}{5} - \dfrac{3\sqrt{12}}{4} + \dfrac{\sqrt{12}}{2}$ **36.** $\dfrac{x^{1/2}y^{1/3}}{5} - \dfrac{2x^{1/2}y^{1/3}}{3}$ **37.** $\sqrt{2} + \sqrt{8}$ **38.** $3\sqrt{5} - \sqrt{20}$

39. $5\sqrt{7} + \sqrt{63} - \sqrt{28}$ **40.** $3\sqrt{98} - (\sqrt{32} + 4\sqrt{18})$ **41.** $\sqrt{a^5b^2} - a\sqrt{a^3b^2}$

42. $\sqrt{100x^2y^4z^5} - \sqrt{25y^4z}$ **43.** $\dfrac{\sqrt{25a^2b^4c}}{2} - \dfrac{\sqrt{9a^2b^2c}}{3}$ **44.** $\dfrac{\sqrt{xyz^2}}{5} + \dfrac{\sqrt{4xyz^4}}{4}$

In Exercises 45–52 multiply as indicated.

45. $\sqrt{3}(3 + \sqrt{3} - \sqrt{2})$ **46.** $(\sqrt{a} + \sqrt{b})(\sqrt{a} - \sqrt{b})$ **47.** $\left(\sqrt{x} + 2\sqrt{y}\right)^2$

48. $\left(\sqrt{x+y} - \sqrt{y}\right)\left(\sqrt{x+y} + \sqrt{y}\right)$ **49.** $x^{1/2}(x^{1/2} + 1)$ **50.** $y^{1/6}(y - y^{1/2} + y^{1/3})$

51. $(x^{1/4} - y^{1/4})^2$ **52.** $(x^{2/3} + x^{1/3})(x^{1/2} - x^{1/3})$

In Exercises 53–58, simplify.

53. $\dfrac{8 + 10\sqrt{3}}{2}$ **54.** $\dfrac{3a\sqrt{x} + 6b\sqrt{x}}{3\sqrt{x}}$ **55.** $\dfrac{\sqrt{18a^2b^3} - \sqrt{9a^4b^3}}{3a^4b}$

56. $\dfrac{\sqrt{125x^2y^4} - \sqrt{49x^3y^2}}{x^2y}$ **57.** $\dfrac{8a^{1/2}b^{3/2} - 2a^{3/2}b^{1/2}}{4a^{1/2}b^{1/2}}$ **58.** $\dfrac{(9ab^3)^{1/2} - (36ab^3)^{1/2}}{3a^{1/2}b^{3/2}}$

59. Let $f(x) = \sqrt{x}$, $x \geq 0$, and let $g(x) = 2x - 1$ for all x.

(a) Find $(f \circ g)(x)$. What is the domain of $f \circ g$? **(b)** Find $(g \circ f)(x)$. What is the domain of $g \circ f$?

(c) Find $(f \circ g)(5)$. **(d)** Find $(g \circ f)(5)$.

60. For all x, let $f(x) = \sqrt[3]{x}$ and $g(x) = x^2 + x + 2$.

(a) Find $(f \circ g)(x)$. **(b)** Find $(g \circ f)(x)$. **(c)** Find $(f \circ g)(2)$. **(d)** Find $(g \circ f)(2)$.

61. Let $h(x) = \sqrt{1 + \sqrt{x}}$, $x \geq 0$.

Find: **(a)** $h(0)$ **(b)** $h(1)$ **(c)** $h(4)$ **(d)** For which value of x does $h(x) = 3$?

62. Find the area of an equilateral triangle of side length 4 centimeters.

5.7 Rationalizing the Denominator

Fractions are usually easiest to work with when their denominators are as simple as possible. For example, $\frac{4}{6}$ can be reduced to $\frac{2}{3}$. Similarly, it is better to write $\frac{4}{-3}$ as $\frac{-4}{3}$ (with *positive* denominator).

When the denominator of a fraction or rational expression involves radicals (or non-integral powers), try to convert to an equivalent expression whose denominator is clear of radicals (or non-integral powers). This process is called **rationalizing the denominator.** For example,

$$\frac{1}{\sqrt{2}} = \frac{1 \cdot \sqrt{2}}{\sqrt{2} \cdot \sqrt{2}} = \frac{\sqrt{2}}{2}$$

The denominator of $\frac{\sqrt{2}}{2}$ is rational. To see that this is the simpler form, consider the *rational approximation* of $\sqrt{2}$:

$$\sqrt{2} \approx 1.414$$

Read: $\sqrt{2}$ is approximately equal to 1.414.

Thus

$$\frac{\sqrt{2}}{2} \approx \frac{1.414}{2} = .707, \qquad \text{whereas} \qquad \frac{1}{\sqrt{2}} \approx \frac{1}{1.414} = \frac{1000}{1414}$$

It is much harder to convert $\frac{1000}{1414}$ to a decimal than it was to convert $\frac{1.414}{2}$.

EXAMPLE 1 Let a be any number, let $x \neq 0$, and let $y > 0$. Rationalize the denominator of each:

(a) $\frac{3}{\sqrt{7}}$ (b) $\frac{a}{x^{1/3}}$ (c) $\frac{-1}{y^{5/4}}$

Solution.

(a) Multiply numerator and denominator by $\sqrt{7}$:

$$\frac{3}{\sqrt{7}} = \frac{3 \cdot \sqrt{7}}{\sqrt{7} \cdot \sqrt{7}} = \frac{3\sqrt{7}}{7}$$

(b) To clear the denominator of fractional powers, multiply the numerator and denominator by $x^{2/3}$. Thus

$$\frac{a}{x^{1/3}} = \frac{a \cdot x^{2/3}}{x^{1/3} \cdot x^{2/3}} = \frac{ax^{2/3}}{x}$$

(c) To obtain an integral power of y in the denominator, multiply numerator and denominator by $y^{3/4}$:

$$\frac{-1}{y^{5/4}} = \frac{-1 \cdot y^{3/4}}{y^{5/4} \cdot y^{3/4}} = \frac{-y^{3/4}}{y^2}$$

□

conjugates

Your next goal is to rationalize the denominator of an expression such as

$$\frac{5}{1 - \sqrt{x}}.$$

Suppose you tried to multiply numerator and denominator by $\sqrt{x}$; then

$$\frac{5}{1 - \sqrt{x}} = \frac{5 \cdot \sqrt{x}}{(1 - \sqrt{x}) \cdot \sqrt{x}} = \frac{5\sqrt{x}}{\sqrt{x} - x}$$

The denominator is still not clear of radicals. Another method must be found. To this end, the expressions

$$A + B\sqrt{C} \qquad \text{and} \qquad A - B\sqrt{C}$$

are called **conjugates.** Each is the **conjugate of** the other. (A may or may not involve a radical. If $B = 1$, it is usually not written, as in parts (a), (b), and (c) of Example 2. For the present, assume $C \geqslant 0$.)

EXAMPLE 2 Let $x > 0$, $y > 0$, and $a > 0$.

(a) $1 + \sqrt{3}$ and $1 - \sqrt{3}$ are conjugates.
(b) $1 - x^{1/2}$ and $1 + x^{1/2}$ are conjugates.

(c) $\sqrt{x} + \sqrt{y}$ and $\sqrt{x} - \sqrt{y}$ are conjugates.
(d) $x + 2\sqrt{a}$ and $x - 2\sqrt{a}$ are conjugates. □

multiplying conjugates

When you multiply conjugates, radicals are cleared. The cross-terms, $AB\sqrt{C}$ and $-AB\sqrt{C}$, are eliminated because of the difference of signs.

EXAMPLE 3 Let $x > 0$, $y > 0$, $a > 0$.

(a) $(1 + \sqrt{3})(1 - \sqrt{3}) = 1 + \cancel{\sqrt{3}} - \cancel{\sqrt{3}} - \sqrt{3}\sqrt{3} = 1 - 3 = -2$

(b) $(1 - x^{1/2})(1 + x^{1/2}) = 1 + \cancel{x^{1/2}} - \cancel{x^{1/2}} - x^{1/2}x^{1/2} = 1 - x$

(c) $(\sqrt{x} + \sqrt{y})(\sqrt{x} - \sqrt{y}) = x + \cancel{\sqrt{x}\sqrt{y}} - \cancel{\sqrt{x}\sqrt{y}} - y$
$= x - y$

(d) $(x + 2\sqrt{a})(x - 2\sqrt{a}) = x^2 + \cancel{2x\sqrt{a}} - \cancel{2x\sqrt{a}} - 4a$
$= x^2 - 4a$ □

To rationalize a denominator of the form $A + B\sqrt{C}$ *or* $A - B\sqrt{C}$, *multiply numerator and denominator by the conjugate of the denominator.*

EXAMPLE 4 Let x ($\neq 1$) be positive, let a be any nonzero number, and let b and c be different positive numbers. Rationalize the denominator of each expression.

(a) $\dfrac{5}{1 - \sqrt{x}}$ (b) $\dfrac{a}{\sqrt{b} + \sqrt{c}}$ (c) $\dfrac{\sqrt{3} + \sqrt{2}}{\sqrt{3} - \sqrt{2}}$

Solution.

(a) $$\frac{5}{1 - \sqrt{x}} = \frac{5 \cdot (1 + \sqrt{x})}{(1 - \sqrt{x}) \cdot (1 + \sqrt{x})} = \frac{5(1 + \sqrt{x})}{1 - x}$$

(b) $$\frac{a}{\sqrt{b} + \sqrt{c}} = \frac{a \cdot (\sqrt{b} - \sqrt{c})}{(\sqrt{b} + \sqrt{c}) \cdot (\sqrt{b} - \sqrt{c})} = \frac{a(\sqrt{b} - \sqrt{c})}{b - c}$$

(c) $$\frac{\sqrt{3} + \sqrt{2}}{\sqrt{3} - \sqrt{2}} = \frac{(\sqrt{3} + \sqrt{2}) \cdot (\sqrt{3} + \sqrt{2})}{(\sqrt{3} - \sqrt{2}) \cdot (\sqrt{3} + \sqrt{2})} = \frac{3 + 2\sqrt{6} + 2}{3 - 2}$$
$$= 5 + 2\sqrt{6}$$

Observe that in (a) and (b), the numerator is left in factored form. There was nothing to be gained by multiplying out. In part (c) the numerator is simpler after multiplying out. □

rationalizing the numerator

In calculus it is sometimes necessary to *rationalize a numerator,* as in Example 5.

EXAMPLE 5 Rationalize the numerator of the difference quotient

$$\frac{f(x + h) - f(x)}{h}, \quad h \neq 0,$$

where (a) $f(x) = \sqrt{x}$, (b) $f(x) = x^{1/3}$

Solution.

(a) $$\frac{f(x+h)-f(x)}{h} = \frac{\sqrt{x+h}-\sqrt{x}}{h}$$
$$= \frac{(\sqrt{x+h}-\sqrt{x})\cdot(\sqrt{x+h}+\sqrt{x})}{h(\sqrt{x+h}+\sqrt{x})}$$
$$= \frac{(\cancel{x}+h)-\cancel{x}}{h(\sqrt{x+h}+\sqrt{x})}$$
$$= \frac{h}{h(\sqrt{x+h}+\sqrt{x})}$$
$$= \frac{1}{\sqrt{x+h}+\sqrt{x}}$$

(b) $$\frac{f(x+h)-f(x)}{h} = \frac{(x+h)^{1/3}-x^{1/3}}{h}$$

To simplify the numerator, recall that
$$a^3 - b^3 = (a-b)(a^2+ab+b^2)$$
Let $a = (x+h)^{1/3}$, $b = x^{1/3}$. Then
$$a^2+ab+b^2 = (x+h)^{2/3} + x^{1/3}(x+h)^{1/3} + x^{2/3}$$
Thus
$$\frac{(x+h)^{1/3}-x^{1/3}}{h}$$
$$= \frac{[(x+h)^{1/3}-x^{1/3}][(x+h)^{2/3}+x^{1/3}(x+h)^{1/3}+x^{2/3}]}{h[(x+h)^{2/3}+x^{1/3}(x+h)^{1/3}+x^{2/3}]}$$
$$= \frac{(\cancel{x}+h)-\cancel{x}}{h[(x+h)^{2/3}+x^{1/3}(x+h)^{1/3}+x^{2/3}]}$$
$$= \frac{1}{(x+h)^{2/3}+x^{1/3}(x+h)^{1/3}+x^{2/3}}$$ □

EXERCISES

In Exercises 1–52 assume that all letters represent positive numbers. Assume, further, that neither the denominator nor the conjugate of the denominator is zero. Rationalize the denominator of each expression.

1. $\frac{2}{\sqrt{3}}$ **2.** $\frac{5}{\sqrt{2}}$ **3.** $\frac{7}{5^{1/2}}$ **4.** $\frac{-1}{\sqrt{6}}$ **5.** $\frac{3}{2\sqrt{2}}$

6. $\frac{-2}{3\sqrt{5}}$ **7.** $\frac{\sqrt{5}}{3\sqrt{2}}$ **8.** $\frac{-3\sqrt{7}}{2\sqrt{5}}$ **9.** $\frac{1}{\sqrt{a}}$ **10.** $\frac{-1}{\sqrt{2x}}$

11. $\frac{x}{\sqrt{ab}}$ **12.** $\frac{3}{\sqrt{x-1}}$, $x > 1$ **13.** $\frac{-5}{2\sqrt{x}}$ **14.** $\frac{-1}{\sqrt{x}\sqrt{y}}$ **15.** $\frac{1+2x}{x^{1/2}}$

16. $\frac{\sqrt{x}+\sqrt{y}}{\sqrt{x^3y}}$ **17.** $\frac{1-\sqrt{2y}}{\sqrt{4y}}$ **18.** $\frac{\sqrt{a}-\sqrt{b}}{\sqrt{8a^3b^3}}$ **19.** $\frac{1}{3^{1/3}}$ **20.** $\frac{-2}{5^{1/4}}$

21. $\frac{3}{\sqrt[3]{x^2}}$ 22. $\frac{-7}{a^{4/5}}$ 23. $\frac{x^{1/2}}{(x-a)^{1/3}}$ 24. $\frac{2}{y^{7/5}}$ 25. $\frac{1}{1+\sqrt{2}}$

26. $\frac{-2}{2+\sqrt{3}}$ 27. $\frac{5}{3-\sqrt{2}}$ 28. $\frac{-3}{7-\sqrt{3}}$ 29. $\frac{4}{1+\sqrt{a}}$ 30. $\frac{-5}{2-\sqrt{b}}$

31. $\frac{\sqrt{x}}{1+\sqrt{x}}$ 32. $\frac{-x\sqrt{x}}{2-\sqrt{x}}$ 33. $\frac{-3}{\sqrt{x+1}}$ 34. $\frac{-5}{\sqrt{x}-y}$ 35. $\frac{-5}{\sqrt{x-y}}$, $x > y$

36. $\frac{-5}{\sqrt{x}-\sqrt{y}}$ 37. $\frac{3}{a-\sqrt{b+1}}$ 38. $\frac{-2ab}{\sqrt{a}-\sqrt{b}}$ 39. $\frac{\sqrt{x}}{\sqrt{x}+\sqrt{y}}$ 40. $\frac{1+\sqrt{x}}{\sqrt{3}+\sqrt{x}}$

41. $\frac{6}{3+2\sqrt{x}}$ 42. $\frac{\sqrt{3}}{5-7\sqrt{a}}$ 43. $\frac{-2}{5+7\sqrt{2}}$ 44. $\frac{9}{3-\sqrt{3x}}$ 45. $\frac{2}{7\sqrt{2}+5\sqrt{3}}$

46. $\frac{-1}{a\sqrt{x}+b\sqrt{y}}$ 47. $\frac{a+b}{a\sqrt{x}-b\sqrt{y}}$ 48. $\frac{7-3x}{3\sqrt{x}-2\sqrt{3}}$ 49. $\frac{1}{\sqrt{x+y}}$ 50. $\frac{1}{\sqrt{x}+\sqrt{y}}$

51. $\frac{a}{\sqrt{a}\,(\sqrt{x}-\sqrt{y})}$ 52. $\frac{y}{\sqrt[3]{x-y}}$

In Exercises 53–58, rationalize the numerator of the difference quotient

$$\frac{f(x+h)-f(x)}{h}, \quad h \neq 0$$

53. $f(x) = \sqrt{x+1}$, $x \geqslant -1$ 54. $f(x) = \sqrt{2x}$, $x \geqslant 0$ 55. $f(x) = \sqrt{1+9x}$, $x \geqslant \frac{-1}{9}$

56. $f(x) = \sqrt{x^2+2}$ 57. $f(x) = x^{1/3} + 2$ 58. $f(x) = x^{2/3}$

59. Rationalize the denominator of $\frac{1}{\sqrt{x+\sqrt{x}}}$, $x > 0$.

5.8 Distance

Definition. The **distance between two points** P_1 and P_2 on the plane is defined to be the length of the line segment $\overline{P_1P_2}$.

Write $\quad \text{dist}(P_1, P_2) \quad$ for the distance between P_1 and P_2.

Because the length of the segment $\overline{P_1P_2}$ is the same as that of $\overline{P_2P_1}$,

$$\text{dist}(P_1, P_2) = \text{dist}(P_2, P_1).$$

horizontal and vertical lines

First consider distance along both horizontal and vertical lines. If the line segment $\overline{P_1P_2}$ is *horizontal*, P_1 and P_2 have the same y-coordinate. Let $P_1 = (x_1, y_1)$ and $P_2 = (x_2, y_1)$. Consequently,

$$\text{dist}(P_1, P_2) = \text{length } \overline{P_1P_2} = |x_2 - x_1|$$

[If P_2 lies to the right of P_1, as in Figure 5.5(a), this length is $x_2 - x_1$. If P_2 lies to the left of P_1, as in Figure 5.5(b), this length is $x_1 - x_2$, which equals $-(x_2 - x_1)$. Note that $|x_2 - x_1|$ covers both cases.]

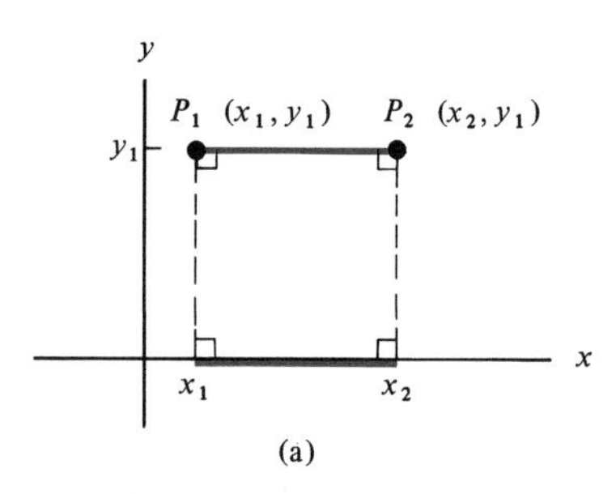

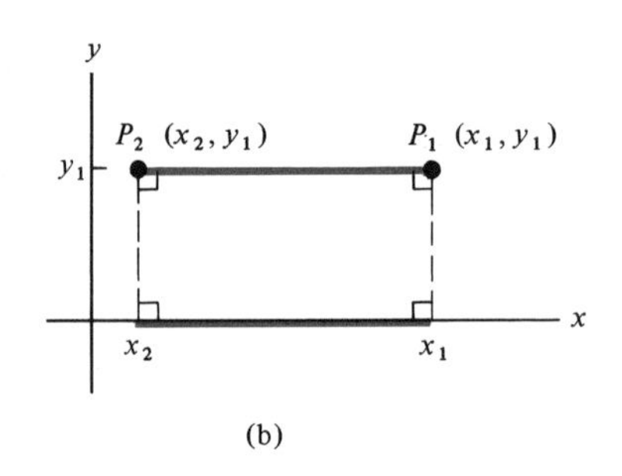

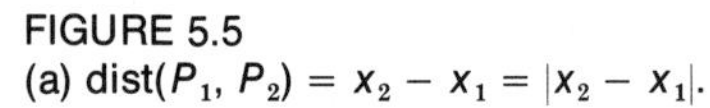
FIGURE 5.5
(a) $\text{dist}(P_1, P_2) = x_2 - x_1 = |x_2 - x_1|$.

(b) $\text{dist}(P_1, P_2) = x_1 - x_2 = |x_2 - x_1|$

If the line segment $\overline{P_1P_2}$ is *vertical,* P_1 and P_2 have the same x-coordinate. Let $P_1 = (x_1, y_1)$ and $P_2 = (x_1, y_2)$. Then

$$\text{dist}(P_1, P_2) = \text{length } \overline{P_1P_2} = |y_2 - y_1|$$

[See Figure 5.6 .]

EXAMPLE 1 [Refer to Figure 5.7 .]

(a) Let $P_1 = (3, 7)$, $P_2 = (8, 7)$. Then $\overline{P_1P_2}$ is horizontal. Both points have the same y-coordinate.
$$\text{dist}(P_1,P_2) = |8 - 3| = 5$$

(b) Let $P_3 = \left(2, \frac{3}{2}\right)$, $P_4 = \left(0, \frac{3}{2}\right)$. Then
$$\text{dist}(P_3, P_4) = |0 - 2| = 2$$

(c) Let $P_5 = (-2, -4)$, $P_6 = (-2, 5)$. Then $\overline{P_5P_6}$ is vertical. Both points have the same x-coordinate.
$$\text{dist}(P_5, P_6) = |5 - (-4)| = 9$$ □

(a)

FIGURE 5.6
(a) $\text{dist}(P_1, P_2) = y_2 - y_1 = |y_2 - y_1|$

Definition. The **midpoint** of a line segment $\overline{P_1P_2}$ is the point M on this segment such that

$$\text{dist}(P_1, M) = \text{dist}(P_2, M).$$

Thus, the midpoint divides the line segment into two segments of equal length. [See Figure 5.8 .]

For a *horizontal* line segment $\overline{P_1P_2}$, let $P_1 = (x_1, y_1)$ and $P_2 = (x_2, y_1)$. Then *the midpoint of* $\overline{P_1P_2}$ *is given by*

$$M = \left(\frac{x_1 + x_2}{2}, y_1\right).$$

Here, $\frac{x_1 + x_2}{2}$ is the *average* of the numbers x_1 and x_2. In fact,

$$\text{dist}(P_1, M) = \left|\frac{x_1 + x_2}{2} - x_1\right| = \left|\frac{x_1 + x_2 - 2x_1}{2}\right| = \frac{|x_2 - x_1|}{2}$$

(b)

FIGURE 5.6
(b) $\text{dist}(P_1, P_2) = y_1 - y_2 = |y_2 - y_1|$

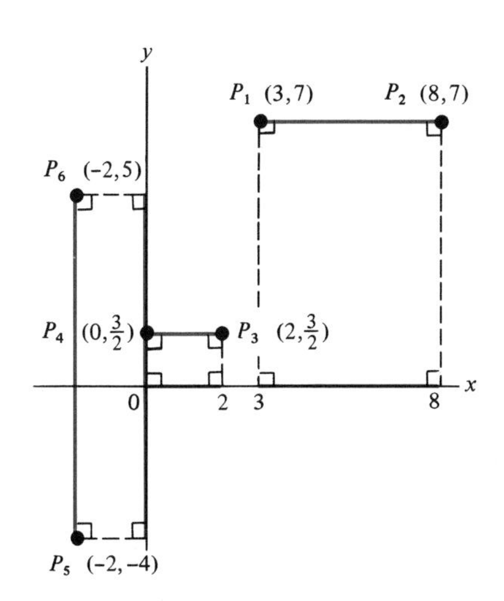

FIGURE 5.7

Also,

$$\text{dist}(P_2, M) = \left|\frac{x_1 + x_2}{2} - x_2\right| = \frac{|x_1 + x_2 - 2x_2|}{2} = \frac{|x_1 - x_2|}{2}$$
$$= \frac{|x_2 - x_1|}{2}$$

Thus

$$\text{dist}(P_1, M) = \text{dist}(P_2, M)$$

Similarly, for a *vertical* line $\overline{P_1P_2}$, let $P_1 = (x_1, y_1)$ and $P_2 = (x_1, y_2)$. Then *the midpoint of* $\overline{P_1P_2}$ *is given by*

$$M = \left(x_1, \frac{y_1 + y_2}{2}\right)$$

$\frac{y_1 + y_2}{2}$ is the *average* of y_1 and y_2.

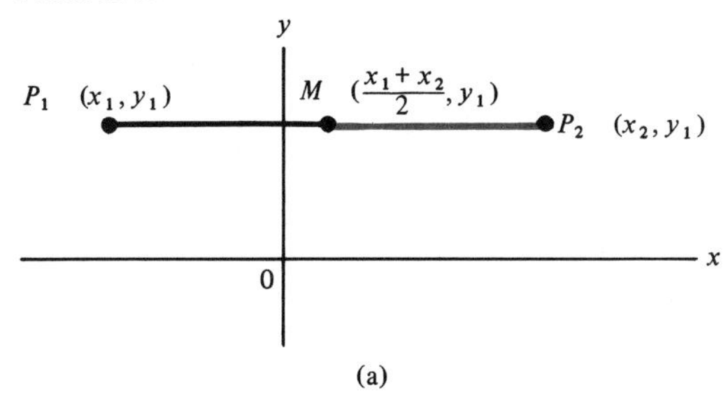

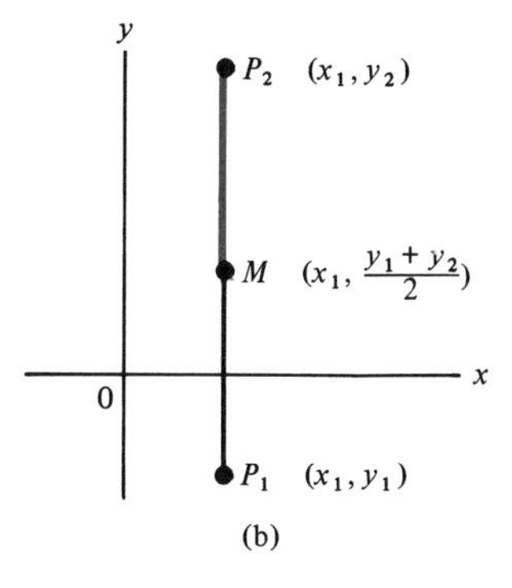

FIGURE 5.8
(a) The midpoint of the horizontal line segment $\overline{P_1P_2}$ is M, where

$$M = \left(\frac{x_1 + x_2}{2}, y_1\right)$$

(b) The midpoint of the vertical line segment $\overline{P_1P_2}$ is M, where

$$M = \left(x_1, \frac{y_1 + y_2}{2}\right)$$

EXAMPLE 2 (a) Let $P_1 = (2, 4)$ and $P_2 = (8, 4)$. Then $\frac{2 + 8}{2} = 5$. Thus the midpoint of $\overline{P_1P_2}$ is (5, 4). [See Figure 5.9 .]

(b) Let $P_3 = (1, -3)$ and $P_4 = (1, 1)$. Then $\frac{-3 + 1}{2} = -1$. Thus the midpoint of $\overline{P_3P_4}$ is $(1, -1)$. [See Figure 5.9 .] □

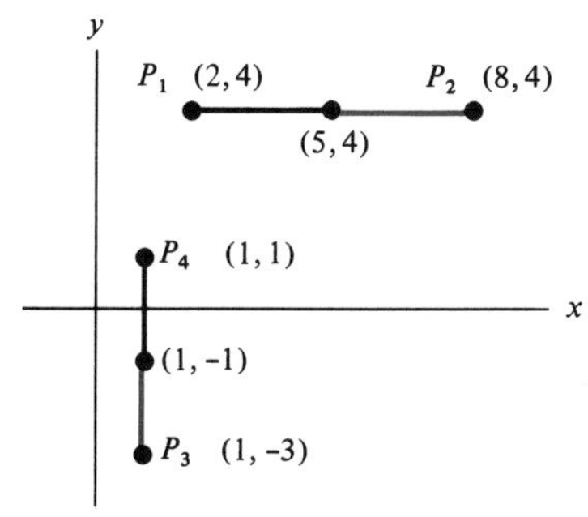

FIGURE 5.9
The midpoint of $\overline{P_1P_2}$ is (5, 4).
The midpoint of $\overline{P_3P_4}$ is $(1, -1)$.

Pythagorean theorem

How do you express $\text{dist}(P_1, P_2)$ when the line segment $\overline{P_1P_2}$ is neither horizontal nor vertical? Let $P_1 = (x_1, y_1)$ and $P_2 = (x_2, y_2)$.

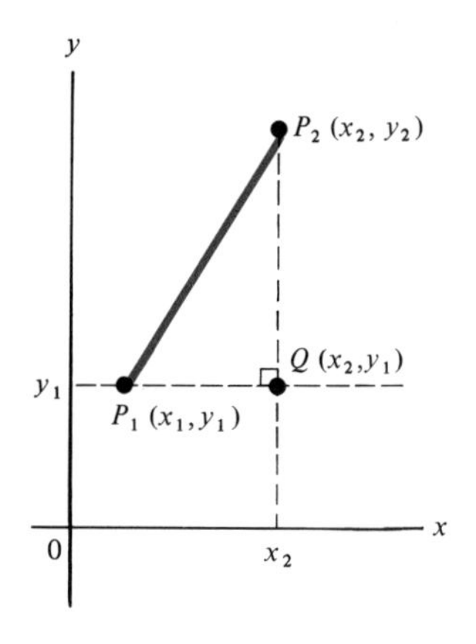

FIGURE 5.10 $\overline{P_1P_2}$ is neither horizontal nor vertical.
$\text{dist}(P_1, P_2) = \sqrt{(x_2 - x_1)^2 + (y_2 - y_1)^2}$

Draw the horizontal line through P_1, and the vertical line through P_2, as in Figure 5.10 . Let Q be the intersection of these lines. Observe that $Q = (x_2, y_1)$. Also, $\overline{P_1P_2}$ is the *hypotenuse of the right triangle* P_1QP_2. According to the *Pythagorean Theorem,*

$$(\text{length } \overline{P_1P_2})^2 = (\text{length } \overline{P_1Q})^2 + (\text{length } \overline{P_2Q})^2$$

or

$$[\text{dist}(P_1, P_2)]^2 = [\text{dist}(P_1, Q)]^2 + [\text{dist}(P_2, Q)]^2$$

But $\overline{P_1Q}$ is horizontal and $\overline{P_2Q}$ is vertical. Therefore

$$[\text{dist}(P_1, P_2)]^2 = (x_2 - x_1)^2 + (y_2 - y_1)^2$$

and

$$\text{dist}(P_1, P_2) = \sqrt{(x_2 - x_1)^2 + (y_2 - y_1)^2}$$

EXAMPLE 3 Let $P_1 = (-2, 1)$, $P_2 = (2, -1)$.

$$\text{dist}(P_1, P_2) = \sqrt{(2 - (-2))^2 + (-1 - 1)^2} = \sqrt{20} = 2\sqrt{5}$$

[See Figure 5.11 .] □

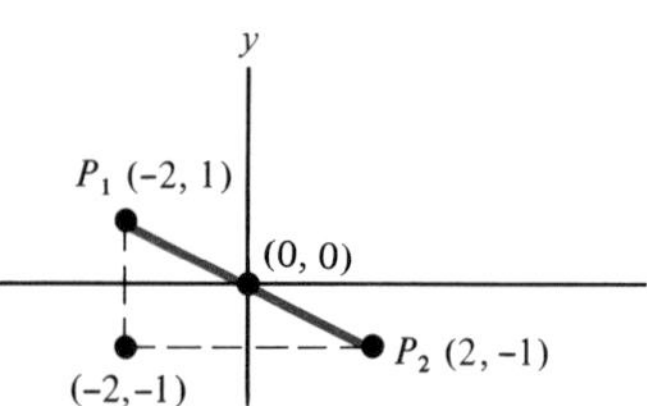

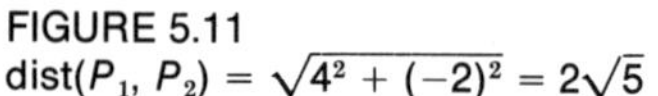

FIGURE 5.11
$\text{dist}(P_1, P_2) = \sqrt{4^2 + (-2)^2} = 2\sqrt{5}$

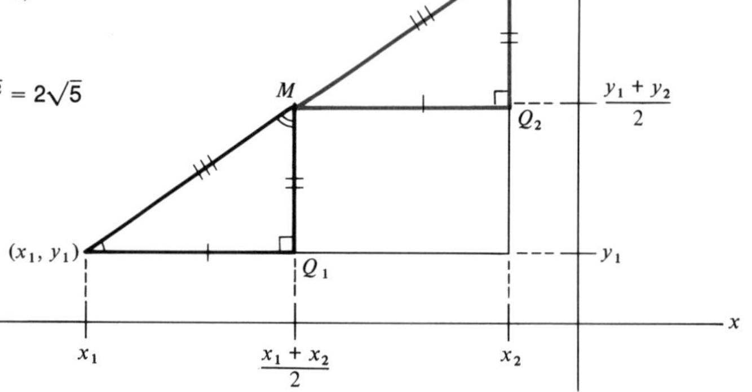

FIGURE 5.12

To find the midpoint of $\overline{P_1P_2}$ when this line segment is neither horizontal nor vertical, observe that in Figure 5.12, right triangles P_1Q_1M and P_2Q_2M are congruent because

$$\text{length } \overline{P_1Q_1} = \text{length } \overline{MQ_2} \text{ and length } \overline{MQ_1} = \text{length } \overline{P_2Q_2}.$$

Thus

$$\text{length } \overline{P_1M} = \text{length } \overline{P_2M}$$

and consequently, M, the midpoint of $\overline{P_1P_2}$, is given by

$$\left(\frac{x_1 + x_2}{2}, \frac{y_1 + y_2}{2}\right).$$ □

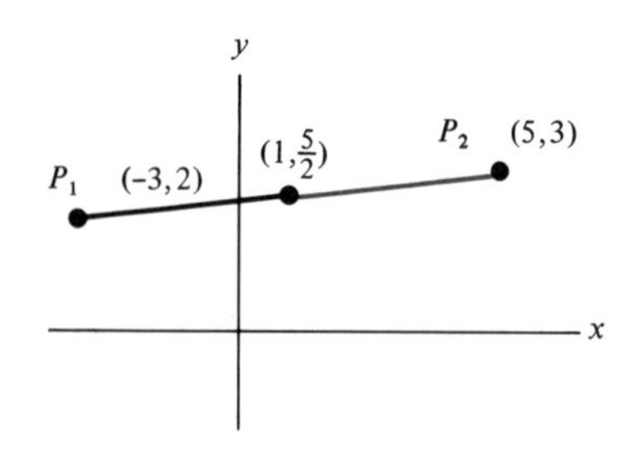

FIGURE 5.13

Thus *each coordinate of the midpoint is the average of the corresponding coordinates of the given points.*

EXAMPLE 4 Let $P_1 = (-3, 2)$ and $P_2 = (5, 3)$. The midpoint of $\overline{P_1P_2}$ is

$$\left(\frac{-3+5}{2}, \frac{2+3}{2}\right) \text{ or } \left(1, \frac{5}{2}\right). \qquad \text{[See Figure 5.13 .]} \qquad \square$$

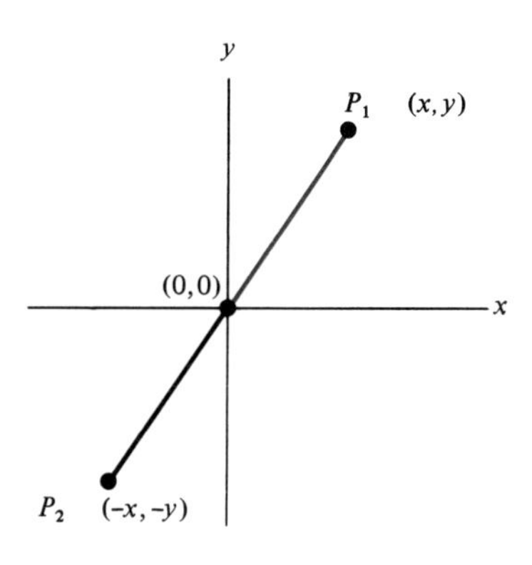

FIGURE 5.14

When $P_1 = (x, y)$ and $P_2 = (-x, -y)$, then the midpoint of $\overline{P_1P_2}$ is

$$\left(\frac{x+(-x)}{2}, \frac{y+(-y)}{2}\right) \qquad \text{or} \qquad (0, 0).$$

[See Figure 5.14 .] Thus, two points P_1 and P_2 are symmetrically located with respect to the origin if the origin is the midpoint of $\overline{P_1P_2}$. This is the case in Figure 5.11, where $P_1 = (-2, 1)$ and $P_2 = (2, -1)$.

EXERCISES

In Exercises 1–14 determine $\text{dist}(P_1, P_2)$.

1. $P_1 = (2, 5), P_2 = (4, 5)$

2. $P_1 = (1, 2), P_2 = (1, 6)$

3. $P_1 = (7, -1), P_2 = (7, -4)$

4. $P_1 = (-3, -3), P_2 = \left(\frac{-1}{2}, -3\right)$

5. $P_1 = (1, 1), P_2 = (4, 5)$

6. $P_1 = (2, 2), P_2 = (4, 4)$

7. $P_1 = (6, 3), P_2 = (3, 6)$

8. $P_1 = (1, 4), P_2 = (8, 5)$

9. $P_1 = (-10, -8), P_2 = (-7, -12)$

10. $P_1 = \left(\frac{1}{2}, 1\right), P_2 = \left(0, \frac{-1}{2}\right)$

11. $P_1 = (8, -3), P_2 = (2, 1)$

12. $P_1 = (9, -5), P_2 = (10, 0)$

13. $P_1 = (1, 1), P_2 = (1.2, .8)$

14. $P_1 = \left(\frac{1}{4}, \frac{1}{2}\right), P_2 = \left(\frac{3}{4}, 0\right)$

In Exercises 15–24 determine the midpoint of $\overline{P_1P_2}$.

15. $P_1 = (4, 3), P_2 = (4, 7)$

16. $P_1 = (-1, 2), P_2 = (9, 2)$

17. $P_1 = \left(0, \frac{1}{2}\right), P_2 = \left(0, \frac{1}{3}\right)$

18. $P_1 = \left(\frac{2}{3}, \frac{1}{3}\right), P_2 = \left(\frac{-3}{2}, \frac{1}{3}\right)$

19. $P_1 = (5, 1), P_2 = (1, 5)$

20. $P_1 = (-2, 5), P_2 = (2, -3)$

21. $P_1 = (7, -4), P_2 = (-7, 4)$

22. $P_1 = \left(6, \frac{1}{4}\right), P_2 = \left(-3, \frac{1}{8}\right)$

23. $P_1 = (9, -9), P_2 = (-3, -13)$

24. $P_1 = (.1, .01), P_2 = (0, 0)$

25. Describe the points of the plane that are the same distance from $(4, 3)$ as from $(-4, 3)$.

26. Find an equation for the points of the plane that are the same distance from (5, 5) as from (5, 1).

27. Let $P = (x_0, y_0)$. Assume $x_0 \neq x_1$, $y_0 \neq y_1$.

(a) Find the (perpendicular) distance from P to the vertical line given by $x = x_1$.
(b) Find the (perpendicular) distance from P to the horizontal line given by $y = y_1$.

28. A motorcycle is traveling along a path described by the equation

$$y = x^3 - 3x^2.$$

(a) What is the motorcycle's distance from the origin when $x = 3$?
(b) Find the distance between the points along its path with x-coordinates 1 and 2, respectively.

29. Show that the points (2, 1), (6, 4), (9, 0), and (5, −3) are the vertices of a square. (Be sure to explain why a pair of adjacent sides is perpendicular.)

30. Let $P = (2, 5)$ and $Q = (4, 9)$. Find the equation of the perpendicular bisector of $\overline{PQ}$.

Review Exercises for Chapter 5

In Exercises 1–12 assume $a > 0$, $b > 0$, $c > 0$, $x > 0$, $y > 0$.

1. Evaluate:

(a) 4^{-2} (b) $\left(\frac{3}{5}\right)^{-1}$ (c) $\frac{2^{-4}}{7^{-1}}$

2. Determine the exponent m.

(a) $\frac{1}{a^3} = a^m$ (b) $\frac{1}{36} = 6^m$ (c) $\frac{1}{16x^4} = (2x)^m$

3. Simplify and write your answers using only positive exponents.

(a) $(a^{-2})^3$ (b) $\frac{a^4a^5}{a^{-2}a^{10}}$ (c) $\frac{(2a^{-1}b)^2}{(4ab^{-2})^{-2}}$

4. Express each in scientific notation.

(a) 6049.4 (b) .1018 (c) .000 639

5. Convert scientific notation back to the usual decimal notation.

(a) 7.19×10^2 (b) 6.38×10^{-1} (c) 1.21×10^{-4}

6. Simplify.

(a) $32^{3/5}$ (b) $100^{-3/2}$ (c) $(64a^{1/4}bc^2)^{1/2}$ (d) $\frac{(2a^{1/3}b^{1/6}c)^6}{8a^2bc^3}$

7. Express in radical notation:

(a) $x^{1/2}y^{1/3}$ (b) $(x + y)^{3/4}$

8. Express in terms of rational exponents:

(a) $\sqrt[5]{(xyz)^3}$ (b) $\frac{\sqrt[3]{x + y}}{\sqrt{x}}$

9. Simplify.

(a) $\sqrt{64a^2b^4}$ (b) $\sqrt[3]{27a^6\sqrt{b^2c^8}}$ (c) $\sqrt{3^2 + 4^2} - \sqrt{a^2} - \sqrt{4c^2}$

10. Combine and simplify:

(a) $\frac{\sqrt{12}}{2} + \frac{\sqrt{12}}{3}$ (b) $x^{1/4}(x + x^{1/2} - x^{3/4})$ (c) $\sqrt{50} + \sqrt{72} - \sqrt{18}$

11. Simplify:

(a) $\dfrac{\sqrt{8} + \sqrt{20}}{2}$ **(b)** $\dfrac{\sqrt{9xy} - \sqrt{25x^3y} + \sqrt{2x^3y^5}}{\sqrt{xy}}$

12. Rationalize the denominator:

(a) $\dfrac{5}{\sqrt{3x}}$ **(b)** $\dfrac{-1}{1 + \sqrt{3}}$ **(c)** $\dfrac{x}{\sqrt{x} - 2\sqrt{y}}$, $x \neq 4y$

13. Let $f(x) = (x - 2)^{1/2} + 1$. Find:

(a) the largest possible domain of this function **(b)** the corresponding range

14. The base of a right triangle is of length 10 inches and the hypotenuse is of length 20 inches. Find the length of the altitude.

15. Let $P_1 = (1, 5)$ and $P_2 = (3, -2)$.

(a) Determine dist (P_1, P_2). **(b)** Determine the midpoint of $\overline{P_1P_2}$.

6 Quadratic Equations and Inequalities

6.1 Solving by Factoring

quadratic equations

A **quadratic,** or **second-degree, equation** (*in a single variable, x*) is an equation that can be written in the form

$$Ax^2 + Bx + C = 0, \qquad \text{where } A \neq 0.$$

The quadratic equation

$$7x^2 + 5x - 31 = 2x - x^2$$

can be transformed into

$$8x^2 + 3x - 31 = 0.$$

Here $A = 8$, $B = 3$, $C = -31$. However, the equation

$$2x^2 + 3x + 2 = 3x^2 - 1 - x^2$$

reduces to

$$\cancel{0x^2} + 3x + 3 = 0$$

and is thus a first-degree, or linear, equation.

$ab = 0$

Some quadratic equations can be solved by means of a very simple principle.

If the product of two numbers is zero, then at least one of these numbers must be zero. And if at least one of the two factors a or b is zero, then the product ab is zero.

Thus you can solve an equation

$$ab = 0$$

by considering two *simpler* equations separately:

$$a = 0, \qquad b = 0$$

EXAMPLE 1 Solve the equation: $(t + 7)(2t - 3) = 0$

Solution. This is the form

$$ab = 0, \qquad \text{where} \qquad a = t + 7, \qquad b = 2t - 3.$$

Obtain two simpler equations:

$$t + 7 = 0, \qquad 2t - 3 = 0$$

Solve each of these equations separately.

$$t + 7 = 0 \qquad t = -7$$

$$2t - 3 = 0 \qquad 2t = 3 \qquad t = \frac{3}{2}$$

The roots of the given equation are -7 and $\frac{3}{2}$. Note that

$$(t + 7)(2t - 3) = 2t^2 + 11t - 21.$$

Thus the roots of the quadratic equation

$$2t^2 + 11t - 21 = 0$$

are -7 and $\frac{3}{2}$. □

Consider a quadratic equation

$$Ax^2 + Bx + C = 0, \qquad A \neq 0.$$

If you can factor the polynomial on the left side into two first-degree polynomials, you can then apply the preceding method.

EXAMPLE 2 Solve and check: $x^2 - 6x = 0$

Solution. Note that $x^2 - 6x = x(x - 6)$. Thus

$$x(x - 6) = 0$$

$$x = 0 \qquad\Big|\qquad x - 6 = 0, \quad x = 6$$

The roots are 0 and 6.

Check. Replace each occurrence of x in the given equation by 0, then by 6.

$x = 0$:

$$0^2 - 6 \cdot 0 \stackrel{?}{=} 0$$

$$0 \stackrel{\checkmark}{=} 0$$

$x = 6$:

$$6^2 - 6 \cdot 6 \stackrel{?}{=} 0$$

$$36 - 36 \stackrel{\checkmark}{=} 0$$

□

EXAMPLE 3 Solve: $x^2 + 12x + 36 = 0$

Solution. $x^2 + 12x + 36 = (x + 6)^2$. Thus solve: $(x + 6)^2 = 0$

If $a^2 = 0$, then $a = 0$, and if $a = 0$, then $a^2 = 0$. Thus

$$x + 6 = 0$$

$$x = -6$$

The only root is -6. □

In the next example, after you clear fractions, a quadratic equation emerges. You will be able to solve this by factoring.

EXAMPLE 4 Assume $x \neq 0$, $x \neq -1$. Solve:

$$\frac{4}{x+1} - \frac{1}{x} = 1$$

Solution. Multiply both sides by the LCD, $x(x + 1)$.

$$4x - (x + 1) = x(x + 1)$$
$$3x - 1 = x^2 + x$$
$$x^2 - 2x + 1 = 0$$
$$(x - 1)^2 = 0$$
$$x = 1$$

□

The same methods apply when the product of more than two factors equals 0.

EXAMPLE 5 Solve: $x^3 - 4x^2 + 3x = 0$

Solution.

$$x^3 - 4x^2 + 3x = x(x^2 - 4x + 3) = x(x - 1)(x - 3)$$

Thus

$$x(x - 1)(x - 3) = 0$$

You obtain three simpler equations:

$x = 0$	$x - 1 = 0$	$x - 3 = 0$
	$x = 1$	$x = 3$

The roots of the given equation are 0, 1, and 3. □

EXAMPLE 6 Find a quadratic equation whose roots are 2 and 4.

Solution. If 2 and 4 are roots of the equation $Ax^2 + Bx + C = 0$, then when you replace x by each of the numbers 2 and 4 in the quadratic equation, you obtain a true statement in each case. Thus,

$$x = 2 \quad \text{or} \quad x = 4$$

Therefore

$$x - 2 = 0 \quad \text{or} \quad x - 4 = 0, \quad \text{so that}$$

$$(x - 2)(x - 4) = 0$$

Consequently,

$$x^2 - 6x + 8 = 0$$

is a quadratic equation whose roots are 2 and 4. If you multiply both sides of this equation by 2, 3, . . . , you obtain other quadratic equations,

$$2x^2 - 12x + 16 = 0,$$
$$3x^2 - 18x + 24 = 0, \qquad \text{and so on}$$

whose roots are 2 and 4. □

quadratic functions

A **quadratic function** is a function of the form

$$f(x) = Ax^2 + Bx + C, \qquad \text{where } A \neq 0.$$

Thus

$$f(x) = x^2 - 6x + 8$$

is a quadratic function which, by Example 6, takes on the value 0 when $x = 2$ and when $x = 4$.

EXERCISES

In Exercises 1–40 assume $A \neq 0$. Determine all roots of each equation. Check where indicated.

1. $x(x - 2) = 0$

2. $3x(2x + 1) = 0$

3. $(x - 1)(x - 2) = 0$

4. $\left(x + \frac{1}{4}\right)\left(x - \frac{1}{3}\right) = 0$

5. $(3x - 2)(2x - 4) = 0$

6. $(7x + 5)(2x - 3) = 0$

7. $(x - A)(x - B) = 0$

8. $Ax(x + B) = 0$

9. $(Ax - B)(Ax + B) = 0$

10. $(3Ax - C)\left(\frac{x}{A} + B\right) = 0$

11. $(x - 2)(x - 3)(x - 4) = 0$

12. $(x + 3)(x - 1)(2x + 1)(3x + 4)(2x - 5) = 0$

13. $x^2 - 3x = 0$

14. $x^2 + \frac{x}{2} = 0$

15. $4x^2 = 9x$

16. $Ax^2 = Bx$

17. $x^2 - 8x + 7 = 0$ (Check.)

18. $x^2 + 4x + 4 = 0$ (Check.)

19. $x^2 - 5x + 6 = 0$ (Check.)

20. $x^2 - 2x - 8 = 0$ (Check.)

21. $t^2 - 4t - 21 = 0$

22. $y^2 - 3y - 18 = 0$

23. $r^2 - 8r + 16 = 0$

24. $s^2 - s - 90 - 0$

25. $2x^2 + 5x - 3 = 0$

26. $4t^2 - 11t - 3 = 0$

27. $5u^2 - 4u - 1 = 0$

28. $6z^2 - 7z - 3 = 0$

29. $s^2 - \frac{s}{4} - \frac{1}{8} = 0$

30. $t^2 - \frac{2t}{3} + \frac{1}{9} = 0$

31. $1 + \frac{1}{u^2} = \frac{2}{u}, \quad u \neq 0$

32. $\frac{4}{x} + x = 5, \quad x \neq 0$

33. $\frac{2}{x^2} - \frac{3}{x} = -1, \quad x \neq 0$

34. $1 + \frac{6}{t^2} = \frac{5}{t}, \quad t \neq 0$

35. $\frac{2}{1 - u} + \frac{u}{2 + u} = 0, \quad u \neq 1, -2$

36. $\frac{1}{y - 4} + \frac{1}{y + 2} = \frac{-1}{4}, \; y \neq 4, -2$

37. $\frac{30}{x^2 - 9} + 2 = \frac{5}{x - 3}, \quad x \neq 3, -3$

38. $x^3 - 6x^2 + 5x = 0$

39. $x^3 + 2x^2 = 15x$ **40.** $x^3 + 20x = 9x^2$

In Exercises 41–46 determine a quadratic equation whose roots are as indicated.

41. 3 and -4 **42.** -3 and 4 **43.** $\frac{2}{3}$ and 1 **44.** 0 and $\frac{1}{2}$ **45.** only 3 **46.** only $\frac{-1}{4}$

In Exercises 47–50 find each x-value for which $f(x) = 0$.

47. $f(x) = (x - 4)(x + 1)$ **48.** $f(x) = x^2 + x - 12$ **49.** $f(x) = x^2 + 18x + 81$ **50.** $f(x) = 4x^2 + 4x - 3$

6.2 Equations of the Form $x^2 = a$

A quadratic equation of the form $x^2 = a$ is easily handled.

$x^2 = a, a > 0$

If a is positive, there are exactly two numbers whose square is a—namely,

$$\pm\sqrt{a}$$

Read: plus and minus the square root of a

You can also solve the equation $x^2 = a$, for $a > 0$ by factoring "over the *real numbers*." Note that

$x^2 - a = 0$ and thus

$$x^2 - (\sqrt{a})^2 = 0.$$

$$(x - \sqrt{a})(x + \sqrt{a}) = 0$$

$$x - \sqrt{a} = 0 \qquad\qquad x + \sqrt{a} = 0$$

$$x = \sqrt{a} \qquad\qquad x = -\sqrt{a}$$

$$x = \pm\sqrt{a}$$

EXAMPLE 1 Solve: $x^2 = 18$

Solution. The roots are given by

$$x = \pm\sqrt{18} = \pm\sqrt{9\cdot 2} = \pm\sqrt{9}\cdot\sqrt{2} = \pm 3\sqrt{2}.$$ □

(The answer is generally left in radical form, rather than approximated.)

EXAMPLE 2 Solve for x: $9x^2 - 25a^2b^4 = 0$

Solution.

$$9x^2 - 25a^2b^4 = 0$$

$$9x^2 = 25a^2b^4$$

$$x^2 = \frac{25a^2b^4}{9}$$

$$x^2 = \left(\frac{5ab^2}{3}\right)^2$$

$$x = \pm \frac{5ab^2}{3}$$ □

EXAMPLE 3 Solve for y: $(2y + 1)^2 = 9$

Solution. This is of the form $x^2 = 9$ with $x = 2y + 1$. Thus

$$2y + 1 = \pm 3$$

Two first-degree equations are obtained. Each of these is solved for y.

$2y + 1 = 3$	$2y + 1 = -3$
$2y = 2$	$2y = -4$
$y = 1$	$y = -2$

The roots are 1 and -2. □

EXAMPLE 4 Let $A \neq 0$. Solve for x: $(Ax - B)^2 = 7$

Solution. $Ax - B = \pm\sqrt{7}$

$Ax - B = \sqrt{7}$	$Ax - B = -\sqrt{7}$
$Ax = B + \sqrt{7}$	$Ax = B - \sqrt{7}$
$x = \dfrac{B + \sqrt{7}}{A}$	$x = \dfrac{B - \sqrt{7}}{A}$

The 2 roots differ only in the sign of $\sqrt{7}$. Thus, the roots can be written as

$$\frac{B \pm \sqrt{7}}{A}$$ □

$x^2 = 0$ *The equation* $x^2 = 0$ *has* 0 *as its only root.*

EXAMPLE 5 Solve for y: $(3y - C)^2 = 0$

Solution. This is of the form

$$x^2 = 0$$

with $x = 3y - C$. The only root of $x^2 = 0$ is 0. Thus

$$3y - C = 0$$

$$3y = C$$

$$y = \frac{C}{3}$$

The given equation has only 1 root, $\frac{C}{3}$. □

Equations of the form $x^2 = -a$, *where* $a > 0$, will be discussed in Section 6.3 .

substituting u for x^2

In Example 6, when you substitute u for x^2, you obtain a quadratic equation. Solve this for u; then substitute for u and solve for x.

EXAMPLE 6 Solve: $x^4 - 3x^2 + 2 = 0$

Solution. Let $u = x^2$ in the given equation.

$$(x^2)^2 - 3x^2 + 2 = 0$$

$$u^2 - 3u + 2 = 0$$

$$(u - 1)(u - 2) = 0$$

$u - 1 = 0$	$u - 2 = 0$
$u = 1$	$u = 2$

Now substitute x^2 for u:

$x^2 = u = 1$	$x^2 = u = 2$
$x = \pm 1$	$x = \pm\sqrt{2}$

The 4 roots of the given equation are ± 1, $\pm\sqrt{2}$. □

EXERCISES

Assume $A > 0$, $B \neq 0$, $C \neq 0$.

In Exercises 1–26 solve for x. Check the ones so indicated.

1. $x^2 = 1$ **2.** $x^2 = 19$ **3.** $x^2 = 8$ **4.** $x^2 = 24$

5. $5x^2 = 20$ **6.** $100x^2 = 1$ **7.** $\frac{x^2}{4} = 7$ **8.** $16x^2 = 21$

9. $x^2 = 16B^2$ **10.** $x^2 = 5A^4$ **11.** $x^2 = 81A^2B^4C^6$ **12.** $4x^2 = 9A^4B^4$

13. $(x + 3)^2 = 25$ (Check.) **14.** $(3x)^2 = 100$ (Check.)

15. $(2x - 5)^2 = 1$ (Check.) **16.** $\left(\frac{x}{3} + 1\right)^2 = 0$

17. $\left(\frac{x + 2}{2}\right)^2 = 4$ **18.** $\left(\frac{2x - 1}{3}\right)^2 = 3$ **19.** $x^2 = A^2$ **20.** $x^2 = A$

21. $A^2x^2 - B^2 = 0$ **22.** $A^2(x - B)^2 = C^2$ **23.** $4x^2 - 49A^2B^2 = 0$ **24.** $\left(\frac{Ax + B}{C}\right)^2 = 0$

25. $\frac{A^2B^2}{16}(x - C)^2 = 1$ **26.** $16(x - 1)^2 = 25$ (Check.)

In Exercises 27–32 solve for the indicated variable.

27. $(2z + 2)^2 = 4$, for z **28.** $A^2u^2 = 5B^2$, for u **29.** $(t - 1)^2 = 12$, for t

30. $3A^2y^2 = 27B^2$, for y **31.** $A\left(\frac{y-B}{C}\right)^2 = A^3$, for y **32.** $A^2\left(\frac{z+5}{2}\right)^2 = 144$, for z

In Exercises 33 and 34 determine all possible values of x in each proportion.

33. $\frac{x}{2} = \frac{8}{x}$ **34.** $\frac{x}{5} = \frac{4}{x}$

In Exercises 35–38 use an appropriate substitution to obtain a quadratic equation. Then find all roots of the given equation.

35. $x^4 - x^2 = 0$ **36.** $y^4 - 3y^2 = 0$ **37.** $x^4 - 10x^2 + 9 = 0$ **38.** $4t^4 - 13t^2 + 9 = 0$

39. Find all points on the x-axis that are 4 units from (2, 1).

40. Find all points on the y-axis that are 10 units from (−3, 4).

41. Let $f(x) = x^2 - 2$. For which x-values does $f(x) = 7$?

42. Let $g(x) = (2x - 1)^2$. For which x-values does $g(x) = 4$?

43. Let $F(x) = (3x + 5)^2$. For which x-values does $F(x) = 0$?

44. Let $G(x) = (4x - 3)^2$. For which x-values does $G(x) = 5$?

6.3 Complex Roots

The square roots of some integers are also integers. For example,

$$\sqrt{4} = 2, \quad \sqrt{25} = 5, \quad \sqrt{100} = 10$$

You cannot go very far in algebra without meeting square roots of other integers. You consider expressions such as

$$\sqrt{3}, \quad \sqrt{5}, \quad \sqrt{8} \left(\text{or } 2\sqrt{2}\right).$$

These square roots are not integers; in fact, they cannot be expressed as quotients of integers. Thus you consider *irrational numbers. The rational and irrational numbers together form the real numbers.* Up until now, real numbers were all you needed to know about.

In order to solve many problems in advanced mathematics, engineering, and science, you must consider *square roots of negative numbers.* As you know, negative numbers do *not* have square roots *within the real number system* (because for every real number x, $x^2 \geq 0$). To get around this difficulty, it is necessary to consider a *new type of number,* known as a *complex number.*

i ***Definition.*** Call i "the square root of −1".

Write $i = \sqrt{-1}$

Note that i is a *new number,* and *not* a real number. Thus, i does not correspond to a point on the real line. Because $i = \sqrt{-1}$,

i^2 will have to be -1.

You will also have to consider such square roots as

$$\sqrt{-3} \qquad \text{and} \qquad \sqrt{-4}.$$

They are also examples of this *new type of number,* and can be defined in terms of i by extending the rule

$$\sqrt{ab} = \sqrt{a}\,\sqrt{b}.$$

Multiplication in the new number system will be defined in Chapter 10 so that

$$\sqrt{-3} = \sqrt{3(-1)} = \sqrt{3}\,\sqrt{-1} = \sqrt{3}\,i,$$
$$\sqrt{-4} = \sqrt{4(-1)} = \sqrt{4}\,\sqrt{-1} = 2i.$$

Now you will be able to find a root of some second-degree equations, such as

$$x^2 = -3 \qquad \text{and} \qquad x^2 = -4.$$

Extend the rule $(ab)^2 = a^2b^2$ to obtain

$$(2i)^2 = 2^2i^2 = 4(-1) = -4.$$

Therefore, $2i$ is a root of the equation $x^2 = -4$. Furthermore,

$$(-2i)^2 = (-2)^2i^2 = 4(-1) = -4,$$

so that $-2i$ is also a root of the equation $x^2 = -4$. Similarly,

$$(\sqrt{3}\,i)^2 = (\sqrt{3})^2i^2 = 3(-1) = -3 \qquad \text{and}$$
$$(-\sqrt{3}\,i)^2 = (-\sqrt{3})^2i^2 = 3(-1) = -3$$

Thus, $\sqrt{3}\,i$ and $-\sqrt{3}\,i$ are each roots of the equation $x^2 = -3$.

$x^2 = -a,\ a > 0$

Consider an equation of the form

$$x^2 = -a, \quad \text{where } a > 0. \text{ Then}$$
$$x^2 = a(-1),$$

so that

$$x = \pm\sqrt{a}\,\sqrt{-1} = \pm\sqrt{a}\,i$$

EXAMPLE 1 Solve for x:

$$x^2 = -12$$

Solution.

$$x = \pm\sqrt{12}\,i = \pm 2\sqrt{3}\,i$$

□

$x + yi$

As you have seen, you must consider numbers of the form

$$yi, \qquad \text{where } y \text{ is real.}$$

To solve certain quadratic equations, such as the one in Example 5, you must also work with numbers of the form

$$x + yi, \qquad \text{where } x \text{ and } y \text{ are both real.}$$

Definition. A **complex number** is an expression of the form

$$x + yi,$$

where x and y are real numbers, and $i = \sqrt{-1}$.

EXAMPLE 2 Each of the following is a complex number:

(a) $2 + 3i$ (b) $3 - 2i$ (or $3 + (-2)i$)

(c) $\dfrac{\sqrt{3} + i}{4} \left(\text{or } \dfrac{\sqrt{3}}{4} + \dfrac{1}{4} i \right)$ □

The complex number $x + 0i$ is considered to be the same as the real number x. Write

$$x \quad \text{instead of} \quad x + 0i$$

Complex numbers of the form $0 + yi$ are called **imaginary.** Write

$$yi \quad \text{instead of} \quad 0 + yi$$

Thus write

$$6 \text{ for } 6 + 0i \quad \text{and} \quad 7i \text{ for } 0 + 7i$$

Definition. Let $x + yi$ be a complex number. Then x is called the **real part** and y the **imaginary part of** $x + yi$.

Note that *both the real and the imaginary parts of a complex number are real numbers.*

EXAMPLE 3 (a) The real part of $3 + 5i$ is 3; the imaginary part is 5 (and *not* $5i$).

(b) The real part of $\frac{1}{2} - \sqrt{2}\, i$ is $\frac{1}{2}$; the imaginary part is $-\sqrt{2}$.

(c) The real part of $-3i$ is 0; the imaginary part is -3. □

Definition. The **conjugate of a complex number** $x + yi$ is $x - yi$.

EXAMPLE 4 (a) The conjugate of $3 + 2i$ is $3 - 2i$.

(b) The conjugate of $3 - 2i$ is $3 - (-2i)$, or $3 + 2i$.

(c) The conjugate of $\sqrt{2}\, i$ is $-\sqrt{2}\, i$ (because $\sqrt{2}\, i = 0 + \sqrt{2}\, i$).

(d) The conjugate of 7 is 7 (because $7 = 7 + 0i = 7 - 0i$). □

A real number is its own conjugate.

EXAMPLE 5 Solve: $(x - 1)^2 = -9$

Solution.

$$x - 1 = \pm 3i$$

$$x = 1 \pm 3i$$

Thus, the roots are the complex conjugates $1 + 3i$ and $1 - 3i$.

Check. Add the real and imaginary parts separately. (See Section 10.1.)

$((1+3i)-1)^2 \stackrel{?}{=} -9$	$((1-3i)-1)^2 \stackrel{?}{=} -9$
$(3i)^2 \stackrel{?}{=} -9$	$(-3i)^2 \stackrel{?}{=} -9$
$3^2 i^2 \stackrel{?}{=} -9$	$(-3)^2 i^2 \stackrel{?}{=} -9$
$9(-1) \stackrel{\checkmark}{=} -9$	$9(-1) \stackrel{\checkmark}{=} -9$

□

EXERCISES

In Exercises 1–8 write each expression in the form *bi*.

1. $\sqrt{-9}$ **2.** $\sqrt{-100}$ **3.** $\sqrt{-8}$ **4.** $\sqrt{-7}$ **5.** $\sqrt{-\frac{1}{4}}$ **6.** $\sqrt{-\frac{1}{16}}$ **7.** $\sqrt{\frac{-4}{9}}$
8. $\sqrt{\frac{-2}{5}}$

In Exercises 9–18 determine: (a) the real part (b) the imaginary part and (c) the conjugate of the given complex number

9. $1+i$ **10.** $1-i$ **11.** $-2+4i$ **12.** 2 **13.** $2i$ **14.** $-2i$
15. $\frac{1}{2}+\sqrt{2}i$ **16.** $\frac{3-5i}{2}$ **17.** $-\sqrt{2}+\sqrt{3}i$ **18.** 0

In Exercises 19–36 assume $A > 0$, $B \neq 0$, $C \neq 0$. Solve for x. Check the ones so indicated.

19. $x^2 = -16$ (Check.) **20.** $x^2 = -27$ (Check.) **21.** $x^2 + 20 = 0$
22. $x^2 + 500 = 0$ **23.** $x^2 + A^2 = 0$ **24.** $x^2 + A = 0$
25. $(x-2)^2 = -4$ (Check.) **26.** $(x+1)^2 = -25$ (Check.) **27.** $(x+5)^2 = -3$
28. $(2x-1)^2 = -9$ **29.** $(3x+2)^2 = -7$ **30.** $(5x-3)^2 = -8$ (Check.)
31. $(4x+3)^2 + 24 = 0$ **32.** $Ax^2 + B^2 = 0$ **33.** $\left(\frac{Ax+B}{C}\right)^2 = -1$
34. $25x^2 + 36A^2B^2 = 0$ **35.** $4(x-3)^2 = -8$ **36.** $3 + (x+2)^2 = -2$

In Exercises 37 and 38 determine all possible values of x in each proportion.

37. $\frac{x}{9} = \frac{-4}{x}$ **38.** $\frac{x}{-3} = \frac{6}{x}$

In Exercises 39–42 use an appropriate substitution to obtain a quadratic equation. Then find all roots of the given equation.

39. $x^4 - 16 = 0$ **40.** $x^4 - 49 = 0$ **41.** $x^4 + x^2 = 0$ **42.** $x^4 + 3x^2 - 4 = 0$

6.4 Completing the Square‡

Consider the equation:

$$x^2 + 4x - 9 = 0$$

You cannot factor the polynomial on the left side (at least not by the previous techniques). Add 9 to both sides, and obtain:

$$x^2 + 4x = 9 \tag{6.1}$$

The left side is not a square, as is. But perhaps by adding a constant to both sides, the equation

$$x^2 + 4x + \_\_\_\_ = 9 + \_\_\_\_$$

will be of the form

$$t^2 = a,$$

as in Sections 6.2 and 6.3. Consider $(x + 2)^2$:

$$\begin{array}{r} x + 2 \\ x + 2 \\ \hline x^2 + 2x \\ 2x + 4 \\ \hline x^2 + 4x + 4 \end{array}$$

Thus $(x + 2)^2$ is 4 more than $x^2 + 4x$. Therefore add 4 to both sides of Equation (**6.1**), and obtain:

$$\begin{aligned} x^2 + 4x + 4 &= 9 + 4 \\ (x + 2)^2 &= 13 \\ x + 2 &= \pm\sqrt{13} \\ x &= -2 \pm \sqrt{13} \end{aligned}$$

If you are given the equation

$$x^2 + 10x - 9 = 0,$$

in which the coefficient of x is 10 (instead of 4, as in the first equation) you would then consider:

$$(x + 5)^2 = x^2 + 10x + 25$$

Note that

$$5 = \frac{10}{2} \quad \text{and} \quad 5^2 = \left(\frac{10}{2}\right)^2 = 25.$$

A = 1 In general, for the equation

$$x^2 + Bx + C = 0 \quad \text{or} \quad x^2 + Bx = -C,$$

add $\left(\frac{B}{2}\right)^2$ to both sides to **complete the square**:

‡Those wishing to check *complex* roots of equations in Sections 6.4 and 6.5 should first cover Section 10.1.

$$x^2 + Bx + \left(\frac{B}{2}\right)^2 = \underbrace{\left(\frac{B}{2}\right)^2 - C}_{D}$$

$$\left(x + \frac{B}{2}\right)^2 = D$$

$$\begin{array}{r} x + \frac{B}{2} \\ x + \frac{B}{2} \\ \hline x^2 + \frac{B}{2}x \\ \frac{B}{2}x + \left(\frac{B}{2}\right)^2 \\ \hline x^2 + Bx + \left(\frac{B}{2}\right)^2 \end{array}$$

If $D \geqslant 0$, then

$$x + \frac{B}{2} = \pm\sqrt{D}$$

$$x = -\frac{B}{2} \pm \sqrt{D}$$

If $D < 0$, then $-D > 0$ and

$$\left(x + \frac{B}{2}\right)^2 = (-D)(-1)$$

$$x + \frac{B}{2} = \pm\sqrt{-D}\,i$$

$$x = -\frac{B}{2} \pm \sqrt{-D}\,i$$

EXAMPLE 1 Solve: $x^2 - 8x - 5 = 0$

Solution.

$$x^2 - 8x = 5$$

Here $B = -8$, $\quad \frac{B}{2} = -4$, $\quad \left(\frac{B}{2}\right)^2 = (-4)^2 = 16$

$$x^2 - 8x + 16 = 5 + 16$$

$$(x - 4)^2 = 21$$

$$x - 4 = \pm\sqrt{21}$$

$$x = 4 \pm \sqrt{21}$$

□

EXAMPLE 2 Solve: $y^2 + 7y + 8 = 0$

Solution.

$$y^2 + 7y = -8$$

$B = 7$, $\frac{B}{2} = \frac{7}{2}$, $\left(\frac{B}{2}\right)^2 = \frac{49}{4}$

$$y^2 + 7y + \frac{49}{4} = -8 + \frac{49}{4} = \frac{-32 + 49}{4} = \frac{17}{4}$$

$$\left(y + \frac{7}{2}\right)^2 = \frac{17}{4}$$

$$y + \frac{7}{2} = \pm\frac{\sqrt{17}}{2}$$

$$y = \frac{-7}{2} \pm \frac{\sqrt{17}}{2} = \frac{-7 \pm \sqrt{17}}{2}$$

□

EXAMPLE 3 Solve: $x^2 + 10x + 40 = 0$

Solution.

$$x^2 + 10x = -40 \qquad B = 10, \quad \left(\frac{B}{2}\right)^2 = 25$$

$$x^2 + 10x + 25 = 25 - 40$$

$$(x + 5)^2 = -15$$

$$x + 5 = \pm\sqrt{-15} = \pm\sqrt{15}i$$

$$x = -5 \pm \sqrt{15}i$$

□

A ≠ 1 Up to now, you have considered equations of the form

$$x^2 + Bx + C = 0$$

with leading coefficient 1. For the more general equation

$$Ax^2 + Bx + C = 0,$$

where $A \neq 0$, divide both sides by A.

$$x^2 + \frac{B}{A}x + \frac{C}{A} = 0$$

This is now of the form

$$x^2 + bx + c = 0 \qquad b = \frac{B}{A}, \quad c = \frac{C}{A}$$

that you have already considered.

EXAMPLE 4 Solve: $4x^2 + 3x - 1 = 0$

Solution. Divide both sides by the leading coefficient, 4.

$$x^2 + \frac{3}{4}x - \frac{1}{4} = 0$$

$$x^2 + \frac{3}{4}x = \frac{1}{4} \qquad b = \frac{3}{4}, \frac{b}{2} = \frac{3}{8}, \left(\frac{b}{2}\right)^2 = \frac{9}{64}$$

$$x^2 + \frac{3}{4}x + \frac{9}{64} = \frac{1}{4} + \frac{9}{64} = \frac{25}{64}$$

$$\left(x + \frac{3}{8}\right)^2 = \left(\frac{5}{8}\right)^2$$

$$x + \frac{3}{8} = \pm\frac{5}{8}$$

$$x = \frac{-3}{8} + \frac{5}{8} = \frac{1}{4} \qquad \Bigg| \qquad x = \frac{-3}{8} - \frac{5}{8} = -1$$

The roots are $\frac{1}{4}$ and -1.

□

The technique of completing the square will also be utilized in your study of parabolas (Section 6.8) and circles (Section 6.10).

EXERCISES

Solve each equation by completing the square. Check where indicated.

1. $x^2 - 2x - 1 = 0$
2. $u^2 + 2u + 1 = 0$
3. $x^2 + 4x - 2 = 0$
4. $x^2 + 6x + 2 = 0$
5. $y^2 - 4y - 1 = 0$
6. $x^2 + 4x - 1 = 0$
7. $x^2 - 8x + 13 = 0$ (Check.)
8. $x^2 + 14x + 47 = 0$
9. $z^2 - 2z - 4 = 0$ (Check.)
10. $y^2 - y - 5 = 0$
11. $x^2 + x - 1 = 0$
12. $x^2 + 3x + 1 = 0$
13. $x^2 + 5x + 2 = 0$
14. $x^2 - 3x = 5$
15. $x^2 - 5x = -5$
16. $x^2 - 9x + 19 = 0$
17. $x^2 + \frac{1}{2}x - 1 = 0$
18. $x^2 + \frac{3}{2}x - \frac{1}{2} = 0$
19. $x^2 + \frac{2}{3}x - 2 = 0$
20. $x^2 + 4x + 10 = 0$
21. $t^2 + 4t + 2 = 0$
22. $y^2 + y + 1 = 0$
23. $x^2 + 2x + 2 = 0$
24. $x^2 + 4x + 1 = 0$ (Check.)
25. $x^2 - 3x + 3 = 0$
26. $x^2 + 3x + 3 = 0$
27. $2x^2 + 4x + 1 = 0$
28. $2x^2 + 4x - 1 = 0$ (Check.)
29. $2t^2 - 6t + 1 = 0$
30. $4x^2 + 8x + 1 = 0$
31. $4t^2 + 8t = 1$
32. $2x^2 + 5x - 1 = 0$
33. $4x^2 + 4x + 3 = 0$
34. $2x^2 + 3x + 1 = 0$
35. $\frac{x^2}{4} + \frac{x}{2} - \frac{1}{4} = 0$
36. $5x^2 + 2x + 5 = 0$

6.5 The Quadratic Formula

The **quadratic formula** yields the roots of *any* quadratic equation. Write the equation in the form

$$Ax^2 + Bx + C = 0, \qquad A \neq 0.$$

Then the quadratic formula asserts that

$$x = \frac{-B \pm \sqrt{B^2 - 4AC}}{2A}.$$

Before verifying the quadratic formula, first consider how it is used.

the discriminant

The quantity $B^2 - 4AC$ is known as the **discriminant** of the equation. The quadratic formula calls for the square root of the discriminant. Three cases occur, according to whether the discriminant is positive, zero, or negative.

Case 1. $B^2 - 4AC > 0$. There are 2 distinct roots, and they are *real numbers*.

Case 2. $B^2 - 4AC = 0$. There is exactly 1 root. It is the real number $\frac{-B}{2A}$.

Case 3. $B^2 - 4AC < 0$. There are 2 distinct roots. But they are *complex conjugates*. (Recall that complex conjugates are complex numbers of the form $x + yi$ and $x - yi$.)

EXAMPLE 1 Use the quadratic formula to solve:

$$2x^2 + 5x - 3 = 0$$

Solution. $A = 2, \quad B = 5, \quad C = -3$

$$\begin{aligned} x &= \frac{-B \pm \sqrt{B^2 - 4AC}}{2A} \\ &= \frac{-5 \pm \sqrt{25 - 4(2)(-3)}}{2(2)} \\ &= \frac{-5 \pm \sqrt{49}}{4} \\ &= \frac{-5 \pm 7}{4} \end{aligned}$$

$$x = \frac{-5 + 7}{4} = \frac{1}{2} \qquad \Big| \qquad x = \frac{-5 - 7}{4} = -3$$

The roots are $\frac{1}{2}$ and -3. Note that the discriminant is positive:

$$B^2 - 4AC = 49 > 0$$

There are 2 real roots. □

EXAMPLE 2 Use the quadratic formula to solve:

$$4x^2 + 9 = 12x$$

Solution. First write the equation in the form $Ax^2 + Bx + C = 0$.

$$4x^2 - 12x + 9 = 0 \qquad A = 4, B = -12, C = 9$$

$$x = \frac{12 \pm \sqrt{144 - 4(4)(9)}}{2(4)} = \frac{12 \pm \sqrt{144 - 144}}{8} = \frac{12}{8} = \frac{3}{2}$$

Here the discriminant is zero: $B^2 - 4AC = 0$. There is exactly 1 root, and it is the real number $\frac{3}{2}$. (See Exercise 41.) □

EXAMPLE 3 Use the quadratic formula to solve:

$$x^2 + x + 1 = 0$$

Solution. $A = B = C = 1$

$$x = \frac{-1 \pm \sqrt{1 - 4(1)(1)}}{2(1)} = \frac{-1 \pm \sqrt{-3}}{2}$$

The roots are $\frac{-1}{2} \pm \frac{\sqrt{3}}{2}i$. Note that they are complex conjugates. Here the discriminant is negative:

$$B^2 - 4AC = -3 < 0$$ □

derivation of the quadratic formula

The quadratic formula is obtained by completing the square. Consider the quadratic equation

$$Ax^2 + Bx + C = 0, \qquad A \neq 0.$$

Transform this as follows:

$$x^2 + \frac{B}{A}x + \frac{C}{A} = 0$$

$$x^2 + \frac{B}{A}x = -\frac{C}{A}$$

Complete the square. The coefficient of x is $\frac{B}{A}$.

$$x^2 + \frac{B}{A}x + \left(\frac{B}{2A}\right)^2 = \left(\frac{B}{2A}\right)^2 - \frac{C}{A}$$

$$\left(x + \frac{B}{2A}\right)^2 = \frac{B^2}{4A^2} - \frac{C}{A} = \frac{B^2 - 4AC}{4A^2}$$

$$x + \frac{B}{2A} = \pm\sqrt{\frac{B^2 - 4AC}{(2A)^2}} = \pm\frac{\sqrt{B^2 - 4AC}}{2A}$$

$$x = \frac{-B}{2A} \pm \frac{\sqrt{B^2 - 4AC}}{2A} = \frac{-B \pm \sqrt{B^2 - 4AC}}{2A}$$ △

EXERCISES

In Exercises 1–12: (a) Determine the discriminant. (b) Without solving, state whether the given equation has 2 distinct real roots, exactly 1 real root, or 2 complex conjugate roots.

1. $x^2 + 7x - 1 = 0$ **2.** $x^2 + 2x + 3 = 0$ **3.** $x^2 - x - 1 = 0$ **4.** $x^2 + x + 2 = 0$

5. $x^2 + 1 = 2x$ **6.** $x^2 + x + 5 = 0$ **7.** $2x^2 + 5x + 1 = 0$ **8.** $3x^2 + 1 = x$

9. $9x^2 + 6x + 1 = 0$ **10.** $2x^2 + 4x = 3$ **11.** $3x^2 + 10x - 1 = 0$ **12.** $4x^2 - 20x + 25 = 0$

In Exercises 13–34 solve by means of the quadratic formula. Check where indicated.

13. $x^2 + x + 2 = 0$ **14.** $x^2 + 3x + 1 = 0$ **15.** $x^2 - 2x - 2 = 0$ (Check.)

16. $x^2 + 3x + 3 = 0$ **17.** $x^2 - 5x = 10$ **18.** $x^2 + 4 = 4x$ (Check.)

19. $2x^2 + 1 = 0$ **20.** $3x^2 + 4x = 0$ **21.** $2x^2 + 9x - 2 = 0$

22. $9x^2 - 24x + 16 = 0$ **23.** $5x^2 + 10x + 3 = 0$ **24.** $3x^2 - 7x + 3 = 0$

25. $\frac{x^2}{7} + 2x + 7 = 0$ **26.** $\frac{x^2}{2} - 3x = 5$ **27.** $x^2 + \frac{1}{4} = x$ (Check.)

28. $x^2 + .2x = .4$ **29.** $x^2 + 12x + 36 = 0$ **30.** $3x^2 + 10x - 10 = 0$

31. $\dfrac{x^2 + 3x}{2} + 1 = 0$ **32.** $\dfrac{x^2 - x}{4} + \dfrac{1}{2} = 1$ **33.** $2x^2 + Bx - B^2 = 0$

34. $x^2 + Bx + B = 0$

In Exercises 35–39 determine B so that the given equation has just 1 root. [*Hint:* Consider the discriminant.]

35. $x^2 + Bx + 16 = 0$ **36.** $x^2 - Bx + 4 = 0$ **37.** $x^2 + Bx + 1 = 0$

38. $x^2 + Bx + 9 = 0$ **39.** $x^2 + Bx + 4 = 0$

40. Show that the equation $x^2 + Bx - 2 = 0$ always has real roots (regardless of the coefficient B).

41. Solve the equation $4x^2 - 12x + 9 = 0$, obtained in Example 2 on page 235, by factoring the left side.

42. Let $f(x) = x^2 - x - 1$. For which x-values does $f(x) = 0$?

43. Let $g(x) = 3x^2 + 2x - 2$. For which x-values does $g(x) = 0$?

44. Let $h(x) = x^2 + x - 1$. For which x-values does $h(x) = 3$?

6.6 Applications

number problems

Here is a problem whose solution utilizes the quadratic formula.

EXAMPLE 1 Find a *positive* number that is 2 more than its reciprocal. Check your result.

Solution. Let x be this number. Then $\dfrac{1}{x}$ is its reciprocal.

$$\underbrace{\text{The number}}_{x}\ \ \underset{=}{\text{is}}\ \ \underbrace{\text{2 more than}}_{2\,+}\ \ \underbrace{\text{its reciprocal.}}_{\frac{1}{x}}$$

Multiply both sides by x. (See the beginning of Section 6.7 for what can happen when you multiply both sides of an equation by a *variable*.)

$$x^2 = 2x + 1$$

$$x^2 - 2x - 1 = 0$$

By the quadratic formula,

$$x = \frac{2 \pm \sqrt{(-2)^2 - 4(1)(-1)}}{2} = \frac{2 \pm \sqrt{4 + 4}}{2} = \frac{2 \pm 2\sqrt{2}}{2}$$
$$= 1 \pm \sqrt{2}$$

Because x must be *positive* and $\sqrt{2} \approx 1.4$, choose the root $1 + \sqrt{2}$. (The other root is *negative* because $1 - \sqrt{2} \approx -.4$.)

Check. First note that

$$\frac{1}{1 + \sqrt{2}} = \frac{1\cdot(1 - \sqrt{2})}{(1 + \sqrt{2})\cdot(1 - \sqrt{2})} = \frac{1 - \sqrt{2}}{1 - 2} = \sqrt{2} - 1. \qquad \textbf{(6.2)}$$

Now check:

$$1 + \sqrt{2} \stackrel{?}{=} 2 + \frac{1}{1 + \sqrt{2}}$$

$$1 + \sqrt{2} \stackrel{?}{=} 2 + (\sqrt{2} - 1) \qquad \text{by } (6.2)$$

$$1 + \sqrt{2} \stackrel{\checkmark}{=} 1 + \sqrt{2}$$

□

geometry

Geometric problems can often be reduced to quadratic equations. Whenever possible, solve the associated quadratic equation by factoring.

EXAMPLE 2 The perimeter (boundary) of a rectangle is 42 centimeters. The area of the rectangle is 108 square centimeters. Determine the dimensions of the rectangle.

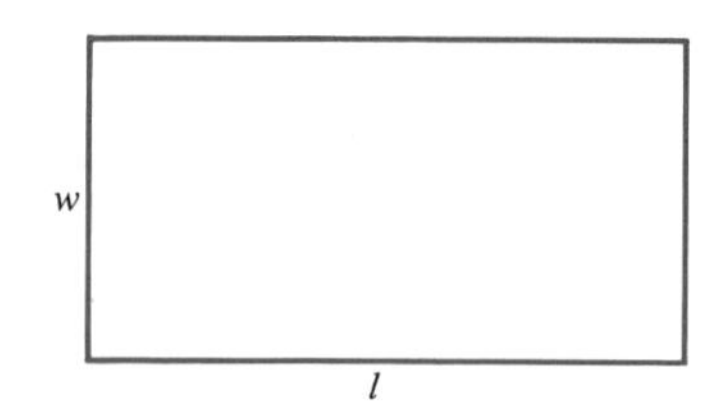

FIGURE 6.1

Solution. Let l and w represent the dimensions of the rectangle. [See Figure 6.1 .] The perimeter is $2(l + w)$. Thus

$$2(l + w) = 42$$
$$l + w = 21$$
$$w = 21 - l$$

The area of the rectangle is $l \cdot w$, or $l(21 - l)$. Thus

$$l(21 - l) = 108$$
$$21l - l^2 = 108$$
$$0 = l^2 - 21l + 108$$
$$0 = (l - 9)(l - 12)$$

$l - 9 = 0$	$l - 12 = 0$
$l = 9$	$l = 12$
$w = 21 - 9 = 12$	$w = 21 - 12 = 9$

The dimensions of the rectangle are 12 centimeters by 9 centimeters. □

distance

In distance problems (Section 2.3), *the distance that an object travels at a constant rate of speed equals its rate multiplied by the time in transit*. If

d = distance, $\quad r$ = rate, $\quad t$ = time,

then $\quad d = r \cdot t$

EXAMPLE 3 A canoeist paddles at a constant rate. She finds that it takes her 2 hours longer to make a 12-kilometer trip upstream than it does downstream. If the current is 3 kilometers per hour, how long would the trip take in still water?

Solution. Let r be the rate (in hours) in still water. The rate is $r + 3$ downstream (with the current) and $r - 3$ upstream (against the current). The distance, 12 kilometers, is the same both ways. [See Table 6.1 .] To find the time t, divide d by r.

$$\underbrace{\frac{12}{r-3}}_{\text{The time upstream}} \underset{\text{is}}{=} \underbrace{2+}_{\text{2 hours more than}} \underbrace{\frac{12}{r+3}}_{\text{the time downstream}}$$

Multiply by the LCD, $(r-3)(r+3)$:

$$12(r+3) = 2(r-3)(r+3) + 12(r-3)$$
$$6(r+3) = (r-3)(r+3) + 6(r-3)$$
$$\cancel{6r} + 18 = r^2 - 9 + \cancel{6r} - 18$$
$$0 = r^2 - 45$$
$$r^2 = 45$$
$$r = \pm\sqrt{45} = \pm 3\sqrt{5}$$

TABLE 6.1

	r	$t = \frac{d}{r}$	d
downstream	$r+3$	$\frac{12}{r+3}$	12
upstream	$r-3$	$\frac{12}{r-3}$	12

The rate is positive; thus $r = 3\sqrt{5}$ (approximately 6.7) kilometers per hour. □

There are distance problems that involve an object traveling at a *variable rate*. Since the rate is *not constant*, the formula $d = r \cdot t$ no longer applies. In fact, the distance is often expressed by means of a quadratic function.

EXAMPLE 4 A skier accelerates as he travels downhill. The distance s traveled in t seconds is given by

$$s = 10t^2 + 10t \text{(feet)}.$$

How long does it take him to go 560 feet downhill?

Solution. Set $s = 560$.

$$560 = 10t^2 + 10t$$
$$56 = t^2 + t$$
$$t^2 + t - 56 = 0$$
$$(t+8)(t-7) = 0$$

$t + 8 = 0$	$t - 7 = 0$
$t = -8$	$t = 7$
(Reject, because t, the number of seconds, must be positive.)	

It takes the skier 7 seconds to travel 560 feet. □

work Work problems were discussed in Section 2.3 . Recall that in work problems,

(the fraction of work done in 1 time unit)$\cdot$(time)
= the fraction of work done

EXAMPLE 5 A father and son working together can paint their house in 8 days. Working alone, it would take the son 12 days longer than his father to paint the house. How long would it take the father, alone, to paint the house? Also, when working together, what fraction of the work does the father do?

Solution. Let x be number of days for the father to paint the house. Then $x + 12$ is the number of days for the son to paint the house.

In 1 day the father paints $\frac{1}{x}$ of the house; the son paints $\frac{1}{x + 12}$ of the house. They work together 8 days. [See Table 6.2 .]

TABLE 6.2

	Fraction of Work Done in 1 Day ·	Days =	Fraction of Work Done
Father	$\frac{1}{x}$	8	$\frac{8}{x}$
Son	$\frac{1}{x + 12}$	8	$\frac{8}{x + 12}$

There is 1 job to be done—namely, to paint the house.

$$\underbrace{\text{The fraction of work done by the father}}_{\frac{8}{x}} + \underbrace{\text{the fraction of work done by the son}}_{\frac{8}{x+12}} = 1$$

Multiply both sides by the LCD, $x(x + 12)$.

$$8(x + 12) + 8x = x(x + 12)$$

$$8x + 96 + 8x = x^2 + 12x$$

$$x^2 - 4x - 96 = 0$$

$$(x - 12)(x + 8) = 0$$

$$x - 12 = 0 \quad\Big|\quad x + 8 = 0$$

$$x = 12 \quad\Big|\quad x = -8 \quad (\textit{Reject.})$$

Clearly, x, the number of days, is positive. Thus $x = 12$ (and $x + 12 = 24$). The father paints $\frac{2}{3}$ of the house because the fraction of work done by him is $\frac{8}{12}$. □

profit Quadratic equations often describe the profit obtained from the sale of goods. As more items are produced, the profit will at first increase. Because of limited plant facilities, a certain number of items produced will result in a maximum profit. Thereafter, as more items are produced the profit decreases. In this case, the "profit curve" is an inverted parabola in which the y-coordinate of the vertex represents the maximum profit. [See Figure 6.2 .]

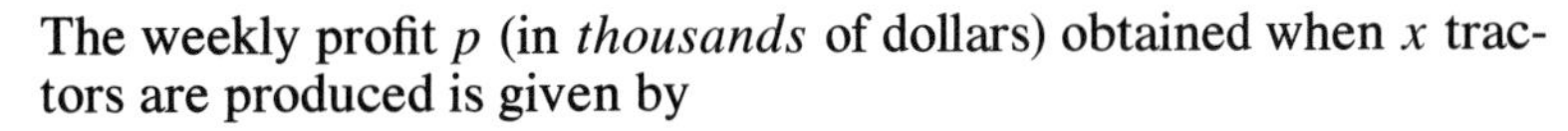

EXAMPLE 6 The weekly profit p (in *thousands* of dollars) obtained when x tractors are produced is given by

$$p = -x^2 + 40x - 150.$$

How many tractors must be produced in a week in order to derive a $250 000 profit?

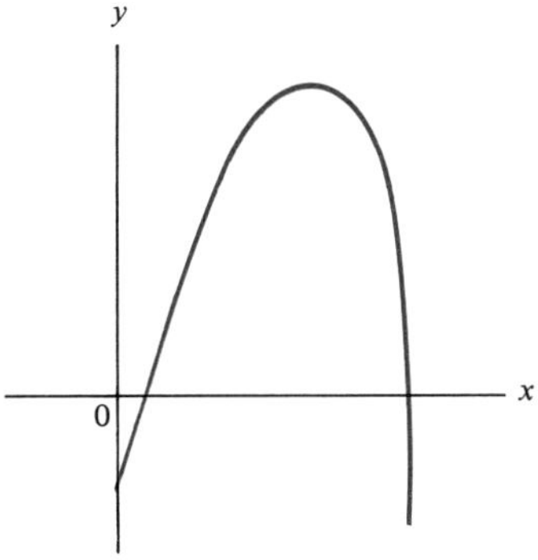

FIGURE 6.2 Profit curve.

Solution. Let $p = 250$ in the given equation.

$$250 = -x^2 + 40x - 150$$

$$x^2 - 40x + 400 = 0$$

$$(x - 20)^2 = 0$$

$$x - 20 = 0$$

$$x = 20$$

Thus, 20 tractors must be produced. □

EXERCISES

1. The sum of two numbers is 12 and the product is 20. Find these numbers.
2. The sum of two numbers is 17 and the product is 72. Find these numbers.
3. The sum of two numbers is 0 and the product is -25. Find these numbers.
4. Twice a certain positive number is five less than its square. Find this number.
5. The square of a negative number is six more than five times the number. Find this number.
6. Show that no real number can equal two more than its square.
7. Determine all nonzero numbers that are such that the sum of the number and its reciprocal is 4.
8. One more than the reciprocal of a negative number is twice that number. Find the number.
9. The product of two consecutive negative integers is 56. Determine these integers.
10. The sum of the squares of two consecutive positive odd integers is 74. Determine these integers.
11. The perimeter of a rectangle is 30 centimeters. The area of the rectangle is 50 square centimeters. Determine the dimensions of the rectangle.
12. The perimeter of a rectangle is 3 feet, 10 inches. The area of the rectangle is 130 square inches. Determine the dimensions of the rectangle.

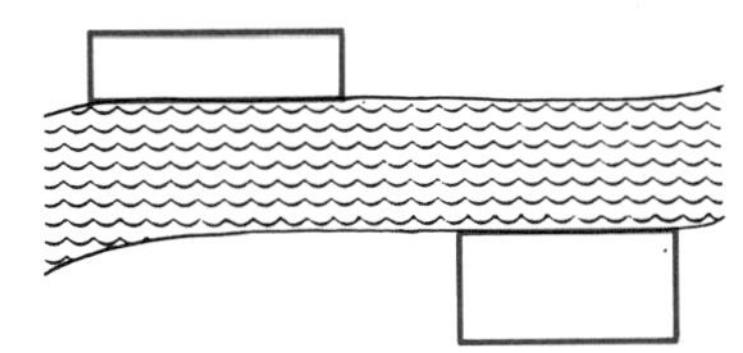

FIGURE 6.3

13. Two rectangular fields with different dimensions border on a river. [See Figure 6.3 .] In each case the field is bounded along one of its longer sides by the river, whereas the other three sides are bounded by a fence whose total length is 90 meters. The area of each field is 1000 square meters. Determine the dimensions of each field.

14. The area of a rectangular Persian rug is 100 square feet. If the length of the rug were decreased by 5 feet and the width increased by 5 feet, the area would be 150 square feet. Determine the dimensions of the rug.

15. The area of a rectangular wood-paneled floor is 160 square feet. If the length were increased by 5 feet and the width increased by 2 feet, the area would be 250 square feet. Determine the dimensions of the floor.

16. A canoe travels at a constant rate in still water. It takes 15 hours longer to make a 40-kilometer trip upstream than it does downstream. If the current is 3 kilometers per hour, how long does the trip take in still water?

17. Two cars, each traveling at a constant rate, leave Boston at the same time, headed for Philadelphia, 300 miles away. One car travels 10 miles per hour faster and arrives 1 hour earlier than the other car. Determine the rate of the faster car.

18. To get to town Jill must jog 3 kilometers through the woods and then ride her scooter the remaining 5 kilometers. She rides 14 kilometers per hour faster than she jogs (at a constant rate). If it takes 45 minutes to make the entire trip, how fast does she jog?

19. An object falls from a window 400 feet high. Its height, h feet, after t seconds, is given by

$$h = 400 - 16t^2 \text{(feet)}.$$

How long does it take to hit the ground?

20. A ball is thrown straight up from ground level. Its height, h feet, after t seconds is given by

$$h = 128t - 16t^2.$$

(a) How long does it take to rise 112 feet?
(b) At what time t is it 112 feet above the ground on its way down?

21. If each works alone, a left-handed clerk takes 6 hours longer than a right-hander to stuff a batch of envelopes. Together they do the job in 4 hours. How long does it take the right-handed clerk, alone, to stuff the envelopes?

22. The hot-water faucet takes 30 minutes longer than the cold-water faucet to fill a tub. When both faucets are turned on, the tub is filled in 8 minutes. How long does it take the hot water faucet, alone, to fill the tub?

23. Sam and Suzy sip a soda together in 6 seconds. On separate sodas, Sam can outsip Suzy by 5 seconds. How long does it take Suzy to sip her soda?

24. The weekly profit p (in *thousands* of dollars) obtained from manufacturing prefabricated garages is given by

$$p = -x^2 + 16x - 24$$

where x is the number of garages produced. How many garages must be produced to derive a weekly profit of $40 000?

25. In Exercise 24, how many garages must be produced to derive a weekly profit of $36 000?

26. In Exercise 24, to the nearest integer, what is the largest number of garages that can be produced before the firm loses money? [*Hint:* Set $p = 0$ in the equation of Exercise 24.]

27. The cost C (in dollars) of building a shelf is given by

$$C = \frac{l^2}{10} - 3l$$

where l is the length of the shelf (in meters). How long a shelf can be made if the cost is kept to $100?

28. Find all points on the line with equation $y = x$ that are 3 units from (2, 1).

6.7 Equations with Square Roots

The Addition Principle (Section 2.1) asserts that you can add the same quantity to both sides of an equation. Similarly, the Multiplication Principle asserts that you can multiply both sides of an equation by the same *nonzero constant*. But if you multiply both sides of an equation by a *variable*, or if you *square* both sides of an equation, you may obtain an equation that has roots that are not solutions of the original equation. The following examples illustrate what can happen.

EXAMPLE 1 Consider the equation:

$$x = 4$$

Multiply both sides by the *variable* x:

$$x^2 = 4x$$
$$x^2 - 4x = 0$$
$$x(x - 4) = 0$$
$$x = 0 \quad \Big| \quad x - 4 = 0$$
$$x = 4$$

The new equation, $x^2 = 4x$, now has an additional root, 0, that is not a root of the original equation, $x = 4$. □

EXAMPLE 2 Consider the equation:

$$x = -5$$

Square both sides and obtain:

$$x^2 = 25$$

$$x = \pm 5$$

The new equation has two roots, $+5$ and -5, whereas the original equation has the single root, -5. □

Equations with square roots are often solved by squaring both sides. In so doing, you do not lose any roots. But Example 2 should convince you that you *may* gain extra roots. *You must check to see which roots are also roots of the original equation.*

EXAMPLE 3 Solve: $\sqrt{3x - 14} = x - 6$

Solution. Square both sides.

$$3x - 14 = (x - 6)^2$$

$$3x - 14 = x^2 - 12x + 36$$

$$x^2 - 15x + 50 = 0$$

$$(x - 5)(x - 10) = 0$$

$$x = 5, 10$$

Now *check* each root to see if it satisfies the original equation

$$\sqrt{3x - 14} = x - 6.$$

For $x = 5$:

$$\sqrt{3(5) - 14} \stackrel{?}{=} 5 - 6$$
$$\sqrt{1} \stackrel{?}{=} -1$$

But $\sqrt{1}$ is *positive*.

$$1 \stackrel{\times}{=} -1$$

This is *false*, so that 5 is *not* a root of the given equation.

For $x = 10$:

$$\sqrt{3(10) - 14} \stackrel{?}{=} 10 - 6$$
$$\sqrt{16} \stackrel{?}{=} 4$$
$$4 \stackrel{\checkmark}{=} 4$$

This is *true*. Therefore 10 is a root, and is the only root of

$$\sqrt{3x - 14} = x - 6.$$ □

In part (a) of the next example, when you square, you get just the root of the original equation. In part (b) when you square, you get a root, although the original equation has no roots.

EXAMPLE 4 Solve:

(a) $\sqrt{x - 3} = 5$ (b) $\sqrt{x - 3} = -5$

Solution.

(a)
$$\sqrt{x - 3} = 5$$
$$x - 3 = 25$$
$$x = 28$$

Check.

$$\sqrt{28 - 3} \stackrel{?}{=} 5$$
$$5 \stackrel{\checkmark}{=} 5$$

Thus 5 is the only root of the given equation.

(b) $\sqrt{t} \geq 0$; thus $\sqrt{x - 3} = -5$ has no root. □

isolating the radical

EXAMPLE 5 Solve: $x + \sqrt{x - 2} = 8$

Solution. When only one radical occurs, isolate this on one side.

$$\sqrt{x - 2} = 8 - x$$

Square both sides.

$$x - 2 = (8 - x)^2$$
$$x - 2 = 64 - 16x + x^2$$
$$x^2 - 17x + 66 = 0$$
$$(x - 6)(x - 11) = 0$$
$$x = 6, 11$$

Check to see if these are roots of the given equation.

$x = 6$:	$x = 11$:
$6 + \sqrt{6-2} \stackrel{?}{=} 8$	$11 + \sqrt{11-2} \stackrel{?}{=} 8$
$6 + \sqrt{4} \stackrel{?}{=} 8$	$11 + \sqrt{9} \stackrel{?}{=} 8$
$6 + 2 \stackrel{\checkmark}{=} 8$	$11 + 3 \stackrel{\times}{=} 8$

Thus 6 is the only root of the given equation. □

two radicals

EXAMPLE 6 Solve: $\sqrt{1+4t} - 1 - \sqrt{2t} = 0$

Solution. Two radicals occur. Bring one of the radicals to the right side:

$$\sqrt{1+4t} - 1 = \sqrt{2t}$$

Square both sides.

$$1 + 4t - 2\sqrt{1+4t} + 1 = 2t$$

Now there is only one radical. Isolate the radical.

$$2t + 2 = 2\sqrt{1+4t}$$

$$t + 1 = \sqrt{1+4t}$$

Again, square both sides to eliminate the radical.

$$t^2 + 2t + 1 = 1 + 4t$$

$$t^2 - 2t = 0$$

$$t(t-2) = 0$$

$$t = 0, 2$$

Check. (the given equation)

$t = 0$:	$t = 2$:
$\sqrt{1+4(0)} - 1 - \sqrt{2(0)} \stackrel{?}{=} 0$	$\sqrt{1+4(2)} - 1 - \sqrt{2(2)} \stackrel{?}{=} 0$
$1 - 1 \stackrel{\checkmark}{=} 0$	$3 - 1 - 2 \stackrel{\checkmark}{=} 0$

Both 0 and 2 are roots of the given equation. □

products of radicals

EXAMPLE 7 Solve: $\sqrt{3x}\,\sqrt{x+1} = 6$

Solution. Square both sides and use $(ab)^2 = a^2b^2$.

$$3x(x+1) = 36$$

$$x(x+1) = 12$$

$$x^2 + x - 12 = 0$$

$$(x+4)(x-3) = 0$$

$$x = -4, 3$$

Check. (the given equation)

$x = -4$:	$x = 3$:
$\sqrt{3(-4)}\,\sqrt{-4+1} \overset{?}{=} 6$	$\sqrt{3(3)}\,\sqrt{3+1} \overset{?}{=} 6$
$\sqrt{4(-3)}\,\sqrt{-3} \overset{?}{=} 6$	$\sqrt{9}\,\sqrt{4} \overset{?}{=} 6$
$2\sqrt{3}i\,\sqrt{3}i \overset{?}{=} 6$	$3(2) \overset{\surd}{=} 6$
$6i^2 \overset{?}{=} 6$	
$-6 \overset{\times}{=} 6$	

Thus 3 is the only root of the given equation. □

EXERCISES

Solve each equation and check the (possible) roots.

1. $\sqrt{x+2} = 2$
2. $\sqrt{x-1} = 3$
3. $\sqrt{2t+2} = 4$
4. $\sqrt{3t+1} = -4$
5. $\sqrt{x+8} - 5 = 0$
6. $\sqrt{2y-5} + 3 = 0$
7. $(13x - 3)^{1/2} - 6 = 0$
8. $\sqrt{x - \frac{3}{4}} = \frac{1}{2}$
9. $\sqrt{2x+2} = 2\sqrt{3}$
10. $\sqrt{5x+3} - 3\sqrt{2} = 0$
11. $x + \sqrt{x-1} = 7$
12. $3x + \sqrt{x-1} = 7$
13. $\sqrt{2u-1} - 8 = -u$
14. $\sqrt{6u+1} - 2u = 1 - u$
15. $\sqrt{1-2x} + x + 1 = 0$
16. $\sqrt{x} + \sqrt{x+5} = 5$
17. $\sqrt{2y} - \sqrt{y-1} = 1$
18. $\sqrt{8-y} + \sqrt{y+1} = 3$
19. $\sqrt{4y+1} + 3 = \sqrt{16+y}$
20. $\sqrt{t-3} - \sqrt{2t-5} + 1 = 0$
21. $\sqrt{1+3y} - 3 = \sqrt{y}$
22. $\sqrt{x+1} - \sqrt{x-2} = 1$
23. $\sqrt{x}\,\sqrt{x-3} = 2$
24. $\sqrt{x}\,\sqrt{x+7} = 12$
25. $\sqrt{z+3}\,\sqrt{z-9} = 8$
26. $\sqrt{3t+1}\,\sqrt{t-7} = 5$
27. $\sqrt{3u}\,\sqrt{9u+1} = 2$
28. $2\sqrt{4x+1} = 3\sqrt{2x}$
29. $5\sqrt{3x+6} = 3\sqrt{9x+10}$
30. $3\sqrt{2x} = 4\sqrt{x+1}$
31. $(x+2)^{1/3} = 2$
32. $x^{2/3} - 4x^{1/3} + 3 = 0$
33. $\sqrt{48x+64} - x = 12$

6.8 Parabolas

The graph of a quadratic equation (*in two variables*)

$$y = Ax^2 + Bx + C, \qquad A \neq 0,$$

is a parabola with a vertical axis of symmetry. Some parabolas were graphed in Section 3.2. Examples 1–5 show you how to cope with various degrees of complexity of such parabolas, beginning with the **basic parabola**

$$y = x^2.$$

EXAMPLE 1 Graph each of the following parabolas.

(a) $y = x^2$ (b) $y = (x + 1)^2$ (c) $y = x^2 + 1$
(d) $y = (x + 1)^2 - 2$

Use these graphs to determine (or to estimate) the roots of the related equations

(a) $x^2 = 0$ (b) $(x + 1)^2 = 0$
(c) $x^2 + 1 = 0$ or $x^2 = -1$
(d) $(x + 1)^2 - 2 = 0$ or $(x + 1)^2 = 2$

obtained by setting $y = 0$ in each given equation.

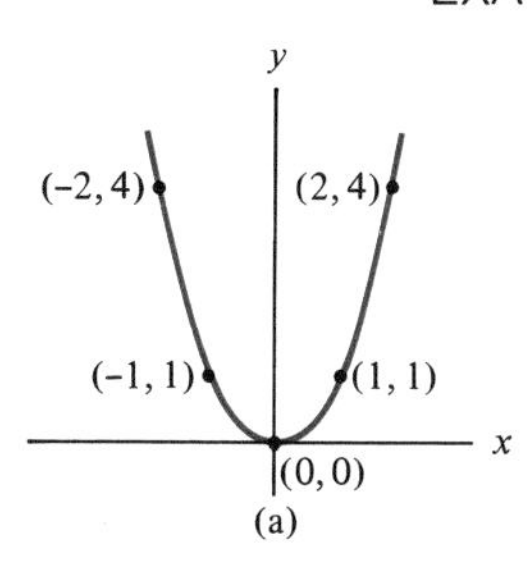

FIGURE 6.4(a) $y = x^2$
The vertex is the origin. The axis of symmetry is the y-axis ($x = 0$).

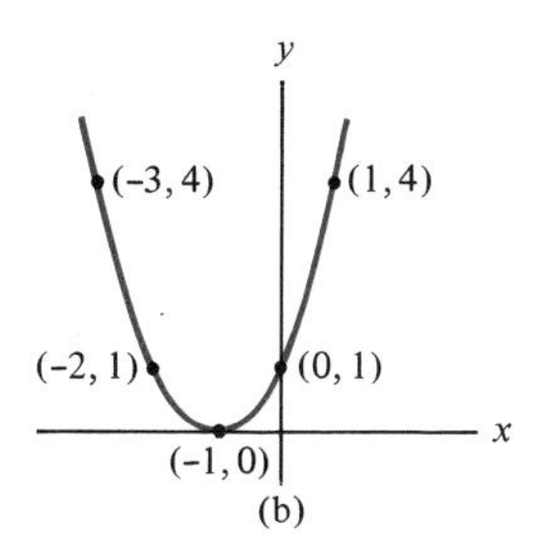

FIGURE 6.4(b)
$y = (x + 1)^2$
The vertex is $(-1, 0)$.
The axis of symmetry is $x = -1$.

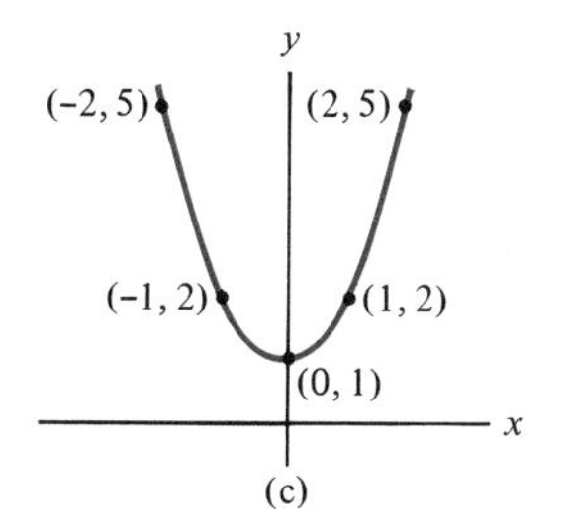

FIGURE 6.4(c) $y = x^2 + 1$
The vertex is $(0, 1)$.
The axis of symmetry is the y-axis.

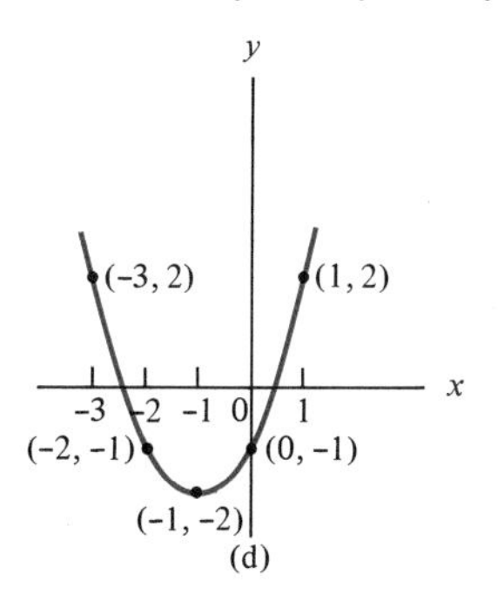

FIGURE 6.4(d) $y = (x + 1)^2 - 2$
The vertex is $(-1, -2)$.
The axis of symmetry is $x = -1$.

Solution. Each of these parabolas opens upward because the coefficient, 1, of x^2 is positive.

(a) See Figure 6.4(a). Observe that $y = 0$ when $x = 0$. Thus replace y by 0 in the equation

$$y = x^2$$

to obtain the related equation

$$x^2 = 0.$$

The only root of this second equation is the x-intercept, 0.

(b) $y = (x + 1)^2$

Thus $y = 0$ when $x = -1$. Shift the graph of the basic parabola 1 unit to the *left*, as in Figure 6.4(b). The only root of the related equation

$$(x + 1)^2 = 0$$

is the x-intercept, -1.

In general, for $y = (x + k)^2$, shift the graph of the basic parabola k units to the left if k is positive, and $|k|$ units to the right if k is negative.

(c) $y = x^2 + 1$

Shift the graph of the basic parabola 1 unit upward, as in Figure 6.4(c). Note that the entire graph is above the x-axis ($y = 0$), and thus has no x-intercept. Therefore, the related equation

$$x^2 + 1 = 0 \quad \text{or} \quad x^2 = -1$$

has no real root.

In general, for $y = x^2 + k$, shift the graph of the basic parabola k units upward if k is positive, and $|k|$ units downward if k is negative.

(d) $y = (x + 1)^2 - 2$

Shift the graph of $y = (x + 1)^2$, of part (b), 2 units downward, as in Figure 6.4(d). The graphical method enables you to *estimate* the roots of

$$(x + 1)^2 - 2 = 0 \quad \text{or} \quad (x + 1)^2 = 2$$

by considering the x-intercepts. One root lies between -3 and -2, and the other root lies between 0 and 1. (The *exact* roots may be obtained by applying the quadratic formula.)

In general, for $y = (x + k)^2 + l$, first shift the graph of the basic parabola $|k|$ units to the left or right, and then shift this new graph $|l|$ units upward or downward. □

In order to graph a parabola whose equation is given in the form

$$y = x^2 + Bx + C, \qquad \text{where } B \neq 0,$$

complete the square. This will make the equation resemble one of the types encountered in Example 1.

EXAMPLE 2 Graph the parabola $y = x^2 + 2x + 3$.

Solution. Half of 2, the middle coefficient, is 1. Thus complete the square on the right side by considering $(x + 1)^2$, which equals $x^2 + 2x + 1$. Add 2 to this to obtain $x^2 + 2x + 3$.

$$y = \underbrace{(x + 1)^2}_{x^2 + 2x + 1} + 2$$

In part (b) of Example 1 you found that $y = (x + 1)^2$ had vertex $(-1, 0)$. By adding 2 to the right side, you shift the graph 2 units upward. Thus the vertex of the parabola given by

$$y = x^2 + 2x + 3 = (x + 1)^2 + 2$$

is $(-1, 2)$. [See Figure 6.5 .] □

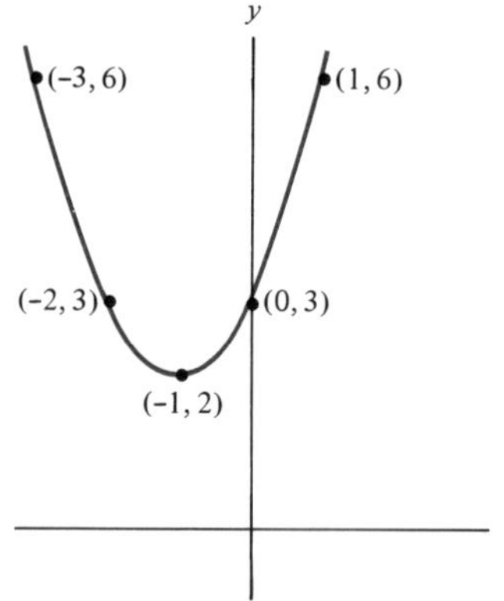

FIGURE 6.5
$y = x^2 + 2x + 3 = (x + 1)^2 + 2$
The vertex is $(-1, 2)$. The axis of symmetry is $x = -1$. The graph lies entirely above the x-axis ($y = 0$). Thus, the related equation $x^2 + 2x + 3 = 0$ has no root.

What happens to the graph of

$y = Ax^2 + Bx + C$ when $A \neq 1$?

EXAMPLE 3 Graph each of the following parabolas.

(a) $y = 2x^2$ (b) $y = -2x^2$
(c) $y = -2(x - 1)^2$ (d) $y = -2(x^2 - 1)$

Solution.

(a) $y = 2x^2$
When $x = 0$, $y = 0$. The vertex is $(0, 0)$. For each value of x, multiply by 2 the corresponding y-value of the basic parabola, $y = x^2$, to obtain $2x^2$. [See Figure 6.6(a).] Here $A = 2$. When A is positive, the parabola opens upward.

(b) $y = -2x^2$
Multiply by -2 each y-value of the basic parabola. Here $A = -2$. When A is negative, the parabola opens downward. [See Figure 6.6(b).]

(c) $y = -2(x - 1)^2$
When $x = 1$, $y = 0$. Shift the graph of part (b) one unit to the right. [See Figure 6.6(c).]

(d) $y = -2(x^2 - 1) = -2x^2 + 2$
Shift the graph of part (b) two units upward. [See Figure 6.6(d).] □

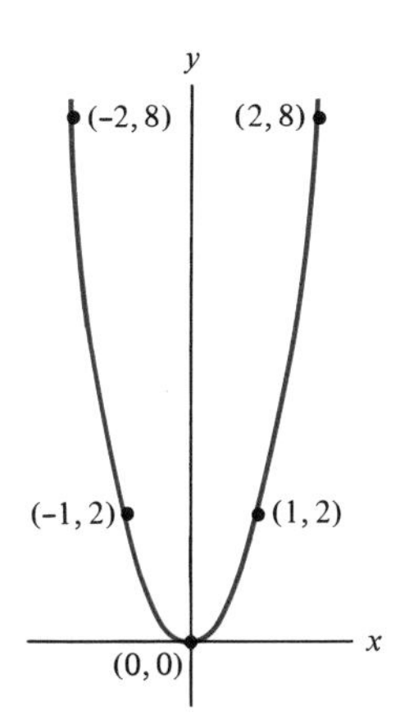

FIGURE 6.6(a)
$y = 2x^2$. The vertex is (0, 0). The axis of symmetry is the y-axis. The related equation $2x^2 = 0$ has 0 as its only root.

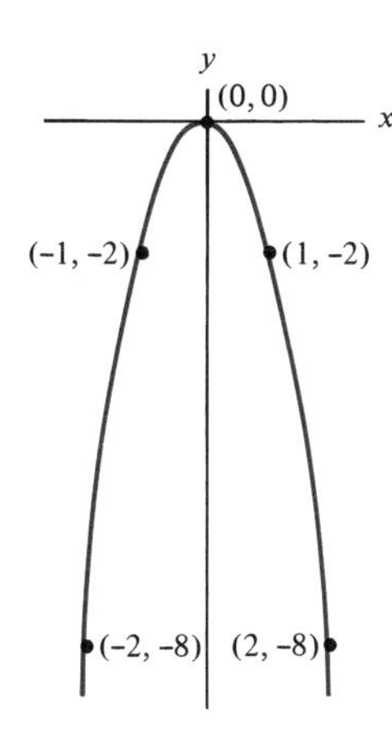

FIGURE 6.6(b)
$y = -2x^2$. The vertex is (0, 0). The axis of symmetry is the y-axis. The related equation $-2x^2 = 0$ has 0 as its only root.

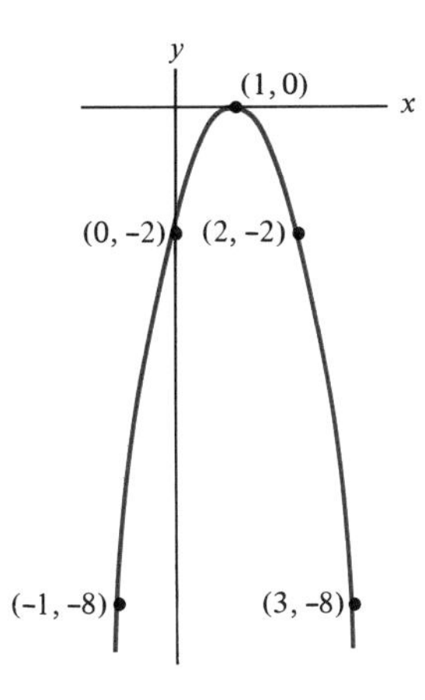

FIGURE 6.6(c)
$y = -2(x - 1)^2$
The vertex is (1, 0). The axis of symmetry is $x = 1$. The related equation $-2(x - 1)^2 = 0$ has 1 as its only root.

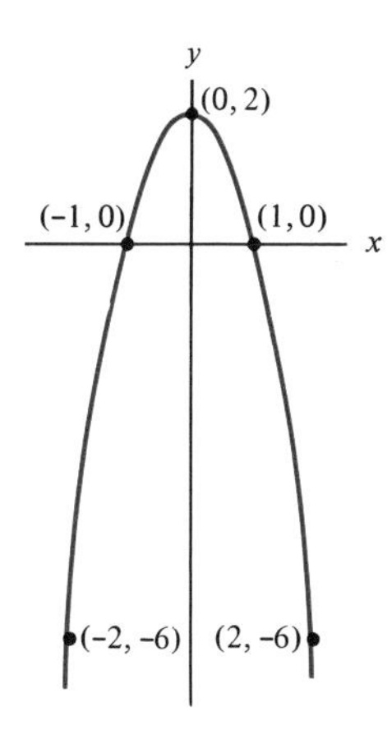

FIGURE 6.6(d)
$y = -2(x^2 - 1) = -2x^2 + 2$
The vertex is (0, 2). The axis of symmetry is the y-axis. The related equation $-2(x^2 - 1) = 0$ has roots -1 and 1.

In Examples 4 and 5, in order to graph a parabola given by

$$y = Ax^2 + Bx + C, \qquad A \neq 1,$$

you must first *complete the square*.

EXAMPLE 4 Graph the parabola whose equation is

$$y = -2x^2 + 4x.$$

Solution.

$$y = -2x^2 + 4x = -2(x^2 - 2x)$$

Complete the square for $x^2 - 2x$. Note that $\frac{-2}{2} = -1$ and $(-1)^2 = 1$. Thus, within the parentheses add 1 and subtract 1.

$$\begin{aligned} y &= -2(x^2 - 2x + 1 - 1) \\ &= -2[(x - 1)^2 - 1] \qquad (-2)(-1) = 2 \\ &= -2(x - 1)^2 + 2 \end{aligned}$$

Thus, shift the graph of part (c) of Example 3 two units upward. [See Figure 6.7 on page 250.] □

EXAMPLE 5 To complete the square for

$$y = -3x^2 + 4x,$$

first "isolate -3".

$$\begin{aligned} y &= -3x^2 + 4x \\ &= -3\left(x^2 - \frac{4}{3}x\right) \end{aligned}$$

$$\frac{-\frac{4}{3}}{2} = \frac{-2}{3} \text{ and } \left(\frac{-2}{3}\right)^2 = \frac{4}{9}$$

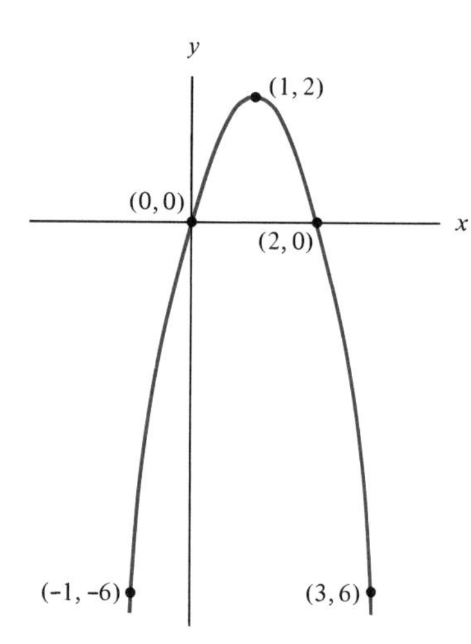

FIGURE 6.7
$y = -2x^2 + 4x = -2(x - 1)^2 + 2$
The vertex is (1, 2). The axis of symmetry is $x = 1$. The related equation $-2x^2 + 4x = 0$ has roots 0 and 2.

$$= -3\left[\left(x^2 - \frac{4}{3}x + \frac{4}{9}\right) - \frac{4}{9}\right]$$

$$= -3\left[\left(x - \frac{2}{3}\right)^2 - \frac{4}{9}\right] \qquad (-3)\left(-\frac{4}{9}\right) = \frac{4}{3}$$

$$= -3\left(x - \frac{2}{3}\right)^2 + \frac{4}{3}$$

The parabola opens downward (because $A = -3$); the vertex is $\left(\frac{2}{3}, \frac{4}{3}\right)$. [See Figure 6.8 .] □

For a parabola given by

$$f(x) = Ax^2 + Bx + C,$$

you complete the square as follows:

$$f(x) = A\left(x^2 + \frac{B}{A}x + \frac{C}{A}\right)$$

$$= A\left[\left(x + \frac{B}{2A}\right)^2 - \frac{B^2}{4A^2} + \frac{C}{A}\right]$$

$$f(x) = A\left[\left(x + \frac{B}{2A}\right)^2 + \frac{-B^2 + 4AC}{4A^2}\right] \tag{6.3}$$

The vertex is located where $x + \frac{B}{2A} = 0$, that is, where

$$x = -\frac{B}{2A} \quad \text{and} \quad y = f\left(-\frac{B}{2A}\right).$$

Setting $x = -\frac{B}{2A}$, in Equation (**6.3**) you obtain

$$f\left(-\frac{B}{2A}\right) = A\left[\underbrace{\left(-\frac{B}{2A} + \frac{B}{2A}\right)^2}_{0} + \frac{-B^2 + 4AC}{4A^2}\right] = \frac{4AC - B^2}{4A}$$

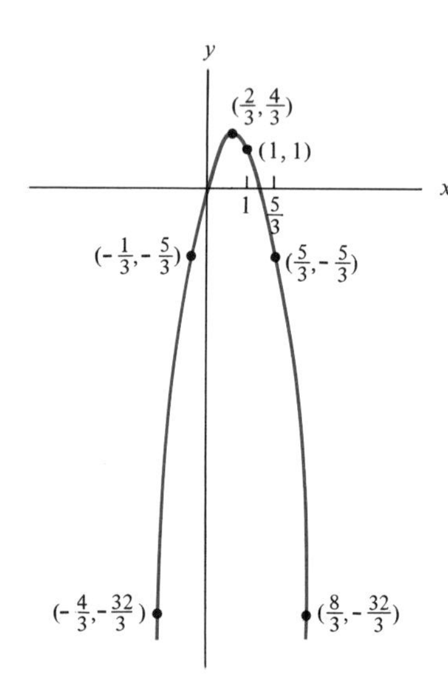

FIGURE 6.8
$y = -3x^2 + 4x = -3\left(x - \frac{2}{3}\right)^2 + \frac{4}{3}$
The vertex is $\left(\frac{2}{3}, \frac{4}{3}\right)$. The axis of symmetry is $x = \frac{2}{3}$. One root of the related equation
$-3x^2 + 4x = 0$ is 0. Because
$-3(1^2) + 4(1) = -3 + 4 = 1 > 0,$
the other root lies between 1 and $\frac{5}{3}$.

Thus the coordinates of the vertex are

$$\left(-\frac{B}{2A}, \frac{4AC - B^2}{4A}\right).$$

For the parabola of Example 5, given by

$$y = -3x^2 + 4x,$$

this yields the vertex previously obtained:

$$\left(-\frac{4}{-6}, \frac{0 - 16}{4(-3)}\right) \quad \text{or} \quad \left(\frac{2}{3}, \frac{4}{3}\right).$$

EXERCISES

In Exercises 1–26: (a) Does the parabola open upward or downward? (b) Find the vertex. (c) Find the equation of the axis of symmetry. (d) Graph the parabola. (e) Use the graph to determine (or to estimate) the roots of the related equation obtained by setting $y = 0$ in the given equation.

1. $y = (x + 2)^2$ **2.** $y = (x - 1)^2$

3. $y = x^2 + 2$ **4.** $y = x^2 - 1$

5. $y = (x - 2)^2 + 3$ **6.** $y = (x + 1)^2 - 2$

7. $y = 3x^2$ **8.** $y = -3x^2$

9. $y = \frac{x^2}{2}$ **10.** $y = 2x^2 + 1$

11. $y = 3x^2 - 4$ **12.** $y = 3 - x^2$

13. $y = 2(x - 3)^2$ **14.** $y = -3(x + 1)^2$

15. $y = 2(x^2 - 2)$ **16.** $y = -4(x^2 + 1)$

17. $y = x^2 + 2x + 5$ **18.** $y = x^2 + 4x + 6$

19. $y = x^2 - 2x + 1$ **20.** $y = x^2 - 4x$

21. $y = x^2 + x + 1$ **22.** $y = x^2 - x$

23. $y = 2x^2 + 8x + 8$ **24.** $y = -4x^2 + 8x$

25. $y = 5 - 5x^2$ **26.** $y = 2x^2 + 3x + 4$

FIGURE 6.9

27. In Figure 6.9 match each parabola with one of the following equations:

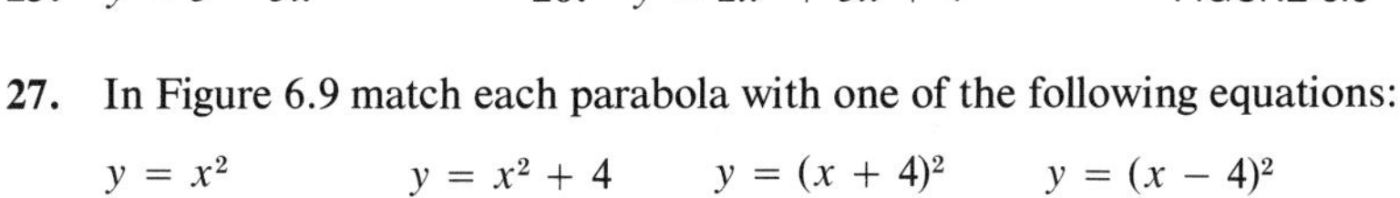

$y = x^2$ $y = x^2 + 4$ $y = (x + 4)^2$ $y = (x - 4)^2$

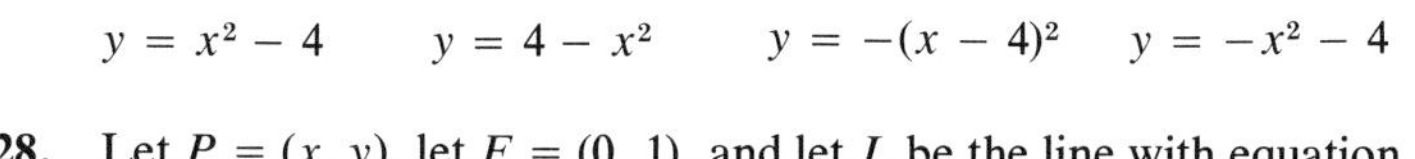

$y = x^2 - 4$ $y = 4 - x^2$ $y = -(x - 4)^2$ $y = -x^2 - 4$

28. Let $P = (x, y)$, let $F = (0, 1)$, and let L be the line with equation $y = -1$, as in Figure 6.10.

(a) Express dist (P, F).

(b) Express the (perpendicular) distance from P to L.

(c) Find an equation for the set of points P that are the same distance from the point F as from the line L.

(d) Show that this equation represents a parabola with vertex at the origin, as indicated in Figure 6.10.

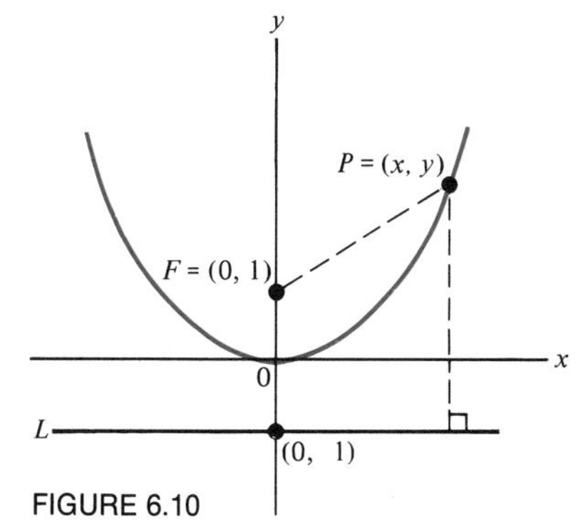

FIGURE 6.10

29. A farmer wishes to fence in a rectangular pasture. One side of the pasture lies along a river, so that fencing is required only for the other three sides. If the farmer has 240 meters of fencing available, what dimensions will yield the maximum area? [*Hint:* Express the area by means of a quadratic function, and find the vertex of the corresponding parabola.]

30. Find the maximum possible area of a rectangle whose perimeter is 40 centimeters.

31. A rectangular playground is to be fenced off along one side of a school. The side bounded by the school requires no fencing. Fencing costs \$8 per (running) foot for the parallel side and \$4 per foot for the other two sides. If \$1200 is appropriated for the fencing, find the maximum area that can be enclosed.

32. A publisher thinks that she can sell 1000 copies of a book on sailing priced at \$20 per copy. For each dollar she lowers the price, she thinks she can sell an additional 500 copies. According to her thinking:

(a) At what price will her total revenue from sales be the maximum?

(b) How many copies will she sell at this price? **(c)** What will be the total revenue?

6.9 Quadratic Inequalities

A **quadratic inequality** is an inequality that can be obtained from a quadratic equation

$$Ax^2 + Bx + C = 0, \qquad A \neq 0,$$

by replacing = by one of the symbols

$$<, \quad \leqslant, \quad >, \quad \geqslant$$

These inequalities frequently arise in solving problems in calculus. Only quadratic inequalities in which the left side can easily be factored will be considered.

In solving a quadratic inequality, you must often consider the sign of a product ab.

$ab > 0,$ if *both* factors are positive or *both* are negative;
$ab = 0,$ if at least one factor is 0;
$ab < 0,$ if one factor is positive and the other negative.

EXAMPLE 1 Solve: $x^2 + 3x > 0$

Solution. Factor the left side to obtain

$$x(x + 3) > 0.$$

Next, observe that

$$x(x + 3) = 0 \quad \text{when } x = 0 \quad \text{and also, when} \quad x = -3.$$

The points 0 and -3 divide the x-axis into three regions. *In each region $x(x + 3)$ is either positive throughout the entire region or else negative*. Figure 6.11 indicates the signs of the factors x and $x + 3$ and, hence, of their product throughout each region. For instance, in the left-most region, that is, in $(-\infty, -3)$, x is always less than -3, which, in turn, is less than 0:

$$x < -3 < 0 \quad \text{and thus} \quad x < 0$$

Also,

$$x + 3 < -3 + 3 = 0 \quad \text{and thus} \quad x + 3 < 0$$

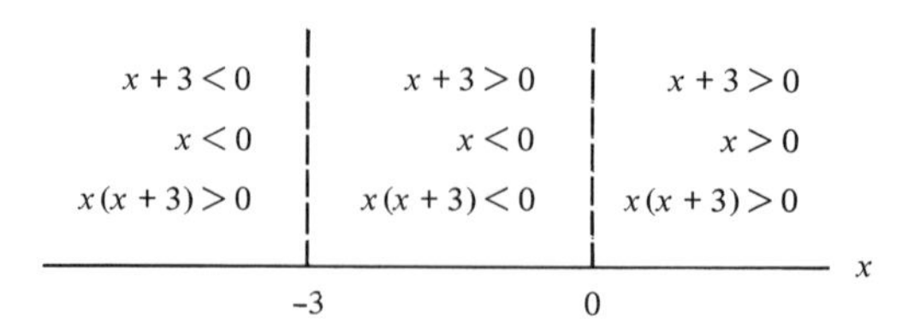

FIGURE 6.11

Each of the factors x and $x + 3$ is negative throughout $(-\infty, -3)$. Consequently,

$$x(x + 3) > 0 \quad \text{throughout } (-\infty, -3)$$

Instead of the preceding analysis, you may pick an arbitrary number in each region. This will indicate the sign of $x(x + 3)$ throughout the region. For example, in the middle region, that is, in $(-3, 0)$, replace x by -1 in the expression $x(x + 3)$:

$$(-1)(-1 + 3) = -2 < 0$$

-3 0 x

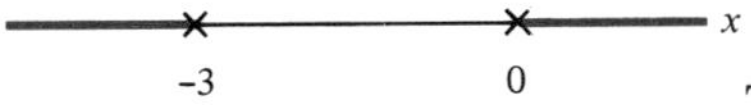

FIGURE 6.12(a) The solution set of $x^2 + 3x > 0$ is $(-\infty, -3) \cup (0, \infty)$.

Thus $x(x + 3) < 0$ throughout the region $(-3, 0)$.

In the right-most region, $(0, \infty)$, both factors, x and $x + 3$, are positive, as, consequently, is the product, $x(x + 3)$.

The solution set of the given inequality consists of all x in

$$(-\infty, -3) \quad \text{or in} \quad (0, \infty).$$

Thus the solution set is

$$(-\infty, -3) \cup (0, \infty).‡$$

[See Figure 6.12(a).] The parabola given by

$$y = x^2 + 3x = \left(x + \frac{3}{2}\right)^2 - \frac{9}{4}$$

is graphed in Figure 6.12(b). Observe that

$$y > 0 \quad \text{when} \quad x < -3 \quad \text{and when} \quad x > 0. \qquad \square$$

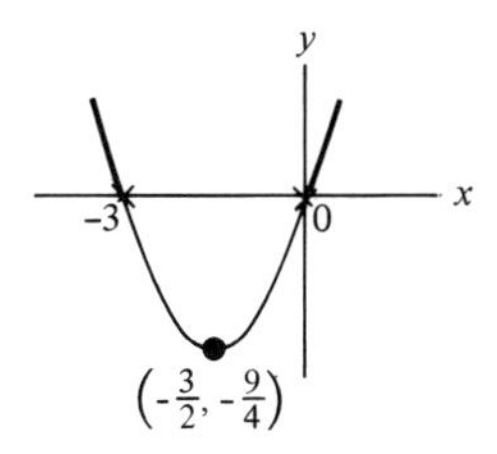

FIGURE 6.12(b)
$y = x^2 + 3x = \left(x + \frac{3}{2}\right)^2 - \frac{9}{4}$.
$y > 0$ when $x < 0$ and when $x > 3$.

EXAMPLE 2 Solve:

$$x^2 + x - 2 \leq 0$$

Solution. Factor the left side:

$$(x + 2)(x - 1) \leq 0$$

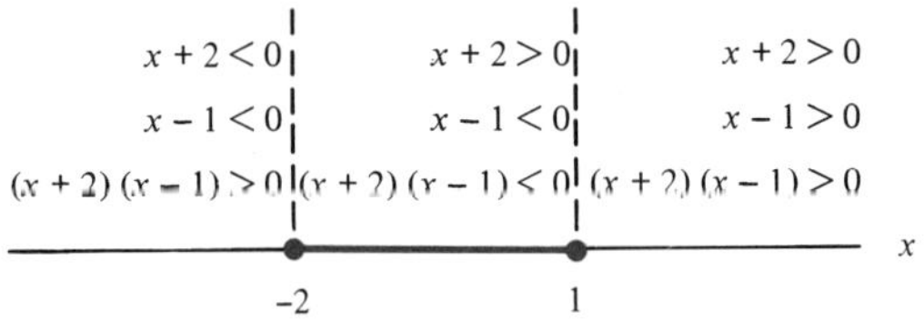

$x + 2 < 0$ | $x + 2 > 0$ | $x + 2 > 0$
$x - 1 < 0$ | $x - 1 < 0$ | $x - 1 > 0$
$(x + 2)(x - 1) > 0$ | $(x + 2)(x - 1) < 0$ | $(x + 2)(x - 1) > 0$

-2 1 x

FIGURE 6.13(a)

Here the points -2 and 1 divide the x-axis into three regions. The possibilities for the factors are shown in Figure 6.13(a). Because $(x + 2)(x - 1) \leq 0$ when $x = -2$ and when $x = 1$, the solution set for the inequality

$$(x + 2)(x - 1) \leq 0,$$

‡See page 95 for the notation $A \cup B$.

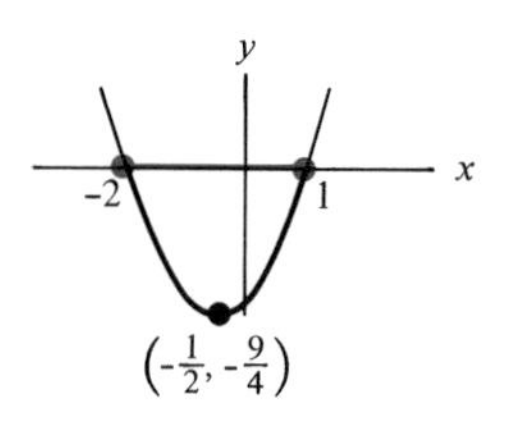

FIGURE 6.13(b)
$y = x^2 + x - 2 = \left(x + \frac{1}{2}\right)^2 - \frac{9}{4}$.
$y \leqslant 0$ when $-2 \leqslant x \leqslant 1$.

and hence for the given inequality,

$$x^2 + x - 2 \leqslant 0$$

is the closed interval $[-2, 1]$. The parabola given by

$$y = x^2 + x - 2 = \left(x + \frac{1}{2}\right)^2 - \frac{9}{4}$$

is graphed in Figure 6.13(b). Observe that

$$y \leqslant 0 \quad \text{when} \quad -2 \leqslant x \leqslant 1. \qquad \square$$

EXAMPLE 3 Solve:

$$x^2 + 25 > 10x$$

$x - 5 < 0$ | $x - 5 > 0$
$(x - 5)^2 > 0$ | $(x - 5)^2 > 0$
5 x

FIGURE 6.14(a)

Solution. Transform this as follows:

$$x^2 + 25 > 10x$$
$$x^2 - 10x + 25 > 0$$
$$(x - 5)^2 > 0$$

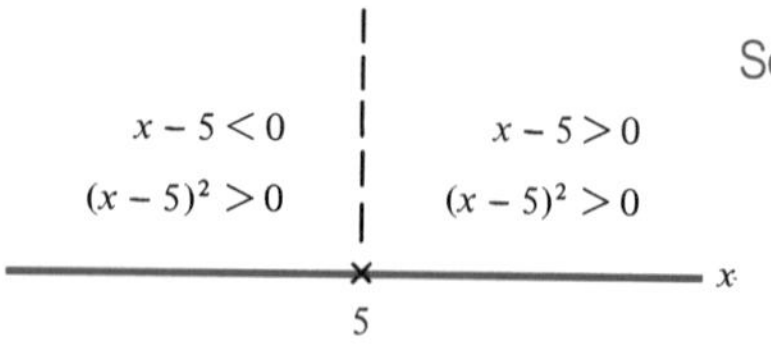

Next, $(x - 5)^2 = 0$ only when $x - 5 = 0$, that is, when $x = 5$. Thus there are only two regions in Figure 6.14(a). Observe that in both regions $(x - 5)^2 > 0$. The solution set of the given inequality consists of all real numbers other than 5.

The parabola given by

$$y = (x - 5)^2$$

is graphed in Figure 6.14(b). Observe that

$$y > 0 \quad \text{when} \quad x < 5 \quad \text{and when} \quad x > 5. \qquad \square$$

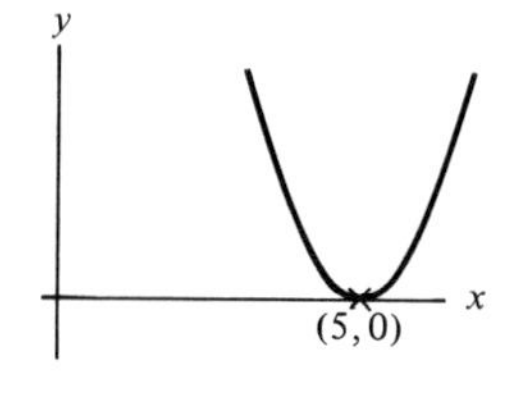

FIGURE 6.14(b) $y = (x - 5)^2$
$y > 0$ when $x < 5$ and when $x > 5$.

You can solve certain inequalities most efficiently by transforming them into quadratic inequalities.

EXAMPLE 4 Solve:

$$\frac{3}{x} < 5$$

Solution. Note that $x \neq 0$ (because division by x is assumed). If you multiply both sides by x, you will have to consider two separate

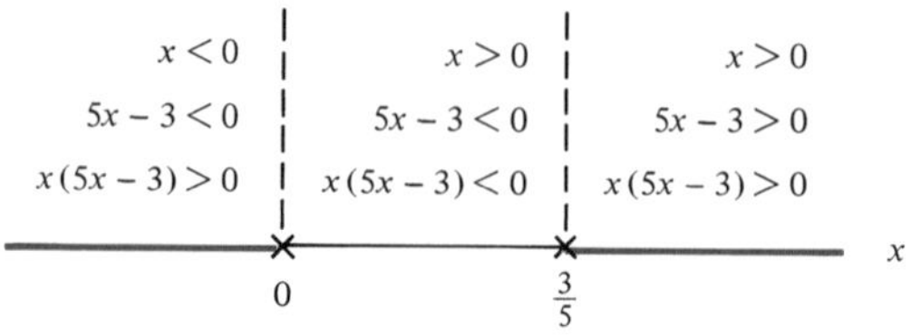

$x < 0$ | $x > 0$ | $x > 0$
$5x - 3 < 0$ | $5x - 3 < 0$ | $5x - 3 > 0$
$x(5x - 3) > 0$ | $x(5x - 3) < 0$ | $x(5x - 3) > 0$
0 $\frac{3}{5}$ x

FIGURE 6.15

cases, depending on whether x is positive or negative. For, multiplication by a *negative* number *reverses* the sense of an inequality. To avoid this difficulty, multiply both sides by x^2, which is *positive* (because $x \neq 0$).

$$\frac{3}{x} < 5$$

$$\frac{3}{x} \cdot x^2 < 5 \cdot x^2$$

$$3x < 5x^2$$

$$0 < 5x^2 - 3x$$

$$0 < x(5x - 3)$$

Note that $5x - 3 = 0$ when $x = \frac{3}{5}$. Figure 6.15 indicates that the solution set is $(-\infty, 0) \cup \left(\frac{3}{5}, \infty\right)$. □

EXAMPLE 5 Solve:

$$\frac{x+3}{x+2} \geq 2$$

Solution. Here $x \neq -2$ because division by $x + 2$ is assumed. If you multiply both sides by $x + 2$, you have to consider two cases, according to whether $x + 2 > 0$ or $x + 2 < 0$. Instead, multiply both sides by $(x + 2)^2$, which is positive.

$$\frac{x+3}{x+2} \cdot (x+2)^2 \geq 2 \cdot (x+2)^2$$

$$(x+3)(x+2) \geq 2(x^2 + 4x + 4)$$

$$x^2 + 5x + 6 \geq 2x^2 + 8x + 8$$

$$0 \geq x^2 + 3x + 2$$

$$0 \geq (x+2)(x+1)$$

or $(x+2)(x+1) \leq 0$

Note that $(x + 2)(x + 1) \leq 0$ (when $x = -2$ and) when $x = -1$. Figure 6.16 indicates that the solution set is the interval $(-2, -1]$. □

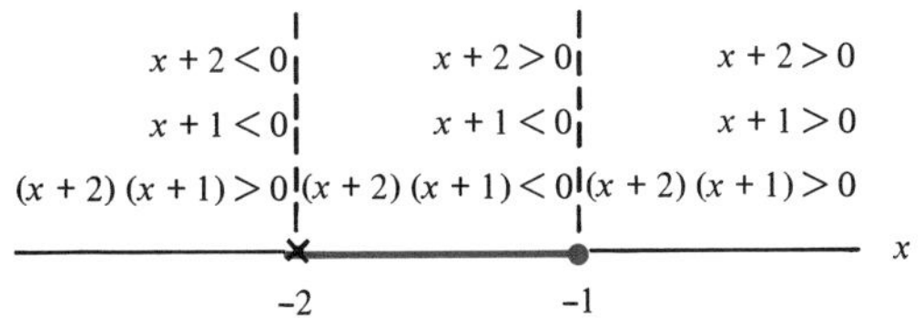

FIGURE 6.16

EXERCISES

In Exercises 1–40 solve each inequality.

1. $(x + 7)(x - 3) < 0$
2. $(x - 5)(x - 9) > 0$
3. $x^2 - 5x < 0$
4. $x^2 > 7x$
5. $t^2 - 4t \leqslant 0$
6. $y^2 + 10y \geqslant 0$
7. $x^2 + 2x + 1 < 0$
8. $x^2 - 2x > -1$
9. $x^2 + 4x + 4 \geqslant 0$
10. $x^2 + 9 \leqslant 6x$
11. $z^2 + 4z + 3 > 0$
12. $x^2 + 5x + 6 \leqslant 0$
13. $x^2 - 7x + 10 < 0$
14. $x^2 + 11x + 24 > 0$
15. $u^2 - 4u - 12 \leqslant 0$
16. $x^2 + 2x > 15$
17. $2x^2 + 3x + 1 < 0$
18. $3x^2 + 7x + 3 > 0$
19. $4t^2 - 8t + 3 > 0$
20. $25x^2 - 1 \leqslant 0$
21. $\frac{3}{x} < 1$
22. $\frac{-2}{x} < 3$
23. $\frac{4}{t} > -1$
24. $1 - \frac{5}{z} \geqslant \frac{1}{2}$
25. $\frac{1}{x + 1} < 1$
26. $\frac{1}{x + 5} > 1$
27. $\frac{1}{x + 4} \leqslant 1$
28. $\frac{1}{x - 2} < 1$
29. $\frac{x}{x + 2} < 1$
30. $\frac{x}{x - 3} \geqslant 1$
31. $\frac{x}{x + 5} < 2$
32. $\frac{5x}{x + 1} > 2$
33. $\frac{t + 3}{t + 2} < 1$
34. $\frac{u + 2}{u + 9} \geqslant 1$
35. $\frac{x + 5}{x + 2} < 2$
36. $\frac{x + 1}{x - 1} > 2$
37. $\frac{x + 2}{x + 4} < 3$
38. $\frac{x + 2}{x - 1} < 2$
39. $(x + 2)^2 + 4 > 0$
40. $(x - 3)^2 + 5 < 0$
41. Find the set of all real numbers whose square is less than twice the number.
42. Find the set of all real numbers whose square is at most three times the number.
43. Let $f(x) = x^2 + 7x + 10$. Find the set of x-values for which $f(x)$ is negative.
44. Let $g(x) = x^2 - 8x$. Find the set of x-values for which $g(x)$ is positive.
45. Let $F(x) = x^2 - x$. Find the set of x-values for which $F(x) < 2$.
46. Let $G(x) = x^2 + 5x$. Find the set of x-values for which $G(x) \geqslant 6$.

6.10 Circles

Definition A **circle** consists of all points of the plane at a fixed distance from a given point. The given point is called the **center** of the circle. Each line segment from the center to a point on the circle is called a **radius** of the circle.

The length of each radius is the same.

EXAMPLE 1 Consider the circle centered at the origin and of radius length 5. This circle consists of all points 5 units away from the origin, P_0. Let $P = (x, y)$ and suppose P is on the circle. Then

$$\text{dist}(P, P_0) = 5$$

Therefore, because $P_0 = (0, 0)$,

$$\sqrt{(x - 0)^2 + (y - 0)^2} = 5$$

Square both sides. Thus

$$x^2 + y^2 = 25$$

TABLE 6.3

x	y
0	± 5
3	± 4
-3	± 4
4	± 3
-4	± 3
5	0
-5	0

This equation describes the circle. The set of (real) solutions of the equation is the set of points on the circle. Some of these are given in Table 6.3 . Thus, if $x = 3$, then y can be either 4 or -4. In fact,

$$3^2 + 4^2 = 9 + 16 = 25 \qquad \text{and} \qquad 3^2 + (-4)^2 = 9 + 16 = 25$$

Similarly, if $x = -3$, then y can be either 4 or -4. The circle is pictured in Figure 6.17 . Observe that there are vertical lines that intersect a circle twice. Thus, a circle is not the graph of a function. □

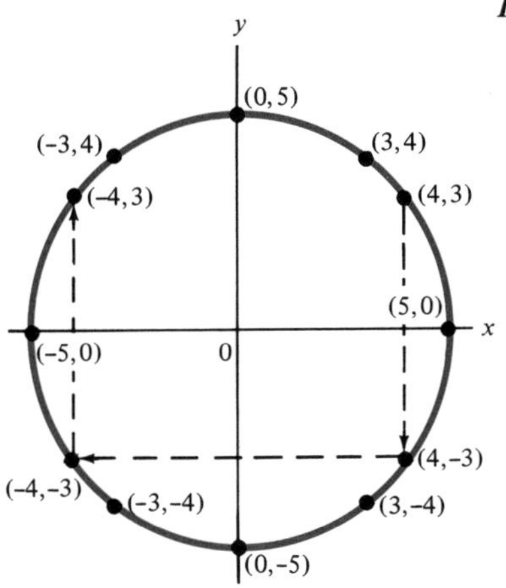

FIGURE 6.17
The graph of the circle with the equation $x^2 + y^2 = 25$

The circle of Example 1 was centered at the origin and had radius length 5. Its equation was

$$x^2 + y^2 = 5^2.$$

Now let $P_0 = (x_0, y_0)$ and let $r > 0$. Consider the circle centered at P_0 and of radius length r. A point P is on this circle if

$$\text{dist}(P, P_0) = r.$$

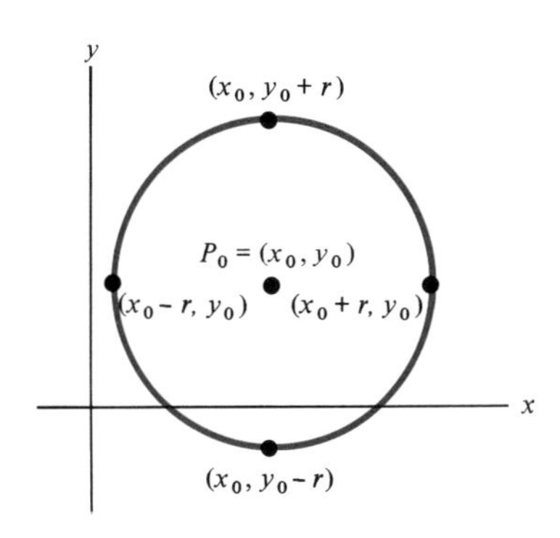

FIGURE 6.18
The circle centered at P_0 and of radius length r. Its equation is $(x - x_0)^2 + (y - y_0)^2 = r^2$

Thus, if $P = (x, y)$, then dist $(P, P_0) = \sqrt{(x - x_0)^2 + (y - y_0)^2}$, and

$$\sqrt{(x - x_0)^2 + (y - y_0)^2} = r$$

Square both sides to obtain:

$$(x - x_0)^2 + (y - y_0)^2 = r^2, \qquad r > 0$$

This is the equation of a circle centered at (x_0, y_0) and of radius length r. [See Figure 6.18 .]

EXAMPLE 2 Determine the equation of the circle consisting of all points that are 3 units from (1, 2).

Solution. Let $(x_0, y_0) = (1, 2)$, $r = 3$. [The center of the circle is (1, 2); the radius length is 3.] Therefore, the equation of the circle is

$$(x - 1)^2 + (y - 2)^2 = 3^2 \qquad \text{or} \qquad (x - 1)^2 + (y - 2)^2 = 9.$$

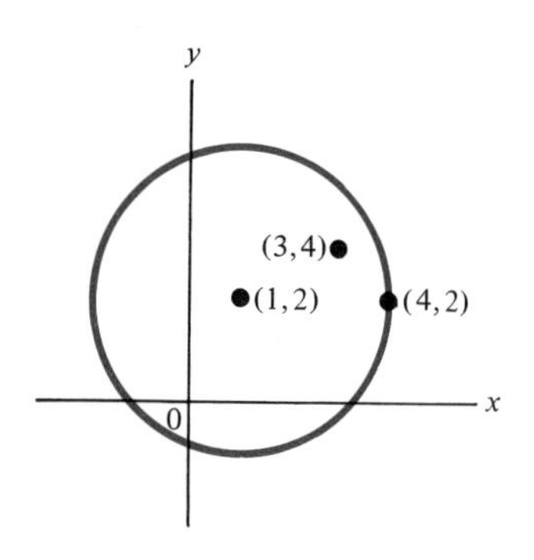

FIGURE 6.19

The point (4, 2) is on this circle because

$$(4 - 1)^2 + (2 - 2)^2 = 3^2 + 0^2 = 9.$$

The point (3, 4) is *not* on the circle because

$$(3 - 1)^2 + (4 - 2)^2 = 2^2 + 2^2 = 8 \neq 9.$$

[See Figure 6.19 .] □

EXAMPLE 3 (a) Show that the equation

$$x^2 + y^2 + 8x - 2y = 0$$

represents a circle.

(b) Determine the center and radius of this circle.

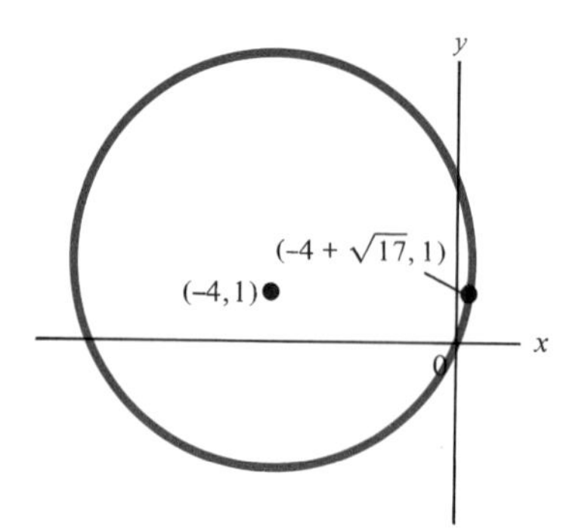

FIGURE 6.20

Solution.

(a) Regroup terms to obtain

$$(x^2 + 8x) + (y^2 - 2y) = 0.$$

Next, *complete the square* within each group. To $x^2 + 8x$, add $\left(\frac{8}{2}\right)^2$ (or 16); to $y^2 - 2y$, add $\left(\frac{-2}{2}\right)^2$ (or 1). Thus

$$(x^2 + 8x + 16) + (y^2 - 2y + 1) = 16 + 1$$

and

$$(x + 4)^2 + (y - 1)^2 = 17$$

This final equation is of the form

$$(x - x_0)^2 + (y - y_0)^2 = r^2, \qquad \text{where}$$

$$x_0 = -4, \qquad y_0 = 1, \qquad r = \sqrt{17} \approx 4.12$$

(b) The center is at $(-4, 1)$. The radius length is $\sqrt{17}$. [See Figure 6.20 .] □

EXERCISES

In Exercises 1–12 determine: (a) the center and (b) the radius length of the given circle.

1. $x^2 + y^2 = 9$

2. $x^2 + y^2 = 100$

3. $x^2 + y^2 = 8$

4. $x^2 + (y - 3)^2 = 1$

5. $(x - 2)^2 + (y - 4)^2 = 49$

6. $(x + 2)^2 + (y + 4)^2 = 64$

7. $(x + 8)^2 + (y - 7)^2 = 144$

8. $\left(x - \frac{1}{2}\right)^2 + (y + 1)^2 = 20$

9. $\left(x - \frac{1}{4}\right)^2 + \left(y - \frac{1}{2}\right)^2 = \frac{1}{4}$

10. $\left(x + \frac{2}{5}\right)^2 + \left(y - \frac{4}{5}\right)^2 = \frac{9}{25}$

11. $(x + 2.7)^2 + (y - 1.7)^2 = .01$

12. $(x + \pi)^2 + (y - \pi)^2 = .09\pi^2$

13. Determine the equation of the circle consisting of all points that are 4 units from the origin.

14. Determine the equation of the circle consisting of all points that are 2 units from (2, 0).

15. Determine the equation of the circle consisting of all points that are 5 units from (2, 3).

16. Determine the equation of the circle consisting of all points that are 3 units from (−3, −3).

17. Determine the equation of the circle consisting of all points that are $\frac{1}{2}$ unit from $\left(\frac{1}{4}, \frac{1}{2}\right)$.

18. Determine the equation of the circle consisting of all points that are .4 unit from (1.2, −.8).

In Exercises 19–22, with the aid of a compass, graph the indicated circle.

19. Center is the origin, radius length 6.

20. Center is (0, 4), radius length 4.

21. $(x - 3)^2 + y^2 = 9$

22. $(x - 3)^2 + (y - 7)^2 = 4$

In Exercises 23–34 complete the square to determine the center and radius length of the corresponding circle.

23. $x^2 + y^2 + 10x = 0$

24. $3x^2 + 3y^2 - 12y = 0$

25. $x^2 + y^2 + 2x + 6y = 0$

26. $x^2 + y^2 + 4x - 4y = 0$

27. $x^2 + y^2 + 6x + 18y = 0$

28. $x^2 + y^2 + 12x - 10y = 0$

29. $x^2 + y^2 + 6x + 5y = 0$

30. $x^2 + y^2 + 3x - 9y = 0$

31. $x^2 + y^2 + 4x - 3y = 1$

32. $x^2 + y^2 + 10x = y + 6$

33. $2x^2 + 2y^2 + 8x + 40y = 0$

34. $4x^2 + 4y^2 + 20x + 10y = 1$

*6.11 Ellipses and Hyperbolas

Let $a > 0$ and $b > 0$. The equation

$$\frac{x^2}{a^2} + \frac{y^2}{a^2} = 1 \quad \text{or} \quad x^2 + y^2 = a^2$$

describes a circle. The equation

$$\frac{x^2}{a^2} + \frac{y^2}{b^2} = 1, \quad \text{where } a \neq b,$$

describes an **ellipse,** or flattened circle, as in Figure 6.21 . The

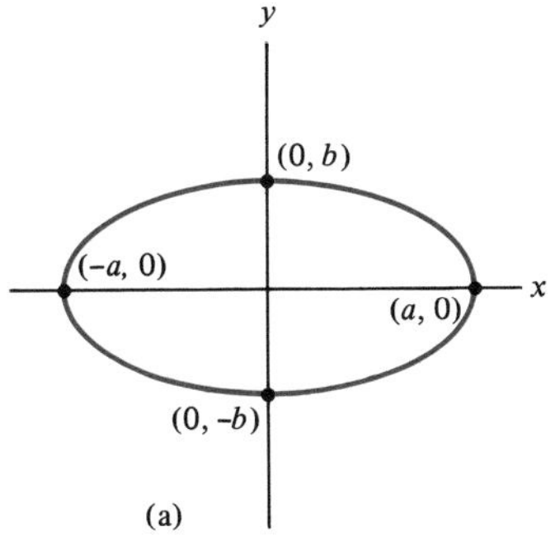

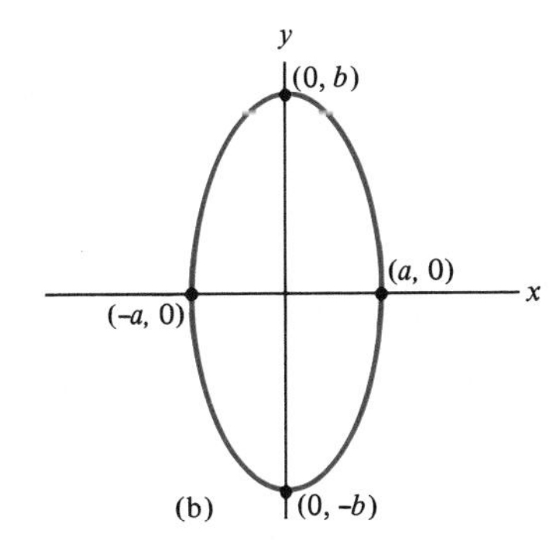

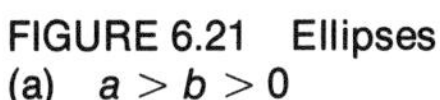
FIGURE 6.21 Ellipses
(a) $a > b > 0$ (b) $b > a > 0$

*Optional topic

equations

$$\frac{x^2}{a^2} - \frac{y^2}{b^2} = 1 \quad \text{and} \quad -\frac{x^2}{a^2} + \frac{y^2}{b^2} = 1$$

describe figures known as **hyperbolas,** as shown in Figure 6.22 .

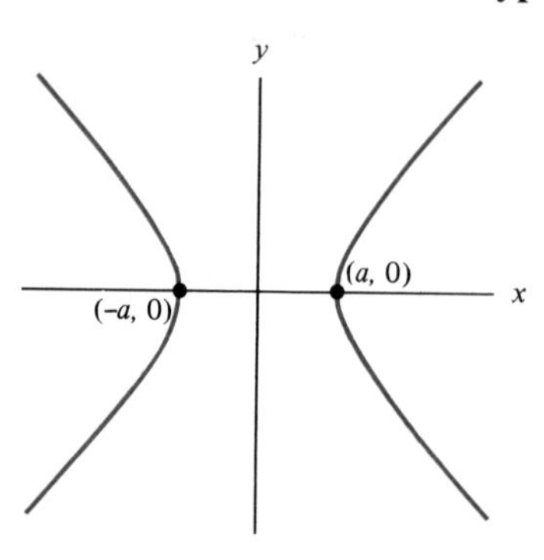

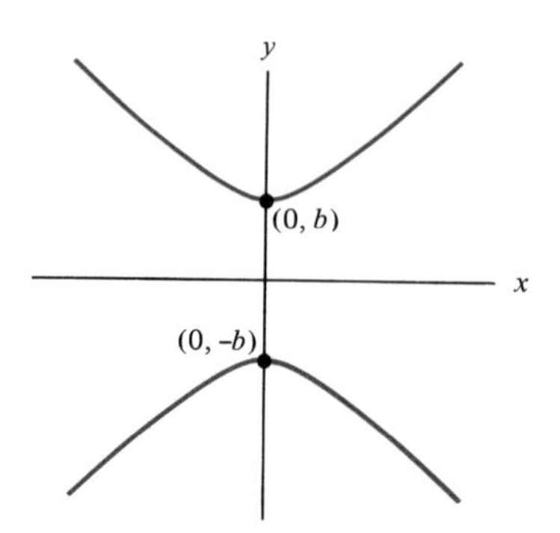

FIGURE 6.22 Hyperbolas
(a) The graph of $\frac{x^2}{a^2} - \frac{y^2}{b^2} = 1$ (b) The graph of $-\frac{x^2}{a^2} + \frac{y^2}{b^2} = 1$

Thus

$$\frac{x^2}{25} + \frac{y^2}{25} = 1 \quad \text{or} \quad x^2 + y^2 = 25 \quad \text{describes a } \textit{circle};$$

$$\frac{x^2}{25} + \frac{y^2}{9} = 1 \quad \text{describes an } \textit{ellipse};$$

$$\frac{x^2}{16} - \frac{y^2}{9} = 1 \quad \text{and} \quad -\frac{x^2}{9} + \frac{y^2}{16} = 1 \text{ describe } \textit{hyperbolas}.$$

In each case the curve is "centered" at the origin. Both x^2 and y^2 appear in the equations, and there are no other occurrences of x or y. Thus, whenever (x, y) is on the graph so are

$$(x, -y), \qquad (-x, y), \qquad \text{and} \qquad (-x, -y).$$

Consequently, the graphs are symmetric with respect to the x-axis, the y-axis, and the origin.

ellipses

The **vertices** of an ellipse given by

$$\frac{x^2}{a^2} + \frac{y^2}{b^2} = 1$$

are the intersections with the coordinate axes. As Example 1 will illustrate, the vertices are the points

$$(a, 0), \qquad (-a, 0), \qquad (0, b), \qquad (0, -b).$$

EXAMPLE 1 Graph the ellipse whose equation is $\frac{x^2}{25} + \frac{y^2}{9} = 1$.

Solution. To graph this ellipse, locate the vertices by first setting $y = 0$ and then setting $x = 0$ in the given equation. Setting $y = 0$,

TABLE 6.4

x	y
0	± 3
3	$\pm\frac{12}{5}$
-3	$\pm\frac{12}{5}$
5	0
-5	0

you obtain:

$$\frac{x^2}{25} = 1$$

$$x^2 = 25$$

$$x = \pm 5$$

Thus (5, 0) and $(-5, 0)$ lie on the curve. Now substitute 0 for x in the given equation.

$$\frac{y^2}{9} = 1$$

$$y = \pm 3$$

Thus (0, 3) and $(0, -3)$ lie on the curve.

To find other points on the curve, multiply both sides of the given equation by 25×9 to obtain

$$9x^2 + 25y^2 = 225. \qquad (6.4)$$

When $x = 3$ and when $x = -3$, $x^2 = 9$. Replace x^2 by 9 in Equation **(6.4)** and obtain:

$$9(9) + 25y^2 = 225$$

$$25y^2 = 144$$

$$y^2 = \frac{144}{25}$$

$$y = \frac{\pm 12}{5}$$

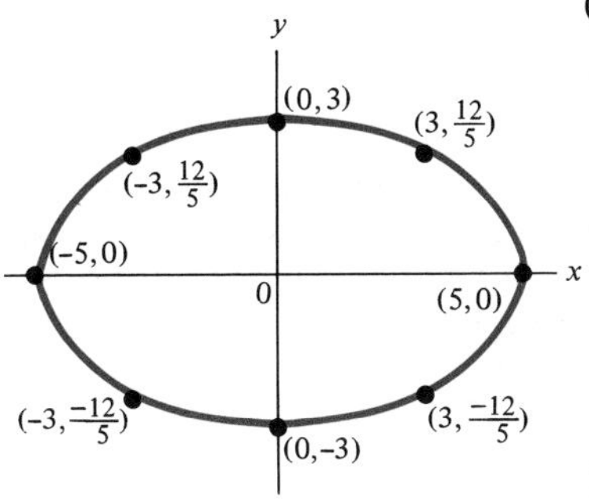

Thus $\left(3, \frac{12}{5}\right)$, $\left(3, \frac{-12}{5}\right)$, $\left(-3, \frac{12}{5}\right)$, and $\left(-3, \frac{-12}{5}\right)$ lie on the curve, so that all the points indicated in Table 6.4 are on the curve. Draw a smooth curve through the vertices, as in Figure 6.23. □

FIGURE 6.23

hyperbolas

EXAMPLE 2 Graph the hyperbola whose equation is $\frac{x^2}{16} - \frac{y^2}{9} = 1$.

Solution. Set $y = 0$ and obtain:

$$\frac{x^2}{16} = 1$$

$$x = \pm 4$$

Thus, the points (4, 0) and $(-4, 0)$ are on the hyperbola. However, when $x = 0$, you do *not* obtain any point on the hyperbola because the equation

$$-\frac{y^2}{9} = 1 \qquad \text{or} \qquad y^2 = -9$$

has no real solution. Therefore, the y-axis ($x = 0$) lies between the two parts, or **branches,** of the hyperbola.

In order to see how to graph the hyperbola, observe that the given equation can be written as

$$\frac{x^2}{4^2} - \frac{y^2}{3^2} = 1 \qquad \text{or} \qquad \frac{y^2}{3^2} = \frac{x^2}{4^2} - 1$$

If you were to drop the -1 in the second form of the equation, you would no longer have an equation of the hyperbola. However, it is instructive to consider the equation

$$\frac{y^2}{3^2} = \frac{x^2}{4^2}$$

which simplifies to

$$\frac{y}{3} = \pm\frac{x}{4} \qquad \text{or} \qquad y = \pm\frac{3}{4}x\,.$$

The lines through the origin

$$y = \frac{3}{4}x \qquad \text{and} \qquad y = -\frac{3}{4}x$$

although not part of the hyperbola, nevertheless play a key role in graphing the hyperbola. These lines are called **asymptotes.** [See Figure 6.24 .] A point moving along the hyperbola gets closer and closer to an asymptote as the point gets further and further from the origin. □

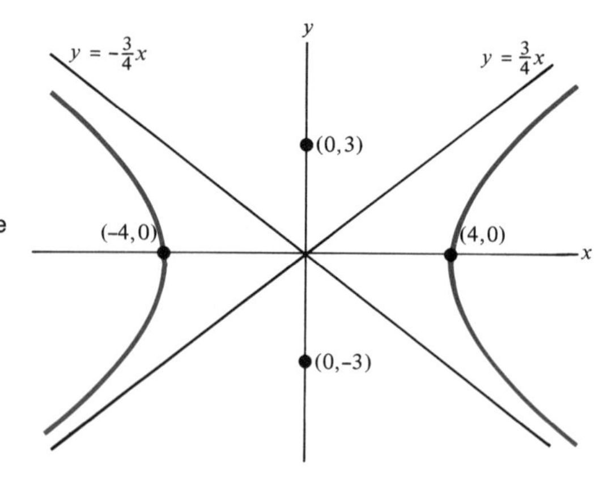

FIGURE 6.24
The hyperbola whose equation is $\frac{x^2}{16} - \frac{y^2}{9} = 1$. Each curve (in color) is a branch of the hyperbola. The vertices are (4, 0) and (−4, 0). The asymptotes are given by $y = \frac{3}{4}x$ and $y = -\frac{3}{4}x$.

The **vertices** of a hyperbola are the intersections with the coordinate axes. For a hyperbola with equation

$$\frac{x^2}{a^2} - \frac{y^2}{b^2} = 1,$$

the vertices are (a, 0) and (−a, 0). For a hyperbola with equation

$$-\frac{x^2}{a^2}+\frac{y^2}{b^2}=1,$$

the vertices are $(0, b)$ and $(0, -b)$. For both types of hyperbolas the asymptotes are given by

$$y=\frac{b}{a}x \quad \text{and} \quad y=-\frac{b}{a}x.$$

To graph a hyperbola:

1. Plot the vertices.
2. Draw the asymptotes.
3. Sketch the hyperbola so that as $|x|$ gets larger, each curve approaches the asymptotes. *The hyperbola never crosses its asymptotes.*

EXAMPLE 3 (a) Show that the equation $16y^2 - x^2 = 16$ represents a hyperbola.

(b) Locate the vertices.

(c) What are the asymptotes?

(d) Sketch the graph.

Solution.

(a) Divide both sides of the given equation by 16.

$$y^2-\frac{x^2}{16}=1$$

This is the equation of a hyperbola centered at the origin and with vertices on the y-axis.

(b) Set $x = 0$. The vertices are (0, 1) and (0, −1).

(c) In the equation

$$-\frac{x^2}{a^2}+\frac{y^2}{b^2}=1,$$

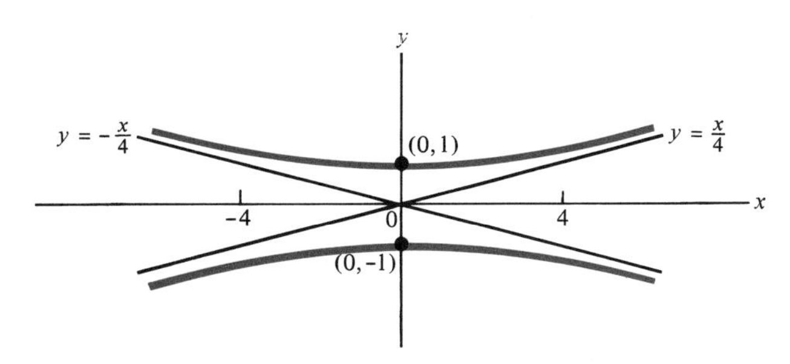

FIGURE 6.25
The hyperbola whose equation is $16y^2 - x^2 = 16$, or $y^2 - \frac{x^2}{16} = 1$. The vertices are (0, 1) and (0, −1). The asymptotes are given by $y = \frac{x}{4}$ and $y = -\frac{x}{4}$.

set $a = 4$ and $b = 1$. The asymptotes are given by $y = \frac{b}{a}x$ and $y = -\frac{b}{a}x$. Thus they are the lines whose equations are

$$y = \frac{x}{4} \quad \text{and} \quad y = -\frac{x}{4}.$$

(d) See Figure 6.25 . □

circles, ellipses, hyperbolas, lines, and points

As was previously indicated,

$$x^2 + y^2 = 1$$

is the equation of a circle. Note that for real numbers x and y, both $x^2 \geqslant 0$ and $y^2 \geqslant 0$. Thus, the equation

$$x^2 + y^2 = 0$$

is only satisfied when $x = 0$ and when $y = 0$, and consequently describes only one point (of the real plane‡), the origin. And the equation

$$x^2 + y^2 = -1$$

does not describe any point. Similarly, the equation

$$x^2 + \frac{y^2}{4} = 1$$

describes an ellipse; the equation

$$x^2 + \frac{y^2}{4} = 0$$

describes a single point; and the equation

$$x^2 + \frac{y^2}{4} = -1$$

does not describe any point.

$$x^2 - y^2 = 1$$

is the equation of a hyperbola, as is the equation

$$x^2 - y^2 = -1, \qquad \text{or equivalently,} \qquad y^2 - x^2 = 1.$$

The equation $x^2 - y^2 = 0$ is equivalent to $x^2 = y^2$, which, in turn means that $y = x$ or $y = -x$. Consequently,

$$x^2 - y^2 = 0$$

describes two lines that intersect at the origin, as in Figure 6.26 .

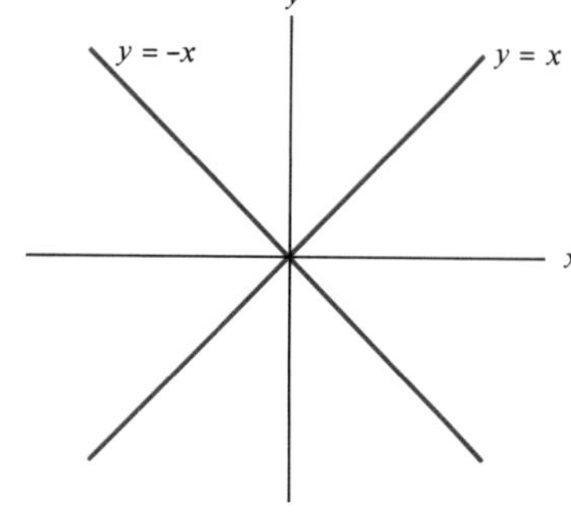

FIGURE 6.26 The graph of $x^2 - y^2 = 0$ or $x^2 = y^2$

quadratic equations in x and y

A **quadratic,** or **second-degree, equation in** x *and* y is an equation that can be written in the form

$$Ax^2 + Bxy + Cy^2 + Dx + Ey + F = 0,$$
where A, B, and C are *not all* 0.

‡The "complex plane" will be discussed in Section 10.2 .

Only equations in which $B = 0$ have been considered here; equations of the form $xy = k$ were treated in Section 4.6 . As you have seen, parabolas, circles, ellipses, and hyperbolas can be described by quadratic equations in x and y. It is interesting to note that these figures can also be obtained as the intersections of planes and cones, as indicated in Figure 6.27 . For this reason, parabolas, circles, ellipses, and hyperbolas are known as **conic sections.**

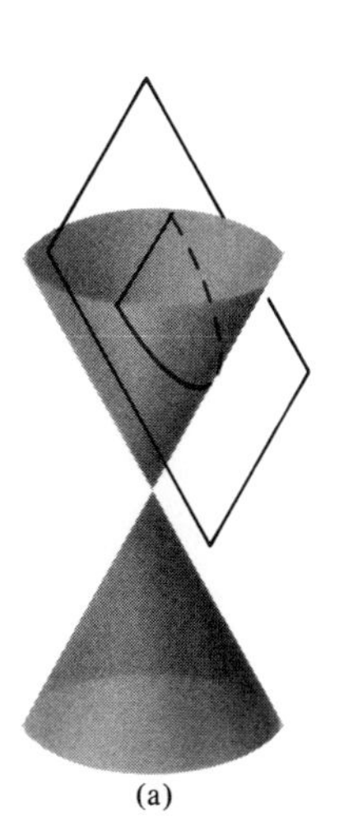

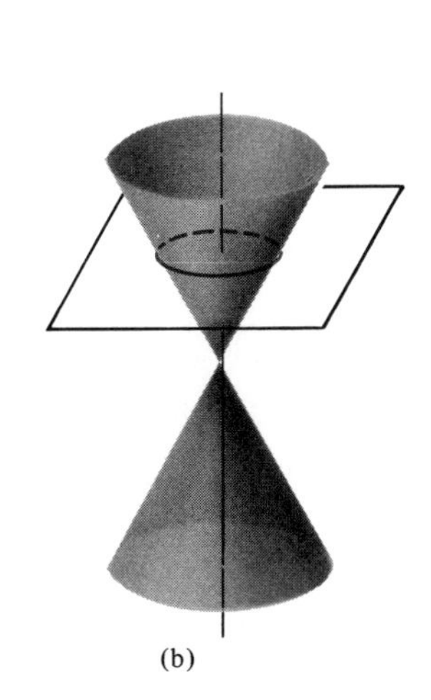

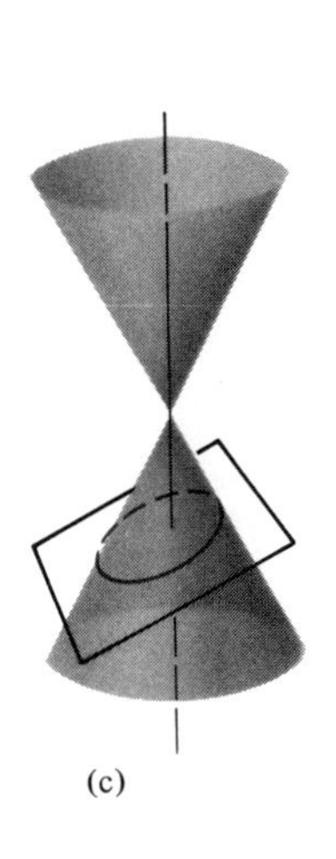

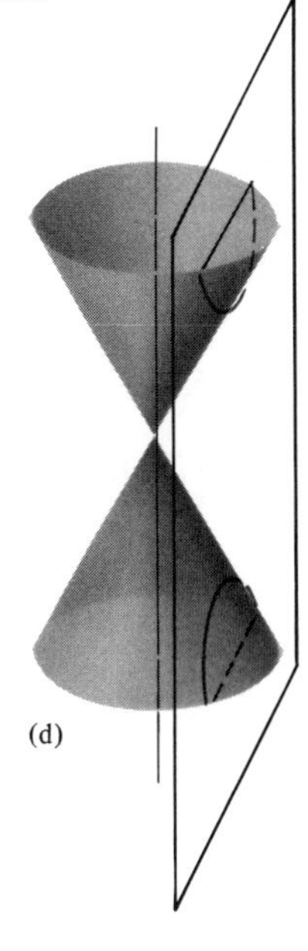

FIGURE 6.27 (a) Parabola (b) Circle (c) Ellipse (d) Hyperbola

EXERCISES

In Exercises 1–14 an equation of an ellipse is given. Locate the vertices. Graph the indicated ellipses.

1. $\frac{x^2}{25} + \frac{y^2}{16} = 1$ (Graph.)

2. $\frac{x^2}{16} + \frac{y^2}{25} = 1$ (Graph.)

3. $\frac{x^2}{9} + \frac{y^2}{4} = 1$

4. $\frac{x^2}{9} + \frac{y^2}{25} = 1$

5. $\frac{x^2}{169} + \frac{y^2}{144} = 1$ (Graph.)

6. $\frac{x^2}{169} + \frac{y^2}{25} = 1$ (Graph.)

7. $\frac{x^2}{100} + \frac{y^2}{36} = 1$ (Graph.)

8. $\frac{x^2}{25} + \frac{y^2}{169} = 1$ (Graph.)

9. $x^2 + \frac{y^2}{4} = 1$

10. $\frac{x^2}{100} + \frac{y^2}{81} = 1$

11. $\frac{x^2}{13} + \frac{y^2}{4} = 1$

12. $\frac{x^2}{64} + \frac{y^2}{15} = 1$

13. $9x^2 + 4y^2 = 36$

14. $25x^2 + y^2 = 25$

In Exercises 15–28 an equation of a hyperbola is given. (**a**) Locate the vertices. (**b**) What are the equations of the asymptotes? (**c**) Graph the indicated hyperbolas.

15. $\frac{x^2}{9} - \frac{y^2}{16} = 1$ (Graph.)

16. $\frac{x^2}{64} - \frac{y^2}{36} = 1$ (Graph.)

17. $x^2 - \frac{y^2}{4} = 1$

18. $\frac{x^2}{9} - y^2 = 1$

19. $\frac{y^2}{16} - \frac{x^2}{9} = 1$ (Graph.)

20. $\frac{y^2}{25} - \frac{x^2}{144} = 1$ (Graph.)

21. $\dfrac{y^2}{9} - x^2 = 1$ **22.** $\dfrac{y^2}{64} - \dfrac{x^2}{36} = 1$ **23.** $4x^2 - 9y^2 = 36$ (Graph.)

24. $4y^2 - 16x^2 = 16$ (Graph.) **25.** $25x^2 - 100y^2 = 100$ **26.** $9y^2 - 36x^2 = 36$

27. $\dfrac{x^2}{25} - \dfrac{y^2}{24} = 1$ **28.** $x^2 - y^2 = 1$

In Exercises 29–40 indicate whether the equation describes: i. a circle ii. an ellipse iii. a hyperbola iv. two intersecting lines v. a single point (of the real plane) vi. no point (of the real plane)

29. $\dfrac{x^2}{12} + \dfrac{y^2}{12} = 1$ **30.** $\dfrac{x^2}{12} - \dfrac{y^2}{12} = 1$ **31.** $\dfrac{x^2}{12} + \dfrac{y^2}{12} = 0$ **32.** $\dfrac{x^2}{12} + \dfrac{y^2}{12} = -1$

33. $\dfrac{x^2}{12} - \dfrac{y^2}{12} = -1$ **34.** $\dfrac{x^2}{12} + y^2 = 1$ **35.** $3x^2 + 4y^2 = 5$ **36.** $3x^2 - 4y^2 = 1$

37. $5x^2 + \dfrac{x^2}{5} = 1$ **38.** $3x^2 + 4y^2 = 0$ **39.** $3x^2 - 4y^2 = 0$ **40.** $3x^2 - 4y^2 = -5$

41. Find the equation of an ellipse with vertices (4, 0), (−4, 0), (0, 6), and (0, −6).

42. Find the equation of an ellipse with vertices (6, 0), (−6, 0), (0, 4), and (0, −4).

43. Find the equation of a hyperbola that has (3, 0) as a vertex and $y = \dfrac{1}{2}x$ as an asymptote.

44. Find the equation of a hyperbola that has (0, 3) as a vertex and $y = \dfrac{1}{2}x$ as an asymptote.

Review Exercises for Chapter 6

1. Determine all roots of each equation.

(a) $(x - 2)(x + 3) = 0$ (b) $x^2 + 5x + 4 = 0$ (c) $1 - \dfrac{1}{t} = \dfrac{6}{t^2}$

2. Determine a quadratic equation whose roots are 4 and −5.

3. Write each expression in the form bi.

(a) $\sqrt{-36}$ (b) $\sqrt{\dfrac{-2}{9}}$

4. Determine: (a) the real part (b) the imaginary part and (c) the conjugate of the given complex number

(a) $3 - 4i$ (b) -8 (c) $\dfrac{4 + i}{2}$

5. Solve for x.

(a) $x^2 = 25A^2, \quad A > 0$ (b) $\left(\dfrac{2x + 1}{3}\right)^2 = 4$ (c) $x^2 + 48 = 0$

6. Solve each equation by completing the square.

(a) $x^2 + 6x = 11$ (b) $t^2 - 5t + 1 = 0$ (c) $2y^2 + 8y + 3 = 0$

7. (a) Determine the discriminant. (b) Without solving, state whether there are 2 distinct real roots, exactly 1 real root, or 2 complex conjugate roots.

i. $x^2 + 5x + 5 = 0$ ii. $3y^2 - 7y + 5 = 0$ iii. $9x^2 + 12x + 4 = 0$

8. Solve by means of the quadratic formula.

(a) $x^2 + 10x + 1 = 0$ (b) $x^2 + 9 = 9x$ (c) $y^2 - 5y + 5 = 0$ (d) $2x^2 - 5x + 5 = 0$

9. Determine all nonzero numbers that are such that the sum of the number and its reciprocal is 9.

10. The perimeter of a rectangle is 32 inches. The area of the rectangle is 63 square inches. Determine the dimensions of the rectangle.

11. Solve each equation and check the (possible) roots.

(a) $\sqrt{x + 3} - 3 = 0$ (b) $\sqrt{t + 1} - \sqrt{t} + 1 = 0$ (c) $\sqrt{y}\,\sqrt{y + 3} = 2$

12. For each parabola: (a) Does the parabola open upward or downward? (b) Find the vertex. (c) Find the equation of the axis of symmetry. (d) Graph the parabola. (e) Use the graph to determine (or to estimate) the roots of the related equation obtained by setting $y = 0$ in the given equation.

i. $y = (x + 2)^2$ ii. $y = 2 - (x - 1)^2$ iii. $y = x^2 - 6x$

13. Solve each inequality.

(a) $x^2 + 5x + 6 > 0$ (b) $16t^2 \leqslant 1$ (c) $\dfrac{x + 4}{x + 1} < 2$

14. Determine: (a) the center and (b) the radius length of each circle

i. $(x + 3)^2 + (y - 5)^2 = 8$ ii. $x^2 + y^2 + 6x - 5y = 0$

15. (a) Locate the vertices of the ellipse

$$\frac{x^2}{100} + \frac{y^2}{36} = 1.$$

(b) Graph the ellipse.

16. (a) Locate the vertices of the hyperbola

$$\frac{x^2}{25} - \frac{y^2}{9} = 1.$$

(b) What are the equations of the asymptotes?

(c) Graph the hyperbola.

17. Indicate whether each equation describes: i. a circle, ii. an ellipse, iii. a hyperbola, iv. two intersecting lines, v. a single point (of the real plane), vi. no point (of the real plane)

(a) $\dfrac{x^2}{4} - \dfrac{y^2}{4} = 1$ (b) $\dfrac{x^2}{4} + \dfrac{y^2}{4} = 1$ (c) $\dfrac{x^2}{4} - \dfrac{y^2}{4} = 0$ (d) $\dfrac{x^2}{4} + \dfrac{y^2}{4} = 0$ (e) $\dfrac{x^2}{4} + \dfrac{y^2}{4} = -1$

7 Logarithms

7.1 Definition of Logarithms

Logarithms, which you will be learning about in this chapter, have important uses in such fields as physics, chemistry, and economics, as well as in advanced mathematics. Some of these applications will be illustrated at the end of this chapter.

Logarithms are exponents. When you say

$$b^m = x$$

b is called the **base,** m is the **exponent,** and x is the **mth power of b.** Often you are asked:

What power of b is x?

The emphasis here is on the *exponent* m, and you want to feature m (alone) on one side of an equation. For this purpose, logarithmic notation is useful.

Definition. Let b and x be positive numbers with $b \neq 1$. If

$$b^m = x,$$

then m is called **the logarithm of x to the base b.**

Write $\quad \log_b x = m \quad$ when $\quad b^m = x$

(You often speak of "the log of x" instead of "the logarithm of x".)

EXAMPLE 1 (a) $\log_2 8 = 3$

↑ ↑ ↑

base power exponent

(the log of 8 to the base 2 is 3) because

$2^3 = 8$

↑ ↖ ↑

base exponent power

(b) $\log_{10} 10\,000 = 4$

↑ base ↑ power ↑ exponent

(the log of 10 000 to the base 10 is 4) because

$10^4 = 10\,000$

↑ base ↖ exponent ↑ power

(c) $\log_{10} .01 = -2$

↑ base ↑ power ↑ exponent

(the log of .01 to the base 10 is -2) because

$$10^{-2} = \frac{1}{10^2} = \frac{1}{100} = .01$$

↑ base ↖ exponent ↑ power □

Observe that $\log_b x$ *is an exponent* because

$$\log_b x = m \qquad \text{means} \qquad b^m = x.$$

Also, if $\log_b x = m$,

then x is the $(\log_b x)$th power of b $[x = b^{\log_b x}]$

because x is the mth power of b. $[x = b^m]$

All powers, b^m, of a positive number b are positive. Thus, $\log_b x$ *is defined only for positive x.* For example, suppose $b = 4$.

$$4^2 = 16 \qquad (\text{or } \log_4 16 = 2)$$

$$4^{1/2} = 2 \qquad \left(\text{or } \log_4 2 = \frac{1}{2}\right)$$

$$4^0 = 1 \qquad (\text{or } \log_4 1 = 0)$$

$$4^{-1} = \frac{1}{4} \qquad \left(\text{or } \log_4 \frac{1}{4} = -1\right)$$

$$4^{-3} = \frac{1}{64} \qquad \left(\text{or } \log_4 \frac{1}{64} = -3\right)$$

The various powers, 4^m, of 4 are all positive. Thus, $\log_4 x$ is defined only for positive x because

$$\log_4 x = m \qquad \text{means} \qquad 4^m = x.$$

In Chapter 5 negative exponents and rational exponents were defined. Logarithms are exponents; here are some examples in which some of the logarithms are *not* positive integers.

EXAMPLE 2 (a) $\log_4 16 = 2$ because $4^2 = 16$

(b) $\log_{16} 4 = \frac{1}{2}$ because $16^{1/2} = 4$

(c) $\log_{16} \frac{1}{4} = \frac{-1}{2}$ because $16^{-1/2} = \frac{1}{16^{1/2}} = \frac{1}{4}$ □

EXAMPLE 3 (a) $\log_4 8 = \frac{3}{2}$ because $4^{3/2} = (4^{1/2})^3 = 2^3 = 8$

(b) $\log_{16} 8 = \frac{3}{4}$ because $16^{3/4} = (16^{1/4})^3 = 2^3 = 8$

(c) $\log_{1/2} 8 = -3$ because $\left(\frac{1}{2}\right)^{-3} = \frac{1}{\left(\frac{1}{2}\right)^3} = \frac{1}{\frac{1}{8}} = 8$

Note that $\log_{1/2} 8$ stands for $\log_{\frac{1}{2}} 8$, which is difficult to read in print. □

EXAMPLE 4 Find the value of:

(a) $\log_3 27$ (b) $\log_6 1$ (c) $\log_{1/8} \frac{1}{2}$

Solution.

(a) What power of 3 is 27? *Answer:* The 3rd power

$$3^3 = 27$$

Therefore

$$\log_3 27 = 3$$

(b) What power of a number, such as 6, is 1? *Answer:* The 0th power

$$6^0 = 1$$

Therefore

$$\log_6 1 = 0$$

(c) $$\left(\frac{1}{8}\right)^{1/3} = \frac{1}{8^{1/3}} = \frac{1}{2}$$

Therefore

$$\log_{1/8} \frac{1}{2} = \frac{1}{3}$$ □

EXAMPLE 5 (a) Find the value of b: $\log_b 125 = 3$

(b) Find the value of x: $\log_9 x = -2$

Solution.

(a) $\log_b 125 = 3$ means $b^3 = 125$

Thus $b = 5$ because $5^3 = 125$

(b) $\log_9 x = -2$ means $9^{-2} = x$

Thus $x = 9^{-2} = \frac{1}{9^2} = \frac{1}{81}$ □

In the definition of $\log_b x$, note that $b \neq 1$. This is because every power of 1 is 1. Thus the base 1 is not useful.

EXERCISES

In Exercises 1–20 express in logarithmic notation.

1. $2^4 = 16$ **2.** $5^3 = 125$ **3.** $10^6 = 1\,000\,000$ **4.** $7^3 = 243$ **5.** $10^{-1} = \frac{1}{10}$

6. $9^{-2} = \frac{1}{81}$ **7.** $4^{1/2} = 2$ **8.** $27^{1/3} = 3$ **9.** $16^{3/4} = 8$ **10.** $125^{2/3} = 25$

11. $9^{-1/2} = \frac{1}{3}$ **12.** $64^{-1/3} = \frac{1}{4}$ **13.** $\left(\frac{1}{3}\right)^{-2} = 9$ **14.** $\left(\frac{2}{5}\right)^3 = \frac{8}{125}$ **15.** $\left(\frac{4}{7}\right)^{-1} = \frac{7}{4}$

16. $\left(\frac{5}{3}\right)^{-2} = \frac{9}{25}$ **17.** $(.1)^4 = .0001$ **18.** $(.2)^3 = .008$ **19.** $.01^{1/2} = .1$ **20.** $.0144^{1/2} = .12$

In Exercises 21–32 express in exponential notation.

21. $\log_2 16 = 4$ **22.** $\log_3 27 = 3$ **23.** $\log_8 2 = \frac{1}{3}$ **24.** $\log_{100} 10 = \frac{1}{2}$ **25.** $\log_7 \frac{1}{7} = -1$

26. $\log_{16} 8 = \frac{3}{4}$ **27.** $\log_{49} 7 = \frac{1}{2}$ **28.** $\log_7 \frac{1}{49} = -2$ **29.** $\log_{10} .01 = -2$ **30.** $\log_{.2} .04 = 2$

31. $\log_{81} 27 = \frac{3}{4}$ **32.** $\log_{2/3} \frac{8}{27} = 3$

In Exercises 33–48 find the value of each logarithm.

33. $\log_5 25$ **34.** $\log_{11} 11$ **35.** $\log_{10} 10\,000$ **36.** $\log_2 64$ **37.** $\log_4 64$ **38.** $\log_8 64$

39. $\log_{16} 64$ **40.** $\log_{64} 8$ **41.** $\log_2 \frac{1}{64}$ **42.** $\log_{1/2} 64$ **43.** $\log_2 1$ **44.** $\log_{.3} .09$

45. $\log_{1/3} \frac{1}{27}$ **46.** $\log_{3/2} \frac{27}{8}$ **47.** $\log_{2/3} \frac{27}{8}$ **48.** $\log_{1/27} 81$

In Exercises 49–64 find the value of x.

49. $\log_5 x = 2$ **50.** $\log_x 8 = 3$ **51.** $\log_7 7 = x$ **52.** $\log_x 29 = 1$

53. $\log_x \frac{1}{3} = -2$ **54.** $\log_5 \sqrt{5} = x$ **55.** $\log_{10} 1 = x$ **56.** $\log_x 121 = 2$

57. $\log_x \frac{1}{4} = -2$ **58.** $\log_{2/5} \frac{4}{25} = x$ **59.** $\log_x .64 = 2$ **60.** $\log_x .64 = -2$

61. $\log_x .0016 = 4$ **62.** $\log_{.09} x = 2$ **63.** $\log_x 100\,000 = -5$ **64.** $\log_{32} 64 = x$

7.2 Properties of Logarithms

exponents and logarithms

Logarithms have been defined in terms of exponents. Thus, it is not surprising that the properties of logarithms are derived from the rules for exponents.

Recall the *rules for exponents:*

(A) $b^{m_1+m_2} = b^{m_1}b^{m_2}$

(B) $b^{m_1-m_2} = \frac{b^{m_1}}{b^{m_2}}, \qquad b \neq 0$

(C) $(b^m)^r = b^{mr}$

The corresponding *properties of logarithms* are as follows:

(A) $\log_b x_1x_2 = \log_b x_1 + \log_b x_2$

(B) $\log_b \frac{x_1}{x_2} = \log_b x_1 - \log_b x_2$

(C) $\log_b x^r = r \log_b x$

Observe that *the base b is held constant in each equation.* Also, x, x_1, x_2, and b are any positive real numbers with $b \neq 1$, and r is any rational number. (With more advanced techniques, irrational exponents can also be considered; see Section 7.5 .) Property (A) asserts:

The log of a product is the sum of the logs.

Property (B) asserts:

The log of a quotient is the difference of the logs.

Property (C) asserts:

The log of the rth *power of x is r times the log of x.*

Before proving these properties, consider some examples.

EXAMPLE 1 $\log_2 (8 \cdot 4) = \log_2 8 + \log_2 4$, according to Property (A). Consider the left side of this equation:

$$\log_2 (8 \cdot 4) = \log_2 32 = 5$$

Now consider the right side and show that this too equals 5:

$$\underbrace{\log_2 8}_{3} + \underbrace{\log_2 4}_{2} = 3 + 2 = 5$$

□

EXAMPLE 2 $\log_3 \frac{27}{9} = \log_3 27 - \log_3 9$, according to Property (B).

$$\log_3 \frac{27}{9} = \log_3 3 = 1$$

$$\underbrace{\log_3 27}_{3} - \underbrace{\log_3 9}_{2} = 3 - 2 = 1$$

□

EXAMPLE 3 $\log_{10} 100^3 = 3 \log_{10} 100$, according to Property (C).

$$\log_{10} 100^3 = \log_{10} 1\,000\,000 = 6$$

because $10^6 = 1\,000\,000$

$$3 \underbrace{\log_{10} 100}_{2} = 3 \cdot 2 = 6$$

□

EXAMPLE 4 $\log_3 \frac{1}{9} = \log_3 1 - \log_3 9 = 0 - 2 = -2$ □

proofs These examples illustrate the properties of logarithms, *but they do not prove them*. Perhaps you might think these properties hold only for specially chosen numbers. You will now see that, in fact, they are true for all numbers (for which exponents or logarithms have been defined). The proofs utilize Rules **(A)**, **(B)**, and **(C)** for exponents.

Assume x, x_1, x_2, and b are any positive real numbers with $b \neq 1$, and r is any rational number.

(A) $\boxed{\log_b x_1 x_2 = \log_b x_1 + \log_b x_2}$

Proof. Let $\log_b x_1 = m_1$, $\log_b x_2 = m_2$.

By the definition of logarithms,

$$b^{m_1} = x_1, \qquad b^{m_2} = x_2$$

Thus

$$\begin{aligned} \log_b x_1 x_2 &= \log_b (b^{m_1} b^{m_2}) && \text{by substitution} \\ &= \log_b (b^{m_1+m_2}) && \text{by Rule (A) for exponents} \\ &= m_1 + m_2 && \text{by the definition of logarithms} \\ &= \log_b x_1 + \log_b x_2 && \text{by substitution} \end{aligned}$$ △

Property **(C)** will be verified next.

(C) $\boxed{\log_b x^r = r \log_b x}$

Proof. Let $\log_b x = m$. By the definition of logarithms,

$$b^m = x$$

Consider the rth power of each side.

$$(b^m)^r = x^r$$

Apply Rule **(C)** of exponents to the left side.

$$b^{mr} = (b^m)^r = x^r$$

Logarithmic notation yields

$$\log_b x^r = mr = rm = r \log_b x.$$ △

Properties **(A)** and **(C)** will be used to prove Property **(B)**.

(B) $\boxed{\log_b \frac{x_1}{x_2} = \log_b x_1 - \log_b x_2}$

Proof.

$$\log_b \frac{x_1}{x_2} = \log_b \left(x_1 \cdot \frac{1}{x_2} \right)$$

$$\begin{aligned} &= \log_b x_1 + \log_b \frac{1}{x_2} && \text{by (A)} \\ &= \log_b x_1 + \log_b x_2^{-1} \\ &= \log_b x_1 + (-1) \log_b x_2 && \text{by (C)} \\ &= \log_b x_1 - \log_b x_2 && \triangle \end{aligned}$$

In Property (**B**), let $x_1 = 1$. Then, as in the proof of Property (**B**),

$$\log_b \frac{1}{x_2} = \log_b 1 - \log_b x_2 = 0 - \log_b x_2 = -\log_b x_2$$

Thus,

$$(\mathbf{B}') \quad \boxed{\log_b \frac{1}{x_2} = -\log_b x_2}$$

further examples

Here are some further examples illustrating the properties of logarithms. (Assume all letters represent positive numbers.)

EXAMPLE 5 Express $\log_2 \dfrac{x^3}{yz}$ in terms of simpler logarithms.

Solution.

$$\begin{aligned} \log_2 \frac{x^3}{yz} &= \log_2 x^3 - \log_2 yz \\ &= 3 \log_2 x - (\log_2 y + \log_2 z) \\ &= 3 \log_2 x - \log_2 y - \log_2 z \end{aligned}$$

□

EXAMPLE 6 Express $\log_{10} \dfrac{\sqrt{ax^3}}{\sqrt[3]{5y^2}}$ in terms of simpler logarithms.

Solution.

$$\begin{aligned} \log_{10} \frac{\sqrt{ax^3}}{\sqrt[3]{5y^2}} &= \log_{10} \sqrt{ax^3} - \log_{10} \sqrt[3]{5y^2} \\ &= \log_{10} (ax^3)^{1/2} - \log_{10} (5y^2)^{1/3} \\ &= \frac{1}{2} \log_{10} (ax^3) - \frac{1}{3} \log_{10} (5y^2) \\ &= \frac{1}{2} (\log_{10} a + 3 \log_{10} x) - \frac{1}{3} (\log_{10} 5 + 2 \log_{10} y) \end{aligned}$$

or

$$= \frac{1}{2} \log_{10} a + \frac{3}{2} \log_{10} x - \frac{1}{3} \log_{10} 5 - \frac{2}{3} \log_{10} y$$

□

EXAMPLE 7 Express $\log_b 48$ in terms of $\log_b 2$ and $\log_b 3$.

Solution.

$$\log_b 48 = \log_b \underbrace{(2^4 \cdot 3)}_{48} = \log_b 2^4 + \log_b 3 = 4 \log_b 2 + \log_b 3 \quad \square$$

EXERCISES

Assume all variables represent positive numbers.

In Exercises 1–10: (a) Verify the following equations by evaluating both sides. (b) Indicate which properties of logarithms are illustrated.

1. $\log_5 125 = \log_5 25 + \log_5 5$ *Answer:* (a) $3 = 2 + 1$
(b) $\log_b x_1x_2 = \log_b x_1 + \log_b x_2$. Here $b = 5$, $x_1 = 25$, $x_2 = 5$
2. $\log_3 81 = \log_3 3 + \log_3 27$
3. $\log_2 64 = \log_2 4 + \log_2 16$
4. $\log_3 9 = \log_3 27 - \log_3 3$
5. $\log_{10} 1000 = \log_{10} 100\,000 - \log_{10} 100$
6. $\log_4 1 = \log_4 64 - \log_4 64$
7. $\log_2 64 = 2 \log_2 8$
8. $\log_{10} 100\,000\,000 = 4 \log_{10} 100$
9. $\log_3 81 = 2 \log_3 9$
10. $\log_2 128 = \log_2 4 + 2 \log_2 8 - \log_2 2$

In Exercises 11–30 express each in terms of simpler logarithms.

11. $\log_b 10x$ *Answer:* $\log_b 10 + \log_b x$
12. $\log_b xyz$
13. $\log_b \dfrac{x}{y}$
14. $\log_b \dfrac{x}{3}$
15. $\log_{10} \dfrac{ab}{c}$
16. $\log_{10} \dfrac{a}{bc}$
17. $\log_{10} x^3$
18. $\log_2 y^{1/2}$
19. $\log_5 xy^4$
20. $\log_b (xy)^4$
21. $\log_3 \dfrac{x^2y^{1/2}}{z}$
22. $\log_b \sqrt{xyz}$
23. $\log_b \sqrt{\dfrac{5x}{3}}$
24. $\log_b \dfrac{\sqrt{5x}}{3}$
25. $\log_{10} \dfrac{a^4b^2}{c^3}$
26. $\log_{10} \dfrac{19a^7b^2}{cd^4}$
27. $\log_b x^{1/2}y^{-1/3}z^{3/4}$
28. $\log_b \dfrac{1}{\sqrt{x^3y^4}}$
29. $\log_b \left(\dfrac{x^{1/2}y^2}{z^3}\right)^{-5}$
30. $\log_b \dfrac{(x^3y^{1/3})^7}{(wz^2)^3}$

In Exercises 31–43, given that $\log_{10} 2 = .301$, $\log_{10} 3 = .477$, and $\log_{10} 5 = .699$, express each of the following to 2 decimal places.

31. $\log_{10} 6$
32. $\log_{10} 15$
33. $\log_{10} 25$
34. $\log_{10} 81$
35. $\log_{10} \dfrac{2}{3}$
36. $\log_{10} \dfrac{3}{4}$
37. $\log_{10} 12$
38. $\log_{10} 24$
39. $\log_{10} \dfrac{48}{25}$
40. $\log_{10} \dfrac{15}{64}$
41. $\log_{10} \dfrac{1}{3}$
42. $\log_{10} \dfrac{1}{20}$
43. Show that $\dfrac{\log_{10} 5}{\log_{10} 2} \neq \log_{10} 5 - \log_{10} 2$.

In Exercises 44–56 express each as a single logarithm. In your answer, do not rationalize the denominator.

44. $\log_b s + \log_b t$
45. $\log_b x - \log_b y$
46. $\log_b r + \log_b s + \log_b t$
47. $\log_b 10 - \log_b a + \log_b x$
48. $\log_b 10 - (\log_b a + \log_b x)$
49. $\log_b 10 - (\log_b a - \log_b x)$

50. $2 \log_b x$

51. $3 \log_b m + \frac{1}{2} \log_b n$

52. $\frac{1}{4} \log_b x - \frac{1}{3} \log_b y^2 + \log_b z$

53. $\frac{1}{5}\left(\log_b r - 2 \log_b s + \frac{1}{3} \log_b t\right)$

54. $\frac{\log_b 10 - \log_b 7}{3}$

55. $\frac{\log_{10} a + 3 \log_{10} b}{5} + \frac{\log_{10} c - \log_{10} d}{3}$

56. $3\left[2 \log_3 9 + \frac{1}{2}(\log_3 5 - \log_3 10)\right]$

57. The pH of a chemical solution is defined to be $-\log_{10}[H^+]$, where $[H^+]$ stands for the hydrogen ion concentration expressed in moles per liter. Find the pH of a solution whose $[H^+]$ is 6×10^{-5}.

7.3 Common Logarithms

Logarithms to the base 10 are known as **common logarithms.** As you will see in Section 7.7, common logarithms simplify arithmetic computations.

powers of 10

When you work with common logs you are considering powers of 10. Thus

$$\log_{10} 100 = 2 \quad \text{because} \quad 10^2 = 100$$

You can immediately find logs of powers of 10. In fact,

$$\log_{10} 10 = 1 \quad \text{because} \quad 10^1 = 10$$

Furthermore, by Property (C) of logarithms,

$$\log_{10} 10^r = r \log_{10} 10 = r \cdot 1 = r$$

Thus

$$\begin{aligned} \log_{10} 100 &= \log_{10} 10^2 = 2 \\ \log_{10} 1000 &= \log_{10} 10^3 = 3 \\ \log_{10} 10\,000 &= \log_{10} 10^4 = 4 \\ \log_{10} 100\,000 &= \log_{10} 10^5 = 5 \end{aligned}$$

and

$$\begin{aligned} \log_{10} 1 & & &= \log_{10} 10^0 = 0 \\ \log_{10} .1 &= \log_{10} \frac{1}{10} &&= \log_{10} 10^{-1} = -1 \\ \log_{10} .01 &= \log_{10} \frac{1}{100} &&= \log_{10} 10^{-2} = -2 \\ \log_{10} .001 &= \log_{10} \frac{1}{1000} &&= \log_{10} 10^{-3} = -3 \\ \log_{10} .0001 &= \log_{10} \frac{1}{10\,000} &&= \log_{10} 10^{-4} = -4 \\ \log_{10} .000\,01 &= \log_{10} \frac{1}{100\,000} &&= \log_{10} 10^{-5} = -5 \end{aligned}$$

You can also determine the logs of *rational powers of* 10. For example,

$$\log_{10} 10^{1/2} = \frac{1}{2} \log_{10} 10 = \frac{1}{2}$$

$$\log_{10} 10^{2/3} = \frac{2}{3} \log_{10} 10 = \frac{2}{3}$$

The trouble is that numbers such as $\log_{10} 5$, $\log_{10} \frac{4}{3}$ and $\log_{10} 1.32$ are *not* rational and are *not* easy to determine. Recall that $\log_{10} x$ is defined for all positive x. *Except for special values of* x (such as 10, 100, $10^{1/2}$) $\log_{10} x$ *is irrational.*

log table

It would be a hopeless task to determine, directly, the common log of most positive numbers (rational or irrational). For instance, to find $\log_{10} 1.32$, you would have to find the power of 10 that yields 1.32 . Fortunately, tables have been compiled (by advanced methods) that indicate the *approximate* log of any number between 1 and 10. Such a table of common logs is found in the appendix. As you will see, this table will enable you to approximate the log of every positive number.

Let $1 \leqslant x < 10$. Furthermore, suppose x is a 3-digit number. For example, consider:

$$x = 1.32$$

The first 2 digits (together with the decimal point) are indicated in the first column in Table 7.1 . The logs of all 3-digit numbers beginning with 1.3 are given in the (horizontal) *row* of 1.3 . Thus, $\log_{10} 1.32$ is in the (vertical) *column* headed by 2.

TABLE 7.1 $\log_{10} 1.32 = .1206$

x	0	1	2	3	4	5	6	7	8	9
1.0	.0000	.0043	.0086	.0128	.0170	.0212	.0253	.0294	.0334	.0374
1.1	.0414	.0453	.0492	.0531	.0569	.0607	.0645	.0682	.0719	.0755
1.2	.0792	.0828	.0864	.0899	.0934	.0969	.1004	.1038	.1072	.1106
1.3	.1139	.1173	.1206	.1239	.1271	.1303	.1335	.1367	.1399	.1430
1.4	.1461	.1492	.1523	.1553	.1584	.1614	.1644	.1673	.1703	.1732

(Strictly speaking, .1206 is only the *approximate* value of log 1.32 .) From now on, the base 10 will usually be omitted in the notation. Thus write:

$$\log 1.32 = .1206$$

EXAMPLE 1 Find the value of: log 4.89

Solution. Consider the row of 4.8 and the column of 9 in Table 7.2 on page 278. □

EXAMPLE 2 Find the value of: log 6.70

Solution. Consider the row of 6.7 and the column headed by 0 in Table 7.3 on page 278. □

TABLE 7.2 log 4.89 = .6893

x	0	1	2	3	4	5	6	7	8	9
4.5	.6532	.6542	.6551	.6561	.6571	.6580	.6590	.6599	.6609	.6618
4.6	.6628	.6637	.6646	.6656	.6665	.6675	.6684	.6693	.6702	.6712
4.7	.6721	.6730	.6739	.6749	.6758	.6767	.6776	.6785	.6794	.6803
4.8	.6812	.6821	.6830	.6839	.6848	.6857	.6866	.6875	.6884	.6893
4.9	.6902	.6911	.6920	.6928	.6937	.6946	.6955	.6964	.6972	.6981
5.0	.6990	.6998	.7007	.7016	.7024	.7033	.7042	.7050	.7059	.7067
5.1	.7076	.7084	.7093	.7101	.7110	.7118	.7126	.7135	.7143	.7152
5.2	.7160	.7168	.7177	.7185	.7193	.7202	.7210	.7218	.7226	.7235
5.3	.7243	.7251	.7259	.7267	.7275	.7284	.7292	.7300	.7308	.7316
5.4	.7324	.7332	.7340	.7348	.7356	.7364	.7372	.7380	.7388	.7396

TABLE 7.3 log 6.70 = .8261

x	0	1	2	3	4	5	6	7	8	9
6.5	.8129	.8136	.8142	.8149	.8156	.8162	.8169	.8176	.8182	.8189
6.6	.8195	.8202	.8209	.8215	.8222	.8228	.8235	.8241	.8248	.8254
6.7	.8261	.8267	.8274	.8280	.8287	.8293	.8299	.8306	.8312	.8319
6.8	.8325	.8331	.8338	.8344	.8351	.8357	.8363	.8370	.8376	.8382
6.9	.8388	.8395	.8401	.8407	.8414	.8420	.8426	.8432	.8439	.8445

To find the log of a number x bigger than 10 or the log of a positive decimal less than 1, use **scientific notation** [Section 5.3]. Thus write x as the product of a number between 1 and 10 and a power of 10. Then use the following properties of logarithms:

(A) $\log x_1 x_2 = \log x_1 + \log x_2, \qquad x_1 > 0,\ x_2 > 0$

(C′) $\log_{10} 10^r = r, \qquad r$ **rational**

EXAMPLE 3 Find the value of: log 743

Solution.

$$743 = 7.43 \times 10^2$$

$$\begin{aligned} \log 743 &= \log (7.43 \times 10^2) \\ &= \log 7.43 + \log 10^2 \\ &= \log 7.43 + 2 \end{aligned}$$

Determine log 7.43. [See Table 7.4 .]

$$\log 743 = \underbrace{\log 7.43}_{.8710} + 2 = 2.8710$$ □

TABLE 7.4 log 7.43 = .8710

x	0	1	2	3	4	5	6	7	8	9
7.0	.8451	.8457	.8463	.8470	.8476	.8482	.8488	.8494	.8500	.8506
7.1	.8513	.8519	.8525	.8531	.8537	.8543	.8549	.8555	.8561	.8567
7.2	.8573	.8579	.8585	.8591	.8597	.8603	.8609	.8615	.8621	.8627
7.3	.8633	.8639	.8645	.8651	.8657	.8663	.8669	.8675	.8681	.8686
7.4	.8692	.8698	.8704	.8710	.8716	.8722	.8727	.8733	.8739	.8745

characteristic and mantissa

The log of a number consists of an integral part, called the **characteristic** and a decimal part, called the **mantissa.** In Example 4,

$$\log 743 = 2.8710$$

integral part or characteristic (2); decimal part or mantissa (.8710)

Thus here the characteristic is 2; the mantissa is .8710 . Observe that *the characteristic equals the exponent of the base 10 when the original number is written in scientific notation.* For example,

$$743 = 7.43 \times 10^2$$

(characteristic: the exponent 2)

and the characteristic of log 743 is 2.

The characteristic is always an integer—positive, negative, or zero. The mantissa is the part of the logarithm read from the table. It is important to note that *whenever the table is to be used, the mantissa is always written as a positive decimal—never as a negative decimal.*

EXAMPLE 4 Find the value of: log 936 000

Solution. Write 936 000 = 9.36×10^5. Now log 9.36 can be read from the table. Thus

$$\log 936\,000 = \log (9.36 \times 10^5)$$

TABLE 7.5 log 9.36 = .9713 (mantissa)

x	0	1	2	3	4	5	6	7	8	9
9.0	.9542	.9547	.9552	.9557	.9562	.9566	.9571	.9576	.9581	.9586
9.1	.9590	.9595	.9600	.9605	.9609	.9614	.9619	.9624	.9628	.9633
9.2	.9638	.9643	.9647	.9652	.9657	.9661	.9666	.9671	.9675	.9680
9.3	.9685	.9689	.9694	.9699	.9703	.9708	.9713	.9717	.9722	.9727
9.4	.9731	.9736	.9741	.9745	.9750	.9754	.9759	.9763	.9768	.9773

$$= \underbrace{\log 9.36}_{\text{mantissa}} + \underbrace{\log 10^5}_{}$$

$$= \underbrace{\log 9.36}_{\text{mantissa}} + \underbrace{5}_{\text{characteristic}}$$

Use Table 7.5 on page 279 to find log 9.36 .

$$\log 936\ 000 = 5.9713$$ □

EXAMPLE 5 Find the value of: log .008 36

Solution.

$$.008\ 36 = 8.36 \times 10^{-3}$$

$$\log .008\ 36 = \log (8.36 \times 10^{-3})$$

$$= \log 8.36 + \log 10^{-3}$$

$$= \underbrace{(\log 8.36)}_{\text{mantissa}} \underbrace{- 3}_{\text{characteristic}}$$

Next, log 8.36 can be read from Table 7.6 .

TABLE 7.6 $\log 8.36 = \underbrace{.9222}_{\text{mantissa}}$

x	0	1	2	3	4	5	6	7	8	9
8.0	.9031	.9036	.9042	.9047	.9053	.9058	.9063	.9069	.9074	.9079
8.1	.9085	.9090	.9096	.9101	.9106	.9112	.9117	.9122	.9128	.9133
8.2	.9138	.9143	.9149	.9154	.9159	.9165	.9170	.9175	.9180	.9186
8.3	.9191	.9196	.9201	.9206	.9212	.9217	.9222	.9227	.9232	.9238
8.4	.9243	.9248	.9253	.9258	.9263	.9269	.9274	.9279	.9284	.9289

$$\log .008\ 36 = \underbrace{\log 8.36}_{.9222} - 3 = \underbrace{.9222}_{\text{mantissa}} \underbrace{- 3}_{\text{characteristic}}$$

If you added the positive mantissa .9222 to the negative characteristic -3 you would obtain

$$\begin{array}{r} -3.0000 \\ + \ .9222 \\ \hline -2.0778 \end{array} = -2 - .0778$$

This would change the mantissa from .9222 to the *negative decimal* $-.0078$, and you would no longer be able to locate it *directly* from the table. Thus

$$\log .008\ 36 = .9222 - 3$$ □

EXERCISES

In Exercises 1–14 find the value of the indicated common log.

1. log 1.03 **2.** log 1.92 **3.** log 4.31 **4.** log 9.77 **5.** log 3.00 **6.** log 8.07

7. log 9.40 **8.** log 9.99 **9.** log 6.03 **10.** log 5.11 **11.** log 7.28 **12.** log 6.46

13. log 1.09 **14.** log 9.85

In Exercises 15–30 determine the characteristic of the indicated common log.

15. log 27.4 **16.** log 382 **17.** log 68 700 **18.** log 9.14

19. log .0384 **20.** log .101 **21.** log .000 724 **22.** log .0101

23. log 8 760 000 **24.** log .000 087 6 **25.** $\log (4.82 \times 10^8)$ **26.** $\log (2.84 \times 10^{-7})$

27. $\log (485 \times 10^4)$ **28.** $\log (283 \times 10^{-6})$ **29.** $\log (30 \times 200)$ **30.** $\log (30 \times 40)$

In Exercises 31–50 find the value of the indicated common log.

31. log 872 **32.** log 9430 **33.** log 16.9 **34.** log 148 000 **35.** log 256 000 000

36. log 70 100 **37.** log .935 **38.** log .004 32 **39.** log .0136 **40.** log .652

41. log 6.39 **42.** log 86.5 **43.** log .000 832 **44.** log .0736 **45.** log 62

46. log .38 **47.** log .004 920 **48.** log 830 000 **49.** $\log (8.36 \times 10^{-5})$ **50.** $\log (.413 \times 10^{-2})$

7.4 Antilogs

In addition to determining the common log of a number, you will also want to find the number with a given logarithm. For instance, you may be told that

$$\log x = 2.5717$$

and be asked to find this number x. This can also be expressed by:

$$x = \text{antilog } 2.5717$$

Definition. Let m be any real number. Define:

$$\textbf{antilog } m = x \qquad \text{if} \qquad \log x = m$$

Do you see that antilog m is simply 10^m? In fact, $\log x$ and 10^x (or antilog x) define functions that are inverse to each other. [See Section 7.5 .] Thus

$$\text{antilog } 3 = 1000 \text{ because } \log 1000 = 3 \text{ (or because } 10^3 = 1000)$$

To determine an antilog, reverse the process of finding a log. Before finding an antilog, recall how the log is found.

EXAMPLE 1

$$\log 484 = \log(4.84 \times 10^2) = \underbrace{2 +}_{\text{characteristic}} \underbrace{\log 4.84}_{\text{mantissa}} = 2.\underbrace{6848}_{\text{mantissa}}$$

because log 4.84 = .6848

[See Table 7.7 .]

TABLE 7.7 log 4.84 = .6848 and thus antilog .6848 = 4.84

4.5	.6532	.6542	.6551	.6561	.6571	.6580	.6590	.6599	.6609	.6618
4.6	.6628	.6637	.6646	.6656	.6665	.6675	.6684	.6693	.6702	.6712
4.7	.6721	.6730	.6739	.6749	.6758	.6767	.6776	.6785	.6794	.6803
4.8	.6812	.6821	.6830	.6839	.6848	.6857	.6866	.6875	.6884	.6893
4.9	.6902	.6911	.6920	.6928	.6937	.6946	.6955	.6964	.6972	.6981
x	0	1	2	3	4	5	6	7	8	9

Suppose you are asked to determine:

antilog 2.6848

The mantissa came from the log table. Thus, you now skim through the table to locate this mantissa, .6848. You note the row 4.8 and the column 4 in which the mantissa .6848 is located. Thus this mantissa corresponds to 4.84, and you have determined the *digits* of the antilog. Now *use the characteristic to place the decimal point.* Because the characteristic is 2, multiply by 10^2. This means you must move the decimal point 2 places to the *right.*

antilog .6848 = 4.84

antilog 2.6848 = 484

Thus $10^{2.6848}$ is approximately equal to 484. □

EXAMPLE 2 Find the value of: antilog 3.6665

Solution. You are given that

$$\log n = \underbrace{3}_{\text{characteristic}}.\underbrace{6665}_{\text{mantissa}}$$

and you must find n. First locate the mantissa .6665 in Table 7.8 . Now move the decimal point 3 places to the right because the characteristic is 3. You must add a 0 after the second 4.

log 4640 = 3.6665

4640 = antilog 3.6665 □

EXAMPLE 3 Find the value of: antilog .9917

TABLE 7.8 log 4.64 = .6665

x	0	1	2	3	4	5	6	7	8	9
4.5	.6532	.6542	.6551	.6561	.6571	.6580	.6590	.6599	.6609	.6618
4.6	.6628	.6637	.6646	.6656	.6665	.6675	.6684	.6693	.6702	.6712
4.7	.6721	.6730	.6739	.6749	.6758	.6767	.6776	.6785	.6794	.6803
4.8	.6812	.6821	.6830	.6839	.6848	.6857	.6866	.6875	.6884	.6893
4.9	.6902	.6911	.6920	.6928	.6937	.6946	.6955	.6964	.6972	.6981

Solution. The characteristic is 0. According to Table 7.9,

$$\log 9.81 = .9917$$

Thus

$$9.81 = \text{antilog } .9917$$

Because the characteristic is 0, this is the solution. □

TABLE 7.9

x	0	1	2	3	4	5	6	7	8	9
9.5	.9777	.9782	.9786	.9791	.9795	.9800	.9805	.9809	.9814	.9818
9.6	.9823	.9827	.9832	.9836	.9841	.9845	.9850	.9854	.9859	.9863
9.7	.9868	.9872	.9877	.9881	.9886	.9890	.9894	.9899	.9903	.9908
9.8	.9912	.9917	.9921	.9926	.9930	.9934	.9939	.9943	.9948	.9952
9.9	.9956	.9961	.9965	.9969	.9974	.9978	.9983	.9987	.9991	.9996

EXAMPLE 4 Find the value of: antilog (.8182 − 4)

Solution. Locate the mantissa, .8182, in Table 7.10 . Because the characteristic is −4, move the decimal point 4 places to the left. Add enough 0's.

$$\log .000\,658 = .8182 - 4$$

$$.000\,658 = \text{antilog } (.8182 - 4)$$

□

TABLE 7.10

x	0	1	2	3	4	5	6	7	8	9
6.5	.8129	.8136	.8142	.8149	.8156	.8162	.8169	.8176	.8182	.8189
6.6	.8195	.8202	.8209	.8215	.8222	.8228	.8235	.8241	.8248	.8254
6.7	.8261	.8267	.8274	.8280	.8287	.8293	.8299	.8306	.8312	.8319
6.8	.8325	.8331	.8338	.8344	.8351	.8357	.8363	.8370	.8376	.8382
6.9	.8388	.8395	.8401	.8407	.8414	.8420	.8426	.8432	.8439	.8445

In Section 7.6 you will learn how to approximate the antilog when the given mantissa is *not* located in the table of logs.

EXERCISES

Find the value of each antilog.

1. antilog .2945 **2.** antilog .9355 **3.** antilog .8716 **4.** antilog .8254

5. antilog .4346 **6.** antilog .2504 **7.** antilog .5623 **8.** antilog 1.5623

9. antilog 4.5623 **10.** antilog (.5623 − 1) **11.** antilog (.5623 − 3) **12.** antilog (.5623 − 5)

13. antilog .7846 **14.** antilog 2.7846 **15.** antilog 6.7846 **16.** antilog (.7846 − 1)

17. antilog (.7846 − 2) **18.** antilog (.7846 − 6) **19.** antilog 1.0374 **20.** antilog 4.9809

21. antilog 2.9805 **22.** antilog 6.9708 **23.** antilog (.8488 − 3) **24.** antilog (.9818 − 1)

25. antilog (.9805 − 2) **26.** antilog (.2227 − 4) **27.** antilog .0294 **28.** antilog 5.6010

29. antilog .5623 **30.** antilog 4.7067 **31.** antilog 5.6493 **32.** antilog (.6493 − 5)

33. antilog 8.8089 **34.** antilog (.8089 − 8)

7.5 Logarithmic and Exponential Functions

log function

When the base of a logarithmic equation is fixed, such as in

$$m = \log_{10} x,$$

or when base 10 is understood, as in

$$m = \log x$$

a correspondence is set up between numbers x and m. *To each given* $x > 0$ *there corresponds exactly one value* m. For example,

when $x = 100$, then $m = \log 100 = 2$
when $x = 10\,000$, then $m = \log 10\,000 = 4$

and from the log table,

when $x = 62.8$, then $m = \log 62.8 = 1.7980$
when $x = .628$, then $m = \log .628 = .7980 - 1 = -.2020$

A function

$$f(x) = \log x$$

is thereby defined. You have been able to determine the log of a *positive* 3-digit number by use of a table. Actually, the log of any *positive* number, irrational as well as rational, can be approximated by means of the techniques of calculus. Logs of "nearby" numbers are close to one another. For example, consider the irrational number $\sqrt{101}$. Because

$\sqrt{101}$ is *approximately* $\sqrt{100}$ (or 10)

TABLE 7.11

x	$\log x$
.1	-1
.5	$.6990 - 1 = -.3010$
1	0
5	.6990
10	1
50	1.6990
100	2

and because

$$\log 10 = 1,$$

$\log \sqrt{101}$ is close to (and slightly larger than) 1. Again, $\log x$ is defined for *all positive* numbers. A glance at the log table suggests that $\log x$ is an *increasing function*. This is in fact true, and thus $\log x$ is *one-to-one*.

The graph of the common logarithmic function can now be drawn. Consider the sample values in Table 7.11 .

The graph of $\log x$ is shown in Figure 7.1 . As usual, the vertical axis will be called the y-axis (instead of the m-axis). Note that different scales are used on the x- and y-axes. Log x is negative for $0 < x < 1$. The range of $\log x$ is the set of all real numbers.

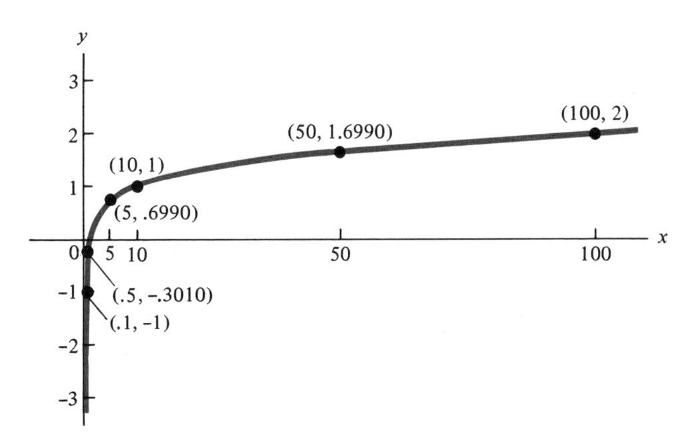

FIGURE 7.1 The graph of $y = \log x$

exponential function

Recall that whenever a function is one-to-one, its inverse is defined and is also a one-to-one function. Let

$$f(x) = \log x.$$

Then $f^{-1}(x)$ denotes the inverse function of $\log x$. Geometrically, the graph of $f^{-1}(x)$ is the reflection of $\log x$ in the line $y = x$ [See Figure 7.2 on page 286.] To obtain $f^{-1}(x)$, first interchange x and y in the defining equation

$$y = \log_{10} x,$$

and obtain

$$x = \log_{10} y.$$

Next, solve for y in terms of x. Because of the definition of the logarithm to the base 10,

$$10^x = y$$

Therefore, f^{-1} is given by

$$f^{-1}(x) = 10^x = \text{antilog } x.$$

In other words, $f^{-1}(x)$, which is an **exponential function,** expresses all powers, 10^x, of 10. Note that the domain of 10^x is the range of

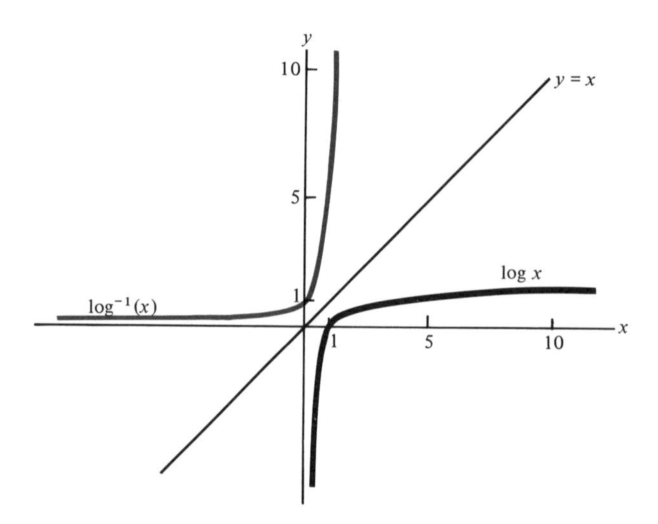

FIGURE 7.2 $\log^{-1}(x)$ is the reflection of $\log x$ in the line $y = x$. $\log^{-1}(x)$ is defined for all real numbers.
$\log^{-1}(x) = 10^x$

$\log x$, that is, the set of *all real numbers,* irrational as well as rational. In this way, irrational powers of 10 can be defined. Thus

$$10^{\sqrt{2}} = y \qquad \text{if} \qquad \log y = \sqrt{2}$$

Finally, 10^x is an increasing function, just as $\log x$ is. And the range of 10^x is the domain of $\log x$, that is, the set of positive numbers.

other exponential and logarithmic functions

Exponential functions describe many growth and decay processes in biology and chemistry. Suppose, for example, that the number of bacteria in a colony doubles each day and that N_0 is the number of bacteria initially present. Then in 1 day the number of bacteria will be $2N_0$, in 2 days $4N_0$, or 2^2N_0, in 3 days $8N_0$, or 2^3N_0, and in x days 2^xN_0. For simplicity, let $N_0 = 1$, and consider the exponential function defined by

$$F(x) = 2^x \quad \text{for } all \text{ real numbers } x.$$

For $x = 0, 1, 2, 3, \ldots$, $F(x)$ describes the number of bacteria present in x days. Some values of $F(x)$ are given in Table 7.12(a) for nonnegative x and in Table 7.12(b) for negative x. As Table 7.12 suggests, $F(x)$ is an increasing function, and therefore has an inverse. The graph of $F(x)$, or 2^x, together with its inverse function, $F^{-1}(x)$, or $\log_2 x$, which is also increasing, are given in Figure 7.3 . Observe that

$$\text{domain } F = \text{range } F^{-1} = \text{set of real numbers;}$$
$$\text{range } F = \text{domain } F^{-1} = \text{set of positive numbers.}$$

Now suppose that a chemical decomposes in such a way that each hour there is only half as much remaining as there was an hour earlier. At the beginning of an experiment suppose there is 1 gram present. In 1 hour there will then be $\frac{1}{2}$ gram present, in 2 hours there

TABLE 7.12 (a)

x	2^x
0	2^0 or 1
$\frac{1}{2}$	$2^{1/2}$ or $\sqrt{2}$
1	2^1 or 2
2	2^2 or 4
3	2^3 or 8
4	2^4 or 16

TABLE 7.12 (b)

x	2^x
$-\frac{1}{2}$	$2^{-(1/2)}$ or $\frac{1}{\sqrt{2}}$ or $\frac{\sqrt{2}}{2}$
-1	2^{-1} or $\frac{1}{2}$
-2	2^{-2} or $\frac{1}{4}$
-3	2^{-3} or $\frac{1}{8}$
-4	2^{-4} or $\frac{1}{16}$

will be $\frac{1}{4}$, or $\left(\frac{1}{2}\right)^2$, gram present, in 3 hours there will be $\frac{1}{8}$, or $\left(\frac{1}{2}\right)^3$, gram present. An hour *before* the experiment began, there were 2 grams present. $\left[\text{Note that } \left(\frac{1}{2}\right)^{-1} = 2.\right]$ And 2 hours before the experiment began, there were 4, or $\left(\frac{1}{2}\right)^{-2}$, grams present.

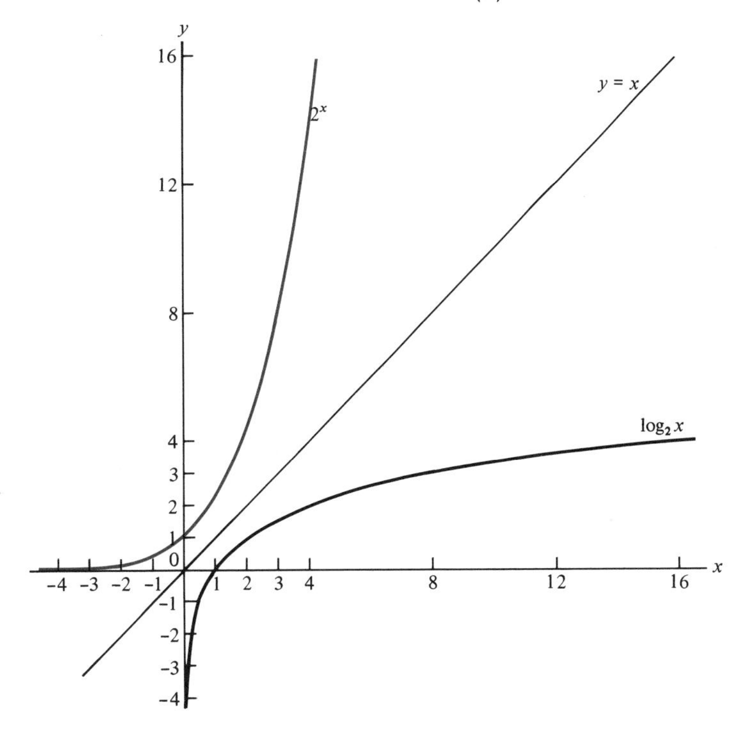

FIGURE 7.3

Consider the function defined by

$$g(x) = \left(\frac{1}{2}\right)^x \quad \text{for } all \text{ real numbers } x.$$

Sample values of $g(x)$ are given in Table 7.13 on page 288. As Table 7.13 suggests, g is a decreasing function, and consequently, has an inverse. The graphs of $g(x)$, or $\left(\frac{1}{2}\right)^x$, and its inverse function, $g^{-1}(x)$, or $\log_{1/2} x$, are given in Figure 7.4 on page 288. Observe that

$$\text{domain } g = \text{range } g^{-1} = \text{set of real numbers;}$$
$$\text{range } g = \text{domain } g^{-1} = \text{set of positive numbers.}$$

In general, for $a > 1$, both a^x and $\log_a x$ are *increasing functions,* whereas for $0 < a < 1$, both a^x and $\log_a x$ are *decreasing functions.*

EXAMPLE 1 Let $f(x) = 9^x$. Then

(a) $f(1) = 9^1 = 9$ (b) $f(2) = 9^2 = 81$

(c) $f\left(\frac{1}{2}\right) = 9^{1/2} = 3$ (d) $f(-1) = 9^{-1} = \frac{1}{9}$ □

TABLE 7.13 (a)

x	$\left(\frac{1}{2}\right)^x$
0	$\left(\frac{1}{2}\right)^0$ or 1
1	$\left(\frac{1}{2}\right)^1$ or $\frac{1}{2}$
2	$\left(\frac{1}{2}\right)^2$ or $\frac{1}{4}$
3	$\left(\frac{1}{2}\right)^3$ or $\frac{1}{8}$
4	$\left(\frac{1}{2}\right)^4$ or $\frac{1}{16}$

TABLE 7.13 (b)

x	$\left(\frac{1}{2}\right)^x$
−1	$\left(\frac{1}{2}\right)^{-1}$ or 2
−2	$\left(\frac{1}{2}\right)^{-2}$ or 4
−3	$\left(\frac{1}{2}\right)^{-3}$ or 8
−4	$\left(\frac{1}{2}\right)^{-4}$ or 16

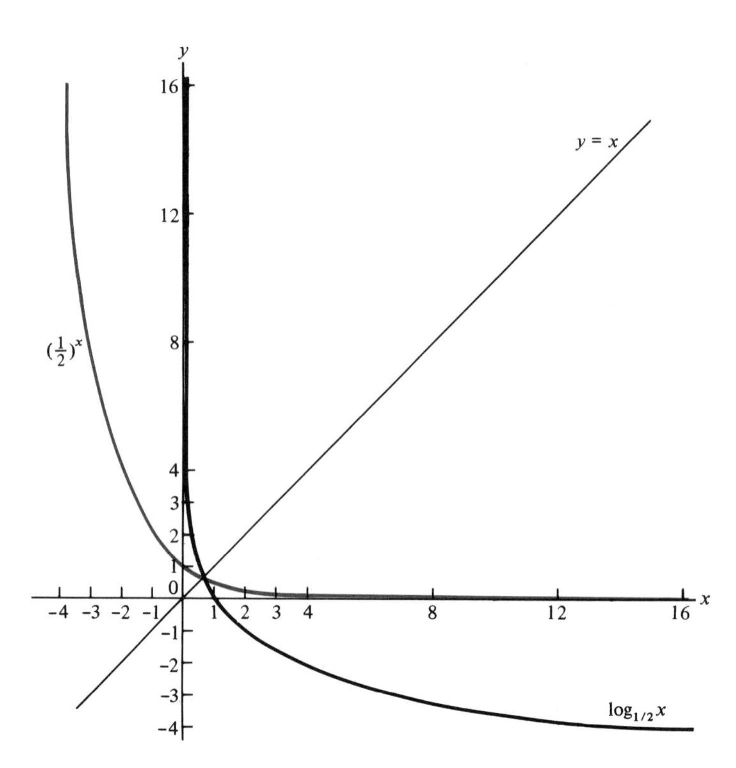

FIGURE 7.4

EXAMPLE 2 Let $f(x) = 2^x$, $g(x) = x + 1$. Then

(a) $(f \circ g)(x) = f(g(x)) = 2^{g(x)} = 2^{x+1}$
(b) $(f \circ g)(0) = 2^{0+1} = 2^1 = 2$
(c) $(f \circ g)(1) = 2^{1+1} = 2^2 = 4$
(d) $(g \circ f)(x) = g(f(x)) = f(x) + 1 = 2^x + 1$
(e) $(g \circ f)(0) = 2^0 + 1 = 1 + 1 = 2$
(f) $(g \circ f)(1) = 2^1 + 1 = 3$ □

natural logarithm

In calculus the logarithmic function of primary importance is the *natural logarithm,* whose base is an irrational number e, which is *approximately* equal to 2.718 . The very definition of e requires a knowledge of calculus. The graphs of this natural logarithmic function and of its inverse function e^x are shaped the same as the graphs in Figure 7.2 and 7.3 .

EXERCISES

In Exercises 1–12 find the indicated values of $f(x)$.

1. $f(x) = 3^x$ **(a)** $f(1)$ **(b)** $f(2)$ **(c)** $f(3)$ **(d)** $f(-1)$

2. $f(x) = 10(5^x)$ **(a)** $f(2)$ **(b)** $f(-2)$ **(c)** $f(3)$ **(d)** $f\left(\frac{1}{2}\right)$

3. $f(x) = 100^x$ **(a)** $f(1)$ **(b)** $f(-1)$ **(c)** $f\left(\frac{1}{2}\right)$ **(d)** $f\left(-\frac{1}{2}\right)$

4. $f(x) = \frac{16^x}{2}$ **(a)** $f\left(\frac{1}{2}\right)$ **(b)** $f\left(\frac{1}{4}\right)$ **(c)** $f\left(-\frac{1}{4}\right)$ **(d)** $f\left(\frac{1}{8}\right)$

5. $f(x) = \left(\frac{1}{4}\right)^x$ **(a)** $f(1)$ **(b)** $f(2)$ **(c)** $f(-1)$ **(d)** $f(-2)$

6. $f(x) = \left(\frac{2}{3}\right)^x$ **(a)** $f(2)$ **(b)** $f(3)$ **(c)** $f(-1)$ **(d)** $f(-2)$

7. $f(x) = \log_2 x$ **(a)** $f(1)$ **(b)** $f(2)$ **(c)** $f(4)$ **(d)** $f(32)$

8. $f(x) = 2 \log_3 x$ **(a)** $f(3)$ **(b)** $f\left(\frac{1}{3}\right)$ **(c)** $f(27)$ **(d)** $f(\sqrt{3})$

9. $f(x) = \log_4 x$ **(a)** $f(16)$ **(b)** $f(2)$ **(c)** $f(64)$ **(d)** $f(8)$

10. $f(x) = \log_9 x$ **(a)** $f(3)$ **(b)** $f\left(\frac{1}{3}\right)$ **(c)** $f\left(\frac{1}{81}\right)$ **(d)** $f(27)$

11. $f(x) = \log_{100} x$ **(a)** $f(10)$ **(b)** $f(.01)$ **(c)** $f(.1)$ **(d)** $f(1000)$

12. $f(x) = \log_{1/10} x$ **(a)** $f\left(\frac{1}{100}\right)$ **(b)** $f(10)$ **(c)** $f(100)$ **(d)** $f(10^6)$

13. Let $f(x) = 2^x$, $g(x) = 3x$. Find:

(a) $(f \circ g)(x)$ **(b)** $(f \circ g)(1)$ **(c)** $(f \circ g)(2)$ **(d)** $(g \circ f)(x)$ **(e)** $(g \circ f)(1)$ **(f)** $(g \circ f)(2)$

14. Let $f(x) = \log_{10} x$, $g(x) = x + 2$. Find:

(a) $(f \circ g)(x)$ **(b)** $(f \circ g)(8)$ **(c)** $(g \circ f)(x)$ **(d)** $(g \circ f)(100)$

(e) What is the largest possible domain of f? **(f)** What is the largest possible domain of $f \circ g$?

15. Let $f(x) = 2^x$, $g(x) = x^2$. Find:

(a) $(f \circ g)(x)$ **(b)** $(f \circ g)(2)$ **(c)** $(f \circ g)(3)$ **(d)** $(g \circ f)(x)$ **(e)** $(g \circ f)(2)$ **(f)** $(g \circ f)(3)$

16. Let $f(x) = 5^x$, $g(x) = \log_5 x$. Find:

(a) $(f \circ g)(x)$ **(b)** $(f \circ g)(1000)$ **(c)** $(g \circ f)(x)$ **(d)** $(g \circ f)\left(\frac{3}{10}\right)$

17. Let $f(x) = \log_6 x$. Suppose $(g \circ f)(x) = x$ for all $x > 0$. Find $g(x)$.

18. Let $f(x) = 2^x$ and $g(x) = 8^x$. Which of the following are true?

(a) For all x, $g(x) > f(x)$ **(b)** For all x, $g(x) = (4f)(x)$

(c) For all x, $g(x) = (f(x))^3$ **(d)** For all x, $\left(\frac{g}{f}\right)(x) = 4^x$

19. On the same coordinate system, graph

(a) 3^x, for $-2 \leqslant x \leqslant 2$, and **(b)** $\log_3 x$, for $\frac{1}{9} \leqslant x \leqslant 9$.

20. On the same coordinate system, graph

(a) $\left(\frac{1}{3}\right)^x$, for $-2 \leqslant x \leqslant 2$, and (b) $\log_{1/3} x$, for $\frac{1}{9} \leqslant x \leqslant 9$

In Exercises 21–30 for each function: (a) Indicate whether the function is increasing, decreasing, or neither. (b) Find the largest possible domain. (c) Find the corresponding range.

21. 7^x **22.** $\log_7 x$ **23.** $\left(\frac{1}{7}\right)^x$ **24.** $\log_{1/7} x$ **25.** -3^x

26. $\frac{1}{\log_3 x}$ **27.** 3^{-x} **28.** $\log_3(-x)$ **29.** $5^{(x^2)}$ **30.** $-\frac{1}{2}\log_{10} x$

31. The number of bacteria in a culture after t hours is given by $1000\,(4^t)$. What is the number of bacteria in the culture (a) after $\frac{1}{2}$ hour? (b) after 2 hours? (c) How many hours will it take to obtain 64 000 bacteria?

32. Each hour the amount of chemical solution in a beaker triples. If it takes 24 hours to fill the beaker, how long does it take to fill one-third of the beaker?

*7.6 Interpolating

linear interpolation

The log of a 4-digit number, such as 2.215, does not appear in the log table. To obtain log 2.215, you *approximate* the value by a method known as **linear interpolation.** This method is based upon the assumption that for a small change in x, the graph of $\log x$ can be *approximated* by a line segment. (See Figure 7.5, which has been distorted to make it readable.) Thus, you can find the corresponding change in y along this line segment by solving a proportion. Note the similar triangles in Figure 7.5 .

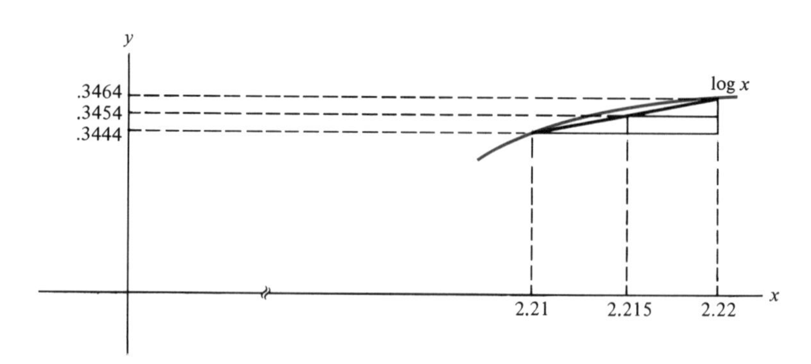

FIGURE 7.5

logs

Throughout this section, base 10 is understood.

EXAMPLE 1

Find the value of: log 2.215

*Optional topic

Solution. First observe that

$$2.210 < 2.215 < 2.220$$

The logs of the 3-digit numbers

$$2.21\ (= 2.210) \qquad \text{and} \qquad 2.22\ (= 2.220)$$

are given in Table 7.14 .

TABLE 7.14 log 2.21 = .3444, log 2.22 = .3464

x	0	1	2	3	4	5	6	7	8	9
2.0	.3010	.3032	.3054	.3075	.3096	.3118	.3139	.3160	.3181	.3201
2.1	.3222	.3243	.3263	.3284	.3304	.3324	.3345	.3365	.3385	.3404
2.2	.3424	.3444	.3464	.3483	.3502	.3522	.3541	.3560	.3579	.3598
2.3	.3617	.3636	.3655	.3674	.3692	.3711	.3729	.3747	.3766	.3784
2.4	.3802	.3820	.3838	.3856	.3874	.3892	.3909	.3927	.3945	.3962

TABLE 7.15

	x	$\log x$	
.010	2.220	.3464	.0020
.005	2.215	$.3444 + h$	h
	2.210	.3444	

Clearly, 2.215 lies exactly halfway between 2.210 and 2.220, as you can see in Table 7.15 .

In Table 7.15 the left-hand column indicates the values of x. The right-hand column indicates the values of $\log x$. Observe the differences between the *largest* and *smallest* values in *each column*.

$$\begin{array}{r} 2.220 \\ -2.210 \\ \hline .010 \end{array} \qquad \begin{array}{r} .3464 \\ -.3444 \\ \hline .0020 \end{array}$$

You seek the log of the intermediate number, 2.215. Therefore, also note the differences:

$$\begin{array}{r} 2.215 \\ -2.210 \\ \hline .005 \end{array} \qquad \text{and} \qquad \begin{array}{r} .3444 + h \\ -.3444 \\ \hline h \end{array}$$

You need the *proportional increase, h*, on the right side. Thus set up the proportion:

$$\frac{.005}{.010} = \frac{h}{.0020}$$

Multiply the numerator and denominator on the left side by 1000, and reduce to lowest terms:

$$\frac{h}{.0020} = \frac{5}{10} = \frac{1}{2}$$

$$h = \frac{.0020}{2} = .0010$$

Thus add:

$$\begin{array}{r} .3444 \\ .0010 \\ \hline .3454 \end{array}$$

and obtain:

$$\log 2.215 = .3454$$

This is a reasonably good *approximation* to the value of log 2.215 .

In practice it is often unnecessary to be so careful about the decimal points in setting up the proportion. You could let d stand for the number of units (ten-thousandths) in the log column [See Table 7.16 .]

TABLE 7.16

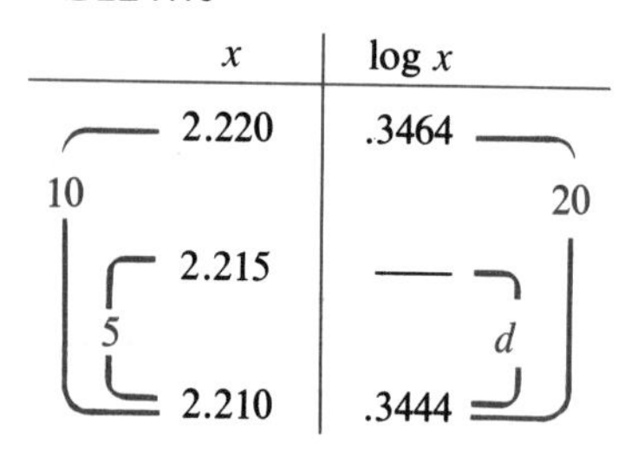

	x	$\log x$	
10	2.220	.3464	20
5	2.215	—	d
	2.210	.3444	

$$\frac{d}{20} = \frac{5}{10}$$

$$d = \frac{5 \cdot 2\not{0}}{1\not{0}} = 10$$

Thus add 10 "units" (.0010) to .3444 and obtain

$$\begin{array}{r} .3444 \\ \underline{.0010} \\ \log 2.215 = .3454, \end{array}$$

as shown previously. □

EXAMPLE 2 Find the value of: log .089 11

Solution.

$$\log .089\ 11 = \log (8.911 \times 10^{-2}) = \log 8.911 - 2$$

[See Table 7.17 .]

Set up the proportion:

TABLE 7.17

	x	$\log x$	
10	8.920	.9504	5
1	8.911	—	d
	8.910	.9499	

$$\frac{d}{5} = \frac{1}{10}$$

$$d = \frac{5}{10} = \frac{1}{2}$$

Because this is $\frac{1}{2}$ (or more), approximate d as 1. Add:

$$\begin{array}{r} .9499 \\ \underline{.0001} \\ \log 8.911 = .9500 \end{array}$$

Thus

$$\log .089\ 11 = .9500 - 2$$

□

antilogs

Interpolation can also be used to determine the *antilog* of a 4-placed decimal that is not in the log table.

EXAMPLE 3 Find the value of: antilog .9276

Solution. .9276 is not found in the log table. Write down the closest numbers to .9276 that are in the table—one smaller, the other larger than .9276 . These values are .9274 and .9279 .

TABLE 7.18

x	$\log x$
8.470	.9279
8.46d	.9276
8.460	.9274

(Brackets: 10 spans 8.460 to 8.470; d spans 8.460 to 8.46d; 5 spans .9274 to .9279; 2 spans .9274 to .9276.)

$$\log 8.46 = .9274 \qquad \text{and} \qquad \log 8.47 = .9279$$

Consider Table 7.18 .

Because .9276 is $\frac{2}{5}$ of the way from .9274 to .9279, antilog .9276 should be about $\frac{2}{5}$ of the way from 8.460 to 8.470.

Now d represents the increase in units (thousandths) in the *left-hand* column. Set up the proportion:

$$\frac{d}{10} = \frac{2}{5}$$

$$d = \frac{2 \cdot 10}{5} = 4$$

Add:

$$\begin{array}{r} 8.460 \\ \underline{.004} \\ \log 8.464 = .9276 \end{array}$$

Therefore

$$8.464 = \text{antilog } .9276$$ □

EXAMPLE 4 Find the value of: antilog 3.5253

Solution. As in Section 7.4 first consider the mantissa, .5253 . Find x such that

$$\log x = .5253$$

[See Table 7.19 .]

TABLE 7.19

x	$\log x$
3.360	.5263
3.35d	.5253
3.350	.5250

(Brackets: 10 spans 3.350 to 3.360; d spans 3.350 to 3.35d; 13 spans .5250 to .5263; 3 spans .5250 to .5253.)

$$\frac{d}{10} = \frac{3}{13}$$

$$d = \frac{3 \cdot 10}{13} = \frac{30}{13} = 2\frac{4}{13}$$

Because $\frac{4}{13} < \frac{1}{2}$, take $d = 2$. Therefore

$$3.352 = \text{antilog } .5253$$

To find antilog 3.5253 (with characteristic 3), move the decimal point 3 places to the right.

$$3352. = \text{antilog } 3.5253$$ □

EXERCISES

In Exercises 1–30 find the value of each common log by interpolating.

1. log 4.105 **2.** log 2.246 **3.** log 8.124 **4.** log 9.908 **5.** log 41.37

6. log 309.3
7. log .7035
8. log 99 480
9. log 103 600
10. log .085 52
11. log .007 699
12. log 9301
13. log 7.002
14. log 70 020
15. log .7002
16. log .070 02
17. log 3.589
18. log 35 890
19. log .003 589
20. log .000 035 89
21. log 8585
22. log 89.37
23. log .008 511
24. log .000 060 06
25. log (3.947×10^7)
26. log (5.053×10^{-4})
27. log (8366×10^{-8})
28. log (8888×10^2)
29. log (98.01×10^8)
30. log (6329×10^{-12})

In Exercises 31–60 find the value of each antilog by interpolating.

31. antilog .6022
32. antilog .6587
33. antilog .9544
34. antilog .1538
35. antilog 1.9796
36. antilog 3.9215
37. antilog 4.9876
38. antilog (.9967 − 1)
39. antilog (.9984 − 2)
40. antilog 5.8390
41. antilog .8230
42. antilog 1.8230
43. antilog 4.8230
44. antilog (.8230 − 1)
45. antilog .8240
46. antilog 6.8240
47. antilog (.8240 − 2)
48. antilog (.8240 − 5)
49. antilog 6.0270
50. antilog (.3432 − 4)
51. antilog (.3931 − 1)
52. antilog (.3950 − 3)
53. antilog .1900
54. antilog .0005
55. antilog (.9999 − 1)
56. antilog 3.8247
57. antilog 6.6666
58. antilog 4.4343
59. antilog (.9224 − 2)
60. antilog (.9580 − 7)

*7.7 Computing with Logarithms

Computations involving products, quotients, and powers (or roots) can be greatly simplified by the use of log tables. (Throughout this section, base 10 is understood.) Of course, the results you obtain will usually be approximations.

In this era of the pocket calculator, you may wonder why these computations should be done. The answer is that computing with logarithms is probably the best way to come to grips with the properties of logarithms. And you will need to understand logarithms in your further studies of the natural sciences and mathematics.

First recall the basic properties of logarithms that will be used. Let x, x_1, and x_2 be positive and let r be a rational number.

(A) $\log x_1 x_2 = \log x_1 + \log x_2$

(B) $\log \dfrac{x_1}{x_2} = \log x_1 - \log x_2$

(C) $\log x^r = r \log x$

Thus when taking the log of a number, instead of multiplying or dividing, you now add or subtract. Instead of finding a power or root, you now multiply or divide. The operations you perform are therefore simpler.

*Optional topic

products

The first example could easily be done directly. However, it illustrates the method.

EXAMPLE 1 Find the value of: (8.39) (.002 64)

Solution. Let $x = (8.39)(.002\,64)$. Instead of finding x directly, consider $\log x$ and use Property (**A**).

$$\begin{aligned}\log x &= \log (8.39)(.002\,64)\\ &= \log (8.39) + \log (.002\,64)\\ &= \log 8.39 + \log (2.64 \times 10^{-3})\\ &= .9238 + .4216 - 3\end{aligned}$$

Thus add:

$$\begin{array}{r} .9238 \\ +\ .4216 - 3 \\ \hline 1.3454 - 3 = .3454 - 2 \end{array}$$

$\llcorner 1 - 3 = -2 \lrcorner$

Therefore

$$\begin{aligned}\log x &= .3454 - 2\\ x &= \text{antilog}\,(.3454 - 2)\end{aligned}$$

In other words, to find the original product x, determine the indicated antilog. [See Table 7.20 .]

TABLE 7.20

	x	$\log x$	
10	2.220	.3464	20
d	2.21d	.3454	10
	2.210	.3444	

$$\frac{d}{10} = \frac{1\not{0}}{2\not{0}}$$

$$d = \frac{10}{2} = 5$$

$$x = 2.215 \times 10^{-2} = .022\,15$$

Compare this with the *exact* answer you obtain by multiplying:

$$\begin{array}{r} 8.3\,9 \\ \times .0\,0\,2\,6\,4 \\ \hline 3\,3\,5\,6 \\ 5\,0\,3\,4 \\ 1\,6\,7\,8 \\ \hline .0\,2\,2\,1\,4\,9\,6 \end{array}$$

If you round off the last 2 digits, you obtain .022 15, as in the logarithmic method.

quotients

EXAMPLE 2 Find the value of: $\dfrac{1.78}{9.35}$

Solution. Let $x = \dfrac{1.78}{9.35}$. Then

$$\log x = \log \left(\frac{1.78}{9.35}\right) = \log 1.78 - \log 9.35 = .2504 - .9708$$

If you subtracted directly, you would obtain the *negative mantissa*, $-.7204$, which you could not locate in your table of *positive mantissas*. To avoid this, write

$$.2504 = 1.2504 - 1$$

Thus you obtain:

$$\begin{array}{r} 1.2504 - 1 \\ -\ \ .9708 \quad\ \ \\ \hline .2796 - 1 \end{array}$$

$\underbrace{.2796}_{\text{positive mantissa}}$

$$\log x = .2796 - 1$$

$$x = \text{antilog}\,(.2796 - 1)$$

TABLE 7.21

x	$\log x$
1.910	.2810
1.90d	.2796
1.900	.2788

(Brackets: 10 spans 1.910 to 1.900; d spans 1.90d to 1.900; 22 spans .2810 to .2788; 8 spans .2796 to .2788.)

Interpolate as in Table 7.21 .

$$\frac{d}{10} = \frac{8}{22}$$

$$d = \frac{10 \cdot 8}{22} = \frac{80}{22} = 3\frac{14}{22} \approx 4$$

$$x = 1.904 \times 10^{-1} = .1904$$

□

roots

EXAMPLE 3 Find the value of: $\sqrt[5]{.308}$

Solution. Let

$$x = \sqrt[5]{.308} = (.308)^{1/5}. \qquad \text{Then}$$

$$\log x = \log\,(.308)^{1/5} = \frac{1}{5}\log .308 = \frac{\log\,(3.08 \times 10^{-1})}{5} = \frac{.4886 - 1}{5}$$

If you were to divide at this point, you would obtain:

$$\frac{.4886 - 1}{5} = .9772 - \frac{1}{5}$$

This is not in proper form because $-\frac{1}{5}$ *is not an integer*. Recall that *the characteristic must be an integer* in order to use the log table. When you divide by 5, you must obtain a negative integer as characteristic, in addition to a positive decimal mantissa. Thus write

$$.4886 - 1 = \underbrace{4.4886 - 5}_{4 - 5 = -1},$$

and obtain:

$$\log x = \frac{4.4886 - 5}{5} = .8977 - 1$$

TABLE 7.22

x	$\log x$
7.910	.8982
7.90d	.8977
7.900	.8976

(Brackets: 10 spans 7.910 to 7.900; d spans 7.90d to 7.900; 6 spans .8982 to .8976; 1 spans .8977 to .8976.)

Now antilog $(.8977 - 1)$ can be determined from Table 7.22 .

$$\frac{d}{10} = \frac{1}{6}$$

$$d = \frac{10}{6} = 1\frac{4}{6} \approx 2$$

$$x = 7.902 \times 10^{-1} = .7902$$ □

several operations

EXAMPLE 4 Find the value of: $\left[\frac{(9.37)^4 \times \sqrt{.008\,46}}{947}\right]^{3/4}$

Solution. Let $x = \left[\frac{(9.37)^4 \times (.008\,46)^{1/2}}{947}\right]^{3/4}$

$$\log x = \log \left[\frac{(9.37)^4(.008\,46)^{1/2}}{947}\right]^{3/4}$$

$$= \frac{3}{4} \log \left[\frac{(9.37)^4(.008\,46)^{1/2}}{947}\right]$$ by property (C)

$$= \frac{3}{4}[\log (9.37)^4 + \log (.008\,46)^{1/2} - \log 947]$$ by properties (A) and (B)

$$= \frac{3}{4}\left[4 \log 9.37 + \frac{1}{2}\log (.008\,46) - \log 947\right]$$

$$= 3 \log 9.37 + \frac{3}{8}\log .008\,46 - \frac{3}{4}\log 947$$

$$= 3 \log 9.37 + \frac{3}{8}\log (8.46 \times 10^{-3}) - \frac{3}{4}\log (9.47 \times 10^{2})$$

$$= 3\,(.9717) + \frac{3}{8}(.9274 - 3) - \frac{3}{4}(2.9763)$$

$$= 2.9151 + \frac{3}{8}(.9274 - 3) - 2.2322$$

The middle expression requires further discussion.

$$\frac{3}{8}(.9274 - 3) = \frac{2.7822 - 9}{8} = \frac{1.7822 - 8}{8}$$

Here $2 - 9 = 1 - 8\ (= -7)$. But now the negative part, -8, *is divisible by* 8. Upon dividing you obtain

$$\underbrace{.2228}_{\text{mantissa}} \quad \underbrace{-\ 1}_{\text{characteristic}}$$

(with an *integer*, -1, as the characteristic).

Thus

$$\begin{array}{r} 2.9151 \\ +\ .2228 - 1 \\ \hline 3.1379 - 1 \\ -2.2322 \\ \hline \end{array}$$

$$\log x = .9057 - 1$$

$$x = \text{antilog}\,(.9057 - 1)$$

TABLE 7.23

	x	$\log x$	
10	8.050	.9058	5
d	$8.04d$	.9057	4
	8.040	.9053	

Interpolate, as in Table 7.23 .

$$\frac{d}{10} = \frac{4}{5}$$

$$d = \frac{10 \cdot 4}{5} = \frac{40}{5} = 8$$

$$x = 8.048 \times 10^{-1} = .8048$$

EXERCISES

In Exercises 1–32 compute using common logarithms.

1. $(8.93)(6.16)$
2. $(107)(84.8)$
3. $(.0926)(9840)$
4. $(891)(6.32)(.007\ 41)$
5. $\dfrac{842}{1.73}$
6. $\dfrac{1.95}{9.83}$
7. $\dfrac{80\ 800}{1.78}$
8. $\dfrac{.009\ 36}{.000\ 74}$
9. $(34.3)^5$
10. $(.832)^{12}$
11. $107^{1/3}$
12. $\sqrt{99\ 600}$
13. $\dfrac{(44.9)(8.85)}{73\ 600}$
14. $\dfrac{(.003\ 96)(.91)}{.004\ 35}$
15. $\dfrac{.731}{(903)(.835)}$
16. $\dfrac{(.828)(6.14)}{(.003\ 62)(.941)}$
17. $(9.84)^3(7.35)^5$
18. $\dfrac{(62.1)^7}{(.0386)^4}$
19. $\sqrt{184}\ \sqrt[3]{1840}$
20. $\sqrt{\dfrac{89\ 200}{1.05}}$
21. $\dfrac{\sqrt{89.3}\ (996)}{804^5}$
22. $\dfrac{(692\ 000^{1/10})(386\ 000\ 000^{1/5})}{(2.04)^{10}}$
23. $\sqrt{\dfrac{888(.333)}{77.7}}$
24. $\dfrac{\sqrt{888(.333)}}{77.7}$
25. $\dfrac{\sqrt{193}\ \sqrt[4]{8.83}}{(92.4)^5(.0083)^4}$
26. $\sqrt{\dfrac{(93.2)^5(.171)^{1/3}}{(8.09)^{3/4}}}$
27. $(9.935)(.083\ 47)$
28. $\dfrac{83.82}{(.004\ 962)(.6396)}$
29. $\sqrt{\dfrac{(9.936)(.8706)}{.024\ 71}}$
30. $\dfrac{(1.444)^5(.083\ 02)^3}{(.1628)^{3/4}}$
31. $\sqrt{\dfrac{(.9909)^5(.1131)^3}{7.707}}$
32. $\dfrac{(9413^4)(18\ 820\ 000^{1/2})}{(.007\ 352)^9(.008\ 214)^{1/3}}$
33. The amount A of money in a bank after n years is given by
$$A = P(1 + r)^n,$$
where P is the original principal and r is the annual interest rate. How much money do you have in the bank after 20 years if you deposit \$1000 at 5%?
34. In Exercise 33, how much money do you have if you deposit \$2500 at 6% for 12 years?
35. The kinetic energy K of a body is given by $K = \frac{1}{2} mv^2$, where m is the body's mass and v is its velocity. Determine the kinetic energy of a body when m = 676 000 lb and v = 1980 ft/sec.
36. The mechanical efficiency e of a machine is defined by:
$$e = \frac{wh}{Fd}$$
Here a weight w is lifted to a height h by a force F that when applied to the machine, acts for a distance d. Determine e when: w = 485 pounds, h = 52.8 feet, F = 312 pounds, d = 101 feet

7.8 Exponential Equations; Changing the Base

If nothing is written to the contrary, base 10 will be understood.

exponential equations

In an **exponential equation** a variable (or a nonconstant polynomial) appears as an exponent.

$$7^t = 15 \quad \text{and} \quad 3^{2x-1} = .145$$

are each exponential equations.

Property (**C**) of logarithms states:

$$\log x^r = r \log x, \qquad \text{where } x > 0,\ r \text{ rational}$$

This property enables you to solve many exponential equations.

EXAMPLE 1 Solve for t: $7^t = 15$

Solution. Because the log function is one-to-one, consider the log of each side of the given equation.

$$\log 7^t = \log 15$$

$$t \log 7 = \log 15 \qquad \text{by Property (C)}$$

$$t = \frac{\log 15}{\log 7}$$

By use of the log table,

$$t = \frac{1.1761}{.8451} \approx 1.39 \quad \text{(to 2 decimal places)}$$

Make sure you understand that

$$\frac{\log 15}{\log 7} \neq \log 15 - \log 7.$$

$\left(\text{Rather, } \log \frac{15}{7} = \log 15 - \log 7. \text{ But } \log \frac{15}{7} \text{ is } \textit{not} \text{ the same as } \frac{\log 15}{\log 7}.\right)$ □

EXAMPLE 2 Solve for x: $3^{2x-1} = .145$

Solution.

$$3^{2x-1} = .145$$

$$\log 3^{2x-1} = \log .145$$

$$(2x - 1) \log 3 = \log (1.45 \times 10^{-1})$$

$$2x - 1 = \frac{\log (1.45 \times 10^{-1})}{\log 3}$$

$$= \frac{.1614 - 1}{.4771}$$

$$= \frac{-.8386}{.4771}$$

Note that here a negative decimal is permissible. Only when the mantissa table is to be used, are negative decimals avoided.

$$2x - 1 \approx -1.76$$
$$2x \approx -.76$$
$$x \approx -.38$$
□

EXAMPLE 3 A chemical substance increases according to the formula

$$y = 25 \cdot 10^t,$$

where y is the number of grams present after t hours. Determine how long it takes to obtain 1000 grams of this substance.

Solution. Substitute 1000 for y in the given equation,

$$y = 25 \cdot 10^t$$

Thus

$$1000 = 25 \cdot 10^t$$
$$40 = 10^t$$
$$\log 40 = \log 10^t$$
$$\log 40 = t \log 10$$
$$\frac{\log 40}{\log 10} = t$$
$$\frac{1.60}{1} \approx t$$

It takes about 1 hour and 36 minutes to obtain 1000 grams. □

EXAMPLE 4 Solve the exponential equation: $10^x - 10^{-x} = 2$

Solution. To eliminate the negative exponent, multiply both sides by 10^x.

$$(10^x)^2 - 1 = 2\,(10^x)$$
$$(10^x)^2 - 2\,(10^x) - 1 = 0 \tag{7.1}$$

Let $u = 10^x$; then u is *positive,* and $u^2 = (10^x)^2$. Use this substitution to obtain a quadratic equation.

$$u^2 - 2u - 1 = 0$$
$$u = \frac{2 \pm \sqrt{4 + 4}}{2} = \frac{2 \pm 2\sqrt{2}}{2} = 1 \pm \sqrt{2}$$

Observe that $1 - \sqrt{2} < 0$, but $u = 10^x > 0$. Reject $1 - \sqrt{2}$. Thus

$$10^x = u = 1 + \sqrt{2} \qquad \text{and}$$

$$x = \log_{10}(1 + \sqrt{2})$$

Note that in Equation (**7.1**), $(10^x)^2 = 10^{2x}$; thus the equation

$$10^{2x} - 2(10^x) + 1 = 0$$

is solved in the same way. □

logarithmic equations

A **logarithmic equation** is an equation in which the logarithm of a variable (or of a nonconstant polynomial) appears. Properties of logarithms are often useful in solving logarithmic equations.

EXAMPLE 5 Solve for x: $\log 2 + \log x = 4$

Solution. The log of a product is the sum of the logs. Thus

$$\log_{10} 2 + \log_{10} x = \log_{10} 2x$$

and the given equation becomes:

$$\log_{10} 2x = 4 \qquad \text{Therefore}$$
$$10^4 = 2x$$
$$10\,000 = 2x$$
$$5000 = x$$

□

change-of-base formula

A logarithmic equation enables you to change the base of a given logarithm.

$$\mathbf{\log_a x = \frac{\log x}{\log a}, \qquad x > 0,\ a > 0}$$

On the right side, the common log is assumed, although, as you will see, any (positive) base can be used. Before proving this formula, consider the following example.

EXAMPLE 6 Show that $\log_2 93 = \dfrac{\log 93}{\log 2}$. Apppproximate $\log_2 93$ to 2 decimal places.

Solution.

$$\text{Let } y = \log_2 93. \qquad \text{Then}$$
$$2^y = 93$$
$$\log 2^y = \log 93$$
$$y \log 2 = \log 93$$
$$y = \frac{\log 93}{\log 2}$$

Note that the *approximate* numerical value of y can be obtained by dividing these common logs.

$$y = \frac{1.9685}{.3010} \approx 6.54$$

□

The proof of the change-of-base formula is similar to Example 6.

Proof. Let $x > 0$, $a > 0$, and let $y = \log_a x$. Then

$$a^y = x$$
$$\log a^y = \log x$$

$$y \log a = \log x$$

$$y = \frac{\log x}{\log a}$$

Because $y = \log_a x$, it follows that

$$\log_a x = \frac{\log x}{\log a}$$

△

Note that the base 10 was not essential in this proof. Thus, the formula can be generalized.

general change-of-base formula

$$\log_a x = \frac{\log_b x}{\log_b a}, \qquad x > 0,\ a > 0,\ b > 0$$

EXAMPLE 7 Use the General Change-of-Base Formula to express $\log_{17} 536$ in terms of base 3.

Solution. In the General Change-of-Base Formula, let $x = 536$, $a = 17$, $b = 3$.

$$\log_{17} 536 = \frac{\log_3 536}{\log_3 17}$$

□

EXERCISES

In Exercises 1–12 solve the indicated exponential equations to 2 decimal places.

1. $5^x = 35$ **2.** $8^t = 100$ **3.** $7^{2y} = 93$ **4.** $3^{x+1} = 148$

5. $12^{5x} = 685$ **6.** $2^{-2t} = 100$ **7.** $(14.8)^{4t-1} = 17.3$ **8.** $9^{-x} = 73$

9. $6^{3-x} = 27$ **10.** $4^{1-2t} = 30$ **11.** $100^{3t+1} = .01$ **12.** $600^{1-3t} = 9000$

In Exercises 13–22 solve the indicated logarithmic equations. (Base 10 is understood.)

13. $\log x = 4$ **14.** $\log 5x = 2$ **15.** $\log(-2x) = 1$ **16.** $\log(x + 1) = -3$

17. $\log 4x = -1$ **18.** $\log(x + 1) = -3$ **19.** $\log \frac{x}{2} = 3$ **20.** $\log(4x + 1) = 0$

21. $\log 3 + \log x = 1$ **22.** $\log 2 + \log(x + 1) = 3$

23. A chemical substance increases according to the formula

$$s = 10\,000 \cdot 5^t,$$

where s is the number of grams present after t hours. Determine how long it takes to obtain 5 000 000 grams of the substance.

24. A radioactive substance s decays according to the formula

$$s = s_0(2.7)^{-t/4},$$

where s_0 is the amount originally present and t is the time in hours. After how many hours will half of the original substance be left?

25. The amount A of money in a bank after n years is given by
$$A = P(1 + r)^n,$$
where P is the original principal and r is the annual interest rate. If you invest \$2000 at 5%, how many years will it take for this to grow to over \$3000?

26. Use the formula of Exercise 25 to determine how long it will take to double your original principal **(a)** at 4%, **(b)** at 8%.

27. A colony of bacteria grows according to the formula
$$N = N_0 \cdot 2^{t/10},$$
where N_0 is the number of bacteria originally present, t is the time in hours, and N is the number present after t hours. After how many hours will the colony quadruple in size? [*Hint:* Find t when $N = 4N_0$.]

28. A colony of bacteria grows according to the formula
$$N = N_0 \cdot 2^{t/k},$$
where N, N_0 and t are as in Exercise 27, and where k is the number of hours for the colony to double in size. Suppose there are 1 000 000 bacteria initially present and 800 000 000 after 12 hours. In how many hours did the original colony double in size?

29. The expected population of a country is given by the formula
$$P = P_0 3^{kt},$$
where P_0 is the initial population, t is the number of years under discussion, and P is the population after t years. Suppose the population of the country was 160 000 000 in 1960 and 200 000 000 in 1970. Use the formula to arrive at an expected population (to the nearest million) in 2000.

30. The temperature T of an object is given by the formula
$$T = T_1 + (T_0 - T_1)10^{-.01t},$$
where T_0 is the object's initial temperature, T_1 is air temperature, and t is the number of minutes. Suppose the temperature of a cup of coffee is initially 150°F and the air temperature holds steady at 70°F. How long will it take the cup of coffee to cool to 100°F?

In Exercises 31–38 find the value of the indicated logarithm to 2 decimal places.

31. $\log_2 3$ **32.** $\log_3 2$ **33.** $\log_5 19$ **34.** $\log_7 138$ **35.** $\log_6 84$

36. $\log_4 60\,000$ **37.** $\log_{100} 2800$ **38.** $\log_{.001} 934\,000$

In Exercises 39–46 use the General Change-of-Base Formula to express the given logarithm to the indicated base.

39. $\log_2 5$, base 3 *Answer:* $\dfrac{\log_3 5}{\log_3 2}$

40. $\log_3 10$, base 5 **41.** $\log_4 29$, base 2 **42.** $\log_7 364$, base 8 **43.** $\log_{10} 93$, base 2

44. $\log_{64} 843$, base 4 **45.** $\log_7 9.02$, base 12 **46.** $\log_5 .001$, base 3

In Exercises 47–48 assume $x > 0$. Simplify each expression.

47. $10^{\log_{10} x}$ **48.** $\log_{10} 10^x$

In Exercises 49–58 solve for x. In Exercises 49–56 answers may be left in terms of common logarithms.

49. $3^x = 2^{x+1}$ **50.** $10^x = 5^{2x+1}$ **51.** $10^x - 10^{-x} = 5$ **52.** $10^{2x} + 10^x = 2$

53. $\dfrac{10^x + 10^{-x}}{2} = 4$ **54.** $10^{2x} - 10^{-2x} = 3$ **55.** $2^x + 2^{-x} = 4$ **56.** $3^x - 3^{-x} = 10$

57. $\log_{10} x + \log_{10}(x + 2) = 1$ **58.** $\log_{10} 2x + \log_{10}(x - 1) = 0$

Review Exercises for Chapter 7

1. Express in logarithmic notation.
(a) $10^3 = 1000$ (b) $\left(\frac{1}{2}\right)^2 = \frac{1}{4}$ (c) $16^{1/2} = 4$

2. Find the value of each indicated logarithm.
(a) $\log_6 36$ (b) $\log_3 81$ (c) $\log_{.2} .04$

3. Find the value of x.
(a) $\log_3 9 = x$ (b) $\log_x 25 = 2$ (c) $\log_3 \sqrt{3} = x$

4. Express $\log_b 60$ in terms of $\log_b 2$, $\log_b 3$, and $\log_b 5$.

5. Express $\log_b x + 3 \log_b y - \log_b z$ as a single logarithm.

In Exercises 6–13 common logs are understood.

6. Find the value of: $\log 8.42$

7. Determine the characteristic of the indicated logarithms:
(a) $\log 485$ (b) $\log .000\,142$

8. Find the value of each log.
(a) $\log 614$ (b) $\log .0631$

9. Find the value of each antilog.
(a) antilog 3.6618 (b) antilog (.5955 − 1)

10. Let $f(x) = 8^x$. Find:
(a) $f(1)$ (b) $f(2)$ (c) $f(-2)$ (d) $f\left(\frac{1}{3}\right)$

11. Let $g(x) = \log_{25} x$. Find:
(a) $g(25)$ (b) $g(5)$ (c) $g\left(\frac{1}{5}\right)$ (d) $g(125)$

12. Find the value of each log by interpolating.
(a) $\log 8.206$ (b) $\log .073\,14$

13. Find the value of each antilog by interpolating.
(a) antilog .6994 (b) antilog 1.9797

14. Compute using logarithms.
(a) $(6.42)(9.09)$ (b) $\dfrac{396^{1/10}}{(1.71)^2}$ (c) $\dfrac{\sqrt{291(.444)}}{(28.4)^2}$

15. Solve the logarithmic equation:

$$\log(5x) = 6$$

16. Solve to 2 decimal places:

$$3^{x-1} = 259$$

17. Find the value of $\log_3 5000$ to 2 decimal places.

18. Use the General Change-of-Base Formula to express $\log_2 7$ to the base 3.

8 Trigonometric Functions

8.1 Right Triangles

Trigonometry treats relations among the sides and angles of a triangle.

angles

An **angle** is the union of two closed rays originating from the same point, that is, an angle is the set of points making up these two rays. The intersecting rays are called the **sides** of the angle, and the point of intersection is called the **vertex.** To identify an angle you often use points on their sides. Thus, in Figure 8.1(a), the angle is written

$$\angle ABC \qquad \text{or} \qquad \angle CBA,$$

or, if confusion seems unlikely, simply $\angle B$. Note that B represents the vertex. Sometimes, $\angle B$ is indicated as in Figure 8.1(b).

An angle is often measured in **degrees;** finer measurements employ degrees and **minutes.** *Each degree equals 60 minutes.* Write

$30°$ for 30 degrees

Write $\angle A = 30°$ when $\angle A$ *measures* $30°$, and
write $\angle A < 30°$ when $\angle A$ *measures less than* $30°$.

Also, write $47°15'$ for 47 degrees, 15 minutes.

Normally, an angle measures between $0°$ and $360°$. A **right angle** measures $90°$. An **acute angle** measures between $0°$ and $90°$, whereas an **obtuse angle** measures between $90°$ and $180°$. More precisely,

$\angle A$ is acute if $0° < \angle A < 90°$;

$\angle B$ is obtuse if $90° < \angle B < 180°$.

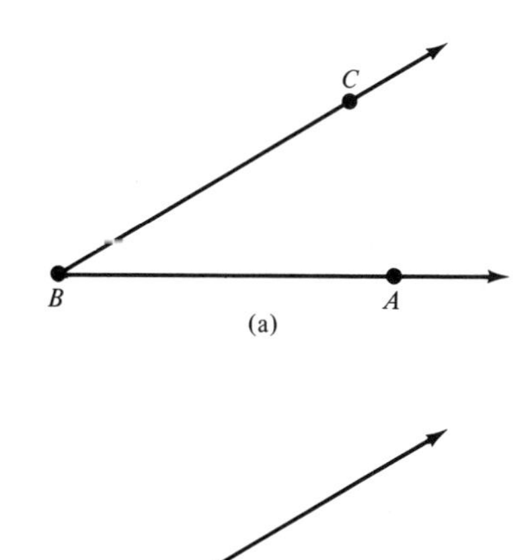

FIGURE 8.1

[See Figure 8.2 on page 306.]

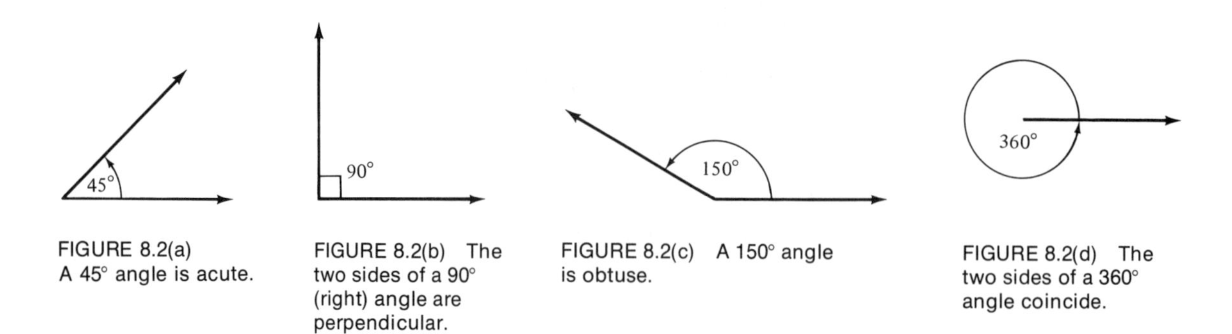

FIGURE 8.2(a) A 45° angle is acute.

FIGURE 8.2(b) The two sides of a 90° (right) angle are perpendicular.

FIGURE 8.2(c) A 150° angle is obtuse.

FIGURE 8.2(d) The two sides of a 360° angle coincide.

right triangles

A **triangle** is a closed figure in the plane composed of three line segments. The line segment from any vertex drawn perpendicular to the opposite side is called an **altitude** of the triangle. [See Figures 8.3(a), (b), (c).] A **right triangle** is one in which one of the angles is a right angle.[See Figure 8.3(c).]

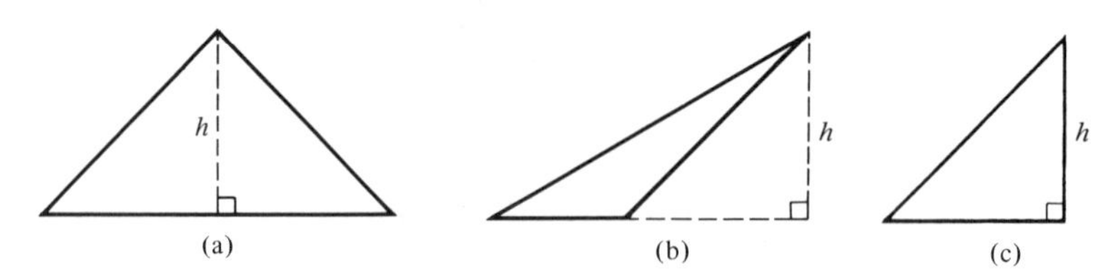

FIGURE 8.3 In each triangle, an altitude, h, is indicated.

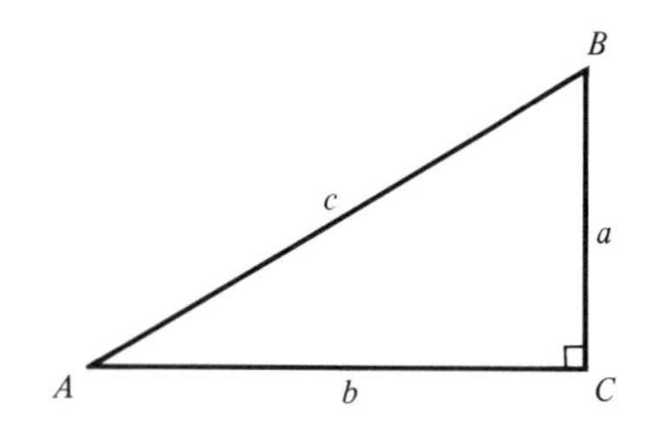

FIGURE 8.4 $a^2 + b^2 = c^2$

Let ΔABC be a right triangle with right angle C, as in Figure 8.4 . The side (of length c) opposite the right angle is known as the **hypotenuse** of the right triangle. According to the *Pythagorean Theorem,* pages 215–216,

$$a^2 + b^2 = c^2$$

sine, cosine, tangent

Now consider $\angle A$ of Figure 8.4 . The side of length a is its **opposite side** and the side of length b is its **adjacent side.** The three most important trigonometric functions of an *acute angle* are defined in terms of $\angle A$.

$$\text{Sine } A = \frac{a}{c} \qquad \text{Think of } \frac{\text{Opposite}}{\text{Hypotenuse}}.$$

$$\text{Cosine } A = \frac{b}{c} \qquad \text{Think of } \frac{\text{Adjacent}}{\text{Hypotenuse}}.$$

$$\text{Tangent } A = \frac{a}{b} \qquad \text{Think of } \frac{\text{Opposite}}{\text{Adjacent}}.$$

To remember these relationships, think of

SOH – CAH – TOA

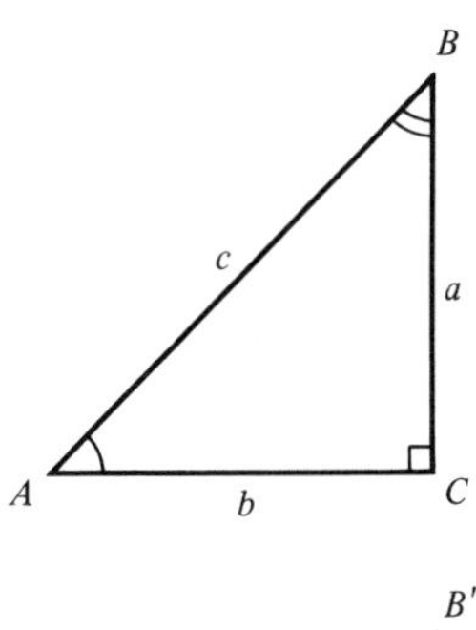

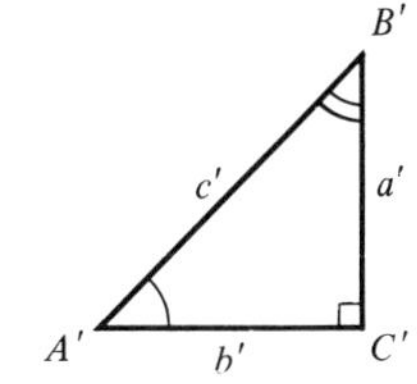

FIGURE 8.5

Abbreviate by using:

$\sin A$	for	sine A
$\cos A$	for	cosine A
$\tan A$	for	tangent A

Recall that *similar triangles* are triangles in which the lengths of corresponding sides are proportional, or alternatively, in which corresponding angles are equal. If ΔABC and $\Delta A'B'C'$ are similar right triangles with corresponding angles as indicated in Figure 8.5, then:

$$\sin A = \frac{a}{c} = \frac{a'}{c'}, \qquad \cos A = \frac{b}{c} = \frac{b'}{c'}, \qquad \tan A = \frac{a}{b} = \frac{a'}{b'}$$

Thus, the values of these trigonometric functions depend only upon the measurements of the angles and not upon the particular right triangles used.

EXAMPLE 1 In Figure 8.6 determine

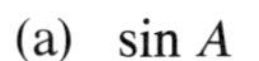

(a) $\sin A$ (b) $\cos A$ (c) $\tan A$
(d) $\sin B$ (e) $\cos B$ (f) $\tan B$

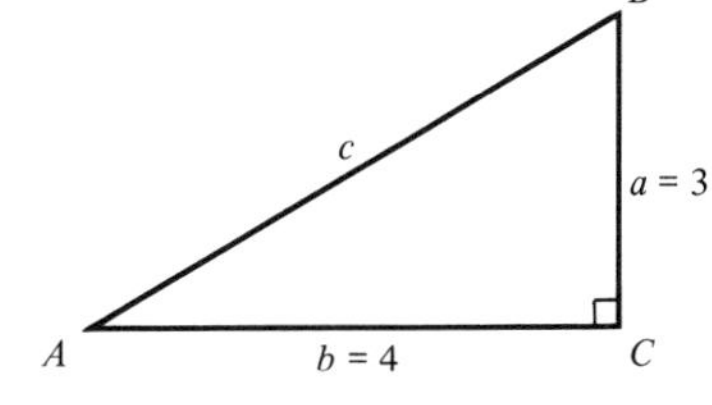

FIGURE 8.6

Solution. By the Pythagorean Theorem,

$$c^2 = a^2 + b^2 = 3^2 + 4^2 = 9 + 16 = 25$$

Thus $c = 5$

(a) $\sin A = \frac{a}{c} = \frac{3}{5}$ (b) $\cos A = \frac{b}{c} = \frac{4}{5}$ (c) $\tan A = \frac{a}{c} = \frac{3}{4}$

To determine the functions of $\angle B$, rather than of $\angle A$, note that now b is the opposite side and a the adjacent side. Thus

(d) $\sin B = \frac{b}{c} = \frac{4}{5}$ (e) $\cos B = \frac{a}{c} = \frac{3}{5}$ (f) $\tan B = \frac{b}{a} = \frac{4}{3}$

□

Observe that in Example 1,

$$\sin A = \cos B \quad \text{and} \quad \cos A = \sin B$$

The sum of the measures of the angles of a triangle equals 180°. Because $\angle C = 90°$, it follows that in a *right* triangle,

$$\angle A + \angle B = 90°$$

and thus

$$\angle A = 90° - \angle B \quad \text{and} \quad \angle B = 90° - \angle A.$$

In this case $\angle A$ and $\angle B$ are each called the **complement of** the other. Furthermore,

$$\sin A = \cos(90° - A) \quad \text{and} \quad \cos A = \sin(90° - A)$$

For example, $\sin 42° = \cos 48°$; $\cos 66° = \sin 24°$. "Cosine" means "Complement's sine".

30°, 45°, 60°

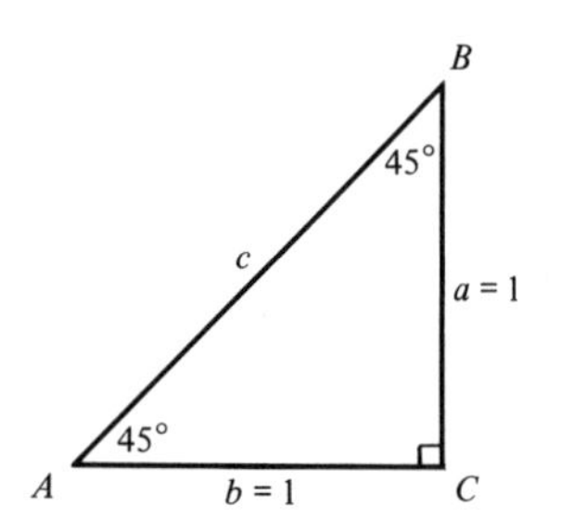

FIGURE 8.7

Generally, to find the trig functions of an acute angle, a trig table must be used. (See Section 8.2 .) However, the trig functions of 30°, 45°, and 60° angles can be determined *without* a table. To study these angles, consider both *isosceles and equilateral triangles.*

An **isosceles triangle** is a triangle in which exactly two sides are of equal length. The angles opposite the "equal sides" are also equal (in measure). In an **isosceles right triangle**, the equal angles each measure 45°. Thus *an isosceles right triangle is a 45°, 45°, 90°-triangle*. Consider the isosceles right triangle of Figure 8.7, in which the equal sides are each of length 1. By the Pythagorean Theorem,

$$c^2 = 1^2 + 1^2 = 2$$

$$c = \sqrt{2}$$

In general, in any isosceles right triangle, the side lengths are in the ratio of

$$1 : 1 : \sqrt{2}$$

Thus,

$$\sin 45° = \frac{1}{\sqrt{2}} = \frac{1 \cdot \sqrt{2}}{\sqrt{2} \cdot \sqrt{2}} = \frac{\sqrt{2}}{2}$$

$$\cos 45° = \frac{1}{\sqrt{2}} = \frac{\sqrt{2}}{2}$$

$$\tan 45° = \frac{1}{1} = 1$$

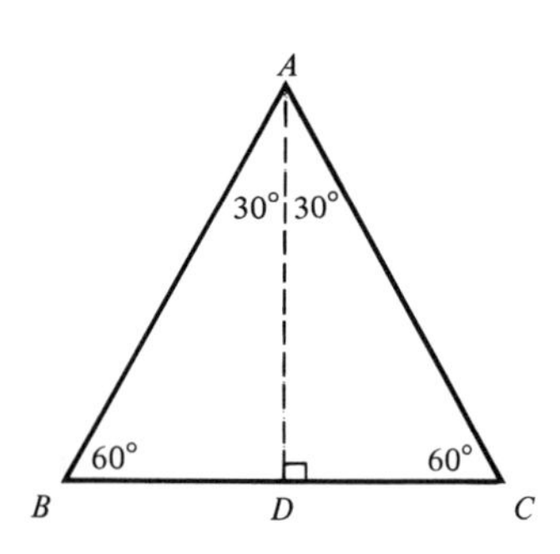

FIGURE 8.8(a)

Next, an **equilateral triangle** is one in which all three side lengths (and all three angle measurements) are equal. Each angle of an equilateral triangle measures 60°. It can be shown that each altitude bisects (cuts in half), the corresponding perpendicular side as well as the corresponding vertex. [See Figure 8.8(a).] In Figure 8.8(b), ΔABC is an equilateral triangle each of whose sides is of length 2. Altitude $\overline{AD}$ bisects side $\overline{BC}$ so that $\overline{BD} = \overline{DC} = 1$. And $\overline{AD}$ bisects $\angle BAC$ so that

$$\angle BAD = \angle DAC = 30°.$$

To determine the length of $\overline{AD}$, use the Pythagorean Theorem.

$$\begin{aligned} (\text{length } \overline{AD})^2 + 1^2 &= 2^2 \\ (\text{length } \overline{AD})^2 &= 3 \\ \text{length } \overline{AD} &= \sqrt{3} \end{aligned}$$

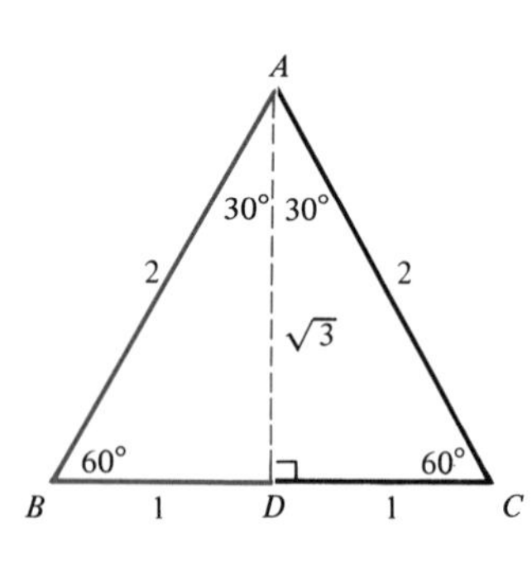

FIGURE 8.8(b)

Observe that ΔABD (or ΔADC) is a "30°, 60°, 90°-triangle".

In any 30°, 60°, 90°-triangle, the side lengths are in the ratio of

1	:	2	:	$\sqrt{3}$
side opposite 30°		hypotenuse		side opposite 60°

Read off the values of the trigonometric functions of 30° and 60° angles from Figure 8.8(b).

$$\sin 30° = \frac{1}{2} \qquad \sin 60° = \frac{\sqrt{3}}{2}$$

$$\cos 30° = \frac{\sqrt{3}}{2} \qquad \cos 60° = \frac{1}{2}$$

$$\tan 30° = \frac{1}{\sqrt{3}} = \frac{\sqrt{3}}{3} \qquad \tan 60° = \frac{\sqrt{3}}{1} = \sqrt{3}$$

EXAMPLE 2 A 12-foot ladder leaning against a wall makes a 60° angle with the ground. How far is the foot of the ladder from the base of the wall? And how high is the top of the ladder? Approximate to 2 decimal places, using $\sqrt{3} = 1.732$.

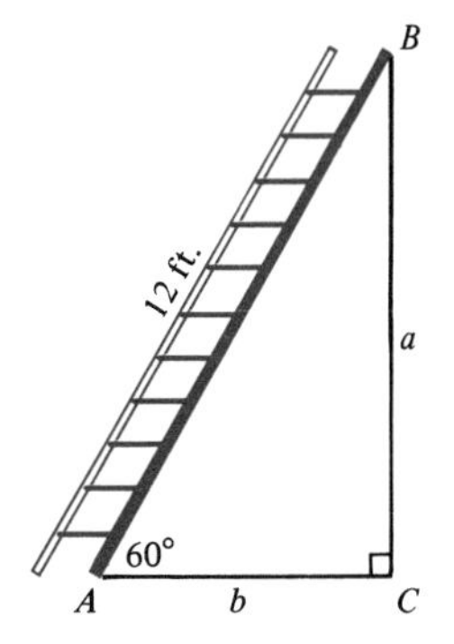

FIGURE 8.9

Solution. Consider Figure 8.9.

b = distance from foot of ladder to bottom of wall

a = height of top of ladder

$\angle B = 180° - (60° + 90°) = 30°$

In a 30°, 60°, 90°-triangle the sides are of the ratio $1 : 2 : \sqrt{3}$.

$$b = \frac{1}{2} \cdot 12 \text{ ft.} = 6 \text{ ft.}$$

$$a = 6\sqrt{3} \text{ ft.} \approx 6 \times 1.732 \text{ ft.} = 10.39 \text{ ft.}$$

□

secant, cosecant, cotangent

The three other trigonometric functions of an *acute angle* can be defined in terms of the sine, cosine, and tangent. [Refer to Figure 8.10.]

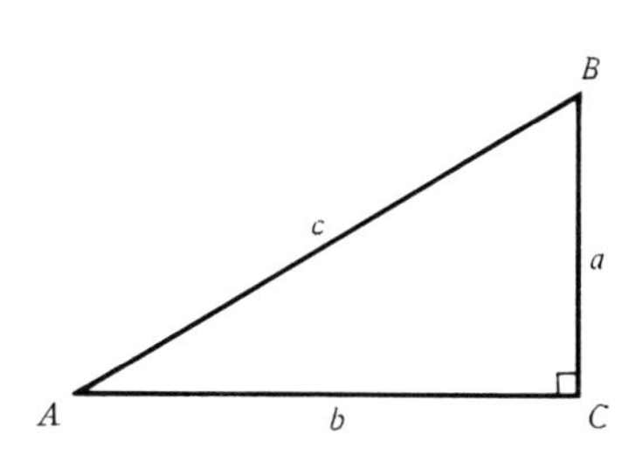

FIGURE 8.10

$$\text{Secant } A = \frac{1}{\cos A} = \frac{c}{b} = \frac{\text{Hypotenuse}}{\text{Adjacent}}$$

$$\text{Cosecant } A = \frac{1}{\sin A} = \frac{c}{a} = \frac{\text{Hypotenuse}}{\text{Opposite}}$$

$$\text{Cotangent } A = \frac{1}{\tan A} = \frac{b}{a} = \frac{\text{Adjacent}}{\text{Opposite}}$$

Abbreviate by using:

sec A for secant A
csc A for cosecant A
cot A for cotangent A

To find values of the secant, cosecant, or cotangent, use the definition in terms of cosine, sine, and tangent.

EXAMPLE 3

(a) $$\sec 45° = \frac{1}{\cos 45°} = \frac{1}{\frac{\sqrt{2}}{2}} = \frac{2}{\sqrt{2}} = \frac{2 \cdot \sqrt{2}}{\sqrt{2} \cdot \sqrt{2}} = \frac{2\sqrt{2}}{2} = \sqrt{2}$$

(b) $$\csc 30° = \frac{1}{\sin 30°} = \frac{1}{\frac{1}{2}} = 2$$

(c) $$\cot 60° = \frac{1}{\tan 60} = \frac{1}{\sqrt{3}} = \frac{1 \cdot \sqrt{3}}{\sqrt{3} \cdot \sqrt{3}} = \frac{\sqrt{3}}{3}$$

□

$$\sec A = \csc (90^\circ - A) \quad \text{and} \quad \csc A = \sec (90^\circ - A)$$

$$\tan A = \cot (90^\circ - A) \quad \text{and} \quad \cot A = \tan (90^\circ - A)$$

For example,

$$\sec 60^\circ = \csc 30^\circ; \qquad \tan 30^\circ = \cot 60^\circ$$

"Cosecant" means "complement's secant" and "cotangent" means "complement's tangent".

Given any one trigonometric function-value of an *(acute) angle*, the other five trigonometric function-values for that angle can be found.

EXAMPLE 4 Suppose $\angle A$ is acute and $\sin A = \frac{3}{4}$. Find the other trigonometric function-values for $\angle A$.

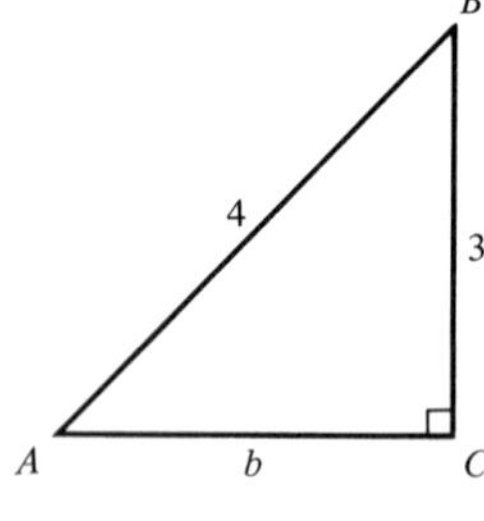

FIGURE 8.11

Solution. Use the triangle in Figure 8.11 . By the Pythagorean Theorem,

$$b^2 = c^2 - a^2 = 4^2 - 3^2 = 7$$

$$b = \sqrt{7}$$

$$\sin A = \frac{3}{4} \qquad \csc A = \frac{4}{3}$$

$$\cos A = \frac{\sqrt{7}}{4} \qquad \sec A = \frac{4}{\sqrt{7}} = \frac{4\sqrt{7}}{7}$$

$$\tan A = \frac{3}{\sqrt{7}} = \frac{3\sqrt{7}}{7} \qquad \cot A = \frac{\sqrt{7}}{3}$$

□

EXERCISES

In Exercises 1–4 convert to degrees and minutes.

1. 35.2° **2.** 56.5° **3.** 100.75° **4.** 14.05°

In Exercises 5–8 write in decimal notation.

5. 42°30′ **6.** 13°24′ **7.** 75°15′ **8.** 60°9′

In Exercises 9–12 find the measure of the third angle of the given triangle.

9.

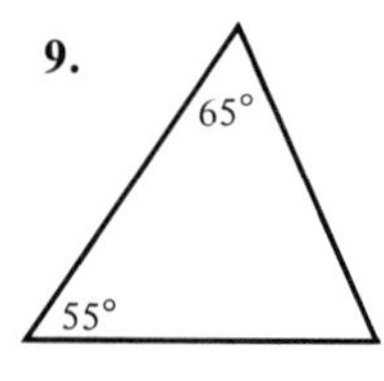

FIGURE 8.12

10.

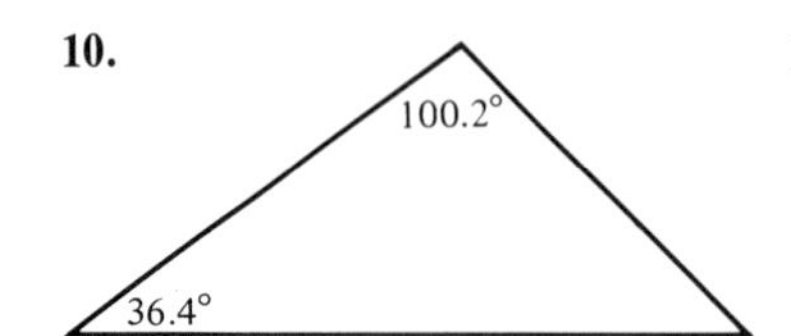

FIGURE 8.13

11.

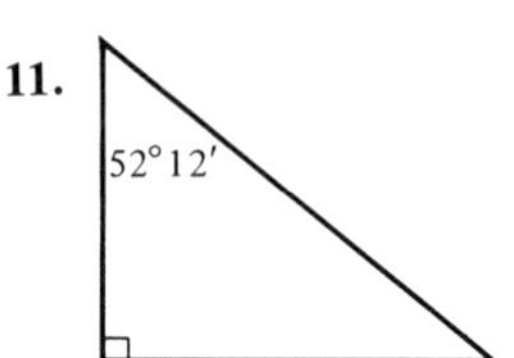

FIGURE 8.14

12.

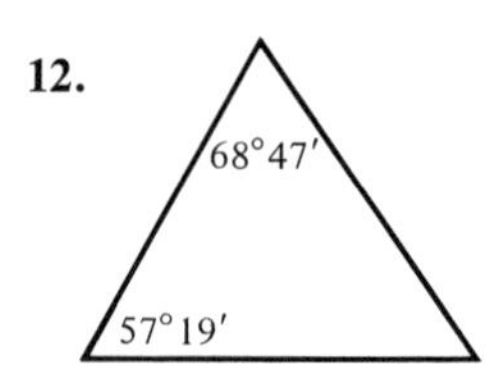

FIGURE 8.15

In Exercises 13–20 consider Figure 8.16. Determine: (a) $\sin A$ (b) $\cos A$ (c) $\tan A$ (d) $\sin B$ (e) $\cos B$ (f) $\tan B$

13. $a = 9,\ b = 12$ **14.** $a = 12,\ c = 20$ **15.** $a = 2,\ b = 2$ **16.** $b = 7,\ c = 25$

17. $a = 2,\ b = 4$ **18.** $a = 5,\ c = 7$ **19.** $a = \sqrt{2},\ b = \sqrt{3}$ **20.** $b = 1,\ c = \sqrt{5}$

In Exercises 21–30 consider Figure 8.16. Determine the other five trigonometric function-values for $\angle A$.

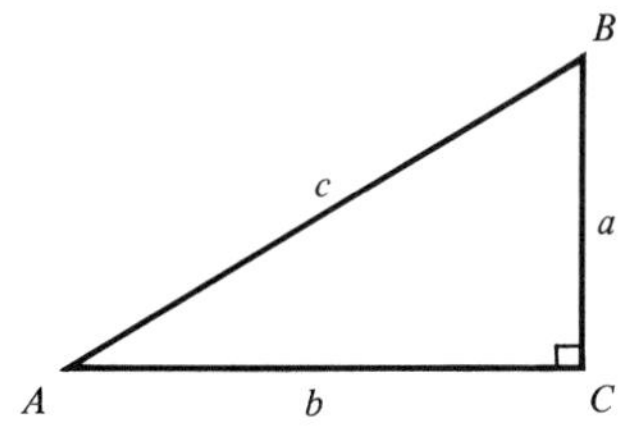

FIGURE 8.16

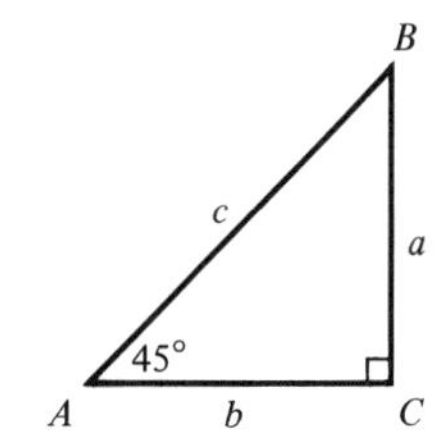

FIGURE 8.17

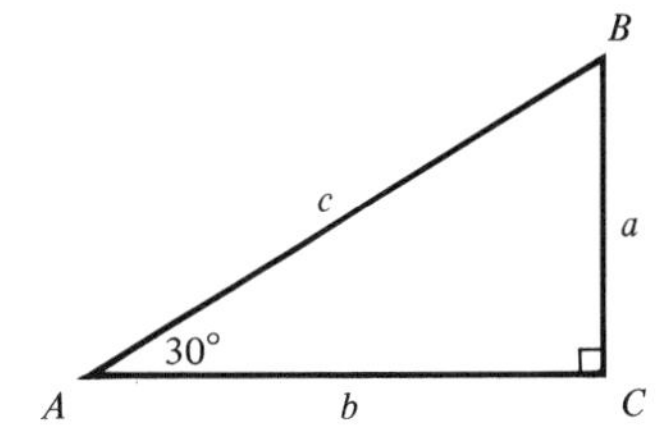

FIGURE 8.18

21. $\sin A = \frac{4}{5}$ **22.** $\cos A = \frac{1}{2}$ **23.** $\tan A = \frac{4}{3}$ **24.** $\cot A = 2$ **25.** $\sin A = \frac{1}{4}$

26. $\sec A = \frac{5}{4}$ **27.** $\csc A = 5$ **28.** $\tan A = \frac{7}{24}$ **29.** $\sec A = \frac{10}{3}$ **30.** $\cot A = 10$

In Exercises 31–33 consider Figure 8.17.

31. Suppose $a = 5$. Determine b and c. **32.** Suppose $b = \sqrt{2}$. Determine a and c.

33. Suppose $c = 10$. Determine a and b.

In Exercises 34–36 consider Figure 8.18.

34. Suppose $a = 4$. Determine b and c. **35.** Suppose $c = 18$. Determine a and b.

36. Suppose $b = 8$. Determine a and c.

In Exercises 37–46 determine each trigonometric function-value.

37. $\sin 30°$ **38.** $\tan 45°$ **39.** $\cos 60°$ **40.** $\cos 30°$ **41.** $\tan 60°$

42. $\cot 60°$ **43.** $\sec 30°$ **44.** $\csc 45°$ **45.** $\cot 45°$ **46.** $\sec 60°$

In Exercises 47–54, when approximations are called for, use $\sqrt{2} \approx 1.414$ and $\sqrt{3} \approx 1.732$. Approximate to 1 decimal place.

47. A 20-foot ladder leaning against a wall makes a 45° angle with the ground. (a) How far is the foot of the ladder from the base of the wall? (b) Approximately how high is the top of the ladder?

48. As an airplane ascends, its path makes a 60° angle with the runway. Approximately how many feet does the plane rise while traveling 1200 feet in the air?

49. Two surveyors stand at points A and B that are 250 feet apart along a straight river bank. Point C lies on the opposite bank directly across the river from B. The surveyor at A finds that $\angle BAC$ measures 60°. Find the approximate width of the river at B, that is, the distance from B to C. [See Figure 8.19 on page 312.]

50. One plane leaves an airport flying due north at an average of 400 miles per hour. At the same time a second plane leaves the airport traveling northeastward. The second plane is always due east of the first. Approximately what is the average rate of the second plane?

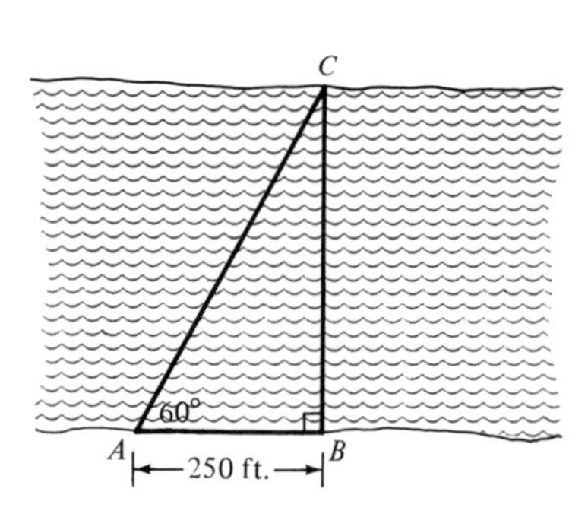

FIGURE 8.19

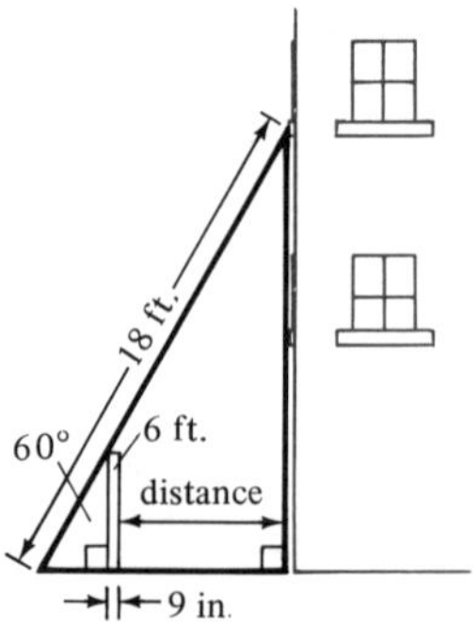

FIGURE 8.20

51. Approximately how high above ground level does a 90-foot ramp rise if it is inclined at a 45° angle with the ground?

52. The top of an 18-foot ladder rests against a window ledge of a house. The ladder makes a 60° angle with the ground and touches the top of a wall parallel to the side of the house. If the wall is 6 feet high and 9 inches thick, find its distance from the house. (See Figure 8.20 .)

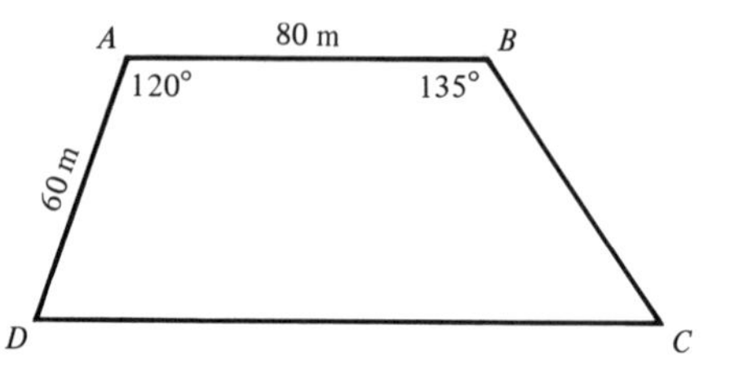

FIGURE 8.21 $\overline{AB} \parallel \overline{CD}$

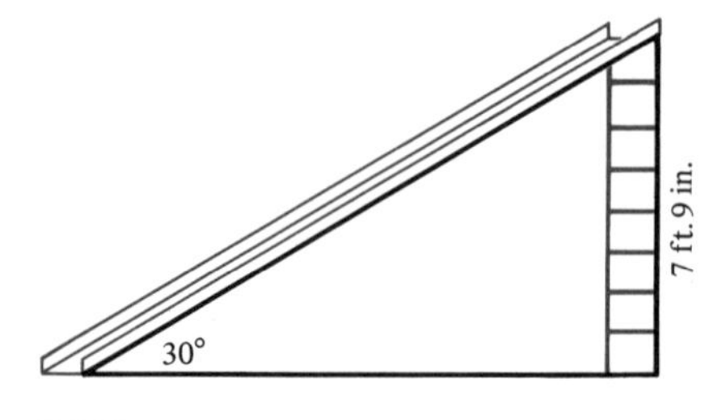

FIGURE 8.22

53. Find the area of the trapezoid in Figure 8.21 .

54. In building the slide pictured in Figure 8.22, a carpenter uses a 16-foot long board. How much must he cut away?

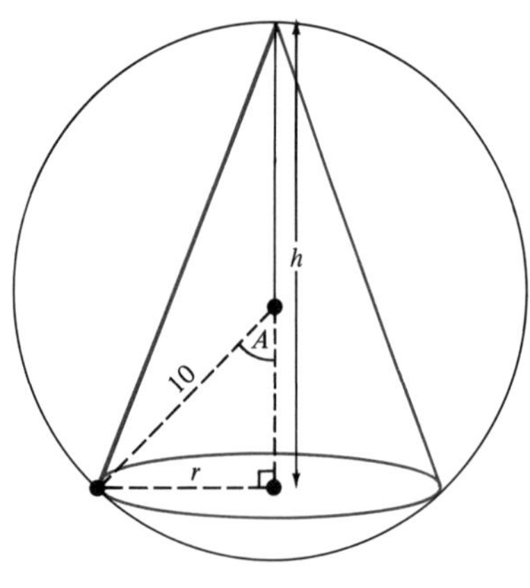

FIGURE 8.23

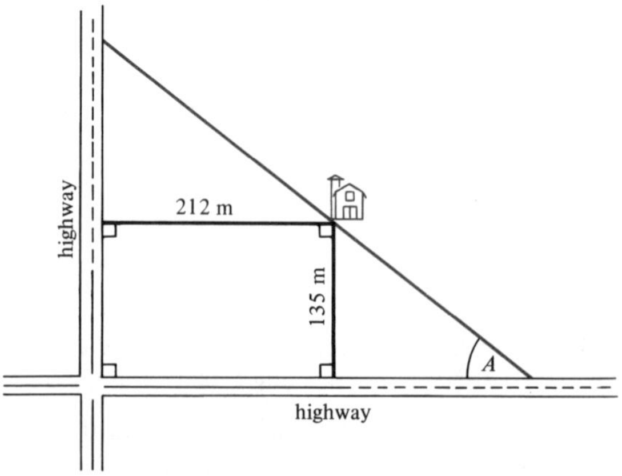

FIGURE 8.24

55. A right circular cone is inscribed in a sphere of radius 10, as in Figure 8.23 . Express: (a) r, the length of the base radius of the cone, (b) h, the length of the altitude of the cone, and (c) V, the volume of the cone as functions of $\angle A$.

56. Two highways intersect at right angles. A barn is located on a straight path that cuts across from one highway to the other, as in Figure 8.24. If the barn is 212 meters from one highway and 135 meters from the other highway, express the length of the path as a function of $\angle A$.

8.2 Use of Table

Values of the sine, cosine, tangent, and cotangent are listed at intervals of 10 minutes in the trig table (Table 3) in the appendix. For convenience, excerpts from the table are reprinted here.

$0^\circ < \angle A < 45^\circ$

First consider an angle between 0° and 45°. Find the function in question at the top of the column; then read down and locate the angle at the left.

EXAMPLE 1 Determine tan 37°20′.

Solution. See Table 8.1.

TABLE 8.1 tan 37°20′ = .7627

Degrees	Sine	Tangent	Cotangent	Cosine	
36°00′	.5878	.7265	1.3764	.8090	**54°00′**
10′	.5901	.7310	1.3680	.8073	50′
20′	.5925	.7355	1.3597	.8056	40′
30′	.5948	.7400	1.3514	.8039	30′
40′	.5972	.7445	1.3432	.8021	20′
50′	.5995	.7490	1.3351	.8004	10′
37°00′	.6018	.7536	1.3270	.7986	**53°00′**
10′	.6041	.7581	1.3190	.7969	50′
20′	.6065	.7627	1.3111	.7951	40′
30′	.6088	.7673	1.3032	.7934	30′
40′	.6111	.7720	1.2954	.7916	20′
50′	.6134	.7766	1.2876	.7898	10′

□

$45^\circ < \angle A < 90^\circ$

Recall that

$$\sin A = \cos (90^\circ - A) \qquad \text{and} \qquad \cos A = \sin (90^\circ - A)$$

Thus, $\sin 50^\circ = \cos 40^\circ$ and $\cos 80^\circ = \sin 10^\circ$. Similarly, it can be shown that

$$\tan A = \cot (90^\circ - A) \qquad \text{and} \qquad \cot A = \tan (90^\circ - A)$$

For an angle between 45° and 90°, the function in question is listed at the bottom of the column. Read up and locate the angle at the right.

EXAMPLE 2 Determine cos 74°40′.

Solution. See Table 8.2 on page 314. Note that

$$74^\circ 40' = 90^\circ - 15^\circ 20'$$

so

$$\cos 74°40' = \sin 15°20' = .2644$$

$$\begin{array}{r} 89°60' \\ \not{90°00'} \\ -15°20' \\ \hline 74°40' \end{array}$$

TABLE 8.2 cos 74°40′ = .2644

Degrees	Sine	Tangent	Cotangent	Cosine	
9°00′	.1564	.1584	6.3138	.9877	**81°00′**
10′	.1593	.1614	6.1970	.9872	50′
20′	.1622	.1644	6.0844	.9868	40′
30′	.1650	.1673	5.9758	.9863	30′
40′	.1679	.1703	5.8708	.9858	20′
50′	.1708	.1733	5.7694	.9853	10′
15°00′	.2588	.2679	3.7321	.9659	**75°00′**
10′	.2616	.2711	3.6891	.9652	50′
20′	.2644	.2742	3.6470	.9644	40′
30′	.2672	.2773	3.6059	.9636	30′
40′	.2700	.2805	3.5656	.9628	20′
50′	.2728	.2836	3.5261	.9621	10′
16°00′	.2756	.2867	3.4874	.9613	**74°00′**
10′	.2784	.2899	3.4495	.9605	50′
20′	.2812	.2931	3.4124	.9596	40′
30′	.2840	.2962	3.3759	.9588	30′
40′	.2868	.2994	3.3402	.9580	20′
50′	.2896	.3026	3.3052	.9572	10′
17°00′	.2924	.3057	3.2709	.9563	**73°00′**
10′	.2952	.3089	3.2371	.9555	50′
20′	.2979	.3121	3.2041	.9546	40′
30′	.3007	.3153	3.1716	.9537	30′
40′	.3035	.3185	3.1397	.9528	20′
50′	.3062	.3217	3.1084	.9520	10′
18°00′	.3090	.3249	3.0777	.9511	**72°00′**
	Cosine	**Cotangent**	**Tangent**	**Sine**	**Degrees**

□

finding $\angle A$, given sin A

In Chapter 7, you sometimes had to find the antilog of a number. Thus when given log N, you used the log table to find N. To find the acute angle with a given sine, locate the value of the sine in the trig table, and then read across to find the angle.

EXAMPLE 3 Find $\angle A$ if $\sin A = .3173$.

Solution. For *acute* angles, the sine increases as the angle measure increases. Because

$$\sin 45° = .7071 \qquad \text{and} \qquad .3173 < .7071$$

it follows that

$$\angle A < 45°.$$

Read down the sine column in Table 3 to locate .3173 . See also Table 8.3 . At the *left*, note that

$$\sin 18°30' = .3173$$

TABLE 8.3 sin 18°30′ = .3173

Degrees	Sine	Tangent	Cotangent	Cosine	
18°00′	.3090	.3249	3.0777	.9511	**72°00′**
10′	.3118	.3281	3.0475	.9502	50′
20′	.3145	.3314	3.0178	.9492	40′
30′	.3173	.3346	2.9887	.9483	30′
40′	.3201	.3378	2.9600	.9474	20′
50′	.3228	.3411	2.9319	.9465	10′

□

solving right triangles

To **solve a right triangle** means to determine (the lengths of) all unknown sides and (the measures of) all unknown angles. (Sometimes "side b" will be used for "the side of length b".

EXAMPLE 4 Solve the right triangle in Figure 8.25 .Approximate each side length to one decimal place and each angle measurement to the nearest 10 minutes.

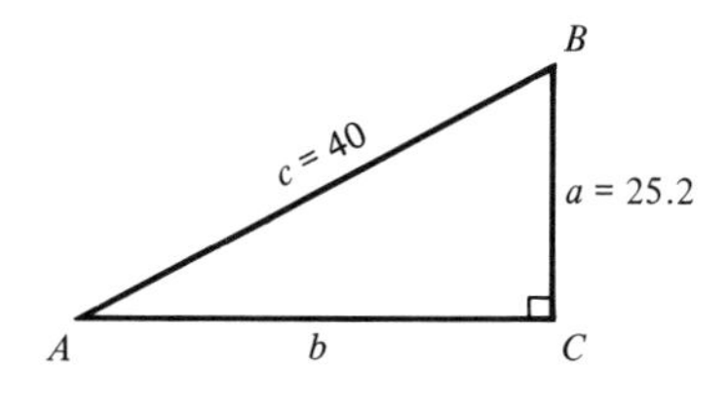

FIGURE 8.25

Solution. You must find $\angle A$, $\angle B$, and side b. To find $\angle A$, first consider sin A.

$$\sin A = \frac{a}{c} = \frac{25.2}{40} = .63 = .6300$$

Because

$$\sin 45° = .7071 \qquad \text{and} \qquad .6300 < .7071,$$

it follows that

$$\angle A < 45°.$$

Read down the sine column in the table to determine the angle to the nearest 10 minutes. See also Table 8.4 .

TABLE 8.4 sin 39°00′ = .6293; sin 39°10′ = .6316; cos 39° = .7771

Degrees	Sine	Tangent	Cotangent	Cosine	
39°00′	.6293	.8098	1.2349	.7771	**51°00′**
10′	.6316	.8146	1.2276	.7753	50′
20′	.6338	.8195	1.2203	.7735	40′
30′	.6361	.8243	1.2131	.7716	30′
40′	.6383	.8292	1.2059	.7698	20′
50′	.6406	.8342	1.1988	.7679	10′

Observe:

$$\begin{aligned} \sin 39°00' &= .6293 \\ &\ .6300 \\ \sin 39°10' &= .6316 \end{aligned} \quad \begin{matrix} \} \ .0007 \\ \} \ .0016 \end{matrix}$$

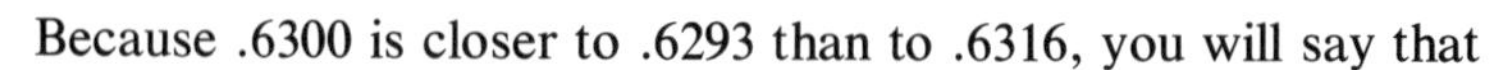

Because .6300 is closer to .6293 than to .6316, you will say that

$$\angle A = 39°$$

It follows that

$$\angle B = 90° - 39° = 51°.$$

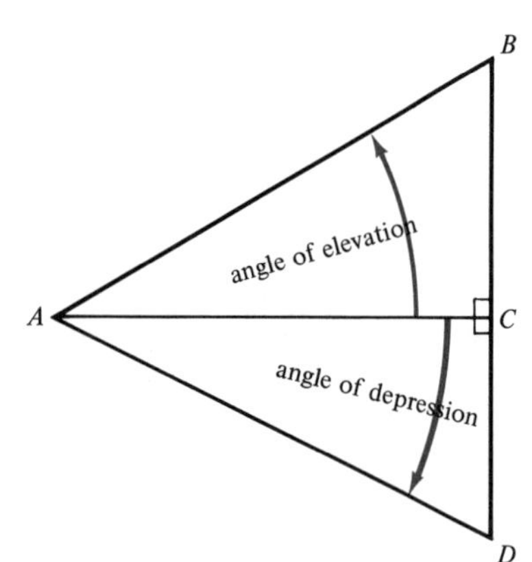

FIGURE 8.26

To find side b, you could use the Pythagorean Theorem. However, it is easier to use Table 8.4 on page 315.

$$\cos 39° = \frac{b}{40}$$

$$40 \cos 39° = b$$

$$40 \times .7771 = b$$

To one decimal place, $b = 31.1$ □

angle of elevation

Suppose an observer is at point A and horizontal eye level is given by $\overline{AC}$, as in Figure 8.26 . Then the **angle of elevation** of point B is given by $\angle CAB$ and the **angle of depression** of point D by $\angle CAD$.

EXAMPLE 5 The angle of elevation of the top of a building from a point at street level 200 feet from the base of the building is 48°. To the nearest foot, how high is the building? [See Figure 8.27 .]

Solution. Let a be the height of the building (in feet).

$$\tan 48° = \frac{a}{200}$$

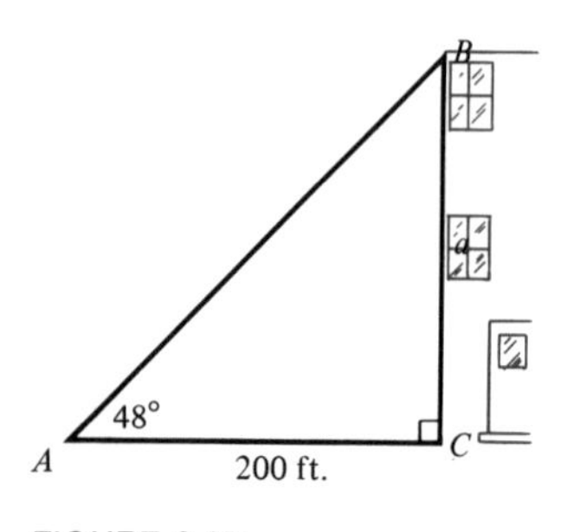

FIGURE 8.27

Because

$$45° < 48° < 90°,$$

read *up* Table 3 and locate the angle at the *right*.

$$1.1106 = \frac{a}{200}$$

$$200 \times 1.1106 = a$$

$$222.12 = a$$

The building is approximately 222 feet high. □

EXAMPLE 6 A man on a 30-foot telephone pole sights his car 20 feet from the base of the pole. To the nearest 10 minutes, what is the angle of depression of the car?

Solution. In Figure 8.28, $\angle CAD$ is the angle of depression.

$$\tan \angle CAD = \frac{30}{20} = 1.5000$$

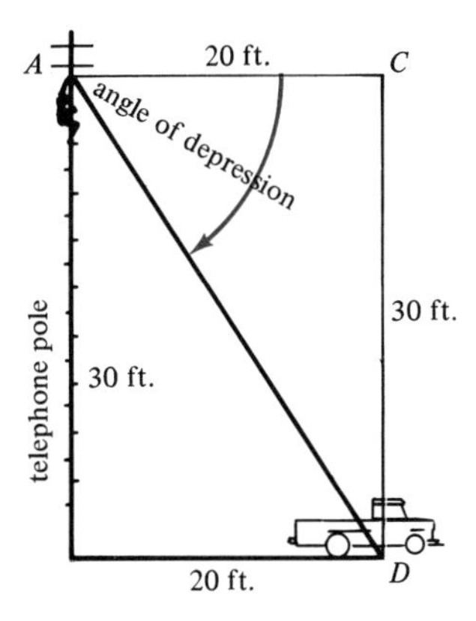

FIGURE 8.28

Because tan 45° = 1.0000, the angle of depression must be more than 45°. Thus, read *up* the table.

$$\left.\begin{array}{l}\tan 56°20' = 1.5013 \\ \,1.5000 \\ \tan 56°10' = 1.4919\end{array}\right\} \begin{array}{l}.0013 \\ .0081\end{array}$$

Take

$$\angle CAD = 56°\ 20'$$

□

TABLE 8.5 tan 56°10′ = 1.4919; tan 56°20′ = 1.5013

33°00′	.5446	.6494	1.5399	.8387	**57°00′**
10′	.5471	.6536	1.5301	.8371	50′
20′	.5495	.6577	1.5204	.8355	40′
30′	.5519	.6619	1.5108	.8339	30′
40′	.5544	.6661	1.5013	.8323	20′
50′	.5568	.6703	1.4919	.8307	10′
34°00′	.5592	.6745	1.4826	.8290	**56°00′**
	Cosine	**Cotangent**	**Tangent**	**Sine**	**Degrees**

interpolation

To obtain greater accuracy when using the trig table, *linear interpolation* can be employed, as it was when using the log table. [See Section 7.6 .] The error that occurs in interpolation is negligible for most purposes.

EXAMPLE 7 Determine sin 14°25′.

Solution. 14°25′ lies halfway between 14°20′ and 14°30′, which are both listed in the trig table.

TABLE 8.6 sin 14°20′ = .2476; sin 14°30′ = .2504

Degrees	Sine	Tangent	Cotangent	Cosine	
14°00′	.2419	.2493	4.0108	.9703	**76°00′**
10′	.2447	.2524	3.9617	.9696	50′
20′	.2476	.2555	3.9136	.9689	40′
30′	.2504	.2586	3.8667	.9681	30′
40′	.2532	.2617	3.8208	.9674	20′
50′	.2560	.2648	3.7760	.9667	10′

In Table 8.7 on page 318, the left-hand column indicates degrees and minutes, the right-hand column, the value of the sine function.

TABLE 8.7

degrees and minutes	sin
14°20′	.2476
14°25′	.2476 + x
14°30′	.2504

(Brackets: 5′ between 14°20′ and 14°25′; 10′ between 14°20′ and 14°30′; x between .2476 and .2476 + x; .0028 between .2476 and .2504.)

Recall that (for acute angles) an increase in the angle measure means an *increase* in the value of the sine.

Note the differences between the *largest* and *smallest* values in *each column*.

$$\begin{array}{r} 14°30' \\ -14°20 \\ \hline 10' \end{array} \quad \text{and} \quad \begin{array}{r} .2504 \\ -.2476 \\ \hline .0028 \end{array}$$

To obtain the sine of the intermediate number of degrees and minutes, 14°25′, note the differences:

$$\begin{array}{r} 14°25' \\ -14°20' \\ \hline 05' \end{array} \quad \text{and} \quad \begin{array}{r} .2476 + x \\ -.2476 \\ \hline x \end{array}$$

You want the *proportional increase, x,* on the right side. As was the case with logarithmic interpolation, you need not be so careful about the decimal points in setting up the proportion. Let d be the number of units (ten thousandths) in the sine column. [See Table 8.8 .]

TABLE 8.8

degrees and minutes	sin
14°20′	.2476
14°25′	___
14°30′	.2504

(Brackets: 5 between 14°20′ and 14°25′; 10 between 14°20′ and 14°30′; d between .2476 and ___; 28 between .2476 and .2504.)

$$\frac{d}{28} = \frac{5}{10}$$

$$d = \frac{\overset{1}{\cancel{5}} \cdot 28}{\underset{2}{\cancel{10}}} = 14$$

Add 14 "units" (.0014) to .2476 to obtain

$$\begin{array}{r} .2476 \\ .0014 \\ \hline .2490 \end{array}$$

Thus,

sin 14°25′ = .2490 □

EXAMPLE 8 Determine cos 26°13′.

Solution. See Table 3 at the back, as well as Table 8.9 .

For acute angles, an increase in the angle measure means a *decrease* in the value of the cosine. Thus d "units" are to be subtracted. Set up the proportion:

TABLE 8.9

degrees and minutes	cos
26°10′	.8975
26°13′	___
26°20′	.8962

(Brackets: 3 between 26°10′ and 26°13′; 10 between 26°10′ and 26°20′; d between .8975 and ___; 13 between .8975 and .8962.)

$$\frac{d}{13} = \frac{3}{10}$$

$$d = \frac{3 \cdot 13}{10} = \frac{39}{10} \approx 4$$

Subtract:

$$\begin{array}{r} .8975 \\ -.0004 \\ \hline .8971 \end{array}$$

Thus

cos 26°13′ = .8971 □

EXAMPLE 9 Determine $\angle A$ to the nearest minute if $\tan A = .5535$

Solution. .5535 is not found in the tangent column of Table 3. Write down the closest numbers to .5535 that are in the table—one smaller than, the other larger than, .5535 .

$$\tan 28°50' = .5505 \qquad \text{and} \qquad \tan 29°00' = .5543$$

The tangent increases as the measure of an acute angle increases. Now d will be the number of *minutes added to* 28°50′. See Table 8.10 .

TABLE 8.10

degrees and minutes	tan
28°50′	.5505
28°(50 + d)′	.5535
29°00′	.5543

(Brackets: d between 28°50′ and 28°(50 + d)′; 10 between 28°50′ and 29°00′; 30 between .5505 and .5535; 38 between .5505 and .5543.)

Set up the proportion:

$$\frac{d}{10} = \frac{30}{38}$$

$$d = \frac{30 \cdot 10}{38} = 7\frac{34}{38} \approx 8$$

Add, to determine $\angle A$.

$$\begin{array}{r} 28°50' \\ 08' \\ \hline \angle A = 28°58' \end{array}$$

□

EXAMPLE 10 Determine $\angle A$ to the nearest minute if $\cos A = .1321$.

Solution. The cosine *decreases* as the measure of an acute angle increases. Because

$$\cos 45° = .7071 \qquad \text{and} \qquad .1321 < .7071$$

it follows that

$$45° < \angle A < 90°$$

Read *up* Table 3 and locate the angle at the *right*. See also Table 8.11 .

TABLE 8.11

degrees and minutes	cos
.1305	82°30′
.1321	82°(20 + d)′
.1334	82°20′

(Brackets: 29 between .1305 and .1334; 13 between .1321 and .1334; 10 between 82°30′ and 82°20′; d between 82°(20 + d)′ and 82°20′.)

Here d represents the increase in the angle.

$$\frac{d}{10} = \frac{13}{29}$$

$$d = \frac{13 \cdot 10}{29} = 4\frac{14}{29}$$

Because $\frac{14}{29} < \frac{1}{2}$, take $d \approx 4$. Add:

$$\begin{array}{r} 82°20' \\ 04' \\ \hline \angle A = 82°24' \end{array}$$

□

parallel lines In Chapter 4, on page 160, it was noted that

> if two distinct nonvertical lines are parallel, then their slopes are equal.
>
> And if their slopes are equal, then two such lines are parallel.

For horizontal lines, this is immediate. Otherwise, to see why this is true, consider Figures 8.29(a) and (b). Let $m_1 =$ slope L_1, $m_2 =$ slope L_2.

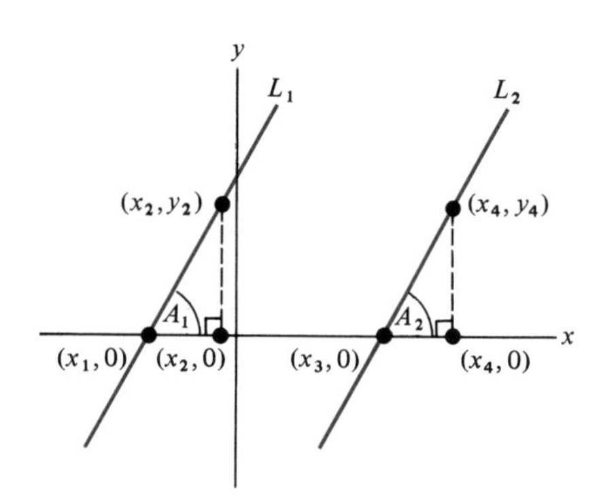

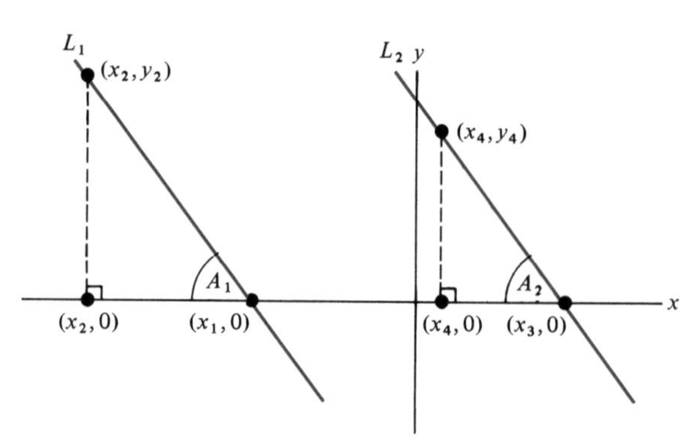

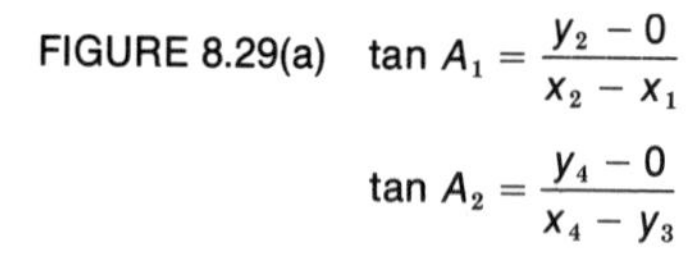
FIGURE 8.29(a) $\tan A_1 = \dfrac{y_2 - 0}{x_2 - x_1}$

$\tan A_2 = \dfrac{y_4 - 0}{x_4 - y_3}$

FIGURE 8.29(b) $\tan A_1 = \dfrac{y_2 - 0}{x_1 - x_2} = -\dfrac{y_2 - 0}{x_2 - x_1}$

$\tan A_2 = \dfrac{y_4 - 0}{x_3 - x_4} = -\dfrac{y_4 - 0}{x_4 - x_3}$

Suppose $L_1 \parallel L_2$. Then

$$\angle A_1 = \angle A_2$$

because corresponding angles of parallel lines are equal in measure. Therefore

$$\tan A_1 = \tan A_2$$

Thus

$$\frac{y_2 - 0}{x_2 - x_1} = \frac{y_4 - 0}{x_4 - x_3}$$

and

$$m_1 = m_2$$

To show that if $m_1 = m_2$, then $L_1 \parallel L_2$, you essentially reverse this argument. Thus suppose $m_1 = m_2$. Then

$$\frac{y_2 - 0}{x_2 - x_1} = \frac{y_4 - 0}{x_4 - x_3} \quad \text{and therefore} \quad \tan A_1 = \tan A_2$$

The tangent of an acute angle *increases* as the angle increases. Thus if their tangents are *the same,*

$$\angle A_1 = \angle A_2, \quad \text{and} \quad L_1 \parallel L_2$$

because when two lines are cut by a transversal (the x-axis), if corresponding angles are equal, the lines are parallel. △

perpendicular lines

Let L_1 and L_2 be distinct lines that are neither vertical nor horizontal. Let $m_1 =$ slope L_1 and $m_2 =$ slope L_2. Clearly, $m_1 \neq 0$, $m_2 \neq 0$.

If $L_1 \perp L_2$, then $m_2 = -\dfrac{1}{m_1}$

And if $m_2 = -\dfrac{1}{m_1}$, then $L_1 \perp L_2$

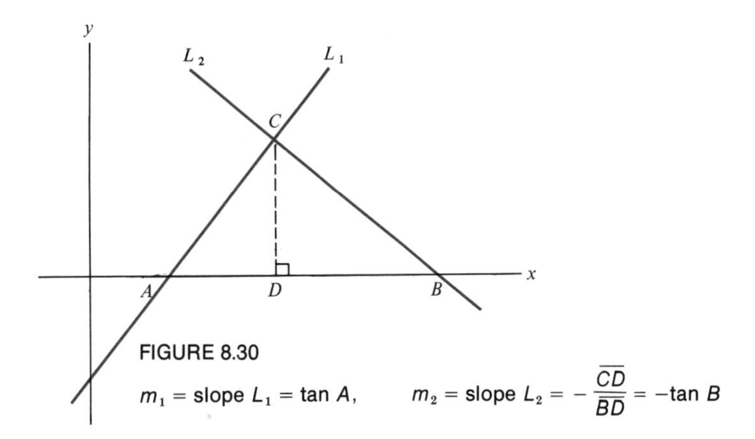

FIGURE 8.30

m_1 = slope L_1 = tan A, m_2 = slope $L_2 = -\dfrac{\overline{CD}}{\overline{BD}} = -\tan B$

[See page 162.] To see why this is true, consider $\triangle ABC$ in Figure 8.30 .

Suppose $L_1 \perp L_2$. Then

$$\angle C = 90^\circ \qquad \text{and} \qquad \angle B = 90^\circ - \angle A$$

Thus

$$m_2 = -\tan B = -\tan(90^\circ - A) = -\cot A = -\frac{1}{\tan A} = -\frac{1}{m_1}$$

Now suppose that $m_2 = \dfrac{-1}{m_1}$. Then $-\tan B = -\dfrac{1}{\tan A}$ or

$$\tan B = \frac{1}{\tan A} = \cot A = \tan(90^\circ - A)$$

Because the tangent of an *acute* angle increases as the measure of the angle increases, two acute angles with the *same* tangent must be equal. Thus

$$\angle B = 90^\circ - \angle A,$$
$$\angle A + \angle B = 90^\circ,$$

and therefore

$$\angle C = 180^\circ - (\angle A + \angle B) = 90^\circ$$

Consequently,

$$L_1 \perp L_2$$

△

EXERCISES

Refer to Table 3 in the appendix.
In Exercises 1–12 determine each value.

1. sin 25° **2.** cos 38° **3.** tan 9° **4.** sin 52° **5.** cos 80° **6.** tan 89°

7. sin 14°10′ **8.** cos 20°50′ **9.** tan 44.5° **10.** sin 74°50′ **11.** cos 88°40′ **12.** cot 36°20′

In Exercises 13–24 assume $\angle A$ is acute. Determine $\angle A$.

13. $\sin A = .1908$ **14.** $\tan A = .4877$ **15.** $\cos A = .8290$ **16.** $\sin A = .8746$

17. $\tan A = 2.4751$ **18.** $\cos A = .9994$ **19.** $\sin A = .1421$ **20.** $\cos A = .9100$

21. $\tan A = .2095$ **22.** $\sin A = .9872$ **23.** $\tan A = 8.3450$ **24.** $\cot A = 42.964$

In Exercises 25–32 solve each right triangle. Refer to Figure 8.31 . Do not interpolate. Approximate each side length to 1 decimal place and each angle measurement to the nearest 10 minutes.

25. $a = 7, b = 10$ **26.** $a = 22, c = 30$

27. $b = 16, c = 40$ **28.** $\angle A = 36°, a = 12$

29. $\angle A = 42°20', c = 25$ **30.** $\angle B = 63°10', a = 9$

31. $\angle B = 6°50', b = 18$ **32.** $\angle A = 88°10', c = 20.25$

FIGURE 8.31

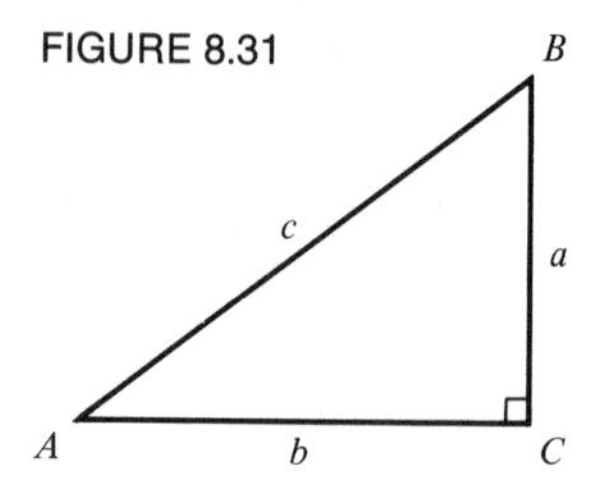

In Exercises 33–46 do not interpolate. Approximate each side length to 1 decimal place and each angle measurement to the nearest 10 minutes.

33. Find the angle of elevation of the sun if a 45-foot monument casts a 36-foot shadow.

34. Find the height of a man who casts a 6-foot shadow when the angle of elevation of the sun is 43°40′.

35. A 24-foot ladder leaning against a wall makes a 36° angle with the ground.
(a) Approximately how far is the foot of the ladder from the base of the wall?
(b) Approximately how high is the top of the ladder?

36. From on top of a tower 432 feet high a man measures the angle of depression of his motorcycle to be 48°40′. Approximately how far is the motorcycle from the base of the tower? [See Figure 8.32 .]

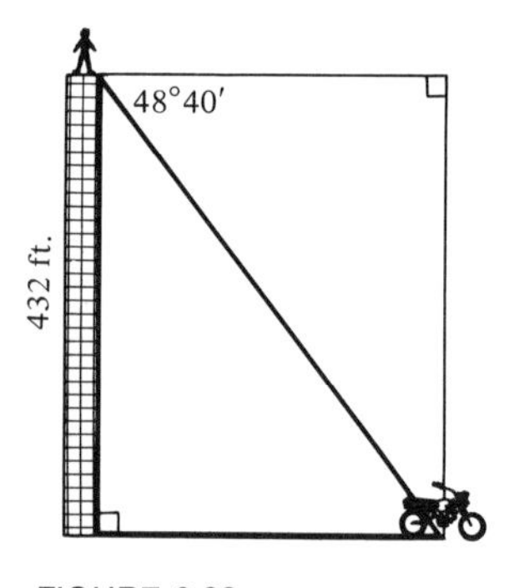

FIGURE 8.32

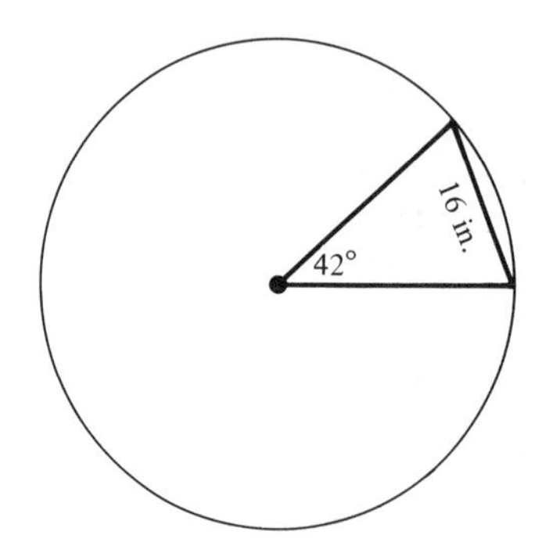

FIGURE 8.33

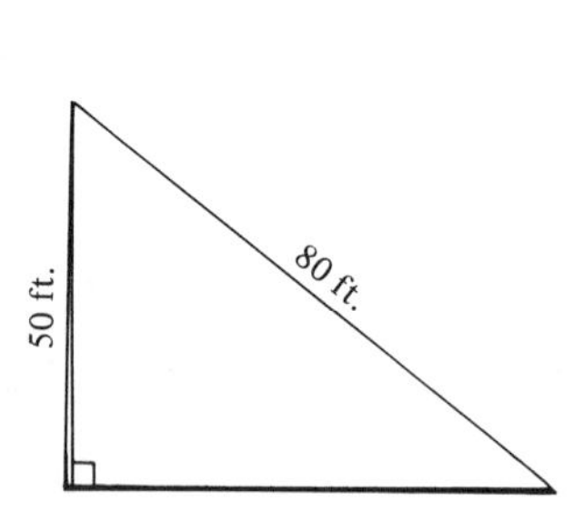

FIGURE 8.34

37. A straight 10-mile road begins at sea level and makes a 6° angle with the horizontal. How many feet above sea level is the end of the road?

38. A woman sits at a window 18 feet above street level. The angle of depression of her daughter, who is playing in the backyard, is 40°20′. How far is her daughter from the base of the house?

39. Find the length of the radius of a circle in which a 16-inch chord subtends (cuts off) a 42° central angle. [See Figure 8.33 .]

40. An airplane flying horizontally at an altitude of 15 000 feet approaches a 100-foot tower. When the angle of depression of the top of the tower is 20°50′, how far must the plane fly to be directly above the tower?

41. A guy wire 80 feet long is attached to the top of a pole 50 feet high. What angle does the wire make with the pole? [See Figure 8.34 .]

42. As an airplane ascends, its path makes a 12°20′ angle with the runway. How many feet does the plane rise while traveling 1200 feet in the air?

43. From the top of a lighthouse 62 feet above sea level, the angle of depression of a small craft is 13°40′. How far from the foot of the lighthouse is the craft?

44. In order to measure the height of a cloud, a spotlight is beamed vertically upward. A woman stands 250 meters from the spotlight. From her point of view, the angle of elevation of the spot on the cloud is 70°. How high is the cloud? [See Figure 8.35 .]

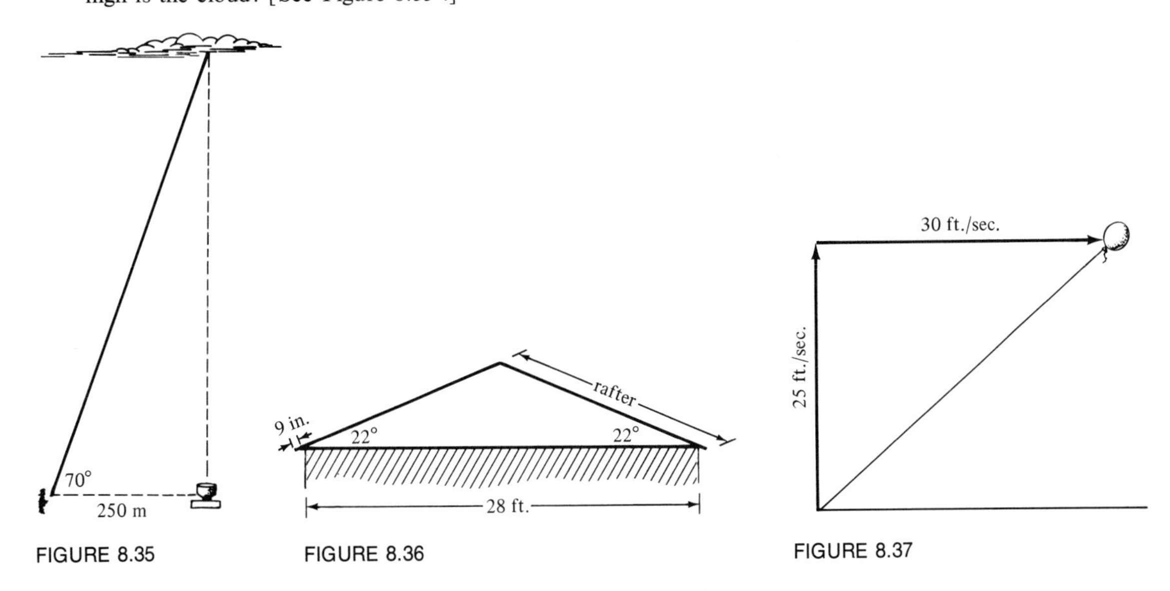

FIGURE 8.35 FIGURE 8.36 FIGURE 8.37

45. The roof of a bungalow slopes 22°, as in Figure 8.36 . How long a rafter should be used in order to obtain a 9-inch overhang?

46. As a helium balloon rises vertically at the rate of 25 feet per second, it is blown horizontally by a wind whose velocity is 30 feet per second. Find the angle of elevation of the balloon. [See Figure 8.37 .]

In Exercises 47–54 interpolate to determine each value.

47. sin 19°35′ **48.** cos 23°41′ **49.** tan 40°07′ **50.** sin 58°12′ **51.** cos 85°05′

52. tan 12°59′ **53.** sin 1°01′ **54.** cot 36°36′

In Exercises 55–60 assume $\angle A$ is acute. Interpolate to determine $\angle A$ to the nearest minute.

55. $\sin A = .4160$ **56.** $\cos A = .8554$ **57.** $\tan A = .8752$ **58.** $\cos A = .6600$

59. $\tan A = 4.501$ **60.** $\cot A = .1773$

In Exercises 61–64 interpolate in solving each right triangle. Refer to Figure 8.31 . Approximate each side length to one decimal place. Determine each angle to the nearest minute.

61. $a = 19.2, b = 17.5$ **62.** $\angle A = 38°24', b = 28.1$ **63.** $\angle B = 72°06', c = 49.5$

64. $\angle A = 41.3°, a = 17.6$

8.3 The Unit Circle and Radian Measure

unit circle

Thus far, only the sine, cosine, etc., of *acute* angles have been discussed. And angles have been measured in terms of *degrees and*

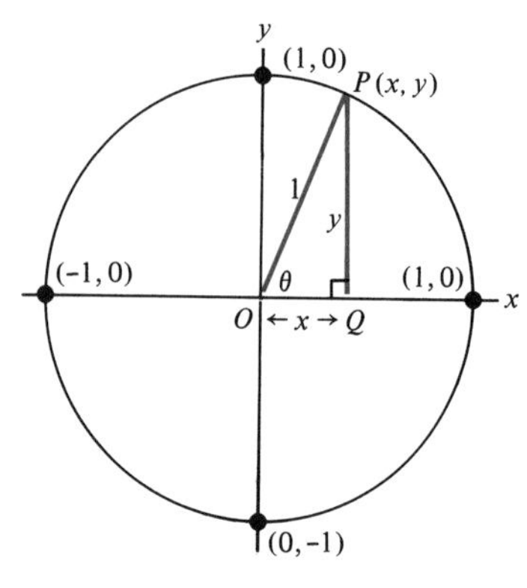

FIGURE 8.38
$\theta = \angle QOP$
$P = (x, y) = (\cos\theta, \sin\theta)$

minutes. For the further development of mathematics, it is preferable to define the trigonometric functions so that the domain of each is the set of *all (possible) real numbers*. To this end, consider the circle centered at the origin and of radius 1, known as the **unit circle.**

The equation of the unit circle is

$$x^2 + y^2 = 1$$

Observe that in Figure 8.38, for a point P on the unit circle and in the first quadrant,

$$\cos \angle QOP = \frac{x}{1} = x$$

$$\sin \angle QOP = \frac{y}{1} = y$$

Let θ (theta) $= \angle QOP$. Then

$$P = (x, y) = (\cos\theta, \sin\theta)$$

In other words, the coordinates of P are the cosine and sine of angle θ.

radians

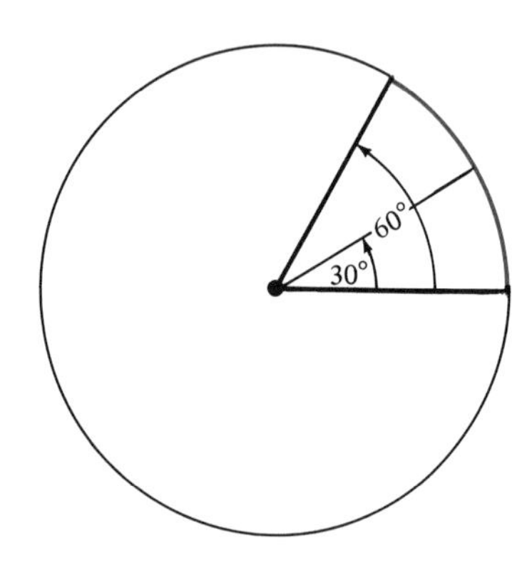

FIGURE 8.39 The length of the arc subtended by a 60° angle is twice the length of the arc subtended by a 30° angle.

Angle θ is a **central angle,** that is, an angle formed by two radii (plural of "radius") of the circle. The measure of a central angle is proportional to the length of the arc of the circle it subtends (cuts off). [See Figure 8.39 .] This observation leads to a key definition. One **radian** is defined to be the measure of the central angle that subtends an arc whose length is that of the radius. Note that the length of the radius of the unit circle is one unit. [See Figure 8.40 .] The circumference, C, of the unit circle is 2π (because the radius length is 1 and $C = 2\pi r$); thus, a central angle of 360° has radian measure 2π. [See Figure 8.41(a).] By proportionality,

$$\pi \text{ radians} = 180°;$$

this is the basic relationship between radian and degree measurement. A right angle, 90°, has radian measure $\frac{\pi}{2}$. [See Figure 8.41(b).]

Because π radians = 180 degrees, it follows that

$$1 \text{ radian} = \frac{\pi}{\pi} \text{ radian} = \frac{180}{\pi} \text{ degrees}$$

$$1 \text{ radian} \approx \frac{180}{3.1416} \text{ degrees} \approx 57.296°$$

To convert from

degrees to radians, multiply by $\frac{\pi}{180°}$;

radians to degrees, multiply by $\frac{180°}{\pi}$.

EXAMPLE 1 Express in radian measure.

(a) 60° (b) 72° (c) 135°

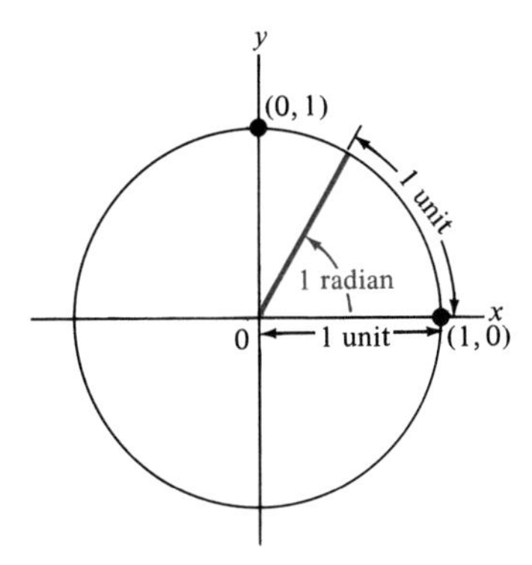

FIGURE 8.40 The length of the subtended arc equals the length of the radius, which, in the unit circle is one unit.

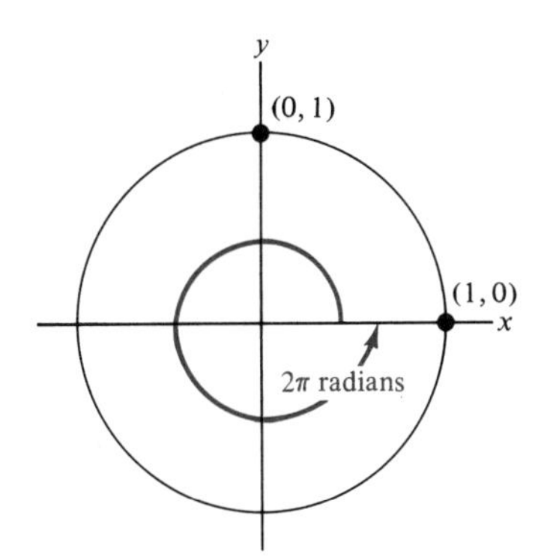

FIGURE 8.41(a) A central angle of 360° has radian measure 2π, the circumference of the unit circle.

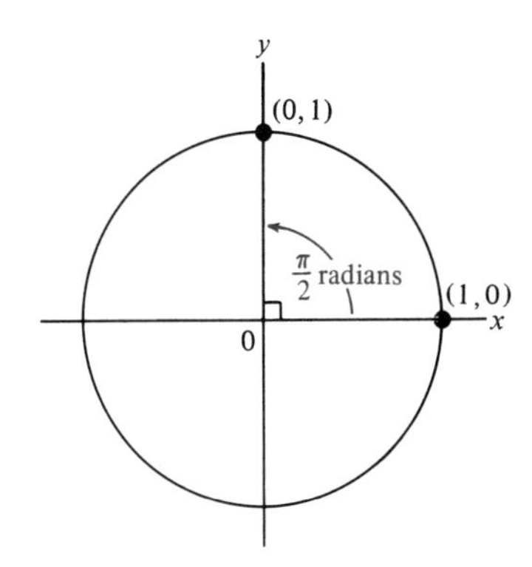

FIGURE 8.41(b) A right angle has radian measure $\frac{\pi}{2}$.

Solution.

(a) $60° = 60° \cdot \dfrac{\pi}{180°} = \dfrac{\pi}{3}$ ("Radians" is usually omitted.)

(b) $72° = \overset{2}{\cancel{72}}° \cdot \dfrac{\pi}{\underset{5}{\cancel{180}}°} = \dfrac{2\pi}{5}$

(c) $135° = 135° \cdot \dfrac{\pi}{180°} = \dfrac{3\pi}{4}$

(Here 135 and 180 were each divided by 45.) □

EXAMPLE 2 Express each radian measure in terms of degrees.

(a) $\dfrac{\pi}{6}$ (b) $\dfrac{3\pi}{10}$ (c) $\dfrac{5\pi}{8}$

Solution.

(a) $\dfrac{\pi}{6} = \dfrac{\cancel{\pi}}{\underset{1}{\cancel{6}}} \cdot \dfrac{\overset{30°}{\cancel{180}}°}{\cancel{\pi}} = 30°$

(b) $\dfrac{3\pi}{10} = \dfrac{3\cancel{\pi}}{\underset{1}{\cancel{10}}} \cdot \dfrac{\overset{18°}{\cancel{180}}°}{\cancel{\pi}} = 54°$

(c) $\dfrac{5\pi}{8} = \dfrac{5\cancel{\pi}}{\underset{2}{\cancel{8}}} \cdot \dfrac{\overset{45°}{\cancel{180}}°}{\cancel{\pi}} = \dfrac{225°}{2} = 112.5°$, or $112°30'$ □

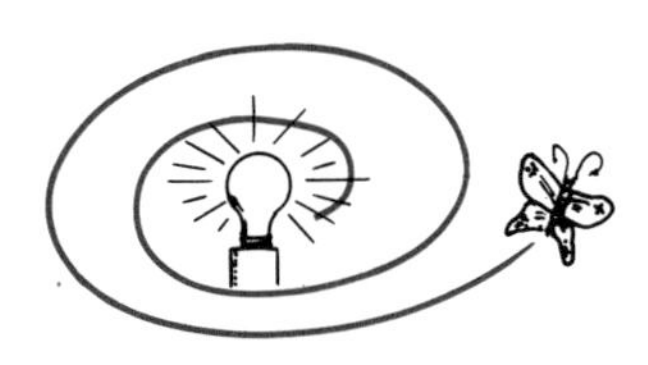

FIGURE 8.42
An angle of 4π radians

At times, it is useful to measure angles of more than 2π radians. For example, if a moth encircles a light bulb twice, as in Figure 8.42, you would measure its flight by an angle of 4π radians (or 720°). Note that the moth's flight is *counterclockwise*.

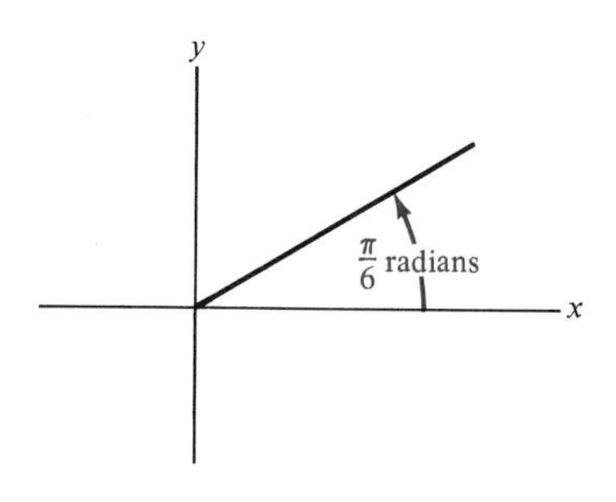

FIGURE 8.43(a)

It is convenient to think of an angle in **standard position** with the vertex at the origin and one side (**the initial side**) of the angle on the *positive* x-axis. The angle is considered to be **positive** if its other side (or **terminal side**) is measured in a *counterclockwise* direction and **negative** if measured in a *clockwise* direction. Thus, Figure 8.43(a) depicts an angle of $\frac{\pi}{6}$ (or 30°) and Figure 8.43(b) depicts an angle of $-\frac{\pi}{6}$ (or $-30°$).

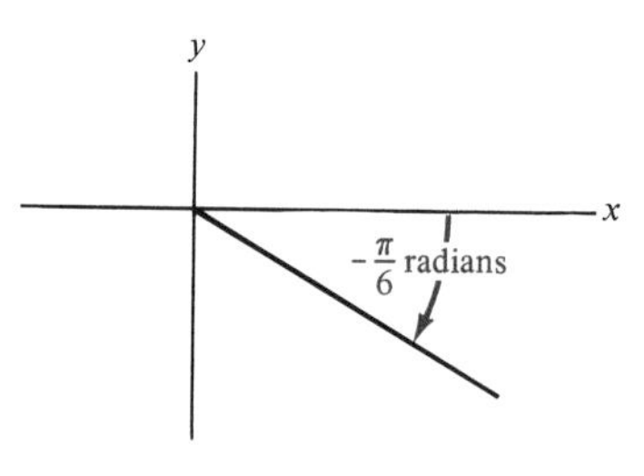

FIGURE 8.43(b)

To every real number θ—positive, negative, or 0—there corresponds a unique angle in standard position that measures θ radians. Thus the number 3, which is slightly less than π, corresponds to the angle in Figure 8.44(a); the number -5 corresponds to the angle in Figure 8.44(b). In contrast, to every angle there correspond *infinitely many* real numbers that measure the angle. Any two of these numbers differ by a multiple of 2π. Thus in Figure 8.44(c), the right angle corresponds to the following numbers:

$$\frac{\pi}{2}, \quad \frac{\pi}{2} + 2\pi\left(\text{or } \frac{5\pi}{2}\right), \quad \frac{\pi}{2} + 4\pi\left(\text{or } \frac{7\pi}{2}\right), \ldots$$

$$\frac{\pi}{2} - 2\pi\left(\text{or } -\frac{3\pi}{2}\right), \quad \frac{\pi}{2} - 4\pi\left(\text{or } -\frac{7\pi}{2}\right), \ldots$$

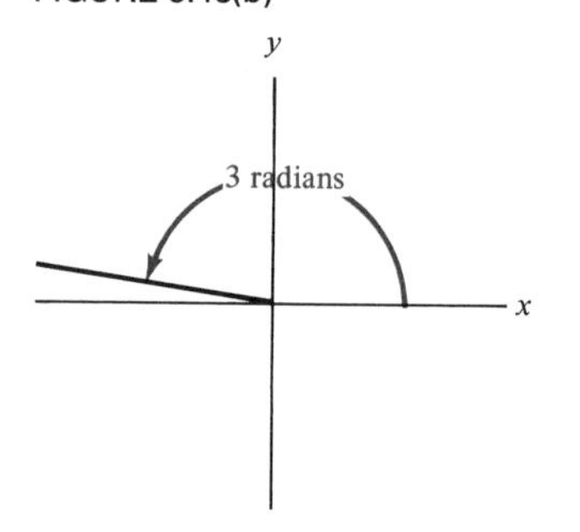

FIGURE 8.44(a)
An angle of 3 radians

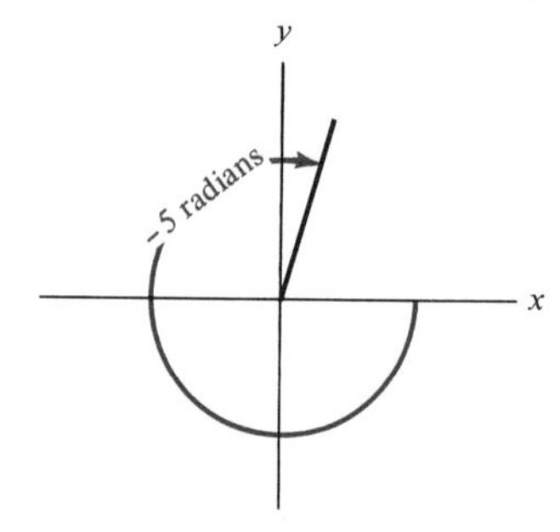

FIGURE 8.44(b)
An angle of -5 radians

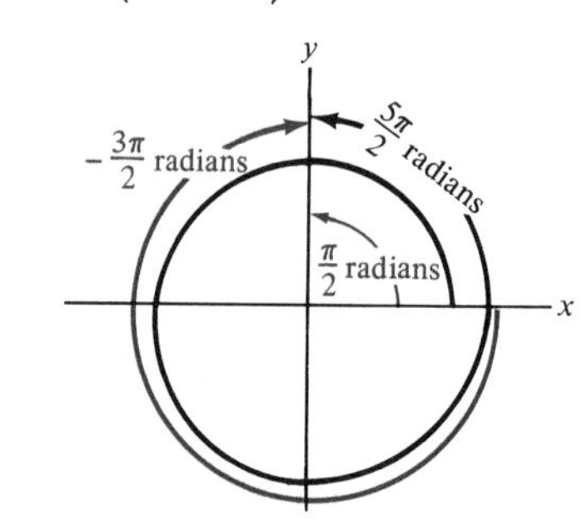

FIGURE 8.44(c)

EXAMPLE 3 What other numbers correspond to an angle of $\frac{\pi}{6}$ radians?

Solution.

$$\frac{\pi}{6} + 2\pi\left(\text{or } \frac{13\pi}{6}\right), \quad \frac{\pi}{6} + 4\pi\left(\text{or } \frac{25\pi}{6}\right), \quad \frac{\pi}{6} + 6\pi\left(\text{or } \frac{37\pi}{6}\right), \ldots$$

$$\frac{\pi}{6} - 2\pi\left(\text{or } -\frac{11\pi}{6}\right), \quad \frac{\pi}{6} - 4\pi\left(\text{or } -\frac{23\pi}{6}\right),$$

$$\frac{\pi}{6} - 6\pi\left(\text{or } -\frac{35\pi}{6}\right), \ldots$$

□

extending the trig functions

The trigonometric functions can be extended so as to apply to *any number* (of radians).

Let θ be *any real number,* and consider the (unique) corresponding (central) angle of θ radians, in standard position. Suppose the terminal side of the angle intersects the unit circle at P, with coordinates (x, y). [See Figure 8.45 .] Then $\cos\theta$ is defined to be x, and $\sin\theta$ is defined to be y. Thus, a major improvement has been made in the definition in that the cosine and sine of *any number* have now been defined. In physical applications there is often a need to speak of $\cos t$ and $\sin t$, where t is a number of seconds.

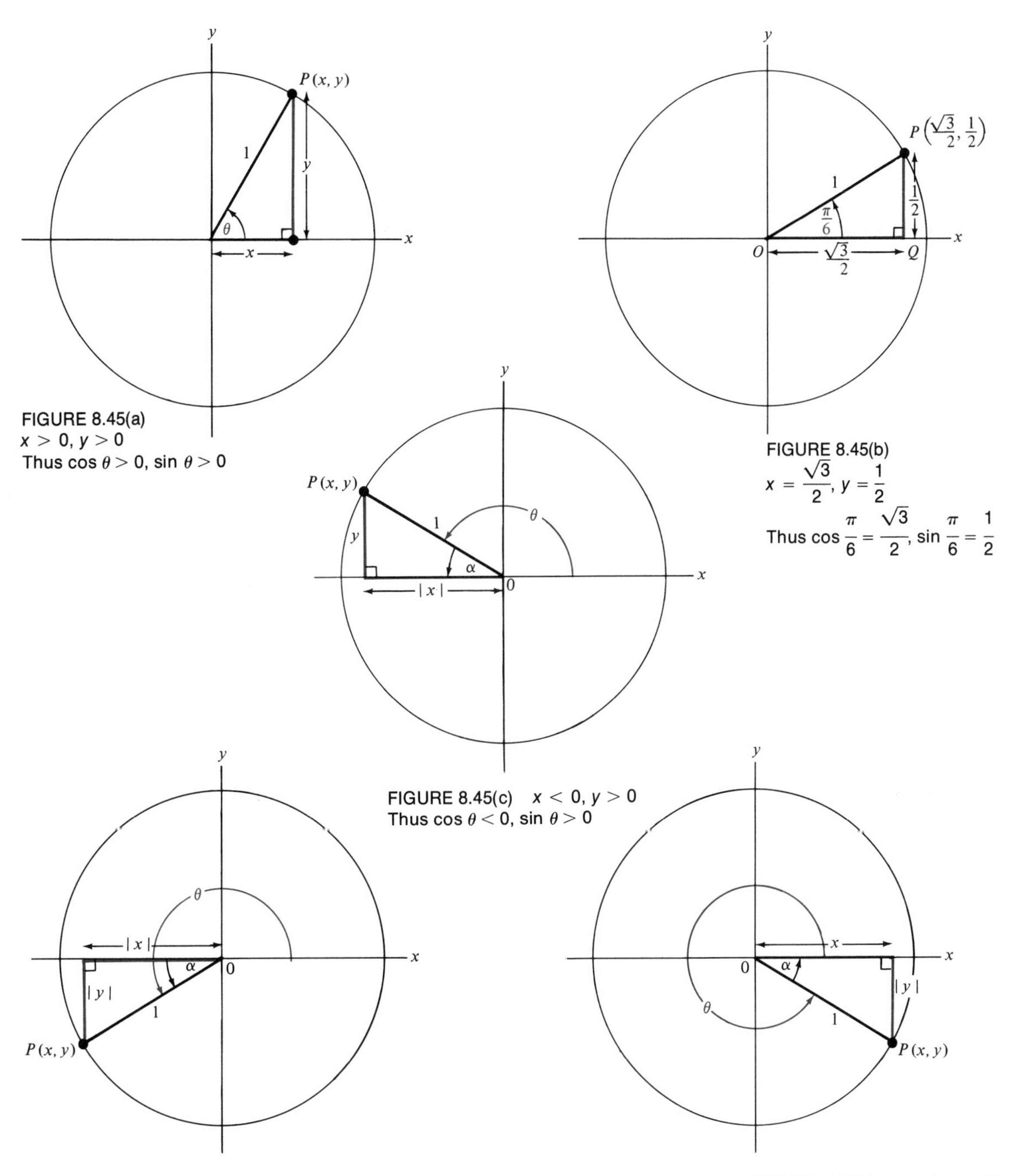

FIGURE 8.45(a)
$x > 0, y > 0$
Thus $\cos\theta > 0$, $\sin\theta > 0$

FIGURE 8.45(b)
$x = \frac{\sqrt{3}}{2}, y = \frac{1}{2}$
Thus $\cos\frac{\pi}{6} = \frac{\sqrt{3}}{2}$, $\sin\frac{\pi}{6} = \frac{1}{2}$

FIGURE 8.45(c) $x < 0, y > 0$
Thus $\cos\theta < 0$, $\sin\theta > 0$

FIGURE 8.45(d) $x < 0, y < 0$
Thus $\cos\theta < 0$, $\sin\theta < 0$

FIGURE 8.45(e) $x > 0, y < 0$
Thus $\cos\theta > 0$, $\sin\theta < 0$

For an acute angle θ, this new definition preserves the definitions of cos θ and sin θ previously given. [See Figure 8.45(a).] For example, when $\theta = \frac{\pi}{6}$ (or 30°), as in Figure 8.45(b), then ΔOPQ is a 30°, 60°, 90°-triangle with the length of hypotenuse $\overline{OP}$ equal to 1, so that length $\overline{PQ} = \frac{1}{2}$ and length $\overline{OQ} = \frac{\sqrt{3}}{2}$. Thus, using the *previous* (SOH-CAH-TOA) *definition,*

$$\cos \underbrace{\frac{\pi}{6}}_{30^\circ} = \frac{\text{length } \overline{OQ}}{\text{length } \overline{OP}} = \frac{\frac{\sqrt{3}}{2}}{1} = \frac{\sqrt{3}}{2}$$

$$\sin \underbrace{\frac{\pi}{6}}_{30^\circ} = \frac{\text{length } \overline{PQ}}{\text{length } \overline{OP}} = \frac{\frac{1}{2}}{1} = \frac{1}{2}$$

Using the *new definition,*

$$\cos \frac{\pi}{6} = x = \frac{\sqrt{3}}{2}, \qquad \sin \frac{\pi}{6} = y = \frac{1}{2}$$

When the terminal side of an angle is in Quadrant II, cos θ is negative (because x is negative) and sin θ is positive. [See Figure 8.45(c).]

When the terminal side of an angle is in Quadrant III, cos θ and sin θ are both negative (because both x and y are negative). [See Figure 8.45(d).]

When the terminal side of an angle is in Quadrant IV, cos θ is positive but sin θ is negative. [See Figure 8.45(e).]

An angle is said to be *in* Quadrant I when its terminal side is in Quadrant I, and similarly for angles *in* Quadrants II, III, and IV.

The other trigonometric functions are defined in terms of cos θ and sin θ:

$$\tan\theta = \frac{\sin\theta}{\cos\theta}, \quad \text{provided } \cos\theta \neq 0$$

$$\sec\theta = \frac{1}{\cos\theta}, \quad \text{provided } \cos\theta \neq 0$$

$$\csc\theta = \frac{1}{\sin\theta}, \quad \text{provided } \sin\theta \neq 0$$

$$\cot\theta = \frac{\cos\theta}{\sin\theta}, \quad \text{provided } \sin\theta \neq 0$$

Whenever tan θ *is defined and is nonzero,* you can use

$$\cot\theta = \frac{1}{\tan\theta}$$

[See Example 7 and the remark that follows it.]

Table 8.12 indicates which of the six trigonometric functions are positive in each quadrant.

In order to evaluate the trigonometric functions of an angle whose terminal side is in Quadrants II, III, or IV, you can refer to a right triangle with hypotenuse $\overline{OP}$ and with base on the x-axis. [See

TABLE 8.12 Moving counterclockwise from Quadrant IV, read CAST.

II Sin and csc are positive.	I All are positive.
III Tan and cot are positive.	IV Cos and sec are positive.

Figure 8.45 .] *Except for sign,* the acute angle α depicted will have the same cosine and sine as the angle θ in question. Moreover, $\sin\theta = \sin\alpha$ in Quadrants I and II, where $\sin\theta$ is positive; $\sin\theta = -\sin\alpha$ in Quadrants III and IV, where $\sin\theta$ is negative. Similarly,

$$\cos\theta = \begin{cases} \cos\alpha, & \text{in Quadrants I and IV} \\ -\cos\alpha, & \text{in Quadrants II and III} \end{cases}$$

[Refer to Table 8.12 .]

trig functions of special angles

EXAMPLE 4 Determine the six trigonometric function values at $\frac{3\pi}{4}$.

Solution. Recall that $\frac{\pi}{4} = 45°$, so that $\frac{3\pi}{4} = 3\left(\frac{\pi}{4}\right) = 135°$. The terminal side of an angle measuring $\frac{3\pi}{4}$ radians lies in Quadrant II. Thus $\cos\frac{3\pi}{4}$ is negative and $\sin\frac{3\pi}{4}$ is positive. Consider Figure 8.46 and recall the ratio of the side lengths of an isosceles right triangle.

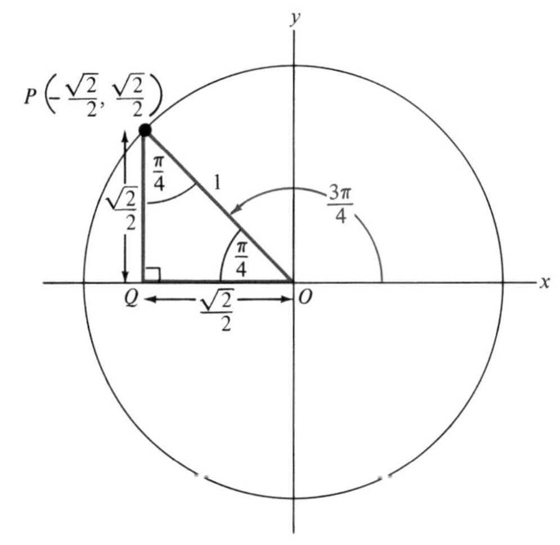

FIGURE 8.46
The length of the hypotenuse of this isosceles triangle is 1. Thus the length of each perpendicular side is $\frac{\sqrt{2}}{2}$ because

$$x^2 + x^2 = 1$$
$$x^2 = \frac{1}{2}$$
$$x = \frac{\sqrt{2}}{2}$$

Note that $P = (x, y) = \left(-\frac{\sqrt{2}}{2}, \frac{\sqrt{2}}{2}\right)$.

$$\cos\frac{3\pi}{4} = x = -\frac{\sqrt{2}}{2}$$

$$\sin\frac{3\pi}{4} = y = \frac{\sqrt{2}}{2}$$

$$\tan\frac{3\pi}{4} = \frac{\sin\frac{3\pi}{4}}{\cos\frac{3\pi}{4}} = \frac{\frac{\sqrt{2}}{2}}{-\frac{\sqrt{2}}{2}} = -1$$

$$\sec\frac{3\pi}{4} = \frac{1}{\cos\frac{3\pi}{4}} = \frac{1}{-\frac{\sqrt{2}}{2}} = -\sqrt{2}$$

$$\csc\frac{3\pi}{4} = \frac{1}{\sin\frac{3\pi}{4}} = \sqrt{2}$$

$$\cot\frac{3\pi}{4} = \frac{1}{\tan\frac{3\pi}{4}} = \frac{1}{-1} = -1$$

□

EXAMPLE 5 Find the six trigonometric function-values at $\frac{7\pi}{6}$.

Solution. Recall that $\frac{\pi}{6} = 30°$, so that $\frac{7\pi}{6} = 210°$. The terminal side of an angle measuring $\frac{7\pi}{6}$ radians lies in Quadrant III. Cos $\frac{7\pi}{6}$ and sin $\frac{7\pi}{6}$ are both negative. Consider Figure 8.47 and recall the basic relationships in a 30°, 60°, 90°-triangle.

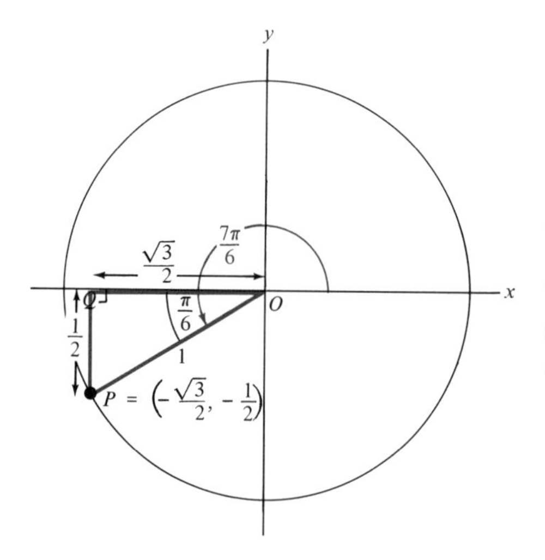

FIGURE 8.47
The length of the hypotenuse is 1. Thus, the side opposite the 30° angle $\left(\frac{\pi}{6}\right)$ has length $\frac{1}{2}$ and the remaining side has length $\frac{\sqrt{3}}{2}$.

$$\cos\frac{7\pi}{6} = x = -\frac{\sqrt{3}}{2}$$

$$\sin\frac{7\pi}{6} = y = -\frac{1}{2}$$

$$\tan\frac{7\pi}{6} = \frac{\sin\frac{7\pi}{6}}{\cos\frac{7\pi}{6}} = \frac{-\frac{1}{2}}{-\frac{\sqrt{3}}{2}} = \frac{1}{\sqrt{3}} = \frac{\sqrt{3}}{3}$$

$$\sec\frac{7\pi}{6} = \frac{1}{\cos\frac{7\pi}{6}} = -\frac{2\sqrt{3}}{3}$$

$$\csc\frac{7\pi}{6} = \frac{1}{\sin\frac{7\pi}{6}} = -2$$

$$\cot\frac{7\pi}{6} = \frac{1}{\tan\frac{7\pi}{6}} = \sqrt{3}$$ □

EXAMPLE 6 Find:

$$\sin\left(-\frac{\pi}{5}\right), \quad \cos\left(-\frac{\pi}{5}\right), \quad \tan\left(-\frac{\pi}{5}\right) \quad \cot\left(-\frac{\pi}{5}\right)$$

Solution. The terminal side of an angle measuring $-\frac{\pi}{5}$ lies in Quadrant IV. Cos $\left(-\frac{\pi}{5}\right)$ is positive and sin $\left(-\frac{\pi}{5}\right)$ is negative. Consider Figure 8.48 . $\frac{\pi}{5}$ corresponds to 36°; use the trig table at the back of the book to evaluate the trigonometric funtions of 36°.

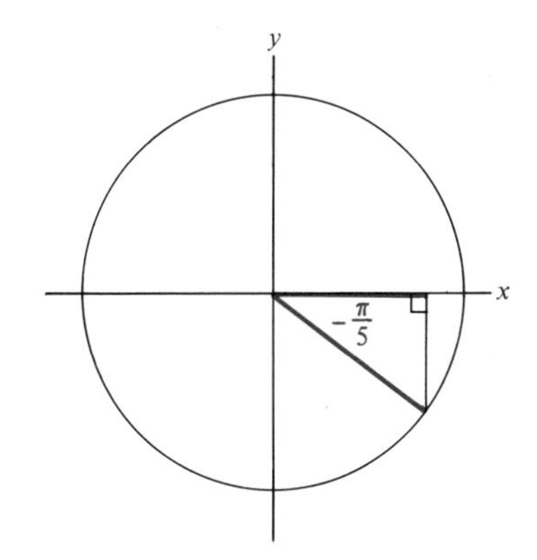

FIGURE 8.48

$$\sin\left(-\frac{\pi}{5}\right) = -\sin\frac{\pi}{5} = -\sin 36° = -\ .5878$$

$$\cos\left(-\frac{\pi}{5}\right) = \cos\frac{\pi}{5} = \cos 36° = .8090$$

$$\tan\left(-\frac{\pi}{5}\right) = -\tan\frac{\pi}{5} = -\tan 36° = -\ .7265$$

$$\cot\left(-\frac{\pi}{5}\right) = -\cot\frac{\pi}{5} = -\cot 36° = -1.3764$$ □

When the terminal side of an angle lies on a coordinate axis, there is no triangle involved. But in this case the coordinates of the point P are easily determined.

EXAMPLE 7 Find the trigonometric functions of π.

Solution. In Figure 8.49 on page 332, $P = (-1, 0)$. Thus,

$$\cos\pi = -1 \qquad \sec\pi = \frac{1}{\cos\pi} = \frac{1}{-1} = -1$$

$$\sin\pi = 0 \qquad \csc\pi \ \frac{1}{\underbrace{\sin\pi}_{0}} \text{ (undefined)}$$

$$\tan\pi = \frac{\sin\pi}{\cos\pi} = \frac{0}{-1} = 0 \qquad \cot\pi \ \frac{1}{\underbrace{\tan\pi}_{0}} \text{ (undefined)}$$ □

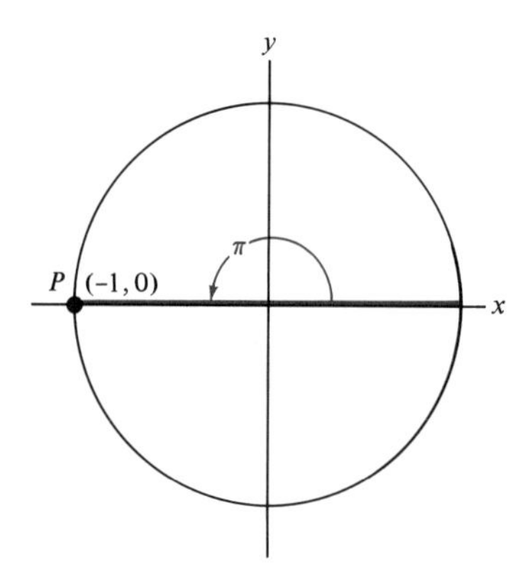

FIGURE 8.49

TABLE 8.13 Trigonometric functions of special angles

quadrant	radians	degrees	sine	cosine	tangent
	0	$0°$	0	1	0
I	$\frac{\pi}{6}$	$30°$	$\frac{1}{2}$	$\frac{\sqrt{3}}{2}$	$\frac{\sqrt{3}}{3}$
I	$\frac{\pi}{4}$	$45°$	$\frac{\sqrt{2}}{2}$	$\frac{\sqrt{2}}{2}$	1
I	$\frac{\pi}{3}$	$60°$	$\frac{\sqrt{3}}{2}$	$\frac{1}{2}$	$\sqrt{3}$
	$\frac{\pi}{2}$	$90°$	1	0	undefined
II	$\frac{2\pi}{3}$	$120°$	$\frac{\sqrt{3}}{2}$	$-\frac{1}{2}$	$-\sqrt{3}$
II	$\frac{3\pi}{4}$	$135°$	$\frac{\sqrt{2}}{2}$	$-\frac{\sqrt{2}}{2}$	-1
II	$\frac{5\pi}{6}$	$150°$	$\frac{1}{2}$	$-\frac{\sqrt{3}}{2}$	$-\frac{\sqrt{3}}{3}$
	π	$180°$	0	-1	0
III	$\frac{7\pi}{6}$	$210°$	$-\frac{1}{2}$	$-\frac{\sqrt{3}}{2}$	$\frac{\sqrt{3}}{3}$
III	$\frac{5\pi}{4}$	$225°$	$-\frac{\sqrt{2}}{2}$	$-\frac{\sqrt{2}}{2}$	1
III	$\frac{4\pi}{3}$	$240°$	$-\frac{\sqrt{3}}{2}$	$-\frac{1}{2}$	$\sqrt{3}$
	$\frac{3\pi}{2}$	$270°$	-1	0	undefined
IV	$\frac{5\pi}{3}$	$300°$	$-\frac{\sqrt{3}}{2}$	$\frac{1}{2}$	$-\sqrt{3}$
IV	$\frac{7\pi}{4}$	$315°$	$-\frac{\sqrt{2}}{2}$	$\frac{\sqrt{2}}{2}$	-1
IV	$\frac{11\pi}{6}$	$330°$	$-\frac{1}{2}$	$\frac{\sqrt{3}}{2}$	$-\frac{\sqrt{3}}{3}$
	2π	$360°$	0	1	0

Note that

$$\cot\frac{\pi}{2} = \frac{\cos\frac{\pi}{2}}{\sin\frac{\pi}{2}} = \frac{0}{1} = 0$$

and that

$$\cot\frac{3\pi}{2} = \frac{\cos\frac{3\pi}{2}}{\sin\frac{3\pi}{2}} = \frac{0}{-1} = 0.$$

However, $\tan\theta$ is *undefined* at both $\frac{\pi}{2}$ and at $\frac{3\pi}{2}$. Thus, you *cannot* use the formula $\cot\theta = \frac{1}{\tan\theta}$ in these cases.

Table 8.13, which lists the sine, cosine, and tangent of various special angles, is useful for reference.

EXERCISES

In Exercises 1–8 indicate whether the point lies on the unit circle.

1. $(-1, 0)$ **2.** $(1, -1)$ **3.** $\left(\frac{1}{2}, -\frac{1}{2}\right)$ **4.** $(-\sqrt{2}, \sqrt{2})$ **5.** $\left(-\frac{\sqrt{2}}{2}, -\frac{\sqrt{2}}{2}\right)$

6. $\left(\frac{1}{2}, -\frac{\sqrt{3}}{2}\right)$ **7.** $\left(\frac{\sqrt{5}}{3}, \frac{2}{3}\right)$ **8.** $\left(\frac{\sqrt{6}}{3}, -\frac{\sqrt{3}}{3}\right)$

In Exercises 9–18 express in radian measure.

9. $45°$ **10.** $120°$ **11.** $144°$ **12.** $135°$ **13.** $270°$ **14.** $330°$ **15.** $900°$

16. $-90°$ **17.** $-150°$ **18.** $-240°$

In Exercises 19–28 express in degrees.

19. $\frac{\pi}{3}$ **20.** $\frac{\pi}{5}$ **21.** $\frac{\pi}{4}$ **22.** $\frac{5\pi}{6}$ **23.** $\frac{7\pi}{4}$ **24.** $\frac{3\pi}{2}$ **25.** $\frac{7\pi}{10}$

26. 3π **27.** -6π **28.** $-\frac{5\pi}{4}$

In Exercises 29–36 find a number θ, satisfying $0 \leq \theta < 2\pi$, that measures the same angle as the given number.

29. 3π **30.** $\frac{5\pi}{2}$ **31.** $\frac{13\pi}{6}$ **32.** 12π **33.** $-\frac{\pi}{4}$ **34.** $-\frac{5\pi}{6}$ **35.** $-\frac{7\pi}{4}$

36. $-\frac{11\pi}{6}$

In Exercises 37–48 determine the six trigonometric function-values at the indicated angle.

37. $\frac{2\pi}{3}$ **38.** $\frac{5\pi}{6}$ **39.** $\frac{3\pi}{2}$ **40.** $\frac{4\pi}{3}$ **41.** $-\frac{3\pi}{2}$ **42.** $\frac{5\pi}{3}$ **43.** $\frac{7\pi}{4}$

44. 2π **45.** $-\pi$ **46.** $-\frac{\pi}{4}$ **47.** $\frac{9\pi}{4}$ **48.** 5π

In Exercises 49–54, for each indicated angle, use Table 3 in the appendix to determine its (a) sine (b) cosine (c) tangent (d) cotangent.

49. $\frac{\pi}{12}$ **50.** $\frac{2\pi}{5}$ **51.** $\frac{5\pi}{8}$ **52.** $-\frac{\pi}{12}$ **53.** $-\frac{7\pi}{12}$ **54.** $-\frac{2\pi}{9}$

55. In three hours through how many radians does (a) the hour hand of a clock rotate? (b) the minute hand rotate?

56. Express in radians the acute angle made by the hands of a clock at 4:00. [See Figure 8.50 .]

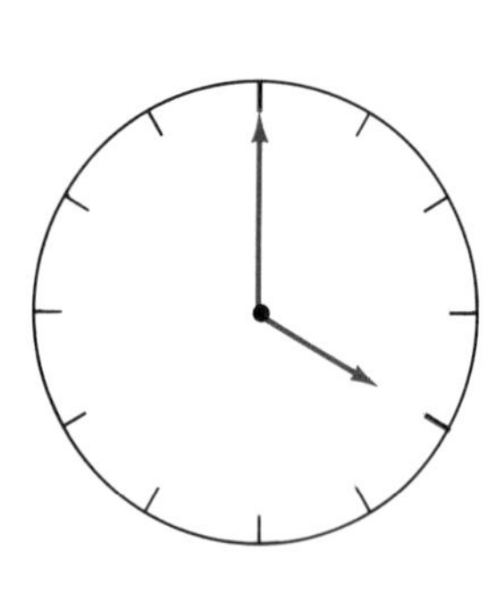

FIGURE 8.50

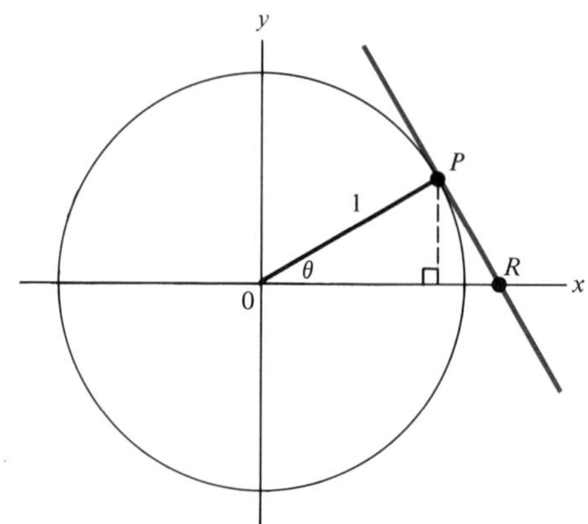

FIGURE 8.51

57. In Figure 8.51, suppose $\overline{PR}$ is tangent to the circle. Express tan θ in terms of length $\overline{PR}$.

8.4 Graphs of Trigonometric Functions

periodic functions

Some functions, such as the one pictured in Figure 8.52, repeat their values at regular intervals. Consider a function f whose domain is a set of real numbers. Then f is called **periodic** if there is a positive number p such that for *every* real number x, whenever x is in the domain of f, then so is $x + p$ and

$$f(x + p) = f(x).$$

The smallest such positive number p is then called the **period** of the function. The function f depicted in Figure 8.52 has period 1. All six trigonometric functions are periodic. To discover the period of both the cosine and sine, recall that each central angle determines a point on the unit circle. The central angle of the entire circle is 2π. Thus

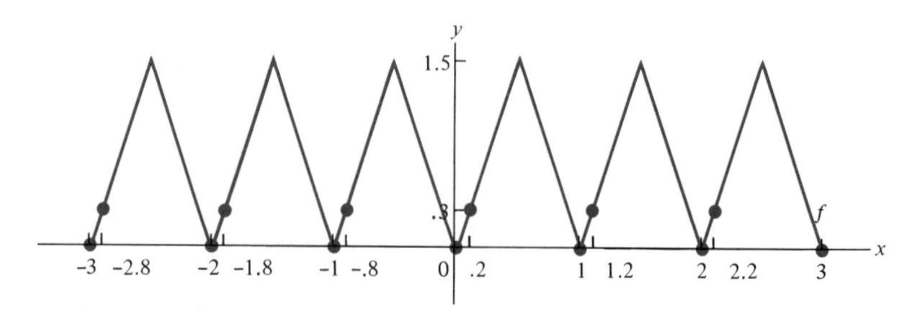

FIGURE 8.52 A periodic function, f.
$f(-3) = f(-2) = f(-1) = f(0) = f(1) = f(2) = f(3) = 0$
Also, $f(-2.8) = f(-1.8) = f(-.8) = f(.2) = f(1.2) = f(2.2) = .3$ The period is 1.

when the central angle is increased by 2π, the same point P on the unit circle is determined by the new and by the original central angles. [See Figure 8.53 .] Because the coordinates of P are $(\cos\theta, \sin\theta)$, it follows that

$$\cos(\theta + 2\pi) = \cos\theta, \qquad \sin(\theta + 2\pi) = \sin\theta$$

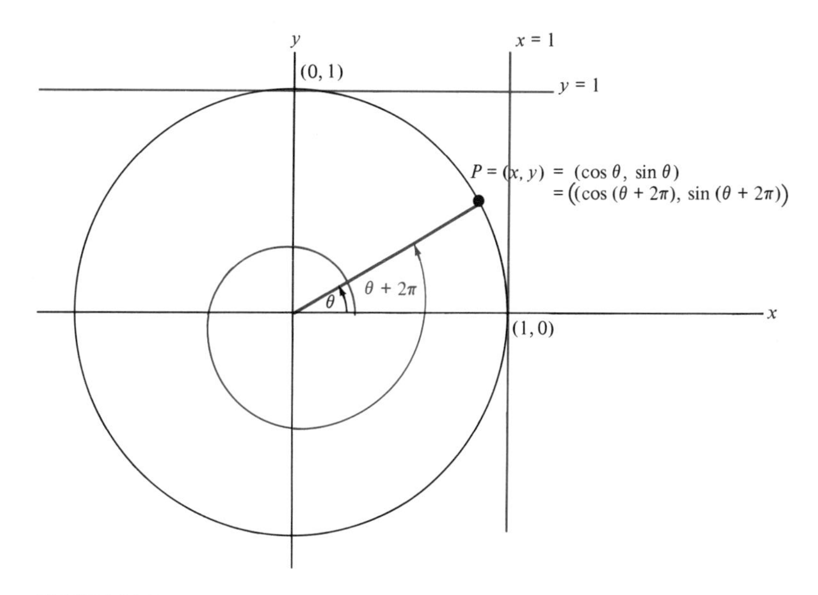

FIGURE 8.53

To see that 2π is the *smallest* positive number p for which $\sin(\theta + p) = \sin\theta$, observe that $\sin\theta = 1$ only when $y = 1$ and hence only when $P = (0, 1)$. This occurs when $\theta = \frac{\pi}{2}$ or when $\theta = \frac{\pi}{2}$ + (an integral multiple of 2π). [See Figure 8.53 .] It follows that the sine has period 2π. Similarly, $\cos\theta = 1$ only when $x = 1$ and thus only when $P = (1, 0)$. Thus $\cos 0 = \cos 2\pi = 1$, but $\cos\theta < 1$ for $0 < \theta < 2\pi$. Thus the cosine also has period 2π.

When graphing both the sine and cosine, once you have drawn the graph for an interval of 2π units, the rest of the graph repeats. It is therefore easy to extend the graph to other values of θ. You will first graph the sine function.

graph of sine

In Figures 8.54(a) and (b) on page 336, note that the points on the unit circle corresponding to θ and to $-\theta$ have y-coordinates that are (additive) inverses of each other. Because each y-coordinate is the sine of the angle in question,

$$\sin(-\theta) = -\sin\theta, \text{ for every } \theta$$

Thus *the sine is an odd function.* This means that once you have found the values of $\sin\theta$ for $0 \leqslant \theta \leqslant \pi$, observe that $-\theta$ satisfies $-\pi \leqslant -\theta \leqslant 0$, and use $\sin(-\theta) = -\sin\theta$. For example,

$$\sin\left(-\frac{\pi}{6}\right) = -\sin\frac{\pi}{6} = -\frac{1}{2}$$

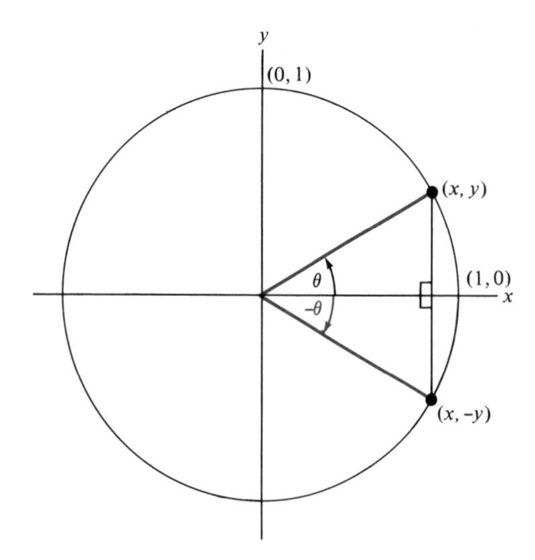

FIGURE 8.54(a)

FIGURE 8.54(b)

In fact, it turns out that once you have drawn the graph of sin θ for the interval

$$0 \leqslant \theta \leqslant \frac{\pi}{2}$$

it is easy to extend the graph to all other values of θ. In Figure 8.55, observe that

$$\sin \theta = \frac{y}{1} = \sin (\pi - \theta)$$

In fact, this relationship holds for *all* values of θ. [See Section 9.2 .] Thus if you know

$$\underbrace{\sin \frac{\pi}{5}}_{36^\circ},$$

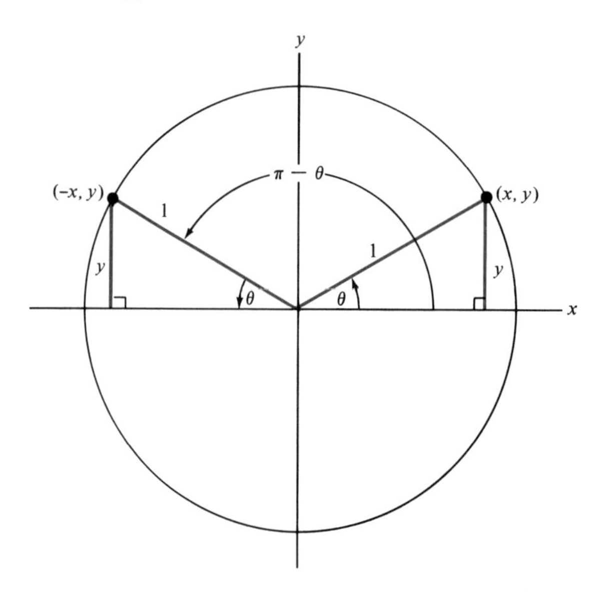

FIGURE 8.55 sin θ = sin ($\pi - \theta$)

for instance, you also know the sine of the supplement, that is $\sin\left(\pi - \frac{\pi}{5}\right)$ or $\sin\frac{4\pi}{5}$.
$\lfloor 180° - 36° \rfloor$ $\lfloor 144° \rfloor$

Table 8.14 indicates values of the sine function at intervals of $\frac{\pi}{12}$ (15°). The values at $\frac{\pi}{12}$ and $\frac{5\pi}{12}$ are obtained from Table 3, the trig table, at the back of the book; the other values you already know by considering the unit circle or the appropriate right triangle.

TABLE 8.14

radians	degrees	sin	degrees	radians
0	0°	0 = .0000	180°	π
$\frac{\pi}{12}$	15°	.2588	165°	$\frac{11\pi}{12}$
$\frac{\pi}{6}$	30°	$\frac{1}{2} = .5000$	150°	$\frac{5\pi}{6}$
$\frac{\pi}{4}$	45°	$\frac{\sqrt{2}}{2} \approx .7071$	135°	$\frac{3\pi}{4}$
$\frac{\pi}{3}$	60°	$\frac{\sqrt{3}}{2} \approx .8660$	120°	$\frac{2\pi}{3}$
$\frac{5\pi}{12}$	75°	.9659	105°	$\frac{7\pi}{12}$
$\frac{\pi}{2}$	90°	1 = 1.0000	90°	$\frac{\pi}{2}$

In Figure 8.56(a), the preceding values of the sine are plotted. Note that the horizontal axis is now called the θ-axis. A glance at Table 8.14 suggests that *the sine is an increasing function for* $0 \leqslant \theta \leqslant \frac{\pi}{2}$ *and is a decreasing function for* $\frac{\pi}{2} \leqslant \theta \leqslant \pi$. [See also Table 3.] In Figure 8.56(b), a smooth curve is drawn for the interval $0 \leqslant \theta \leqslant \pi$. In Figure 8.56(c) on page 338, the relation

$$\sin(-\theta) = -\sin\theta$$

is used to extend the graph to the interval $-\pi \leqslant \theta \leqslant \pi$. In Figure 8.56(d), the periodicity of the sine is used to indicate how the graph extends over the entire real line.

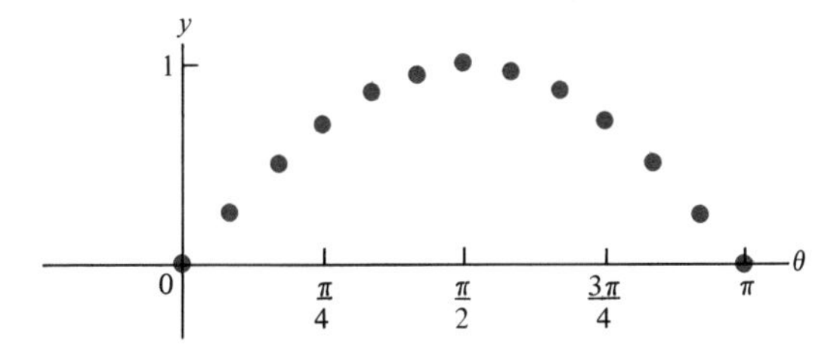

FIGURE 8.56(a)

FIGURE 8.56(b)

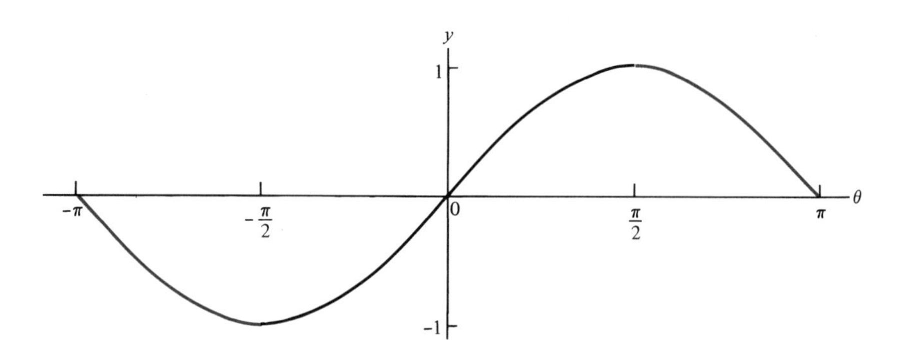

FIGURE 8.56(c)

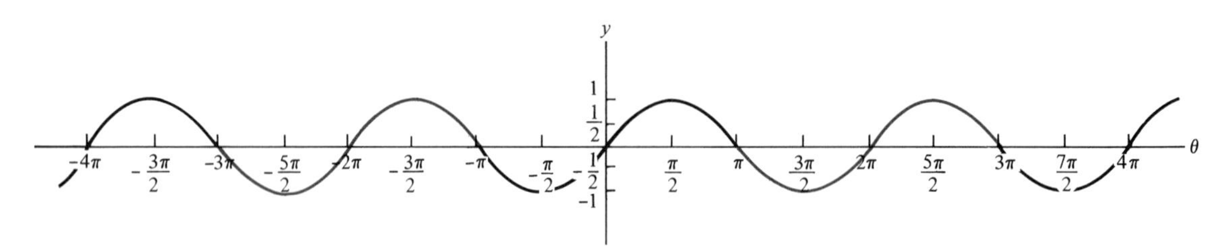

FIGURE 8.56(d)

From Figure 8.56(d), note the following properties of the sine function:

1. **The domain is the set of all real numbers.**
2. **The range is the closed interval $[-1, 1]$.**
3. **The y-intercept is 0.**
4. **The x-intercepts are $0, \pm\pi, \pm 2\pi, \pm 3\pi, \ldots$**
5. **$\sin(\theta + \pi) = -\sin\theta$, for all θ** (This will be *proved* in Section 9.2 .)

   For example,

$$\sin\left(\frac{\pi}{2} + \pi\right) = \sin\frac{3\pi}{2} = -1 = -\sin\frac{\pi}{2}$$

   You already know Properties **6–8**:

6. **$\sin(-\theta) = -\sin\theta$, for all θ.** **(The sine is an *odd* function.)**
7. **$\sin(\pi - \theta) = \sin\theta$. (This holds for *all* θ, *not* just for $0 \leqslant \theta \leqslant \pi$.)** [See Section 9.2 .]
8. **The sine function is periodic with period 2π.**

graph of cosine

In order to graph the cosine function, observe the following pattern:

$$\cos 0° = 1 = \sin 90° = \cos(0 + \mathbf{90})°$$

$$\cos 30° = \frac{\sqrt{3}}{2} = \sin 60° = \sin 120° = \sin(30 + \mathbf{90})°$$

$$\cos 45^\circ = \frac{\sqrt{2}}{2} = \sin 45^\circ = \sin 135^\circ = \sin (45 + 90)^\circ$$

$$\cos 60^\circ = \frac{1}{2} = \sin 30^\circ = \sin 150^\circ = \sin (60 + 90)^\circ$$

$$\cos 90^\circ = 0 = \sin 180^\circ = \sin (90 + 90)^\circ$$

These values of the cosine and the corresponding values of the sine are plotted in Figure 8.57 . In each case,

$$\cos A^\circ = \sin (A + 90)^\circ,$$

or in radian measure,

$$\cos \theta = \sin \left(\theta + \frac{\pi}{2}\right).$$

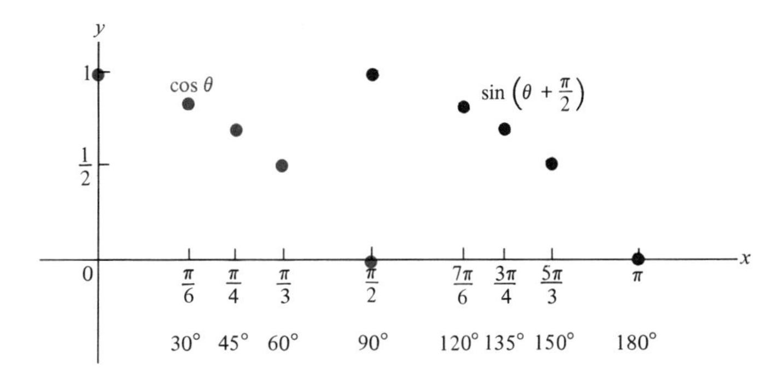

FIGURE 8.57 Each of these values of the cosine can be obtained by shifting the sine graph $\frac{\pi}{2}$ units to the left.

To see that this relationship holds for *all* θ satisfying $0 \leq \theta \leq \frac{\pi}{2}$, consider Figure 8.58 on page 340. Let $\angle\theta = \angle QOP$ and $\angle\alpha = \angle QPO$. In right triangle OPQ, $\angle\alpha$ is the *complement* of $\angle\theta$. Next observe that

$$\angle\theta + \angle POR + \angle ROS = 180^\circ.$$

Because $\angle POR$ is a right angle, it follows that $\angle ROS$ is the *complement* of $\angle\theta$. Thus $\angle ROS = \angle\alpha$.

In right triangle ORS, $\angle ORS$ is the complement of $\angle\alpha$; therefore, $\angle ORS = \angle\theta$. Because $\angle\theta$ and $\angle\alpha$ are complements,

$$\cos \theta = \sin \alpha$$
$$= \sin (\pi - \alpha) \qquad \text{by Property 7 of the sine function.}$$
$$= \sin \left(\theta + \frac{\pi}{2}\right) \qquad \triangle$$

In fact, this relationship,

$$\cos \theta = \sin \left(\theta + \frac{\pi}{2}\right)$$

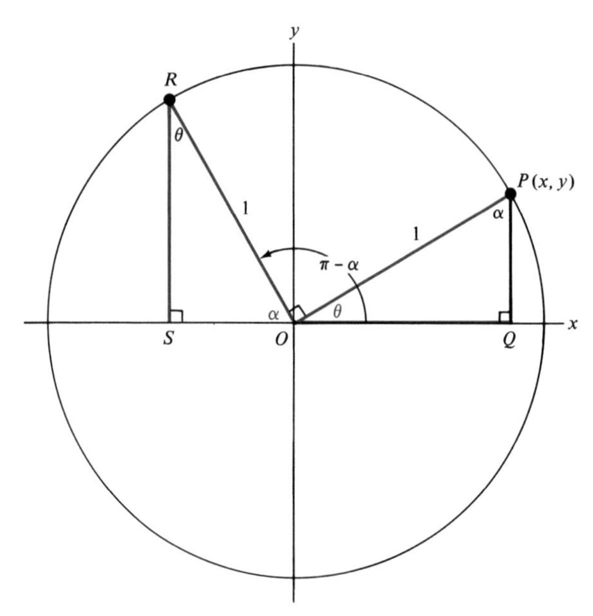

FIGURE 8.58 $\cos\theta = \sin\alpha = \sin(\pi - \alpha) = \sin\left(\theta + \frac{\pi}{2}\right)$

holds for *all* θ; this will be proved in Section 9.2. For example,

$$\cos\frac{2\pi}{3} = \sin\underbrace{\left(\frac{2\pi}{3} + \frac{\pi}{2}\right)}_{\frac{7\pi}{6}} = \frac{-1}{2}$$

$$\cos\frac{3\pi}{4} = \sin\underbrace{\left(\frac{3\pi}{4} + \frac{\pi}{2}\right)}_{\frac{5\pi}{4}} = -\frac{\sqrt{2}}{2}$$

$$\cos\pi = \sin\underbrace{\left(\pi + \frac{\pi}{2}\right)}_{\frac{3\pi}{2}} = -1$$

$$\cos\frac{3\pi}{2} = \sin\underbrace{\left(\frac{3\pi}{2} + \frac{\pi}{2}\right)}_{2\pi} = 0$$

This relationship between the cosine and sine enables you to draw the cosine graph by "shifting" the sine graph $\frac{\pi}{2}$ units to the *left*, as indicated in Figure 8.59(a).

Here is a summary of important properties of the cosine function that can be seen from its graph [Figure 8.59(b)].

1. **The domain is the set of all real numbers.**
2. **The range is [−1, 1].**

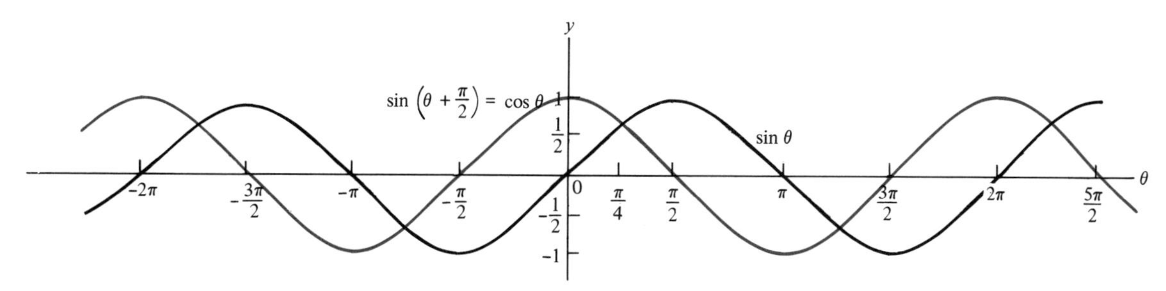

FIGURE 8.59(a) $\cos\theta = \sin\left(\theta + \frac{\pi}{2}\right)$

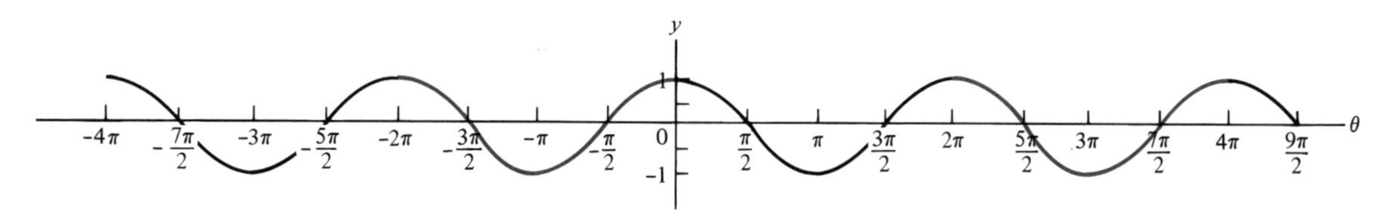

FIGURE 8.59(b) $\cos\theta$. Note that $\cos\theta$ has period 2π.

3. **The y-intercept is 1.**

4. **The x-intercepts are $\pm\frac{\pi}{2}, \pm\frac{3\pi}{2}, \pm\frac{5\pi}{2}, \pm\frac{7\pi}{2}, \ldots$**

5. **$\cos(\pi + \theta) = \cos(\pi - \theta) = -\cos\theta$ for all θ.** (This will be *proved* in Section 9.2.) For example,

$$\cos(0 + \pi) = \cos\pi = -1 = -\cos 0$$

6. **$\cos(-\theta) = \cos\theta$ for all θ. (The cosine is an *even* function.)**

7. **The cosine function is periodic with period 2π.**

graph of tangent

In order to draw the graph of the tangent function, recall that

$$\tan\theta = \frac{\sin\theta}{\cos\theta}, \qquad \cos\theta \neq 0.$$

But $\cos\theta = 0$ for $\theta = \pm\frac{\pi}{2}, \pm\frac{3\pi}{2}, \pm\frac{5\pi}{2}, \ldots$ Thus, *the tangent is not defined for these values of θ.* Also,

$$\tan\theta = 0 \text{ when } \sin\theta = 0, \text{ that is, when } \theta = 0, \pm\pi, \pm2\pi, \ldots \tag{8.1}$$

Because the sine is an odd function and the cosine an even function, the tangent is an odd function. $\left[\frac{\text{odd function}}{\text{even function}} = \text{odd function}\right]$

TABLE 8.15

radians	degrees	tan
0	$0°$	$0 = 0.0000$
$\frac{\pi}{12}$	$15°$	$.2679$
$\frac{\pi}{6}$	$30°$	$\frac{\sqrt{3}}{3} \approx .5774$
$\frac{\pi}{4}$	$45°$	$1 = 1.000$
$\frac{\pi}{3}$	$60°$	$\sqrt{3} \approx 1.732$
$\frac{5\pi}{12}$	$75°$	3.732
	$85°$	11.43
	$88°$	28.64
	$89°$	57.29
	$89°50'$	343.77
	$89°59'$	3437.75
$\frac{\pi}{2}$	$90°$	undefined

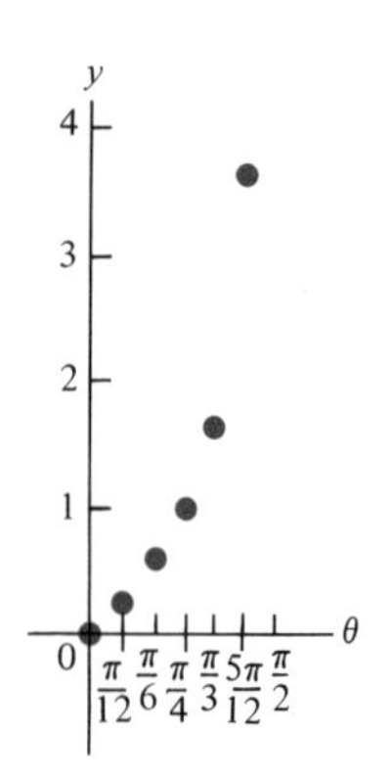

FIGURE 8.60(a)

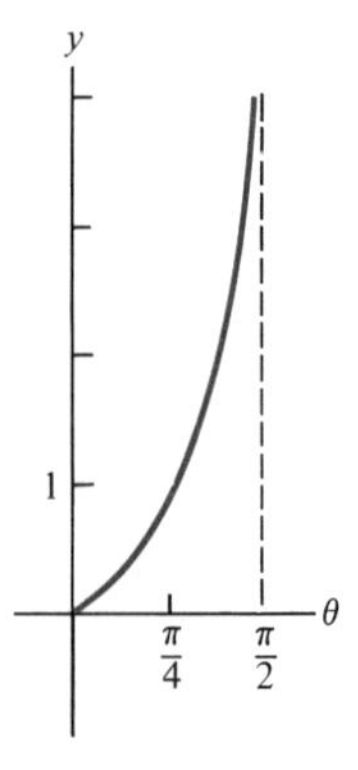

FIGURE 8.60(b)

In fact,

$$\tan(-\theta) = \frac{\sin(-\theta)}{\cos(-\theta)} = \frac{-\sin\theta}{\cos\theta} = -\tan\theta \tag{8.2}$$

for all θ in the domain of the tangent.

For all θ in the domain of the tangent,

$$\tan(\theta + \pi) = \frac{\sin(\theta + \pi)}{\cos(\theta + \pi)} = \frac{-\sin\theta}{-\cos\theta} = \tan\theta$$

And π is the *smallest* positive number p for which $\tan(\theta + p) = \tan\theta$ for all θ because $\tan 0 = \tan\pi = 0$, whereas, by Statement (**8.1**), $\tan\theta \neq 0$ for $0 < \theta < \pi$. Therefore, the tangent function has period π, and thus its graph repeats every π units. In fact, if you draw the graph of $\tan\theta$ for the interval $0 \leqslant \theta < \frac{\pi}{2}$, it is easy to extend the graph to all other values for which $\tan\theta$ is defined. For, you can use the fact that the tangent is an odd function to extend the graph to $-\frac{\pi}{2} < \theta < \frac{\pi}{2}$, and you can use the periodicity to extend the graph to its entire domain. Values of the tangent for $0 \leqslant \theta < \frac{\pi}{2}$ are given in Table 8.15 .

The tangent is undefined at $\frac{\pi}{2}$ (or 90°). In terms of degree measure, as θ *approaches* 90°, through values such as 89°, 89°50′, 89°59′, . . . , the tangent gets larger and larger "beyond all bounds". (We could say that the tangent "approaches infinity" as θ approaches $\frac{\pi}{2}$ through arguments *less than* $\frac{\pi}{2}$.) In Figure 8.60(a) some of the preceding values of the tangent are plotted. In Figure 8.60(b) a smooth curve is drawn for the interval $0 \leqslant \theta < \frac{\pi}{2}$. In Figure 8.60(c) relation (**8.2**),

$$\tan(-\theta) = -\tan\theta,$$

is used to extend the graph to the interval $-\frac{\pi}{2} < \theta < \frac{\pi}{2}$. In Figure 8.60(d), the periodicity of the tangent is used to illustrate how the graph extends to the entire domain of the tangent.

Here is a summary of some important properties of the tangent function.

1. **The domain is the set of real numbers *excluding***

   $$\pm\frac{\pi}{2}, \pm\frac{3\pi}{2}, \pm\frac{5\pi}{2}, \pm\frac{7\pi}{2}, \ldots$$

2. **The range is the set of *all* real numbers.**
3. **The y-intercept is 0.**
4. **The x-intercepts are $0, \pm\pi, \pm 2\pi, \pm 3\pi, \ldots$**
5. **$\tan(\theta + \pi) = \tan\theta$ for all θ in the domain of the tangent, that is, the tangent is periodic with period π.**
6. **$\tan(-\theta) = -\tan\theta$ for all θ in the domain of the tangent. (The tangent is an *odd* function.)**

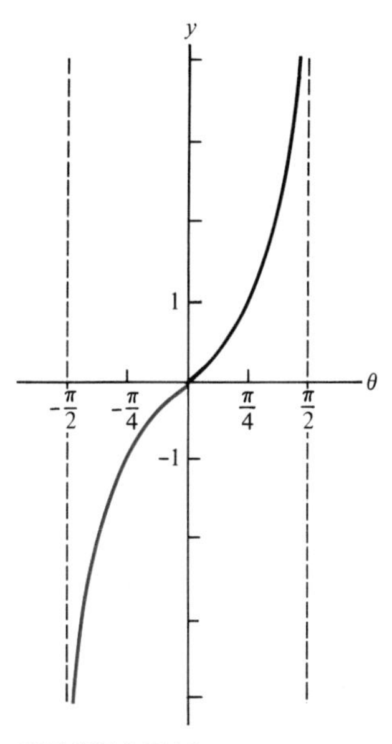

FIGURE 8.60(c)

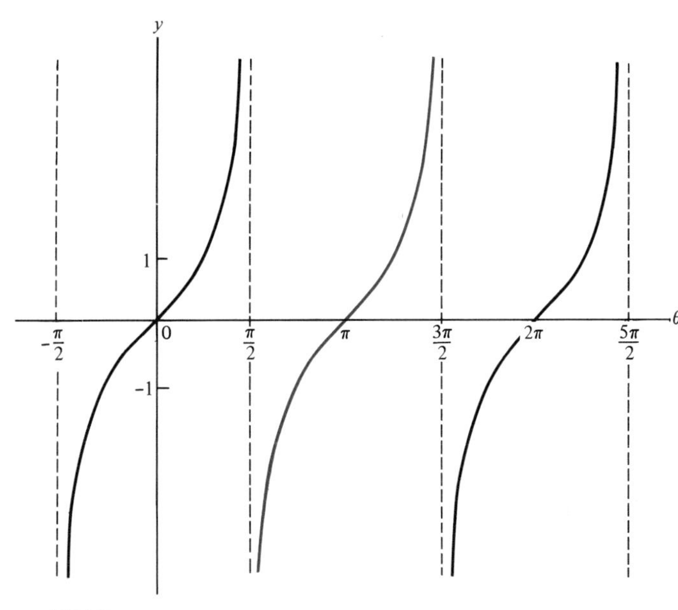

FIGURE 8.60(d)

7. tan θ "approaches infinity" as θ approaches $\frac{\pi}{2}$ through arguments less than $\frac{\pi}{2}$.

graphs of secant, cosecant, cotangent

The graphs of the remaining functions are drawn in Figure 8.61 on pages 344 and 345. Recall that

$$\sec \theta = \frac{1}{\cos \theta} \quad \text{for} \quad \cos \theta \neq 0$$

$$\csc \theta = \frac{1}{\sin \theta} \quad \text{for} \quad \sin \theta \neq 0$$

$$\cot \theta = \begin{cases} \dfrac{1}{\tan \theta} & \text{for } \theta \text{ in the domain of the tangent and } \tan \theta \neq 0 \\ 0 & \text{for} \quad \theta = \pm\frac{\pi}{2}, \pm\frac{3\pi}{2}, \pm\frac{5\pi}{2}, \ldots \\ & \text{(where } \tan \theta \text{ is undefined)} \end{cases}$$

Thus secant θ is not defined at

$$\pm\frac{\pi}{2}, \pm\frac{3\pi}{2}, \pm\frac{5\pi}{2}, \pm\frac{7\pi}{2}, \ldots,$$

where the cosine equals 0. And neither cosecant θ nor cotangent θ is defined at

$$0, \pm\pi, \pm 2\pi, \pm 3\pi, \ldots,$$

where both the sine and tangent equal 0.

The secant is an *even* function $\left(\frac{\text{even function}}{\text{even function}}\right)$, whereas the cosecant and cotangent are *odd* functions. $\left(\frac{\text{even function}}{\text{odd function}}\right)$

Both the secant and cosecant have period 2π, whereas the cotangent has period π.

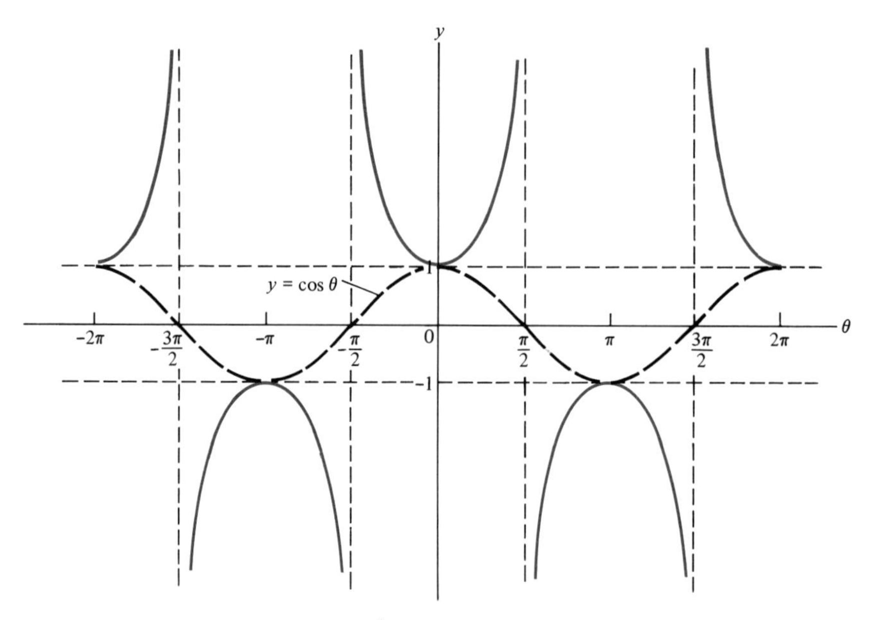

FIGURE 8.61(a) sec θ (in color) $= \frac{1}{\cos\theta}$, for $\cos\theta \neq 0$

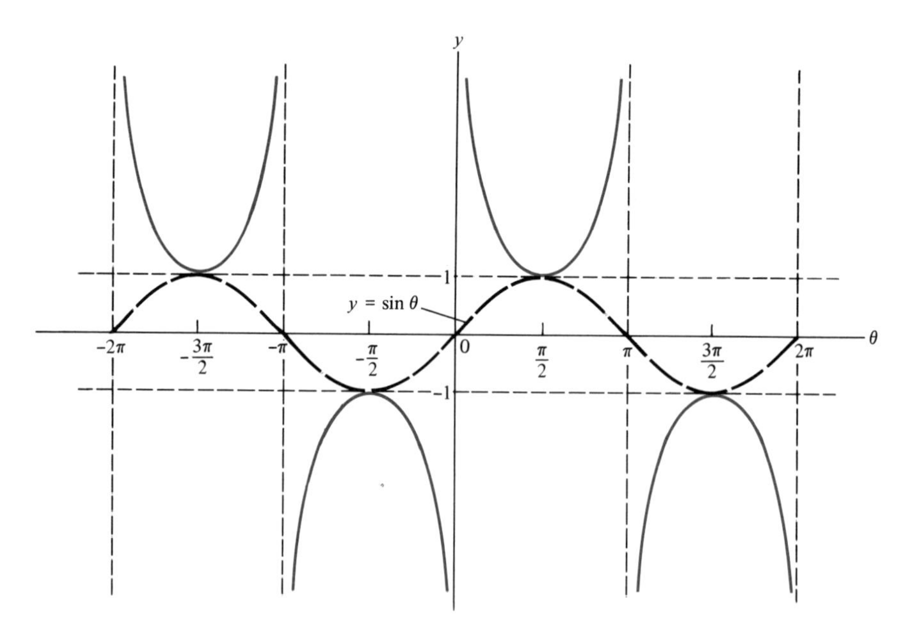

FIGURE 8.61(b) csc θ (in color) $= \frac{1}{\sin\theta}$, for $\sin\theta \neq 0$

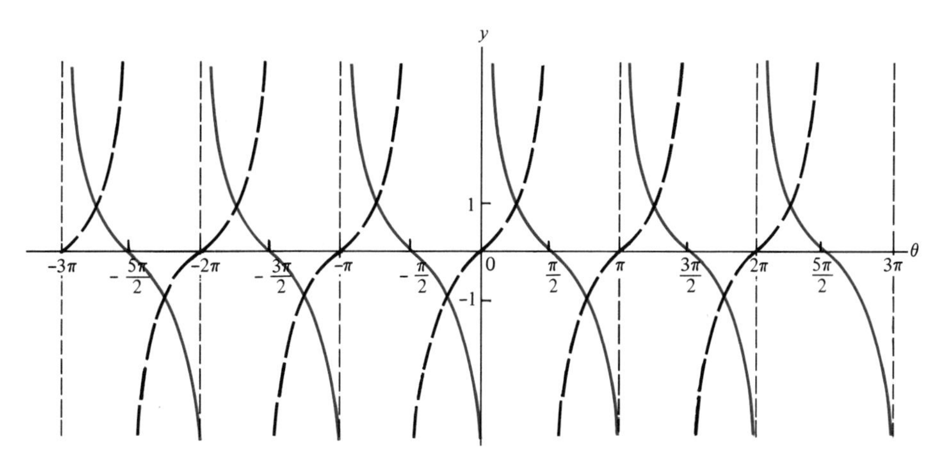

FIGURE 8.61(c) $\cot\theta$ (in color) $= \begin{cases} \dfrac{1}{\tan\theta}, & \text{for } \tan\theta \neq 0 \\ 0, & \text{for } \theta = \pm\dfrac{\pi}{2}, \pm\dfrac{3\pi}{2}, \pm\dfrac{5\pi}{2}, \ldots \end{cases}$

EXERCISES

In Exercises 1–10 graph each function for the indicated interval.

1. $\sin\theta,\ 0 \leqslant \theta \leqslant 2\pi$
2. $\cos\theta,\ 0 \leqslant \theta \leqslant 2\pi$
3. $\tan\theta,\ -\frac{\pi}{2} < \theta < \frac{\pi}{2}$
4. $\sin\theta,\ -\frac{\pi}{2} \leqslant \theta \leqslant \frac{3\pi}{2}$

5. $\cos\theta,\ \pi \leqslant \theta \leqslant 3\pi$
6. $\tan\theta,\ \frac{\pi}{2} < \theta < \frac{3\pi}{2}$
7. $\cot\theta,\ 0 < \theta < \pi$
8. $\sec\theta,\ -\frac{\pi}{2} < \theta < \frac{\pi}{2}$

9. $\sec\theta,\ \frac{\pi}{2} < \theta < \frac{3\pi}{2}$
10. $\csc\theta,\ 0 < \theta < \pi$

11. Draw the graphs of $\sin\theta$ and $\cos\theta$, $-2\pi \leqslant \theta \leqslant 2\pi$, on the same coordinate system.

12. Draw the graphs of $\cos\theta$ and $\sec\theta$, $-\frac{3\pi}{2} \leqslant \theta \leqslant \frac{3\pi}{2}$, on the same coordinate system.

13. Draw the graphs of $\sin\theta$ and $\csc\theta$, $-2\pi \leqslant \theta \leqslant \pi$, on the same coordinate system.

14. Draw the graphs of $\tan\theta$ and $\cot\theta$, $-\frac{3\pi}{2} < \theta < \frac{3\pi}{2}$, on the same coordinate system.

In Exercises 15–32 which trigonometric functions (sine, cosine, tangent, secant, cosecant, cotangent) have the following properties?

15. The domain is the set of all real numbers.

16. The domain is the set of real numbers excluding $0, \pm\pi, \pm 2\pi, \pm 3\pi, \ldots$

17. The domain is the set of real numbers excluding $\pm\frac{\pi}{2}, \pm\frac{3\pi}{2}, \pm\frac{5\pi}{2}, \ldots$

18. The range is the set of all real numbers.

19. The range is $[-1, 1]$.

20. The range is the set of real numbers excluding the open interval $(-1, 1)$.

21. The y-intercept is 0.

22. The y-intercept is 1.

23. There is no y-intercept.

24. The x-intercepts are $0, \pm\pi, \pm2\pi, \pm3\pi, \ldots$

25. The x-intercepts are $\pm\frac{\pi}{2}, \pm\frac{3\pi}{2}, \pm\frac{5\pi}{2}, \ldots$

26. There is no x-intercept.

27. The function has period π.

28. The function has period 2π.

29. 0 is in the range of the function.

30. $\frac{1}{2}$ is in the range of the function.

31. 1 is in the range of the function.

32. 2 is in the range of the function.

In Exercises 33–38 which trigonometric functions f have the following properties? Here, θ varies over *all* numbers in the domain of f.

33. $f(\theta + \pi) = f(\theta)$ **34.** $f(\theta + \pi) = -f(\theta)$ **35.** $f(\pi - \theta) = f(\theta)$ **36.** $f(-\theta) = f(\theta)$

37. $f(-\theta) = -f(\theta)$ **38.** $-1 \leq f(\theta) \leq 1$

In Exercises 39–42 which pairs of trigonometric functions, f and g, have the following relationships? Here θ varies over *all* numbers that are in the domain of *both* f and g.

39. $f(\theta) = \dfrac{1}{g(\theta)}$ **40.** $f\left(\dfrac{\pi}{2} - \theta\right) = g(\theta)$ **41.** $f(\theta) = g\left(\theta + \dfrac{\pi}{2}\right)$ **42.** $(f(\theta))^2 + (g(\theta))^2 = 1$

8.5 Graphs of Composite Trigonometric Functions

The graphs of the sine and cosine functions can be easily modified to obtain the graphs of such functions as

$$\sin 2\theta, \quad 2 \sin \theta, \quad \text{and} \quad -3 \cos \frac{\theta}{2}.$$

Waves and vibrations are often periodic, and are described by such composite functions.

EXAMPLE 1 Draw the graphs of (a) $\sin 2\theta$ (b) $2 \sin \theta$.

Solution.

(a) First note that $\sin \theta$ *increases* from 0 to 1 as θ increases from 0 to $\frac{\pi}{2}$. And $\sin \theta$ *decreases* from 1 to 0 as θ increases from $\frac{\pi}{2}$ to π. [See Figure 8.56(b) on page 337.] To obtain the corresponding increase and decrease in $\sin 2\theta$, note that

$$\text{if} \quad \theta = \frac{\pi}{4}, \quad \text{then} \quad 2\theta = 2\left(\frac{\pi}{4}\right) = \frac{\pi}{2},$$

$$\text{and if} \quad \theta = \frac{\pi}{2}, \quad \text{then} \quad 2\theta = 2\left(\frac{\pi}{2}\right) = \pi.$$

Thus, $\sin 2\theta$ *increases* from 0 to 1 as θ increases from 0 to $\frac{\pi}{4}$. And $\sin 2\theta$ *decreases* from 1 to 0 as θ increases from $\frac{\pi}{4}$ to $\frac{\pi}{2}$. In general, $\sin 2\theta$ runs through its values *twice* as fast as does $\sin \theta$. The period of $\sin 2\theta$ is $\frac{2\pi}{2}$, or π, which is *half* the period of $\sin \theta$. [See Table 8.16 and Figure 8.62(a).]

TABLE 8.16

θ	2θ	$\sin 2\theta$	2θ	θ
0	0	0	π	$\frac{\pi}{2}$
$\frac{\pi}{12}$	$\frac{\pi}{6}$	$\frac{1}{2}$	$\frac{5\pi}{6}$	$\frac{5\pi}{12}$
$\frac{\pi}{8}$	$\frac{\pi}{4}$	$\frac{\sqrt{2}}{2} \approx .71$	$\frac{3\pi}{4}$	$\frac{3\pi}{8}$
$\frac{\pi}{6}$	$\frac{\pi}{3}$	$\frac{\sqrt{3}}{2} \approx .87$	$\frac{2\pi}{3}$	$\frac{\pi}{3}$
$\frac{\pi}{4}$	$\frac{\pi}{2}$	1		

(b) To obtain the graph of $2 \sin \theta$ from that of $\sin \theta$, for each θ, simply double the y-value, as illustrated in Figure 8.62(b). □

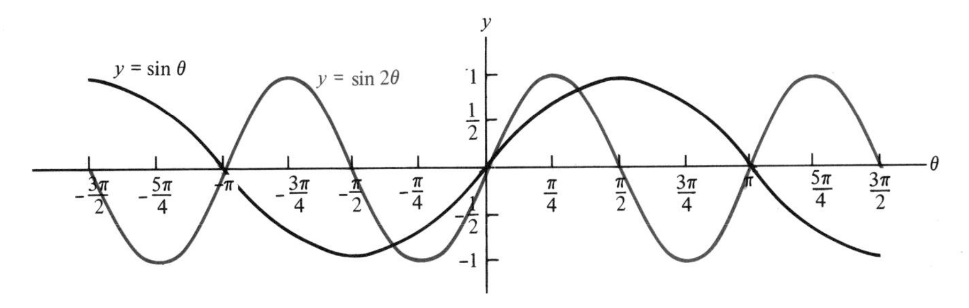

FIGURE 8.62(a)

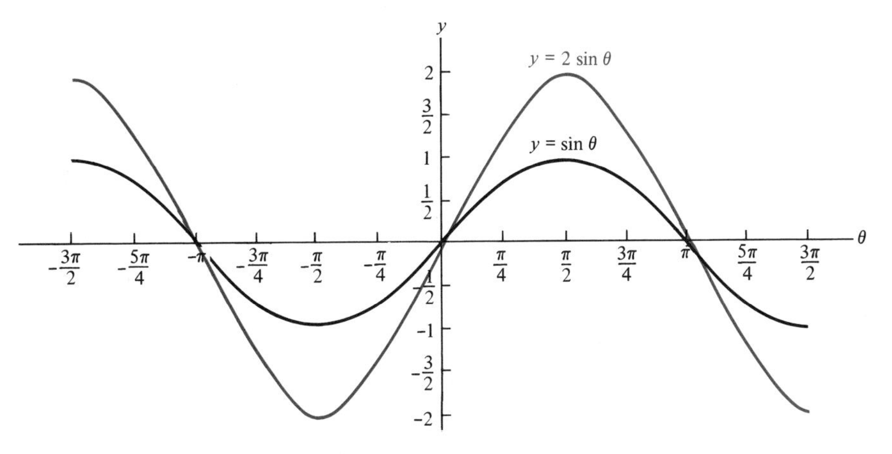

FIGURE 8.62(b)

amplitude and period

The **amplitude** of a function of the form $a \sin b\theta$ or $a \cos b\theta$, $a \neq 0$, $b \neq 0$, is the maximum y-value of its graph.† Thus the amplitude of $\sin \theta$ is 1, whereas the amplitude of $2 \sin \theta$ is 2. See Table 8.17. *The amplitude is always positive.*

TABLE 8.17

$f(\theta)$	period	amplitude
$\sin \theta$	2π	1
$\sin 2\theta$	π	1
$2 \sin \theta$	2π	2

EXAMPLE 2 Draw the graph of $-3 \cos \frac{\theta}{2}$.

Solution. In Figure 8.63, first consider the (solid black) graph of $\cos \frac{\theta}{2}$, which runs through its values half as fast as does $\cos \theta$ (dotted). The period of $\cos \frac{\theta}{2}$ is therefore $\frac{2\pi}{\frac{1}{2}}$, or 4π. For each value of θ, multiply the y-value of $\cos \frac{\theta}{2}$ by -3 to obtain the y-value of $-3 \cos \frac{\theta}{2}$. The graph of $-3 \cos \frac{\theta}{2}$ is in color. Note that the amplitude

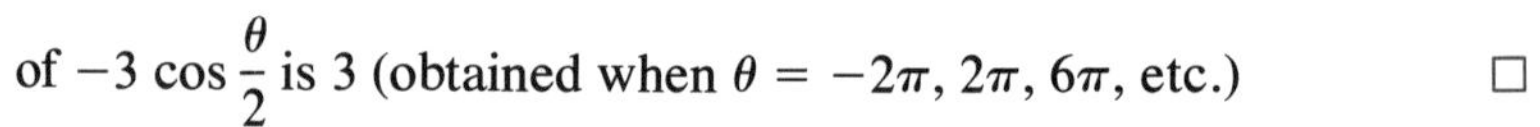
of $-3 \cos \frac{\theta}{2}$ is 3 (obtained when $\theta = -2\pi, 2\pi, 6\pi$, etc.) □

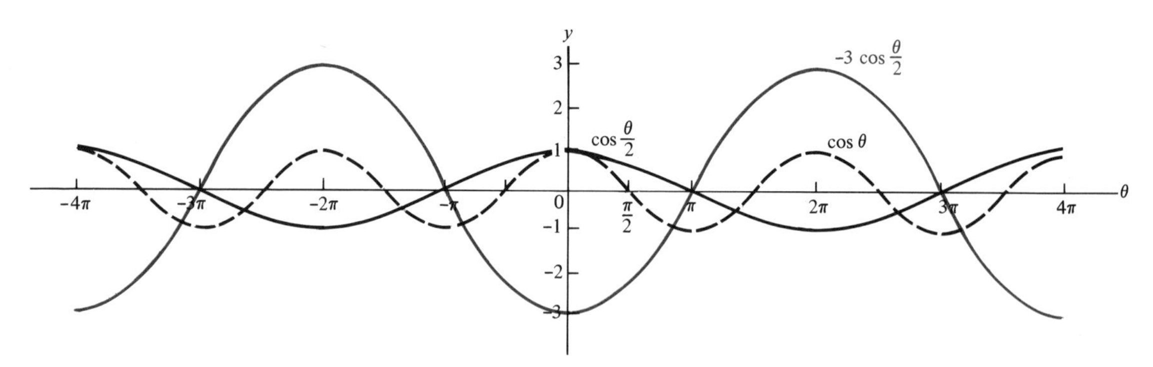

FIGURE 8.63

In general, for a function of the form

$$a \sin b\theta \qquad \text{or} \qquad a \cos b\theta,$$

where $a \neq 0$ and $b > 0$, the period is $\frac{2\pi}{b}$ and the amplitude is $|a|$.

†More generally, the amplitude can be defined as

$$\frac{1}{2}(\text{maximum } y\text{-value} - \text{minimum } y\text{-value}),$$

so that $1 + \sin \theta$ has amplitude $\frac{1}{2}(2 - 0)$, or 1.

EXAMPLE 3 TABLE 8.18

function	period	amplitude
$-\sin 4\theta$	$\frac{2\pi}{4}$ or $\frac{\pi}{2}$	$\|-1\|$ or 1
$\frac{1}{2}\cos 3\theta$	$\frac{2\pi}{3}$	$\frac{1}{2}$
$-5\sin\frac{\theta}{3}$	$\frac{2\pi}{\frac{1}{3}}$ or 6π	$\|-5\|$ or 5
$-\frac{1}{4}\cos \pi\theta$	$\frac{2\pi}{\pi}$ or 2	$\left\|-\frac{1}{4}\right\|$ or $\frac{1}{4}$

□

phase shift

EXAMPLE 4 Draw the graph of $\tan\left(\theta - \frac{\pi}{4}\right)$.

Solution. To obtain the graph of $\tan\left(\theta - \frac{\pi}{4}\right)$, let

$f(\theta) = \tan\left(\theta - \frac{\pi}{4}\right)$. Note that

$$\theta - \frac{\pi}{4} = 0 \quad \text{when} \quad \theta = \frac{\pi}{4}, \quad \text{so that} \quad f\left(\frac{\pi}{4}\right) = 0$$

And

$$\theta - \frac{\pi}{4} = \frac{\pi}{2} \quad \text{when} \quad \theta = \frac{\pi}{2} + \frac{\pi}{4}\left(\text{or } \frac{3\pi}{4}\right),$$

so that $f\left(\frac{3\pi}{4}\right)$ is undefined.

Thus, to obtain the graph of $\tan\left(\theta - \frac{\pi}{4}\right)$, shift the graph of $\tan\theta$ $\frac{\pi}{4}$ units to the *right*, as illustrated in Figure 8.64 on page 350. □

In Example 4, $\frac{\pi}{4}$ is called the *phase shift*. For a function of the form

$$\sin(\theta - k),$$

the **phase shift** is k. Here, k can be any number—positive, negative, or 0. The graph is obtained by shifting $\sin\theta$ k units to the *right* if k is *positive* and $|k|$ units to the *left* if k is *negative*. Similar remarks apply to

$$\cos(\theta - k) \quad \text{and to} \quad \tan(\theta - k).$$

Suppose $f(\theta) = \cos\left(\theta + \frac{\pi}{3}\right)$. Write $f(\theta)$ in the form $\cos\left(\theta - \left(-\frac{\pi}{3}\right)\right)$ to see that the phase shift is $-\frac{\pi}{3}$. The graph is obtained by shifting $\cos\theta$ $\frac{\pi}{3}$ units to the *left*.

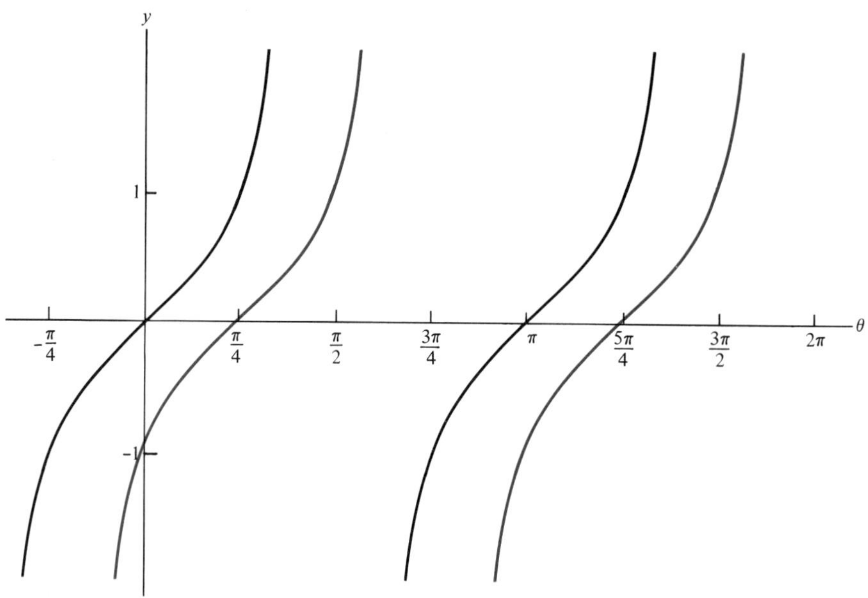

FIGURE 8.64 $\tan \theta$ is in black. $\tan\left(\theta - \frac{\pi}{4}\right)$ is in color.

The phase shift, period, and amplitude can also be defined for more complicated functions.

EXAMPLE 5 Draw graphs of:

(a) $2 \sin\left(\theta - \frac{\pi}{2}\right)$ (b) $3 \sin\left(2\theta - \frac{\pi}{2}\right)$

Solution.

(a) In Figure 8.65(a), first shift the (dotted) graph of $\sin \theta$ $\frac{\pi}{2}$ units to the *right* to obtain the (solid black) graph of $\sin\left(\theta - \frac{\pi}{2}\right)$. Then double each y-value of $\sin\left(\theta - \frac{\pi}{2}\right)$ to obtain the y-value of $2 \sin\left(\theta - \frac{\pi}{2}\right)$. The graph of $2 \sin\left(\theta - \frac{\pi}{2}\right)$ is in color.

(b) To find the phase shift here, observe that

$$2\theta - \frac{\pi}{2} = 2\left(\theta - \frac{\pi}{4}\right).$$

In Figure 8.65(b), first shift the (dotted black) graph of $\sin \theta$ $\frac{\pi}{4}$ units *right* to obtain the (solid black) graph of $\sin\left(\theta - \frac{\pi}{4}\right)$. Next, the period of $\sin\left(2\theta - \frac{\pi}{2}\right)$, or $\sin 2\left(\theta - \frac{\pi}{4}\right)$ is $\frac{2\pi}{2}$, or π. The graph

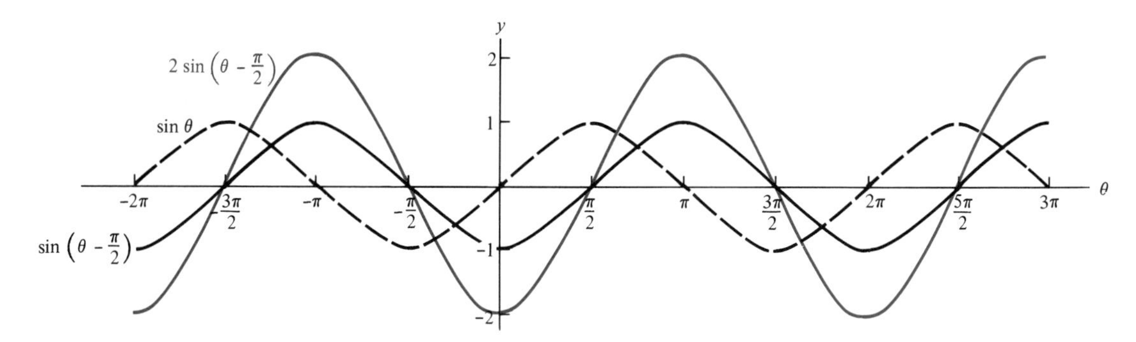

FIGURE 8.65(a)

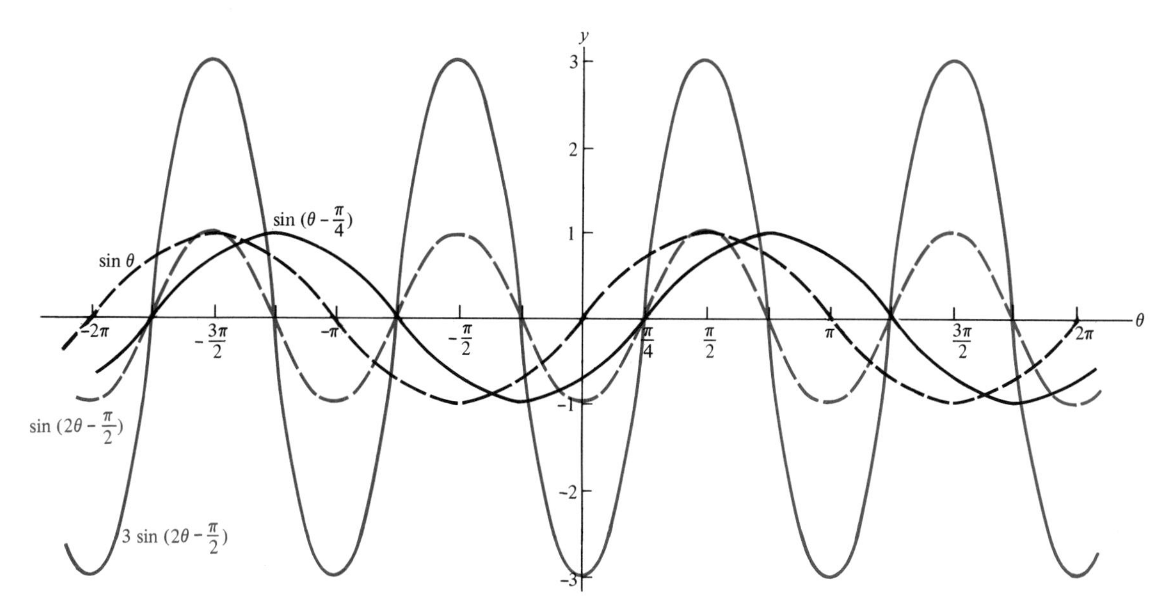

FIGURE 8.65(b)

of $\sin\left(2\theta - \frac{\pi}{2}\right)$ is dotted in color. Finally, triple each y-value of $\sin\left(2\theta - \frac{\pi}{2}\right)$. This yields the (solid, colored) graph of $3\sin\left(2\theta - \frac{\pi}{2}\right)$ □

Let $a \neq 0$ and $b > 0$. For a function of the form

$a \sin(b\theta - k)$, that is, $a \sin b\left(\theta - \frac{k}{b}\right)$,

or of the form

$a \cos(b\theta - k)$,

the phase shift is $\frac{k}{b}$, the period is $\frac{2\pi}{b}$, and the amplitude can be defined as $|a|$. Thus in part (a) of Example 5, for $2\sin\left(\theta - \frac{\pi}{2}\right)$, the phase shift is $\frac{\pi}{2}$, the period is 2π, and the amplitude is 2. In part (b) of Example 5, for $3 \sin\left(2\theta - \frac{\pi}{2}\right)$, the phase shift is $\frac{\frac{\pi}{2}}{2}$, or $\frac{\pi}{4}$, the period is π, and the amplitude is 3.

For a function of the form

$$a \tan (b\theta - k), \qquad a \neq 0, b > 0,$$

the phase shift is $\frac{k}{b}$, the period is $\frac{\pi}{b}$, but the amplitude is not defined. (Why not?)

Recall that

$$(f + g)(\theta) \text{ is defined to be } f(\theta) + g(\theta), \quad \text{for all } \theta.$$

For example, if

$$f(1) = 3 \text{ and } g(1) = 5, \text{ then } (f + g)(1) = 3 + 5 = 8$$

EXAMPLE 6 Let $h(\theta) = \sin\theta + \cos\theta$. Draw the graph.

Solution. Here, let $f(\theta) = \sin\theta$, $g(\theta) = \cos\theta$, so that

$$h(\theta) = (f + g)(\theta) = f(\theta) + g(\theta) = \sin\theta + \cos\theta, \text{ for all } \theta.$$

To obtain the graph of h from the graphs of the sine and cosine, add the corresponding y-values of $\sin\theta$ and $\cos\theta$, as indicated in Figure 8.66. Observe that

$$h(\theta) = 0 \quad \text{means} \quad \sin\theta + \cos\theta = 0,$$

and thus,

$$\sin\theta = -\cos\theta.$$

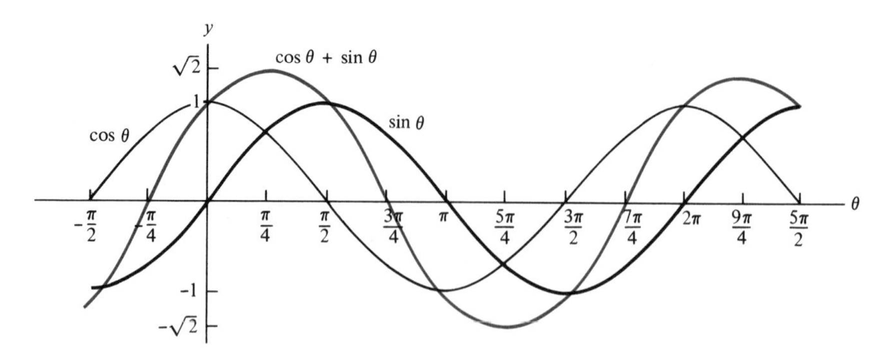

FIGURE 8.66 $y = \cos\theta + \sin\theta$

For $\cos\theta \neq 0$, divide both sides by $\cos\theta$ to obtain

$$\tan\theta = -1.$$

This occurs when $\theta = -\frac{\pi}{4}, \frac{3\pi}{4}, \frac{7\pi}{4}$, etc. □

EXERCISES

In Exercises 1–10 determine (a) the period and (b) the amplitude of each function.

1. $\cos 3\theta$ **2.** $\sin \frac{\theta}{4}$ **3.** $2 \cos \theta$ **4.** $-2 \sin \theta$ **5.** $\frac{1}{2} \sin 4\theta$

6. $\frac{\sin 5\theta}{5}$ **7.** $-6 \cos \frac{\theta}{10}$ **8.** $-\frac{2}{3} \sin \frac{2}{3} \theta$ **9.** $\sin 2\pi\theta$ **10.** $-\pi \sin \frac{\pi}{2} \theta$

In Exercises 11–24 draw the graph of each function.

11. $\cos 2\theta$ **12.** $2 \cos \theta$ **13.** $\sin 4\theta$ **14.** $4 \sin \theta$ **15.** $\sin \frac{\theta}{4}$ **16.** $\frac{1}{4} \sin \theta$

17. $4 \sin 2\theta$ **18.** $4 \sin \frac{\theta}{4}$ **19.** $-\sin \theta$ **20.** $-2 \cos \theta$ **21.** $2 \sin \frac{\theta}{2}$ **22.** $2 \tan \theta$

23. $\tan 2\theta$ **24.** $\tan \frac{\theta}{2}$

In Exercises 25–30: (a) Determine the phase shift of each function. (b) Is the shift to the left or to the right?

25. $\tan \left(\theta - \frac{\pi}{3}\right)$ **26.** $\sin \left(\theta - \frac{\pi}{6}\right)$ **27.** $\cos (\theta + \pi)$ **28.** $2 \sin \left(\theta + \frac{\pi}{4}\right)$

29. $-\cot (\theta + 1)$ **30.** $\sin \left(\frac{2\theta - \pi}{2}\right)$

In Exercises 31–46 for each function: (a) Determine the period (b) determine the phase shift (c) determine the amplitude, or state that it is undefined (d) draw the graph.

31. $\sin \left(\theta - \frac{\pi}{2}\right)$ **32.** $\cos (\theta - \pi)$ **33.** $\sin \left(\theta - \frac{\pi}{6}\right)$ **34.** $\cos \left(\theta - \frac{\pi}{2}\right)$

35. $\cos \left(\theta - \frac{\pi}{3}\right)$ **36.** $\tan \left(\theta + \frac{\pi}{2}\right)$ **37.** $2 \sin \left(\theta - \frac{\pi}{4}\right)$ **38.** $-\cos \left(\theta + \frac{\pi}{4}\right)$

39. $\sin \left(2\theta - \frac{\pi}{4}\right)$ **40.** $2 \sin \left(\frac{\theta}{2} - \pi\right)$ **41.** $3 \cos \left(\theta - \frac{\pi}{2}\right)$ **42.** $\frac{1}{2} \cos \left(3\theta - \frac{\pi}{2}\right)$

43. $-\cos (2\theta + \pi)$ **44.** $-\sin \left(\theta - \frac{\pi}{4}\right)$ **45.** $2 \tan \left(\theta - \frac{\pi}{2}\right)$ **46.** $-2 \tan (\theta + \pi)$

In Exercises 47–52 draw the graph of each function. In Exercises 47–51 draw this for $0 \leqslant \theta \leqslant 2\pi$; in Exercise 52 draw this for $-2\pi \leqslant \theta \leqslant 2\pi$.

47. $\cos \theta - \sin \theta$ **48.** $1 + \cos \theta$ **49.** $2 \cos \theta + 3 \sin \theta$ **50.** $\sin \theta - 2 \cos \theta$

51. $\cos 2\theta + \sin 2\theta$ **52.** $2 \sin \frac{\theta}{2} + \cos \frac{\theta}{2}$

53. Find an interval on which $\sin 3\theta$ increases from -1 to 1.

54. Find an interval on which $\cos 4\theta$ decreases from 1 to -1.

55. Find an interval on which $2 \sin \frac{\theta}{2}$ increases from -1 to 1.

56. Find an interval on which $3 \tan 2\theta$ increases from -3 to 3.

8.6 The Law of Sines

solving oblique triangles

An **oblique triangle** is one that does not contain a right angle. In an oblique triangle all three angles may be acute, as in the **acute triangle** of Figure 8.67(a), or one angle may be obtuse (and the other two acute), as in the **obtuse triangle** of Figure 8.67(b).

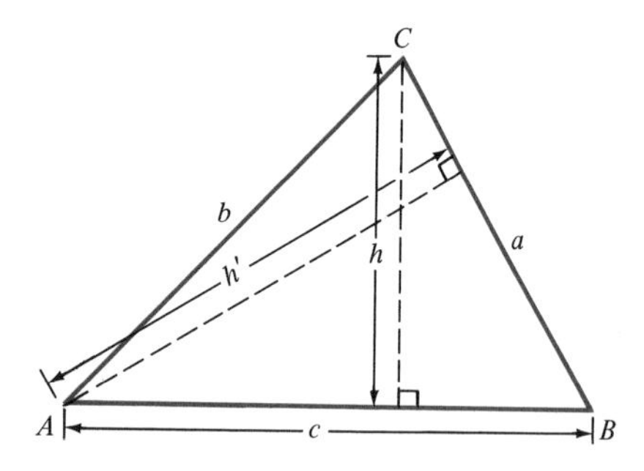

FIGURE 8.67(a) $\triangle ABC$ is an acute triangle.

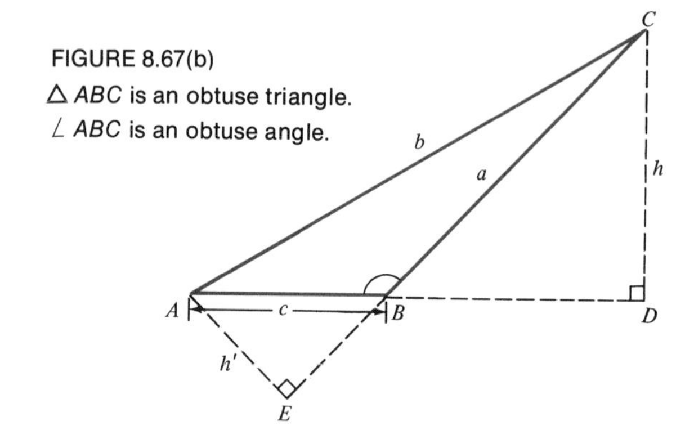

FIGURE 8.67(b)
$\triangle ABC$ is an obtuse triangle.
$\angle ABC$ is an obtuse angle.

In Section 8.2 you *solved right triangles*. When given the lengths of two sides of the right triangle, or the length of one side together with the measure of one acute angle, you determined all three sides and all three angles. Actually, you had *three* pieces of information to work with because one angle was always a right angle.

Now consider the problem of *solving an oblique triangle,* given three pieces of information:

AAS: Two angles and the side opposite one of them (Angle, Angle, Side), as in Figure 8.68(a)

ASA: Two angles and the *included* side (Angle, Side, Angle), as in Figure 8.68(b)

SSA: Two sides and the angle opposite one of them (Side, Side, Angle), as in Figure 8.68(c)

SAS: Two sides and the *included* angle (Side, Angle, Side), as in Figure 8.68(d)

SSS: Three sides (Side, Side, Side), as in Figure 8.68(e)

AAA: Three angles (Angle, Angle, Angle), as in Figure 8.68(f)

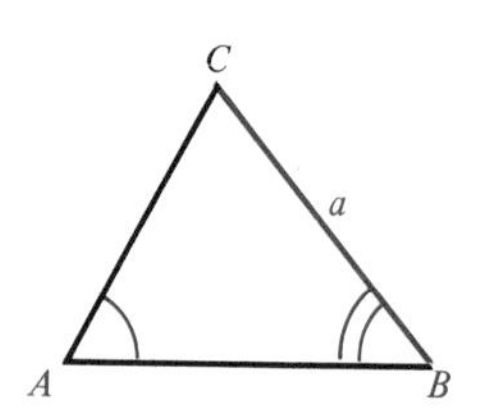

FIGURE 8.68(a) AAS
$\angle C = 180° - (\angle A + \angle B)$

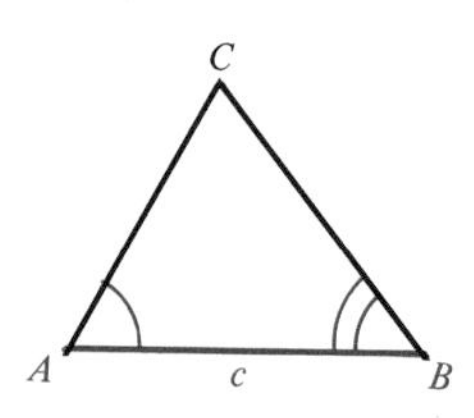

FIGURE 8.68(b) ASA
$\angle C = 180° - (\angle A + \angle B)$

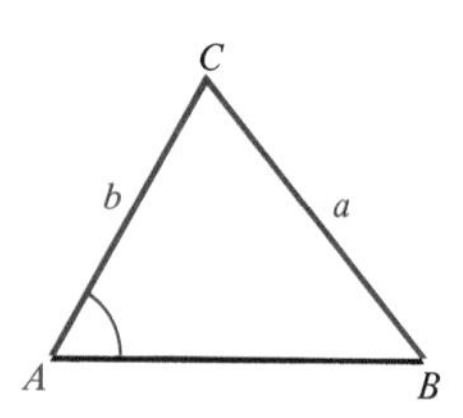

FIGURE 8.68(c) SSA

The AAS, ASA, and SSA cases are solved by means of the *Law of Sines*. The SAS and SSS cases will be treated by means of the *Law of Cosines* in Section 8.7 . The AAA case is ambiguous because three angles determine a triangle only up to similarity, as illustrated in Figure 8.68(f).

The Law of Sines states:

In any triangle the lengths of the sides are proportional to the sines of the opposite angles:

$$\frac{a}{\sin A} = \frac{b}{\sin B} = \frac{c}{\sin C}$$

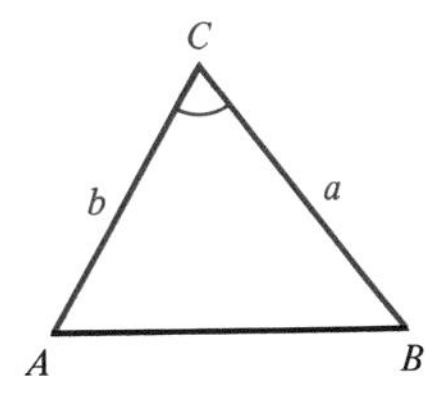

FIGURE 8.68(d) SAS

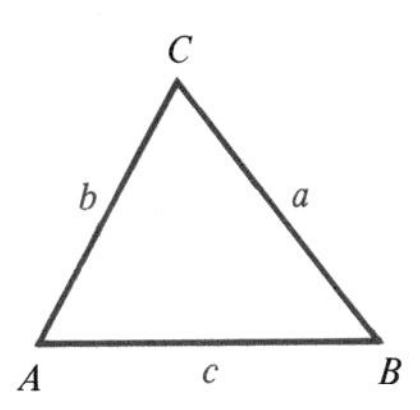

FIGURE 8.68(e) SSS

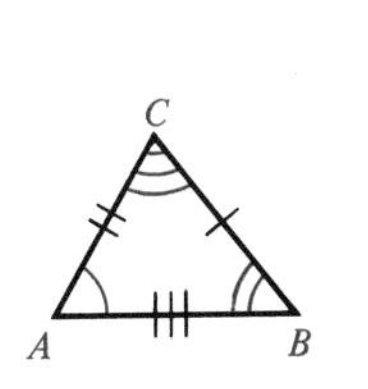

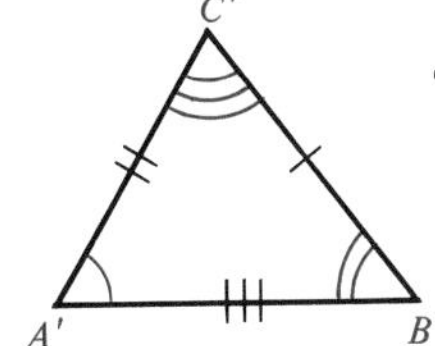

FIGURE 8.68(f)
AAA The two triangles are similar. Corresponding angles and corresponding sides are indicated.

[Refer to Figure 8.67(a) and (b). In Figure 8.67(b), $\angle B$ refers to $\angle ABC$.]

To see that $\dfrac{a}{\sin A} = \dfrac{b}{\sin B}$, observe that in both Figure 8.67(a) and Figure 8.67(b),

$$\frac{h}{b} = \sin A$$

Moreover, in Figure 8.67(a),

$$\frac{h}{a} = \sin B$$

whereas in Figure 8.67(b)

$$\frac{h}{a} = \sin \angle CBD = \sin(\pi - B) = \sin B$$

Thus in both cases,

$$h = b \sin A \text{ and } h = a \sin B$$

Therefore,

$$b \sin A = a \sin B \qquad \text{Divide both sides by } \sin A \sin B.$$

$$\frac{b}{\sin B} = \frac{a}{\sin A}$$

To see that $\dfrac{b}{\sin B} = \dfrac{c}{\sin C}$, use $\dfrac{h'}{b} = \sin C$ and $\dfrac{h'}{c} = \sin B$ to obtain

$$h' = b \sin C = c \sin B$$

and therefore

$$\frac{b}{\sin B} = \frac{c}{\sin C}$$

△

Because tables will be used in computations, it is more convenient to measure angles in degrees, rather than in radians.

First apply the Law of Sines when one side and *any* two angles are known (AAS and ASA). See Figures 8.68(a) and (b). The third angle is then the supplement of the sum of the two known angles. The use of logarithms greatly simplifies computations† because cumbersome products and quotients are usually involved and

$$\log AB = \log A + \log B; \qquad \log \frac{A}{B} = \log A - \log B,$$

$$A > 0, B > 0$$

The log of the sine (or *log sin*) can be obtained directly from Table 4 in the appendix.

†It might be interesting to work these computations with the aid of a hand calculator and without using logarithms. The results could then be compared with those obtained in the text by the logarithmic method.

AAS We illustrate with the AAS case.

EXAMPLE 1 Solve the triangle in Figure 8.69.

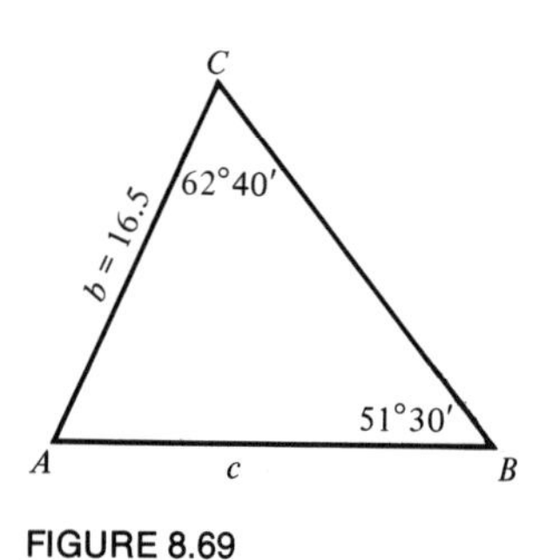

FIGURE 8.69

Solution.

$$\angle A = 180° - (51°30' + 62°40')$$
$$= 180° - 114°10'$$
$$= 65°50'$$

To find a:

$$\frac{a}{\sin A} = \frac{b}{\sin B}$$

$$a = \frac{b \cdot \sin A}{\sin B} = \frac{16.5 \sin 65°50'}{\sin 51°30'}$$

$$\log a = \log\left(\frac{16.5 \sin 65°50'}{\sin 51°30'}\right)$$
$$= \log 16.5 + \log \sin 65°50' - \log \sin 51°30'$$

Use Tables 2 and 4; in Table 4, because both angles measure more than 45°, read up and locate the angle at the right

$$\log a = 1.2175 + (.9602 - 1) - (.8935 - 1) = 1.2842$$

From Table 2 (without interpolating):

$$a = \text{antilog } 1.2842 = 1.92 \times 10^1 = 19.2$$

To find c:

(It is better to use b, the *given* side, rather than a, because a may involve an approximation error.)

$$\frac{c}{\sin C} = \frac{b}{\sin B}$$

$$c = \frac{b \sin C}{\sin B} = \frac{16.5 \sin 62°40'}{\sin 51°30'}$$

$$\log c = \log 16.5 + \log \sin 62°40' - \log \sin 51°30'$$
$$= 1.2175 + (.9486 - 1) - (.8935 - 1)$$
$$= 1.2726$$

$$c = \text{antilog } 1.2726 = 18.7$$

□

SSA SSA is often called the *ambiguous case,* for there are several possibilities. Suppose you know the lengths a and b of two sides of a triangle and the measure of $\angle A$, the angle opposite side a. At first, assume $\angle A$ is acute. You may suppose that $\angle A$ is in standard position, as in Figure 8.70. One of the vertices, B, is then on the *positive* x-axis and the third vertex, C, is in the first quadrant. Note that $b \sin A$ is the distance from C to the x-axis. Because a is known, consider an arc (of a circle) of length a centered at C. There are four possibilities, depending on a, as indicated in Figure 8.71.

Now suppose $\angle A$ is obtuse. Again, vertex B must lie on the

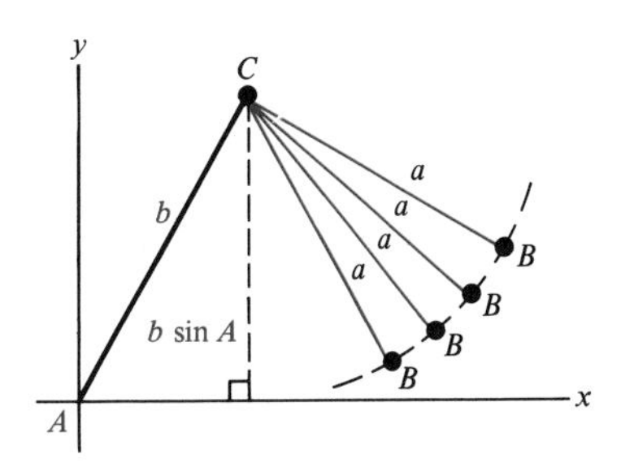

FIGURE 8.70

positive x-axis. Figure 8.72 on page 358 illustrates that a (unique) triangle exists if $a > b$, but *not* if $a \leq b$.

In practice, it is unnecessary to recognize which of these figures applies. In the course of solving a problem, the situation becomes apparent, as illustrated in Examples 2, 3, and 4.

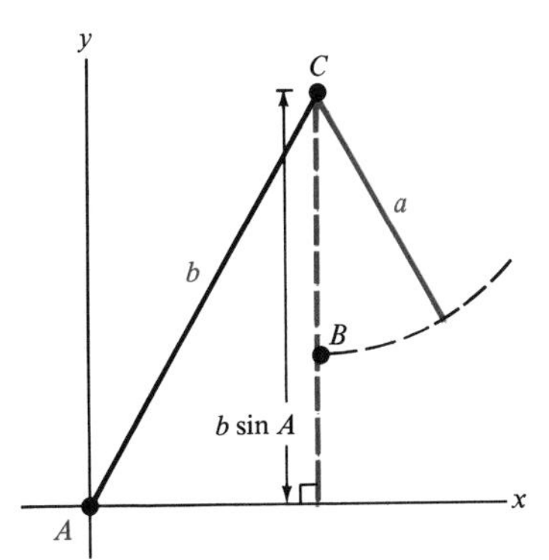

FIGURE 8.71(a).
$a < b \sin A$. (Side a is too short!)
No triangle is possible.

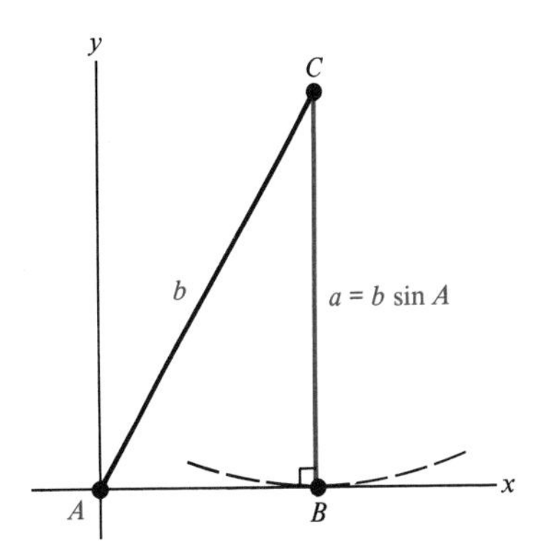

FIGURE 8.71(b).
$a = b \sin A$. This is the case of a unique *right* triangle.

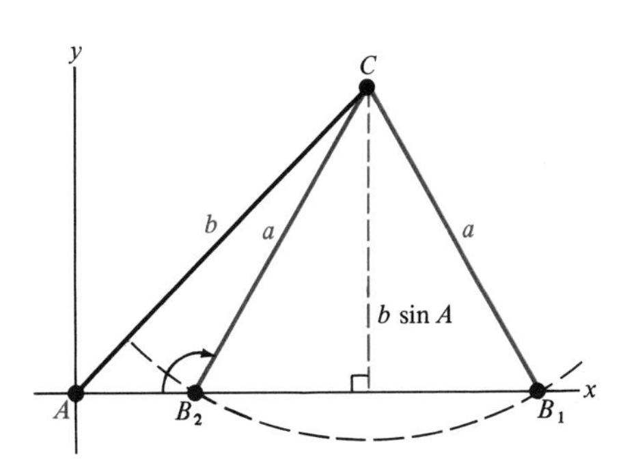

FIGURE 8.71(c) $b \sin A < a < b$.
Side a intersects the *positive* x-axis twice. There are *two* triangles—ΔAB_1C (acute) and ΔAB_2C (obtuse). Note that $\angle B_1B_2C = \quad AB_1C$ and therefore $\angle AB_2C = \pi - \angle AB_1C$

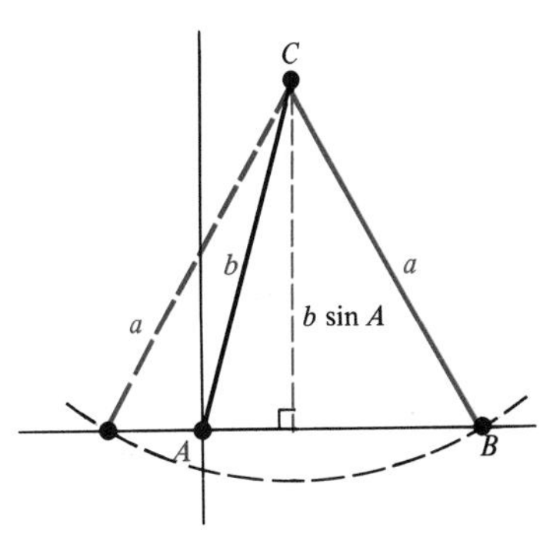

FIGURE 8.71(d) $b \leq a$ and therefore $\underbrace{b \sin A}_{\angle 1} < b \leq a$. Side a intersects the *positive* x-axis, forming a legitimate triangle. Side a also intersects the *negative* x-axis (or coincides with b), contradicting the original assumption. There is *one* (legitimate) triangle.

EXAMPLE 2 Solve ΔABC if $a = 12$, $b = 20$, and $\angle A = 78°$.

Solution. Use the Law of Sines in the form

$$\frac{\sin B}{b} = \frac{\sin A}{a},$$

so that

$$\sin B = \frac{b \sin A}{a} = \frac{20\,(.9781)}{12} \approx 1.6$$

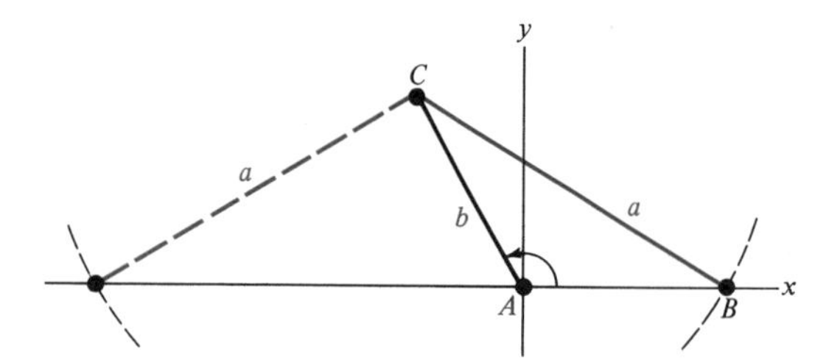

FIGURE 8.72(a) $a > b$. Side a intersects the x-axis twice. As in Figure 8.71(d), only the intersection with the *positive* x-axis yields a legitimate triangle.

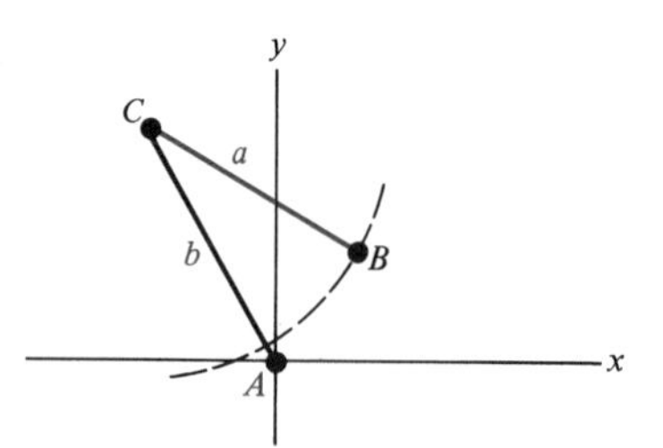

FIGURE 8.72(b) $a \leqslant b$ (Side a is too short!) No legitimate triangle is formed.

Because $\sin B > 1$, no triangle is possible. Note that

$$\angle A \text{ is acute} \quad \text{and} \quad a < b \sin A, \qquad 12 < 20\,(.9781)$$

so that Figure 8.71(a) on page 357 applies. □

EXAMPLE 3 Solve ΔABC if $a = 49$, $b = 44$, and $\angle A = 38°10'$.

Solution.

$$\sin B = \frac{b \sin A}{a} = \frac{44 \sin 38°10'}{49}$$

Use Tables 2 and 4 at the back.

$$\begin{aligned} \log \sin B &= \log\left(\frac{44 \sin 38°10'}{49}\right) \\ &= \log 44 + \log \sin 38°10' - \log 49 \\ &= 1.6435 + (.7910 - 1) - 1.6902 \\ &= .7443 - 1 \end{aligned}$$

$$\angle B \approx 33°40'$$

$$\begin{array}{r} 1.6435 \\ +.7910 - 1 \\ \hline 2.4345 - 1 \\ -1.6902 \\ \hline .7443 - 1 \end{array}$$

Because $\sin(180° - B) = \sin B$, you might also consider $\angle B \approx 146°20'$. But then

$$\angle A + \angle B \approx 38°10' + 146°20' = 184°30' > 180°.$$

This cannot be, so that $\angle B \approx 33°40'$, and only one triangle is possible. [Figure 8.71(d) on page 357.]

Next,

$$\angle C = 180° - (38°10' + 33°40') = 108°10'$$

To solve for c, use

$$\frac{c}{\sin C} = \frac{a}{\sin A}$$

$$\begin{aligned} c &= \frac{a \sin C}{\sin A} \\ &= \frac{49 \sin 108°10'}{\sin 38°10'} \\ &= \frac{49 \sin 71°50'}{\sin 38°10'} \end{aligned}$$

$$\sin 108°10' = \sin \underbrace{(180° - 108°10')}_{71°50'}$$

$$\begin{aligned}\log c &= \log 49 + \log \sin 71°50' - \log \sin 38°10' \\ &= 1.6902 + (.9778 - 1) - (.7910 - 1) \\ &= 1.8770\end{aligned}$$

$$c = 75.3$$

□

EXAMPLE 4 Solve for $\angle A$ if $a = 12$, $c = 10$, $\angle C = 30°$.

Here, use

$$\sin A = \frac{a \sin C}{c} = \frac{12(.5)}{10} = .6000$$

Use Table 3 to obtain

$$A_1 \approx 36°50' \quad \text{or} \quad A_2 \approx 180° - (36°50') = 143°10'$$

Note that for each choice of $\angle A$,

$$\angle A + \angle C < 180°$$

so that *two triangles are possible*. [Figure 8.71(c) on page 357 with C at the origin and A_1 and A_2 on the positive x-axis] □

EXERCISES

Refer to the tables at the back of the book. Express side lengths to 1 decimal place, and angle measurements to the nearest 10 minutes.

In Exercises 1–24 find all possible solutions for ΔABC.

1. $\angle A = 26°$, $\angle B = 48°$, $c = 18$

2. $\angle B = 49°$, $\angle C = 45°$, $a = 26$

3. $\angle A = 19°$, $\angle C = 120°$, $b = 40$

4. $\angle B = 35°$, $\angle C = 61°$, $b = 120.8$

5. $\angle A = 45°$, $a = 48$, $b = 80$

6. $\angle A = 52°$, $a = 10$, $b = 10$

7. $\angle A = 30°$, $a = 19.1$, $b = 38.2$

8. $\angle C = 39°$, $b = 180$, $c = 109$

9. $\angle A = 31°$, $a = 10$, $b = 25$

10. $\angle A = 64°$, $\angle B = 101°$, $c = 59$

11. $\angle B = 8°$, $a = 15$, $b = 18$

12. $\angle C = 70°$, $b = 10$, $c = 9$

13. $\angle C = 47°10'$, $b = 14.8$, $c = 17.2$

14. $\angle B = 12°30'$, $a = 38.1$, $b = 27.9$

15. $\angle A = 100°$, $a = 40$, $b = 53$

16. $\angle A = 95°$, $a = 28$, $b = 28$

17. $\angle B = 42°20'$, $\angle C = 27°50'$, $c = 19.2$

18. $\angle C = 106°$, $a = 15.2$, $c = 17.5$

19. $\angle A = 27.5°$, $a = 12$, $b = 21$

20. $\angle A = 81°$, $a = 24.5$, $c = 18.1$

21. $\angle B = 135°$, $a = 12$, $b = 16$

22. $\angle B = 132°$, $a = 39$, $b = 37$

23. $\angle C = 141°30'$, $a = 15.1$, $c = 12.9$

24. $\angle A = 78°$, $a = 15.3$, $c = 24$

25. Two drivers at points A and B, 4 miles apart along a straight highway, see a monument at point C some distance from the road. If $\angle ABC = 44°$ and $\angle BAC = 37°$, how far is the monument from each driver? [See Figure 8.73 on page 360.]

26. Two points A and B lie on opposite banks of a river. At 190 meters further up, where the river narrows, a point C is chosen on the same bank as A. If $\angle BAC$ measures 68°40′ and $\angle ACB$ measures 57°10′, find the distance from A to B.

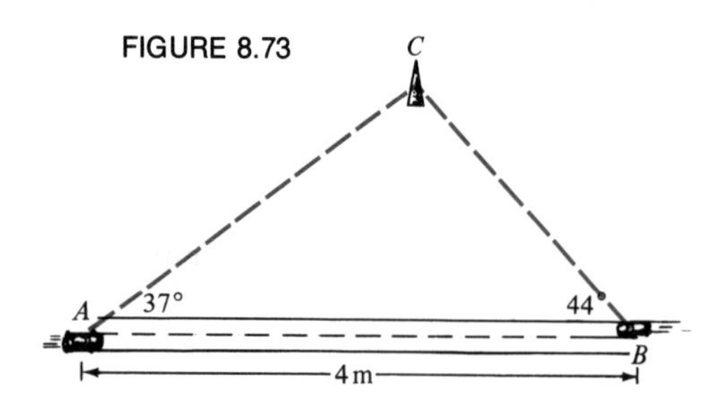

FIGURE 8.73

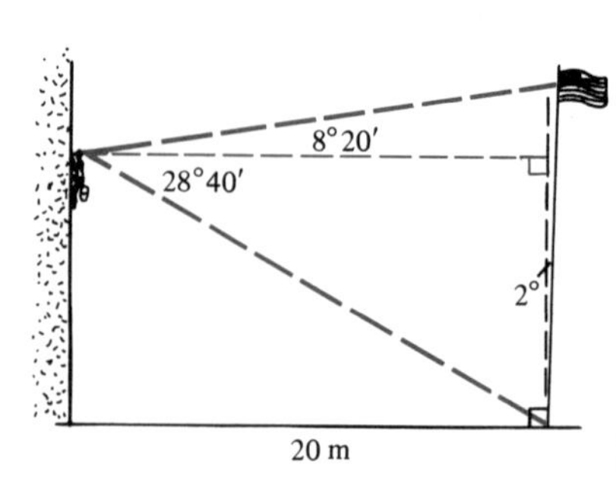

FIGURE 8.74

27. Two angles of a triangular flag measure 46° and 52°. The side included between them is 22 centimeters long. Find the lengths of the other two sides.

28. A tower that tilts at a 5° angle away from the sun casts a 50-meter shadow when the angle of elevation of the sun is 58°. How high is the tower?

29. A flagpole tilts at a 2° angle away from the side of a building on which a window washer is working. [See Figure 8.74.] The base of the flagpole is 20 meters from the base of the building. From the window washer's prospective, the angle of elevation of the top of the flag is 8°20′ and the angle of depression of the base of the pole is 28°40′. (a) How high is the window washer's eye level? (b) How high is the top of the flag?

8.7 The Law of Cosines

The **Law of Cosines** asserts:

In any triangle ABC, with sides of lengths a, b, and c,

$$c^2 = a^2 + b^2 - 2ab\cos C$$

The proof for acute triangles is given in the caption to Figure 8.75(a) and for obtuse triangles in the caption to Figure 8.75(b). Note that if $\angle C$ is a right angle, then $\cos C = 0$, so that the formula then reduces to that of the Pythagorean Theorem.

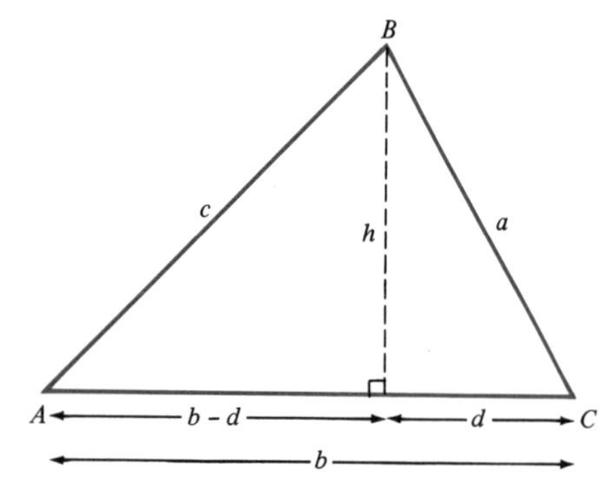

FIGURE 8.75(a) Triangle ABC is acute:

$$\begin{aligned} c^2 &= (b-d)^2 + h^2 \\ &= b^2 - 2bd + d^2 + h^2 \\ &= b^2 - 2ba\cos C + a^2 \\ &= a^2 + b^2 - 2ab\cos C \end{aligned}$$

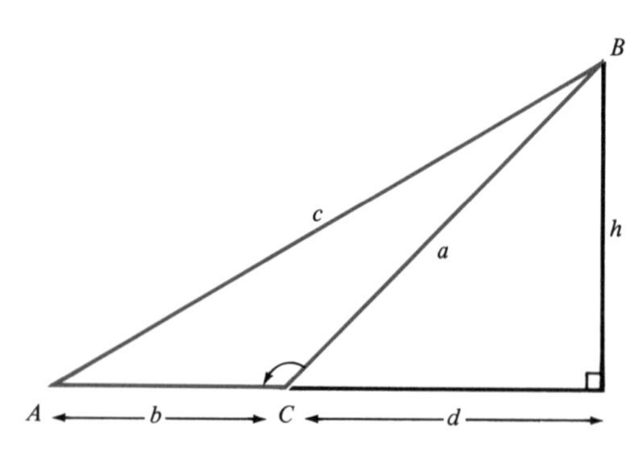

FIGURE 8.75(b) Triangle ABC is obtuse:

$$\begin{aligned} c^2 &= (b+d)^2 + h^2 \\ &= b^2 + 2bd + d^2 + h^2 \\ &= b^2 + 2ba\underbrace{\cos(\pi - C)}_{-\cos C} + a^2 \\ &= a^2 + b^2 - 2ab\cos C \end{aligned}$$

△

The Law of Cosines involves three sides and one angle of a triangle. It can therefore be used in case SAS to find the third side and in case SSS to find an angle. (See page 354.)

SAS

In Examples 1 and 2 find side lengths to the nearest integer and angle measurements to the nearest 10 minutes.

EXAMPLE 1 Let $a = 70$, $b = 80$, $C = 76°$. Find c.

Solution.

$$\begin{aligned} c^2 &= a^2 + b^2 - 2ab\cos C \\ &= 70^2 + 80^2 - 2(70)(80)\cos 76° \\ &= 4900 + 6400 - 11\,200\,(.2419) \\ &\approx 8591 \end{aligned}$$

$c \approx 93$ See Table 1 in the appendix. □

If you are given a, c, and $\angle B$, you would solve for b by using the Law of Cosines in the form

$$b^2 = a^2 + c^2 - 2ac\cos B$$

Here the roles of sides b and c and of angles B and C have been exchanged. And given b, c, and $\angle A$, you would use

$$a^2 = b^2 + c^2 - 2bc\cos A$$

SSS

EXAMPLE 2 Solve ΔABC if $a = 15$, $b = 11$, $c = 25$.

Solution. First solve for $\angle C$.

$$\begin{aligned} c^2 &= a^2 + b^2 - 2\,ab\cos C \\ 25^2 &= 15^2 + 11^2 - 2(15)(11)\cos C \\ 625 &= 225 + 121 - 330\cos C \\ -\frac{279}{330} &= \cos C \\ -.8455 &= \cos C \qquad (\angle C \text{ is obtuse}) \\ .8455 &= \cos(180° - C) \\ 32°20' &\approx 180° - C \\ \angle C &\approx 147°40' \end{aligned}$$

To solve for $\angle A$ at this point, it is probably easier to use the Law of Sines.†

$$\sin A = \frac{a\sin C}{c}$$

†At times, a diagram is helpful if the Law of Sines (rather than the Law of Cosines) is used at this point. See Exercise 19 on page 362.)

$$= \frac{15 \sin 147°40'}{25}$$

$\sin 147°40' = \sin \underbrace{(180° - 147°40')}_{32°20'}$

$$= (.6) \sin 32°20'$$
$$= (.6)(.5348)$$
$$= .3209$$

$$\angle A = 18°40' \qquad (\angle A \text{ is acute.})$$

Finally,

$$\angle B = 180° - (\angle A + \angle C)$$
$$= 180° - (18°40' + 147°40')$$
$$= 13°40'$$

□

EXERCISES

Refer to the tables at the back of the book. Find side lengths to the nearest integer (or unit) and angle measurements to the nearest degree.

In Exercises 1–10 solve for the indicated side length or angle of ΔABC.

1. $a = 12$, $b = 10$, $\angle C = 78°$. Find c.

2. $a = 40$, $c = 80$, $\angle B = 42°$. Find b.

3. $a = 8$, $b = 9$, $c = 10$. Find $\angle C$.

4. $a = b = 12$, $c = 15$. Find $\angle A$.

5. $b = 16$, $c = 28$, $\angle A = 110°$. Find a.

6. $a = 10$, $b = 9$, $c = 16$. Find $\angle C$.

7. $a = 24$, $b = 28$, $c = 30$. Find $\angle A$.

8. $a = 100$, $c = 38$, $\angle B = 55°$. Find b.

9. $a = 40$, $b = 50$, $c = 60$. Find $\angle B$.

10. $b = 38.2$, $c = 72$, $\angle A = 16°$. Find a.

In Exercises 11–16 use logarithms to solve for the indicated side length or angle of ΔABC.

11. $a = 38$, $b = 45$, $\angle C = 80°$. Find c.

12. $a = 94$, $b = 49$, $c = 72$. Find $\angle C$.

13. $a = 136$, $b = 152$, $c = 176$. Find $\angle A$.

14. $a = 9.6$, $b = 7.8$, $c = 8.7$. Find $\angle B$.

15. $a = 94$, $c = 49$, $\angle B = 115°$. Find b.

16. $b = c = 208$, $\angle A = 29°30'$. Find a.

In Exercises 17–22 solve ΔABC.

17. $a = 5$, $b = 6$, $c = 7$

18. $a = 27$, $c = 29$, $\angle B = 85°$

19. $a = 20$, $b = 15$, $\angle C = 20°$ (*Sketch.*)

20. $a = b = 10$, $c = 13$

21. $b = 39$, $c = 55$, $\angle A = 18°$

22. $a = 47$, $b = 27$, $\angle C = 125°$

23. Two adjacent sides of a parallelogram are 16 inches long and 18 inches long, respectively. If the included angle is 40°, find the length of each diagonal.

24. One attic wall of an A-shaped house is in the form of an isosceles triangle. The equal legs are each 9 meters long and the included angle is 54°. Find the length of the base of the wall. [See Figure 8.76 .]

25. A plot of land in the form of a parallelogram has sides of length 45 meters and 65 meters. If one diagonal measures 70 meters, find the angles of the parallelogram.

26. A gasoline station lies at the intersection of two straight roads. Two cars leave the gas station at the same time, one traveling at 65 miles per hour along one road and the other at 60 miles per hour along the other road. If a 68° angle is formed by the cars' paths, how far apart are the cars 2 hours later? [See Figure 8.77 .]

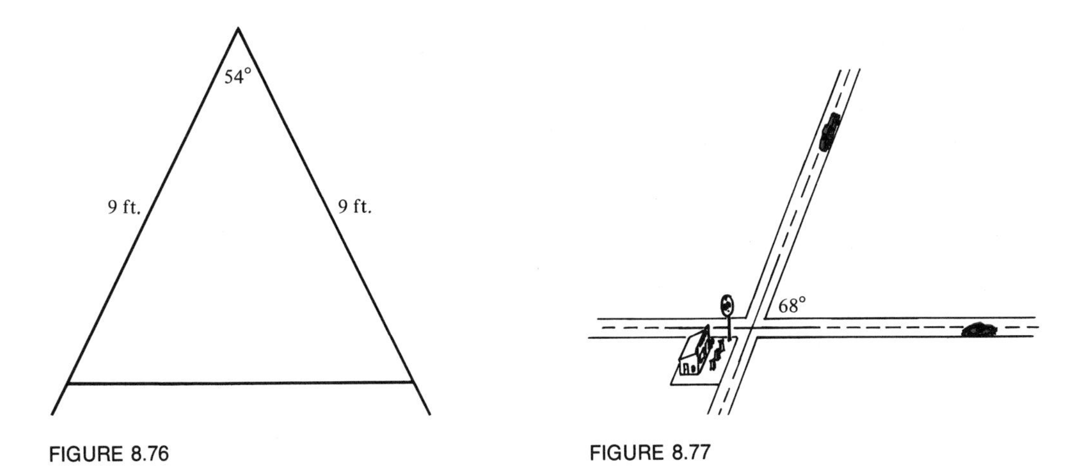

FIGURE 8.76

FIGURE 8.77

27. To find the distance between points P and Q, a third point, R, is chosen that is 30 meters from P and 35 meters from Q. If $\angle PRQ$ measures 40°, find the distance from P to Q.

28. The (air) distance from Boston to New York is 210 miles, from New York to Philadelphia is 90 miles, and from Philadelphia to Boston is 250 miles. Find the measures of the angles of the triangle formed by these cities.

Review Exercises for Chapter 8

1. Consider Figure 8.78 . Determine: **(a)** sin A **(b)** cos A **(c)** tan A **(d)** sin B **(e)** cos B **(f)** tan B

FIGURE 8.78

FIGURE 8.79

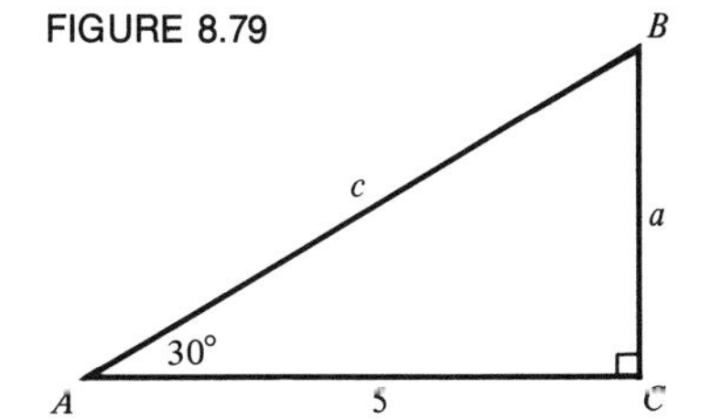

FIGURE 8.80

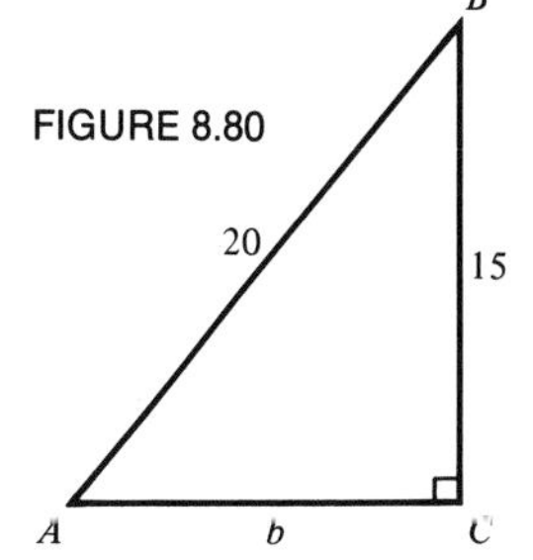

2. Suppose $\sin A = \frac{3}{5}$ and $\angle A$ is acute. Determine the other trigonometric function-values for $\angle A$.

3. In Figure 8.79, determine side lengths a and c.

4. Determine each trigonometric function-value. **(a)** cos 30° **(b)** sin 45° **(c)** tan 60° **(d)** sec 30°

5. A woman standing 42 meters from the base of a tree measures the angle of elevation of the top of the tree to be 60°. How high is the tree?

6. Use the trig table to determine each value. **(a)** sin 27° **(b)** cos 44°40′ **(c)** tan 75°10′

7. Assume $\angle A$ is acute. Use the trig table to determine $\angle A$ in each case.

   (a) $\sin A = .5471$ **(b)** $\tan A = 1.2131$ **(c)** $\cos A = .9674$

8. In Figure 8.80 on page 363, solve the right triangle. Do not interpolate. Approximate side length b to 1 decimal place, and $\angle A$ and $\angle B$ to the nearest 10 minutes.

9. Interpolate to determine

(a) $\sin 57°46'$ **(b)** $\cos 18°12'$ **(c)** $\angle A$ (to the nearest minute), if $\tan A = 1.538$.

10. Express in radian measure: **(a)** $75°$ **(b)** $225°$ **(c)** $-120°$

11. Express in degrees: **(a)** $\frac{2\pi}{3}$ **(b)** $\frac{3\pi}{10}$ **(c)** $\frac{3\pi}{2}$

12. Determine the six trigonometric function-values at each angle. **(a)** $\frac{\pi}{4}$ **(b)** $\frac{7\pi}{6}$ **(c)** $-\frac{\pi}{2}$

13. Graph $\sin \theta$, for $-\pi \leqslant \theta \leqslant \pi$. **14.** Graph $\cos \theta$, for $-\frac{\pi}{2} \leqslant \theta \leqslant \frac{3\pi}{2}$.

15. Consider $3 \sin \frac{\theta}{2}$.

(a) Determine the period. **(b)** Determine the amplitude. **(c)** Graph the function for $0 \leqslant \theta \leqslant 4\pi$.

16. Consider $\cos\left(\theta + \frac{\pi}{2}\right)$.

(a) Determine the phase shift. **(b)** Is the shift to the left or to the right?

(c) Determine the period. **(d)** Graph the function for $0 \leqslant \theta \leqslant 2\pi$.

In Exercises 17–20 refer to Tables 3 and 4 at the back of the book. In Exercises 17–19 find side lengths to one decimal place, and angle measurements to the nearest 10 minutes.

17. Find all possible solutions for ΔABC if $\angle B = 42°30'$, $\angle C = 67°10'$ and $b = 13.8$.

18. Find all possible solutions for ΔABC if $a = 12.1$, $b = 8.8$, and $\angle B = 36°30'$.

19. Solve ΔABC if $a = 9$, $b = 10$, $c = 13$.

20. Two adjacent sides of a parallelogram are 20 centimeters long and 45 centimeters long, respectively. If the included angle is 36°, find the length of each diagonal to the nearest centimeter.

Trigonometric Formulas

9

9.1 Trigonometric Identities

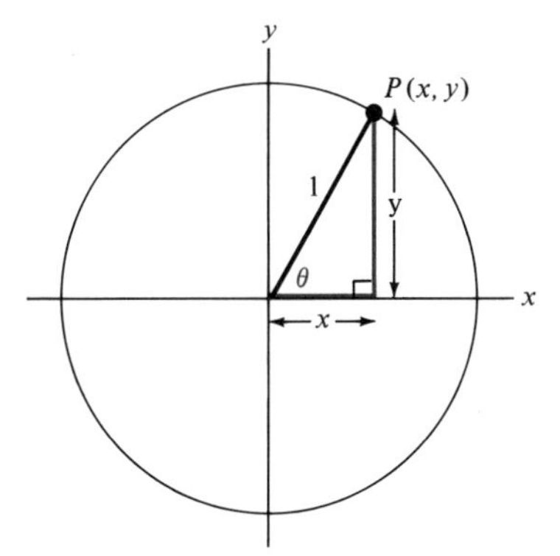

FIGURE 9.1 $x > 0$, $y > 0$
Thus $\cos\theta > 0$, $\sin\theta > 0$

The six trigonometric functions

cosine, sine, tangent, secant, cosecant, cotangent

were defined (in Section 8.3) in terms of the unit circle and in terms of radian measure. Recall that if θ is any real number, there corresponds to it a (unique) angle of θ radians in standard position. Let the terminal side of the angle intersect the unit circle at point P, with coordinates (x, y). [See Figure 9.1 for points P in the first quadrant and Figures 8.45(c), (d), and (e) on page 328 for points in the other quadrants.] Then

cosine θ was defined to be x and sine θ was defined to be y.

The four other trigonometric functions were defined in terms of the cosine and sine:

$$\tan\theta = \frac{\sin\theta}{\cos\theta}, \text{ provided } \cos\theta \neq 0 \quad \left(\text{that is, for } \theta \neq \pm\frac{\pi}{2}, \pm\frac{3\pi}{2}, \ldots\right) \tag{9.1}$$

$$\sec\theta = \frac{1}{\cos\theta}, \text{ provided } \cos\theta \neq 0 \tag{9.2}$$

$$\csc\theta = \frac{1}{\sin\theta}, \text{ provided } \sin\theta \neq 0 \quad (\text{that is, for } \theta \neq 0, \pm\pi, \pm 2\pi, \ldots) \tag{9.3}$$

$$\cot\theta = \frac{\cos\theta}{\sin\theta} \text{ provided } \sin\theta \neq 0 = \begin{cases} \dfrac{1}{\tan\theta} & \text{for } \theta \text{ in the domain of the tangent and } \tan\theta \neq 0 \\ 0 & \text{for } \theta = \pm\dfrac{\pi}{2}, \pm\dfrac{3\pi}{2}, \pm\dfrac{5\pi}{2}, \ldots, \text{ where } \tan\theta \text{ is undefined} \end{cases} \tag{9.4}$$

The preceding definitions relate the six trigonometric functions. You will now consider other relationships among them.

An **identity** is an equation for which every *permissible* real number is a root. Thus, because every real number other than 1 and -1 (for which a denominator would be 0) is a root of

$$\frac{1}{x-1}\cdot\frac{1}{x+1}=\frac{1}{x^2-1},$$

this equation is an identity. In a *trigonometric identity* every real number in the (common) domain of each occurring trigonometric function is a root. Thus in Formulas (**9.1**), (**9.2**), (**9.3**), and (**9.4**), the equations are trigonometric identities. Formula (**9.5**), which you next consider, contains another basic identity from which many others follow.

Let θ be any real number. Consider the point (x, y) on the unit circle determined by the angle of θ radians in standard position. [See Figure 9.1 .] Then by the Pythagorean Theorem

$$x^2 + y^2 = 1$$

But $x = \cos\theta$ and $y = \sin\theta$. Thus

$$\cos^2\theta + \sin^2\theta = 1 \qquad \text{for all } \theta \tag{9.5}$$

$\cos^2\theta$ stands for $(\cos\theta)^2$ and $\sin^2\theta$ stands for $(\sin\theta)^2$.

Other useful forms of Formula (**9.5**) are:

$$\cos^2\theta = 1 - \sin^2\theta \qquad \text{for all } \theta \tag{9.5'}$$

$$\sin^2\theta = 1 - \cos^2\theta \qquad \text{for all } \theta \tag{9.5''}$$

Next, when $\cos\theta \neq 0$, divide both sides of (**9.5**) by $\cos^2\theta$ to obtain

$$1+\frac{\sin^2\theta}{\cos^2\theta}=\frac{1}{\cos^2\theta}.$$

This yields:

For $\theta \neq \pm\frac{\pi}{2}, \pm\frac{3\pi}{2}, \ldots,$

$$1+\tan^2\theta = \sec^2\theta \quad \text{or equivalently,} \quad \sec^2\theta - \tan^2\theta = 1 \tag{9.6}$$

And when $\sin\theta \neq 0$, divide both sides of (**9.5**) by $\sin^2\theta$, to obtain

$$\frac{\cos^2\theta}{\sin^2\theta}+1=\frac{1}{\sin^2\theta}.$$

Thus

For $\theta \neq 0, \pm\pi, \pm 2\pi, \ldots,$

$$\cot^2\theta + 1 = \csc^2\theta, \quad \text{or equivalently,} \quad \csc^2\theta - \cot^2\theta = 1 \tag{9.7}$$

For the most part, *identities will be stated without explicitly mentioning the values for which one of the functions is undefined.*

Formulas **(9.1)**–**(9.5)** are crucial for all further work involving trigonometry. (You should know them in your sleep.) Formulas **(9.6)** and **(9.7)**, though easily derived from **(9.5)**, come up frequently, so that you should remember them. DO NOT MEMORIZE most of the hundred or so other identities that will be discussed, but instead, understand the *method of deriving them* from *identities* **(9.1)** – **(9.5)**. The most useful identities that *should be memorized* are printed in color. The examples that follow illustrate some of the most important ways of verifying, or discovering, an identity.

EXAMPLE 1 Verify the identity: $(1 - \sin^2\theta)\sec^2\theta = 1$

Solution. Recall that

$$1 - \sin^2\theta = \cos^2\theta \qquad \text{and} \qquad \sec\theta = \frac{1}{\cos\theta}$$

Thus

$$(1 - \sin^2\theta)\sec^2\theta = \cos^2\theta\cdot\frac{1}{\cos^2\theta} = 1$$ □

EXAMPLE 2 Verify the identity:

$$\frac{1}{\sec\theta + \tan\theta} = \sec\theta - \tan\theta$$

Solution. Recall that to rationalize the denominator (Section 5.7) of $\dfrac{1}{a + \sqrt{b}}$, you multiply numerator and denominator by $a - \sqrt{b}$. Here, given $\dfrac{1}{\sec\theta + \tan\theta}$, to simplify the denominator, multiply numerator and denominator by $\sec\theta - \tan\theta$, and use Formula **(9.6)**.

$$\frac{1\cdot(\sec\theta - \tan\theta)}{(\sec\theta + \tan\theta)\cdot(\sec\theta - \tan\theta)} = \frac{\sec\theta - \tan\theta}{\underbrace{\sec^2\theta - \tan^2\theta}_{1}} = \sec\theta - \tan\theta$$ □

EXAMPLE 3 Verify the identity:

$$\frac{\sin\theta}{1 + \cos\theta} = \frac{1 - \cos\theta}{\sin\theta}$$

Solution.

$$\frac{\sin\theta\cdot(1 - \cos\theta)}{(1 + \cos\theta)\cdot(1 - \cos\theta)} = \frac{\sin\theta\,(1 - \cos\theta)}{1 - \cos^2\theta} = \frac{\sin\theta\,(1 - \cos\theta)}{\sin^2\theta} = \frac{1 - \cos\theta}{\sin\theta}$$ □

From now on, letters such as A, B, C, x, y, as well as θ will be used to indicate the number of radians of the angles in question.

In Examples 1 and 2, you worked with the left side of an equation and "transformed" it into the right side. It is often easiest to work with *both* sides and to transform them until you arrive at the same trigonometric expression on each side. Also, converting to sines and cosines is often helpful.

EXAMPLE 4 Verify the identity: $\sin^2 A\,(\cot A + 1)^2 = \cos^2 A\,(\tan A + 1)^2$

Solution. Transform each side as indicated.

$\sin^2 A\,(\cot A + 1)^2$	$\cos^2 A\,(\tan A + 1)^2$
$= \sin^2 A\,(\cot^2 A + 2\cot A + 1)$	$= \cos^2 A\,(\tan^2 A + 2\tan A + 1)$
$= \sin^2 A\left(\dfrac{\cos^2 A}{\sin^2 A} + 2\dfrac{\cos A}{\sin A} + 1\right)$	$= \cos^2 A\left(\dfrac{\sin^2 A}{\cos^2 A} + 2\dfrac{\sin A}{\cos A} + 1\right)$
$= \cos^2 A + 2\sin A\cos A + \sin^2 A$	$= \sin^2 A + 2\cos A\sin A + \cos^2 A$

The bottom expressions are clearly equal. □

EXAMPLE 5 Verify the identity: $\sec B - \cos B = \sin B\tan B$

Solution. Transform each side as indicated.

$\sec B - \cos B$	$\sin B\tan B$
$= \dfrac{1}{\cos B} - \cos B$	$= \sin B \cdot \dfrac{\sin B}{\cos B}$
$= \dfrac{1 - \cos^2 B}{\cos B}$	$= \dfrac{\sin^2 B}{\cos B}$
$= \dfrac{\sin^2 B}{\cos B}$	

□

EXAMPLE 6 Assume $0 \leqslant \theta < 2\pi$. Express $\sin\theta$ in terms of $\cos\theta$.

Solution. By **(9.5″)**,

$$\sin^2\theta = 1 - \cos^2\theta \qquad \text{for all } \theta$$

The sine is *positive* in the first and second quadrants and *negative* in the third and fourth quadrants. Also $\sin 0 = \sin\pi = 0$; $\sin\dfrac{\pi}{2} = 1$; $\sin\dfrac{3\pi}{2} = -1$. Therefore, to obtain the correct square root, take:

$$\sin\theta = \sqrt{1 - \cos^2\theta}, \quad 0 \leqslant \theta \leqslant \pi$$
$$\sin\theta = -\sqrt{1 - \cos^2\theta}, \quad \pi < \theta < 2\pi$$

□

EXERCISES

In Exercises 1–10 simplify each expression.

1. $1 - \sin^2\theta$
2. $\tan^2 x - \sec^2 x$
3. $\dfrac{1+\cot^2 A}{1+\tan^2 A}$
4. $(1-\cos^2 B)\cot^2 B$
5. $\tan y \cot y$
6. $\dfrac{\sin\theta}{\csc\theta}$
7. $\dfrac{\sin^2 A - 1}{\sec A}$
8. $\dfrac{\tan B}{\cot B} + \sin B \csc B$
9. $\dfrac{\cos\theta}{1-\sin\theta} + \dfrac{\cos\theta}{1+\sin\theta}$
10. $\dfrac{1+\tan\theta}{\sec\theta}$

In Exercises 11–38 verify the following identities.

11. $\cos\theta\cdot\sec\theta = 1$
12. $\dfrac{\cos A}{\cot A} = \sin A$
13. $\cot^2 x - \csc^2 x + 1 = 0$
14. $(\cos B + 1)(\cos B - 1) = -\sin^2 B$
15. $\dfrac{\cos^2 A}{\sin A} = \dfrac{\cot A}{\sec A}$
16. $\tan^2 C + \sin^2 C = \sec^2 C - \cos^2 C$
17. $\tan y + \cot y = \dfrac{1}{\cos y \sin y}$
18. $\cos\theta \tan\theta \csc\theta = 1$
19. $(\sin A - \cos A)^2 = 1 - 2\sin A\cos A$
20. $\csc^4 B - \cot^4 B = \csc^2 B + \cot^2 B$
21. $(\sin^2\theta - \cos^2\theta)(\tan^2\theta + 1) = \tan^2\theta - 1$
22. $\dfrac{\cos\theta}{1-\sin\theta} + \dfrac{\cos\theta}{1+\sin\theta} = 2\sec\theta$
23. $\dfrac{\sin A + \cos A}{2} = \dfrac{\sin^2 A - \frac{1}{2}}{\sin A - \cos A}$
24. $\dfrac{\tan B}{\sec B - \cos B} = \dfrac{\cot B}{\cos B}$
25. $\dfrac{\sin A}{1+\sin A} = \dfrac{1}{1+\csc A}$
26. $\tan B - \sec B = \sin B \sec B - \tan B \csc B$
27. $\csc C - \sin C = \cot C \cos C$
28. $\sin\theta + \cos\theta\cot\theta = \csc\theta$
29. $\dfrac{-1}{\cot A + \csc A} = \cot A - \csc A$
30. $\dfrac{\sec A - \cos A}{\tan A - \sin A} = \cot A + \csc A$
31. $2\csc^2\theta = \dfrac{1}{1-\cos\theta} + \dfrac{1}{1+\cos\theta}$
32. $\sec^4 y - \tan^4 y = 2\sec^2 y - 1$
33. $\dfrac{1+\cot\theta}{1+\tan\theta} = \dfrac{\cot\theta - 1}{1-\tan\theta}$
34. $\cos^3 A - \sin^3 A = \cos^2 A \sin A - \sin^2 A \cos A + \cos A - \sin A$
35. $\cos^4 B - \sin^4 B = \cos^2 B - \sin^2 B$
36. $\dfrac{\sec C - \cos C}{\sin C + \tan C} = \csc C - \cot C$
37. $\dfrac{\cos^2 x + 1}{\cos x + \sec x} = \dfrac{\sin^2 x}{\sec x - \cos x}$
38. $\cos\theta\sin^3\theta + \sin\theta\cos^3\theta = \cos\theta\sin\theta$
39. For $0 \le \theta < 2\pi$, express $\cos\theta$ in terms of $\sin\theta$.
40. Express $\tan\theta$ in terms of $\sec\theta$ for:
    (a) $0 \le \theta < \dfrac{\pi}{2}$ (b) $\dfrac{\pi}{2} < \theta \le \pi$
    (c) $\pi \le \theta < \dfrac{3\pi}{2}$ (d) $\dfrac{3\pi}{2} < \theta \le 2\pi$
41. Express $\csc\theta$ in terms of $\cot\theta$ for:
    (a) $0 < \theta \le \dfrac{\pi}{2}$ (b) $\dfrac{\pi}{2} \le \theta < \pi$
    (c) $\pi < \theta \le \dfrac{3\pi}{2}$ (d) $\dfrac{3\pi}{2} \le \theta < 2\pi$
42. Express $\tan\theta$ in terms of $\cos\theta$ for $0 \le \theta < \dfrac{\pi}{2}$.

43. Show that for $\theta \neq 0$, and for $\theta \neq \pm\pi, \pm 3\pi, \pm 5\pi, \ldots,$

$$\frac{1 - \cos\theta}{\theta} = \frac{\sin\theta}{\theta} \cdot \frac{\sin\theta}{1 + \cos\theta}$$

(This relationship is important in calculus.)

In Exercises 44–46 you will consider *trigonometric substitutions* which are useful in calculus. In each case, suppose $a > 0$ and $0 < \theta < \frac{\pi}{2}$.

44. Let $x = a \sin\theta$. Show that $\sqrt{a^2 - x^2} = a \cos\theta$.

45. Let $x = a \sec\theta$. Show that $\sqrt{x^2 - a^2} = a \tan\theta$.

46. Let $x = a \tan\theta$. Show that $\sqrt{x^2 + a^2} = a \sec\theta$.

9.2 Addition Formulas

You will now establish important identities for the cosine, sine, and tangent of the sum or difference of two angles, commonly called **Addition Formulas.** In establishing the Addition Formulas you will use Formulas **(9.1)**–**(9.7)** of Section 9.1, as well as

$$\cos(-\theta) = \cos\theta \qquad \text{for all } \theta$$

$$\sin(-\theta) = -\sin\theta \qquad \text{for all } \theta$$

$$\tan(-\theta) = -\tan\theta \qquad \text{for } \theta \neq \pm\frac{\pi}{2}, \pm\frac{3\pi}{2}, \pm\frac{5\pi}{2}, \ldots$$

These identities for $\cos(-\theta)$ and for $\sin(-\theta)$ follow immediately from the definition of $\sin\theta$ and of $\cos\theta$ in terms of points on the unit circle. The identity for $\tan(-\theta)$ follows immediately from the definition of $\tan\theta$ in terms of $\sin\theta$ and of $\cos\theta$. [See Sections 8.3 and 8.4 .] No other formulas from Chapter 8 will be used here without first being established.

$\cos(A - B)$, $\cos(A + B)$

First consider $\cos(A - B)$.

In the unit circle of Figure 9.2, $\angle A < \angle B$ and central angles A and B are in standard position, as is central angle $(A - B)$. The initial side of central angle $(B - A)$ is the terminal side of $\angle A$. Moreover,

$$\angle(A - B) = -\angle(B - A)$$

and therefore

$$|\angle(A - B)| = |\angle(B - A)|$$

Thus, both central angles subtend chords of the same length:

$$\text{dist}(P_1, P_4) = \text{dist}(P_2, P_3)$$

Thus

$$[\text{dist}(P_1, P_4)]^2 = [\text{dist}(P_2, P_3)]^2$$

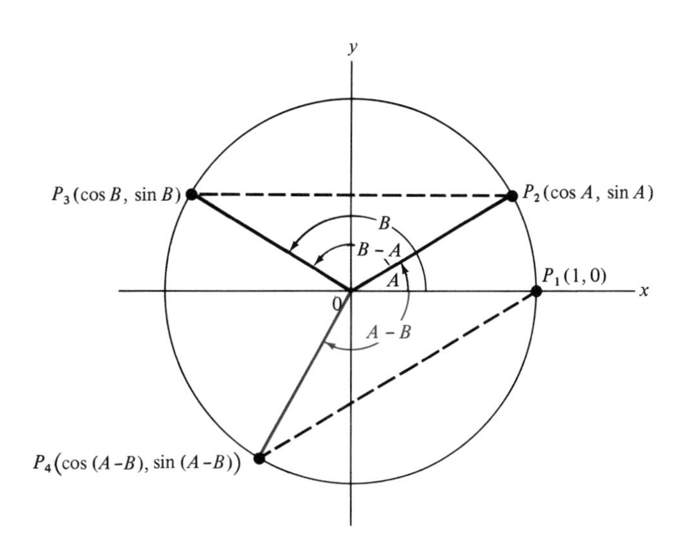

FIGURE 9.2

By the distance formula, p. 216,

$$(\cos (A - B) - 1)^2 + (\sin (A - B) - 0)^2$$
$$= (\cos B - \cos A)^2 + (\sin B - \sin A)^2$$

$$\cos^2 (A - B) - 2 \cos (A - B) + 1 + \sin^2 (A - B)$$
$$= (\cos^2 B - 2 \cos A \cos B + \cos^2 A)$$
$$+ (\sin^2 B - 2 \sin A \sin B + \sin^2 A)$$

Rearranging terms on each side, you obtain

$$\underbrace{\cos^2 (A - B) + \sin^2 (A - B)}_{1} + 1 - 2 \cos (A - B)$$
$$= \underbrace{\cos^2 B + \sin^2 B}_{1} + \underbrace{\cos^2 A + \sin^2 A}_{1}$$
$$- 2 \cos A \cos B - 2 \sin A \sin B$$
$$2 - 2 \cos (A - B) = 2 - 2(\cos A \cos B + \sin A \sin B)$$

$$\boxed{\cos (A - B) = \cos A \cos B + \sin A \sin B} \qquad \textbf{(9.8)}$$

This formula holds for all numbers (of radians) A and B, as do Formulas **(9.9)**–**(9.13)**, which follow.

To obtain the formula for the cosine of a *sum,* let

$x = A, y = -B.$

Then using Formula **(9.8)** for x and y, observe that

$$\cos (A + B) = \cos (x - y)$$
$$= \cos x \cos y + \sin x \sin y$$
$$= \cos A \underbrace{\cos (-B)}_{\cos B} + \sin A \underbrace{\sin (-B)}_{-\sin B}$$
$$= \cos A \cos B - \sin A \sin B$$

$$\boxed{\cos(A+B) = \cos A \cos B - \sin A \sin B} \qquad (9.9)$$

sin (A + B), sin (A − B)

In Section 8.1 you saw that for *acute angles A*,

$$\sin A = \cos\left(\frac{\pi}{2} - A\right) \qquad \text{and} \qquad \cos A = \sin\left(\frac{\pi}{2} - A\right).$$

In fact, these relationships hold for *all numbers* θ. For, by Formula **(9.8)**,

$$\cos\left(\frac{\pi}{2} - \theta\right) = \underbrace{\cos\frac{\pi}{2}}_{0}\cos\theta + \underbrace{\sin\frac{\pi}{2}}_{1}\sin\theta = \sin\theta$$

And replacing θ by $\frac{\pi}{2} - \theta$ in the preceding equation, note that

$$\underbrace{\cos\left(\frac{\pi}{2} - \left(\frac{\pi}{2} - \theta\right)\right)}_{\theta} = \sin\left(\frac{\pi}{2} - \theta\right),$$

$$\cos\theta = \sin\left(\frac{\pi}{2} - \theta\right)$$

$$\boxed{\sin\theta = \cos\left(\frac{\pi}{2} - \theta\right)} \qquad (9.10)$$

$$\boxed{\cos\theta = \sin\left(\frac{\pi}{2} - \theta\right)} \qquad (9.11)$$

You will use these together with Formula **(9.9)** with $\frac{\pi}{2} - A$ (in place of A) and $-B$ (in place of B) to obtain:

$$\begin{aligned}
\sin(A+B) &= \cos\left[\frac{\pi}{2} - (A+B)\right] \qquad \text{by (9.10)}\\
&= \cos\left[\left(\frac{\pi}{2} - A\right) + (-B)\right]\\
&= \underbrace{\cos\left(\frac{\pi}{2} - A\right)}_{\sin A}\underbrace{\cos(-B)}_{\cos B} - \underbrace{\sin\left(\frac{\pi}{2} - A\right)}_{\cos A}\underbrace{\sin(-B)}_{(-\sin B)}\\
&= \sin A \cos B + \cos A \sin B
\end{aligned}$$

$$\boxed{\sin(A+B) = \sin A \cos B + \cos A \sin B} \qquad (9.12)$$

To obtain the formula for $\sin(A - B)$, let $x = A$, $y = -B$. Then use Formula **(9.12)** for x and y:

$$\begin{aligned}
\sin(A - B) &= \sin(x + y)\\
&= \sin x \cos y + \cos x \sin y
\end{aligned}$$

$$= \sin A \underbrace{\cos(-B)}_{\cos B} + \cos A \underbrace{\sin(-B)}_{-\sin B}$$
$$= \sin A \cos B - \cos A \sin B$$

$$\boxed{\sin(A - B) = \sin A \cos B - \cos A \sin B} \qquad (9.13)$$

EXAMPLE 1 Compute (a) $\cos 75°$ (b) $\sin 75°$ without using a table.

Solution. $75° = 30° + 45°$

(a) $\cos 75° = \cos(30° + 45°) = \cos 30° \cos 45° - \sin 30° \sin 45°$

$$= \frac{\sqrt{3}}{2}\cdot\frac{\sqrt{2}}{2} - \frac{1}{2}\cdot\frac{\sqrt{2}}{2} = \frac{\sqrt{2}}{4}(\sqrt{3} - 1)$$

(b) $\sin 75° = \sin 30° \cos 45° + \cos 30° \sin 45°$

$$= \frac{1}{2}\cdot\frac{\sqrt{2}}{2} + \frac{\sqrt{3}}{2}\cdot\frac{\sqrt{2}}{2} = \frac{\sqrt{2}}{4}(1 + \sqrt{3})$$

□

You are now able to establish the properties of the sine and cosine "observed from their graphs" in Section 8.4. In establishing these properties you use only facts that have been verified. As an example,

$$\sin(\theta + \pi) = \sin\theta \underbrace{\cos\pi}_{-1} + \cos\theta \underbrace{\sin\pi}_{0} = -\sin\theta$$

Other properties are verified in the Exercises.

$\tan(A + B)$, $\tan(A - B)$

Formulas for $\tan(A + B)$ and $\tan(A - B)$ can be obtained from the preceding Addition Formulas. Let A and B be any numbers *for which A, B, and $A + B$ are not odd integral multiples of* $\frac{\pi}{2}$, so that $\tan A$, $\tan B$, and $\tan(A + B)$ are all defined.

$$\tan(A + B) = \frac{\sin(A + B)}{\cos(A + B)}$$
$$= \frac{\sin A \cos B + \cos A \sin B}{\cos A \cos B - \sin A \sin B}$$
$$= \frac{\dfrac{\sin A}{\cos A} + \dfrac{\sin B}{\cos B}}{1 - \dfrac{\sin A}{\cos A}\cdot\dfrac{\sin B}{\cos B}}$$
$$= \frac{\tan A + \tan B}{1 - \tan A \tan B}$$

Divide numerator and denominator by $\cos A \cos B$, which is nonzero.

$$\tan(A + B) = \frac{\tan A + \tan B}{1 - \tan A \tan B}, \qquad \left.\begin{matrix} A \\ B \\ A + B \end{matrix}\right\} \neq \pm\frac{\pi}{2}, \pm\frac{3\pi}{2}, \pm\frac{5\pi}{2}, \ldots \tag{9.14}$$

In Exercise 29, you will use Formula **(9.14)** together with

$\tan(-B) = -\tan B$

to establish

$$\tan(A - B) = \frac{\tan A - \tan B}{1 + \tan A \tan B}, \qquad \left.\begin{matrix} A \\ B \\ A - B \end{matrix}\right\} \neq \pm\frac{\pi}{2}, \pm\frac{3\pi}{2}, \pm\frac{5\pi}{2}, \ldots \tag{9.15}$$

EXAMPLE 2 Compute tan 15° without using a table.

Solution.

$$\tan 15° = \tan(45° - 30°) = \frac{\tan 45° - \tan 30°}{1 + \tan 45° \tan 30°}$$

$$= \frac{1 - \dfrac{\sqrt{3}}{3}}{1 + \dfrac{\sqrt{3}}{3}} = \frac{\dfrac{3 - \sqrt{3}}{\not{3}}}{\dfrac{3 + \sqrt{3}}{\not{3}}}$$

$$= \frac{(3 - \sqrt{3}) \cdot (3 - \sqrt{3})}{(3 + \sqrt{3}) \cdot (3 - \sqrt{3})} = \frac{(3 - \sqrt{3})^2}{6}$$

□

EXAMPLE 3 Suppose

$$\sin A = \frac{3}{5}, \text{ where } \angle A \text{ is in Quadrant I;}$$

$$\cos B = \frac{-12}{13}, \text{ where } \angle B \text{ is in Quadrant II.}$$

Find: (a) $\sin(A + B)$ (b) $\cos(A + B)$ (c) $\tan(A + B)$
(d) In which quadrant or on which axis is $\angle(A + B)$?

Solution. Angles A and B can be represented in terms of 3, 4, 5- and 5, 12, 13- right triangles, as in Figure 9.3 . Clearly,

$$\cos A = \frac{4}{5}, \qquad \tan A = \frac{3}{4}$$

In Quadrant II, the sine is positive and the tangent is negative:

$$\sin B = \frac{5}{13}, \qquad \tan B = \frac{-5}{12}$$

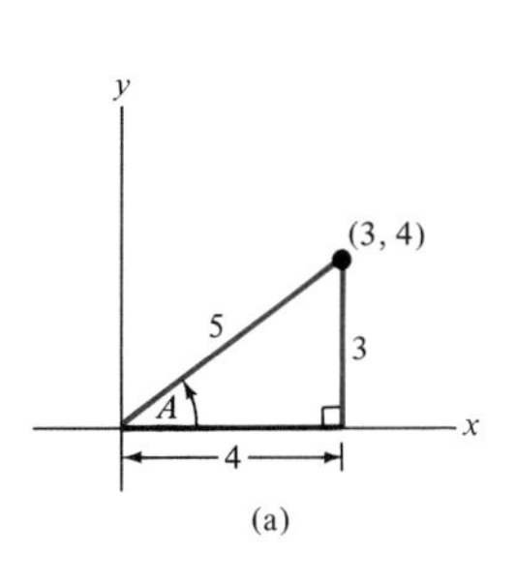

FIGURE 9.3

(a) $\sin A = \dfrac{3}{5}$

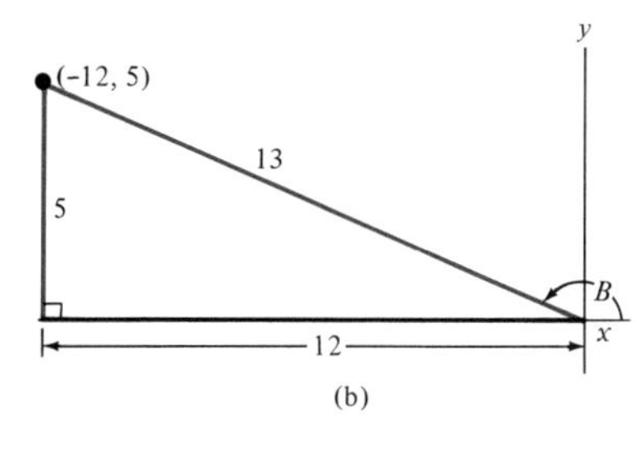

(b) $\cos B = \dfrac{-12}{13}$

(a) $$\sin(A+B) = \sin A \cos B + \cos A \sin B = \frac{3}{5}\left(\frac{-12}{13}\right) + \frac{4}{5}\cdot\frac{5}{13} = \frac{-16}{65}$$

(b) $$\cos(A+B) = \cos A \cos B - \sin A \sin B = \frac{4}{5}\left(\frac{-12}{13}\right) - \frac{3}{5}\cdot\frac{5}{13} = \frac{-63}{65}$$

(c) $$\tan(A+B) = \frac{\tan A + \tan B}{1 - \tan A \tan B} = \frac{\frac{3}{4} + \frac{-5}{12}}{1 - \frac{3}{4}\left(\frac{-5}{12}\right)} = \frac{\frac{9-5}{12}}{\frac{48+15}{48}}$$

$$= \frac{4}{12}\cdot\frac{48}{63} = \frac{16}{63}, \quad \text{or } \tan(A+B) = \frac{\sin(A+B)}{\cos(A+B)}$$

(d) $\angle A$ is in the first quadrant and $\angle B$ is in the second quadrant. Thus

$$0° < \angle A < 90°,$$
$$90° < \angle B < 180°,$$

and therefore, by adding corresponding parts of these inequalities,

$$90° < \angle(A+B) < 270°$$

Thus $\angle(A+B)$ is in either the second or the third quadrant or on the (negative) x-axis. However, $\tan(A+B)$ is positive; consequently, $\angle(A+B)$ must be in the third quadrant. □

EXERCISES

In Exercises 1–10 find: (a) the cosine, (b) the sine, and (c) the tangent of the indicated angle *without* using a table.

1. 105° **2.** 165° **3.** −15° **4.** −75° **5.** 315° **6.** 255° **7.** 195° **8.** 285° **9.** 345° **10.** 375°

In Exercises 11–16 find: **(a)** $\cos(A+B)$ **(b)** $\cos(A-B)$ **(c)** $\sin(A+B)$ **(d)** $\sin(A-B)$ **(e)** $\tan(A+B)$ **(f)** $\tan(A-B)$ **(g)** In which quadrant or on which axis is $\angle(A+B)$? **(h)** In which quadrant or on which axis is $\angle(A-B)$?

11. $\cos A = \dfrac{-3}{5}$, $\angle A$ in Quadrant II;
$\sin B = \dfrac{4}{5}$, $\angle B$ in Quadrant I

12. $\tan A = \dfrac{-4}{3}$, $\angle A$ in Quadrant II;
$\tan B = \dfrac{4}{3}$, $\angle B$ in Quadrant I

13. $\cos A = \dfrac{5}{13}$, $\angle A$ in Quadrant IV;
$\sin B = \dfrac{12}{13}$, $\angle B$ in Quadrant II

14. $\sin A = \dfrac{7}{25}$, $\angle A$ in Quadrant II;
$\cos B = \dfrac{-24}{25}$, $\angle B$ in Quadrant II

15. $\sin A = \dfrac{-\sqrt{5}}{5}$, $\angle A$ in Quadrant III;
$\tan B = -2$, $\angle B$ in Quadrant II

16. $\sin A = \cos B = \dfrac{3}{5}$;
$\angle A$ in Quadrant II, $\angle B$ in Quadrant IV

In Exercises 17–22 use an Addition Formula to establish each of the following identities. (Some of these were originally given in Section 8.4 .)

17. $\cos(\theta + \pi) = -\cos\theta$ **18.** $\cos(\pi - \theta) = -\cos\theta$ **19.** $\sin(\pi - \theta) = \sin\theta$

20. $\sin\left(\theta + \frac{\pi}{2}\right) = \cos\theta$ **21.** $\tan(\theta + \pi) = \tan\theta$ **22.** $\sin(\theta + 2\pi) = \sin\theta$

In Exercises 23–28 use an Addition Formula to find each value.

23. $\cos 50° \cos 40° - \sin 50° \sin 40°$ **24.** $\sin 35° \cos 10° + \cos 35° \sin 10°$

25. $\cos 110° \cos 80° + \sin 110° \sin 80°$ **26.** $\sin 160° \cos 25° - \cos 160° \sin 25°$

27. $\dfrac{\tan 36° + \tan 9°}{1 - \tan 36° \tan 9°}$ **28.** $\dfrac{\tan 105° - 1}{1 + \tan 105°}$

29. Verify the identity: $\tan(A - B) = \dfrac{\tan A - \tan B}{1 + \tan A \tan B}$. For which numbers A and B does it hold?

30. Verify the identity: $\cot(A + B) = \dfrac{\cot A \cot B - 1}{\cot A + \cot B}$. For which numbers A and B does it hold?

31. Verify the identity: $\tan\left(\theta + \frac{\pi}{2}\right) = \cot\theta$. For which numbers θ does it hold?

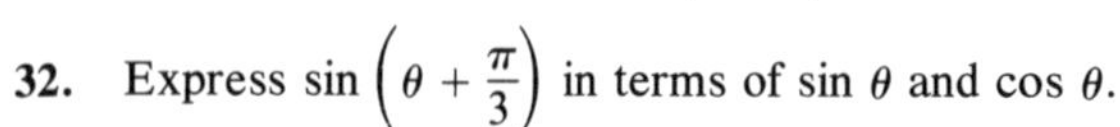

32. Express $\sin\left(\theta + \frac{\pi}{3}\right)$ in terms of $\sin\theta$ and $\cos\theta$.

33. Express $\cos\left(\theta - \frac{\pi}{4}\right)$ in terms of $\sin\theta$ and $\cos\theta$.

34. Express $\tan\left(\theta + \frac{\pi}{4}\right)$ in terms of $\tan\theta$. For which numbers θ does this hold?

35. Let $f(x) = \sin x$. Show that for $h \neq 0$,

$$\frac{f(x+h) - f(x)}{h} = \sin x\left(\frac{\cos h - 1}{h}\right) + \cos x\left(\frac{\sin h}{h}\right)$$

(This relationship is important in calculus.)

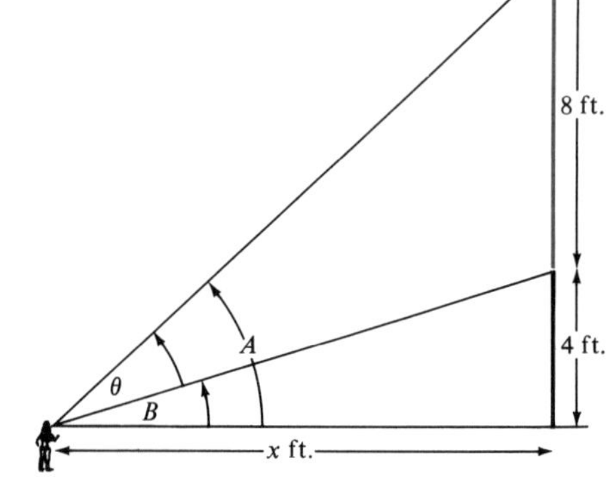

FIGURE 9.4

36. The lower edge of an 8-foot-high painting is 4 feet above an observer's eye level. [See Figure 9.4 .] The observer stands x feet from the wall. Let θ denote the angle subtended by the painting at the observer's eye level. Let B denote the angle subtended, at her eye level, by the portion of the wall above eye level but below the painting. Let $\angle A = \angle B + \angle \theta$, Express $\tan\theta$ as a function of x.

9.3 Multiple-Angle, Product, and Sum Formulas

The purpose of this section is to show how a number of other useful formulas can be derived from the Addition Formulas. Do NOT memorize them. Instead:

1. You should know that these Double-Angle, Half-Angle, Product and Sum Formulas exist. (They are boxed here for easy reference, and are also listed on the inside back cover.)
2. You should know how to use them, as indicated by the Illustrative Examples and Exercises.

3. You should have a sense of how to obtain each from previously derived formulas.

Double-Angle Formulas

Let A be any number. From the Addition Formulas,

$$\cos(A + A) = \cos A \cos A - \sin A \sin A,$$
$$\sin(A + A) = \sin A \cos A + \cos A \sin A,$$

and if A is not an integral multiple of $\frac{\pi}{4}$,

$$\tan(A + A) = \frac{\tan A + \tan A}{1 - \tan A \tan A}$$

Thus, you obtain the **Double-Angle Formulas:**

$$\boxed{\cos 2A = \cos^2 A - \sin^2 A} \qquad \textbf{(9.16)}$$

You can establish alternative forms of **(9.16)** as follows:

$$\cos 2A = \cos^2 A - (1 - \cos^2 A) = 2\cos^2 A - 1$$

Similarly,

$$\cos 2A = (1 - \sin^2 A) - \sin^2 A = 1 - 2\sin^2 A$$

$$\boxed{\cos 2A = 2\cos^2 A - 1} \qquad \textbf{(9.16')}$$

$$\boxed{\cos 2A = 1 - 2\sin^2 A} \qquad \textbf{(9.16'')}$$

Moreover, from the formulas for $\sin(A + A)$ and $\tan(A + A)$, you obtain:

$$\boxed{\sin 2A = 2\sin A \cos A} \qquad \textbf{(9.17)}$$

$$\boxed{\tan 2A = \frac{2\tan A}{1 - \tan^2 A}, \quad A \neq \pm\frac{\pi}{4}, \pm\frac{\pi}{2}, \pm\frac{3\pi}{4}, \pm\pi, \ldots} \qquad \textbf{(9.18)}$$

EXAMPLE 1 Suppose $\cos A = \frac{4}{5}$ and $\angle A$ is in Quadrant I. Find: (a) $\cos 2A$ (b) $\sin 2A$ (c) $\tan 2A$ (d) In which quadrant is $2A$?

Solution. Clearly, $\sin A = \frac{3}{5}$, $\tan A = \frac{3}{4}$

(a) $\cos 2A = \cos^2 A - \sin^2 A = \frac{16}{25} - \frac{9}{25} = \frac{7}{25}$

(b) $\sin 2A = 2\sin A \cos A = 2 \cdot \frac{3}{5} \cdot \frac{4}{5} = \frac{24}{25}$

(c) $\tan 2A = \frac{2\tan A}{1 - \tan^2 A} = \frac{2 \cdot \frac{3}{4}}{1 - \frac{9}{16}} = \frac{3}{2} \cdot \frac{16}{7} = \frac{24}{7}$

Note that you could have also obtained

$$\tan 2A = \frac{\sin 2A}{\cos 2A} = \frac{\frac{24}{25}}{\frac{7}{25}} = \frac{24}{7}$$

(d) The sine, cosine, and tangent of $2A$ are all positive. Therefore $2A$ is in Quadrant I. □

Half-Angle Formulas

By Formula (**9.16′**), $\cos 2\theta = 2\cos^2\theta - 1$. Thus

$$\frac{1+\cos 2\theta}{2} = \cos^2\theta$$

and

$$\pm\sqrt{\frac{1+\cos 2\theta}{2}} = \cos\theta \tag{9.19}$$

Also, by Formula (**9.16″**),

$$\cos 2\theta = 1 - 2\sin^2\theta,$$

$$\sin^2\theta = \frac{1-\cos 2\theta}{2},$$

$$\sin\theta = \pm\sqrt{\frac{1-\cos 2\theta}{2}} \tag{9.20}$$

Let $A = 2\theta$ in (**9.19**) and (**9.20**) and obtain the **Half-Angle Formulas:**

$$\boxed{\cos\frac{A}{2} = \pm\sqrt{\frac{1+\cos A}{2}}} \tag{9.21}$$

$$\boxed{\sin\frac{A}{2} = \pm\sqrt{\frac{1-\cos A}{2}}} \tag{9.22}$$

In each Formula (**9.19**)–(**9.22**), the choice of sign depends upon the quadrant of angle $\frac{A}{2}$. For example, in Formula (**9.22**),

$\sin\frac{A}{2} \geqslant 0$ if $0 \leqslant \frac{A}{2} \leqslant \pi$ that is, if $0 \leqslant A \leqslant 2\pi$. And

$\sin\frac{A}{2} < 0$ if $\pi < \frac{A}{2} < 2\pi$, that is, if $2\pi < A < 4\pi$.

EXAMPLE 2 Simplify the expression $\sqrt{\frac{1+\cos 200°}{2}}$.

Solution. In Formula (**9.21**), let $A = 200°$. Because the cosine is negative in the second quadrant,

$$\cos 100° = -\sqrt{\frac{1+\cos 200°}{2}}$$

$$-\cos 100° = \sqrt{\frac{1 + \cos 200°}{2}}$$ □

To obtain a formula for $\tan \frac{A}{2}$, where A is not an integral multiple of π, use

$$\tan^2 \frac{A}{2} = \frac{\sin^2 \frac{A}{2}}{\cos^2 \frac{A}{2}} = \frac{\frac{1 - \cos A}{2}}{\frac{1 + \cos A}{2}} = \frac{(1 - \cos A)\cdot(1 - \cos A)}{(1 + \cos A)\cdot(1 - \cos A)}$$

$$= \frac{(1 - \cos A)^2}{\underbrace{1 - \cos^2 A}_{\sin^2 A}} = \left(\frac{1 - \cos A}{\sin A}\right)^2$$

When you take square roots, you obtain

$$\tan \frac{A}{2} = \frac{1 - \cos A}{\sin A}, \qquad A \neq 0, \pm\pi, \pm 2\pi, \ldots \tag{9.23}$$

It is unnecessary to write $\pm$ because for all A,

$1 - \cos A \geq 0$ and

(if $\tan \frac{A}{2}$ is defined) $\tan \frac{A}{2}$ and $\sin A$ agree in sign.

(See Exercise 53.)

EXAMPLE 3 Find: (a) $\cos 22.5°$ (b) $\tan 22.5°$

Solution. $22.5° = \frac{1}{2} \cdot 45°$

(a) $$\cos 22.5° = \sqrt{\frac{1 + \cos 45°}{2}} = \sqrt{\frac{1 + \frac{\sqrt{2}}{2}}{2}} = \sqrt{\frac{\frac{2 + \sqrt{2}}{2}}{2}}$$

$$= \sqrt{\frac{2 + \sqrt{2}}{4}} = \frac{1}{2}\sqrt{2 + \sqrt{2}}$$

(b) $$\tan 22.5° = \frac{1 - \cos 45°}{\sin 45°} = \frac{1 - \frac{\sqrt{2}}{2}}{\frac{\sqrt{2}}{2}} = \frac{\frac{2 - \sqrt{2}}{2}}{\frac{\sqrt{2}}{2}}$$

$$= \frac{(2 - \sqrt{2})\cdot\sqrt{2}}{\sqrt{2}\cdot\sqrt{2}} = \frac{2\sqrt{2} - 2}{2} = \frac{2(\sqrt{2} - 1)}{2} = \sqrt{2} - 1$$ □

Formula (**9.23**) expresses $\tan \frac{A}{2}$ in terms of $\cos A$ and $\sin A$. In calculus it is sometimes useful to write each of the expressions

$\sin A$ and $\cos A$ in terms of $\tan \frac{A}{2}$. Thus if $\theta \neq \pm\frac{\pi}{2}, \pm\frac{3\pi}{2}, \pm\frac{5\pi}{2}, \ldots,$

$$\sin 2\theta = 2 \sin \theta \cos \theta = \frac{2 \sin \theta \cos \theta \cdot \frac{1}{\cos^2 \theta}}{1 \cdot \frac{1}{\cos^2 \theta}} = \frac{2 \tan \theta}{\sec^2 \theta}$$

$$= \frac{2 \tan \theta}{1 + \tan^2 \theta}$$

Replace θ by $\frac{A}{2}$:

$$\boxed{\sin A = \frac{2 \tan \frac{A}{2}}{1 + \tan^2 \frac{A}{2}}, \qquad A \neq \pm\pi, \pm 3\pi, \pm 5\pi, \ldots} \tag{9.24}$$

And, under the preceding condition on θ,

$$\cos 2\theta = \cos^2 \theta - \sin^2 \theta = \frac{(\cos^2 \theta - \sin^2 \theta) \cdot \frac{1}{\cos^2 \theta}}{1 \cdot \frac{1}{\cos^2 \theta}}$$

$$= \frac{\frac{\cos^2 \theta}{\cos^2 \theta} - \frac{\sin^2 \theta}{\cos^2 \theta}}{\sec^2 \theta} = \frac{1 - \tan^2 \theta}{\sec^2 \theta} = \frac{1 - \tan^2 \theta}{1 + \tan^2 \theta}$$

Replace θ by $\frac{A}{2}$:

$$\boxed{\cos A = \frac{1 - \tan^2 \frac{A}{2}}{1 + \tan^2 \frac{A}{2}}, \qquad A \neq \pm\pi, \pm 3\pi, \pm 5\pi, \ldots} \tag{9.25}$$

Product Formulas

Product Formulas involving cosines and sines can be derived from the Addition Formulas. These Product Formulas, **(9.26)–(9.29)**, hold for all A and B.

$$\cos (A + B) + \cos (A - B) = (\cos A \cos B - \sin A \sin B) + (\cos A \cos B + \sin A \sin B) = 2 \cos A \cos B$$

Thus

$$\boxed{\cos A \cos B = \frac{1}{2}[\cos (A + B) + \cos (A - B)]} \tag{9.26}$$

Also

$$-\cos(A+B)+\cos(A-B) = -(\cos A\cos B - \sin A\sin B) + (\cos A\cos B + \sin A\sin B) = 2\sin A\sin B$$

Therefore

$$\sin A\sin B = \frac{1}{2}[-\cos(A+B)+\cos(A-B)] \tag{9.27}$$

And

$$\sin(A+B)+\sin(A-B) = (\sin A\cos B + \cos A\sin B) + (\sin A\cos B - \cos A\sin B) = 2\sin A\cos B$$

Thus

$$\sin A\cos B = \frac{1}{2}[\sin(A+B)+\sin(A-B)] \tag{9.28}$$

Reversing the roles of A and B in Formula **(9.28)**, you obtain

$$\sin B\cos A = \frac{1}{2}[\sin(B+A)+\sin(B-A)]$$

Moreover, for all A and B,

$$\sin(B-A) = \sin(-(A-B)) = -\sin(A-B)$$

Thus

$$\cos A\sin B = \frac{1}{2}[\sin(A+B)-\sin(A-B)] \tag{9.29}$$

EXAMPLE 4 Express $\cos 50°\cos 40°$ in terms of $\cos 10°$.

Solution. Let $A = 50°$, $B = 40°$ in Formula **(9.26)**.

$$\cos 50°\cos 40° = \frac{1}{2}[\underbrace{\cos(50°+40°)}_{0} + \cos(50°-40°)]$$

$$= \frac{1}{2}\cos 10°$$

□

Sum Formulas

At times it is convenient to express the sum or difference of two cosines or of two sines in terms of products. Thus, in Formulas **(9.26)–(9.29)**, reverse sides, multiply both sides by 2, and let

$$x = A+B, \qquad y = A-B, \qquad \text{so that}$$

$$x+y = 2A \qquad \text{or} \qquad A = \frac{x+y}{2},$$

$$x-y = 2B \qquad \text{or} \qquad B = \frac{x-y}{2}.$$

You obtain the following **Sum Formulas:**

$$\cos x + \cos y = 2 \cos\left(\frac{x+y}{2}\right) \cos\left(\frac{x-y}{2}\right) \tag{9.30}$$

$$-\cos x + \cos y = 2 \sin\left(\frac{x+y}{2}\right) \sin\left(\frac{x-y}{2}\right) \tag{9.31}$$

$$\sin x + \sin y = 2 \sin\left(\frac{x+y}{2}\right) \cos\left(\frac{x-y}{2}\right) \tag{9.32}$$

$$\sin x - \sin y = 2 \cos\left(\frac{x+y}{2}\right) \sin\left(\frac{x-y}{2}\right) \tag{9.33}$$

EXAMPLE 5 Simplify: $\dfrac{\cos 4\theta - \cos 2\theta}{\sin 2\theta + \sin 4\theta}$

Solution. Apply Formulas **(9.31)** and **(9.32)** with $x = 2\theta$, $y = 4\theta$, so that

$$\frac{x+y}{2} = \frac{2\theta + 4\theta}{2} = 3\theta, \qquad \frac{x-y}{2} = \frac{2\theta - 4\theta}{2} = -\theta$$

$$\frac{\cos 4\theta - \cos 2\theta}{\sin 2\theta + \sin 4\theta} = \frac{-\cos 2\theta + \cos 4\theta}{\sin 2\theta + \sin 4\theta} = \frac{2 \sin 3\theta \sin(-\theta)}{2 \sin 3\theta \cos(-\theta)}$$
$$= \tan(-\theta) = -\tan \theta$$

□

EXERCISES

In Exercises 1–8 simplify each expression by means of a Double-Angle Formula.

1. $\cos^2 40° - \sin^2 40°$ **2.** $\sin 3\theta \cos 3\theta$ **3.** $\dfrac{\tan 17°}{1 - \tan^2 17°}$ **4.** $\cos^2 \dfrac{\pi}{12} - \sin^2 \dfrac{\pi}{12}$

5. $\sin \dfrac{3\pi}{8} \cos \dfrac{3\pi}{8}$ **6.** $\dfrac{\tan \frac{\pi}{8}}{1 - \tan^2 \frac{\pi}{8}}$ **7.** $1 - 2 \sin^2 \dfrac{5\pi}{8}$ **8.** $\cos^2 4A - \dfrac{1}{2}$

In Exercises 9–14 simplify each expression by means of a Half-Angle Formula.

9. $\sqrt{\dfrac{1 + \cos 42°}{2}}$ **10.** $1 - \cos 28°$ **11.** $\dfrac{1 - \cos 4A}{\sin 4A}$ **12.** $\cos \dfrac{\pi}{8}$ **13.** $\sin \dfrac{5\pi}{12}$ **14.** $\tan \dfrac{7\pi}{12}$

15. Express $\cos^2 5\theta$ in terms of $\cos 10\theta$.

16. Express $\sin^2 \dfrac{\theta}{2}$ in terms of $\cos \theta$.

In Exercises 17–22 find: **(a)** $\cos 2A$ **(b)** $\sin 2A$ **(c)** $\tan 2A$ **(d)** $\cos \dfrac{A}{2}$ **(e)** $\sin \dfrac{A}{2}$ **(f)** $\tan \dfrac{A}{2}$

17. $\cos A = \frac{3}{5}$ and $\angle A$ is in Quadrant I.

18. $\sin A = \frac{12}{13}$ and $\angle A$ is in Quadrant II.

19. $\tan A = 4$ and $\angle A$ is in Quadrant I.

20. $\cos A = \frac{-1}{5}$ and $\angle A$ is in Quadrant III.

21. $\tan A = \frac{-4}{3}$ and $\angle A$ is in Quadrant IV.

22. $\sin A = \frac{-1}{10}$ and $\angle A$ is in Quadrant III.

In Exercises 23–26 express each product in terms of a sum (or difference).

23. $\cos 50° \cos 10°$

24. $2 \sin 20° \sin 70°$

25. $8 \sin \frac{3\pi}{5} \cos \frac{\pi}{5}$

26. $\cos \frac{\pi}{7} \sin \frac{\pi}{10}$

In Exercises 27–30 express each sum (or difference) as a product. Simplify.

27. $\cos 48° + \cos 40°$

28. $\sin \frac{7\pi}{10} + \sin \frac{\pi}{10}$

29. $\cos 26° - \cos 12°$

30. $\sin \frac{\pi}{10} - \sin \frac{\pi}{20}$

In Exercises 31–39 simplify each expression by using Formulas **(9.16)–(9.33)**.

31. $\sin^2 15° - \cos^2 15°$

32. $\dfrac{\tan \frac{3\pi}{8}}{1 - \tan^2 \frac{3\pi}{8}}$

33. $\cos 75° + \cos 15°$

34. $\sin 140° - \sin 40°$

35. $\sin 29° \cos 16°$

36. $\cos 36° + \cos 9°$

37. $\sin 97° + \sin 7°$

38. $\cos 20° \sin 10°$

39. $\dfrac{\tan \frac{7\pi}{12}}{\sec^2 \frac{7\pi}{12}}$

40. Simplify $\sin \frac{\pi}{12}$ by using **(a)** a Half-Angle Formula **(b)** an Addition Formula [for $\sin (A - B)$].

(c) Show directly that both results are equal.

In Exercises 41–52 verify each identity.

41. $\sec 2A = \dfrac{\sec^2 A}{1 - \tan^2 A}$

42. $\cos A = \dfrac{\sin 2A}{2 \sin A}$

43. $2 \sin^2 2A = 1 - \cos 4A$

44. $\dfrac{1 + \cos 2A + \sin 2A}{2 (\cos A + \sin A)} = \cos A$

45. $\dfrac{\tan A}{2} = \dfrac{\tan \frac{A}{2}}{1 - \tan^2 \frac{A}{2}}$

46. $\tan \frac{A}{2} = \dfrac{\sin A}{1 + \cos A}$

47. $\sin 3A = 3 \sin A - 4 \sin^3 A$

48. $\sin (A + \frac{\pi}{3}) + \sin (A - \frac{\pi}{3}) = \sin A$

49. $\tan A \tan B = \dfrac{\cos (A - B) - \cos (A + B)}{\cos (A + B) + \cos (A - B)}$

50. $\dfrac{\sec^2 A}{\sec 2A} = 1 - \tan^2 A$

51. $8 \sin^4 A = \cos 4A - 4 \cos 2A + 3$

52. $\cot A = \dfrac{1 - \tan^2 \frac{A}{2}}{2 \tan \frac{A}{2}}$

53. **(a)** Show that for all A, $1 - \cos A \geqslant 0$.

(b) Show that if A is not an integral multiple of π, then $\tan \frac{A}{2}$ and $\sin A$ agree in sign.

(c) Given that $\tan^2 \frac{A}{2} = \left(\frac{1 - \cos A}{\sin A}\right)^2$, $A \neq 0, \pm\pi, \pm 2\pi, \ldots,$ show that $\tan \frac{A}{2} = \frac{1 - \cos A}{\sin A}$.

9.4 Trigonometric Equations

An equation involving trigonometric functions, such as

$$\sin \theta = \frac{1}{2} \qquad \text{or} \qquad 2\cos^2 \theta - \sin \theta = 1,$$

is known as a **trigonometric equation.** You want to know all values of θ in the domain of each function that satisfy the given equation.

Here is a relatively simple example that illustrates the present problem.

EXAMPLE 1 Suppose $0 \leq \theta < 2\pi$. Solve each equation.

(a) $\sin \theta = 1$ (b) $\sin \theta = \frac{1}{2}$ (c) $\sin \theta = 2$

Solution.

(a) $\sin \frac{\pi}{2} = 1$

$\frac{\pi}{2}$ is the only number in this interval whose sine is 1. The (only) root of the equation is $\frac{\pi}{2}$.

(b) The sine is positive in Quadrants I and II.

$$\sin \frac{\pi}{6} = \frac{1}{2} \qquad \text{and} \qquad \sin \frac{5\pi}{6} = \frac{1}{2}$$

The roots of the equation are $\frac{\pi}{6}$ and $\frac{5\pi}{6}$.

(c) For all θ,

$$-1 \leq \sin \theta \leq 1$$

Thus $\sin \theta$ is *never* 2. Therefore the equation

$$\sin \theta = 2$$

has no solution. □

Next, recall that the sine and cosine are each periodic with period 2π, whereas the tangent is periodic with period π. Thus if you want *all* roots of the equations in Example 1, not just those θ satisfying $0 \leq \theta < 2\pi$, you must add *multiples* of 2π to the ones obtained.

EXAMPLE 2 Let θ be *any* real number.

(a) The roots of $\sin \theta = 1$ are all numbers of the form

$$\frac{\pi}{2} + 2n\pi, \quad \text{where } n \text{ is any integer.}$$

Among these numbers are:

$$\frac{\pi}{2} \text{ (when } n = 0), \quad \frac{5\pi}{2} \text{ (when } n = 1), \quad \frac{9\pi}{2} \text{ (when } n = 2),$$

$$-\frac{3\pi}{2} \text{ (when } n = -1), \quad -\frac{7\pi}{2} \text{ (when } n = -2)$$

(b) The roots of $\sin\theta = \frac{1}{2}$ are all numbers of either of the forms:

$$\frac{\pi}{6} + 2n\pi \quad \text{or} \quad \frac{5\pi}{6} + 2n\pi, \quad \text{where } n \text{ is any integer.} \quad \square$$

In Example 3, a *composite trigonometric function* occurs in the equation.

EXAMPLE 3 Solve the equation: $\cos 2\theta = \frac{\sqrt{2}}{2}$

Solution. The cosine is positive in Quadrants I and IV.

$$\cos\frac{\pi}{4} = \frac{\sqrt{2}}{2} \quad \text{and} \quad \cos\frac{7\pi}{4} = \frac{\sqrt{2}}{2}$$

Thus solve:

$2\theta = \frac{\pi}{4}$	$2\theta = \frac{7\pi}{4}$
$\theta = \frac{\pi}{8}$	$\theta = \frac{7\pi}{8}$

The period of $\cos 2\theta$ is $\frac{2\pi}{2}$, or π. Thus add multiples of π to each of the values obtained. The roots are all numbers of either of the forms

$$\frac{\pi}{8} + n\pi \quad \text{or} \quad \frac{7\pi}{8} + n\pi, \quad \text{where } n \text{ is any integer.} \quad \square$$

EXAMPLE 4 Solve the equation:

$$\sin\theta\tan\theta = \sin\theta, \quad \text{where } \theta \neq \frac{\pi}{2} + n\pi,\ n = 0, \pm 1, \pm 2, \ldots$$

Solution.

$$\sin\theta\tan\theta - \sin\theta = 0 \qquad \text{Factor the left side.}$$

$$\sin\theta(\tan\theta - 1) = 0$$

$\sin\theta = 0$	$\tan\theta - 1 = 0$
	$\tan\theta = 1$
Note that	Note that
$\sin 0 = 0$ and $\sin\pi = 0$.	$\tan\frac{\pi}{4} = 1$.

Add multiples of 2π to 0 and to π, thus obtaining:

$$\ldots, -2\pi, 0, 2\pi, 4\pi, \ldots$$

and

$$\ldots, -3\pi, -\pi, \pi, 3\pi, \ldots$$

or together,

$$\ldots, -3\pi, -2\pi, -\pi, 0, \pi, 2\pi, 3\pi, \ldots$$

Because $\tan\theta$ has period π, add multiples of π to $\frac{\pi}{4}$. Thus, the roots of the given equation are all numbers of either of the forms

$$n\pi \quad \text{or} \quad \frac{\pi}{4} + n\pi, \quad \text{where } n \text{ is any integer.}$$ □

quadratic equations

EXAMPLE 5 Suppose $0 \leqslant \theta < 2\pi$. Solve the equation: $2\cos^2\theta - \sin\theta = 1$

Solution. Use $\cos^2\theta = 1 - \sin^2\theta$ to write this as a "quadratic equation" in $\sin\theta$ (instead of in x).

$$2(1 - \sin^2\theta) - \sin\theta = 1$$
$$2 - 2\sin^2\theta - \sin\theta = 1$$
$$2\sin^2\theta + \sin\theta - 1 = 0$$

This equation is of the form

$$2x^2 + x - 1 = 0.$$

Factor the left side

$$(2x - 1)(x + 1) = 0$$

Now replace x by $\sin\theta$ to obtain

$$(2\sin\theta - 1)(\sin\theta + 1) = 0$$

$2\sin\theta - 1 = 0$	$\sin\theta + 1 = 0$
$2\sin\theta = 1$	$\sin\theta = -1$
$\sin\theta = \frac{1}{2}$	$\theta = \frac{3\pi}{2}$
$\theta = \frac{\pi}{6}, \frac{5\pi}{6}$	

The roots in this interval are $\frac{\pi}{6}$, $\frac{5\pi}{6}$, and $\frac{3\pi}{2}$. □

EXAMPLE 6 Suppose $0 \leqslant \theta < 2\pi$. Solve the equation: $\sin^2\theta - 3\cos\theta = 2$

Solution. Write this as a "quadratic equation" in terms of $\cos\theta$.

$$1 - \cos^2\theta - 3\cos\theta = 2$$
$$\underbrace{\cos^2\theta}_{x^2} + \underbrace{3\cos\theta}_{3x} + 1 = 0$$

The left side of the equation

$$x^2 + 3x + 1 = 0$$

does not readily factor; apply the quadratic formula to the corresponding trigonometric equation and solve for $\cos\theta$.

$$\cos\theta = \frac{-3 \pm \sqrt{3^2 - 4(1)(1)}}{2} = \frac{-3 \pm \sqrt{5}}{2}$$

From Table 1, in the appendix, use $\sqrt{5} \approx 2.236$

$$\cos\theta \approx \frac{-3 + 2.236}{2} = \frac{-.764}{2} = -.382$$

From Table 3, to the nearest *degree*,

$$\cos 68^\circ \approx .382$$

The cosine is negative in Quadrants II and III. Thus

$$\theta \approx 180^\circ - 68^\circ = 112^\circ$$

and

$$\theta \approx 180^\circ + 68^\circ = 248^\circ$$

$$\cos\theta \approx \frac{-3 - 2.236}{2} = \frac{-5.236}{2} = -2.618$$

But $-1 \leqslant \cos\theta \leqslant 1$. (No root.)

In radian measure, the roots are (approximately)

$$112\cdot\frac{\pi}{180} \qquad \text{and} \qquad 248\cdot\frac{\pi}{180}$$ □

EXAMPLE 7 Solve the equation: $\cos 2A + \sin^2 A = 0$

Solution. Use the identity $\cos 2A = \cos^2 A - \sin^2 A$ to obtain

$$\cos^2 A - \sin^2 A + \sin^2 A = 0$$
$$\cos^2 A = 0$$
$$\cos A = 0$$

Recall that

$$\cos\frac{\pi}{2} = 0 \qquad \text{and} \qquad \cos\underbrace{\frac{3\pi}{2}}_{\frac{\pi}{2}+\pi} = 0$$

Thus the roots are all numbers of the form

$$\frac{\pi}{2} + n\pi, \qquad \text{where } n \text{ is any integer.}$$ □

EXAMPLE 8 Suppose $0 \leqslant \theta < 2\pi$ $\left(\text{and } \theta \neq \frac{\pi}{2}, \frac{3\pi}{2}\right)$. Solve the equation

$$\sec^2\theta - \sec\theta - 2 = 0$$

Solution. Factor the left side.

$$(\sec\theta - 2)(\sec\theta + 1) = 0$$

$$\begin{array}{rl|rl} \sec\theta - 2 = 0 & & \sec\theta + 1 = 0 & \\ \sec\theta = 2 & & \sec\theta = -1 & \\ \cos\theta = \frac{1}{2} & & \cos\theta = -1 & \qquad \cos\theta = \frac{1}{\sec\theta} \\ \cos\frac{\pi}{3} = \frac{1}{2} & \text{and} & \cos\pi = -1 & \\ \cos\frac{5\pi}{3} = \frac{1}{2} & & & \end{array}$$

The roots in this domain are $\frac{\pi}{3}$, π, and $\frac{5\pi}{3}$. □

EXERCISES

In Exercises 1–24 suppose $0 \leq \theta < 2\pi$ and θ is in the domain of each function that occurs in the equation. Solve the equation. Do *not* use a table.

1. $\sin\theta = 0$

2. $\cos\theta = \frac{\sqrt{2}}{2}$

3. $\tan\theta = \sqrt{3}$

4. $\cos\theta = \sqrt{3}$

5. $\sin 2\theta = \frac{\sqrt{3}}{2}$

6. $\tan 2\theta = 1$

7. $\sqrt{2}\sin\theta = 1$

8. $\cos\frac{\theta}{2} = -\frac{1}{2}$

9. $\sec 3\theta = 2$

10. $\sin\left(\theta + \frac{\pi}{6}\right) = 1$

11. $\sin\theta\,(1 + \cos\theta) = 0$

12. $\left(\sin\theta - \frac{1}{2}\right)\left(\cos\theta + \frac{1}{2}\right) = 0$

13. $2\sin^2\theta - 3\sin\theta + 1 = 0$

14. $\tan^2\theta = \tan\theta$

15. $4\cos^2\theta = 1$

16. $3\tan^2\theta - 1 = 0$

17. $4\sin^2\theta = 3$

18. $\csc^2\theta = 2$

19. $(2\sin\theta - \sqrt{2})(2\cos\theta - \sqrt{3}) = 0$

20. $2\cos^2\theta = 3(1 + \sin\theta)$

21. $\sin 2\theta - \sin\theta = 0$

22. $\sec^2\theta + \tan\theta = 1$

23. $\sin^2\theta + 1 = \frac{5}{2}\sin\theta$

24. $\sin^4\theta = \sin^2\theta$

In Exercises 25–40 find *all* roots of the given equation. Do *not* use a table.

25. $\sin\theta + 1 = 0$

26. $2\cos 3\theta - 1 = 0$

27. $(1 + \tan\theta)(1 - \tan\theta) = 0$

28. $\sin\theta\tan\theta = \sin\theta$

29. $(2\cos\theta - 1)(\sqrt{2}\sin\theta + 2) = 0$

30. $4\sin^2\theta - 3 = 0$

31. $\cos^2\theta + 2\cos\theta = 3$

32. $3 - 2\cos^2\theta = 3\sin\theta$

33. $\sec^2\theta - 3\sec\theta + 2 = 0$

34. $\tan\theta = \sin\theta$

35. $\sin 2\theta = 2\sin\theta$

36. $\sin\frac{\theta}{2} = \cos\theta$

37. $2\cos^2\theta - 9\cos\theta - 5 = 0$

38. $2\csc^2\theta - 5\csc\theta + 2 = 0$

39. $\cot^2\theta + \cot\theta = 0$

40. $5\sin\theta - 6\cos^2\theta + 2 = 0$

In Exercises 41–50 suppose $0 \leq \theta < 2\pi$ and θ is in the domain of each function that occurs in the equation. Use Table 3 to solve each equation. Express your results to the nearest degree; then convert to radian measure.

41. $\sin\theta = \frac{3}{4}$

42. $\cos 2\theta = .9703$

43. $\tan\theta - 1 = .192$

44. $\tan^2 \theta = 9$

45. $(4 \cos \theta - 1)(5 \sin \theta + 1) = 0$

46. $4 \cos^2 \theta + 7 \cos \theta - 2 = 0$

47. $4 \tan^2 \theta - 12 \tan \theta + 9 = 0$

48. $\sin^2 \theta - 2 \sin \theta = 2$

49. $\cos^2 \theta + \cos \theta + 1 = 0$

50. $\tan^2 \theta - \tan \theta - 1 = 0$

In Exercises 51–54 suppose $0 \leqslant \theta < 2\pi$. Solve each inequality.

51. $\sin \theta > \frac{1}{2}$

52. $\cos \theta < \frac{1}{2}$

53. $\tan \theta \geqslant 1$, where $\theta \neq \frac{\pi}{2}, \frac{3\pi}{2}$

54. $4 \sin^2 \theta \geqslant 1$

9.5 Inverse Trigonometric Functions

Inverse functions were introduced in Section 3.5, and were further discussed in Section 5.4 and in Chapter 7. *Inverse trigonometric functions* have properties that are useful in calculus and in applications. Instead of finding the value of y when given x in the equation

$$y = \sin x,$$

you will now ask:

The sine of what number equals y?

arc sin

The sine function takes on all values between -1 and 1 infinitely many times. [See Figure 9.5(a).] Recall that the inverse of a function is only defined for one-to-one functions (Section 3.5). Thus if you simply interchanged the coordinates, the resulting correspondence, $x = \sin y$, [illustrated in color in Figure 9.5(b)] would *not* represent a function. Note that there are vertical lines (such as the y-axis) that intersect the (colored) graph more than once.

A similar difficulty arose when you considered square roots (Section 5.4) because the function $g(x)$ defined by

$$g(x) = x^2 \qquad \text{for } all \text{ real numbers } x$$

is not one-to-one. You got around this by considering, instead, the *restricted* function $G(x)$ defined by

$$G(x) = x^2 \qquad \text{for} \qquad x \geqslant 0.$$

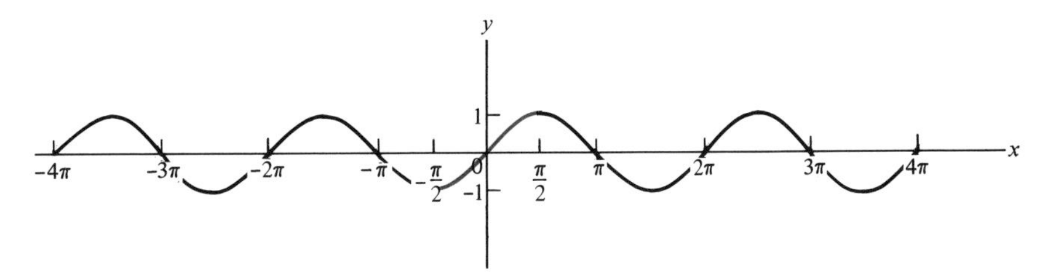

FIGURE 9.5(a) $y = \sin x$

Note that on the interval $\left[-\frac{\pi}{2}, \frac{\pi}{2}\right]$, the sine takes on all values between -1 and 1.

As x varies over the real numbers, the sine takes on each value between -1 and 1 infinitely many times.

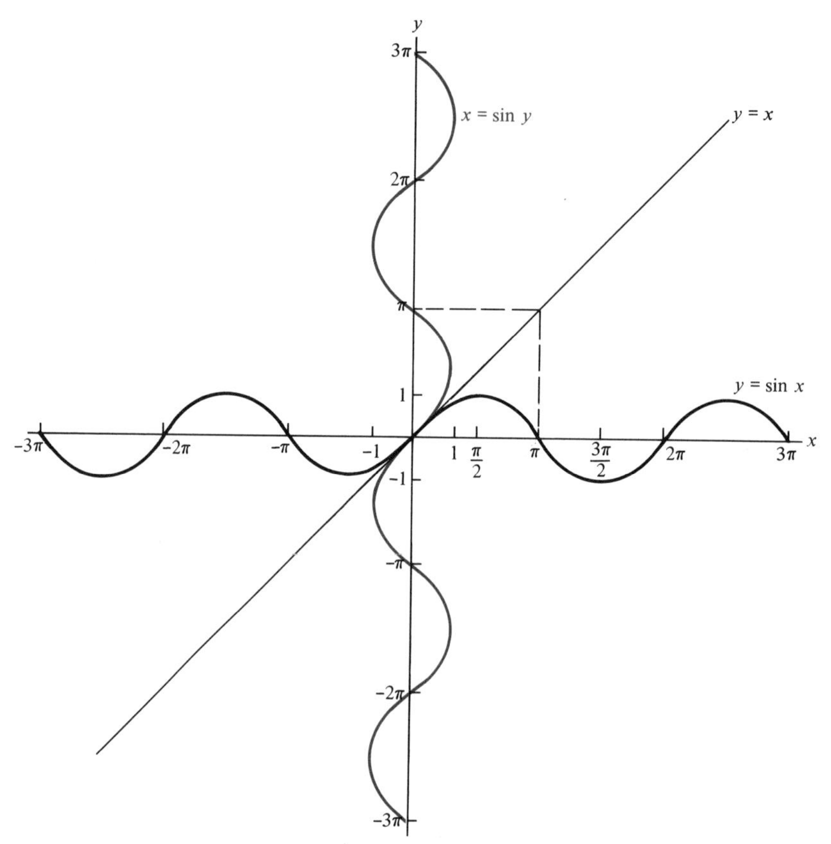

FIGURE 9.5(b) The graph of $x = \sin y$ is obtained by reflecting the graph of $y = \sin x$ in the line $y = x$.

This restricted function is one-to-one, and therefore has an inverse, given by

$$G^{-1}(x) = x^{1/2} \qquad \text{for} \qquad x \geqslant 0$$

[See Figure 9.6 .]

Figure 9.5(a) suggests that the sine function be restricted to some closed interval of length π on which it is increasing (or on which it is decreasing). On any such interval the sine takes on all its values. Because of its centrality, the interval $\left[-\frac{\pi}{2}, \frac{\pi}{2}\right]$ is commonly used. Thus let

$$F(x) = \sin x, \qquad -\frac{\pi}{2} \leqslant x \leqslant \frac{\pi}{2}.$$

Then F is an increasing function on this interval, and is therefore one-to-one. As x increases from $-\frac{\pi}{2}$ to $\frac{\pi}{2}$, $F(x)$ increases from -1 to 1. F^{-1} is therefore defined, and is called *arc sin*.† Each point

†We use *arc sin* x rather than $\sin^{-1}(x)$ because the latter suggests $\frac{1}{\sin x}$, which can be confused with $\csc x$. However, $\sin^{-1}(x)$ is also in use.

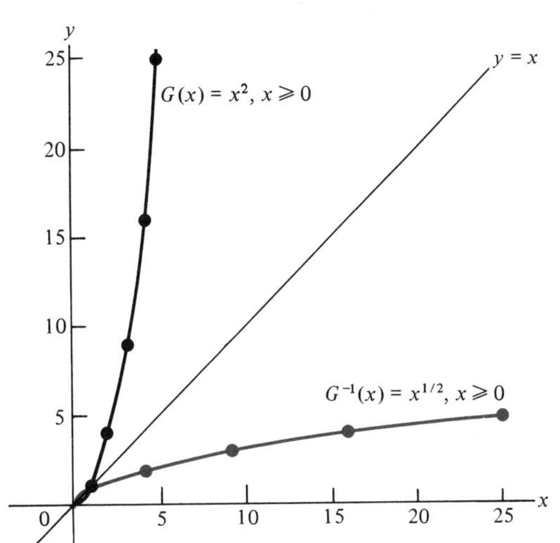

FIGURE 9.6 Note that $G(5) = 25$ $(5^2 = 25)$, whereas $G^{-1}(25) = 5$ $(25^{1/2} = 5)$.

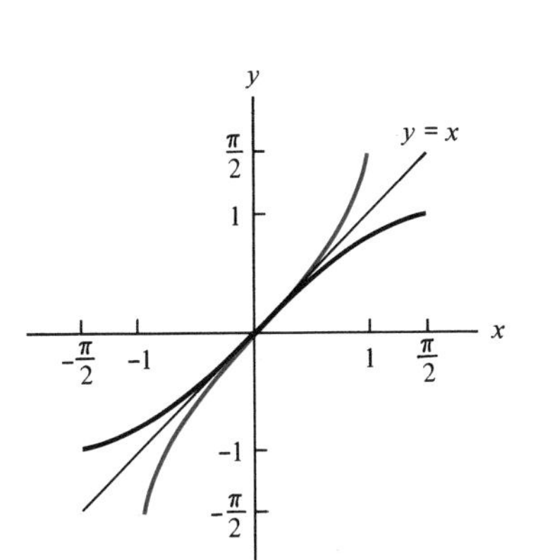

FIGURE 9.7
$F(x) = \sin(x)$, $-\frac{\pi}{2} \leqslant x \leqslant \frac{\pi}{2}$
[in black]

$F^{-1}(x) = \text{arc sin } x$, $-1 \leqslant x \leqslant 1$
[in color]

Note that the sine curve "bends" toward the x-axis whereas the arc sine curve "bends" toward the y-axis.

(b, a) of the arc sin graph is obtained from the point (a, b) of the *restricted* sine graph by reflection in the line $y = x$. [See Figure 9.7.]

The domain of F^{-1} (arc sin) is the range of F, namely, $[-1, 1]$. The range of the arc sin is the domain of F, or $\left[-\frac{\pi}{2}, \frac{\pi}{2}\right]$. And arc sin x is also an increasing function—hence, a one-to-one function. Its inverse is $F(x)$, the *restricted sine function*.

$$\text{If } -\frac{\pi}{2} \leqslant x \leqslant \frac{\pi}{2}, \quad \text{arc sin}(\sin x) = x$$
$$\text{If } -1 \leqslant x \leqslant 1, \quad \sin(\text{arc sin } x) = x$$

EXAMPLE 1 Find: (a) $\text{arc sin}\frac{1}{2}$ (b) $\text{arc sin}\frac{\sqrt{2}}{2}$ (c) $\text{arc sin}(-1)$

Solution. In each case find an x satisfying

$$-\frac{\pi}{2} \leqslant x \leqslant \frac{\pi}{2},$$

whose sine is the number in question.

(a) $\text{arc sin}\left(\frac{1}{2}\right) = x$ when $\sin x = \frac{1}{2}$

Thus take $x = \frac{\pi}{6}$, and therefore $\text{arc sin}\left(\frac{1}{2}\right) = \frac{\pi}{6}$

(b) $\sin\frac{\pi}{4} = \frac{\sqrt{2}}{2}$ and thus $\text{arc sin}\left(\frac{\sqrt{2}}{2}\right) = \frac{\pi}{4}$

(c) $\sin\left(-\frac{\pi}{2}\right) = -1$ and thus $\text{arc sin}(-1) = -\frac{\pi}{2}$ □

arc cos The cosine function is not one-to-one. Nor is it one-to-one on $\left[-\frac{\pi}{2}, \frac{\pi}{2}\right]$, the interval chosen for the restricted sine. [See Figure 8.59(b) on page 341.] To define an inverse function for the cosine, restrict the domain to $[0, \pi]$. On this interval the cosine *decreases* from 1 to -1, and thus it is one-to-one. Also, it takes on all its values on $[0, \pi]$. Let

$$G(x) = \cos x, \qquad 0 \leqslant x \leqslant \pi,$$

and let

$$G^{-1}(x) = \text{arc cos } x, \qquad -1 \leqslant x \leqslant 1. \qquad \text{[See Figure 9.8 .]}$$

Thus the range of G^{-1}, the arc cosine, is $[0, \pi]$, the domain of G.

If $0 \leqslant x \leqslant \pi$, $\quad$ arc cos (cos x) $= x$
If $-1 \leqslant x \leqslant 1$, $\quad$ cos (arc cos x) $= x$

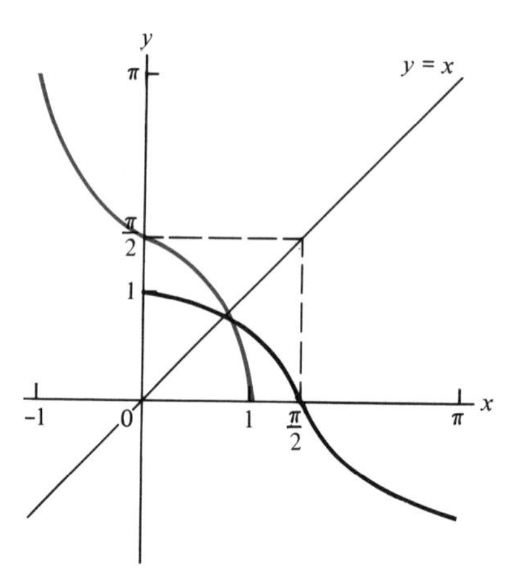

FIGURE 9.8 $G(x) = \cos x, 0 \leqslant x \leqslant \pi$ [in black]
$G^{-1}(x) = \text{arc cos } x, -1 \leqslant x \leqslant 1$ [in color]
The graph of G^{-1} is obtained by reflecting the graph of G in the line $y = x$. Note that the cosine curve "bends" toward the x-axis, whereas the arc cosine curve "bends" toward the y-axis.

EXAMPLE 2 Find: (a) $\text{arc cos } \frac{\sqrt{3}}{2}$ (b) arc cos 0 (c) $\text{arc cos}\left(-\frac{1}{2}\right)$

Solution. In each case find an x satisfying $0 \leqslant x \leqslant \pi$ whose cosine is the number in question.

(a) $\text{arc cos } \frac{\sqrt{3}}{2} = x$ when $\cos x = \frac{\sqrt{3}}{2}$

Thus take $x = \frac{\pi}{6}$, and therefore $\text{arc cos } \frac{\sqrt{3}}{2} = \frac{\pi}{6}$

(b) $\cos \frac{\pi}{2} = 0$ and thus $\text{arc cos } 0 = \frac{\pi}{2}$

(c) $\cos\frac{2\pi}{3} = -\frac{1}{2}$ and thus $\text{arc cos}\left(-\frac{1}{2}\right) = \frac{2\pi}{3}$ □

arc tan

To define an inverse for the tangent, restrict x to the *open* interval

$$-\frac{\pi}{2} < x < \frac{\pi}{2},$$

as illustrated in Figure 9.9 . In this interval the tangent takes on *all* real values. Recall that the tangent is not defined at either $-\frac{\pi}{2}$ or $\frac{\pi}{2}$.

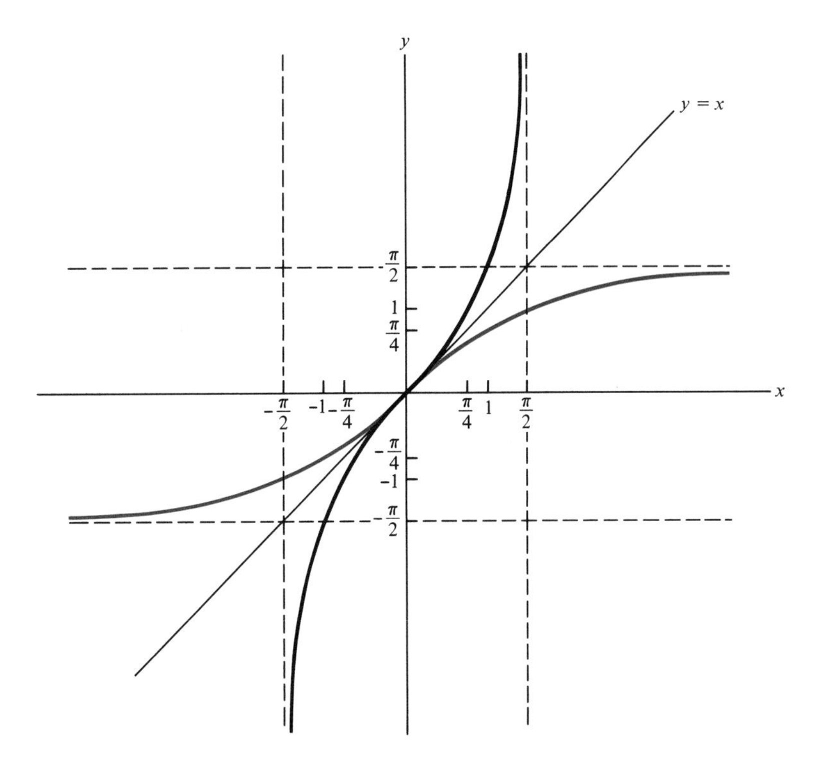

FIGURE 9.9 $H(x) = \tan x,\ -\frac{\pi}{2} < x < \frac{\pi}{2}$ [in black]
$H^{-1}(x) = \text{arc tan } x$, for all x [in color]
The graph of H^{-1} is obtained by reflecting the graph of H in the line $y = x$. Note that the tangent curve "bends" toward the y-axis, whereas the arc tangent curve "bends" toward the x-axis.

Let

$$H(x) = \tan x, \qquad -\frac{\pi}{2} < x < \frac{\pi}{2},$$

and let

$$H^{-1}(x) = \text{arc tan } x, \quad \text{for } all\ x.$$

Then

domain (arc tan x) = range tan x = set of all real numbers

$$\text{range (arc tan } x) = \text{domain tan } x = \left(-\frac{\pi}{2}, \frac{\pi}{2}\right)$$

$$\text{If} \quad -\frac{\pi}{2} < x < \frac{\pi}{2}, \qquad \text{arc tan (tan } x) = x$$

$$\text{For all } x, \qquad \tan(\text{arc tan } x) = x$$

EXAMPLE 3 Find: (a) arc tan 1 (b) arc tan $(-\sqrt{3})$ (c) arc tan 8

Solution. In each case find an x satisfying

$$-\frac{\pi}{2} < x < \frac{\pi}{2}$$

whose tangent is the number in question.

(a) $\quad \text{arc tan } 1 = x \quad \text{when} \quad \tan x = 1$

Thus take $x = \frac{\pi}{4}$, and therefore $\text{arc tan } 1 = \frac{\pi}{4}$

(b) $\quad \tan\left(-\frac{\pi}{3}\right) = -\sqrt{3}$. Thus $\text{arc tan }(-\sqrt{3}) = -\frac{\pi}{3}$

(c) Use Table 3 to find that, to the nearest degree, $\tan 83° \approx 8$. Thus (in radian measure)

$$\tan\frac{83\pi}{180} \approx 8 \quad \text{and} \quad \text{arc tan } 8 \approx \frac{83\pi}{180}$$

□

EXAMPLE 4 Find: sin (arc tan (−2))

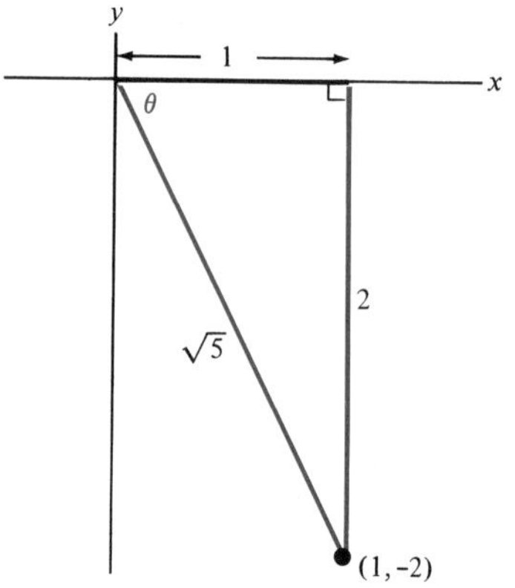

Solution. Let $\theta = \text{arc tan }(-2)$. Then

$$-\frac{\pi}{2} < \theta < \frac{\pi}{2}$$

Because $\tan\theta = -2$ (negative), the angle that measures θ radians is in Quadrant IV, rather than in Quadrant I. Thus, consider Figure 9.10 and note that the length of the hypotenuse is $\sqrt{5}$. Thus

$$\sin(\text{arc tan }(-2)) = \sin\theta = \frac{-2}{\sqrt{5}} = \frac{-2\sqrt{5}}{5}$$

□

FIGURE 9.10

EXAMPLE 5 Find: $\sin\left(2 \text{ arc cos } \frac{4}{5}\right)$.

Solution. Let $\theta = \text{arc cos } \frac{4}{5}$. Then $0 \leqslant \theta \leqslant \pi$. Use $\sin 2\theta = 2\sin\theta\cos\theta$. Then note that because $\cos\theta = \frac{4}{5}$ (positive), the

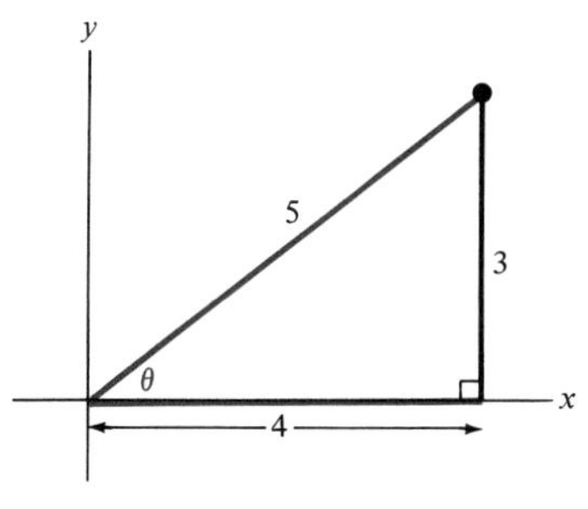

FIGURE 9.11

angle that measures θ radians is in Quadrant I, rather than in Quadrant II. Use Figure 9.11 .

$$\sin\Big(\underbrace{2 \text{ arc cos } \frac{4}{5}}_{\theta}\Big) = \sin 2\theta = 2 \sin\theta \cos\theta = 2\cdot\frac{3}{5}\cdot\frac{4}{5} = \frac{24}{25} \qquad \square$$

Although inverse functions for the cotangent, secant, and cosecant can be defined, they are of less importance than the preceding inverse functions.

EXERCISES

In Exercises 1–12 find each value. (Do *not* use a table.)

1. $\text{arc sin } \frac{\sqrt{3}}{2}$ **2.** $\text{arc sin } \left(-\frac{1}{2}\right)$ **3.** $\text{arc cos } \frac{\sqrt{2}}{2}$ **4.** $\text{arc cos } (-1)$

5. $\text{arc tan } \frac{\sqrt{3}}{3}$ **6.** $\text{arc tan } (-1)$ **7.** $\text{arc cos } 1$ **8.** $\text{arc sin } \left(-\frac{\sqrt{2}}{2}\right)$

9. $\text{arc sin } 0$ **10.** $\text{arc tan } 0$ **11.** $\text{arc tan } (-\sqrt{3})$ **12.** $\text{arc cos } \left(-\frac{\sqrt{3}}{2}\right)$

In Exercises 13–18 use Table 3 to find each value. Express your answer in terms of radians.

13. $\text{arc sin } .2756$ **14.** $\text{arc tan } .7002$ **15.** $\text{arc cos } .5150$ **16.** $\text{arc sin } (-.4067)$

17. $\text{arc tan } (-1.376)$ **18.** $\text{arc cos } (-.0349)$

In Exercises 19–38 find each value. (Do *not* use a table.)

19. $\sin\left(\text{arc sin } \frac{1}{3}\right)$ **20.** $\text{arc cos } \left(\cos \frac{\pi}{12}\right)$ **21.** $\sin\left(\text{arc cos } \frac{3}{5}\right)$ **22.** $\cos\,(\text{arc tan } 3)$

23. $\cos\left(\text{arc sin } \frac{3}{4}\right)$ **24.** $\sin\left(\text{arc tan } (-1)\right)$ **25.** $\tan\left(\text{arc cos } \frac{\sqrt{2}}{2}\right)$ **26.** $\sin\left(\text{arc cos } \left(-\frac{1}{2}\right)\right)$

27. $\text{arc sin } \left(\cos \frac{\pi}{3}\right)$ **28.** $\text{arc cos } \left(\tan \frac{\pi}{4}\right)$ **29.** $\text{arc tan } (\cos 0)$ **30.** $\text{arc cos}\left(\sin\left(-\frac{\pi}{6}\right)\right)$

31. $\cos\left(2 \text{ arc sin } \frac{4}{5}\right)$ **32.** $\sin\left(2 \text{ arc tan } \frac{4}{3}\right)$ **33.** $\tan\left(\frac{1}{2} \text{ arc cos } \frac{3}{5}\right)$

34. $\sin\left(\text{arc tan } 2 + \text{arc cos } \frac{1}{2}\right)$ **35.** $\text{arc sin } \left(\sin \frac{3\pi}{4}\right)$ **36.** $\text{arc cos}\left(\cos\left(-\frac{\pi}{3}\right)\right)$

37. $\text{arc tan } (\tan \pi)$ **38.** $\text{arc sin } \left(\sin \frac{7\pi}{6}\right)$

In Exercises 39–42 simplify each expression.

39. $\text{arc cos } x + \text{arc sin } x$, where $0 \leqslant x \leqslant 1$ **40.** $\text{arc sin } (\cos \theta)$, where $0 \leqslant \theta \leqslant \pi$

41. $\cot\,(\text{arc tan } x)$ **42.** $\sqrt{1 + x^2}\cdot \sin\,(\text{arc tan } x)$

In Exercises 43–49 sketch the following graphs.

43. $2 \text{ arc sin } x, -1 \leq x \leq 1$ **44.** $\text{arc sin } 2x, -\frac{1}{2} \leq x \leq \frac{1}{2}$ **45.** $\frac{1}{2} \text{ arc cos } x, -1 \leq x \leq 1$

46. $\text{arc cos } \frac{x}{2}, -2 \leq x \leq 2$ **47.** $\text{arc tan } 4x$ **48.** $2 \text{ arc tan } \frac{x}{2}$ **49.** $\sin(\text{arc sin } x), -1 \leq x \leq 1$

50. A 15-foot ladder leans against a wall. Express the angle between the ladder and the ground as a function of

(a) the distance x from the foot of the ladder to the bottom of the wall.
(b) the height y of the top of the ladder.

In Exercises 51 and 52 solve for x.

51. $\text{arc sin } 2x = \frac{\pi}{4}$ **52.** $\text{arc cos } (x + 1) = \frac{\pi}{3}$

53. In Figure 9.12 match each curve with one of the following.

$y = \sin x$ $y = \cos x$ $y = \tan x$ $y = \text{arc sin } x$ $y = \text{arc cos } x$ $y = \text{arc tan } x$

FIGURE 9.12

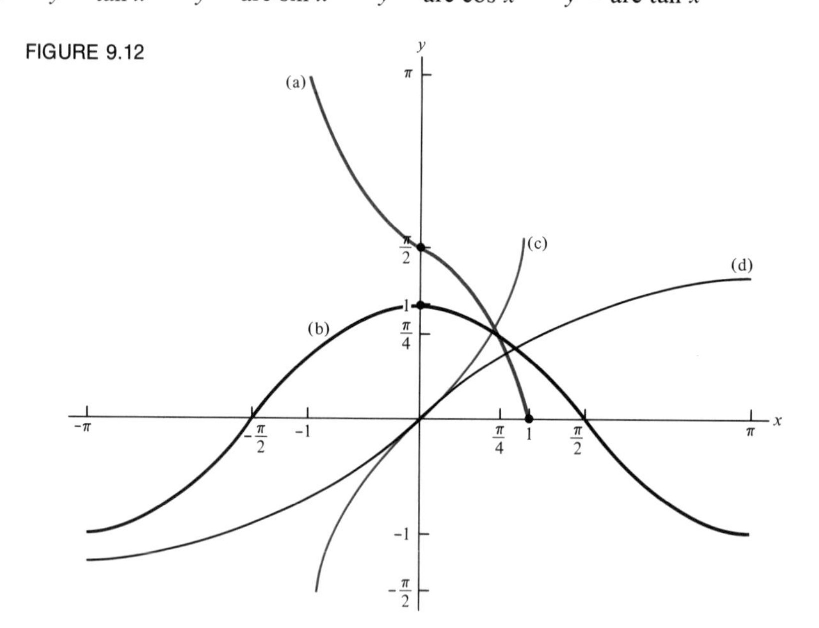

*9.6 Polar Coordinates

In Figure 9.13 the curve can be traced by a point winding around the origin. Curves such as this one are more conveniently represented in *polar coordinates* than in rectangular coordinates.

In polar coordinates the origin, O, is called the **pole**. Let A be an arbitrary point on the positive x-axis. As in Figure 9.14, any point P in the plane can be described by the ordered pair (r, θ), where $r = \text{dist}(O, P)$ and $\theta = \angle AOP$. Then *one pair of* **polar coordinates** for P is (r, θ).

*Optional topic

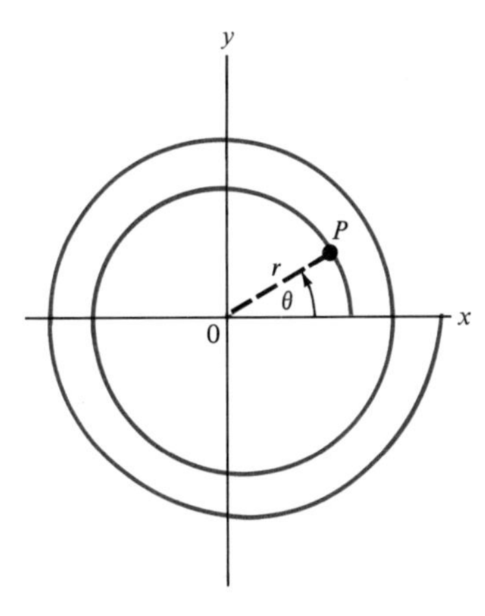

FIGURE 9.13

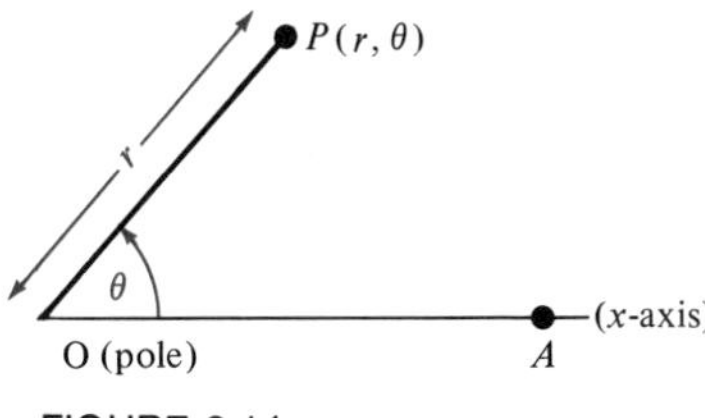

FIGURE 9.14

In Figure 9.15, *rays* determining various angles in standard position are drawn. Circles centered about the pole with radius lengths, r, equal to 1, 2, 3, and 4 are also drawn. Thus, a pair of polar coordinates for A is $\left(3, \frac{\pi}{3}\right)$ and a pair of polar coordinates for B is $\left(2, \frac{4\pi}{3}\right)$. But recall that the measure of an angle is determined only up to a multiple of 2π. Therefore, add multiples of 2π $\left(\text{or } \frac{6\pi}{3}\right)$ to $\frac{\pi}{3}$ and note that A can also be represented by $\left(3, \frac{7\pi}{3}, \right), \left(3, \frac{13\pi}{3}\right)$, etc., as well as by $\left(3, -\frac{5\pi}{3}\right), \left(3, -\frac{11\pi}{3}\right)$, etc.

Observe that in Figure 9.15, A and B lie on the same *line* through the origin. This line is determined both by $\theta = \frac{\pi}{3}$ and by $\theta = \frac{4\pi}{3}$. It proves useful to consider point A as being in the "*negative direction*" given by $\theta = \frac{4\pi}{3}$ and to identify A by means of the polar coordinates $\left(-3, \frac{4\pi}{3}\right)$ as well. Here -3 represents the "*directed distance*" of A (the "wrong way") along the line given by $\theta = \frac{4\pi}{3}$. And by adding multiples of 2π to θ, you obtain still other

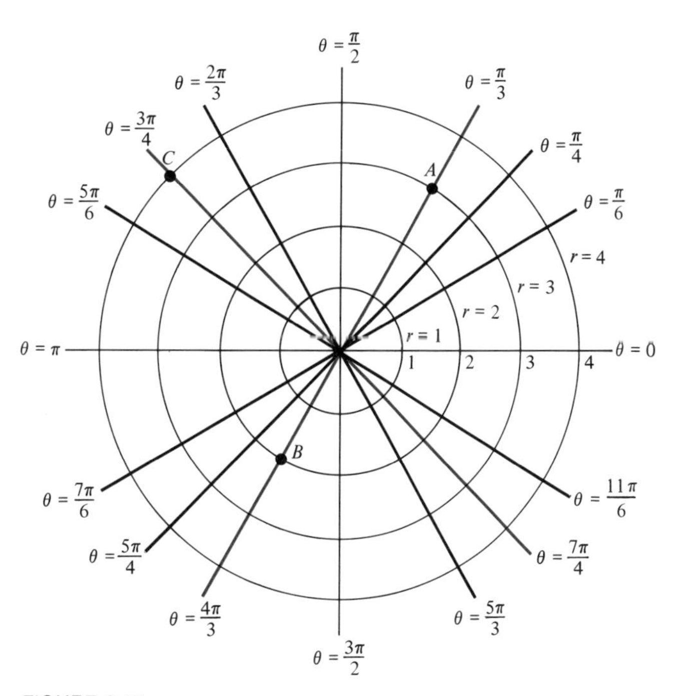

FIGURE 9.15

pairs of polar coordinates of A:

$$\left(-3, \frac{10\pi}{3}\right), \quad \left(-3, \frac{16\pi}{3}\right), \ldots$$
$$\left(-3, -\frac{2\pi}{3}\right), \quad \left(-3, -\frac{8\pi}{3}\right), \ldots$$

Similarly, B has polar coordinates:

$$\left(-2, \frac{\pi}{3}\right), \quad \left(-2, \frac{7\pi}{3}\right), \ldots$$
$$\left(-2, -\frac{5\pi}{3}\right), \quad \left(-2, -\frac{11\pi}{3}\right), \ldots$$

Note that unlike the unique representation of every point P by its rectangular coordinates (x, y), *in polar coordinates, every point P can be represented by infinitely many ordered pairs* (r, θ).

Consider the point C in Figure 9.15 . It lies at the intersection of the *ray* described by $\theta = \frac{3\pi}{4}$ and by the circle with radius 4. Thus, one pair of polar coordinates of C is $\left(4, \frac{3\pi}{4}\right)$. By adding multiples of 2π to θ, you obtain other pairs of polar coordinates:

$$\left(4, \frac{11\pi}{4}\right), \quad \left(4, \frac{19\pi}{4}\right), \ldots$$
$$\left(4, -\frac{5\pi}{4}\right), \quad \left(4, -\frac{13\pi}{4}\right), \ldots$$

But the *line* through the pole and C can also be described by $\theta = \frac{7\pi}{4}$. Thus other pairs of polar coordinates of P are:

$$\left(-4, \frac{7\pi}{4}\right), \quad \left(-4, \frac{15\pi}{4}\right), \ldots$$
$$\left(-4, -\frac{\pi}{4}\right), \quad \left(-4, -\frac{9\pi}{4}\right), \ldots$$

The pole (by itself) does not determine any particular angle. It is convenient to consider the polar coordinates of the pole to be $(0, \theta)$, where θ is *any* number. Thus, among the polar coordinates of the pole are:

$$\left(0, \frac{\pi}{2}\right), \quad \left(0, \frac{3\pi}{4}\right), \quad (0, -\pi), \quad (0, 0)$$

Note that if r is positive, $(r, 0)$ determines a unique point on the positive x-axis, the same point that is determined by the rectangular coordinates $(r, 0)$.

polar and rectangular coordinates

To convert from one coordinate system to the other, consider Figure 9.16, in which r is assumed to be *positive*. Then

$$\frac{x}{r} = \cos \theta, \qquad \text{so that} \qquad x = r \cos \theta,$$

$$\frac{y}{r} = \sin\theta, \qquad \text{so that} \qquad y = r\sin\theta,$$

and

$$r = \sqrt{x^2 + y^2}$$

The following chart is useful:

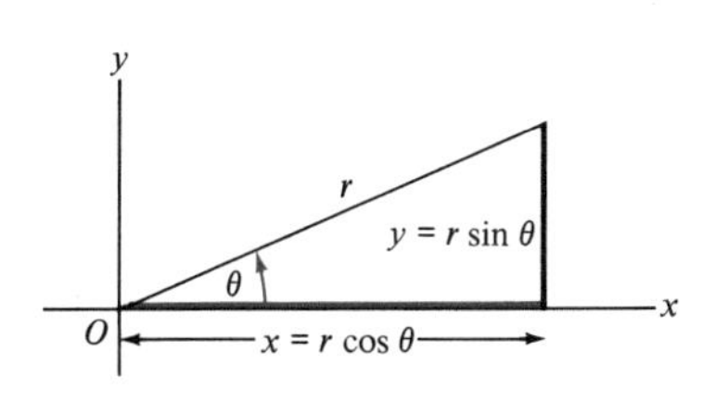

FIGURE 9.16

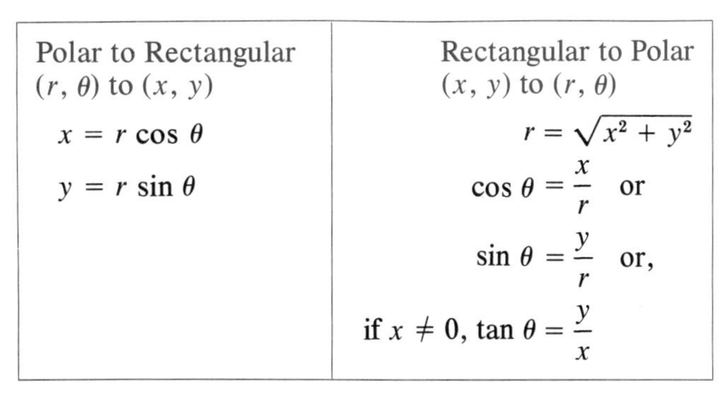

Polar to Rectangular (r, θ) to (x, y)	Rectangular to Polar (x, y) to (r, θ)
$x = r\cos\theta$ $y = r\sin\theta$	$r = \sqrt{x^2 + y^2}$ $\cos\theta = \frac{x}{r}$ or $\sin\theta = \frac{y}{r}$ or, if $x \neq 0$, $\tan\theta = \frac{y}{x}$

Observe that, if $-\frac{\pi}{2} < \theta < \frac{\pi}{2}$, then $\theta = \arctan\frac{y}{x}$.

EXAMPLE 1 Polar coordinates of each point are given. Convert to rectangular coordinates.

(a) $P = \left(4, \frac{\pi}{6}\right)$ (b) $Q = \left(1, \frac{3\pi}{4}\right)$

(c) $R = \left(2, -\frac{\pi}{2}\right)$ (d) $S = (0, \pi)$

(e) $T = \left(-\sqrt{2}, \frac{\pi}{4}\right)$

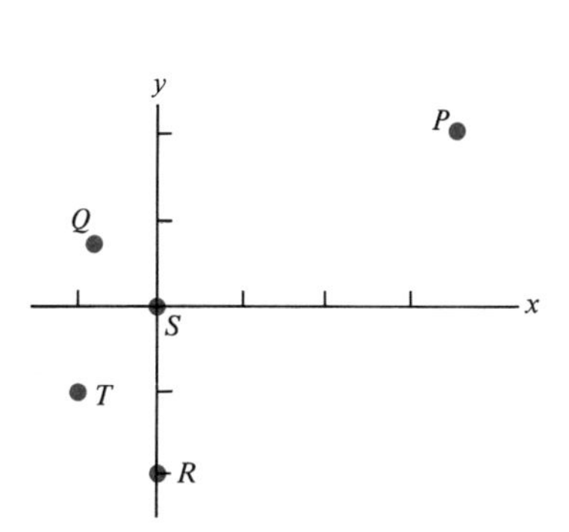

FIGURE 9.17 Polar coordinates are in color. Rectangular coordinates are in black.

$P = \left(4, \frac{\pi}{6}\right) = (2\sqrt{3}, 2)$

$Q = \left(1, \frac{3\pi}{4}\right) = \left(-\frac{\sqrt{2}}{2}, \frac{\sqrt{2}}{2}\right)$

$R = \left(2, -\frac{\pi}{2}\right) = (0, -2)$

$S = (0, \pi) = (0, 0)$

$T = \left(-\sqrt{2}, \frac{\pi}{4}\right) = \left(\sqrt{2}, \frac{5\pi}{4}\right) = (-1, -1)$

Solution. [See Figure 9.17 .]

(a)
$$r = 4, \quad \theta = \frac{\pi}{6}$$

$$x = 4\cos\frac{\pi}{6} = \frac{4\sqrt{3}}{2} = 2\sqrt{3}$$

$$y = 4\sin\frac{\pi}{6} = 4\cdot\frac{1}{2} = 2$$

The rectangular coordinates of P are $(2\sqrt{3}, 2)$.

(b)
$$r = 1, \quad \theta = \frac{3\pi}{4}$$

$$x = 1\cdot\cos\frac{3\pi}{4} = -\frac{\sqrt{2}}{2}$$

$$y = 1\cdot\sin\frac{3\pi}{4} = \frac{\sqrt{2}}{2}$$

The rectangular coordinates of Q are $\left(-\frac{\sqrt{2}}{2}, \frac{\sqrt{2}}{2}\right)$.

(c) $r = 2, \quad \theta = -\frac{\pi}{2}$

$$x = 2 \cos\left(-\frac{\pi}{2}\right) = 2 \cdot 0 = 0$$

$$y = 2 \sin\left(-\frac{\pi}{2}\right) = 2(-1) = -2$$

The rectangular coordinates of R are $(0, -2)$.

(d) $r = 0, \quad \theta = \pi$

Because $r = 0$, S is the pole. Note that

$$x = 0 \cos \pi = 0$$
$$y = 0 \sin \pi = 0$$

The rectangular coordinates of S are $(0, 0)$.

(e) First express $\left(-\sqrt{2}, \frac{\pi}{4}\right)$ with *positive r*:

$$\left(-\sqrt{2}, \frac{\pi}{4}\right) = \left(\sqrt{2}, \frac{\pi}{4} + \pi\right) = \left(\sqrt{2}, \frac{5\pi}{4}\right)$$

Now convert $\left(\sqrt{2}, \frac{5\pi}{4}\right)$ to rectangular coordinates.

$$r = \sqrt{2}, \quad \theta = \frac{5\pi}{4}$$

$$x = r \cos \theta = \sqrt{2} \cos \frac{5\pi}{4} = \sqrt{2}\left(-\frac{\sqrt{2}}{2}\right) = -1$$

$$y = r \sin \theta = \sqrt{2} \sin \frac{5\pi}{4} = \sqrt{2}\left(-\frac{\sqrt{2}}{2}\right) = -1$$

The rectangular coordinates of T are $(-1, -1)$. □

In converting from rectangular to polar coordinates, to obtain the correct value of θ, you must consider the quadrant or axis of the given point (x, y). For example, if you find that $\cos \theta = \frac{1}{2}$, then θ could be $\frac{\pi}{3}$ or $-\frac{\pi}{3}$.

EXAMPLE 2 Rectangular coordinates of each point are given. Convert to polar coordinates.

(a) $P = (4, 4)$ (b) $Q = (0, 3)$ (c) $R = (-5, 0)$
(d) $S = (1, -\sqrt{3})$ (e) $T = (3, 4)$

Solution. [See Figure 9.18 .]

(a) $x = 4, \quad y = 4 \quad$ (Quadrant I)

$$r = \sqrt{4^2 + 4^2} = \sqrt{32} = \sqrt{16 \cdot 2} = 4\sqrt{2}$$

$$\cos \theta = \frac{x}{r} = \frac{4}{4\sqrt{2}} = \frac{1}{\sqrt{2}} = \frac{\sqrt{2}}{2}, \qquad \text{or easier,}$$

$$\tan \theta = \frac{y}{x} = \frac{4}{4} = 1$$

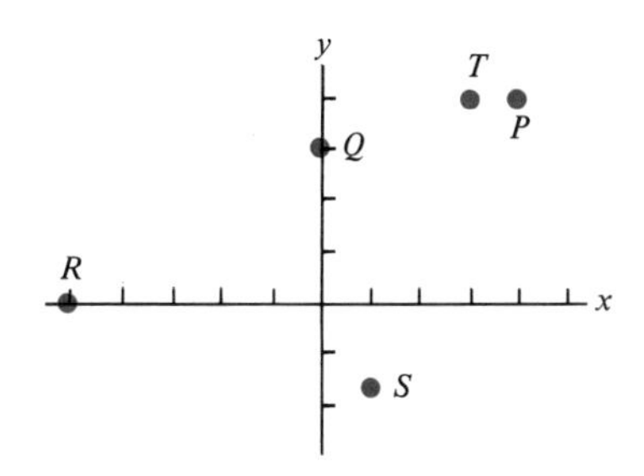

FIGURE 9.18 Rectangular coordinates are in black. Polar coordinates are in color.

$P = (4, 4) = \left(4\sqrt{2}, \frac{\pi}{4}\right)$

$Q = (0, 3) = \left(3, \frac{\pi}{2}\right)$

$R = (-5, 0) = (5, \pi)$

$S = (1, -\sqrt{3}) = \left(2, -\frac{\pi}{3}\right)$

$T = (3, 4) = \left(5, \frac{53\pi}{180}\right)$

Because P is in Quadrant I, $\theta = \frac{\pi}{4}$. Polar coordinates for P are $\left(4\sqrt{2}, \frac{\pi}{4}\right)$.

(b) $x = 0, \quad y = 3$ (*positive* portion of the y-axis)

$r = \sqrt{0^2 + 3^2} = 3$

Because Q is on the *positive* y-axis, $\theta = \frac{\pi}{2}$. (Note also that $\cos\theta = \frac{0}{3} = 0$.) Polar coordinates for Q are $\left(3, \frac{\pi}{2}\right)$.

(c) $x = -5, \quad y = 0$ (*negative* portion of the x-axis)

$r = \sqrt{(-5)^2 + 0^2} = 5, \quad \theta = \pi$

Polar coordinates for R are $(5, \pi)$.

(d) $x = 1, \quad y = -\sqrt{3}$ (S is in Quadrant IV.)

$r = \sqrt{1^2 + (-\sqrt{3})^2} = \sqrt{1 + 3} = \sqrt{4} = 2$

$\cos\theta = \frac{1}{2}$

Because S is in Quadrant IV, $\theta = -\frac{\pi}{3}\left(\text{or } \frac{5\pi}{3}\right)$. Polar coordinates for S are $\left(2, -\frac{\pi}{3}\right)$.

(e) $x = 3, \quad y = 4$ (T is in Quadrant I.)

$r = \sqrt{3^2 + 4^2} = 5$

$\cos\theta = \frac{3}{5} = .6$

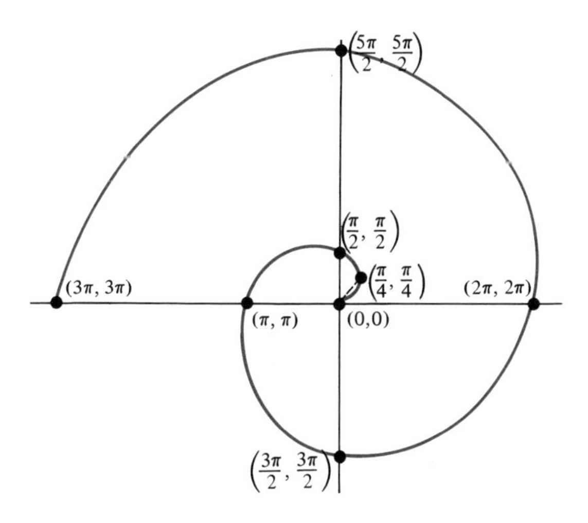

FIGURE 9.19
The spiral given by $r = \theta$, for $0 \leq \theta \leq 3\pi$

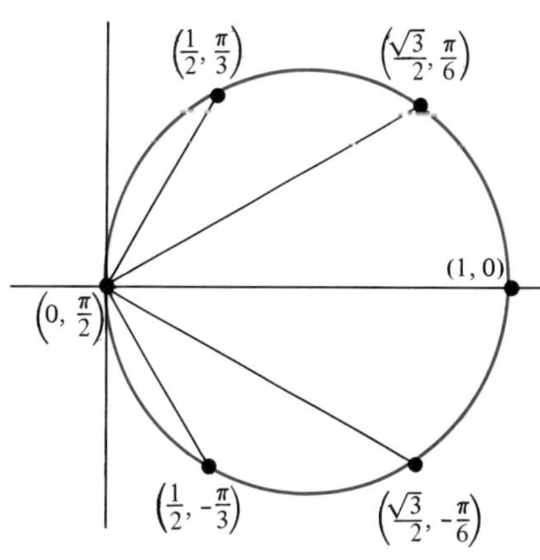

FIGURE 9.20
The circle given by $r = \cos\theta$, for $-\frac{\pi}{2} \leq \theta \leq \frac{\pi}{2}$

Using Table 3 and the quadrant of T, θ is *approximately* 53°, or in radian measure, $\frac{53\pi}{180}$. Thus, polar coordinates for T are $\left(5, \frac{53\pi}{180}\right)$. □

curve tracing

Although curve tracing in polar coordinates is beyond the intended level of this course, Figures 9.19–9.22 illustrate how some curves appear on a polar coordinate system. Note, in particular, that in Figure 9.22, when determining the intersections of two curves, it is necessary to consider various forms of the polar coordinates of the intersection points.

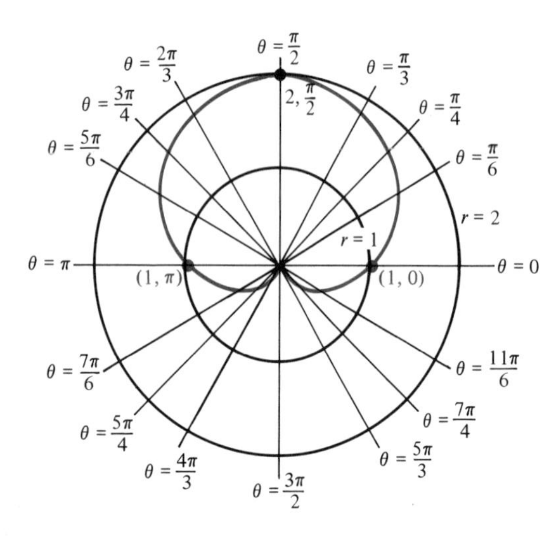

FIGURE 9.21 The cardiod (in color) given by $r = 1 + \sin\theta$ and the circles (in black) given by $r = 1$ and by $r = 2$. The cardiod intersects the inner circle at $(1, 0)$ and at $(1, \pi)$ and the outer circle at $\left(2, \frac{\pi}{2}\right)$.

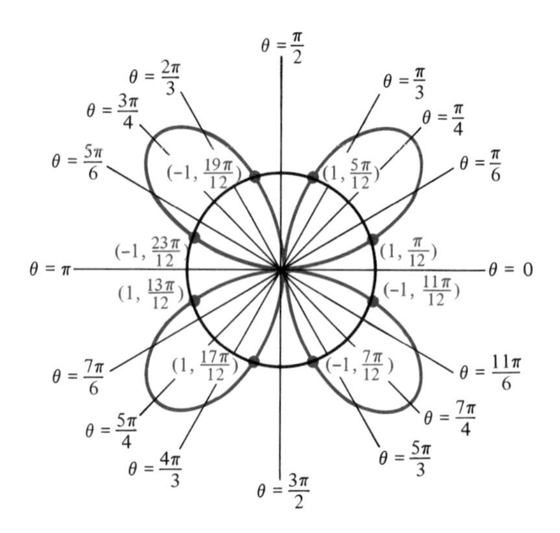

FIGURE 9.22 The circle (in black) given by $r = 1$ and the four-leafed rose (in color) given by $r = 2\sin 2\theta$. There are 8 intersection points. Those intersection points in Quadrants I and III satisfy both of the preceding equations. For example, the point $\left(1, \frac{\pi}{12}\right)$ satisfies the equation of the circle because $r = 1$. It also satisfies the equation of the rose because $2\sin 2\left(\frac{\pi}{12}\right) = 2\sin\frac{\pi}{6} = 2\cdot\frac{1}{2} = 1$.

Every point on the circle has polar coordinates $(-1, \theta + \pi)$, so that the circle is also given by the equation $r = -1$. The intersection points in Quadrants II and IV satisfy this alternate equation of the circle as well as the equation of the rose.

EXERCISES

In Exercises 1–12 polar coordinates of various points are given. In which quadrant or on which axis is each point?

1. $\left(2, \frac{\pi}{4}\right)$ **2.** $\left(1, \frac{3\pi}{2}\right)$ **3.** $\left(\frac{1}{2}, -\frac{\pi}{4}\right)$ **4.** $\left(4, \frac{\pi}{2}\right)$ **5.** $\left(7, -\frac{5\pi}{6}\right)$ **6.** $\left(5, \frac{5\pi}{2}\right)$

7. $(3, 2\pi)$ **8.** $\left(9, \frac{9\pi}{4}\right)$ **9.** $\left(-1, \frac{\pi}{6}\right)$ **10.** $\left(-4, \frac{3\pi}{2}\right)$ **11.** $\left(-2, \frac{5\pi}{4}\right)$ **12.** $\left(0, \frac{\pi}{3}\right)$

In Exercises 13–18, one pair of polar coordinates for P is given. Which other pairs are polar coordinates for P?

13. $P = \left(2, \frac{\pi}{3}\right)$ **(a)** $\left(\frac{1}{2}, \frac{\pi}{3}\right)$ **(b)** $\left(\frac{\pi}{3}, 2\right)$ **(c)** $\left(2, \frac{7\pi}{3}\right)$ **(d)** $\left(-2, \frac{\pi}{3}\right)$

14. $P = \left(4, \frac{2\pi}{3}\right)$ **(a)** $\left(4, \frac{5\pi}{3}\right)$ **(b)** $\left(-4, \frac{5\pi}{3}\right)$ **(c)** $\left(4, \frac{8\pi}{3}\right)$ **(d)** $\left(-4, -\frac{2\pi}{3}\right)$

15. $P = \left(6, \frac{\pi}{2}\right)$ **(a)** $\left(6, \frac{3\pi}{2}\right)$ **(b)** $\left(6, \frac{5\pi}{2}\right)$ **(c)** $\left(-6, \frac{3\pi}{2}\right)$ **(d)** $\left(6, -\frac{\pi}{2}\right)$

16. $P = (5, 0)$ **(a)** $(5, 2\pi)$ **(b)** $(-5, \pi)$ **(c)** $(5, 4\pi)$ **(d)** $(-5, 5\pi)$

17. $\left(-2, \frac{\pi}{2}\right)$ **(a)** $\left(2, \frac{\pi}{2}\right)$ **(b)** $\left(2, \frac{3\pi}{2}\right)$ **(c)** $\left(-2, \frac{3\pi}{2}\right)$ **(d)** $\left(-2, \frac{5\pi}{2}\right)$

18. $\left(0, \frac{3\pi}{4}\right)$ **(a)** $\left(0, \frac{\pi}{4}\right)$ **(b)** $(0, 0)$ **(c)** $\left(0, -\frac{3\pi}{4}\right)$ **(d)** $\left(\frac{3\pi}{4}, 0\right)$

In Exercises 19–22 consider the point with the indicated polar coordinates. Find another pair of polar coordinates (r, θ) for the point, with $r > 0$.

19. $\left(-2, \frac{\pi}{6}\right)$ **20.** $\left(-\frac{1}{2}, \frac{\pi}{2}\right)$ **21.** $\left(-3, -\frac{\pi}{3}\right)$ **22.** $(-5, 0)$

In Exercises 23–26 consider the point with the indicated polar coordinates. Find another pair of polar coordinates (r, θ) for the point with $0 \leqslant \theta < 2\pi$.

23. $\left(2, \frac{5\pi}{2}\right)$ **24.** $\left(0, \frac{10\pi}{3}\right)$ **25.** $\left(1, \frac{-\pi}{4}\right)$ **26.** $\left(-3, \frac{11\pi}{4}\right)$

In Exercises 27–38 convert from polar to rectangular coordinates.

27. $\left(4, \frac{\pi}{6}\right)$ **28.** $\left(8, \frac{5\pi}{6}\right)$ **29.** $\left(2\sqrt{2}, \frac{\pi}{4}\right)$ **30.** $\left(5\sqrt{2}, \frac{5\pi}{4}\right)$ **31.** $(6, 0)$ **32.** $(0, 3\pi)$

33. $\left(6, \frac{\pi}{2}\right)$ **34.** $\left(\pi, \frac{\pi}{2}\right)$ **35.** $(-1, 0)$ **36.** $\left(-8, \frac{\pi}{3}\right)$ **37.** $(10, 5\pi)$ **38.** $(-10, -\pi)$

In Exercises 39–50 convert from rectangular to polar coordinates.

39. $(1, 1)$ **40.** $(1, \sqrt{3})$ **41.** $(\sqrt{3}, 1)$ **42.** $(-4, 4)$ **43.** $(5, -5)$ **44.** $(-2, -2)$

45. $(1, 2)$ **46.** $(2, 1)$ **47.** $(8, 0)$ **48.** $(0, 8)$ **49.** $(-8, 0)$ **50.** $(0, -8)$

In Exercises 51–55 find an equation in polar coordinates for the indicated line or circle:

51. The line with equation $y = x$ **52.** The line with equation $y = -x$ **53.** The x-axis

54. The y-axis **55.** The unit circle

56. For which points P are the rectangular coordinates for P also a pair of polar coordinates for P?

In Exercises 57 and 58 let P have rectangular coordinates (x, y).

57. Which points P have polar coordinates $\left(\sqrt{x^2 + y^2}, \text{arc cos } \frac{x}{\sqrt{x^2 + y^2}}\right)$?

58. Which points P have polar coordinates $\left(\sqrt{x^2 + y^2}, \text{arc tan } \frac{y}{x}\right)$?

Just for fun, in Exercises 59–64, see if you can graph the curves given by the following equations. Begin by plotting the points (r, θ) with $\theta = 0, \frac{\pi}{6}, \frac{\pi}{4}, \frac{\pi}{3}, \frac{\pi}{2}$. Then take values of θ in other quadrants.

59. $r = 2$ **60.** $r = \frac{3\theta}{2}, 0 \leqslant \theta \leqslant 3\pi$ **61.** $r = -\theta, 0 \leqslant \theta \leqslant 3\pi$ **62.** $r = \sin\theta, 0 \leqslant \theta \leqslant \pi$

63. $r = 2\cos\theta, -\frac{\pi}{2} \leqslant \theta \leqslant \frac{\pi}{2}$ **64.** $r = 1 + \cos\theta, -\frac{\pi}{2} \leqslant \theta \leqslant \frac{\pi}{2}$

Review Exercises for Chapter 9

1. Simplify: **(a)** $\sec^2\theta - \tan^2\theta$ **(b)** $\frac{\cos A}{\sec A} + \cos^2 A \tan^2 A$

2. Verify the following identities: **(a)** $\frac{\cos^2\theta}{1 + \sin\theta} = 1 - \sin\theta$ **(b)** $\cos^2 x - \sin^2 x = \frac{1 - \tan^2 x}{1 + \tan^2 x}$

(c) $\frac{1 + \tan A}{\sec A + \csc A} = \sin A$ **(d)** $\cot^2 B \sec^2 B - 1 = \cot^2 B$

3. Compute without using a table: **(a)** $\cos 15°$ **(b)** $\tan 75°$

4. Suppose $\cos A = \frac{4}{5}$, $\angle A$ is in Quadrant I, $\sin B = \frac{3}{5}$, and $\angle B$ is in Quadrant II. Find:

(a) $\cos(A + B)$ **(b)** $\sin(A - B)$ **(c)** $\tan(A + B)$

(d) In which quadrant or on which axis is $\angle(A + B)$?

5. Use an Addition Formula to evaluate: $\frac{\tan 28° + \tan 17°}{1 - \tan 28° \tan 17°}$

6. Simplify: $\cos^2 15° - \sin^2 15°$

7. Express $\cos 34° - \cos 13°$ as a product.

8. Verify the identity: $\sin 4\theta = 4\sin 2\theta \cos^2\theta - 2\sin 2\theta$

9. Suppose $0 \leqslant \theta < 2\pi$. Solve each equation.

(a) $\sin\theta = -\frac{1}{2}$ **(b)** $4\cos^2\theta = 3$ **(c)** $2\sin^2\theta + \sin\theta = 1$

10. Find all roots of the given equation. Do not use a table.

(a) $3\tan 2\theta = \sqrt{3}$ **(b)** $\cos^2\theta + \cos\theta = 0$ **(c)** $\sin^2\theta - \sin\theta - 2 = 0$

11. Find each value without using a table. **(a)** $\arcsin\frac{\sqrt{2}}{2}$ **(b)** $\arccos(-1)$ **(c)** $\arctan(-1)$

12. Use a table to find: **(a)** $\arcsin(-.3090)$ **(b)** $\arctan(3.7321)$

Express your answers in terms of π.

13. Find each value without using a table.

(a) $\cos\left(\arccos\frac{2}{5}\right)$ **(b)** $\sin(\arctan\sqrt{3})$ **(c)** $\arcsin\left(\sin\left(-\frac{\pi}{6}\right)\right)$

14. Sketch the graph of $\arcsin\frac{x}{2}, -2 \leqslant x \leqslant 2$.

15. Consider the point with the indicated polar coordinates. Find another pair of polar coordinates (r, θ) for the point: **(a)** $(3, 0)$ **(b)** $\left(-1, \frac{\pi}{4}\right)$ **(c)** $\left(1, \frac{5\pi}{2}\right)$ **(d)** $(0, \pi)$

16. Convert from polar to rectangular coordinates.

(a) $\left(3, \frac{\pi}{3}\right)$ **(b)** $\left(5, \frac{3\pi}{4}\right)$ **(c)** $(-1, \pi)$ **(d)** $\left(3, \frac{3\pi}{2}\right)$

17. Convert from rectangular to polar coordinates.

(a) $(4, 4)$ **(b)** $(2, 0)$ **(c)** $(-1, \sqrt{3})$ **(d)** $(0, -5)$

10 Complex Numbers

10.1 Arithmetic of Complex Numbers

$x + yi$

Complex numbers were introduced in Section 6.3 in order that equations such as

$$x^2 + 1 = 0 \qquad \text{and} \qquad x^2 + x + 1 = 0$$

may have roots. Thus

$$i \text{ was defined to be } \sqrt{-1}$$

so that $i^2 = -1$, and thus i and $-i$ are roots of $x^2 + 1 = 0$. A *complex number* was defined as an expression of the form

$$x + yi, \quad \text{where } x \text{ and } y \text{ are real numbers.}$$

Here x is called the *real part* and y the *imaginary part* of $x + yi$. Both the real and the imaginary parts of a complex number are real numbers. The complex number $x + 0i$ is considered to be the same as the real number x, and is thus written as x. The complex number $0 + yi$ is called *imaginary* and is written as yi. The complex numbers $x + yi$ and $x - yi$ are called *conjugates*. Thus $-\frac{1}{2} + \frac{\sqrt{3}}{2}i$ and $-\frac{1}{2} - \frac{\sqrt{3}}{2}i$ are conjugates. If you apply the quadratic formula, you will see that these conjugates are the roots of $x^2 + x + 1 = 0$. A real number is its own conjugate. (For further discussion of these matters, see pages 227–230.)

Addition, subtraction, multiplication, and division can be defined for complex numbers. Simply treat complex numbers $x + yi$ as you would treat binomials in x and y and set $i^2 = -1$.

addition and subtraction

To add or subtract complex numbers, add or subtract their real parts and their imaginary parts, separately.

EXAMPLE 1 Compute the following sums and difference:

(a) $(3 + 4i) + (7 - 2i)$ (b) $(3 + 4i) - (7 - 2i)$

(c) $(8 - \sqrt{2}\,i) + (8 + \sqrt{2}\,i)$ (d) $(5 + 2i) + (-5 - 2\,i)$

Solution.

(a) Method 1: $(3 + 4i) + (7 - 2i) = (3 + 7) + [4 + (-2)]i$
$= 10 + 2i$

Method 2: *Add:*
$$\begin{array}{r} 3 + 4i \\ \underline{7 - 2i} \\ 10 + 2i \end{array}$$

(b) Method 1: $(3 + 4i) - (7 - 2i) = (3 - 7) + [4 - (-2)]i$
$= -4 + 6i$

Method 2: *Subtract:*
$$\begin{array}{r} 3 + 4i \\ \underline{\mp 7 \pm 2i} \\ -4 + 6i \end{array}$$

(c) *Add:*
$$\begin{array}{l} 8 - \sqrt{2}\,i \\ \underline{8 + \sqrt{2}\,i} \\ 16 \end{array}$$

(d) *Add:*
$$\begin{array}{r} 5 + 2i \\ \underline{-5 - 2i} \\ 0 \end{array}$$

□

Observe that

$$(x + yi) + (0 + 0i) = x + yi.$$

Thus 0, that is, $0 + 0i$, plays the same role in addition of complex numbers as in addition of real numbers. Moreover,

$$(x + yi) + (-x - yi) = 0$$

In general, the **additive inverse** of a complex number z is defined to be the number added to z to obtain 0, just as the additive inverse of a real number a is the number added to a to obtain 0. Thus, *the additive inverse of* $x + yi$ *is* $-x - yi$. In part (d) of Example 1, the additive inverse of $5 + 2i$ is $-5 - 2i$.

multiplication

To multiply complex numbers, extend the Associative, Commutative, and Distributive Laws from real numbers to complex.

EXAMPLE 2

(a) $5(2 + 5i) = 5 \cdot 2 + 5 \cdot 5i = 10 + 25i$

(b) $2i(3 + 2i) = (2i)3 + (2i)(2i)$
$= 6i + 4i^2$ $\quad i^2 = -1$
$= -4 + 6i$

(c) *Multiply:* $(6 + 3i)(2 + i)$
$$\begin{array}{r} 6 + 3i \\ \underline{2 + i} \\ 12 + 6i \\ \underline{6i - 3} \\ 9 + 12i \end{array}$$
$(3i)i = 3i^2 = -3$

Thus $(6 + 3i)(2 + i) = 9 + 12i$ □

The product of a complex number and its conjugate is a real number. In fact,

$$(x + yi)(x - yi) = x^2 + xyi - xyi - y^2i^2 = x^2 + y^2$$

EXAMPLE 3 Find the product of $2 + 5i$ and its conjugate.

Solution. The conjugate of $2 + 5i$ is $2 - 5i$. *Multiply:*

$$\begin{array}{l} 2 + 5i \\ 2 - 5i \\ \hline 4 + 10i \\ \quad - 10i + 25 \\ \hline 29 \end{array}$$

$(5i)(-5i) = -25i^2 = 25$

Thus $(2 + 5i)(2 - 5i) = 29$ □

division To divide two complex numbers, first express the quotient as a "fraction". Then multiply both numerator and denominator by the conjugate of the denominator. This will yield a fraction whose denominator is real. (Recall a similar use of "conjugates" in rationalizing a denominator.)

EXAMPLE 4

$$\frac{6 + 5i}{3 + 2i} = \frac{(6 + 5i) \cdot (3 - 2i)}{(3 + 2i) \cdot (3 - 2i)} = \frac{28 + 3i}{3^2 + 2^2} = \frac{28}{13} + \frac{3}{13} i$$

$$\begin{array}{l} 6 + 5i \\ 3 - 2i \\ \hline 18 + 15i \\ \quad -12i + 10 \\ \hline 28 + 3i \end{array}$$

□

Note that when the denominator of the fraction is an integer, as in Example 4, the given complex number can easily be put in the form $x + yi$. In Example 4, $x = \frac{28}{13}$, $y = \frac{3}{13}$.

powers Powers of complex numbers are defined as are powers of real numbers. Thus

$$(2 + i)^2 = (2 + i)(2 + i) = 3 + 4i$$

$$(2 + i)^{10} = \underbrace{(2 + i)(2 + i) \ldots (2 + i)}_{10 \text{ times}}$$

$$\begin{array}{l} 2 + i \\ 2 + i \\ \hline 4 + 2i \\ \quad 2i - 1 \\ \hline 3 + 4i \end{array}$$

$$(2 + i)^{-1} = \frac{1}{2 + i}$$

Simplify the denominator.

$$= \frac{1 \cdot (2 - i)}{(2 + i) \cdot (2 - i)} = \frac{2 - i}{4 + 1} = \frac{2}{5} - \frac{1}{5} i$$

The powers of i are particularly important. But after the 4th power, they repeat:

$i^1 = i$

$i^2 = -1$

$i^3 = i^2 \cdot i = -i$

$i^4 = i^2 \cdot i^2 = (-1)(-1) = 1$

$i^5 = i^4 \cdot i = 1 \cdot i = i$

$i^6 = i^4 \cdot i^2 = i^2 = -1$

$i^7 = i^4 \cdot i^3 = i^3 = -i$

$i^8 = i^4 \cdot i^4 = 1$

and so on. Thus

$$i^{10} = i^4 \cdot i^4 \cdot i^2 = -1$$
$$i^{16} = i^4 \cdot i^4 \cdot i^4 \cdot i^4 = 1$$

$\sqrt{ab}$ For *positive* real numbers x and y,

$$\sqrt{-x} = \sqrt{x}\,i \qquad \text{and} \qquad \sqrt{-xy} = \sqrt{x}\sqrt{y}\,i$$

Thus

$$\sqrt{-4} = \sqrt{4}\,i \quad (\text{or } 2i) \qquad \text{and}$$
$$\sqrt{-(4)(9)} = \sqrt{4}\sqrt{9}\,i \quad (= 2 \cdot 3i = 6i)$$

For the most part, the rules that hold for real numbers are also true for complex numbers. However, there is one important exception. The rule

$$\sqrt{a}\sqrt{b} = \sqrt{ab}$$

is true when $a \geqslant 0$ and $b \geqslant 0$. However, what happens when both a and b are negative? Let $a = -4$ and $b = -4$. Both $\sqrt{a}$ and $\sqrt{b}$ are complex numbers. Then

$$\sqrt{a}\sqrt{b} = \sqrt{-4} \cdot \sqrt{-4} = 2i \cdot 2i = 4i^2 = -4$$

but

$$\sqrt{ab} = \sqrt{(-4)(-4)} = \sqrt{16} = 4$$

Thus

$$\sqrt{-4} \cdot \sqrt{-4} \neq \sqrt{(-4)(-4)}$$

The rule

$$\sqrt{a}\sqrt{b} = \sqrt{ab},$$

which is true for $a \geqslant 0$ and $b \geqslant 0$, is *false* when both a and b are negative.

Complex numbers enable you to "factor" previously irreducible polynomials. As a simple example,

$$z^2 + 1 = (z + i)(z - i)$$

rules for conjugates Finally, here are some useful rules relating complex conjugates and arithmetic operations. Let z, z_1, and z_2 be complex numbers and let n be a positive integer.

$$\overline{z_1 \pm z_2} = \overline{z_1} \pm \overline{z_2}$$

The conjugate of a sum (or difference) is the sum (or difference) of the conjugates.

$$\overline{z_1 \cdot z_2} = \overline{z_1} \cdot \overline{z_2}$$

The conjugate of a product is the product of the conjugates.

$$\overline{\left(\frac{z_1}{z_2}\right)} = \frac{\overline{z_1}}{\overline{z_2}}, \text{ where } z_2 \neq 0$$

The conjugate of a quotient is the quotient of the conjugates.

$$\overline{z^n} = (\overline{z})^n$$

The conjugate of an nth power is the nth power of the conjugate.

EXAMPLE 5 Let $z_1 = 5 + 3i$, $z_2 = -2 + i$.

$\overline{z_1} = 5 - 3i$, $\overline{z_2} = -2 - i$

(a) $z_1 + z_2 = 3 + 4i$, so that

$\overline{z_1 + z_2} = 3 - 4i = \overline{z_1} + \overline{z_2}$

(b) $z_1 - z_2 = 7 + 2i$, so that

$\overline{z_1 - z_2} = 7 - 2i = \overline{z_1} - \overline{z_2}$

(c)
$$\begin{aligned} z_1 \cdot z_2 &= (5 + 3i)\cdot(-2 + i) \\ &= -13 - i, \end{aligned}$$

$$\begin{array}{r} 5 + 3i \\ \underline{-2 + i} \\ -10 - 6i \\ \underline{+ 5i - 3} \\ -13 - i \end{array}$$

$$\begin{aligned} \overline{z_1}\cdot\overline{z_2} &= (5 - 3i)\cdot(-2 - i) \\ &= -13 + i, \quad \text{so that} \\ \overline{z_1 \cdot z_2} &= -13 + i = \overline{z_1}\cdot\overline{z_2} \end{aligned}$$

$$\begin{array}{r} 5 - 3i \\ \underline{-2 - i} \\ -10 + 6i \\ \underline{- 5i - 3} \\ -13 + i \end{array}$$

(d)
$$\begin{aligned} \frac{z_1}{z_2} &= \frac{(5 + 3i)\cdot(-2 - i)}{(-2 + i)\cdot(-2 - i)} \\ &= \frac{-7 - 11i}{(-2)^2 + 1} \\ &= \frac{-7}{5} - \frac{11}{5} i; \end{aligned}$$

$$\begin{array}{r} 5 + 3i \\ \underline{-2 - i} \\ -10 - 6i \\ \underline{- 5i + 3} \\ -7 - 11i \end{array}$$

$$\begin{aligned} \frac{\overline{z_1}}{\overline{z_2}} &= \frac{(5 - 3i)\cdot(-2 + i)}{(-2 - i)\cdot(-2 + i)} \\ &= \frac{-7}{5} + \frac{11}{5} i \quad \text{so that} \end{aligned}$$

$$\begin{array}{r} 5 - 3i \\ \underline{-2 + i} \\ -10 + 6i \\ \underline{+ 5i + 3} \\ -7 + 11i \end{array}$$

$$\overline{\left(\frac{z_1}{z_2}\right)} = \frac{-7}{5} + \frac{11}{5} i = \frac{\overline{z_1}}{\overline{z_2}}$$

(e)
$$\begin{aligned} (z_1)^2 &= (5 + 3i)\cdot(5 + 3i) \\ &= 16 + 30i, \\ (\overline{z_1})^2 &= (5 - 3i)(5 - 3i) \\ &= 16 - 30i, \quad \text{so that} \\ (\overline{z_1})^2 &= 16 - 30i = \overline{(z_1)^2} \end{aligned}$$

$$\begin{array}{r} 5 + 3i \\ \underline{5 + 3i} \\ 25 + 15i \\ \underline{+ 15i - 9} \\ 16 + 30i \end{array}$$

$$\begin{array}{r} 5 - 3i \\ \underline{5 - 3i} \\ 25 - 15i \\ \underline{- 15i - 9} \\ 16 - 30i \end{array}$$ □

EXERCISES

In Exercises 1–35 perform the indicated operations and express the resulting complex number in the form $a + bi$.

1. $(5i) + (3i)$ **2.** $(2 + 4i) + (1 + 2i)$ **3.** $(2 - 3i) + (5 - 9i)$ **4.** $6 + (3 + 7i)$

5. $\left(\frac{1}{3} - \frac{1}{2}i\right) + \left(\frac{1}{6} - \frac{3}{4}i\right)$ **6.** $(4 + 7i) - (3 + 4i)$ **7.** $(6 + 5i) - (7 - i)$ **8.** $i - (4 - 2i)$

9. $\left(\frac{1}{2} + \frac{1}{4}i\right) - \left(\frac{1}{4} + \frac{1}{2}i\right)$ **10.** $(2i)(3i)$ **11.** $(5i)(-4i)$ **12.** $-4(2 + 5i)$

13. $3i(7 - 2i)$ **14.** $(1 + i)(1 - i)$ **15.** $(2 + 3i)(1 - i)$ **16.** $(4 + 3i)(2 + i)$

17. $(10 - 3i)(6 + i)$ **18.** $(3i)^2$ **19.** $(-5i)^2$ **20.** $(1 + 2i)^2$

21. $(1 + 2i)^3$ **22.** $(3 - i)^2$ **23.** $(3 - i)^3$ **24.** $\frac{1 + i}{1 - i}$

25. $\frac{2 + i}{1 + i}$ **26.** $\frac{3 - 5i}{i}$ **27.** $\frac{4 - 3i}{2}$ **28.** $\frac{2 + 5i}{1 - 2i}$

29. $\frac{3 + i}{3 - i}$ **30.** $(5 + i)^{-1}$ **31.** $(5 + i)^{-2}$ **32.** $i(2 + i) - (1 - i)^2$

33. $(1 + i)\left(\frac{i}{2 - i}\right)$ **34.** $\frac{(2 + i)^3}{3 - i}$ **35.** $\left(\frac{2 + i}{1 - 2i}\right)^2$

In Exercises 36–44 simplify:

36. i^7 **37.** i^9 **38.** i^{12} **39.** $i^{34} + i^{35}$ **40.** $i^{61} - i^{62}$ **41.** i^{-1}

42. i^{-2} **43.** i^{-3} **44.** i^{-4}

45. Does $\sqrt{(-1)(-1)} = \sqrt{-1}\sqrt{-1}$?

46. **(a)** Factor: $z^2 + 9$ **(b)** Solve the equation: $z^2 + 9 = 0$

47. **(a)** Factor: $z^2 + 6$ **(b)** Solve the equation: $z^2 + 6 = 0$

In Exercises 48–50 show *directly* that

(a) $\overline{z_1 + z_2} = \overline{z_1} + \overline{z_2}$ (b) $\overline{z_1 - z_2} = \overline{z_1} - \overline{z_2}$ (c) $\overline{z_1 \cdot z_2} = \overline{z_1} \cdot \overline{z_2}$

(d) $\overline{\left(\frac{z_1}{z_2}\right)} = \frac{\overline{z_1}}{\overline{z_2}}$ (e) $\overline{(z_2)^2} = (\overline{z_2})^2$.

48. $z_1 = 4, \quad z_2 = 3i$ **49.** $z_1 = 1 + i, \quad z_2 = 4 - 2i$ **50.** $z_1 = 7 + 2i, \quad z_2 = 7 - 2i$

51. Let z $(= x + yi)$ be any complex number. Fill in: (a) $\frac{z + \overline{z}}{2} =$ ▇ (b) $\frac{z - \overline{z}}{2i} =$ ▇

52. Find all complex numbers z such that $\frac{1}{z} = -z$.

10.2 Rectangular Coordinates of Complex Numbers

Complex numbers can be represented by means of a rectangular coordinate system. Instead of an x-axis, you consider a **real axis;** instead of a y-axis there is an **imaginary axis.** The complex number $x + yi$ corresponds to the point whose coordinates are (x, y).

$$x + yi \longleftrightarrow (x, y)$$

EXAMPLE 1 Represent the following complex numbers on a rectangular coordinate system.

(a) $5 + 2i$ (b) $3 - 4i$ (c) $-4 + 3i$
(d) 4 (e) $4i$ (f) $-4i$

Solution. See Figure 10.1 . □

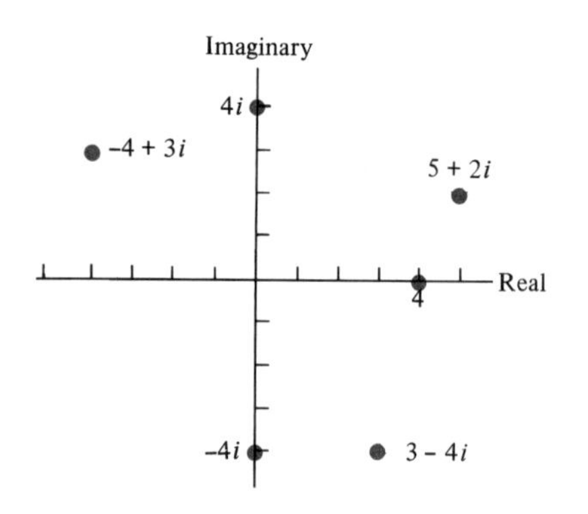

FIGURE 10.1

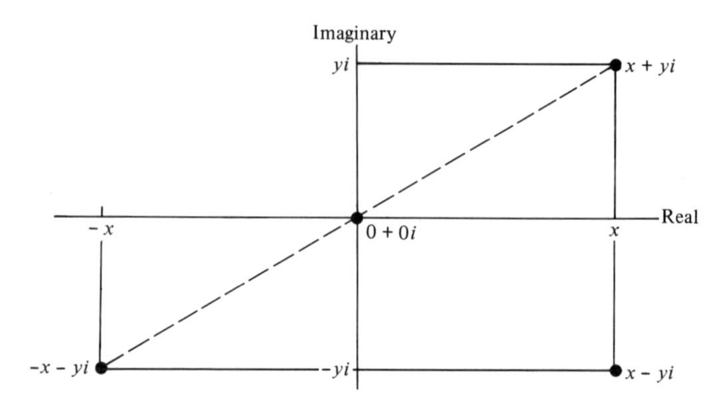

FIGURE 10.2

In Figure 10.2 observe that $x - yi$, the conjugate of $x + yi$, corresponds to the point whose coordinates are $(x, -y)$. Thus, a complex number and its conjugate are symmetrically located with respect to the real axis. Also, $0 + 0i$ corresponds to the origin. Moreover, $x + yi$ and its additive inverse, $-x - yi$, are symmetrically located with respect to the origin. The real number x, or $x + 0i$, corresponds to $(x, 0)$ on the real axis. The imaginary number yi corresponds to $(0, y)$ on the imaginary axis.

parallelogram rule

Addition and subtraction of complex numbers can be represented, geometrically, by means of the **Parallelogram Rule,** as indicated in Figure 10.3 . To add

$$(x_1 + y_1 i) + (x_2 + y_2 i),$$

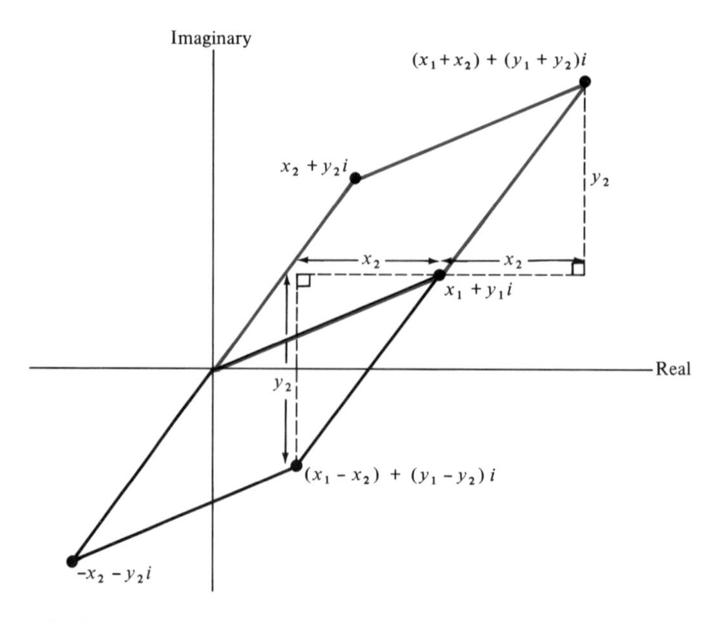

FIGURE 10.3 The Parallelogram Rule for addition (in color) and for subtraction (in black)

increase x_1, the real part of $x_1 + y_1 i$, by x_2; and increase y_1, the imaginary part, by y_2. Thus a parallelogram is formed with a vertex at the origin, a pair of opposite vertices representing $x_1 + y_1 i$ and $x_2 + y_2 i$, and the fourth vertex representing the sum, $(x_1 + x_2) + (y_1 + y_2)i$.

To subtract

$$(x_1 + y_1 i) - (x_2 + y_2 i),$$

increase the real part of $x_1 + y_1 i$ by $-x_2$ and the imaginary part by $-y_2$, and form a parallelogram. [See Figure 10.3 .]

EXAMPLE 2 Use the Parallelogram Rule to represent:

(a) $(2 + 3i) + (-1 + 2i)$
(b) $(2 + 3i) - (-1 + 2i)$

Solution.

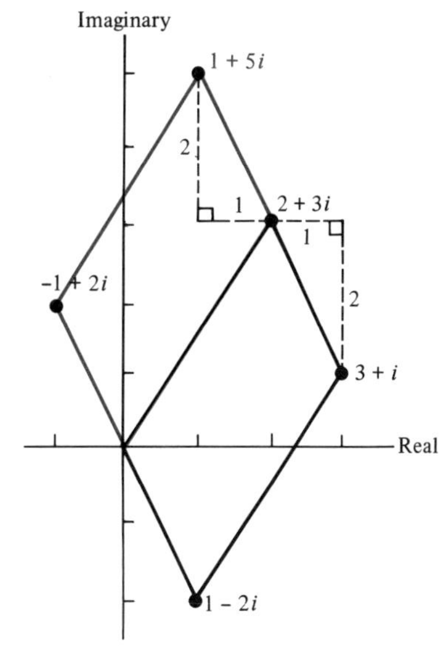

FIGURE 10.4
(a) $(2 + 3i) + (-1 + 2i) = 1 + 5i$
(b) $(2 + 3i) - (-1 + 2i) = 3 + i$

□

EXERCISES

1. In Figure 10.5 on page 414 which complex number is represented by each point?

2. In Figure 10.6 on page 414: **(a)** Which pairs of points represent conjugates? **(b)** Which pairs of points represent additive inverses?

For Exercises 3–12: On a rectangular coordinate system, represent the following complex numbers.

3. $6 + 2i$ **4.** $2 + 6i$ **5.** $3 - 3i$ **6.** $-3 + 2i$ **7.** 1 **8.** $4i$ **9.** $-2i$ **10.** 0
11. $-6 + 6i$ **12.** $-6 - 6i$

In Exercises 13–24 use the Parallelogram Rule to represent each sum or difference, as in Examples 1 and 2 of the text.

13. $(4 + 2i) + (3 + i)$ **14.** $(6 + 3i) - (2 + i)$ **15.** $(7 + i) + (-2 + 4i)$ **16.** $(-2 + i) - (2 - 4i)$

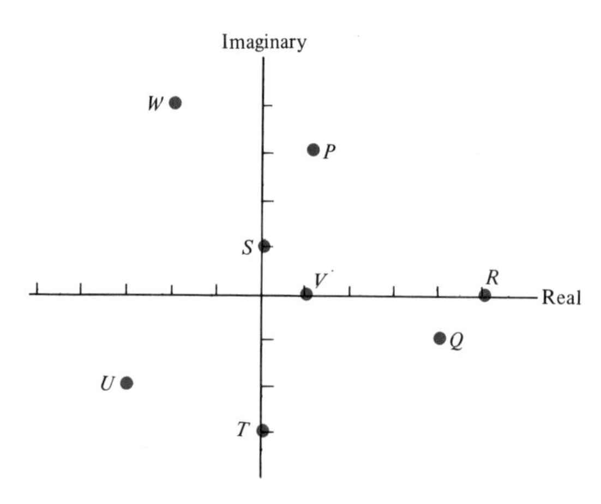

FIGURE 10.5

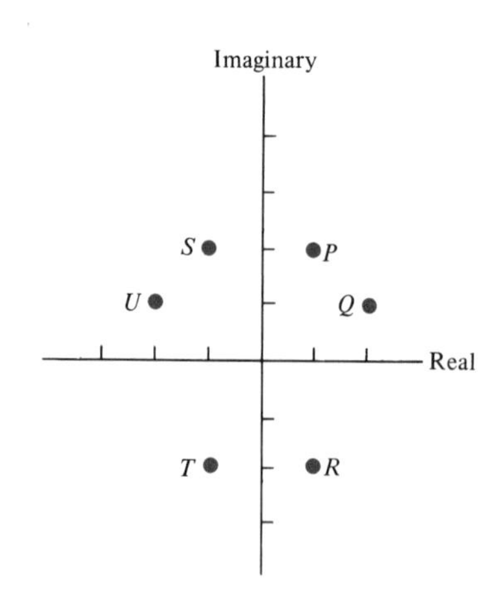

FIGURE 10.6

17. $(-3 + 2i) + (5 + 2i)$ **18.** $(-3 - 4i) - (-2 - i)$ **19.** $(-3) + i$ **20.** $6 - (-5i)$

21. $(-2 + 4i) + (-2 + 4i)$ **22.** $(6 + 3i) + (-6 + 3i)$ **23.** $(2 + 4i) - (2 + 4i)$ **24.** $(5 - 3i) - (5 + 3i)$

*10.3 Polar Coordinates of Complex Numbers

polar form

The complex number $x + yi$ corresponds to the point P with rectangular coordinates (x, y). Recall that in polar coordinates,

$$r = \sqrt{x^2 + y^2}, \qquad x = r\cos\theta, \qquad y = r\sin\theta,$$

and

$$\text{if } x \neq 0, \quad \tan\theta = \frac{y}{x}.$$

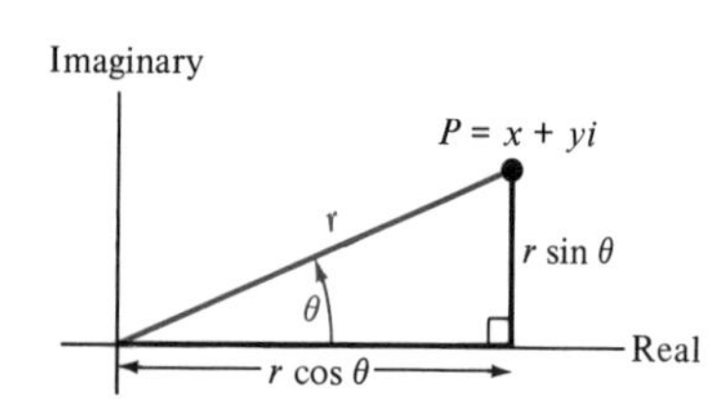

FIGURE 10.7

$x = r\cos\theta,\ y = r\sin\theta$
$x + yi = r\cos\theta + r\sin\theta \cdot i$
$= r(\cos\theta + i\sin\theta)$

(See Figure 10.7 .) Thus

$$x + yi = r\cos\theta + r\sin\theta \cdot i$$

or, isolating the common factor r and writing

$$i\sin\theta \quad \text{instead of} \quad \sin\theta \cdot i,$$

you obtain:

$$x + yi = r(\cos\theta + i\sin\theta),$$

the **polar** (or **trigonometric**) **form** of $x + yi$. Here $r = \text{dist}(O, P)$ and is thus a unique nonnegative (real) number; r is called the **modulus** of the given complex number. (In contrast, negative values for r were allowed in polar representation on the real plane.) And θ is called the **argument** (or **amplitude**) of the complex number; θ is *not* unique. In fact, when you add a multiple of 2π to θ, you obtain another representation of the same point. For example, $\frac{9\pi}{4}$ can be written $\left(\frac{\pi}{4} + 2\pi\right)$, and because $\cos\left(\frac{\pi}{4} + 2\pi\right) = \cos\frac{\pi}{4}$ and

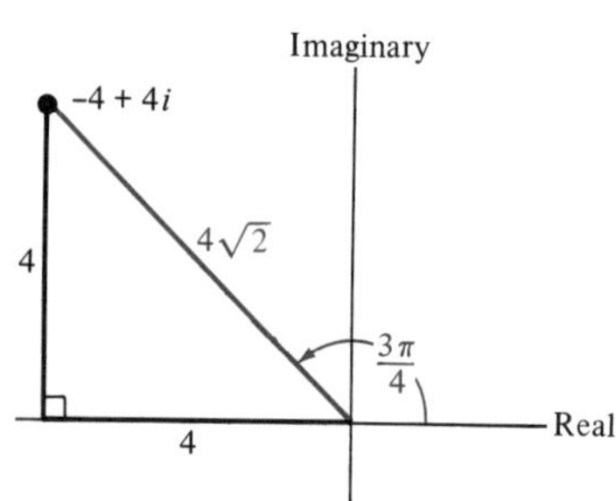

FIGURE 10.8

* Optional topic. Section 9.6 is prerequisite.

$$\sin\left(\frac{\pi}{4} + 2\pi\right) = \sin\frac{\pi}{4}, \quad \text{it follows that}$$

$$3\left(\cos\frac{9\pi}{4} + i\sin\frac{9\pi}{4}\right) = 3\left(\cos\frac{\pi}{4} + i\sin\frac{\pi}{4}\right)$$

EXAMPLE 1 Convert $-4 + 4i$ to polar form.

Solution. First plot $-4 + 4i$, as in Figure 10.8 . Observe that

$$r = \sqrt{(-4)^2 + 4^2} = \sqrt{32} = 4\sqrt{2} \quad \text{and} \quad \tan\theta = \frac{4}{-4} = -1.$$

Because θ is in Quadrant II, it follows that $\theta = \frac{3\pi}{4}$. [See Figure 10.8 .] The polar form is $4\sqrt{2}\left(\cos\frac{3\pi}{4} + i\sin\frac{3\pi}{4}\right)$. □

EXAMPLE 2 Convert $4\left(\cos\frac{\pi}{6} + i\sin\frac{\pi}{6}\right)$ to rectangular form.

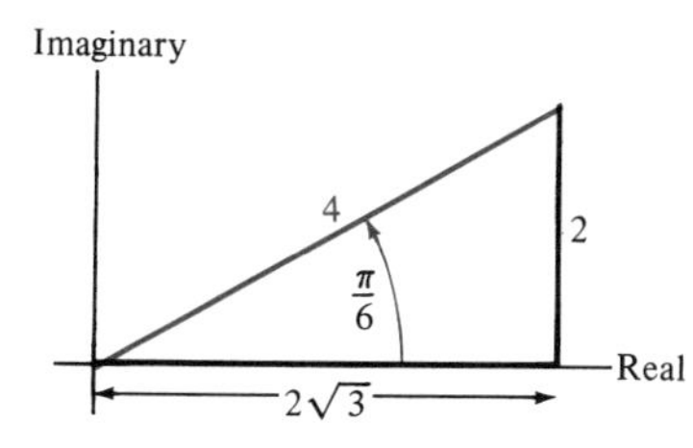

FIGURE 10.9

Solution.

$$x = 4\cos\frac{\pi}{6} = 4\cdot\frac{\sqrt{3}}{2} = 2\sqrt{3}, \qquad y = 4\sin\frac{\pi}{6} = 4\cdot\frac{1}{2} = 2$$

The rectangular coordinates are $(2\sqrt{3}, 2)$, and the rectangular form is $2\sqrt{3} + 2i$. [See Figure 10.9 .] □

$\bar{z}$, $-z$, $\frac{1}{z}$

Let $z = x + yi$ in rectangular form
$= r(\cos\theta + i\sin\theta)$ in polar form.

Then $\bar{z}$, the conjugate of z, has rectangular form $x - yi$. Clearly, $\bar{z}$ has modulus r and argument $-\theta$. [See Figure 10.10 on page 416.] The polar form of $\bar{z}$ is

$$r\,(\underbrace{\cos(-\theta)}_{\cos\theta} + i\,\underbrace{\sin(-\theta)}_{-\sin\theta}), \quad \text{or} \quad r(\cos\theta - i\sin\theta)$$

Also, $-z$, the *additive inverse of* z, has modulus r and rectangular form $-x - yi$. Thus, $-z$ has modulus r and argument $\theta + \pi$. Its polar form is

$$r\,(\underbrace{\cos(\theta+\pi)}_{-\cos\theta} + i\,\underbrace{\sin(\theta+\pi)}_{-\sin\theta}) \quad \text{or} \quad r(-\cos\theta - i\sin\theta).$$

For $z \neq 0$, the polar form of $\frac{1}{z}$, the **reciprocal** or **multiplicative inverse of** z, is obtained as follows:

$$\frac{1}{z} = \frac{1\cdot(x - yi)}{(x + yi)\cdot(x - yi)} = \frac{1}{x^2 + y^2}\cdot(x - yi) = \frac{1}{r^2}\cdot\bar{z}$$

$$= \frac{1}{r^2} \cdot r (\cos \theta - i \sin \theta) = \frac{1}{r} (\cos \theta - i \sin \theta)$$

To summarize, if $z = r (\cos \theta + i \sin \theta)$, then

$$\bar{z} = r (\cos \theta - i \sin \theta)$$

$$-z = r (-\cos \theta - i \sin \theta) = r (\cos (\theta + \pi) + i \sin (\theta + \pi))$$

$$\frac{1}{z} = \frac{1}{r} (\cos \theta - i \sin \theta), \qquad z \neq 0$$

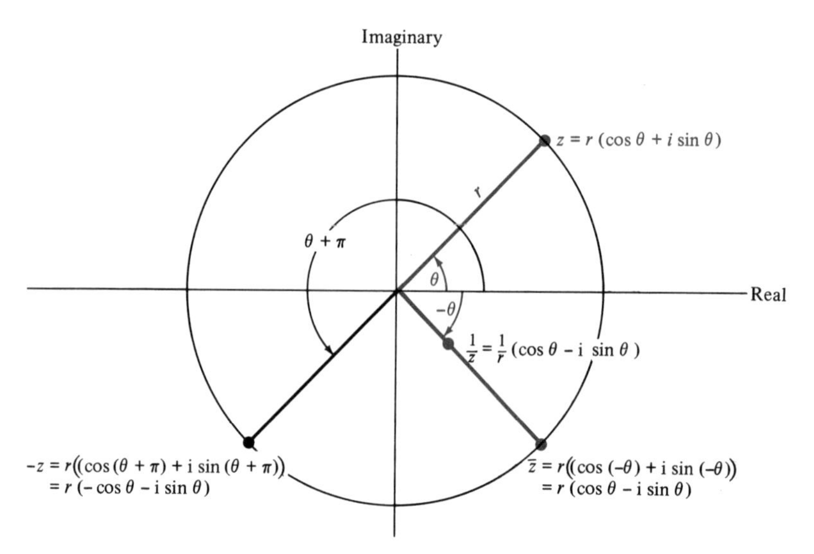

FIGURE 10.10

EXAMPLE 3 Let $z = 5\left(\cos \frac{\pi}{10} + i \sin \frac{\pi}{10}\right)$. In polar form:

(a) $\bar{z} = 5\left(\cos \frac{\pi}{10} - i \sin \frac{\pi}{10}\right)$

(b) $-z = 5\left(-\cos \frac{\pi}{10} - i \sin \frac{\pi}{10}\right) = 5\left(\cos \frac{11\pi}{10} + \text{i} \sin \frac{11\pi}{10}\right)$

(c) $\frac{1}{z} = \frac{1}{5}\left(\cos \frac{\pi}{10} - i \sin \frac{\pi}{10}\right)$ □

multiplication Polar form provides a convenient way to multiply and divide complex numbers. Observe that

$$r_1 (\cos \theta_1 + i \sin \theta_1) \cdot r_2 (\cos \theta_2 + i \sin \theta_2)$$

$$= r_1 r_2 (\cos \theta_1 \cos \theta_2 + i (\sin \theta_1 \cos \theta_2 + \cos \theta_1 \sin \theta_2)$$

$$+ \underbrace{i^2}_{-1} \sin \theta_1 \sin \theta_2)$$

$$= r_1 r_2 ((\cos\theta_1 \cos\theta_2 - \sin\theta_1 \sin\theta_2) + i(\cos\theta_1 \sin\theta_2 + \sin\theta_1 \sin\theta_2))$$

$$= r_1 r_2 (\cos(\theta_1 + \theta_2) + i \sin(\theta_1 + \theta_2)) \quad \text{by an Addition Formula}$$

$$r_1(\cos\theta_1 + i\sin\theta_1)\cdot r_2(\cos\theta_2 + i\sin\theta_2) = r_1 r_2(\cos(\theta_1+\theta_2) + i\sin(\theta_1+\theta_2))$$

Thus to multiply two complex numbers in polar form:

1. Multiply their moduli, (plural of *modulus*)
2. Add their arguments.

[See Figure 10.11 .] In particular,

$$[r(\cos\theta + i\sin\theta)]^2 = r(\cos\theta + i\sin\theta)\cdot r(\cos\theta + i\sin\theta) = r^2(\cos 2\theta + i\sin 2\theta)$$

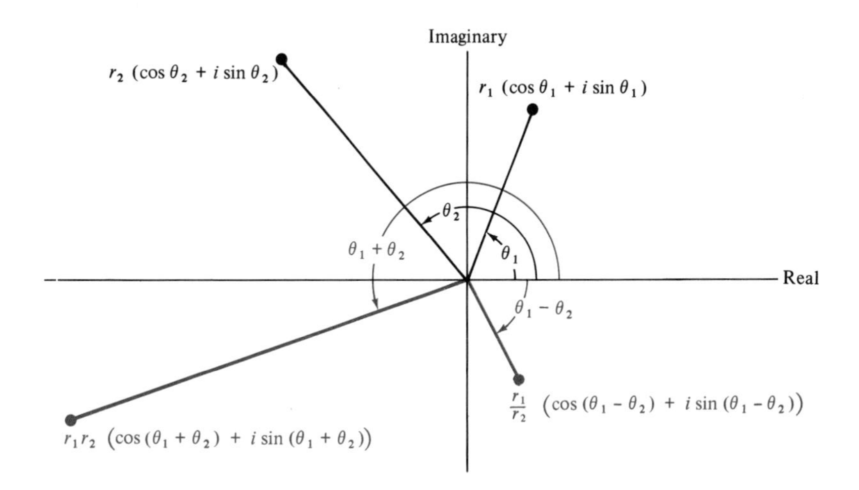

FIGURE 10.11

division To divide, use the preceding rule for multiplication, the formula for $\frac{1}{z}$, and the formulas $\cos\theta = \cos(-\theta)$ and $-\sin\theta = \sin(-\theta)$ to obtain

$$\frac{r_1(\cos\theta_1 + i\sin\theta_1)}{r_2(\cos\theta_2 + i\sin\theta_2)} = r_1(\cos\theta_1 + i\sin\theta_1)\cdot\frac{1}{r_2(\cos\theta_2 + i\sin\theta_2)}$$

$$= r_1(\cos\theta_1 + i\sin\theta_1)\cdot\frac{1}{r_2}(\cos\theta_2 - i\sin\theta_2)$$

$$= r_1 (\cos \theta_1 + i \sin \theta_1) \cdot \frac{1}{r_2} (\cos (-\theta_2) + i \sin (-\theta_2))$$

$$= \frac{r_1}{r_2} (\cos (\theta_1 - \theta_2) + i \sin (\theta_1 - \theta_2))$$

$$\boxed{\frac{r_1 (\cos \theta_1 + i \sin \theta_1)}{r_2 (\cos \theta_2 + i \sin \theta_2)} = \frac{r_1}{r_2} (\cos (\theta_1 - \theta_2) + i \sin (\theta_1 - \theta_2))}$$

Thus to divide two complex numbers in polar form:

1. Divide their moduli.
2. Subtract their arguments.

[See Figure 10.11 .]

EXAMPLE 4 (a) $2\left(\cos \frac{\pi}{8} + i \sin \frac{\pi}{8}\right) \cdot 4\left(\cos \frac{3\pi}{8} + i \sin \frac{3\pi}{8}\right)$

$$= 8\left(\cos\left(\frac{\pi}{8} + \frac{3\pi}{8}\right) + i \sin\left(\frac{\pi}{8} + \frac{3\pi}{8}\right)\right)$$

$$= 8\left(\underbrace{\cos \frac{\pi}{2}}_{0} + i \underbrace{\sin \frac{\pi}{2}}_{1}\right)$$

or in rectangular form,
$= 8i$

(b) $$\frac{2\left(\cos \frac{\pi}{8} + i \sin \frac{\pi}{8}\right)}{4\left(\cos \frac{3\pi}{8} + i \sin \frac{3\pi}{8}\right)} = \frac{2}{4} \cos\left(\frac{\pi}{8} - \frac{3\pi}{8}\right) + i \sin\left(\frac{\pi}{8} - \frac{3\pi}{8}\right)$$

$$= \frac{1}{2}\left(\cos\left(-\frac{\pi}{4}\right) + i \sin\left(-\frac{\pi}{4}\right)\right)$$

or in rectangular form,

$$= \frac{1}{2}\left(\frac{\sqrt{2}}{2} - \frac{\sqrt{2}}{2} i\right) = \frac{\sqrt{2}}{4} - \frac{\sqrt{2}}{4} i \qquad \square$$

EXERCISES

In Exercises 1–8 find: (a) the modulus and (b) the argument of the given complex number. (c) Plot this complex number.

1. $6\left(\cos \frac{\pi}{4} + i \sin \frac{\pi}{4}\right)$ **2.** $4\left(\cos \frac{7\pi}{4} + i \sin \frac{7\pi}{4}\right)$ **3.** $2\left(\cos \frac{\pi}{5} + i \sin \frac{\pi}{5}\right)$

4. $\cos 140° + i \sin 140°$ **5.** $\cos\left(-\frac{3\pi}{5}\right) + i \sin\left(-\frac{3\pi}{5}\right)$ **6.** $2 (\cos 0 + i \sin 0)$

7. $7(\cos\pi + i\sin\pi)$ **8.** $3\left(\cos\left(-\frac{\pi}{2}\right) + i\sin\left(-\frac{\pi}{2}\right)\right)$

In Exercises 9–18: (a) Convert to polar form. (b) Find the modulus. (c) Find the argument.

9. $1+i$ **10.** $1-i$ **11.** $-2+2i$ **12.** $1+\sqrt{3}i$ **13.** $5-5\sqrt{3}i$ **14.** 8 **15.** -8

16. $2i$ **17.** $\frac{i}{2}$ **18.** $-\frac{i}{2}$

In Exercises 19–28: (a) Convert to rectangular form. (b) Find the real part. (c) Find the imaginary part.

19. $3\left(\cos\frac{\pi}{4} + i\sin\frac{\pi}{4}\right)$ **20.** $5\left(\cos\frac{3\pi}{4} + i\sin\frac{3\pi}{4}\right)$ **21.** $9\left(\cos\frac{\pi}{6} + i\sin\frac{\pi}{6}\right)$

22. $\frac{1}{3}\left(\cos\left(-\frac{\pi}{4}\right) + i\sin\left(-\frac{\pi}{4}\right)\right)$ **23.** $12\left(\cos\frac{5\pi}{6} + i\sin\frac{5\pi}{6}\right)$ **24.** $\cos\frac{5\pi}{3} + i\sin\frac{5\pi}{3}$

25. $8(\cos\pi + i\sin\pi)$ **26.** $6\left(\cos\frac{\pi}{2} + i\sin\frac{\pi}{2}\right)$ **27.** $\cos 0 + i\sin 0$

28. $\frac{3}{4}\left(\cos\left(-\frac{\pi}{2}\right) + i\sin\left(-\frac{\pi}{2}\right)\right)$

In Exercises 29–36, for each complex number z, find: (a) $\bar{z}$, (b) $-z$, (c) $\frac{1}{z}$. Leave these in polar form.

29. $4\left(\cos\frac{\pi}{6} + i\sin\frac{\pi}{6}\right)$ **30.** $3\left(\cos\frac{2\pi}{3} + \text{i}\sin\frac{2\pi}{3}\right)$ **31.** $12\left(\cos\frac{3\pi}{4} + i\sin\frac{3\pi}{4}\right)$

32. $20\left(\cos\left(-\frac{\pi}{5}\right) + i\sin\left(-\frac{\pi}{5}\right)\right)$ **33.** $10\left(\cos\frac{\pi}{2} + i\sin\frac{\pi}{2}\right)$ **34.** $\frac{1}{2}(\cos(-\pi) + i\sin(-\pi))$

35. $\pi\left(\cos\frac{7\pi}{8} + i\sin\frac{7\pi}{8}\right)$ **36.** $\frac{\sqrt{2}}{3}\cos\left(\frac{3\pi}{10} + i\sin\frac{3\pi}{10}\right)$

In Exercises 37–42 multiply or divide, and express the result in polar form.

37. $2\left(\cos\frac{\pi}{9} + i\sin\frac{\pi}{9}\right)\cdot 5\left(\cos\frac{4\pi}{9} + i\sin\frac{4\pi}{9}\right)$ **38.** $3\left(\cos\frac{3\pi}{10} + i\sin\frac{3\pi}{10}\right)\cdot 4\left(\cos\frac{\pi}{10} + i\sin\frac{\pi}{10}\right)$

39. $7\left(\cos\frac{\pi}{5} + i\sin\frac{\pi}{5}\right)\cdot 2\left(\cos\frac{\pi}{3} + i\sin\frac{\pi}{3}\right)$ **40.** $\dfrac{6\left(\cos\frac{4\pi}{7} + i\sin\frac{4\pi}{7}\right)}{3\left(\cos\frac{\pi}{7} + i\sin\frac{\pi}{7}\right)}$

41. $\dfrac{12\left(\cos\frac{3\pi}{5} + i\sin\frac{3\pi}{5}\right)}{4\left(\cos\frac{4\pi}{5} + i\sin\frac{4\pi}{5}\right)}$ **42.** $\left(\cos\frac{2\pi}{3} + i\sin\frac{2\pi}{3}\right) \div 5\left(\cos\frac{\pi}{4} + i\sin\frac{\pi}{4}\right)$

In Exercises 43–48, multiply or divide, and express the result in rectangular form.

43. $4\left(\cos\frac{\pi}{4} + i\sin\frac{\pi}{4}\right)\cdot 5\left(\cos\frac{3\pi}{4} + i\sin\frac{3\pi}{4}\right)$ **44.** $2\left(\cos\left(-\frac{\pi}{6}\right) + i\sin\left(-\frac{\pi}{6}\right)\right)\cdot\left(\cos\frac{\pi}{3} + i\sin\frac{\pi}{3}\right)$

45. $8\left(\cos\frac{\pi}{5} + i\sin\frac{\pi}{5}\right)\cdot\frac{3}{4}\left(\cos\frac{3\pi}{10} + i\sin\frac{3\pi}{10}\right)$ **46.** $18\left(\cos\frac{\pi}{2} + i\sin\frac{\pi}{2}\right) \div 9\left(\cos\frac{\pi}{4} + i\sin\frac{\pi}{4}\right)$

47. $\dfrac{20\left(\cos\dfrac{3\pi}{8} + i\sin\dfrac{3\pi}{8}\right)}{4\left(\cos\dfrac{5\pi}{8} + i\sin\dfrac{5\pi}{8}\right)}$ **48.** $\dfrac{9(\cos 5\pi + i\sin 5\pi)}{12(\cos(-\pi) + i\sin(-\pi))}$

In Exercises 49–54: (a) Multiply or divide in rectangular form. (b) Convert the given complex numbers to polar form and then multiply or divide. (c) Show that both results agree.

49. $(1 + i)(1 + \sqrt{3}i)$ **50.** $(4 - 4i)(3 + 3i)$ **51.** $5(3 - 3\sqrt{3}i)$ **52.** $-2i(8 + 8i)$

53. $\dfrac{6 + 6\sqrt{3}i}{2\sqrt{3} - 2i}$ **54.** $\dfrac{3 - 3i}{-2i}$

55. Fill in: Multiplication of a complex number by i corresponds to a ____________ rotation of _____ degrees.

56. Let $z_1 = x_1 + y_1 i$ and $z_2 = x_2 + y_2 i$. Let r_1 be the modulus of z_1, let r_2 be the modulus of z_2, and let r_3 be the modulus of $z_1 + z_2$. Use the Parallelogram Rule (Figure 10.3 on p. 412) to show (graphically) the following **Triangle Inequality:** $r_3 \leq r_1 + r_2$

*10.4 De Moivre's Theorem

powers

Powers and roots of complex numbers can most easily be obtained in polar form. First consider positive integral powers. Recall that

$$[r(\cos\theta + i\sin\theta)]^2 = r^2(\cos 2\theta + i\sin 2\theta).$$

De Moivre's Theorem:†

Let n be *any positive integer*. Then

$$[r(\cos\theta + i\sin\theta)]^n = r^n(\cos n\theta + i\sin n\theta)$$

EXAMPLE 1 Find $(\sqrt{3} + i)^6$.

Solution. Convert to polar form. [See Figure 10.12 .]

$$\sqrt{3} + i = 2\left(\cos\frac{\pi}{6} + i\sin\frac{\pi}{6}\right)$$

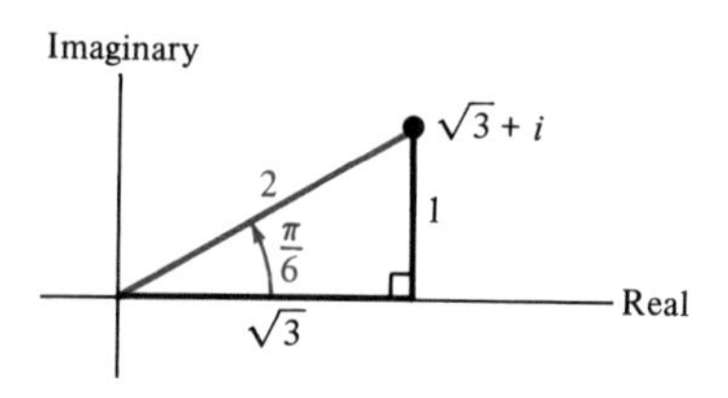

FIGURE 10.12

By De Moivre's Theorem,

$$\left[2\left(\cos\frac{\pi}{6} + i\sin\frac{\pi}{6}\right)\right]^6 = 2^6\left(\cos 6\cdot\frac{\pi}{6} + i\sin 6\cdot\frac{\pi}{6}\right)$$
$$= 64(\underbrace{\cos\pi}_{-1} + i\underbrace{\sin\pi}_{0})$$
$$= -64$$

□

roots

Now consider how to find the nth *roots* of a complex number.

EXAMPLE 2 Find the 4th roots of 16.

* Optional topic

† A proof requires Mathematical Induction. See Section 13.6 .

Solution. Clearly, $2^4 = 16$ and $(-2)^4 = 16$. Moreover,

$$(2i)^4 = 2^4 \cdot i^4 = 16 \cdot 1 = 16$$

and

$$(-2i)^4 = (-2)^4 \cdot i^4 = 16 \cdot 1 = 16$$

Here it was not difficult to find the 4th roots. In order to discover the general pattern, write these numbers in polar form.

Rectangular Form *Polar Form*

$$2 = 2\,(\underbrace{\cos 0}_{1} + i\,\underbrace{\sin 0}_{0})$$

$$2i = 2\left(\underbrace{\cos \frac{\pi}{2}}_{0} + i\,\underbrace{\sin \frac{\pi}{2}}_{1}\right)$$

$$-2 = 2\,(\underbrace{\cos \pi}_{-1} + i\,\underbrace{\sin \pi}_{0})$$

$$-2i = 2\left(\underbrace{\cos \frac{3\pi}{2}}_{0} + i\,\underbrace{\sin \frac{3\pi}{2}}_{-1}\right)$$

[See Figure 10.13 .] Observe that in polar form,

$$16 = 16\,(\cos 0 + i \sin 0)$$

There are four 4th roots of 16. Each 4th root has modulus $16^{1/4}$, or 2, and thus lies on the circle of radius length 2 centered about the pole. The arguments of these 4th roots differ by $\frac{\pi}{2}$, which equals $\frac{2\pi}{4}$. □

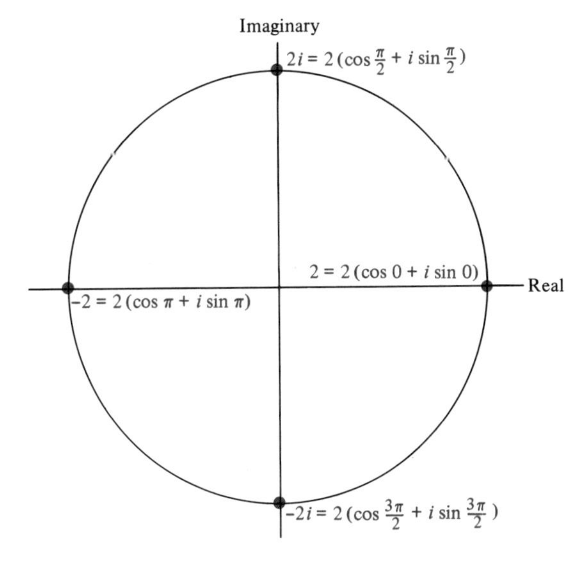

FIGURE 10.13

To find the nth roots of a nonzero complex number

$R\ (\cos\theta_0 + i\sin\theta_0)$,

write these roots in the form $r\ (\cos\theta + i\sin\theta)$. Thus for each such root,

$$[r\ (\cos\theta + i\sin\theta)]^n = R\ (\cos\theta_0 + i\sin\theta_0)$$

[In Example 2, for the 4th roots of 16, $n = 4$, $R = 16$, $R^{1/4} = 2$, $\theta_0 = 0$.] Apply De Moivre's Theorem.

$$r^n\ (\cos n\theta + i\sin n\theta) = R\ (\cos\theta_0 + i\sin\theta_0)$$

Because the modulus of a complex number is unique, each nth root must satisfy:

$$r^n = R \quad \text{or} \quad r = R^{1/n}$$

Furthermore, two arguments of a complex number can differ only by a multiple of 2π. Thus

$$n\theta = \theta_0 + 2k\pi, \quad \text{where } k \text{ is an integer}$$
$$\theta = \frac{\theta_0}{n} + k\cdot\frac{2\pi}{n}$$

As k takes on the values 0, 1, 2, 3, . . . , $n - 1$, you obtain n different complex numbers, each with modulus $R^{1/n}$. [See Figure 10.13, in which $n = 4$, $R = 16$, $R^{1/4} = 2$, and $\theta_0 = 0$.] As k takes on the values n, $n + 1$, . . . , the numbers begin repeating. For example, for $k = n$,

$$\frac{\theta_0}{n} + n\cdot\frac{2\pi}{n} = \frac{\theta_0}{n} + 2\pi,$$

and because

$$\cos\left(\frac{\theta_0}{n} + 2\pi\right) = \cos\frac{\theta_0}{n} \quad \text{and} \quad \sin\left(\frac{\theta_0}{n} + 2\pi\right) = \sin\frac{\theta_0}{n},$$

you obtain the same complex number as when $k = 0$. Geometrically, you have completed one cycle about the circle with radius $R^{1/n}$, and are now repeating the points previously obtained. Thus,
the nth roots of $R\ (\cos\theta_0 + i\sin\theta_0)$ are the n complex numbers:

$$R^{1/n}\left(\cos\frac{\theta_0}{n} + i\sin\frac{\theta_0}{n}\right),$$

$$R^{1/n}\left(\cos\left(\frac{\theta_0}{n} + \frac{2\pi}{n}\right) + i\sin\left(\frac{\theta_0}{n} + \frac{2\pi}{n}\right)\right),$$

$$R^{1/n}\left(\cos\left(\frac{\theta_0}{n} + 2\cdot\frac{2\pi}{n}\right) + i\sin\left(\frac{\theta_0}{n} + 2\cdot\frac{2\pi}{n}\right)\right),$$

. . . .

$$R^{1/n}\left(\cos\left(\frac{\theta_0}{n} + (n-1)\frac{2\pi}{n}\right) + i\sin\left(\frac{\theta_0}{n} + (n-1)\frac{2\pi}{n}\right)\right)$$

EXAMPLE 3 Find the cube roots of $4 - 4\sqrt{3}i$.

Solution. The given number lies in Quadrant IV because its real part is positive and its imaginery part is negative. Convert to polar form, $R\,(\cos\theta_0 + i\sin\theta_0)$.

$$R = \sqrt{4^2 + (-4\sqrt{3})^2} = \sqrt{16 + 16\cdot 3} = \sqrt{64} = 8$$

$$\tan\theta_0 = \frac{-4\sqrt{3}}{4} = -\sqrt{3}$$

Because θ_0 is in Quadrant IV, $\theta_0 = \frac{5\pi}{3}$. Thus

$$4 - 4\sqrt{3}i = 8\left(\cos\frac{5\pi}{3} + i\sin\frac{5\pi}{3}\right)$$

Here $n = 3,\quad R = 8,\quad R^{1/3} = 2,\quad \theta_0 = \frac{5\pi}{3}.$ The cube roots of $4 - 4\sqrt{3}\,i$ are

$$2\left(\cos\frac{5\pi}{9} + i\sin\frac{5\pi}{9}\right),$$

$$2\left(\cos\frac{11\pi}{9} + i\sin\frac{11\pi}{9}\right), \qquad \frac{11\pi}{9} = \frac{5\pi}{9} + \frac{2\pi}{3}$$

$$2\left(\cos\frac{17\pi}{9} + i\sin\frac{17\pi}{9}\right). \qquad \frac{17\pi}{9} = \frac{5\pi}{9} + 2\cdot\frac{2\pi}{3}$$

[See Figure 10.14 .] □

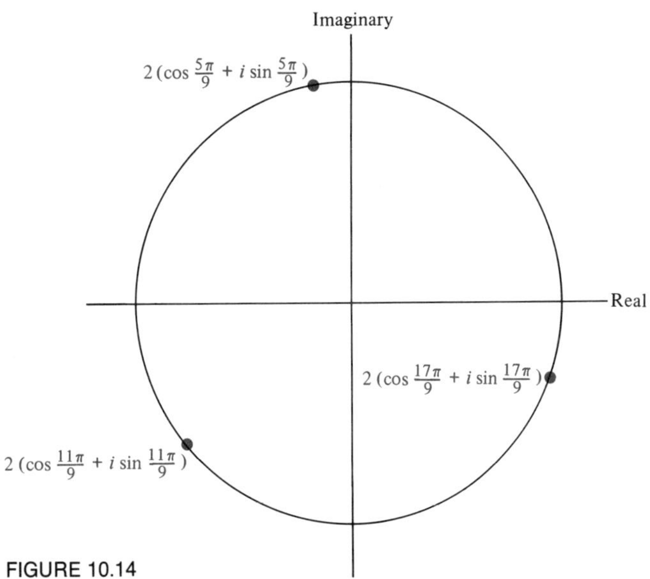

FIGURE 10.14

roots of unity The nth roots of 1 are known as the ***n*th roots of unity.**

EXAMPLE 4 (a) Find the 6th roots of unity. Express these both in polar form and in rectangular form.

(b) Find all complex roots of the equation $z^6 - 1 = 0$.

Solution.

(a) $$1 = 1(\cos 0 + i \sin 0)$$

Here $n = 6$, $R = 1$, $R^{1/6} = 1$, $\theta_0 = 0$. The 6th roots of unity are:

$$\cos 0 + i \sin \theta, \quad \text{or } 1$$

$$\cos \frac{\pi}{3} + i \sin \frac{\pi}{3}, \quad \text{or } \frac{1}{2} + \frac{\sqrt{3}}{2} i$$

$$\cos \frac{2\pi}{3} + i \sin \frac{2\pi}{3}, \quad \text{or } -\frac{1}{2} + \frac{\sqrt{3}}{2} i$$

$$\cos \pi + i \sin \pi, \quad \text{or } -1$$

$$\cos \frac{4\pi}{3} + i \sin \frac{4\pi}{3}, \quad \text{or } -\frac{1}{2} - \frac{\sqrt{3}}{2} i$$

$$\cos \frac{5\pi}{3} + i \sin \frac{5\pi}{3}, \quad \text{or } \frac{1}{2} - \frac{\sqrt{3}}{2} i$$

[See Figure 10.15 .]

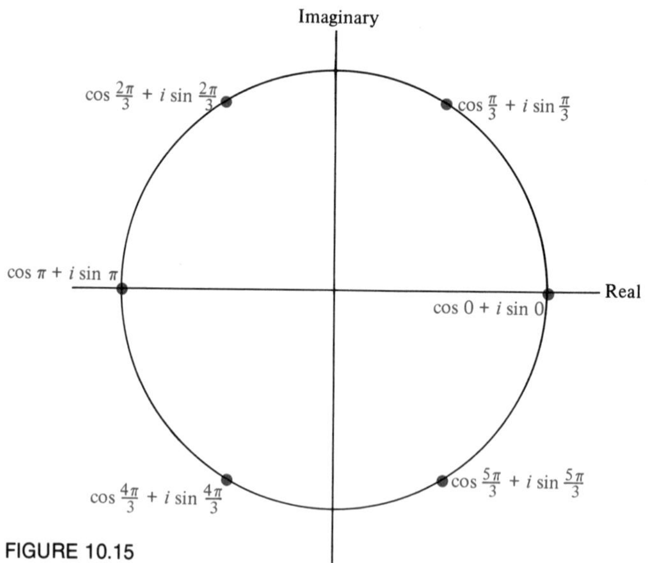

FIGURE 10.15

(b) The (complex) roots of the equation

$$z^6 - 1 = 0 \qquad \text{or} \qquad z^6 = 1$$

are the 6th roots of unity:

$$1, \quad \frac{1}{2} + \frac{\sqrt{3}}{2} i, \quad -\frac{1}{2} + \frac{\sqrt{3}}{2} i, \quad -1, \quad -\frac{1}{2} - \frac{\sqrt{3}}{2} i, \quad \frac{1}{2} - \frac{\sqrt{3}}{2} i$$ □

EXERCISES

1. Find $(3 + 3i)^3$: **(a)** by multiplication **(b)** by using De Moivre's Theorem

2. Find $(2 - 2\sqrt{3}\,i)^4$: **(a)** by multiplication **(b)** by using De Moivre's Theorem

In Exercises 3–12 express the indicated powers in polar form.

3. $\left[3\left(\cos\frac{\pi}{3} + i\sin\frac{\pi}{3}\right)\right]^2$ **4.** $\left[4\left(\cos\frac{3\pi}{4} + i\sin\frac{3\pi}{4}\right)\right]^4$ **5.** $\left[10\left(\cos\left(-\frac{\pi}{8}\right) + i\sin\left(-\frac{\pi}{8}\right)\right)\right]^4$

6. $\left[\frac{1}{2}\left(\cos\frac{\pi}{10} + i\sin\frac{\pi}{10}\right)\right]^5$ **7.** $(1 + i)^3$ **8.** $(\sqrt{3} - i)^6$

9. $(-2 + 2\sqrt{3}\,i)^4$ **10.** $(1 - i)^{10}$ **11.** $\left(\frac{\sqrt{2}}{2} + \frac{\sqrt{2}}{2}i\right)^{12}$

12. $\left(\frac{1}{2} - \frac{\sqrt{3}}{2}i\right)^{15}$

In Exercises 13–18 express the indicated powers in rectangular form.

13. $(1 + i)^2$ **14.** $(\sqrt{3} + i)^3$ **15.** $(1 - i)^6$ **16.** $\left(\frac{\sqrt{2}}{2} + \frac{\sqrt{2}}{2}i\right)^{10}$ **17.** $\left(\frac{1}{2} + \frac{\sqrt{3}}{2}i\right)^{12}$

18. $(-\sqrt{3} + i)^9$

In Exercises 19–24 find *all* of the indicated roots. Express these in polar form. In Exercises 19–22 show the distribution of the roots about the appropriate circle.

19. The square roots of $\frac{\sqrt{3}}{2} + \frac{1}{2}i$ **20.** The 4th roots of -16 **21.** The 6th roots of $-i$

22. The cube roots of $1 - i$ **23.** The 10th roots of i **24.** The 6th roots of $\frac{1}{64}\cos\left(\frac{\pi}{3} + i\sin\frac{\pi}{3}\right)$

In Exercises 25–30 find *all* of the indicated roots. Express these in rectangular form.

25. The square roots of i **26.** The cube roots of -2 **27.** The square roots of $-4 + 4i$

28. The 4th roots of $-8 + 8\sqrt{3}\,i$ **29.** The 8th roots of 256 **30.** The 8th roots of 5^{16}

In Exercises 31–34 find *all* of the indicated roots of unity. Express these in polar form.

31. The 5th roots of unity **32.** The 9th roots of unity **33.** The 10th roots of unity

34. The 15th roots of unity

In Exercises 35–38 find *all* of the indicated roots of unity. Express these in rectangular form.

35. The cube roots of unity **36.** The 4th roots of unity **37.** The 8th roots of unity

38. The 12th roots of unity

In Exercises 39–46 find *all* complex roots of each equation. Express these roots in rectangular form.

39. $z^2 + 4 = 0$ **40.** $z^2 - 4i = 0$ **41.** $z^2 - 9i = 0$ **42.** $z^3 - 8i = 0$ **43.** $z^6 + 64 = 0$

44. $z^4 + 16 = 0$ **45.** $z^6 + 8i = 0$ **46.** $z^3 + 1 = i$

47. Use De Moivre's Theorem (with $r = 1$, $n = 2$) to obtain the Double Angle Formulas.

48. Use De Moivre's Theorem to obtain Triple Angle Formulas for $\cos 3\theta$ and for $\sin 3\theta$.

49. **(a)** Find the sum of the 4th roots of unity. **(b)** Find the sum of the 6th roots of unity.

(c) What rule do parts **(a)** and **(b)** suggest?

Review Exercises for Chapter 10

1. Perform the indicated operations and express the resulting complex number in the form of $a + bi$.

   (a) $(3 - 2i) - (1 - i)$ (b) $(2 + i)(3 - 5i)$ (c) $\dfrac{4 + i}{1 - 3i}$ (d) $(2 - i)^2$

2. On a rectangular coordinate system, represent the following complex numbers.

   (a) $2 + 5i$ (b) $3 - i$ (c) $-3i$

3. Use the Parrallelogram Rule to represent the sum $(3 + 4i) + (-2 + i)$.

4. Find: (a) the modulus and (b) the argument of $3\left(\cos\dfrac{3\pi}{4} + i\sin\dfrac{3\pi}{4}\right)$. (c) Plot this complex number.

5. (a) Convert $4 - 4i$ to polar form. (b) Find the modulus. (c) Find the argument.

6. (a) Convert $2\left(\cos\dfrac{4\pi}{3} + i\sin\dfrac{4\pi}{3}\right)$ to rectangular form. (b) Find the real part.

   (c) Find the imaginary part.

7. Let $z = 3\left(\cos\dfrac{\pi}{4} + i\sin\dfrac{\pi}{4}\right)$. Find: (a) $\bar{z}$ (b) $-z$ (c) $\dfrac{1}{z}$. Leave these in polar form.

8. Multiply, and express the product (a) in polar form (b) in rectangular form.

$$3\left(\cos\frac{\pi}{6} + i\sin\frac{\pi}{6}\right)\cdot\frac{1}{2}\left[\cos\left(-\frac{\pi}{3}\right) + i\sin\left(-\frac{\pi}{3}\right)\right].$$

9. Find: $(\sqrt{3} - i)^3$ (a) by multiplication (b) by using De Moivre's Theorem.

10. Express $\left[2\left(\cos\dfrac{\pi}{3} + i\sin\dfrac{\pi}{3}\right)\right]^4$ in polar form.

11. Express $(1 - i)^3$ in rectangular form.

12. Find all the 6th roots of -64. Express these in polar form.

13. Find all the 8th roots of unity. Express these in polar form.

14. Find all complex roots of the equation $z^3 - 27i = 0$. Express these roots in rectangular form.

11 Polynomials

11.1 Division of Polynomials

division of integers

Division is the inverse operation of multiplication, just as subtraction is the inverse operation of addition. In some respects division of polynomials resembles division of integers.

Definition. Let a, b, c be integers, with $b \neq 0$. Then

$$\frac{a}{b} = c \qquad \text{if} \qquad a = bc$$

Here a is called the **dividend,** b the **divisor,** and c the **quotient.**

$$\text{Thus} \qquad \frac{6}{2} = 3 \qquad \text{because} \qquad 6 = 2 \cdot 3$$

In this example, 6 is the dividend, 2 the divisor, and 3 the quotient. But when you divide 7 by 2 there is also a "remainder." Thus

$$\frac{7}{2} = 3 + \frac{1}{2} \qquad \left(\text{or } 3\frac{1}{2}\right)$$

Here 3 is the quotient and 1 is the remainder. In general, the **Division Algorithm for Integers** asserts that for any integers a and b with $b > 0$, there exist *unique* integers q (the **quotient**) and r (the **remainder**) such that

$$a = bq + r, \qquad \text{where} \qquad 0 \leqslant r < b.$$

Thus

$$\begin{aligned} &\text{if} \quad a = 6 \quad \text{and} \quad b = 2, \quad \text{then} \quad q = 3 \quad \text{and} \quad r = 0; \\ &\text{if} \quad a = 7 \quad \text{and} \quad b = 2, \quad \text{then} \quad q = 3 \quad \text{and} \quad r = 1 \end{aligned}$$

Note that in both cases, $0 \leqslant r < b$.

division algorithm for polynomials

Similarly, division of polynomials is defined in terms of multiplication of polynomials. Because much of this Chapter concerns polynomials as functions, polynomials will be denoted by $P(x)$, $Q(x)$, etc.

Definition. Let $P(x)$, $D(x)$, and $Q(x)$ be polynomials and assume that $D(x)$ is not the zero polynomial. Then

$$\frac{P(x)}{D(x)} = Q(x) \quad \text{if} \quad P(x) = D(x) \cdot Q(x)$$

In this case, $P(x)$ is called the **dividend,** $D(x)$ the **divisor,** and $Q(x)$ the **quotient.**

For example, $\frac{6x^3}{2x^2} = 3x$ because $6x^3 = 2x^2 \cdot 3x$

Here $6x^3$ is the dividend, $2x^2$ the divisor, and $3x$ the quotient. But when you divide $6x^3 + x + 1$ by $2x^2$, there is a remainder. Thus

$$\frac{6x^3 + x + 1}{2x^2} = 3x + \frac{x + 1}{2x^2}$$

Here, $3x$ is the quotient and $x + 1$ is the remainder.

The **Division Algorithm for Polynomials** asserts:

For any polynomial $P(x)$ and *nonzero* polynomial $D(x)$, there exist *unique* polynomials $Q(x)$ [the *quotient*] and $R(x)$ [the *remainder*] such that

$$P(x) = D(x) \cdot Q(x) + R(x),$$

where either $R(x)$ is the zero polynomial or else the degree of $R(x)$ is less than the degree of $D(x)$.

Thus

$$\begin{aligned} &\text{if} \quad P(x) = 6x^3 \quad \text{and} \quad D(x) = 2x^2, \\ &\text{then} \quad Q(x) = 3x \quad \text{and} \quad R(x) \equiv 0; \\ &\text{if} \quad P(x) = 6x^3 + x + 1 \quad \text{and} \quad D(x) = 2x^2, \\ &\text{then} \quad Q(x) = 3x \quad \text{and} \quad R(x) = x + 1 \end{aligned}$$

In one case $R(x) \equiv 0$, and in the other case

$$\text{degree } R(x) < \text{degree } D(x)$$

Observe that in the Division Algorithm, if you divide both sides of the equation by $D(x)$, you obtain:

$$\frac{P(x)}{D(x)} = Q(x) + \frac{R(x)}{D(x)}$$

Thus

$$\frac{6x^3 + x + 1}{2x^2} = 3x + \frac{x + 1}{2x^2}$$

factoring the dividend

By factoring the dividend you can sometimes easily obtain the quotient. For example,

$$\frac{2x^4 + 3x^3 - x^2}{x^2} = \frac{x^2(2x^2 + 3x - 1)}{x^2} = 2x^2 + 3x - 1$$

and

$$\frac{x^2 + 3x + 2}{x + 1} = \frac{(x + 1)(x + 2)}{x + 1} = x + 2$$

long division

Often a procedure similar to long division of integers is employed. Only division of polynomials in a single variable will be considered. To this end, consider the following definition.

Definition. A polynomial in a *single variable* is said to be in **standard form** if terms of the same degree are combined and the resulting terms are arranged in order of decreasing degrees. The term of highest degree is called the **leading term;** the coefficient of this leading term is called the **leading coefficient.**

The standard form of

$$x^5 + x^5 - 8x^7 + 4x^2 - 9 + 10x^2 \quad \text{is} \quad -8x^7 + 2x^5 + 14x^2 - 9.$$

The leading term is $-8x^7$, and the leading coefficient is -8. A monomial in a single variable is automatically in standard form. Thus, $9x^5$ is in standard form.

EXAMPLE 1

$$x + 4 \overline{\big)\, 2x^2 + 11x + 12}$$

Solution. Divide $2x^2$, the leading term of the dividend, by x, the leading term of the divisor.

$$\frac{2x^2}{x} = 2x$$

Thus the leading term of the quotient is $2x$. Multiply this by the divisor, $x + 4$, and subtract the product, $2x^2 + 8x$, from $2x^2 + 11x$. The difference is $3x$. Bring down the next term, 12, and obtain the **first difference polynomial,** $3x + 12$.

$$\begin{array}{rrl}
 & 2x & \leftarrow \text{quotient} \\
\text{divisor} \rightarrow x + 4 & \overline{\big)\, 2x^2 + 11x + 12} & \leftarrow \text{dividend} \\
 & \underline{\mp 2x^2 \mp 8x} & \\
 & 3x + 12 & \leftarrow \text{first difference polynomial}
\end{array}$$

Divide $3x$, the leading term of the difference polynomial, by x, the leading term of the divisor.

$$\frac{3x}{x} = 3$$

Thus the second term of the quotient is 3. Multiply this by the divisor, $x + 4$. Subtract the product, $3x + 12$, from $3x + 12$ to obtain the **second difference polynomial.** Because this is 0, $x + 4$ divides $2x^2 + 11x + 12$ (evenly). The quotient is $2x + 3$.

$$\begin{array}{rrl}
 & 2x + 3 & \\
x + 4 & \overline{\big)\, 2x^2 + 11x + 12} & \\
 & \underline{\mp 2x^2 \mp 8x} & \\
 & 3x + 12 & \\
 & \underline{\mp 3x \mp 12} & \\
 & 0 & \leftarrow \text{second difference polynomial}
\end{array}$$

Note that you could also have obtained the quotient by factoring the dividend.

$$\frac{2x^2 + 11x + 12}{x + 4} = \frac{(x + 4)(2x + 3)}{x + 4} = 2x + 3$$

□

checking

You can *check* a division example by multiplying the quotient and the divisor (in either order). If the remainder is 0, the product should be the dividend.

Check. (For Example 1).

$$\begin{array}{rl} 2x + 3 & \leftarrow \text{quotient} \\ x + 4 & \leftarrow \text{divisor} \\ \hline 2x^2 + 3x & \\ 8x + 12 & \\ \hline 2x^2 + 11x + 12 & \leftarrow \text{dividend} \end{array}$$

remainders

In the following examples there is a nonzero remainder. *Continue the division procedure until you obtain either a 0 difference polynomial or one whose degree is less than that of the divisor. This difference polynomial will be the remainder. In case the degree of the dividend is less than the degree of the divisor, the quotient is 0 and the dividend is the remainder.* Thus

$$\frac{x^2 + 1}{x^3 + 1} = 0 + \frac{x^2 + 1}{x^3 + 1}.$$

The quotient is 0 and the remainder is $x^2 + 1$.

Polynomials (in a single variable) should always be put into standard form, as in Example 2. Thus, the terms are arranged in order of decreasing degrees. *Add* 0's *for the missing terms,* as in Example 2.

EXAMPLE 2 Divide $6x^2 + x^3 + 10x$ by $2 + x + x^2$.

Solution. Write these polynomials in standard form. Add a 0 for the missing constant term of the dividend.

$$\begin{array}{rl} x + 5 + \dfrac{3x - 10}{x^2 + x + 2} & \\ \text{divisor} \rightarrow x^2 + x + 2 \,\overline{\big)\, x^3 + 6x^2 + 10x + 0} & \leftarrow \text{dividend} \\ \overline{+}x^3 \overline{+} x^2 \overline{+} 2x & \\ \hline 5x^2 + 8x + 0 & \leftarrow \text{first difference polynomial} \\ \overline{+} 5x^2 \overline{+} 5x \overline{+} 10 & \\ \hline 3x - 10 & \leftarrow \text{second difference polynomial } \textit{(remainder)} \end{array}$$

Note that on the first step, after subtracting $x^3 + x^2 + 2x$ from $x^3 + 6x^2 + 10x$, there are no more nonzero terms of the dividend to bring down. The first difference polynomial is therefore $5x^2 + 8x$. *Because its degree, 2, equals that of the divisor, the division process continues.*

The degree, 1, of the second difference polynomial, $3x - 10$, is less than the degree, 2, of the divisor, $x^2 + x + 2$. Thus, $3x - 10$ is the remainder. The result is

$$x + 5 + \frac{3x - 10}{x^2 + x + 2}.$$

□

To *check* the result *when there is a remainder,* multiply the quotient and the divisor; then add the remainder. The sum should be the dividend.

Check for Example 2.

$$
\begin{array}{rrrrl}
 & x^2 + & x + & 2 & \leftarrow \text{divisor} \\
 & x + & 5 & & \leftarrow \text{quotient} \\
\hline
 & x^3 + & x^2 + & 2x & \\
 & & + 5x^2 + & 5x + 10 & \\
\hline
 & x^3 + & 6x^2 + & 7x + 10 & \\
+ & & & 3x - 10 & \leftarrow \text{remainder} \\
\hline
 & x^3 + & 6x^2 + & 10x & \leftarrow \text{dividend}
\end{array}
$$

EXAMPLE 3 **Determine the quotient and remainder:**

$$(x^6 - x^4 + 2x^3 - x^2) \div (x^2 + 1)$$

Solution. Add "zero terms" in the dividend and divisor.

$$
\begin{array}{rl}
 & x^4 \quad - 2x^2 + 2x + 1 + \dfrac{-2x - 1}{x^2 + 1} \\
x^2 + 0x + 1 & \overline{\big)\, x^6 + 0x^5 - x^4 + 2x^3 - x^2 + 0x + 0} \leftarrow \text{divisor} \\
\uparrow \\
\text{dividend} & \underline{\mp x^6 \qquad \mp x^4} \\
 & \qquad - 2x^4 + 2x^3 - x^2 \quad \leftarrow \text{first difference polynomial} \\
 & \qquad \underline{\pm 2x^4 \qquad \pm 2x^2} \\
 & \qquad\qquad 2x^3 + x^2 + 0x \quad \leftarrow \text{second difference polynomial} \\
 & \qquad\qquad \underline{\mp 2x^3 \qquad \mp 2x} \\
 & \qquad\qquad\qquad x^2 - 2x + 0 \leftarrow \text{third difference polynomial} \\
 & \qquad\qquad\qquad \underline{\mp x^2 \qquad \mp 1} \\
 & \qquad\qquad\qquad\qquad - 2x - 1 \leftarrow \text{fourth difference polynomial } \textit{(remainder)}
\end{array}
$$

Because the divisor has three "terms" (including the "zero term", $0x$), the first difference polynomial must also have three terms. Thus bring down $2x^3 - x^2$, so that the first difference polynomial is $-2x^4 + 2x^3 - x^2$. Even though you run out of terms of the dividend after the second step, the degree, 3, of the second difference polynomial is greater than 2, the degree of the divisor. Therefore continue the process. Similarly, the degree, 2, of the third difference polynomial *equals* the degree of the divisor. Continue the process until you obtain a difference polynomial whose degree is *less than* that of the divisor. This occurs on the fourth step. The fourth difference polynomial, $-2x - 1$, is the remainder. Thus

$$\frac{x^6 - x^4 + 2x^3 - x^2}{x^2 + 1} = x^4 - 2x^2 + 2x + 1 + \frac{-2x - 1}{x^2 + 1} \qquad \square$$

The next example illustrates a very simple, but often misunderstood, point.

EXAMPLE 4 $2x + 2\overline{)x + 2}$

Solution. The degree of the dividend *equals* the degree of the divisor. Even though the *leading coefficient*, 1, of the dividend is less than the *leading coefficient*, 2, of the divisor, you can still divide. Obtain $\frac{1}{2}$ as your quotient; your remainder turns out to be 1.

$$\begin{array}{r} \frac{1}{2} + \frac{1}{2x+2} \\ 2x + 2 \overline{)\; x + 2} \\ \underline{\mp x \mp 1} \\ 1 \end{array}$$

□

copying the divisor Sometimes, the easiest way to divide is to add and subtract the same polynomial in the dividend in order to obtain a copy of the divisor. This technique is used frequently in calculus problems.

EXAMPLE 5 (a) $$\frac{x}{x-1} = \frac{x-1+1}{x-1} = \frac{x-1}{x-1} + \frac{1}{x-1} = 1 + \frac{1}{x-1}$$

(b) $$\frac{x^4 + x^2 - 2}{x^4 + x^3 + x^2 + 1} = \frac{x^4 + x^2 - 2 + (x^3 + 3) - (x^3 + 3)}{x^4 + x^3 + x^2 + 1}$$
$$= 1 - \frac{x^3 + 3}{x^4 + x^3 + x^2 + 1}$$

□

EXERCISES

In Exercises 1–16 divide. Check the ones so indicated.

1. $(30 + 13z + z^2) \div (3 + z)$

2. $(40 + 13y + y^2) \div (8 + y)$

3. $a + 4\overline{)a^3 + 5a^2 + 6a + 8}$ (Check.)

4. $m - 3\overline{)m^3 + 2m^2 - 21m + 18}$

5. $\dfrac{35 - 16x + 4x^2 + x^3}{7 + x}$

6. $2x + 1\overline{)2x^2 + 9x + 4}$

7. $3x - 5\overline{)6x^2 + 2x - 20}$

8. $4y + 3\overline{)12y^3 + 17y^2 + 14y + 6}$

9. $\dfrac{3(t^3 + 21) - 13t(t + 1)}{3t - 7}$

10. $t^2 + t + 1\overline{)4t^3 - t^2 - t - 5}$

11. $a^2 - 3a + 2\overline{)a^3 + 2a^2 - 13a + 10}$

12. $m^2 + 2m - 5\overline{)4m^3 + 5m^2 - 26m + 15}$

13. $2m^2 - 5\overline{)2m^3 + 2m^2 - 5m - 5}$ (Check.)

14. $3s^2 - 2s + 1\overline{)9s^3 - s + 2}$

15. $x^2 + 1\overline{)x^4 - x^3 + 3x^2 - x + 2}$

16. $x^2 - x - 2\overline{)x^4 + x^3 - 3x^2 - 5x - 2}$

In Exercises 17–38 divide as indicated. In some cases there will be a remainder. Check the exercises so indicated.

17. $x + 2\overline{)x^2 + 4x + 5}$ (Check.)

18. $y - 3\overline{)y^2 + y + 2}$

19. $\dfrac{x^2 - 9x + 7}{x}$

20. $\dfrac{z^2 - 3z + 8}{z - 4}$

21. $x + 4\overline{)x^2 + 9x + 16}$

22. $a^2\overline{)a^3 - a^2 + a + 1}$

23. $c^2 + 2\overline{)c^3 + 2c^2 + 2c + 4}$

24. $y^2 + 3\overline{)y^4 - 3y^2 + 3y + 9}$

25. $x^2 + x + 1\overline{)2x^4 + 3x^2 - x + 2}$ (Check.)

26. $z^2 - z + 1\overline{)2z^4 + z^2 - 1}$

27. $y^2 - 2\overline{)y^4 - 4y^2 + 2y - 4}$ (Check.)

28. $a^4 - 1\overline{)a^8 - 3a^4 + a^2 - 1}$

29. $\dfrac{z + 2}{z + 1}$

30. $\dfrac{2 - z}{z}$

31. $\dfrac{10t}{10t + 1}$

32. $\dfrac{2t^2}{t^2 + 1}$

33. $t^2 + 2\overline{)t^3 - 3t^2 + 2t - 6}$

34. $z^2 - 5\overline{)z^3 + 6z - 2}$

35. $2z^2 - z + 4\overline{)2z^4 + 3z^3 - z + 8}$

36. $m^2 - 2m\overline{)m^4 - 3m^2 + m - 1}$

37. $a^3 - a^2 + 1\overline{)a^6 + 4a^4 - 2a^2 + 2}$

38. $c^2 + 3\overline{)4c^4 + 6c^2 - 2c + 9}$

39. Suppose the divisor is $x + 2$ and the quotient is $x + 3$ (with no remainder). Determine the dividend.

40. Suppose the divisor is $x + 5$, the quotient is $x - 2$, and the remainder is 1. Determine the dividend.

11.2 Synthetic Division and the Remainder Theorem

synthetic division

There are many steps in the division process that are repetitious. *When the divisor is of the form $x - a$,* a simplified process, known as **synthetic division,** can be employed. At first a will be an integer. Actually, a can be any real number or any complex number. In order to see which steps can be eliminated, consider an example worked out by the usual division process.

EXAMPLE 1

$$
\begin{array}{rrrrr}
 & x^2 & -\,2x & -\;\;3 & \\
x - 1\,\big) & x^3 - 3x^2 & -\;\;x & +\,3 & \\
 & \mp x^3 \pm x^2 & & & \\
\hline
 & -\,2x^2 & -\;\;x & & \\
 & \pm\,2x^2 & \mp\,2x & & \\
\hline
 & & -\,3x & +\,3 & \\
 & & \pm\,3x & \mp\,3 & \\
\hline
\end{array}
$$

1. Arrange the terms of the dividend according to descending degrees. Place 0's for the missing terms, including those "at the end". If the lowest-degree term is of degree 1, place 1 zero after the last nonzero term; if the lowest-degree term is of degree 2, place 2 zeros, and so on. You need only list the coefficients. The last coefficient is the constant term, the next to last is that of the first-degree term, and so on. Thus if x is the variable, then

$$1 \qquad -3 \qquad -1 \qquad 3$$

indicates the preceding dividend,

$$x^3 \qquad -3x^2 \qquad -\,x \qquad +3$$

Similarly,

$$2 \qquad 0 \qquad -8 \qquad 1 \quad 0$$

indicates

$$2x^4 \qquad\qquad -8x^2 \quad +\,x$$

or $2x^4 + 0x^3 - 8x^2 + x + 0$

and

1	1	3	−2	0	0	0

indicates

$x^6 \quad + x^5 \quad +3x^4 \quad -2x^3$

2. Below the division line, terms of the same degree are in the same column. Thus you need only list the coefficients. Also, the divisor here is always of the form $x - a$. Just write a for the divisor. The division process for the given example can be written as follows:

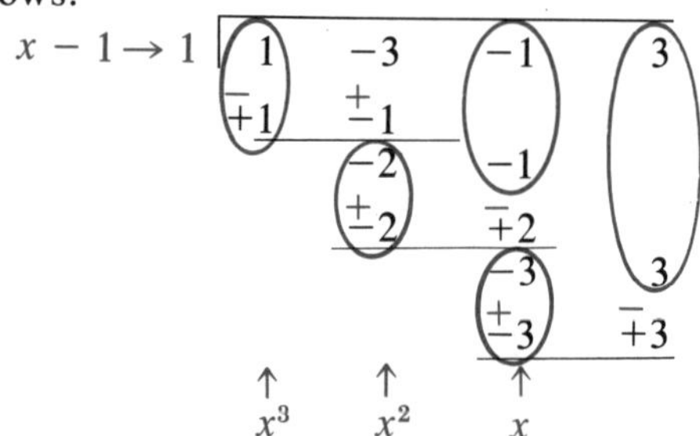

3. Note that in the balloons, the lower numbers repeat the upper numbers. Omit the repetitions:

$$
\begin{array}{r|rrrr}
1 & 1 & -3 & -1 & 3 \\
 & & \pm 1 & \uparrow & \uparrow \\
 & & -2 & & \\
 & & & \mp 2 & \\
 & & & -3 & \\
 & & & & \mp 3
\end{array}
$$

4. Instead of changing signs every time you subtract, immediately list the number subtracted *with its sign changed* on the first available space, as indicated by the arrows above. Bring down the leading coefficient of the dividend, as indicated: *Add* the first two lines, column by column. The quotient is now indicated on the third line.

$$
\begin{array}{r|rrrr}
1 & 1 & -3 & -1 & 3 \\
 & & 1 & -2 & -3 \\
\hline
 & 1 & -2 & -3 &
\end{array}
$$

Read

$1 \quad -2 \quad -3$

as

$x^2 - 2x - 3$ □

EXAMPLE 2 Determine $\dfrac{x^3 + x^2 + x - 3}{x - 1}$ by synthetic division.

Solution. Write

$1 \quad 1 \quad 1 \quad -3$ for the dividend

$x^3 + \ x^2 + \ x - 3$

The divisor is $x - 1$; here $a = 1$. Begin by writing:

$$\underline{1|} \quad 1 \quad 1 \quad 1 \quad -3$$

(The $\underline{1|}$ at the left indicates the divisor, $x - 1$.) After you divide the leading term, x^3, with coefficient 1, by $x - 1$, the resulting quotient must have leading coefficient 1. In synthetic division, the first entry underneath the horizontal line will always be the same as the number lying above it.

$$\begin{array}{r|rrrr} 1 & 1 & 1 & 1 & -3 \\ & \downarrow & & & \\ \hline & 1 & & & \end{array}$$

$$\text{divisor } x - 1 \rightarrow \begin{array}{r|rrrr} 1 & 1 & 1 & 1 & -3 \\ & & 1 & & \\ \hline & (1) & 2 & & \end{array} \leftarrow x^3 + x^2 + x - 3 \text{ dividend}$$

The encircled 1 has been multiplied by the 1 from the divisor. The product is 1, as indicated by the arrow. Adding in the second column, you obtain $1 + 1 = 2$.

Now multiply the encircled 2 by the 1 from the divisor (as shown next) to obtain the product 2 as indicated by the arrow. Adding the numbers in the third column, you obtain $1 + 2 = 3$.

$$\begin{array}{r|rrrr} 1 & 1 & 1 & 1 & -3 \\ & & 1 & 2 & \\ \hline & 1 & (2) & 3 & \end{array}$$

Repeating the process once more, you obtain the final form:

$$\begin{array}{l} \text{divisor } x - 1 \rightarrow \\ \\ \text{quotient } x^2 + 2x + 3 \rightarrow \end{array} \begin{array}{r|rrr|r} 1 & 1 & 1 & 1 & -3 \\ & & 1 & 2 & 3 \\ \hline & 1 & 2 & (3) & 0 \end{array} \begin{array}{l} \leftarrow x^3 + x^2 + x - 3 \text{ dividend} \\ \\ \leftarrow \text{remainder } 0 \end{array}$$

The numbers below the horizontal line indicate the quotient. Bear in mind that when a nonconstant polynomial is divided by $x - a$, its degree is lowered by 1. Because the first line represents a third-degree polynomial, the bottom line represents a second-degree polynomial. In reading this, note that the last digit represents the remainder. Thus

$$1 \quad 2 \quad 3$$

represents the quotient

$$x^2 + 2x + 3;$$

the $\lfloor 0$ at the end indicates that the remainder is 0. Therefore $x - 1$ *is a factor of* $x^3 + x^2 + x - 3$. Observe that if

$$P(x) = x^3 + x^2 + x - 3,$$

then

$$\underset{\substack{\uparrow \\ a}}{P(1)} = 1^3 + 1^2 + 1 - 3$$

Thus $x - 1$ is a factor of $P(x)$, and $P(1) = 0$. □

EXAMPLE 3 (a) Determine $\dfrac{2x^3 - 3x + 10}{x + 2}$ by synthetic division.

(b) Let $P(x) = 2x^3 - 3x + 10$. Determine $P(-2)$.

Solution.

(a) The divisor is $x + 2$, or $x - (-2)$. Here $a = -2$. The dividend, which can be written as $2x^3 + 0x^2 - 3x + 10$, is represented by

2 0 −3 10

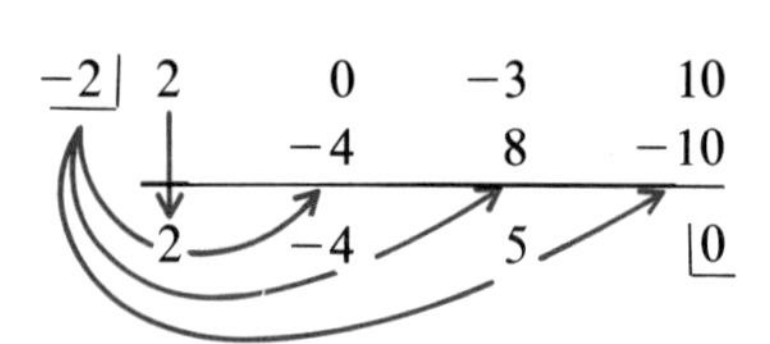

Thus, the quotient is $2x^2 - 4x + 5$; the remainder is 0. Therefore, $x + 2$ is a factor of $2x^3 - 3x + 10$.

(b) $P(-2) = 2(-2)^3 - 3(-2) + 10 = -16 + 6 + 10 = 0$ (with an arrow marking -2 as a)

Thus $x - (-2)$ is a factor of $P(x)$, and $P(-2) = 0$. □

EXAMPLE 4 (a) Determine $\dfrac{x^4 + 3x^2 - x}{x + 1}$ by synthetic division.

(b) Let $P(x) = x^4 + 3x^2 - x$. Determine $P(-1)$.

Solution.

(a) There are no third-degree or degree 0 (constant) terms in the dividend.

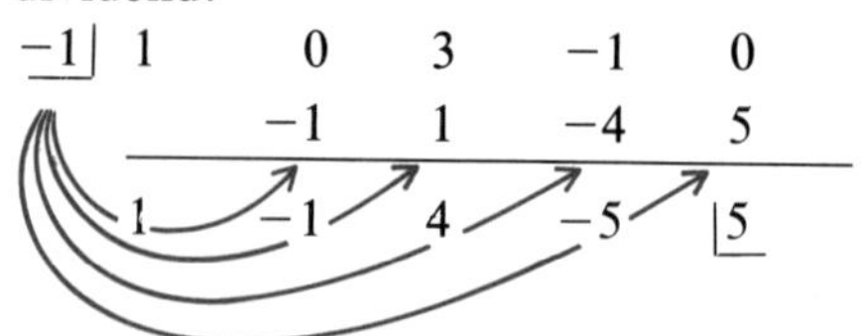

The quotient is $x^3 - x^2 + 4x - 5$, and the remainder is 5. Thus

$$\frac{x^4 + 3x^2 - x}{x + 1} = x^3 - x^2 + 4x - 5 + \frac{5}{x + 1}$$

(b) $P(-1) = (-1)^4 + 3(-1)^2 - (-1) = 1 + 3 + 1 = 5$

Thus the remainder is 5, and $P(-1) = 5$. □

remainder theorem In Examples 2, 3, and 4 a polynomial $P(x)$ is divided by $x - a$. In Examples 2 and 3, $x - a$ is a *factor of* $P(x)$, and $P(a) = 0$. In Example 4 when $P(x)$ is divided by $x - a$, the remainder is 5, and $P(a) = 5$. Let us discover the general rule.

By the Division Algorithm, if $P(x)$ is any polynomial and $D(x)$ is any nonzero polynomial, then

$$\frac{P(x)}{D(x)} = \underset{\text{quotient}}{Q(x)} + \frac{R(x) \leftarrow \text{remainder}}{D(x) \leftarrow \text{divisor}}$$

where either $R(x) \equiv 0$ or else the degree of $R(x)$ is less than that of $D(x)$. Let $D(x) = x - a$. Then

$$\frac{P(x)}{x - a} = Q(x) + \frac{R(x)}{x - a},$$

where either $R(x) \equiv 0$ or else the degree of $R(x)$ is less than that of $x - a$. Because $x - a$ is of degree 1, if $R(x) \not\equiv 0$, then $R(x)$ must be of degree 0. In either case $R(x)$ is a *constant*, r. Thus

$$\frac{P(x)}{x - a} = Q(x) + \frac{r}{x - a}$$

Multiplying both sides by $x - a$, you obtain

$$P(x) = Q(x) \cdot (x - a) + r$$

Set $x = a$. Then

$$P(a) = Q(a) \cdot (\overbrace{a - a}^{0}) + r$$

$$P(a) = r$$

Thus, the remainder equals $P(a)$. △

This demonstrates the **Remainder Theorem:**

Let $P(x)$ be any polynomial and let a be any real number.

When $P(x)$ is divided by $x - a$, the remainder is $P(a)$.

factor theorem

If $P(a) = 0$, then $x - a$ is a factor of $P(x)$ because then

$$P(x) = Q(x) \cdot (x - a)$$

On the other hand, if $x - a$ is a factor of $P(x)$, then

$$P(x) = Q(x) \cdot (x - a),$$

so that

$$P(a) = Q(a) \cdot (a - a) = 0$$

△

This proves the **Factor Theorem:**

If $x - a$ is a factor of $P(x)$, then $P(a) = 0$.

And if $P(a) = 0$, then $x - a$ is a factor of $P(x)$.

EXAMPLE 5 Let $P(x) = x^4 - 3x^3 - 10x^2 + 2x - 10$.

(a) Use the Remainder Theorem to find the remainder when $P(x)$ is divided by $x - 2$.

(b) Use the Factor Theorem to show that $x - 5$ is a factor of $P(x)$.

Solution.

(a)
$$\begin{aligned} P(2) &= 2^4 - 3(2^3) - 10(2^2) + 2(2) - 10 \\ &= 16 - 24 - 40 + 4 - 10 \\ &= -54 \end{aligned}$$

When $P(x)$ is divided by $x - 2$, the remainder is -54.

(b) $$\begin{aligned} P(5) &= 5^4 - 3(5^3) - 10(5^2) + 2(5) - 10 \\ &= 625 - 375 - 250 + 10 - 10 \\ &= 0 \end{aligned}$$

Thus $x - 5$ is a factor of $P(x)$. □

EXERCISES

In Exercises 1–6 write the first line of synthetic division if the divisor is given by (a) and the dividend by (b).

1. (a) $x + 1$ (b) $x^3 + 2x + 4$

2. (a) $x - 5$ (b) $x^3 - 5x^2 + 10x - 5$

3. (a) x (b) $x^4 + 5x^3 + x^2 + x - 1$

4. (a) $x - 10$ (b) $x^4 + 10x^2 - 10$

5. (a) $x + 4$ (b) $x^3 - 3x^2 + 4$

6. (a) $x + 2$ (b) $x^6 + 4x^4 - 2x + 12$

In Exercises 7–12 suppose that the bottom line of a synthetic division example is as indicated. Determine the quotient and if there is one, the remainder.

7. $2 \quad 2 \;\lfloor 1$ **8.** $3 \quad 1 \quad 2 \;\lfloor 0$ **9.** $1 \quad 0 \quad 1 \quad 0 \;\lfloor 0$ **10.** $2 \quad 0 \quad 1 \quad 0 \quad 0 \;\lfloor 0$

11. $1 \quad 0 \quad 0 \quad 1 \quad 0 \quad 2 \;\lfloor 3$ **12.** $2 \quad 0 \quad 1 \quad 0 \quad 0 \quad 1 \quad 0 \quad 1 \;\lfloor -1$

In Exercises 13–25 use synthetic division to determine the quotient.

13. $\dfrac{x^2 + 6x + 9}{x + 3}$ **14.** $\dfrac{x^2 - 12x + 36}{x - 6}$ **15.** $\dfrac{y^2 - 5y + 6}{y - 2}$ **16.** $\dfrac{z^2 + 3z - 18}{z + 6}$

17. $x + 2\overline{)x^3 + 3x^2 + 3x + 2}$ **18.** $x + 1\overline{)x^3 - x}$ **19.** $t - 1\overline{)t^3 - t^2 + t - 1}$

20. $x - 1\overline{)x^4 - 1}$ **21.** $\dfrac{x^4 + x^2 - x - 3}{x + 1}$ **22.** $\dfrac{t^4 - 1}{t + 1}$ **23.** $\dfrac{a^3 + a^2 + a - 3}{a - 1}$

24. $\dfrac{y^6 - y^5 + y^3 - y^2 + y - 1}{y - 1}$ **25.** $\dfrac{y^8 - 1}{y - 1}$

In Exercises 26–32 use synthetic division to determine the quotient and, if there is one, the remainder.

26. $\dfrac{z^2 + 3z + 4}{z + 2}$ **27.** $\dfrac{m^4 + m^2 + 1}{m + 1}$ **28.** $\dfrac{5u^3 - 25u^2 + u - 5}{u - 5}$ **29.** $\dfrac{a^3 + 4a^2 + 5a + 2}{a + 2}$

30. $\dfrac{x^6 + x^4 - x^2 + 2}{x - 1}$ **31.** $\dfrac{z^4 - 9z^2 + 3z}{z - 3}$ **32.** $\dfrac{6z^3 - 6z^2 - 6}{z - 2}$

In Exercises 33–44:
(a) Use the Remainder Theorem to determine the remainder when $P(x)$ is divided by $x - a$. [*Note:* a can be negative, as well as positive.]
(b) Is $x - a$ a factor of $P(x)$?

33. $P(x) = x^2 + 2x + 5, \quad x - 1$

34. $P(x) = x^3 + x^2 - 2x + 1, \quad x + 1$

35. $P(x) = x^3 - x - 2, \quad x - 2$

36. $P(x) = x^3 - 4x^2 + 3x + 1, \quad x - 1$

37. $P(x) = 2x^3 - 9x^2 + 9x - 20, \quad x - 4$

38. $P(x) = 4x^3 + 3x + 1, \quad x + 2$

39. $P(x) = x^4 + x^3 + x^2 + 2x + 1, \quad x + 1$

40. $P(x) = x^4 + x^3 - 2x + 2, \quad x - 1$

41. $P(x) = x^4 - 8x^3 - 2x^2 - x + 8, \quad x - 8$

42. $P(x) = x^4 - 3x^3 + 10x - 100, \quad x - 10$

43. $P(x) = x^8 + 2x - 260, \quad x - 2$

44. $P(x) = x^5 + 3x^4 + 2x^3 + 6x^2 + 3x + 9, \quad x + 3$

In Exercises 45–48 find $P(a)$: **(a)** by direct substitution **(b)** by using the Remainder Theorem.

45. $P(x) = x^3 - 3x^2 + x + 2, \quad a = 3$

46. $P(x) = x^3 - 5x^2 + 3x - 1, \quad a = -1$

47. $P(x) = x^4 + 2x^2 - x + 2, \quad a = -2$

48. $P(x) = x^5 - x^3 + x + 4, \quad a = 2$

49. Determine k so that $x - 2$ is a factor of $x^3 - 2x^2 + 4x - k$.

50. Determine k so that $x + 2$ is a factor of $x^4 - kx^3 + 2x - 20$.

51. Determine k so that $x - 3$ is a factor of $x^3 + kx^2 + kx - 3$.

52. Show that $x^4 + 5x^2 + 1$ has no factor of the form $x - a$ for a real number a.

53. Show that $x - 1$ is a factor of:
(a) $x^4 - 1$ **(b)** $x^6 - 1$ **(c)** $x^{10} - 1$

54. Show that $x - y$ is a factor of:
(a) $x^4 - y^4$ **(b)** $x^6 - y^6$ **(c)** $x^{10} - y^{10}$

11.3 Zeros of Polynomials

The quadratic equation

$$x^2 + 1 = 0$$

has no real root. To remedy this situation, complex numbers were introduced. Every quadratic equation

$$Ax^2 + Bx + C = 0, \qquad A \neq 0,$$

with *real* coefficients has roots: either two distinct real roots, one (repeated) real root, or two complex conjugate roots.

What about **polynomial equations**

$$a_n x^n + a_{n-1} x^{n-1} + \cdot \cdot \cdot + a_1 x + a_0 = 0, \qquad a_n \neq 0,$$

in general? At first, *complex* coefficients will be allowed. Because every real number a can be regarded as the complex number $a + 0i$, when you say "complex coefficients" or "complex roots", this allows the possibility that some of the numbers in question are actually *real* numbers.

zeros and the fundamental theorem of algebra

Definition. A **zero of a polynomial** $P(x)$ is a (complex) number c such that $P(c) = 0$. In other words, a zero of a polynomial $P(x)$ is a root of the polynomial equation $P(x) = 0$.

Thus, 2 is a zero of the polynomial $P(x)$ given by

$$x^3 - 8$$

because

$$P(2) = 2^3 - 8 = 0.$$

The **Fundamental Theorem of Algebra** whose proof involves advanced methods, states:

Every nonconstant polynomial with *complex* coefficients, has at least one *complex* zero.

applying the factor theorem

The Factor Theorem, p. 437, also applies to polynomials with *complex* coefficients.

EXAMPLE 1 To see how the Factor Theorem applies, consider the 4th-degree polynomial,

$$P(x) = 2x^4 - 4x^3 - 2x^2 + 4x.$$

You "peel off" the zeros. First, by isolating the common factor, you obtain:

$$P(x) = 2x(x^3 - 2x^2 - x + 2) = 2(x - 0)\underbrace{(x^3 - 2x^2 - x + 2)}_{P_1(x)}$$

Note that 0 is a zero of $P(x)$ and that $P_1(x)$ is of degree 3, or $4 - 1$. By synthetic division on $P_1(x)$,

$$\begin{array}{r|rrrr} 1 & 1 & -2 & -1 & 2 \\ & & 1 & -1 & -2 \\ \hline & 1 & -1 & -2 & \underline{|0} \end{array}$$

you obtain 1 as a zero of $P_1(x)$, and hence of $P(x)$. Thus

$$P_1(x) = (x - 1)\underbrace{(x^2 - x - 2)}_{P_2(x)}$$

Note that $P_2(x)$ is of degree 2, or $4 - 2$. Furthermore,

$$P_2(x) = (x + 1)(x - 2),$$

so that

$$P(x) = 2(x - 0)\overbrace{(x - 1)\underbrace{(x + 1)(x - 2)}_{P_2(x)}}^{P_1(x)}$$

and 0, 1, -1, and 2 are the zeros of $P(x)$. □

In general, suppose c_1 is a (complex) zero of an nth-degree polynomial $P(x)$, that is, suppose $P(c_1) = 0$. Then $x - c_1$ is a factor of $P(x)$. Therefore

$$P(x) = (x - c_1)P_1(x), \qquad \text{where } P_1(x) \text{ has degree } n - 1.$$

By the Fundamental Theorem, if $n \geqslant 2$, then $P_1(x)$ has a zero c_2, and by the Factor Theorem,

$$P_1(x) = (x - c_2)P_2(x),$$

so that

$$P(x) = (x - c_1)(x - c_2)P_2(x), \text{ where } P_2(x) \text{ has degree } n - 2.$$

After n such steps,† you obtain

$$P(x) = a_n(x - c_1)(x - c_2) \cdot \cdot \cdot (x - c_n)$$

where a_n is the leading coefficient of $P(x)$. △

Thus, if $P(x)$ is a polynomial of degree n, $n > 0$, with complex coefficients and with leading coefficient a_n, then there exist n complex numbers $c_1, c_2, \ldots, c_n$ such that

$$P(x) = a_n(x - c_1)(x - c_2) \cdot \cdot \cdot (x - c_n)$$

† This really utilizes Mathematical Induction (on n, the degree of the polynomial). [See Section 13.6 .]

The zeros $c_1, c_2, \ldots, c_n$ of $P(x)$ may involve repetitions. For example,

$$\begin{aligned} x^5 - 2x^4 + x^3 &= x^3(x^2 - 2x + 1) \\ &= (x - 0)^3(x - 1)^2 \end{aligned}$$

Say that 0 is a zero of **multiplicity** 3 of this polynomial and that 1 is a zero of multiplicity 2. Zeros that do not repeat are called **simple zeros.** Thus if you consider a simple zero as having multiplicity 1, the sum of the multiplicities of the different zeros of an nth-degree polynomial is n.

EXAMPLE 2 Find a polynomial of degree 3 that has 2 as a zero of multiplicity 2 and i as a simple zero.

Solution. Let
$$\begin{aligned} P(x) &= (x - 2)^2(x - i) \\ &= (x^2 - 4x + 4)(x - i) \\ &= x^3 - (4 + i)x^2 + 4(1 + i)x - 4i \end{aligned}$$
□

real coefficients

You will be primarily concerned with polynomials having *real coefficients*. Because a real number a can be regarded as the complex number $a + 0i$, the Fundamental Theorem of Algebra tells you that every polynomial with *real coefficients* has at least one *complex* zero. Furthermore, every such nth-degree polynomial $P(x)$ with *real coefficients* and with leading coefficient a_n factors as follows:

$$P(x) = a_n(x - c_1)(x - c_2) \cdots (x - c_n),$$

where $c_1, c_2, \ldots, c_n$ are complex numbers, some of which may be equal.

rational zeros

It is often a difficult matter to determine the actual zeros of a polynomial. However, you will consider some of the most important ways of finding the more obvious zeros.

EXAMPLE 3 Consider the quadratic polynomial:

$$6x^2 - 7x + 2$$

This factors as $(3x - 2)(2x - 1)$. Thus, the zeros of the quadratic are $\frac{2}{3}$ and $\frac{1}{2}$. Observe that for each of these rational zeros, the *numerator* is a factor of 2, the *constant term* of the quadratic. And each *denominator* is a factor of 6, the *leading coefficient* of the quadratic. □

Rational Zeros Theorem.

Let $P(x) = a_nx^n + a_{n-1}x^{n-1} + \cdots + a_1x + a_0$, and suppose that $P(x)$ has *integral coefficients*. Suppose, further, that $P(x)$ has a rational zero $\frac{c}{d}$, where $d \neq 0$ and where c and d are integers that have no prime factors in common. Then c is a factor of a_0 and d is a factor of a_n.

To prove that c is a factor of a_0, observe that if $c = \pm 1$, then c is obviously a factor of a_0. Otherwise, if $\frac{c}{d}$ is such a rational zero, then

$$a_n \left(\frac{c}{d}\right)^n + a_{n-1}\left(\frac{c}{d}\right)^{n-1} + \cdots + a_1\left(\frac{c}{d}\right) + a_0 = 0.$$

Multiply both sides by d^n:

$$a_n c^n + a_{n-1}c^{n-1}d + \cdots + a_1 cd^{n-1} + a_0 d^n = 0 \qquad (11.1)$$

or

$$a_n c^n + a_{n-1}c^{n-1}d + \cdots + a_1 cd^{n-1} = -a_0 d^n$$

Clearly, c is a common factor on the left side:

$$c(a_n c^{n-1} + a_{n-1}c^{n-2}d + \cdots + a_1 d^{n-1}) = -a_0 d^n$$

Consequently, c is a factor of $-a_0 d^n$ on the right side. Consider the unique prime factorization of c:

$$c = \pm p_1^{k_1} p_2^{k_2} \ldots p_r^{k_r}$$

Each prime power $p_i^{k_i}$ is a factor of c, and hence of $-a_0 d^n$. Because c and d have no factors in common, each prime power $p_i^{k_i}$ must divide a_0. Therefore c, the product of these prime powers, is a factor of a_0.

Similarly, from Equation (**11.1**) it follows that

$$d(a_{n-1}c^{n-1} + \cdots + a_1 cd^{n-2} + a_0 d^{n-1}) = -a_n c^n$$

From this it can be shown that d is a factor of a_n. [See Exercise 53.] △

integral zeros

Next, suppose that $P(x)$ has *integral coefficients* and *leading coefficient* 1, so that $P(x)$ is of the form

$$x^n + a_{n-1}x^{n-1} + \cdots + a_1 x + a_0.$$

Then from the preceding theorem it follows that *every rational zero* $\frac{c}{d}$, where c and d have no prime factors in common, *must be an integer*. This is because d must be a factor of 1 and consequently must be either 1 or -1. Furthermore, each such integral zero equals $\pm c$, and thus must be a factor of a_0.

For example, the only *rational* zeros of a polynomial, such as

$$x^4 - 2x^3 - 7x^2 - 2x - 8,$$

with integral coefficients and with leading coefficient 1, are integers. Every such integral zero must be a factor of -8. Thus, the only *possible* rational zeros are

$$\pm 1, \pm 2, \pm 4, \pm 8$$

EXAMPLE 4 Let $P(x) = x^4 - 2x^3 - 7x^2 - 2x - 8$.

(a) Find all the rational zeros of $P(x)$.

(b) Find all the complex zeros of $P(x)$.

(c) Factor $P(x)$.

Solution.

(a) To see which of the integers ± 1, ± 2, ± 4, ± 8 are actually zeros of $P(x)$, use either synthetic division or the Factor Theorem. Some possibilities can be quickly eliminated by means of the Factor Theorem. For example,

$$P(1) = 1 - 2 - 7 - 2 - 8 < 0$$

However, for the most part, synthetic division is suggested. It turns out that -2 is a zero because when $P(x)$ is divided by $x + 2$, the remainder is 0:

$$\begin{array}{r|rrrrr} x+2 \rightarrow \quad -2 & 1 & -2 & -7 & -2 & -8 \\ & & -2 & +8 & -2 & +8 \\ \hline & 1 & -4 & +1 & -4 & \underline{|0} \end{array}$$

Thus

$$P(x) = (x + 2)(x^3 - 4x^2 + x - 4)$$

One zero of $P(x)$ is -2. The other zeros of $P(x)$ are the zeros of $x^3 - 4x^2 + x - 4$.

Now try the integers *not yet eliminated* with the quotient represented by the *bottom line*. Begin by trying -2 again to check for multiplicity.

$$\begin{array}{r|rrrr} x+2 \rightarrow \quad -2 & 1 & -4 & +1 & -4 \\ & & -2 & +12 & -26 \\ \hline & 1 & -6 & +13 & \underline{|-30} \leftarrow \text{remainder} \end{array}$$

Thus -2 is a *simple* zero of $P(x)$. Also 4 is a zero of $P(x)$:

$$\begin{array}{r|rrrr} x-4 \rightarrow \quad 4 & 1 & -4 & +1 & -4 \\ & & +4 & +0 & +4 \\ \hline & \underbrace{1 \quad +0 \quad +1}_{x^2+1} & & & \underline{|0} \end{array}$$

(b) The zeros of $x^2 + 1$ are $\pm i$. Thus, the 4 zeros of the 4th-degree polynomial $P(x)$ are

$$-2, 4, \pm i$$

(c) $P(x) = (x + 2)(x - 4)(x - i)(x + i)$ □

EXAMPLE 5 Let $P(x) = 4x^3 - 8x^2 + 5x - 1$.

(a) Find all rational zeros of $P(x)$.

(b) Factor $P(x)$.

Solution.

(a) If $\dfrac{c}{d}$ is a rational zero (in lowest terms) of $P(x)$, then c is a factor of -1 and d is a factor of 4. The possibilities for $\dfrac{c}{d}$ are:

$$\pm 1, \pm\frac{1}{2}, \pm\frac{1}{4}$$

Try the integers first.

$$\begin{array}{r|rrrr} x-1\rightarrow\ 1 & 4 & -8 & +5 & -1 \\ & & +4 & -4 & +1 \\ \hline & 4 & -4 & +1 & \lfloor 0 \end{array}$$

Therefore

$$P(x) = (x - 1)(4x^2 - 4x + 1)$$

The quadratic factors as $(2x - 1)^2$, and thus has $\frac{1}{2}$ as a zero of multiplicity 2. (Synthetic division is unnecessary.) The zeros of $P(x)$ are 1 (a simple zero) and $\frac{1}{2}$ (of multiplicity 2). Because $P(x)$ is of degree 3, these are all of the zeros.

(b) The leading coefficient of $P(x)$ is 4. Thus, by part (a),

$$4x^3 - 8x^2 + 5x - 1 = 4(x - 1)\left(x - \frac{1}{2}\right)^2$$ □

Suppose all of the coefficients of a polynomial are rational, but at least one is not an integer. By multiplying by the *lcd* of the coefficients, you will obtain *integral coefficients*, and can apply the preceding method.

EXAMPLE 6 Let

$$Q(x) = \frac{1}{10}x^3 - \frac{1}{5}x^2 + \frac{1}{8}x - \frac{1}{40}.$$

$$\text{LCD}\left(\frac{1}{10}, -\frac{1}{5}, \frac{1}{8}, -\frac{1}{40}\right) = \text{LCM}(10, 5, 8, 40) = 40$$

$$40Q(x) = 4x^3 - 8x^2 + 5x - 1$$

Because $40Q(x) = 0$ exactly when $Q(x) = 0$, you must find the zeros of the polynomial $4x^3 - 8x^2 + 5x - 1$. But this is the polynomial $P(x)$ of Example 5, whose zeros are 1 (simple) and $\frac{1}{2}$ (of multiplicity 2). Recall that

$$P(x) = 4(x - 1)\left(x - \frac{1}{2}\right)^2$$

Thus

$$Q(x) = \frac{P(x)}{40} = \frac{1}{10}(x - 1)\left(x - \frac{1}{2}\right)^2$$ □

real coefficients, complex zeros

Finally, consider an nth-degree polynomial all of whose coefficients are *real numbers*. As in the case of roots of a quadratic equation, *complex zeros of an* n*th-degree polynomial occur in conjugate pairs*.

Before stating the theorem more precisely, recall some useful facts about complex conjugates. [See Section 10.1 .]

If $z = x + yi$, then $\bar{z}$, the conjugate of z, was defined to be $x - yi$. Suppose z, z_1, and z_2 are complex numbers and n is a positive integer. Then, as previously shown,

$$\overline{z_1 + z_2} = \bar{z}_1 + \bar{z}_2 \qquad \textbf{(11.2)}$$

The conjugate of a sum is the sum of the conjugates.

$$\overline{z_1 \cdot z_2} = \overline{z_1} \cdot \overline{z_2} \tag{11.3}$$

The conjugate of a product is the product of the conjugates.

$$\overline{z^n} = (\overline{z})^n \tag{11.4}$$

The conjugate of an nth power is the nth power of the conjugate.

If z is *real*, then $z = \overline{z}$. (11.5)

Theorem (Real coefficients, complex zeros)

Let
$$P(x) = a_n x^n + a_{n-1}x^{n-1} + \cdots + a_1 x + a_0.$$

Suppose all of the coefficients of $P(x)$ are *real*. Then if z is a complex zero of $P(x)$, so is $\overline{z}$, the conjugate of z.

To show that this is true, suppose z is a zero of $P(x)$. Then

$$P(z) = a_n z^n + a_{n-1}z^{n-1} + \cdots + a_1 z + a_0 = 0$$

and

$$\overline{a_n z^n + a_{n-1}z^{n-1} + \cdots + a_1 z + a_0} = \overline{0} = 0 \tag{11.6}$$

By applying rules (**11.2**)–(**11.5**) to the left side†, you obtain

$$\begin{aligned}
&\overline{a_n z^n + a_{n-1}z^{n-1} + \cdots + a_1 z + a_0} \\
&= \overline{a_n z^n} + \overline{a_{n-1}z^{n-1}} + \cdots + \overline{a_1 z} + \overline{a_0} && \text{by (11.2)} \\
&= \overline{a_n}\,\overline{z^n} + \overline{a_{n-1}}\,\overline{z^{n-1}} + \cdots + \overline{a_1}\,\overline{z} + \overline{a_0} && \text{by (11.3)} \\
&= a_n\,\overline{z^n} + a_{n-1}\,\overline{z^{n-1}} + \cdots + a_1\overline{z} + a_0 && \text{by (11.5)} \\
&= a_n(\overline{z})^n + a_{n-1}(\overline{z})^{n-1} + \cdots + a_1\overline{z} + a_0 && \text{by (11.4)}
\end{aligned}$$

But by (**11.6**),

$$\overline{a_n z^n + a_{n-1}z^{n-1} + \cdots + a_1 z + a_0} = 0$$

Thus

$$a_n(\overline{z})^n + a_{n-1}(\overline{z})^{n-1} + \cdots + a_1\overline{z} + a_0 = 0$$

This says that $\overline{z}$ is also a zero of $P(x)$. △

EXAMPLE 7 Find a polynomial of degree 4 with real coefficients that has i and $1 - 2i$ among its zeros.

Solution. By the preceding theorem, $-i$ and $1 + 2i$ must also be zeros. Because the degree of the polynomial is 4, the 4 zeros are $\pm i$, $1 \pm 2i$. Let

$$\begin{aligned}
P(x) &= \underbrace{(x - i)(x + i)}_{(x^2+1)}\underbrace{(x - (1 + 2i))(x - (1 - 2i))}_{(x^2 - 2x + 5)} \\
&= (x^2 + 1)\quad (x^2 - 2x + 5) \\
&= x^4 - 2x^3 + 6x^2 - 2x + 5
\end{aligned}$$

$$\begin{array}{llll}
x - (1 & & + 2i) & \\
x - (1 & & - 2i) & \\
\hline
x^2 & -x & - 2ix & \\
 & -x & + 2ix & + 1 - 4i^2 \\
\hline
x^2 & -2x & & + 1 + 4
\end{array}$$

□

† These steps really utilize Mathematical Induction (on n, the degree of the polynomial.) [See Section 13.6 .]

It follows from the preceding theorem that every (nonconstant) polynomial, $P(x)$, with *real* coefficients can be factored into

1. linear polynomials $x - a$ with *real zeros* and
2. quadratic polynomials with *real coefficients* and *complex conjugate zeros*.

For, if $P(x)$ is of degree n, then by the Fundamental Theorem of Algebra together with the Factor Theorem,

$$P(x) = a_n(x - c_1)(x - c_2) \cdots (x - c_n).$$

By the preceding theorem, whenever $a + bi$ is among the zeros c_j, so is $a - bi$. Thus

$$[x - (a + bi)][x - (a - bi)]$$

or

$$x^2 - 2ax + (a^2 + b^2),$$

$$\begin{array}{l} x - (a + bi) \\ x - (a - bi) \\ \hline x^2 - (a + bi)x \\ \quad - (a - bi)x + (a^2 - b^2i^2) \\ \hline x^2 - 2ax \quad + (a^2 + b^2) \end{array}$$

is a quadratic factor, with real coefficients, of $P(x)$. △

EXAMPLE 8 Factor

$$x^4 - 6x^3 + 9x^2 + 6x - 10$$

into linear polynomials with real zeros and quadratic polynomials with real coefficients and (nonreal) complex conjugate zeros.

Solution. The given polynomial has integral coefficients and leading coefficient 1. Its rational zeros must be among

$$\pm 1, \pm 2, \pm 5, \pm 10.$$

By synthetic division you find that ± 1 are zeros.

$$\begin{array}{rr|rrrrr} x - 1 \rightarrow & 1 & 1 & -6 & +9 & +6 & -10 \\ & & & +1 & -5 & +4 & +10 \\ \hline x + 1 \rightarrow & -1 & 1 & -5 & +4 & +10 & \lfloor 0 \\ & & & -1 & +6 & -10 & \\ \hline x^2 - 6x + 10 \rightarrow & & 1 & -6 & +10 & \lfloor 0 & \end{array}$$

You obtain a quadratic polynomial $x^2 - 6x + 10$, whose discriminant, $36 - 40$, is negative. The zeros of the quadratic polynomial are therefore complex conjugates (which can be obtained by applying the quadratic formula). Thus

$$x^4 - 6x^3 + 9x^2 + 6x - 10 = (x - 1)(x + 1)(x^2 - 6x + 10)$$

□

EXERCISES

In Exercises 1–8 find the zeros of each polynomial. Indicate which are simple zeros. Find the multiplicity of the other zeros.

1. $x(x - 2)(x + 7)$ **2.** $(x - 3)(x + 5)^2(x - i)$ **3.** $(x - 4)^2(x - 1 + i)$

4. $x^3(x + 10)^4(x + i)^2$ **5.** $(3x - 2)(2x - 1)^2$ **6.** $x^2(x^2 + 1)$

7. $x^5 + 4x^3$ **8.** $(x^2 + x + 10)(x^2 + 9)^2$

In Exercises 9–12 find a polynomial that has the indicated degree and zeros. Express the polynomial in the form $a_nx^n + a_{n-1}x^{n-1} + \cdots + a_1x + a_0$.

9. Degree 3; zeros: 1, 2, and 4

10. Degree 4; zeros: 2, -2, $1 + i$, $1 - i$

11. Degree 4; zeros 3 and $4i$, each of multiplicity 2

12. Degree 5; zeros 0 (of multiplicity 3), $1 + 4i$, $1 - 4i$

In Exercises 13–16 find a polynomial that has the indicated degree and zeros. Leave the polynomial in factored form.

13. Degree 4; zeros 0, -2, $1 + i$, $1 - i$

14. Degree 4; zeros 1, $-i$, 5 (of multiplicity 2)

15. Degree 5; zeros 4 (of multiplicity 3), $4 - 3i$ (of multiplicity 2)

16. Degree 6; zeros -3, i, $1 - 4i$, each of multiplicity 2

In Exercises 17–32:
(a) Find all rational zeros of each polynomial.
(b) Find all irrational zeros and all (nonreal) complex zeros of each polynomial.

17. $x^3 + 2x^2 - x - 2$ **18.** $x^3 - x^2 - 5x - 3$ **19.** $x^4 + x^3 - 4x^2 - 4x$ **20.** $x^3 + x^2 + x + 1$

21. $x^4 - 8x^2 - 9$ **22.** $\frac{1}{3}x^4 + \frac{1}{3}x^3 + \frac{1}{3}x^2$ **23.** $\frac{x^3}{10} + \frac{x}{10} - 1$ **24.** $2x^3 + 2x^2 - 4$

25. $2x^3 - x^2 - 2x + 1$ **26.** $4x^3 + 4x^2 - x - 1$ **27.** $6x^4 + 13x^3 + 7x^2 - x - 1$

28. $4x^4 + 3x^2 - 1$ **29.** $5x^4 - 2x^3 - 5x + 2$ **30.** $6x^4 - 7x^3 - 7x^2 + 6x$

31. $\frac{1}{6}x^3 - \frac{1}{2}x + \frac{1}{3}$ **32.** $\frac{x^4}{8} + \frac{x^2}{4} - 1$

33. **(a)** Find a polynomial, $P(x)$, of degree 3 with real coefficients that has both 2 and i among its zeros.
(b) Suppose that, in addition, $P(0) = 4$. Determine $P(x)$.

34. Find a polynomial of degree 4 with real coefficients that has both $-i$ and $1 + i$ among its zeros.

35. Find a polynomial of degree 3 with real coefficients that has $1 + 3i$ among its zeros.

36. Find a polynomial of degree 4 with real coefficients that has $1 + i$ as a zero of multiplicity 2.

37. Suppose $1 + 2i$ is a zero of multiplicity 3 and $1 - i$ is a zero of multiplicity 2 of a polynomial with real coefficients. What is the smallest possible degree of this polynomial?

38. Given that $1 + i$ is a zero of $x^4 - 2x^3 + 6x^2 - 8x + 8$, find all the zeros.

In Exercises 39–44, factor each polynomial into linear polynomials with real zeros and quadratic polynomials with real coefficients and (nonreal) complex conjugate zeros.

39. $x^3 - 5x^2 + x - 5$ **40.** $x^4 + x^3 + 2x^2$ **41.** $x^4 - x^3 + 5x^2 - 5x$

42. $x^4 + 8x^2 + 16$ **43.** $x^4 + 3x^3 - 5x^2 - 6x - 8$ **44.** $4x^4 - 4x^3 + 5x^2 - 4x + 1$

45. How do you know that every polynomial of odd degree with real coefficients has at least one real zero?

46. Find a polynomial of degree 4 with real coefficients that has no real zero.

47. Find a polynomial of degree 3 with nonreal coefficients that has no real zero.

48. Show that $x^4 + x^3 + 2$ has no rational zero.

49. By considering the polynomial $x^2 - 2$, show that $\sqrt{2}$ is irrational.

50. Show that -1 is a *simple* zero of $x^5 - 1$.

51. A 32-cubic-inch box (without a top) is made from a square piece of cardboard of side length 8 inches by cutting a square of side length x inches from each corner, and then folding up the edges. [See Figure 11.1 .] Find x.

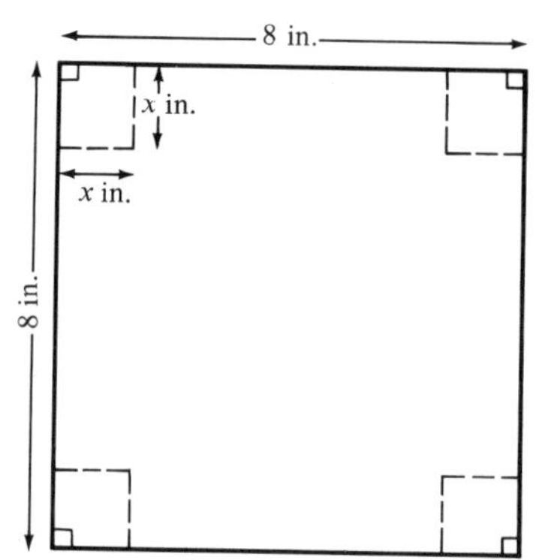

FIGURE 11.1

52. The radius of a cylindrical tank is 3 meters longer than the altitude. If the volume of the tank is 196π cubic meters, find the length of the altitude.

53. In the proof of the Rational Zeros Theorem, show that d is a factor of a_n.

11.4 Graphing Polynomials

From now on, all polynomials considered will have *real coefficients*. You will consider *nth-degree* polynomial functions of the form

$$y = a_n x^n + a_{n-1}x^{n-1} + \cdots + a_1 x + a_0\,, \qquad a_0 \neq 0.$$

The graphs of these polynomial functions of degrees 0 and 1 are lines. [See Sections 4.2 and 4.3 .] The graphs of these polynomial functions of degree 2 are parabolas. [See Section 6.8 .] You will be primarily concerned here with polynomial functions of degree 3 or more.

symmetries

By considering symmetries of a graph, the work is often shortened. Recall that if all the exponents of a polynomial function are *even,* the function is even. This means that its graph is *symmetric with respect to the y-axis*. If all the exponents are *odd,* the function is odd, and its graph is *symmetric with respect to the origin*. If both even and odd powers occur, the function is neither even nor odd, and its graph exhibits neither of these symmetries. [See Sections 3.3 and 3.4 .]

Thus the graphs of

$$y = x^4 \qquad \text{and of} \qquad y = x^6 - 2x^4 + 5$$

are each symmetric with respect to the y-axis. The graphs of

$$y = x^3 \qquad \text{and of} \qquad y = x^5 - 2x^3 + x$$

are each symmetric with respect to the origin. But the graph of

$$y = x^3 + x^2$$

is neither symmetric with respect to the y-axis nor with respect to the origin.

When the graph of a function is symmetric with respect to either the y-axis or with respect to the origin, draw the graph for nonnegative values of x. Then reflect the graph in the y-axis, as in Figure 11.2(a), or through the origin, as in Figure 11.2(b). (The graph of a nonzero function of x is never symmetric with respect to the x-axis.)

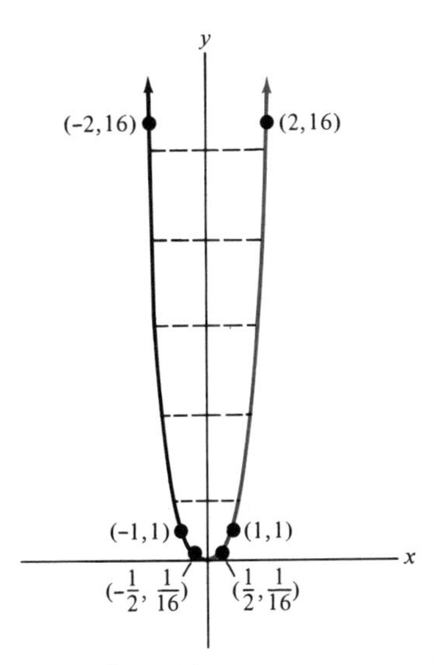

FIGURE 11.2(a)
The graph of $y = x^4$ is symmetric with respect to the y-axis.

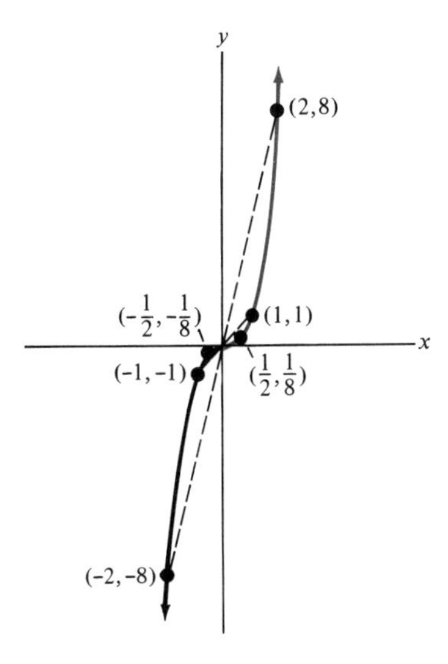

FIGURE 11.2(b)
The graph of $y = x^3$ is symmetric with respect to the origin.

$x \to \infty$, $x \to -\infty$

The graphs of

$$y = x^4, \quad y = x^6, \quad y = x^8, \qquad \text{or in general,} \qquad y = x^{2n},$$

where n is a positive integer, are symmetric with respect to the y-axis, and basically resemble Figure 11.2(a) in shape. As x gets larger and larger, the corresponding values of y increase beyond all bounds (in symbols, as $x \to \infty$, then $y \to \infty$). For example, in the case of $y = x^4$, several values of x and the corresponding values of y are indicated in Table 11.1 .

TABLE 11.1

x	$y = x^4$
1	1
2	16
3	81
4	256
5	625
10	10 000
100	100 000 000

By y-axis symmetry, for *negative* x, as the x-values *decrease* beyond all bounds, the corresponding y-values *increase* beyond all bounds (in symbols, as $x \to -\infty$, then $y \to \infty$).

The graphs of

$$y = x^3, \quad y = x^5, \quad y = x^7, \qquad \text{or in general,} \qquad y = x^{2n+1},$$

where n is a positive integer, are symmetric with respect to the origin and essentially resemble Figure 11.2(b) in shape. As $x \to \infty$, then $y \to \infty$. By origin symmetry, as $x \to -\infty$, then $y \to -\infty$

For all of these functions, when x is close to 0, y is extremely close to 0. For example,

$$\text{if} \quad x = \frac{1}{2}, \quad \text{then} \quad x^3 = \frac{1}{8}; \qquad \text{if} \quad x = \frac{1}{3}, \quad \text{then} \quad x^4 = \frac{1}{81};$$

$$\text{if} \quad x = -\frac{1}{2}, \quad \text{then} \quad x^5 = -\frac{1}{32}$$

continuity, smoothness

A polynomial is defined for *all* real numbers. The graph of a polynomial function is "continuous" and "smooth", that is, there are no jumps, *abrupt* changes, or sharp points in the graph. Nearby values of y correspond to nearby values of x. The functions graphed in Figures 11.3(a) and (b) are continuous; the functions graphed in Figures 11.3(c) and (d) are discontinuous.

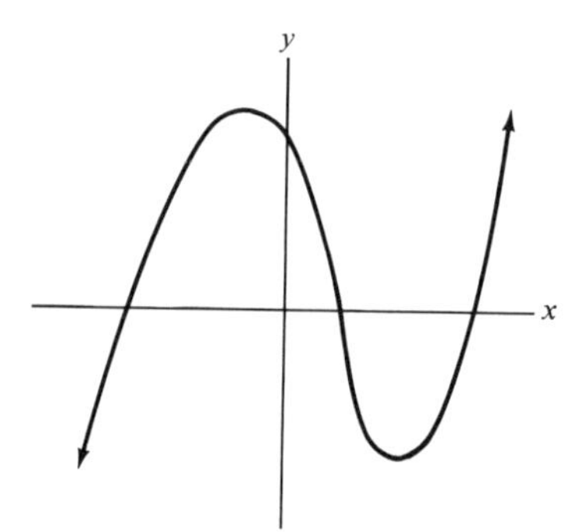

FIGURE 11.3(a) A continuous, smooth function. Nearby values of y correspond to nearby values of x.

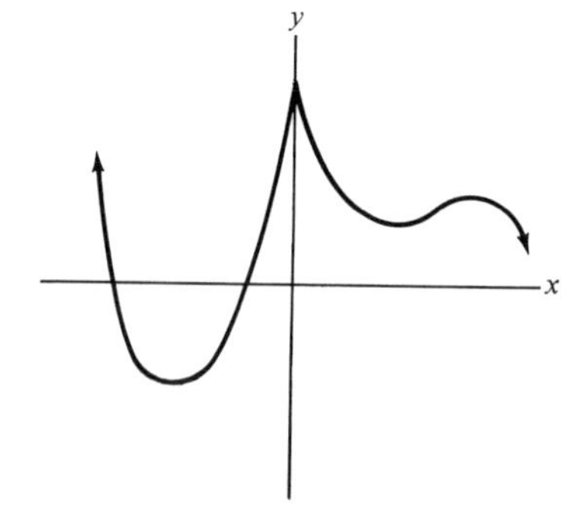

FIGURE 11.3(b) A continuous function that is not smooth. There is a sharp point when $x = 0$.

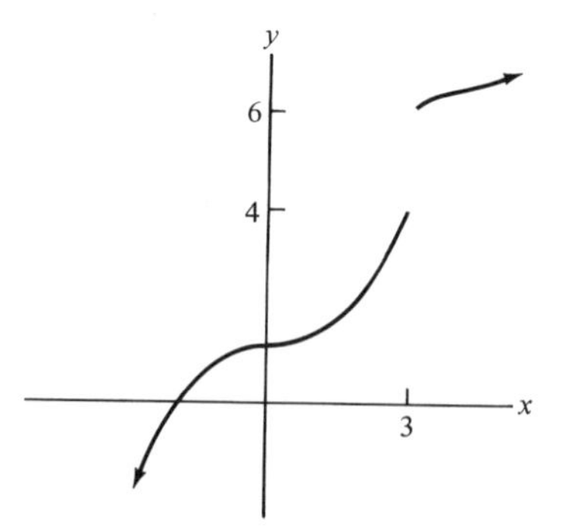

FIGURE 11.3(c) A discontinuous function. As x gets close to 3, the y-values "jump" from about 4 to about 6.

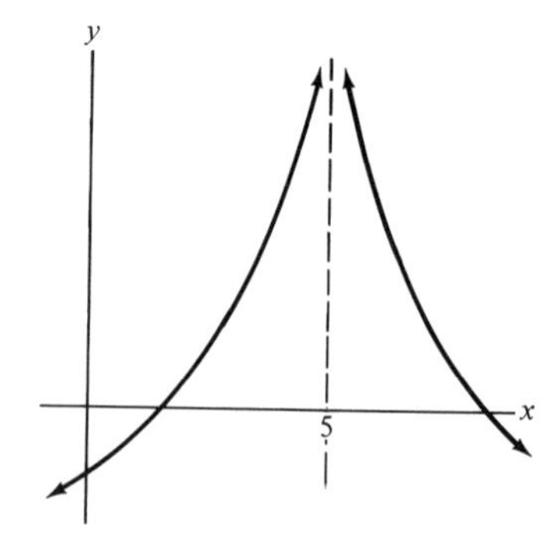

FIGURE 11.3(d) A discontinuous function. As x gets close to 5, the y-values get large beyond all bounds.

$y > 0, \quad y = 0, \quad y < 0$

It is useful to know all values of x for which $y = 0$ (the x-intercepts of the graph), for which $y > 0$, and for which $y < 0$. To find where $y = 0$, determine the real zeros of the polynomial $P(x)$. Then factor $P(x)$ to find where it is positive and where it is negative. Use the fact that a product is *negative* if an *odd* number of factors is negative, and *positive* if an even number of factors is negative (or all factors are positive). For example, if

$$P(x) = x^2 - 2x = x(x - 2),$$

then clearly $P(x) = 0$ when $x = 0$ and when $x = 2$. Figure 11.4 indicates where $P(x)$ is positive and where it is negative. Observe that in the rightmost region, $x > 2$. Therefore $x - 2 > 0$ and $x > 0$. Both factors of $P(x)$, x and $x - 2$, are positive. Thus $P(x)$ is positive.

$x < 0$	$x > 0$	$x > 0$
$x - 2 < 0$	$x - 2 < 0$	$x - 2 > 0$
$x(x-2) > 0$	$x(x-2) < 0$	$x(x-2) > 0$

0 2 x

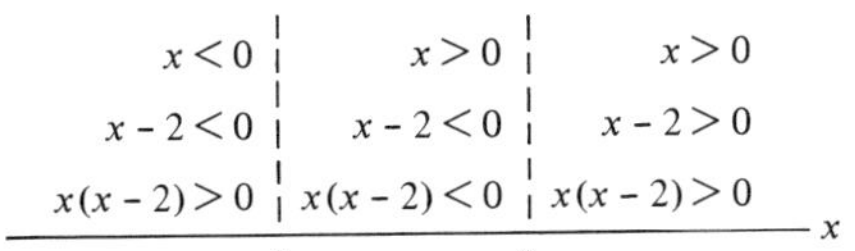

FIGURE 11.4

In the middle region, $0 < x < 2$. Thus, x is positive but $x - 2$ is negative. Consequently, $P(x)$ is negative.

In the leftmost region, $x < 0$. Therefore both factors of $P(x)$, x and $x - 2$, are negative. Thus $P(x)$ is positive.

Alternatively, you could pick a number in each region and evaluate the function at that value. See page 253, beginning with line 4.

leading term

To find out what happens to a polynomial as $x \to \infty$ or as $x \to -\infty$, consider only the leading term. This is because when $|x|$ is large, the remaining terms of the polynomial contribute comparatively little to the total value. For example, if

$$y = x^4 + 2x^3 + 10x^2 - 1, \quad \text{then when} \quad x = 1000,$$

$$x^4 = 1\,000\,000\,000\,000 \text{ but } 2x^3 + 10x^2 - 1 = 2\,009\,999\,999$$

Thus, when $x = 1000$, x^4 is almost 500 times as large as the sum of the remaining terms. Because $x^4 \to \infty$ both as $x \to \infty$ and as $x \to -\infty$, it follows that $y \to \infty$ both as $x \to \infty$ and as $x \to -\infty$.

If $y = -2x^4 + x^3 + 100x + 12$, the leading term is $-2x^4$, which is negative for $x \neq 0$. Thus $y \to -\infty$ both as $x \to \infty$ and as $x \to -\infty$

If $y = x^5 - 2x^4 - 12x^2 - 55$, the leading term, x^5, is positive if x is positive, and negative if x is negative. Thus as $x \to \infty$, then $y \to \infty$, and as $x \to -\infty$, then $y \to -\infty$

EXAMPLE 1 Graph the function given by $y = x^3 - 4x^2$.

Solution. Both odd and even powers of x occur in this polynomial function; thus, neither y-axis nor origin symmetry occurs. Factoring, you obtain:

$$y = x^2(x - 4)$$

TABLE 11.2

x	$y = x^3 - 4x^2$
-2	-24
-1	-5
0	0
1	-3
2	-8
$\frac{8}{3}$	$-\frac{256}{27} (\approx -9.5)$
3	-9
4	0
5	25

Clearly $y = 0$ when $x = 0$ and when $x = 4$. For $x \neq 0$, $x^2 > 0$. Thus for $x \neq 0$ and for $x \neq 4$, the sign of y depends upon the sign of $x - 4$. Clearly

$$x - 4 > 0 \quad \text{when} \quad x > 4$$

and

$$x - 4 < 0 \quad \text{when} \quad x < 4$$

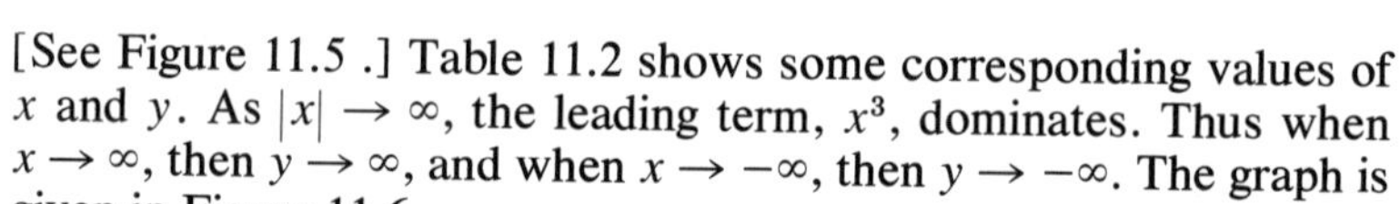

FIGURE 11.5

Therefore

$$y > 0 \quad \text{when} \quad x > 4, \quad \text{and}$$
$$y < 0 \quad \text{when} \quad 0 < x < 4 \quad \text{and when} \quad x < 0$$

[See Figure 11.5 .] Table 11.2 shows some corresponding values of x and y. As $|x| \to \infty$, the leading term, x^3, dominates. Thus when $x \to \infty$, then $y \to \infty$, and when $x \to -\infty$, then $y \to -\infty$. The graph is given in Figure 11.6 .

There is a "turning point" at the origin, because the function increases on one side of the origin and decreases on the other side. There is also a turning point near $x = 3$. A method of calculus is required to show that $x = \frac{8}{3}$ at this second turning point. □

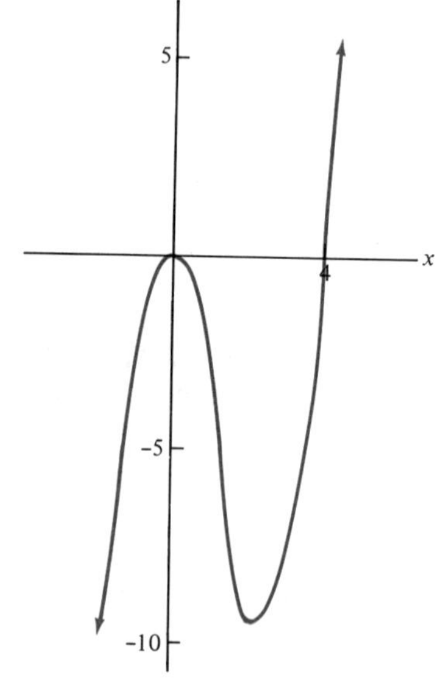

FIGURE 11.6

EXAMPLE 2 Graph the function given by $y = P(x) = x^3 - 9x^2 + 23x - 15$.

Solution. As in Example 1, there are no symmetries. The possible rational zeros are ± 1, ± 3, ± 5, ± 15. In fact, 1 is a zero:

$$\begin{array}{r|rrrr} x - 1 \rightarrow \quad 1 & 1 & -9 & +23 & -15 \\ & & +1 & -8 & +15 \\ \hline & 1 & -8 & +15 & \boxed{0} \end{array}$$

The quotient is $x^2 - 8x + 15$, which factors as $(x - 3)(x - 5)$. Thus

$$y = P(x) = (x - 1)(x - 3)(x - 5)$$

The zeros of $P(x)$ are 1, 3, and 5. When $x > 5$, all three factors are positive, and thus $y > 0$.

If $3 < x < 5$, then $x - 5$ is negative, but the other two factors are positive. Thus y is negative.

If $1 < x < 3$, then $x - 3$ and $x - 5$ are both negative, whereas $x - 1$ is positive. Thus y is positive.

If $x < 1$, all three factors are negative, as is the product y. [See Figure 11.7 .]

$x<1$	$1<x<3$	$3<x<5$	$x>5$
$x-1<0$	$x-1>0$	$x-1>0$	$x-1>0$
$x-3<0$	$x-3<0$	$x-3>0$	$x-3>0$
$x-5<0$	$x-5<0$	$x-5<0$	$x-5>0$
$y<0$	$y>0$	$y<0$	$y>0$

FIGURE 11.7

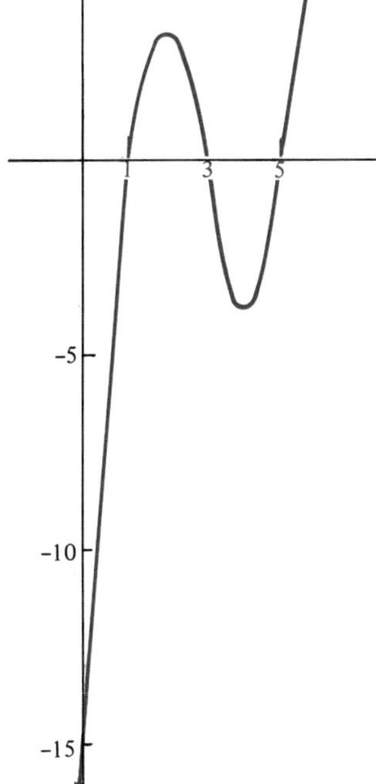

FIGURE 11.8 The graph of $y = (x - 1)(x - 3)(x - 5)$

Table 11.3 indicates some corresponding values of x and y.

As $|x| \to \infty$, the leading term, x^3, dominates. So when $x \to \infty$, then $y \to \infty$, and when $x \to -\infty$, then $y \to -\infty$. The graph is given in Figure 11.8 . Note that the graph "turns" near $x = 2$ and near $x = 4$. □

TABLE 11.3

x	$y = (x-1)(x-3)(x-5)$
-1	$(-2)(-4)(-6) = -48$
0	$(-1)(-3)(-5) = -15$
1	0
2	$1(-1)(-3) = 3$
3	0
4	$3(1)(-1) = -3$
5	0
6	$5 \cdot 3 \cdot 1 = 15$

EXAMPLE 3 Graph the function given by $y = P(x) = x^4 - 3x^2 - 4$.

Solution. All powers of x that occur are even; therefore, there is symmetry about the y-axis. Thus, you need only draw the graph for $x \geqslant 0$, and then reflect the graph in the y-axis.

To find the zeros of $P(x)$, let $u = x^2$. Then set $y = 0$:

$$y = u^2 - 3u - 4 = (u - 4)(u + 1) = 0$$

$u - 4 = 0$	$u + 1 = 0$
$x^2 = u = 4$	$x^2 = u = -1$
$x = \pm 2$	No real roots

Thus $y = (x - 2)(x + 2)\underbrace{(x^2 + 1)}_{\text{positive for all } x}$

$x - 2 < 0$	$x - 2 > 0$
$x + 2 > 0$	$x + 2 > 0$
$x^2 + 1 > 0$	$x^2 + 1 > 0$
$y < 0$	$y > 0$

0 2 x

FIGURE 11.9

TABLE 11.4

x	$y = x^4 - 3x^2 - 4$
0	−4
1	−6
2	0
3	50

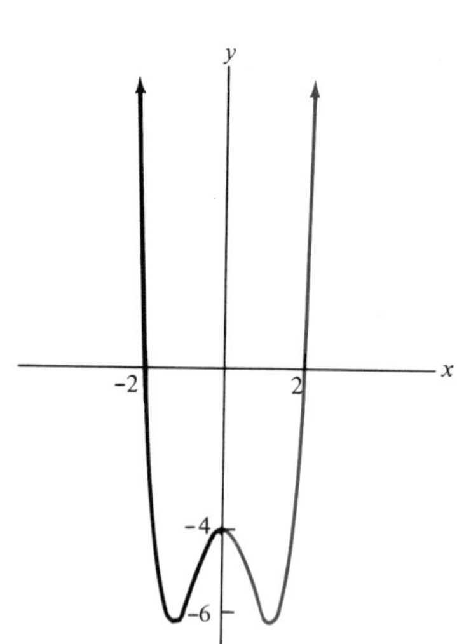

FIGURE 11.10

For $x \geqslant 0$:

$y = 0$ when $x = 2$

$y > 0$ when $x - 2 > 0$, that is, when $x > 2$

$y < 0$ when $0 \leqslant x < 2$

[See Figure 11.9 .]

Table 11.4 shows some corresponding values of x and y.

As $x \to \infty$, the leading term x^4 dominates. Thus when $x \to \infty$, then $y \to \infty$. First draw the graph for $x \geqslant 0$; then reflect this in the y-axis, as in Figure 11.10 . Note the three turning points. □

EXERCISES

In Exercises 1–6 indicate which graphs are symmetric with respect to: **(a)** the y-axis **(b)** the origin.

1. $y = 2x^3 + x$ **2.** $y = x^6 - x^4 + x^2$ **3.** $y = x^3 + x^2 + x$ **4.** $y = x(x + 1)$

5. $y = x^2(x^3 - 3x)$ **6.** $y = 2x(x^4 + x^2)$

In Exercises 7–18 indicate where: **(a)** $y = 0$ **(b)** $y > 0$ **(c)** $y < 0$

7. $y = x^2(x - 5)$ **8.** $y = x^3(x - 5)$ **9.** $y = x(x - 1)(x + 4)$

10. $y = (x + 2)(x - 2)(x - 4)$ **11.** $y = (x - 2)^2(x + 2)$ **12.** $y = (x - 4)^2(x + 1)^3$

13. $y = x(x - 1)(x - 2)(x - 3)$ **14.** $y = x(x - 1)(x - 2)(x - 3)(x - 4)$

15. $y = x^4 - 4x^2$ **16.** $y = x^4 - 5x^2 - 36$

17. $y = 2x^3 - 3x^2 - 3x - 5$ **18.** $y = 6x^4 - x^3 - 7x^2 + x + 1$

In Exercises 19–26 determine the behavior of y: **(a)** as $x \to \infty$ **(b)** as $x \to -\infty$.

19. $y = x^3 + 2x^2 - 10x$ **20.** $y = x^4 - 17x^3$ **21.** $y = 10 - x^5$ **22.** $y = \frac{1}{100}x^6 - 20x^5$

23. $y = (x^2 + 1)(x - 2)$ **24.** $y = (x + 1)(x - 2)^2(x + 3)$

25. $y = (x - 2)^2(3x - 1)(x^2 + 1)$ **26.** $y = (1 - 2x)(3x + 1)^4$

In Exercises 27–50: **(a)** Indicate the symmetries, if any. **(b)** Determine where $y = 0$, $y > 0$, $y < 0$. **(c)** Determine the behavior of y as $x \to \infty$, and also as $x \to -\infty$. **(d)** Graph the function.

27. $y = x^2(x - 3)$ **28.** $y = x(x + 4)^2$ **29.** $y = x(x - 2)(x - 4)$

30. $y = \frac{1}{3}(x + 3)(x - 1)(x - 2)$ **31.** $y = x^2(x - 1)(x - 4)$ **32.** $y = x^2(x - 2)^2$

33. $y = (x + 1)^2(x - 2)(x + 2)$ **34.** $y = (2x - 1)(x - 1)(x - 2)(x - 3)$

35. $y = (x + 2)(x + 1)x(x - 1)(x - 2)$ **36.** $y = 2x^2(x^2 + 2)$

37. $y = x^2(x^2 - 2)$ **38.** $y = x^3(x^2 + 2)$ **39.** $y = x^2(3x - 4)(x + 1)^2$

40. $y = (x^2 + 1)(x^2 + 2)$ **41.** $y = x^3 + x$ **42.** $y = x^4 - 4x^2$

43. $y = x^4 - 4x^3 + 4x^2$ **44.** $y = x^4 + 2x^2 + 2$ **45.** $y = x^3 - 5x^2 + 7x - 3$

46. $y = 1 - x^6$ **47.** $20y = x^4 - 3x^3 - 9x^2 - 3x - 10$

48. $4y = x^4 + 4x^3 - x^2 - 16x - 12$ **49.** $y = 2x^3 - x^2 + 2x - 1$

50. $y = 6x^3 - x^2 - 10x - 3$

51. Find a polynomial function $y = P(x)$ with *all* of the following properties: **i.** $P(x)$ is of degree 4. **ii.** $P(0) = 1$ **iii.** The graph is symmetric with respect to the y-axis. **iv.** As $x \to \infty$, then $y \to \infty$

52. Find a polynomial function $y = P(x)$ with *all* of the following properties. **i.** $P(x)$ is of degree 5. **ii.** $P(x) = 0$ when $x = 0$, 1, and, 3 **iii.** As $x \to \infty$, then $y \to -\infty$ **iv.** As $x \to -\infty$, then $y \to \infty$

Review Exercises for Chapter 11

1. Divide as indicated. In some cases there will be a remainder.

(a) $y - 2\overline{)y^3 - 2y^2 + 3y - 6}$ **(b)** $m^2 + 2\overline{)m^3 + 5m^2 + 2m}$ **(c)** $c^3 + 4\overline{)c^5 + c^3 + 4c^2 + c + 5}$

2. Use synthetic division to determine the quotient. In some cases there will be a remainder.

(a) $\dfrac{x^2 + 10x + 21}{x + 3}$ **(b)** $t - 9\overline{)t^2 - 5t - 30}$ **(c)** $x - 2\overline{)x^3 - 2x^2 + 4x - 8}$ **(d)** $z - 1\overline{)z^3 + 8z^2 - 4z}$

3. **(a)** Use the Remainder Theorem to determine the remainder when $P(x)$ is divided by $x - a$.
(b) Is $x - a$ a factor of $P(x)$?
i. $P(x) = x^3 + 2x^2 + x + 2$, $x - 1$ **ii.** $P(x) = x^3 - 2x + 4$, $x - 2$
iii. $P(x) = x^4 - 3x^2 + x - 4$, $x + 2$

4. Determine k so that $x - 1$ is a factor of $x^3 + kx^2 - 2x + 1$.

5. Find the zeros of each polynomial. Indicate which are simple zeros. Find the multiplicity of the other zeros.

(a) $(x + 3)(x - 1)^2(x - i)^3$ **(b)** $x^6 - 3x^2$ **(c)** $(x^2 + x + 1)(x^2 - 16)^2$

6. Find a polynomial of degree 4 that has -2 as a zero of multiplicity 2 and both $1 + 2i$ and $1 - 2i$ as simple zeros. Express this polynomial in the form $a_nx^n + a_{n-1}x^{n-1} + \cdots + a_1x + a_0$.

7. **(a)** Find all rational zeros of each polynomial. **(b)** Find all (nonreal) complex zeros of each polynomial.
i. $x^3 - 2x^2 + x - 2$ **ii.** $6x^3 - 13x^2 + x + 2$

8. Find a polynomial of degree 4 with real coefficients that has both $1 - i$ and $2 - i$ among its zeros. Express this polynomial in the form $a_nx^n + a_{n-1}x^{n-1} + \cdots + a_1x + a_0$.

9. Factor $x^4 - 2x^3 + x^2 - 8x - 12$ into linear polynomials with real zeros and quadratic polynomials with real coefficients and (nonreal) complex conjugate zeros.

10. One of these polynomials must have at least one real zero. Which one? Do *not* solve.

(a) $x^5 - ix^4 + 5x^3 - 5ix^2 + 4x - 4i$ **(b)** $x^6 + 14x^4 + 49x^2 + 36$

(c) $x^7 - 100x^6 + 202x^5 - 7x^3 + 19x^2 - x + 504$

11. Which graphs are symmetric with respect to: **(a)** the y-axis, **(b)** the origin?

i. $y = x^5 - 4x^3 + 2x$ **ii.** $y = x^8 + 3x^4 - 4x^2 + 5$ **iii.** $y = x^6 - 3x^3 + x^2 - 4$

12. Indicate where: **(a)** $y = 0$ **(b)** $y > 0$ **(c)** $y < 0$

i. $y = (x + 3)(x - 1)(x - 4)$ **ii.** $y = (x - 5)^2(x + 2)^3$ **iii.** $y = x^5 - 16x^3$

13. Determine the behavior of y: **(a)** as $x \to \infty$ **(b)** as $x \to -\infty$

i. $y = x^3 + 2x^2 + x - 10$ **ii.** $y = x^4 - 10x^3 - 20x^2 - 100$ **iii.** $y = -2(x^2 + 1)(x^3 - 1)$

14. **(a)** Indicate the symmetries, if any. **(b)** Determine where $y = 0$, $y > 0$, $y < 0$
(c) Determine the behavior of y as $x \to \infty$, and also as $x \to -\infty$. **(d)** Graph the function.

i. $y = x(x - 1)(x - 4)$ **ii.** $y = (2x - 3)(x - 1)x(x + 1)$ **iii.** $y = x^2(x^2 - 1)$
iv. $y = x^3 - x^5$

Systems of Equations

12

12.1 Linear Systems in Two Variables

There are three possibilities for the intersection of two lines in the plane. Let L_1 and L_2 be these lines.

Case 1. L_1 and L_2 do not intersect. In this case $L_1 \parallel L_2$ [See Figure 12.1(a).]

Case 2. L_1 and L_2 intersect at a single point. [See Figure 12.1(b).]

Case 3. L_1 and L_2 are identical. [See Figure 12.1(c).]

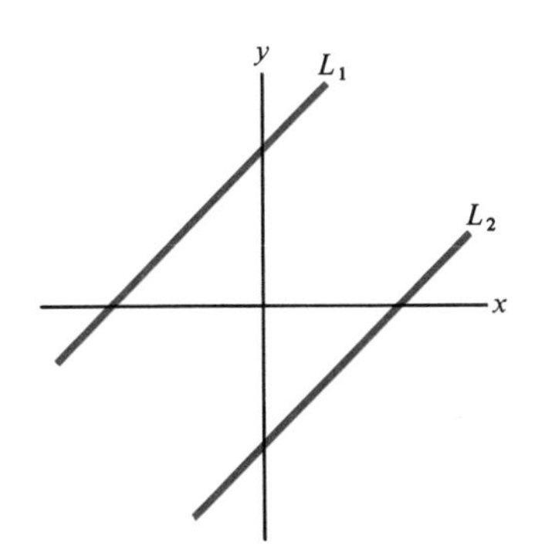

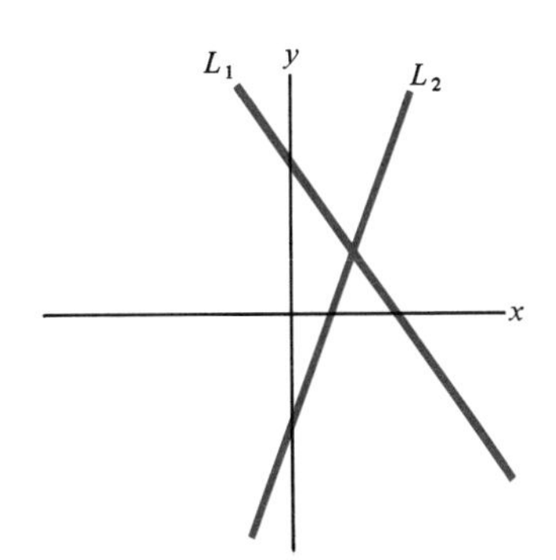

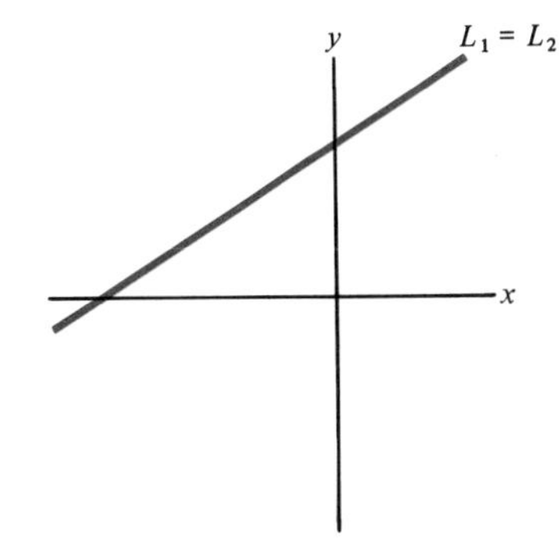

FIGURE 12.1
(a) L_1 and L_2 do not intersect. $L_1 \parallel L_2$

(b) L_1 and L_2 intersect at a single point.

(c) L_1 and L_2 are identical.

Thus, *distinct* lines (Cases 1 and 2) intersect at a single point or not at all. In this section, you will consider Case 2, in which there is a single point of intersection. Case 1, in which there is no intersection (parallel lines) was discussed in Section 4.4 . Case 3 occurs when each of the two equations is "a multiple" of the other.

Suppose neither L_1 nor L_2 is vertical. Let

$$m_1 = \text{slope } L_1; \qquad m_2 = \text{slope } L_2.$$

In Cases 1 and 3, $m_1 = m_2$. *Case 2 is characterized by* $m_1 \neq m_2$.

graphical method One way of determining the *approximate* intersection of two lines is by graphing each of the lines on the same coordinate system to see where they intersect.

EXAMPLE 1 Determine the intersection of

$$L_1\colon 2x + 3y = 8 \qquad \text{and} \qquad L_2\colon 3x - y = 1$$

graphically.

Solution.

$$m_1 = \text{slope } L_1 = \frac{-2}{3}, \qquad m_2 = \text{slope } L_2 = 3$$

There is a single point of intersection because $m_1 \neq m_2$.

Determine two points (for convenience, the x- and y- intercepts) on each line.

TABLE 12.1

L_1: x	L_1: y
0	$\frac{8}{3}$
4	0

L_2: x	L_2: y
0	-1
$\frac{1}{3}$	0

Draw L_1 through points $\left(0, \frac{8}{3}\right)$ and $(4, 0)$ and L_2 through $(0, -1)$ and $\left(\frac{1}{3}, 0\right)$. The point of intersection can be read from the graph as $(1, 2)$. [See Figure 12.2 .] □

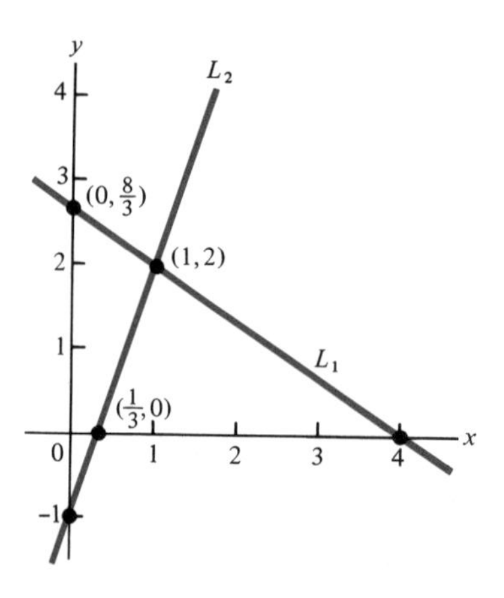

FIGURE 12.2 The single point of intersection is (1, 2).

algebraic methods The preceding graphical method tends to be inaccurate, particularly when the x- or y-coordinate of the intersection point is not an integer. There are also algebraic techniques for determining the intersection point of two given lines L_1 and L_2. The algebraic methods, which give the *exact* intersection, are preferable.

Recall that the general form of the equation of a line is

$$Ax + By + C = 0,$$

where A and B are not *both* 0. You can also write

$$Ax + By = -C$$

Therefore, the equations of lines L_1 and L_2 can be written as

$$a_1x + b_1y = k_1$$
$$a_2x + b_2y = k_2$$

(where a_1 and b_1 are not both 0, and a_2 and b_2 are not both 0).

Together, the equations of L_1 and L_2 form a **system of two linear equations in two variables.** (The word "linear" indicates that each equation represents a line.)

Recall that an ordered pair (x_0, y_0) is a *solution* of (or *satisfies*) an equation in two variables x and y if a true statement results when x_0 replaces x and y_0 replaces y.

Definition. The ordered pair (x_0, y_0) is called a **solution of the system** if (x_0, y_0) is a solution of *both* equations of the system.

Solving the system is the algebraic equivalent of graphing these lines to find their point of intersection. In other words, *a solution of the system corresponds to a point of intersection of the lines.*

Thus in Example 1, you considered the system:

$$2x + 3y = 8$$
$$3x - y = 1$$

The solution of this system represents the point of intersection, (1, 2), of the corresponding lines.

substitution method

In the **substitution method** for solving two equations:

1. Solve one equation for one variable—say y—in terms of the other, x (as you did when you solved literal equations in Section 2.4).
2. Then substitute this expression for y in the other equation and solve for x.
3. Replace x by this value in the expression for y, and solve for y.

EXAMPLE 2 Solve the system:

$$2x + y = 12$$
$$3x - 2y = 11$$

Check the solution.

Solution. From the first equation,

$$y = 12 - 2x$$

Substitute $12 - 2x$ for y in the second equation:

$$3x - 2(12 - 2x) = 11$$
$$3x - 24 + 4x = 11$$
$$7x = 35$$
$$x = 5$$

Replace x by 5 in the expression for y:

$$y = 12 - 2x$$

Thus $y = 12 - 2(5) = 2$

The solution of the system of equations is (5, 2).

Check. Replace x by 5 and y by **2** in each of the original equations:

$2(5) + \mathbf{2} \stackrel{?}{=} 12$	$3(5) - 2\,(\mathbf{2}) \stackrel{?}{=} 11$
$12 \stackrel{\checkmark}{=} 12$	$11 \stackrel{\checkmark}{=} 11$

□

"adding equations"

You can also eliminate a variable by **"adding equations"** (or really, adding the corresponding *sides* of the equations). For if

□ = □

and

▨ = ▨,

then

□ + ▨ = □ + ▨

EXAMPLE 3 Solve the system:

$$4x + 3y = -1$$
$$2x - 3y = 13$$

Solution. Note that $+ 3y$ occurs in the first equation and $-3y$ in the second. Adding, you obtain:

$$4x + 3y = -1$$
$$2x - 3y = 13$$
$$\overline{6x \qquad = 12}$$
$$x = 2$$

Replace x by 2 in, say, the first equation:

$$4(2) + 3y = -1$$
$$8 + 3y = -1$$
$$3y = -9$$
$$y = -3$$

(Had you replaced x by 2 in the second equation, you would have also obtained $y = -3$.) The solution is $(2, -3)$. □

In order to use this method effectively, you first may have to transform one or both of the equations into an equivalent equation.

EXAMPLE 4 Solve the system:

$$\begin{aligned} 5x + 4y &= 8 \\ 3x + 2y &= 6 \end{aligned}$$

Solution. Note that $4y$ occurs in the first equation and $2y$ in the second. Multiply both sides of the second equation by -2:

$$-6x - 4y = -12$$

Now perform the addition to obtain:

$$\begin{array}{rcl} 5x + 4y &=& 8 \\ -6x - 4y &=& -12 \\ \hline -x &=& -4 \end{array} \quad \text{or} \quad x = 4$$

Replace x by 4 in the first equation:

$$\begin{aligned} 5(4) + 4y &= 8 \\ 20 + 4y &= 8 \\ 4y &= -12 \\ y &= -3 \end{aligned}$$

The solution is $(4, -3)$. □

Both equations may have to be transformed so that upon addition one of the variables is eliminated.

EXAMPLE 5 Solve the system:

$$\begin{aligned} 2x - 3y &= 3 \\ 3x - 2y &= 7 \end{aligned}$$

Solution. The coefficients of y are -3 and -2. LCM $(-3, -2) = 6$. Thus multiply, as indicated, in order to eliminate y.

Multiply both sides by 2.→ ← Multiply both sides by -3. Add.

$$\begin{array}{rcl} 2x - 3y &=& 3 \\ 3x - 2y &=& 7 \\ \hdashline 4x - 6y &=& 6 \\ -9x + 6y &=& -21 \\ \hline -5x &=& -15 \\ x &=& 3 \end{array}$$

Replace x by 3 in the first equation:

$$\begin{aligned} 2(3) - 3y &= 3 \\ 6 - 3y &= 3 \\ -3y &= -3 \\ y &= 1 \end{aligned}$$

The solution is $(3, 1)$. □

applications

EXAMPLE 6 Three times a number is two less than another number. Seven times the first number is one more than twice the second number. Find these two numbers.

Solution. Let x represent the first number and y the second. Set up a system of linear equations and solve.

$$\underbrace{\text{Three times a number}}_{3x} \ \underset{=}{\text{is}} \ \underbrace{\text{two less than}}_{-2\,+} \ \underbrace{\text{another number.}}_{y}$$

$$\underbrace{\text{Seven times the first number}}_{7x} \ \underset{=}{\text{is}} \ \underbrace{\text{one more than}}_{1\,+} \ \underbrace{\text{twice the second number.}}_{2y}$$

From the first equation, $3x + 2 = y$

Replace y by $3x + 2$ in the second equation.

$$7x = 1 + 2(3x + 2)$$
$$7x = 1 + 6x + 4$$
$$x = 5$$
$$y = 3x + 2$$
$$y = 17$$

The numbers are 5 and 17. □

EXAMPLE 7 Determine a and b so that the line given by

$$ax + by + 1 = 0$$

goes through the points $(2, -5)$ and $(-1, 1)$.

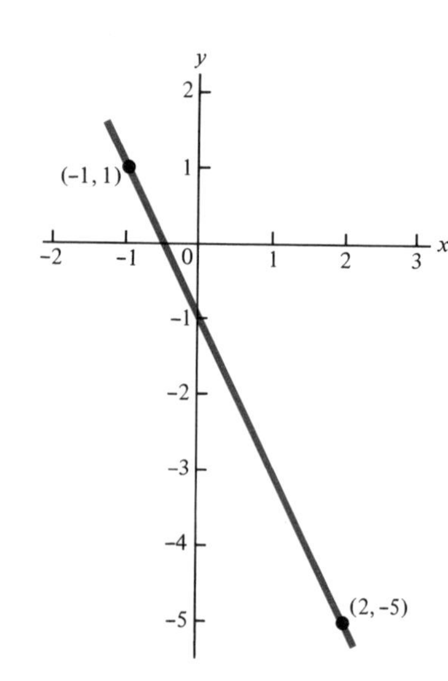

FIGURE 12.3 The line given by $2x + y + 1 = 0$

Solution. Because the *point* $(2, -5)$ is on the line, the *ordered pair* $(2, -5)$ satisfies the equation of the line. Thus replace x by 2 and y by -5. Therefore, one equation *in the variables a and b* is

$$a(2) + b(-5) + 1 = 0 \qquad \text{or} \qquad 2a - 5b = -1.$$

Similarly, replace x by -1 and y by 1 in the equation of the line, and obtain a second equation in a and b:

$$-a + b + 1 = 0 \qquad \text{or} \qquad b = a - 1$$

You now have two equations in a and b:

$$2a - 5b = -1$$
$$b = a - 1$$

Replace b by $a - 1$ in the first equation, and obtain:

$$2a - 5(a - 1) = -1$$

Thus

$$2a - 5a + 5 = -1$$

$$-3a = -6$$
$$a = 2$$

Also,

$$b = a - 1 = 2 - 1 = 1$$

Thus, $a = 2$ and $b = 1$; the equation of the line is then

$$2x + y + 1 = 0$$

[See Figure 12.3 .] □

EXERCISES

In Exercises 1–10 determine the intersection of L_1 and L_2 graphically.

1. L_1: $4x - 3y = 1$, L_2: $x + y = 2$

2. L_1: $y = 3x$, L_2: $y + 1 = 0$

3. L_1: $3x + y = 1$, L_2: $x + y = 1$

4. L_1: $2x - y = 3$, L_2: $x + 2y = -1$

5. L_1: $4x + 2y = 5$, L_2: $x - 2y = 0$

6. L_1: $x - 2y = 1$, L_2: $-x + 3y = -3$

7. L_1: $x - y = 5$, L_2: $3x - 5y = 5$

8. L_1: $x + 2y = 4$, L_2: $2x + 5y = 10$

9. L_1: $5x - 2y = 1$, L_2: $x - 3y = 8$

10. L_1: $x + y = 2$, L_2: $2x - y = 10$

In Exercises 11–18 solve each system by the substitution method.

11. $2x + y = 4$, $3x - y = 1$

12. $3x - 4y = 0$, $x + y = 7$

13. $x + y = 7$, $3x - 5y = 5$

14. $x - 3y = 8$, $3x - 2y = 10$

15. $4u + v = 24$, $u - 3v = 6$

16. $5s + t = 3$, $s - 2t = 5$

17. $4x - 8y = 0$, $x + 2y = 1$

18. $7x - 2y = 1$, $2x + 4y = 14$

In Exercises 19–26 solve each system by "adding equations".

19. $2x + y = 9$, $3x - y = 1$

20. $5x - 2y = 3$, $3x + 2y = 21$

21. $x + y = \frac{\pi}{2}$, $x - y = \frac{\pi}{6}$

22. $2x + 5y = 22$, $4x - 5y = 14$

23. $u + 2v = 3$, $u - 2v = 15$

24. $5x + 4y = -3$, $6x - 2y = 10$

25. $9x - 7y = 15$, $3x + 2y = 18$

26. $3x + 2y = 10$, $2x + 3y = 10$

In Exercises 27–38 solve by either substituting or adding.

27. $2x - 7y = 0$, $x + 4y = 0$

28. $x + 3y = 1$, $5x + 14y = 8$

29. $5x + 3y = 24$, $3x + 2y = 14$

30. $10x + 100y = 2$, $100x - 200y = 8$

31. $2s + 5t = -3$, $5s + 7t = 9$

32. $x + y + 2 = 0$, $3x - 2y + 1 = 0$

33. $2x + 3y = -6$, $7x + 4y = 5$

34. $9x - 8y = 26$, $4x - 3y = 16$

35. $3x + 2y = 5$, $12x + 7y = 25$

36. $-5x + 4y = 1$, $3x - 2y = 0$

37. $3x - 2y = 9$, $5x - 4y = 7$

38. $6x - 3y = 6$, $7x - 4y = -2$

In Exercises 39–51 set up a system of linear equations and solve.

39. The sum of two numbers is 30. The larger number is 4 more than the smaller. Find these numbers.

40. Alice has \$3.40 in dimes and quarters. If the number of dimes and quarters were reversed, she would have \$4.30. How many dimes does she have?

41. Ted is twice as old as his brother. Four years ago he was four times as old as his brother. How old is each boy?

42. Twice a number is one less than three times another number. Five times the first number is one more than seven times the second. Find these two numbers.

43. A bookstore shelves books vertically. A 2-foot shelf holds 12 copies of a dictionary and 3 copies of an encyclopedia. A 3-foot shelf holds 4 copies of the dictionary and 15 copies of the encyclopedia. How thick is each book?

44. The perimeter of an isosceles triangle is 20 inches. If the length of each of the equal sides were increased by 3 inches and the base length were doubled, the perimeter would be 34 inches. Find the length of each side.

45. A shipping clerk finds that 6 cartons of paper and 4 cartons of notebooks together weigh 92 pounds. Later he finds that 3 cartons of paper and 6 cartons of notebooks weigh 90 pounds. How much does a carton of paper weigh?

46. A copy center has two machines. The Xerox in 2 minutes and the IBM in 3 minutes together produce 155 copies. The Xerox in 5 minutes and the IBM in 6 minutes together produce 350 copies. How many copies per minute can each machine produce?

47. Determine a and b so that the line given by $ax + by - 5 = 0$ passes through the points $(2, 1)$ and $(5, 5)$.

48. Determine a and b so that the line given by $ax + by - 2 = 0$ passes through the points $(1, 1)$ and $(4, 6)$.

49. Determine a and b so that the line given by $ax + by - 2 = 0$ passes through the points $(1, 1)$ and $(4, -2)$.

50. Determine a and b so that the line given by $ax + by - 10 = 0$ has x-intercept 2 and y-intercept 5.

51. Determine a and b so that the line given by $ax + by + 2 = 0$ passes through the point $(-1, -1)$ and has slope -1.

In Exercises 52–54 suppose $0 \leq x \leq \frac{\pi}{2}$ and $0 \leq y \leq \frac{\pi}{2}$. Solve for x and y.

52. $\sin x + \cos y = 1$
$\sin x - \cos y = 0$

53. $2 \sin x + 3 \cos y = 4$
$4 \sin x - \cos y = 1$

54. $2 \sin x - 4 \cos y = 1$
$4 \sin x + 3 \cos y = 2$

12.2 Linear Systems in Three Variables; Partial Fractions

solutions The equation

$$2x + y - 3z = 10$$

is in *three* variables. Set $x = 4$, $\mathbf{y = 5}$, $z = 1$, and obtain:

$$2 \cdot 4 + \mathbf{5} - 3 \cdot 1 = 10 \qquad \text{or} \qquad 10 = 10$$

Thus the numbers

$$4, 5, 1 \qquad \text{(in this order)}$$

serve as a "solution" of the given equation.

Definition. Let x_0, y_0, and z_0 be three real numbers. The symbol

$$(x_0, y_0, z_0)$$

is called an **ordered triple,** and represents the given numbers in the order written. The ordered triple (x_0, y_0, z_0) is called a **solution** of an equation

$$Ax + By + Cz = D, \qquad A, B, C \text{ not all } 0,$$

if

$$Ax_0 + By_0 + Cz_0 = D.$$

The equations

$$\begin{aligned} a_1x + b_1y + c_1z &= d_1 \\ a_2x + b_2y + c_2z &= d_2 \\ a_3x + b_3y + c_3z &= d_3 \end{aligned}$$

constitute a **system of three linear equations in three variables** (provided a_1, b_1, and c_1, are not *all* 0; a_2, b_2, and c_2 are not *all* 0; and a_3, b_3, and c_3 are not *all* 0). The ordered triple (x_0, y_0, z_0) is called a **solution of the** preceding **system** if (x_0, y_0, z_0) is a solution of *all three* equations of the system.

solving a system

To solve a system of three linear equations in three variables, either "add the equations" or substitute. In Example 1, you will immediately obtain one linear equation in a single variable. Upon solving for this variable, you then obtain a system of two linear equations in the other two variables. This system can then be solved by "adding" or substituting.

EXAMPLE 1 Solve the system:

$$\begin{aligned} x - y - z &= 0 && \textbf{(A)} \\ x - y + z &= 4 && \textbf{(B)} \\ x + y + z &= 10 && \textbf{(C)} \end{aligned}$$

Solution. Note that the coefficients of x and of z agree in Equations **(B)** and **(C)**. Thus to solve for y, multiply (both sides of) Equation **(B)** by -1 and "add" this to Equation **(C)**.

In symbols: (-1) **(B)** + **(C)**

$$\begin{aligned} -x + y - z &= -4 \\ x + y + z &= 10 \\ \hline 2y &= 6 \\ y &= 3 \end{aligned}$$

To solve for x, "add" **(A)** + **(B)**, and replace y by 3.

$$\begin{aligned} x - y - z &= 0 \\ x - y + z &= 4 \\ \hline 2x - 2y &= 4 \\ 2x - 2(3) &= 4 \\ 2x &= 10 \\ x &= 5 \end{aligned}$$

From **(B)**:

$$5 - 3 + z = 4$$
$$z = 2$$

□

In Example 5 of Section 12.1, you multiplied both sides of one equation by 2 and both sides of another equation by -3, and then "added" the resulting equations. In general, you can multiply both sides of an equation by any nonzero number and both sides of a second equation by another nonzero number, and then "add".

EXAMPLE 2 Solve the system:

$$\begin{aligned} 4x + 3y - 3z &= 2 \quad &\textbf{(A)} \\ 5x - 3y + 2z &= 10 \quad &\textbf{(B)} \\ 2x - 2y + 3z &= 14 \quad &\textbf{(C)} \end{aligned}$$

Solution. First eliminate y so as to obtain two equations, **(D)** and **(E)**, in x and z.

(A) + **(B)**:

$$\begin{aligned} 4x + 3y - 3z &= 2 \quad &\textbf{(A)} \\ 5x - 3y + 2z &= 10 \quad &\textbf{(B)} \\ \hline 9x \qquad\quad - z &= 12 \quad &\textbf{(D)} \end{aligned}$$

2(**A**) + 3(**C**):

$$\begin{aligned} 8x + 6y - 6z &= 4 \quad &2\textbf{(A)} \\ 6x - 6y + 9z &= 42 \quad &3\textbf{(C)} \\ \hline 14x \qquad\quad + 3z &= 46 \quad &\textbf{(E)} \end{aligned}$$

Eliminate z from **(D)** and **(E)**.

3(**D**) + (**E**):

$$\begin{aligned} 27x - 3z &= 36 \quad &3\textbf{(D)} \\ 14x + 3z &= 46 \quad &\textbf{(E)} \\ \hline 41x \qquad &= 82 \\ x \qquad &= 2 \end{aligned}$$

Replace x by 2 in (**D**).

$$\begin{aligned} 9(2) - z &= 12 \\ 6 &= z \end{aligned}$$

Replace x by 2 and z by **6** in (**A**):

$$\begin{aligned} 4(2) + 3y - 3(\mathbf{6}) &= 2 \\ 8 + 3y - 18 &= 2 \\ 3y &= 12 \\ y &= 4 \end{aligned}$$

The solution is (2, 4, 6). □

EXAMPLE 3 The sum of three numbers is 21. Six times the smallest number equals the sum of the others. The largest number is one less than the sum of the others. Find these numbers.

Solution. Let x be the smallest of these numbers, let y be the next smallest, and let z be the largest.

The sum of the three numbers is 21.

$$x + y + z = 21 \qquad \text{(A)}$$

Six times the smallest number equals the sum of the others.

$$6x = y + z \qquad \text{(B)}$$

The largest number is one less than the sum of the others.

$$z = -1 + (x + y) \qquad \text{(C)}$$

From (**A**): $y + z = 21 - x$

From (**B**):

$$6x = 21 - x$$
$$7x = 21$$
$$x = 3$$

From (**B**):

$$z = 18 - y$$

From (**C**):

$$z = 2 + y$$
$$18 - y = 2 + y$$
$$16 = 2y$$
$$8 = y$$

From (**B**):

$$z = 18 - 8 = 10$$

The three numbers are 3, 8, and 10. □

partial fractions

In calculus, it is sometimes useful to rewrite a rational expression as an "equivalent" sum of rational expressions with simpler denominators. In Example 4, $\frac{x^2 + 4}{x^3 + x}$ will be rewritten as $\frac{4}{x} + \frac{-3x}{x^2 + 1}$. The given rational expression is equivalent to the sum in that each expression has the same value for every x. This method of **partial fraction decomposition** involves solving a number of linear equations in the same number of unknowns.

EXAMPLE 4 Find real numbers A, B, and C such that

$$\frac{x^2 + 4}{x^3 + x} = \frac{A}{x} + \frac{Bx + C}{x^2 + 1} \qquad \text{(12.1)}$$

Solution. The denominator on the left side of (**12.1**) factors as indicated:

$$x^3 + x = x(x^2 + 1)$$

Clearly, the LCD of the denominators on the right side of (**12.1**) is $x(x^2 + 1)$.

Suppose Equation (**12.1**) holds. Then adding the expressions on the right side of (**12.1**), you obtain

$$\frac{x^2+4}{x(x^2+1)} = \frac{A}{x} + \frac{Bx+C}{x^2+1} = \frac{A(x^2+1)+(Bx+C)x}{x(x^2+1)}$$

Because the denominators in color are the same, the numerators in color must also be the same. Thus

$$x^2 + 4 = A(x^2+1) + (Bx+C)x = Ax^2 + A + Bx^2 + Cx$$

or

$$1x^2 + \mathbf{0}x + 4 = (A+B)x^2 + \mathbf{C}x + A$$

If two polynomials are equivalent, their coefficients must be the same. Thus

$$A + B = 1, \qquad \mathbf{C = 0}, \qquad A = 4$$

It is easily seen that $B = 1 - 4 = -3$, so that

$$\frac{x^2+4}{x^3+x} = \frac{x^2+4}{x(x^2+1)} = \frac{4}{x} + \frac{-3x}{x^2+1}$$

□

EXERCISES

In Exercises 1–20 solve each system. You may use any method.

1. $\begin{aligned} x + y &= 2 \\ x + z &= 5 \\ y + z &= 5 \end{aligned}$

2. $\begin{aligned} x + y &= 6 \\ y + z &= 5 \\ x + z &= 7 \end{aligned}$

3. $\begin{aligned} x + y &= 0 \\ y + 2z &= 5 \\ x + z &= 4 \end{aligned}$

4. $\begin{aligned} x + 3y &= -2 \\ 2x + z &= 0 \\ x + y + 2z &= 6 \end{aligned}$

5. $\begin{aligned} x + y &= 5 \\ x + 2z &= 2 \\ x + y + z &= 3 \end{aligned}$

6. $\begin{aligned} x + y + z &= 2 \\ 5x - 6z &= 9 \\ 4y + 10z &= 2 \end{aligned}$

7. $\begin{aligned} x - y + z &= 9 \\ 3x - 5y - 4z &= 2 \\ y + z &= 5 \end{aligned}$

8. $\begin{aligned} x + y + 3z &= 10 \\ 2x + 5z &= 3 \\ -y - 3z &= -1 \end{aligned}$

9. $\begin{aligned} 2x + 3y - z &= 6 \\ x + 2y - z &= 1 \\ 4x + y - z &= 0 \end{aligned}$

10. $\begin{aligned} x + 5y - z &= 2 \\ 3x + 3y - 2z &= 4 \\ x - 7y + 5z &= -10 \end{aligned}$

11. $\begin{aligned} 2x - y + z &= 3 \\ 3x + y - 6z &= 7 \\ 5x - 2y - z &= 4 \end{aligned}$

12. $\begin{aligned} \frac{x}{2} + \frac{y}{4} + \frac{z}{4} &= 5 \\ \frac{x}{4} + \frac{y}{2} + \frac{z}{2} &= 4 \\ \frac{x}{4} + \frac{y}{4} + \frac{z}{2} &= 3 \end{aligned}$

13. $\begin{aligned} x + 5y - 10z &= 5 \\ x - 3y - 4z &= 9 \\ x - 7y + z &= 15 \end{aligned}$

14. $\begin{aligned} x + 3y + z &= 0 \\ 5x + y + 3z &= 0 \\ 2x + y + z &= 2 \end{aligned}$

15. $\begin{aligned} 5x - y + z &= -6 \\ 2x + z &= 0 \\ 3x + y + 6z &= 0 \end{aligned}$

16. $\begin{aligned} 4x - 5y - z &= 5 \\ 2x + y - z &= 3 \\ 6x - 9y - 2z &= 13 \end{aligned}$

17. $\begin{aligned} 3x - 10y + 2z &= -4 \\ 2x + 3y - 5z &= 13 \\ 5x - 2y - 2z &= 0 \end{aligned}$

18. $\begin{aligned} 7x - 2y + 2z &= 19 \\ 3x + y - 4z &= -30 \\ 2x + 3y - 3z &= -16 \end{aligned}$

19. $\begin{aligned} x - 2y + z &= 0 \\ 3x + 4y + 3z &= 5 \\ 2x - 2y - z &= 4 \end{aligned}$

20. $\begin{aligned} 2x + 2y + z &= 1 \\ x + 8y + z &= 2 \\ 3x + 2y - 2z &= 3 \end{aligned}$

In Exercises 21–26 set up a system of three linear equations in three variables, and solve.

21. The sum of three numbers is 12. Five times the smallest number equals the sum of the other two. The largest number equals the sum of the other two. Find these numbers.

22. The sum of three numbers is 20. Four times the smallest number is the sum of the other two. Three times the largest number is five more than twice the sum of the other two. Find these numbers.

23. Helen has \$2.60 in nickels, dimes, and quarters. The number of nickels and dimes together is one more than the number of quarters. The value of the nickels is that of the dimes. How many of each coin does she have?

24. With your two dollars you can buy 3 hot dogs, an order of french fries, and two sodas. You can save a dime by having 2 of each item. Or you can save a nickel by ordering 4 hot dogs and 1 soda. How much does a hot dog cost?

25. Find the equation of the circle that passes through the origin, (4, 4), and (1, 5).

26. Let $f(x) = Ax^2 + Bx + C$. Determine f if $f(0) = 4$, $f(1) = 8$, and $f(2) = 14$.

In Exercises 27–38 find real numbers A, B, C, D such that the given rational expression can be rewritten as the indicated sum.

27. $\dfrac{2x}{(x + 1)(x - 1)} = \dfrac{A}{x + 1} + \dfrac{B}{x - 1}$

28. $\dfrac{4x + 7}{(x + 1)(x + 4)} = \dfrac{A}{x + 1} + \dfrac{B}{x + 4}$

29. $\dfrac{-8}{x^2 - 4} = \dfrac{A}{x + 2} + \dfrac{B}{x - 2}$

30. $\dfrac{12x + 14}{x^2 - x - 6} = \dfrac{A}{x - 3} + \dfrac{B}{x + 2}$

31. $\dfrac{4x^2 - 2}{x(x + 1)(x - 1)} = \dfrac{A}{x} + \dfrac{B}{x + 1} + \dfrac{C}{x - 1}$

32. $\dfrac{13x + 19}{(x + 1)(x + 3)(x - 2)} = \dfrac{A}{x + 1} + \dfrac{B}{x + 3} + \dfrac{C}{x - 2}$

33. $\dfrac{2x^2 - 18}{x^3 - 9x} = \dfrac{A}{x} + \dfrac{B}{x + 3} + \dfrac{C}{x - 3}$

34. $\dfrac{x^2 - x}{(x + 1)(x^2 + 1)} = \dfrac{A}{x + 1} + \dfrac{Bx + C}{x^2 + 1}$

35. $\dfrac{-1}{x^2(x + 1)} = \dfrac{A}{x} + \dfrac{B}{x^2} + \dfrac{C}{x + 1}$

36. $\dfrac{4 - 3x}{(x - 1)^2(x - 2)} = \dfrac{A}{x - 1} + \dfrac{B}{(x - 1)^2} + \dfrac{C}{x - 2}$

37. $\dfrac{x^2 + 5x + 2}{x^3 + 2x^2 - 4x - 8} = \dfrac{A}{x + 2} + \dfrac{B}{(x + 2)^2} + \dfrac{C}{x - 2}$

38. $\dfrac{x^2 + x + 1}{x^2(x^2 + 1)} = \dfrac{A}{x} + \dfrac{B}{x^2} + \dfrac{Cx + D}{x^2 + 1}$

12.3 Systems of Linear and Quadratic Equations

solutions and intersections

A linear, or first-degree, equation in x and y represents a line. Recall that a quadratic, or second-degree, equation in x and y is an equation that can be written in the form

$$Ax^2 + Bxy + Cy^2 + Dx + Ey + F = 0,$$

where A, B, and C are not all zero. A quadratic equation of this form *may* represent a parabola, circle, ellipse, or hyperbola.† The

† See pages 264–265 for other possibilities.

intersections of a line with one of these second-degree figures or the intersections of two such second-degree figures can be determined by solving a system of equations. As in systems of two linear equations (Section 12.1), the pair (x_0, y_0) is called a **solution of the system** if (x_0, y_0) is a solution of (or satisfies) *both* equations of the system. *A solution of the system corresponds to a point of intersection of the figures.*

lines, parabolas and circles

EXAMPLE 1 Determine the intersections of the parabola given by

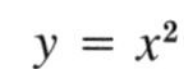

$$y = x^2$$

with the line given by

$$y = 3x.$$

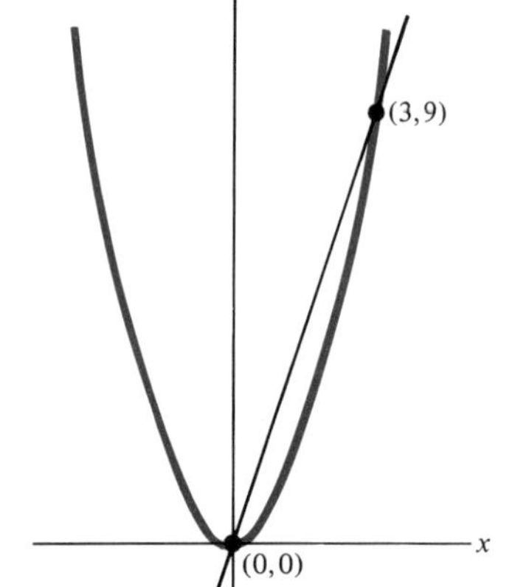

FIGURE 12.4

Solution. Replace y by $3x$ in the equation of the parabola:

$$\begin{aligned} y &= x^2 \\ 3x &= x^2 \\ x^2 - 3x &= 0 \\ x(x-3) &= 0 \\ x = 0 \quad &\text{or} \quad x = 3 \end{aligned}$$

From the equation of the line, when $x = 0$, $y = 3(0) = 0$; when $x = 3$, $y = 3(3) = 9$. The intersection points are these two solutions: (0, 0) and (3, 9). [See Figure 12.4 .]

Check. Because you used the equation of the line to find the y-value corresponding to the x-values 0 and 3, now use the equation of the parabola.

(0, **0**):	$\mathbf{0} \stackrel{?}{=} 0^2$	(3, **9**):	$\mathbf{9} \stackrel{?}{=} 3^2$
	$0 \stackrel{\checkmark}{=} 0$		$9 \stackrel{\checkmark}{=} 9$

□

EXAMPLE 2 Determine the intersections of the circle given by

$$x^2 + y^2 = 16$$

with the line given by

$$y = 2 - x.$$

Solution. From the equation of the line, substitute $2 - x$ for y in the equation of the circle:

$$\begin{aligned} x^2 + y^2 &= 16 \\ x^2 + (2 - x)^2 &= 16 \\ x^2 + 4 - 4x + x^2 &= 16 \\ 2x^2 - 4x - 12 &= 0 \\ x^2 - 2x - 6 &= 0 \end{aligned}$$

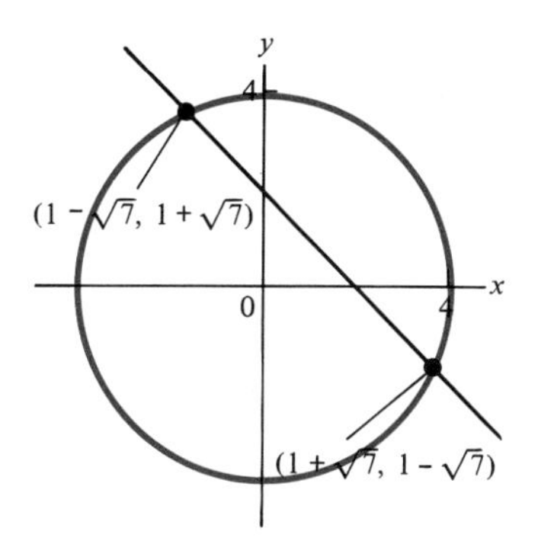

FIGURE 12.5

By the quadratic formula,

$$x = \frac{2 \pm \sqrt{4 - 4(1)(-6)}}{2} = \frac{2 \pm \sqrt{28}}{2} = \frac{2 \pm 2\sqrt{7}}{2} = 1 \pm \sqrt{7}$$

When $x = 1 + \sqrt{7}$,

$$y = 2 - x = 2 - (1 + \sqrt{7}) = 1 - \sqrt{7}$$

When $x = 1 - \sqrt{7}$,

$$y = 2 - x = 2 - (1 - \sqrt{7}) = 1 + \sqrt{7}$$

The solutions are

$$(1 + \sqrt{7}, 1 - \sqrt{7}) \quad \text{and} \quad (1 - \sqrt{7}, 1 + \sqrt{7}).$$

Because $\sqrt{7} \approx 2.6$, the intersection points are *approximately*

$$(3.6, -1.6) \quad \text{and} \quad (-1.6, 3.6).$$

[See Figure 12.5 .] □

EXAMPLE 3 Show that the line given by

$$y = 2x$$

does not intersect the circle given by

$$(x - 2)^2 + (y + 4)^2 = 4$$

(a) by drawing the graphs on the same coordinate plane,
(b) by solving a system of equations.

FIGURE 12.6 The line given by $y = 2x$ does not intersect the circle given by $(x - 2)^2 + (y + 4)^2 = 4$.

Solution.

(a) See Figure 12.6 .

(b) Replace y by $2x$ in the equation of the circle:

$$(x - 2)^2 + (y + 4)^2 = 4$$
$$(x - 2)^2 + (2x + 4)^2 = 4$$
$$x^2 - 4x + 4 + 4x^2 + 16x + 16 = 4$$
$$5x^2 + 12x + 16 = 0$$
$$x = \frac{-12 \pm \sqrt{144 - 4(5)(16)}}{10}$$
$$= \frac{-12 \pm \sqrt{-176}}{10}$$

The solutions are not real because the discriminant, -176, is negative. There is no intersection point. □

ellipses and hyperbolas

EXAMPLE 4† Determine the intersections of the ellipse given by

$$\frac{x^2}{4} + \frac{y^2}{9} = 1$$

†The algebraic solution of Example 4 does *not* depend upon a knowledge of Section 6.11; the graphing of Figure 12.7 does, however.

with the hyperbola given by

$$\frac{x^2}{4} - \frac{y^2}{9} = 1.$$

Solution. "Add these equations":

$$\begin{aligned} \frac{x^2}{4} + \frac{y^2}{9} &= 1 \\ \frac{x^2}{4} - \frac{y^2}{9} &= 1 \\ \hline \frac{2x^2}{4} &= 2 \\ \frac{x^2}{4} &= 1 \\ x^2 &= 4 \\ x &= \pm 2 \end{aligned}$$

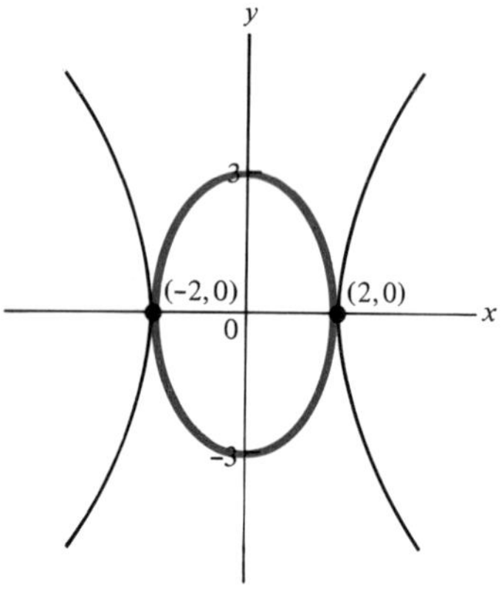

Replace x by 2, and then by -2 in the equation of the ellipse. Because only the second power of x appears, the corresponding y-values are the same.

$$\frac{4}{4} + \frac{y^2}{9} = 1 \qquad \text{Subtract } \frac{4}{4}\text{, or 1, from each side.}$$

$$\frac{y^2}{9} = 0$$

$$y = 0$$

The intersections are (2, 0) and (−2, 0). [See Figure 12.7 .] □

FIGURE 12.7

EXERCISES

In Exercises 1–18:

(a) Determine all intersections.
(b) Indicate what figure (line, parabola, circle) is represented by each equation.
(c) When called for, graph the figures on the same coordinate system to illustrate the points of intersection.

1. $x^2 + y^2 = 25$ and $x = 3$ (Graph.)

2. $y = 3x^2$ and $y = 5$ (Graph.)

3. $y = -8x^2$ and $y = x$

4. $x^2 + y^2 = 81$ and $y = 3x$

5. $x^2 + y^2 = 49$ and $y = 7 - x$ (Graph.)

6. $y = 4x^2$ and $y = 2x$

7. $x = -2y^2$ and $y = -x$

8. $x^2 + y^2 = 9$ and $y = 3 - x$ (Graph.)

9. $x^2 + y^2 = 100$ and $y = 10 - 3x$

10. $y = -x^2$ and $y = 2$ (Graph.)

11. $y = -x^2$ and $x + y = 0$ (Graph.)

12. $x^2 + y^2 = 2$ and $x = y^2$

13. $x^2 + y^2 = 2$ and $x = -y^2$

14. $x^2 + y^2 = 16$ and $x^2 + y^2 = 25$ (Graph.)

15. $(x + 2)^2 + (y - 3)^2 = 1$ and $y = 2x$

16. $(x + 4)^2 + (y + 1)^2 = 25$ and $y = 4 - x$

17. $x^2 + y^2 = 4$ and $(x - 4)^2 + y^2 = 4$ (Graph.) **18.** $x^2 = 5y$ and $x^2 + y^2 = 6$

In Exercises 19–30†

(a) Determine all intersections.
(b) Indicate what figure (line, parabola, circle, ellipse, hyperbola) is represented by each equation.
(c) When called for, graph the figures on the same coordinate system to illustrate the points of intersection.

19. $\frac{x^2}{4} + y^2 = 1$ and $x = 1$ **20.** $x^2 + \frac{y^2}{9} = 1$ and $y = 3$ **21.** $x^2 - y^2 = 1$ and $x = -1$ (Graph.)

22. $x^2 - \frac{y^2}{4} = 1$ and $y = x$ **23.** $\frac{x^2}{16} + \frac{y^2}{9} = 1$ and $y = -x$ **24.** $\frac{x^2}{9} - \frac{y^2}{25} = 1$ and $y = 5x$

25. $x^2 + y^2 = 4$ and $\frac{x^2}{9} + \frac{y^2}{16} = 1$ (Graph.) **26.** $x^2 + y^2 = 16$ and $\frac{x^2}{16} + \frac{y^2}{9} = 1$ (Graph.)

27. $\frac{x^2}{64} - \frac{y^2}{36} = 1$ and $\frac{x^2}{64} + \frac{y^2}{36} = 1$ **28.** $\frac{y^2}{9} - \frac{x^2}{16} = 1$ and $\frac{x^2}{16} + \frac{y^2}{9} = 1$

29. $\frac{x^2}{16} + \frac{y^2}{16} = 1$ and $(x - 2)^2 + y^2 = 36$ (Graph.) **30.** $\frac{x^2}{25} + \frac{y^2}{9} = 1$ and $x^2 - y^2 = 1$

*12.4 2 × 2 Determinants and Cramer's Rule

determinants

"Determinants" enable you to solve systems of *linear* equations mechanically.

Definition. Let $a, b, c,$ and d be any four real numbers. The symbol

$$\begin{vmatrix} a & c \\ b & d \end{vmatrix}$$

is called a **2-by-2 (2 × 2) determinant** with **entries** a, b, c, d; its **value** is

$$ad - bc.$$

EXAMPLE 1 Evaluate each 2 × 2 determinant:

(a) $\begin{vmatrix} 4 & 1 \\ 1 & 2 \end{vmatrix}$ (b) $\begin{vmatrix} 5 & 0 \\ 2 & -1 \end{vmatrix}$ (c) $\begin{vmatrix} 3 & 6 \\ 4 & 8 \end{vmatrix}$

Solution.

(a) Here $a = 4$, $b = c = 1$, $d = 2$. Thus

$$\begin{vmatrix} 4 & 1 \\ 1 & 2 \end{vmatrix} = 4(2) - 1(1) = 7$$

†In Exercises 19–30, part (a) does *not* depend upon a knowledge of Section 6.11; parts (b) and (c) do.

* Optional topic

(b) $$\begin{vmatrix} 5 & 0 \\ 2 & -1 \end{vmatrix} = 5(-1) - 2(0) = -5$$

(c) $$\begin{vmatrix} 3 & 6 \\ 4 & 8 \end{vmatrix} = 3(8) - 4(6) = 0$$

□

EXAMPLE 2 Determine c if $\begin{vmatrix} 5 & c \\ 3 & 4 \end{vmatrix} = 23$.

Solution.

$$\begin{vmatrix} 5 & c \\ 3 & 4 \end{vmatrix} = 5(4) - 3c = 20 - 3c$$

Thus solve $20 - 3c = 23$ and obtain:

$$\begin{aligned} -3c &= 3 \\ c &= -1 \end{aligned}$$

□

Cramer's Rule

In order to determine a mechanical procedure for solving a system of two linear equations in two variables, observe the following.

$$\begin{vmatrix} a_1 & b_1 \\ a_2 & b_2 \end{vmatrix} = a_1b_2 - a_2b_1 \tag{12.2}$$

$$\begin{vmatrix} k_1 & b_1 \\ k_2 & b_2 \end{vmatrix} = k_1b_2 - k_2b_1 \tag{12.3}$$

Now consider two linear equations in two variables:

$$a_1x + b_1y = k_1 \quad \textbf{(A)}$$
$$a_2x + b_2y = k_2 \quad \textbf{(B)}$$

To eliminate y, multiply both sides of **(A)** by b_2, and both sides of **(B)** by $-b_1$.

$$\begin{aligned} a_1b_2x + b_1b_2y &= k_1b_2 \\ -a_2b_1x - b_1b_2y &= -k_2b_1 \end{aligned}$$

Add:

$$(a_1b_2 - a_2b_1)x = k_1b_2 - k_2b_1$$

According to **(12.2)** and **(12.3)**, the preceding equation can be expressed in terms of determinants:

$$\begin{vmatrix} a_1 & b_1 \\ a_2 & b_2 \end{vmatrix} x = \begin{vmatrix} k_1 & b_1 \\ k_2 & b_2 \end{vmatrix}$$

If the left-hand determinant is not zero,

$$x = \frac{\begin{vmatrix} k_1 & b_1 \\ k_2 & b_2 \end{vmatrix}}{\begin{vmatrix} a_1 & b_1 \\ a_2 & b_2 \end{vmatrix}}$$

Observe that the determinant in the *denominator* has as its entries the coefficients of x and y as they appear on the left side of the original equations. Call this determinant D:

$$\begin{matrix} \textcircled{a_1} x + \textcircled{b_1} y \\ a_2 x + b_2 y \end{matrix} \rightarrow \begin{vmatrix} a_1 & b_1 \\ a_2 & b_2 \end{vmatrix} = D$$

The determinant of the *numerator,* called D_x, is obtained from D by replacing the a's (the coefficients of x) by the corresponding k's:

$$D_x = \begin{vmatrix} k_1 & b_1 \\ k_2 & b_2 \end{vmatrix} \qquad \text{so that} \qquad x = \frac{D_x}{D}, \qquad \text{if } D \neq 0$$

Let D_y be the 2 × 2 determinant obtained from D by replacing the b's (the coefficients of y) by the corresponding k's. Thus

$$D_y = \begin{vmatrix} a_1 & k_1 \\ a_2 & k_2 \end{vmatrix}$$

Similarly, it can be shown that

$$y = \frac{\begin{vmatrix} a_1 & k_1 \\ a_2 & k_2 \end{vmatrix}}{\begin{vmatrix} a_1 & b_1 \\ a_2 & b_2 \end{vmatrix}},$$

provided D, the determinant in the denominator, is not zero. The determinant in the numerator is D_y. Thus, when $D \neq 0$,

$$y = \frac{D_y}{D}$$

In this case *the system has the unique solution* (x, y), namely,

$$x = \frac{D_x}{D}, \qquad y = \frac{D_y}{D}$$

This solution by determinants is known as **Cramer's Rule.** △

EXAMPLE 3 Solve by Cramer's Rule:

$$\begin{aligned} 10x - 7y &= 12 \\ 3x - 2y &= 5 \end{aligned}$$

Solution.

$$D = \begin{vmatrix} 10 & -7 \\ 3 & -2 \end{vmatrix} = 10(-2) - 3(-7) = 1$$

$$D_x = \begin{vmatrix} 12 & -7 \\ 5 & -2 \end{vmatrix} = 12(-2) - 5(-7) = 11$$

$$D_y = \begin{vmatrix} 10 & 12 \\ 3 & 5 \end{vmatrix} = 10(5) - 3(12) = 14$$

$D \neq 0$; hence, Cramer's Rule applies:

$$x = \frac{D_x}{D} = \frac{11}{1} = 11 \qquad \text{and} \qquad y = \frac{D_y}{D} = \frac{14}{1} = 14$$

The unique solution is (11, 14). □

If $D = 0$, then the system has either no solution (geometrically, parallel lines, as in Figure 12.1(a) on page 457) or infinitely many solutions (both equations represent the same line, as in Figure 12.1(c)). Cramer's Rule, which involves division by D, does not apply.

EXAMPLE 4 Show that Cramer's Rule does not apply to the system:

$$\begin{aligned} 3x + 5y &= 1 \\ 6x + 10y &= 4 \end{aligned}$$

Solution.

$$D = \begin{vmatrix} 3 & 5 \\ 6 & 10 \end{vmatrix} = 3(10) - 5(6) = 0$$

Cramer's Rule only applies when $D \neq 0$. □

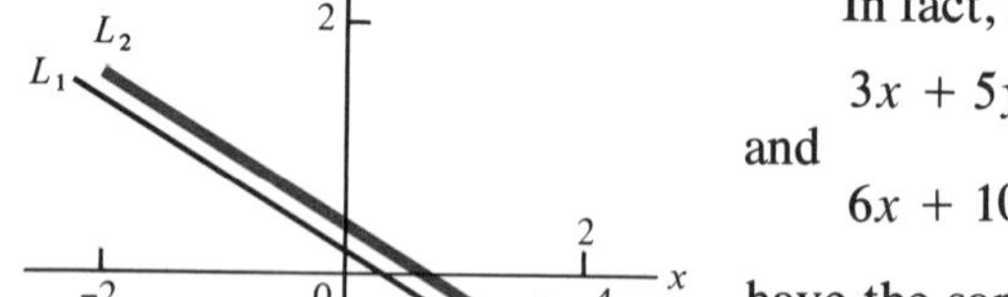

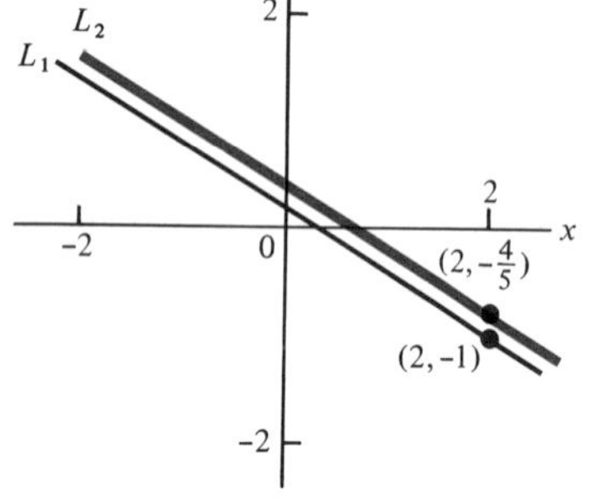

FIGURE 12.8
L_1 is given by $3x + 5y = 1$.
L_2 is given by $6x + 10y = 4$.
$L_1 \parallel L_2$

In fact, the two lines represented by equations

$$3x + 5y = 1 \tag{A}$$

and

$$6x + 10y = 4 \tag{B}$$

have the same slope, $\frac{-3}{5}$, but are not identical. The point $(2, -1)$ is on the line given by (**A**) because

$$3 \cdot 2 + 5(-1) = 1;$$

however, $(2, -1)$ is not on the line given by (**B**) because

$$6 \cdot 2 + 10(-1) \neq 4.$$

Thus the lines given by Equations (**A**) and (**B**) are parallel. [See Figure 12.8 .]

Note that the left side of Equation (**B**) "is a multiple of" the left side of Equation (**A**). In general, the lines given by the equations

$$ax + by = c \qquad \text{and} \qquad kax + kby = kc$$

(a and b not both 0, and $k \neq 0$) are identical. For example,

$$3x + 5y = 1 \qquad \text{and} \qquad 6x + 10y = 2$$

represent the same line. But the lines represented by the equations

$$ax + by = c \quad \text{and} \quad kax + kby = lc, \quad k \neq l, \quad k \neq 0, \quad l \neq 0$$

are parallel. Thus

$$3x + 5y = 1 \qquad \text{and} \qquad 6x + 10y = 4$$

represent parallel lines. [See Figure 12.8.]

EXERCISES

In Exercises 1–14 evaluate each determinant. In Exercises 13 and 14, x can be any real number.

1. $\begin{vmatrix} 1 & 2 \\ 1 & 1 \end{vmatrix}$ **2.** $\begin{vmatrix} 3 & 1 \\ 4 & 1 \end{vmatrix}$ **3.** $\begin{vmatrix} 2 & 0 \\ 1 & 3 \end{vmatrix}$ **4.** $\begin{vmatrix} 5 & 2 \\ 1 & 0 \end{vmatrix}$ **5.** $\begin{vmatrix} 0 & 0 \\ 0 & 1 \end{vmatrix}$ **6.** $\begin{vmatrix} 2 & 4 \\ 4 & -1 \end{vmatrix}$

7. $\begin{vmatrix} 1 & -3 \\ -2 & 4 \end{vmatrix}$ **8.** $\begin{vmatrix} 5 & -9 \\ -7 & -4 \end{vmatrix}$ **9.** $\begin{vmatrix} -2 & -9 \\ 9 & \frac{1}{2} \end{vmatrix}$ **10.** $\begin{vmatrix} 6 & \frac{1}{2} \\ 8 & \frac{1}{3} \end{vmatrix}$ **11.** $\begin{vmatrix} \frac{1}{2} & \frac{2}{5} \\ \frac{1}{4} & \frac{-3}{5} \end{vmatrix}$ **12.** $\begin{vmatrix} .1 & .02 \\ -.4 & .01 \end{vmatrix}$

13. $\begin{vmatrix} \sin x & \cos x \\ -\cos x & \sin x \end{vmatrix}$ **14.** $\begin{vmatrix} \cos x & \sin x \\ 1 & \tan x \end{vmatrix}$

In Exercises 15 and 16, evaluate the following "determinants" with *complex* entries.

15. $\begin{vmatrix} 1 & i \\ i & 1 \end{vmatrix}$ **16.** $\begin{vmatrix} 1+i & 2-i \\ 2+i & 1-i \end{vmatrix}$

In Exercises 17–24 determine x.

17. $\begin{vmatrix} x & 5 \\ 2 & 1 \end{vmatrix} = 0$ **18.** $\begin{vmatrix} 3 & x \\ 8 & 1 \end{vmatrix} = -5$ **19.** $\begin{vmatrix} 2 & x \\ 9 & 1 \end{vmatrix} = 11$ **20.** $\begin{vmatrix} 4 & 3 \\ x & 2 \end{vmatrix} = -13$

21. $\begin{vmatrix} 5 & 2 \\ x & x \end{vmatrix} = 27$ **22.** $\begin{vmatrix} x & -x \\ -4 & 1 \end{vmatrix} = 5$ **23.** $\begin{vmatrix} x & 3 \\ x & x \end{vmatrix} = 0$ **24.** $\begin{vmatrix} x & x \\ -1 & x \end{vmatrix} = 5$

In Exercises 25 and 26, suppose $0 \leq x < 2\pi$. Determine x.

25. $\begin{vmatrix} \sin x & 2 \\ 1 & 2 \end{vmatrix} = 0$ **26.** $\begin{vmatrix} \sin x & -\cos x \\ \sin x & \cos x \end{vmatrix} = 1$

In Exercises 27–30 let a, b, c, d, and k be any real numbers.

27. Evaluate: $\begin{vmatrix} a & a \\ b & b \end{vmatrix}$ **28.** Evaluate: $\begin{vmatrix} a & c \\ ka & kc \end{vmatrix}$ **29.** Compare $\begin{vmatrix} a & c \\ b & d \end{vmatrix}$ with $\begin{vmatrix} c & a \\ d & b \end{vmatrix}$

30. Compare $\begin{vmatrix} a & c \\ b & d \end{vmatrix}$ with $\begin{vmatrix} b & d \\ a & c \end{vmatrix}$.

In Exercises 31–44: (a) Determine whether Cramer's Rule applies. (b) If it applies, use Cramer's Rule to find the solution.

31. $\begin{aligned} x + y &= 5 \\ x - y &= 1 \end{aligned}$ **32.** $\begin{aligned} 2x + y &= 1 \\ 3x + y &= 2 \end{aligned}$ **33.** $\begin{aligned} x - y &= 0 \\ x + 2y &= 1 \end{aligned}$ **34.** $\begin{aligned} 6x - 3y &= 0 \\ x + 2y &= 5 \end{aligned}$

35. $\begin{aligned} x + 3y &= 1 \\ 3x + 9y &= 2 \end{aligned}$ **36.** $\begin{aligned} 5x - 3y &= 10 \\ x - y &= -4 \end{aligned}$ **37.** $\begin{aligned} 2x + 3y &= -10 \\ x + 6y &= 1 \end{aligned}$ **38.** $\begin{aligned} 4x - 6y &= 8 \\ -2x + 3y &= -4 \end{aligned}$

39. $\begin{aligned} 5x + 3y &= 3 \\ 3x + 5y &= 5 \end{aligned}$ **40.** $\begin{aligned} 2x - 7y &= 3 \\ 9x + 2y &= 47 \end{aligned}$ **41.** $\begin{aligned} 15x + 10y &= 25 \\ 3x \quad &= 10 \end{aligned}$ **42.** $\begin{aligned} 9x - 2y &= 3 \\ -4y &= 1 \end{aligned}$

43. $\begin{aligned} \frac{x}{2} + \frac{y}{3} &= \frac{1}{6} \\ 2x + 3y &= 4 \end{aligned}$ **44.** $\begin{aligned} \frac{x}{4} + \frac{y}{2} &= \frac{1}{2} \\ \frac{x}{2} - \frac{y}{4} &= \frac{1}{2} \end{aligned}$

*12.5 3 × 3 Determinants and Cramer's Rule

determinants 3-by-3 (3 × 3) determinants are defined in terms of 2 × 2 determinants.

* Optional topic

Definition. Let $a_1, a_2, a_3, b_1, b_2, b_3, c_1, c_2$, and c_3 be any nine real numbers. The symbol

$$\begin{vmatrix} a_1 & b_1 & c_1 \\ a_2 & b_2 & c_2 \\ a_3 & b_3 & c_3 \end{vmatrix}$$

is called a **3-by-3 (3 × 3) determinants;** its **entries** are the preceding nine numbers; its **value** is

$$a_1\begin{vmatrix} b_2 & c_2 \\ b_3 & c_3 \end{vmatrix} - a_2\begin{vmatrix} b_1 & c_1 \\ b_3 & c_3 \end{vmatrix} + a_3\begin{vmatrix} b_1 & c_1 \\ b_2 & c_2 \end{vmatrix},$$

which equals

$$a_1(b_2c_3 - b_3c_2) - a_2(b_1c_3 - b_3c_1) + a_3(b_1c_2 - b_2c_1).$$

Because the value of each 2×2 determinant is a real number, the resulting expression is a real number. Note that the signs preceding the 2×2 determinants alternate: $+ \quad - \quad +$

The **rows** and **columns** of a 3×3 determinant are as indicated:

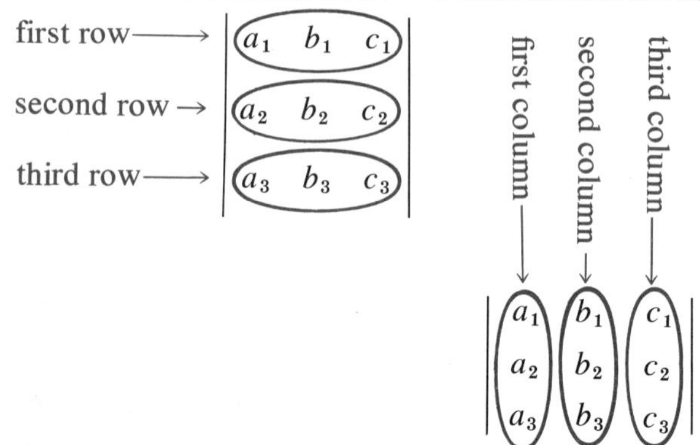

Each 2×2 determinant is obtained from the 3×3 determinant by dropping the appropriate row along with the first column. Thus, to obtain the *second* 2×2 determinant

$$\begin{vmatrix} b_1 & c_1 \\ b_3 & c_3 \end{vmatrix},$$

drop the *second* row and first column:

$$\begin{vmatrix} a_1 & b_1 & c_1 \\ a_2 & b_2 & c_2 \\ a_3 & b_3 & c_3 \end{vmatrix} \rightarrow \begin{vmatrix} b_1 & c_1 \\ b_3 & c_3 \end{vmatrix}$$

EXAMPLE 1

$$\begin{vmatrix} 2 & 1 & 4 \\ 6 & 1 & 2 \\ 1 & 2 & 1 \end{vmatrix} = 2\begin{vmatrix} 2 & 1 & 4 \\ 6 & 1 & 2 \\ 1 & 2 & 1 \end{vmatrix} - 6\begin{vmatrix} 2 & 1 & 4 \\ 6 & 1 & 2 \\ 1 & 2 & 1 \end{vmatrix} + 1\begin{vmatrix} 2 & 1 & 4 \\ 6 & 1 & 2 \\ 1 & 2 & 1 \end{vmatrix}$$

$$= 2\begin{vmatrix} 1 & 2 \\ 2 & 1 \end{vmatrix} - 6\begin{vmatrix} 1 & 4 \\ 2 & 1 \end{vmatrix} + 1\begin{vmatrix} 1 & 4 \\ 1 & 2 \end{vmatrix}$$

$$= 2[1(1) - 2(2)] - 6[1(1) - 2(4)] + 1[1(2) - 1(4)]$$

$$= 2(-3) - 6(-7) + (-2)$$

$$= 34$$

□

EXAMPLE 2

$$\begin{vmatrix} 4 & 1 & 0 \\ 0 & 2 & -8 \\ -2 & -1 & -1 \end{vmatrix} = 4\begin{vmatrix} 2 & -8 \\ -1 & -1 \end{vmatrix} - \underbrace{0\begin{vmatrix} 1 & 0 \\ -1 & -1 \end{vmatrix}}_{0} + (-2)\begin{vmatrix} 1 & 0 \\ 2 & -8 \end{vmatrix}$$

$$= 4[2(-1) - (-1)(-8)] - 2[1(-8) - 2(0)]$$
$$= 4(-10) - 2(-8)$$
$$= -24$$

□

EXAMPLE 3 Determine x:

$$\begin{vmatrix} 2 & 5 & 0 \\ x & 1 & 4 \\ 3 & -1 & 1 \end{vmatrix} = 100$$

Solution.

$$\begin{vmatrix} 2 & 5 & 0 \\ x & 1 & 4 \\ 3 & -1 & 1 \end{vmatrix} = 2\begin{vmatrix} 1 & 4 \\ -1 & 1 \end{vmatrix} - x\begin{vmatrix} 5 & 0 \\ -1 & 1 \end{vmatrix} + 3\begin{vmatrix} 5 & 0 \\ 1 & 4 \end{vmatrix}$$
$$= 2(1 + 4) - x(5 + 0) + 3(20 - 0)$$
$$= 10 - 5x + 60$$
$$= 70 - 5x$$

Thus solve the equation

$$70 - 5x = 100$$

and obtain:

$$-5x = 30$$
$$x = -6$$

□

Cramer's Rule

Just as 2 × 2 determinants were used to solve systems of two linear equations, 3 × 3 determinants are used for systems of three linear equations in three variables.

Consider the system:

$$a_1x + b_1y + c_1z = k_1$$
$$a_2x + b_2y + c_2z = k_2$$
$$a_3x + b_3y + c_3z = k_3$$

Let D be the 3 × 3 determinant whose entries are the coefficients of x, y, and z as they appear on the left sides of these equations:

$$D = \begin{vmatrix} a_1 & b_1 & c_1 \\ a_2 & b_2 & c_2 \\ a_3 & b_3 & c_3 \end{vmatrix}$$

Let D_x be the determinant obtained from D by replacing the a's (the coefficients of x) by the corresponding k's:

$$D_x = \begin{vmatrix} k_1 & b_1 & c_1 \\ k_2 & b_2 & c_2 \\ k_3 & b_3 & c_3 \end{vmatrix}$$

Similarly, replace the b's by the corresponding k's to obtain D_y, and replace the c's by the corresponding k's to obtain D_z:

$$D_y = \begin{vmatrix} a_1 & k_1 & c_1 \\ a_2 & k_2 & c_2 \\ a_3 & k_3 & c_3 \end{vmatrix}, \qquad D_z = \begin{vmatrix} a_1 & b_1 & k_1 \\ a_2 & b_2 & k_2 \\ a_3 & b_3 & k_3 \end{vmatrix}$$

Note that here D, D_x, D_y, and D_z are all 3×3 determinants.

As with systems of two linear equations, formulas for x, y, and z can be obtained in terms of determinants. **Cramer's Rule** *for systems of three linear equations states that if* $D \neq 0$, *then the system has a unique solution* (x, y, z), in which

$$x = \frac{D_x}{D}, \qquad y = \frac{D_y}{D}, \qquad z = \frac{D_z}{D}.$$

If $D = 0$, the system does not have a *unique* solution. In fact, in this case either there are no solutions or there are infinitely many.

EXAMPLE 4 Solve by Cramer's Rule:

$$\begin{aligned} 2x + 4y - 3z &= 1 \\ 5x - y + z &= 6 \\ y + z &= 5 \end{aligned}$$

Check your solution.

Solution.

$$\begin{aligned} D &= \begin{vmatrix} 2 & 4 & -3 \\ 5 & -1 & 1 \\ 0 & 1 & 1 \end{vmatrix} \\ &= 2\begin{vmatrix} -1 & 1 \\ 1 & 1 \end{vmatrix} - 5\begin{vmatrix} 4 & -3 \\ 1 & 1 \end{vmatrix} + 0\begin{vmatrix} 4 & -3 \\ -1 & 1 \end{vmatrix} \\ &= 2[-1(1) - 1(1)] - 5[4(1) - 1(-3)] + 0 \\ &= 2(-2) - 5(7) \\ &= -39 \end{aligned}$$

Because $D \neq 0$, Cramer's Rule applies.

$$\begin{aligned} D_x &= \begin{vmatrix} 1 & 4 & -3 \\ 6 & -1 & 1 \\ 5 & 1 & 1 \end{vmatrix} \\ &= 1\begin{vmatrix} -1 & 1 \\ 1 & 1 \end{vmatrix} - 6\begin{vmatrix} 4 & -3 \\ 1 & 1 \end{vmatrix} + 5\begin{vmatrix} 4 & -3 \\ -1 & 1 \end{vmatrix} \\ &= 1[-1(1) - 1(1)] - 6[4(1) - 1(-3)] + 5[4(1) - (-1)(-3)] \\ &= 1(-2) - 6(7) + 5(1) \\ &= -39 \end{aligned}$$

$$\begin{aligned} D_y &= \begin{vmatrix} 2 & 1 & -3 \\ 5 & 6 & 1 \\ 0 & 5 & 1 \end{vmatrix} \\ &= 2\begin{vmatrix} 6 & 1 \\ 5 & 1 \end{vmatrix} - 5\begin{vmatrix} 1 & -3 \\ 5 & 1 \end{vmatrix} + 0\begin{vmatrix} 1 & -3 \\ 6 & 1 \end{vmatrix} \end{aligned}$$

$$= 2[6(1) - 5(1)] - 5[1(1) - 5(-3)] + 0$$
$$= 2(1) - 5(16)$$
$$= -78$$

$$D_z = \begin{vmatrix} 2 & 4 & 1 \\ 5 & -1 & 6 \\ 0 & 1 & 5 \end{vmatrix}$$
$$= 2\begin{vmatrix} -1 & 6 \\ 1 & 5 \end{vmatrix} - 5\begin{vmatrix} 4 & 1 \\ 1 & 5 \end{vmatrix} + 0\begin{vmatrix} 4 & 1 \\ -1 & 6 \end{vmatrix}$$
$$= 2[-1(5) - 1(6)] - 5[4(5) - 1(1)] + 0$$
$$= 2(-11) - 5(19)$$
$$= -117$$

Thus by Cramer's Rule,

$$x = \frac{D_x}{D} = \frac{-39}{-39} = 1, \qquad y = \frac{D_y}{D} = \frac{-78}{-39} = 2,$$
$$z = \frac{D_z}{D} = \frac{-117}{-39} = 3$$

The solution is (1, 2, 3).

Check. Replace x by **1**, y by **2**, and z by **3** in the original equations:

$2(1) + 4(2) - 3(3) \stackrel{?}{=} 1$	$5(1) - 2 + 3 \stackrel{?}{=} 6$	$2 + 3 \stackrel{?}{=} 5$
$1 \stackrel{\checkmark}{=} 1$	$6 \stackrel{\checkmark}{=} 6$	$5 \stackrel{\checkmark}{=} 5$

All three statements are true; the solution checks. □

EXAMPLE 5 Show that the system

$$\begin{aligned} 2x - \ \ y + 3z &= 1 \\ 5x + 2y - \ \ z &= 0 \\ 7x + \ \ y + 2z &= 4 \end{aligned}$$

does *not* have a unique solution.

Solution.

$$D = \begin{vmatrix} 2 & -1 & 3 \\ 5 & 2 & -1 \\ 7 & 1 & 2 \end{vmatrix}$$
$$= 2\begin{vmatrix} 2 & -1 \\ 1 & 2 \end{vmatrix} - 5\begin{vmatrix} -1 & 3 \\ 1 & 2 \end{vmatrix} + 7\begin{vmatrix} -1 & 3 \\ 2 & -1 \end{vmatrix}$$
$$= 2[2(2) - 1(-1)] - 5[-1(2) - 1(3)] + 7[-1(-1) - 2(3)]$$
$$= 2(5) - 5(-5) + 7(-5)$$
$$= 0$$

When $D = 0$, the system does not have a unique solution. □

As you might expect, 4×4 determinants can be defined in terms of 3×3 determinants, just as 3×3 determinants have been defined in terms of 2×2's. Then 5×5 determinants can be defined in terms of 4×4's, 6×6's in terms of 5×5's, and so on. And Cramer's Rule can be extended to systems of four (or more) linear equations in the same number of variables.

EXERCISES

In Exercises 1–12 evaluate each determinant.

1. $\begin{vmatrix} 5 & 1 & 1 \\ 2 & 0 & 1 \\ 1 & 0 & 1 \end{vmatrix}$ **2.** $\begin{vmatrix} 3 & 8 & 1 \\ 0 & 1 & 0 \\ 1 & 0 & 1 \end{vmatrix}$ **3.** $\begin{vmatrix} 1 & -2 & 1 \\ 2 & 1 & 1 \\ 1 & 0 & 1 \end{vmatrix}$ **4.** $\begin{vmatrix} 1 & 3 & 1 \\ 2 & 1 & 2 \\ -1 & 1 & 0 \end{vmatrix}$ **5.** $\begin{vmatrix} 6 & 1 & 0 \\ 0 & 1 & 0 \\ 1 & 2 & 0 \end{vmatrix}$

6. $\begin{vmatrix} 0 & 0 & 9 \\ 1 & 0 & 2 \\ 1 & 1 & 3 \end{vmatrix}$ **7.** $\begin{vmatrix} 2 & 1 & 4 \\ 1 & 3 & 1 \\ 1 & -2 & 3 \end{vmatrix}$ **8.** $\begin{vmatrix} 1 & -1 & 1 \\ -1 & 1 & -1 \\ 1 & -1 & 1 \end{vmatrix}$ **9.** $\begin{vmatrix} 0 & 2 & 4 \\ 0 & 6 & 2 \\ 3 & 1 & 0 \end{vmatrix}$ **10.** $\begin{vmatrix} 2 & 5 & 3 \\ 1 & 8 & -4 \\ -3 & 2 & 3 \end{vmatrix}$

11. $\begin{vmatrix} 1 & .6 & .2 \\ 0 & .9 & .1 \\ 1 & .5 & .1 \end{vmatrix}$ **12.** $\begin{vmatrix} \frac{1}{2} & \frac{1}{2} & 0 \\ 1 & -1 & \frac{1}{4} \\ 0 & 1 & \frac{1}{2} \end{vmatrix}$

In Exercises 13–16 determine x.

13. $\begin{vmatrix} x & 0 & 0 \\ 0 & 2 & 0 \\ 0 & 0 & 3 \end{vmatrix} = 12$ **14.** $\begin{vmatrix} x & 7 & -1 \\ 0 & 5 & 1 \\ 0 & 0 & 2 \end{vmatrix} = 40$ **15.** $\begin{vmatrix} x & 0 & 0 \\ 4 & 2 & 1 \\ 5 & 2 & 2 \end{vmatrix} = 10$ **16.** $\begin{vmatrix} 0 & 7 & 2 \\ x & 3 & 6 \\ 0 & 4 & 1 \end{vmatrix} = 12$

In Exercises 17–20 let $a_1, a_2, a_3, b_1, b_2, b_3, c_1, c_2, c_3$, and k be any real numbers.

17. Evaluate: $\begin{vmatrix} a_1 & b_1 & b_1 \\ a_2 & b_2 & b_2 \\ a_3 & b_3 & b_3 \end{vmatrix}$ **18.** Evaluate: $\begin{vmatrix} a_1 & b_1 & c_1 \\ 0 & 0 & 0 \\ a_3 & b_3 & c_3 \end{vmatrix}$

19. Compare $\begin{vmatrix} ka_1 & b_1 & c_1 \\ ka_2 & b_2 & c_2 \\ ka_3 & b_3 & c_3 \end{vmatrix}$ with $\begin{vmatrix} a_1 & b_1 & c_1 \\ a_2 & b_2 & c_2 \\ a_3 & b_3 & c_3 \end{vmatrix}$ **20.** Compare $\begin{vmatrix} a_1 & c_1 & b_1 \\ a_2 & c_2 & b_2 \\ a_3 & c_3 & b_3 \end{vmatrix}$ with $\begin{vmatrix} a_1 & b_1 & c_1 \\ a_2 & b_2 & c_2 \\ a_3 & b_3 & c_3 \end{vmatrix}$

In Exercises 21–32: (a) Determine whether Cramer's Rule applies. (b) If it applies, use Cramer's Rule to find the solution.

21. $\begin{aligned} x + y \phantom{{}+z} &= 2 \\ x \phantom{{}+y} - z &= 0 \\ 2x \phantom{{}+y} + z &= 3 \end{aligned}$ **22.** $\begin{aligned} 5x \phantom{{}-3y} + z &= 6 \\ y + z &= 1 \\ x - 3y + z &= 2 \end{aligned}$ **23.** $\begin{aligned} x + 2y \phantom{{}-z} &= 3 \\ y - z &= 2 \\ 5x + 3y \phantom{{}-z} &= 1 \end{aligned}$

24. $\begin{aligned} 3x - y \phantom{{}-z} &= 10 \\ 2y + z &= 9 \\ x + y - z &= 1 \end{aligned}$ **25.** $\begin{aligned} x \phantom{{}-3y} + z &= 0 \\ 2x - 3y + z &= 6 \\ x - y - z &= 12 \end{aligned}$ **26.** $\begin{aligned} 3x + 2y - 3z &= 7 \\ 2x - 5y - 5z &= 5 \\ y + z &= 1 \end{aligned}$

27. $\begin{aligned} 2x + 7y + z &= 0 \\ 5x + 6y \phantom{{}+z} &= 1 \\ 3x - y - z &= 0 \end{aligned}$ **28.** $\begin{aligned} 5x - y + 2z &= 5 \\ 3y - 4z &= 0 \\ 2x \phantom{{}-y} + 4z &= 2 \end{aligned}$ **29.** $\begin{aligned} x + 2y + z &= 1 \\ 5x - y \phantom{{}-z} &= 2 \\ 4x - 3y - z &= 2 \end{aligned}$

30. $\begin{aligned} x + y + 2z &= 0 \\ 3y - z &= 0 \\ 2x - y + 5z &= 0 \end{aligned}$

31. $\begin{aligned} x - y \quad &= 0 \\ 2x - y - z &= 0 \\ x - 3y - 2z &= 4 \end{aligned}$

32. $\begin{aligned} 5x + y - 2z &= 4 \\ -x \quad + z &= -1 \\ -10x - 2y + 4z &= 0 \end{aligned}$

Review Exercises for Chapter 12

1. Determine the intersections of L_1 and L_2 graphically.

L_1: $2x - 3y = -4$
L_2: $x + y = 3$

2. Solve by either substituting or adding:

(a) $\begin{aligned} 4x + 3y &= 1 \\ x + 2y &= -1 \end{aligned}$ **(b)** $\begin{aligned} 3x + 5y &= 4 \\ 2x + 3y &= 2 \end{aligned}$ **(c)** $\begin{aligned} 3x + y &= 4 \\ 3x - 2y &= -8 \end{aligned}$

3. Determine a and b so that the line given by $ax + by + 1 = 0$ passes through the point (1, 1) and has slope $\frac{2}{3}$.

4. Solve each system. You may use any methods.

(a) $\begin{aligned} x \quad + z &= 0 \\ 2x - y \quad &= 0 \\ x + y + z &= 2 \end{aligned}$ **(b)** $\begin{aligned} 2x - y + 4z &= 5 \\ x + y - z &= 4 \\ 3x \quad - z &= 1 \end{aligned}$ **(c)** $\begin{aligned} 7x + y + z &= 2 \\ 3x - 4y + 2z &= 3 \\ 6x + 3y + z &= 7 \end{aligned}$

5. The sum of three numbers is 31. The largest of these numbers is 6 more than the smallest. The sum of the two smaller numbers is 5 more than the largest number. Find these numbers.

6. Find real numbers A, B, C, D so that the given rational expression can be rewritten as the indicated sum.

(a) $\dfrac{3x + 7}{(x + 2)(x + 3)} = \dfrac{A}{x + 2} + \dfrac{B}{x + 3}$ **(b)** $\dfrac{x - 1}{x^3 - x^2 - 2x} = \dfrac{A}{x} + \dfrac{B}{x + 1} + \dfrac{C}{x - 2}$

(c) $\dfrac{x^2}{x^4 - 1} = \dfrac{A}{x + 1} + \dfrac{B}{x - 1} + \dfrac{Cx + D}{x^2 + 1}$

7. **(a)** Determine all intersections.

(b) Indicate what figure (line, circle, parabola, ellipse, hyperbola) is represented by each equation.

(c) When called for, graph the figures on the same coordinate system to illustrate the points of intersection.

i. $x^2 + y^2 = 36$ and $y = x$ (Graph.) **ii.** $y = 4x^2$ and $y = 2x$

iii. $\dfrac{x^2}{9} + \dfrac{y^2}{4} = 1$ and $x^2 + y^2 = 4$ (Graph.)

8. Evaluate each determinant.

(a) $\begin{vmatrix} 2 & 0 \\ 1 & 1 \end{vmatrix}$ **(b)** $\begin{vmatrix} 4 & -1 \\ 2 & 3 \end{vmatrix}$ **(c)** $\begin{vmatrix} 1 & -5 \\ -2 & -1 \end{vmatrix}$

9. Determine x: $\begin{vmatrix} 2 & 1 \\ 1 & x \end{vmatrix} = 3$

10. **(a)** Determine whether Cramer's Rule applies. **(b)** If it applies, use Cramer's Rule to find the solution.

i. $\begin{aligned} 3x + y &= 4 \\ x - 2y &= -8 \end{aligned}$ **ii.** $\begin{aligned} 3x + 2y &= 5 \\ 6x + 4y &= 9 \end{aligned}$ **iii.** $\begin{aligned} 2x - y &= 1 \\ x + y &= 2 \end{aligned}$

11. Evaluate each determinant.

(a) $\begin{vmatrix} 2 & 0 & 1 \\ 0 & 1 & 1 \\ 0 & -1 & 0 \end{vmatrix}$ **(b)** $\begin{vmatrix} 1 & 3 & 0 \\ 1 & 2 & -1 \\ -1 & 1 & 2 \end{vmatrix}$

12. Determine x: $\begin{vmatrix} 0 & 0 & x \\ 7 & 1 & 2 \\ 1 & 1 & -1 \end{vmatrix} = 30$

13. **(a)** Determine whether Cramer's Rule applies.

(b) If it applies, use Cramer's Rule to find the solution.

i. $\begin{aligned} x + y + z &= 1 \\ 2x \quad\quad + z &= 3 \\ y - z &= 1 \end{aligned}$ **ii.** $\begin{aligned} x + y \quad\quad &= 5 \\ y - z &= 0 \\ 2x + y + z &= 10 \end{aligned}$

13 Progressions

13.1 Sequences

A *sequence* can be thought of as an "infinite list". Examples of sequences are

1, 3, 5, 7, 9, 11, . . .
2, 4, 6, 8, 10, 12, . . .
10, 20, 30, 40, 50, 60, . . .
1, 2, 4, 8, 16, 32, . . .
−1, −2, −3, −4, −5, −6, . . .

As usual, the 3 dots are read: "and so on"

Here is a more precise definition.

Definition. A **sequence** is a function whose domain is the set of positive integers.

The letter "a" is usually used to denote a sequence. Instead of writing the function values of a sequence as $a(1)$, $a(2)$, $a(3)$, and so on, it is customary to write:

$$a_1, a_2, a_3, \ldots, a_i, \ldots$$

In each place the three dots signify "and so on". Thus a sequence consists of infinitely many pairs:

$$(1, a_1), \quad (2, a_2), \quad (3, a_3), \ldots$$

Definition. The function values of a sequence,

$$a_1, a_2, \ldots, a_i, \ldots$$

are called the **terms** of the sequence. a_1 is called the **first term,** a_2 is the **second term,** a_i is the ***i*th term.**

Throughout the Chapter (with one exception), *i will refer to the* ith *term of a sequence, and NOT to* $\sqrt{-1}$. The only exception to this is in Exercise 18 on p. 515.

A sequence is often defined by indicating the ith term a_i, as in Example 1.

EXAMPLE 1 Consider the sequence defined by $a_i = i - 4$. Its first 8 terms are given by

$$\begin{array}{lll} a_1 = 1 - 4 = -3 & a_2 = 2 - 4 = -2 & a_3 = 3 - 4 = -1 \\ a_4 = 4 - 4 = 0 & a_5 = 5 - 4 = 1 & a_6 = 6 - 4 = 2 \\ a_7 = 7 - 4 = 3 & a_8 = 8 - 4 = 4 & \end{array}$$

Also,

$$a_{99} = 99 - 4 = 95 \quad \text{and} \quad a_{100} = 100 - 4 = 96$$ □

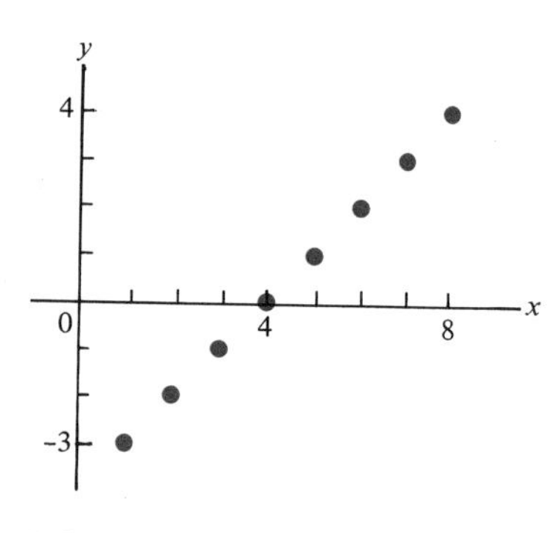

FIGURE 13.1 The graph of the sequence given by $a_i = i - 4$

A sequence is a function and can therefore be graphed. The graph consists of the points

$$(1, a_1), \quad (2, a_2), \quad (3, a_3), \ldots$$

corresponding to the pairs of the sequence. Of course, only finitely many points can be shown in a graph. But often, the general picture can be suggested. Figure 13.1 depicts the graph of the sequence of Example 1, given by $a_i = i - 4$.

EXAMPLE 2 (a) Determine the first 8 terms of the sequence given by

$$a_i = 5 - \frac{i}{2}.$$

(b) Graph this sequence.

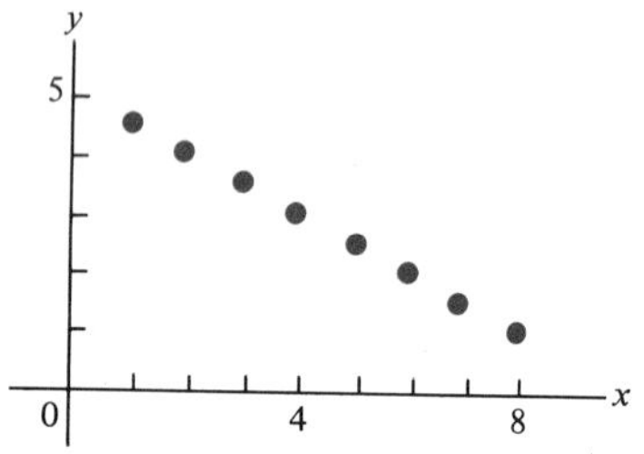

Solution.

(a) $a_1 = 5 - \frac{1}{2} = 4.5 \qquad a_2 = 5 - \frac{2}{2} = 4 \qquad a_3 = 5 - \frac{3}{2} = 3.5$

$a_4 = 5 - \frac{4}{2} = 3 \qquad a_5 = 2.5 \qquad a_6 = 2$

$a_7 = 1.5 \qquad a_8 = 1$

(b) See Figure 13.2 . □

FIGURE 13.2 The graph of the sequence given by $a_i = 5 - \frac{i}{2}$

increasing sequences

Definition. A sequence $a_1, a_2, a_3, \ldots, a_i, \ldots$ is said to be **increasing** if

$$a_i < a_{i+1}$$

for *all* $i = 1, 2, \ldots$, and **decreasing** if

$$a_i > a_{i+1}$$

for *all* $i = 1, 2, \ldots$

Thus for an increasing sequence, the terms get *larger* as the subscripts i get *larger*. For a decreasing sequence, the terms get *smaller* as the subscripts get *larger*.

EXAMPLE 3 (a) The sequences $a_i = i$ and $a_i = \frac{i}{2}$ are increasing.

(b) The sequence $a_i = 5 - \frac{i}{2}$ of Example 2 is decreasing. □

Here is a sequence that is neither increasing nor decreasing.

EXAMPLE 4 Consider the sequence given by $a_i = (-1)^i$.

$$a_1 = (-1)^1 = -1 \qquad a_2 = (-1)^2 = 1$$
$$a_3 = (-1)^3 = -1 \qquad a_4 = (-1)^4 = 1$$
$$a_5 = (-1)^5 = -1 \qquad a_6 = (-1)^6 = 1$$
$$a_7 = (-1)^7 = -1 \qquad a_8 = (-1)^8 = 1$$

□

In general,

$$a_i = -1, \text{ if } i \text{ is } odd; \qquad a_i = 1, \text{ if } i \text{ is } even$$

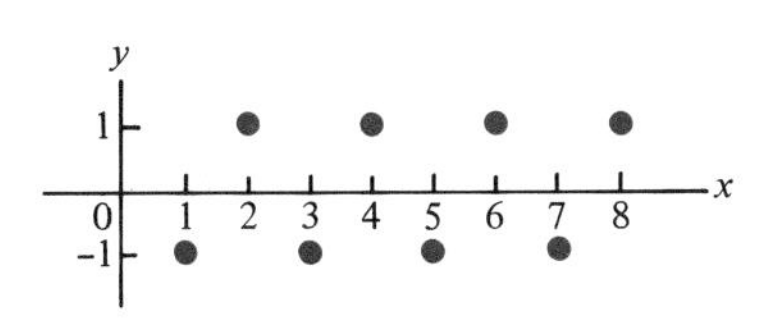

FIGURE 13.3 The graph of the sequence given by $a_i = (-1)^i$

The sequence is also given by:

$$-1, 1, -1, 1, -1, 1, -1, 1, \ldots$$

Observe that $a_3 < a_2$. Therefore the sequence is *not* increasing. Also, $a_2 > a_1$. Therefore the sequence is *not* decreasing. [See Figure 13.3 .]

EXAMPLE 5 Describe the ith term a_i of the suggested sequences.

(a) $5, 10, 15, 20, 25, 30, \ldots$

(b) $4, 3, 2, 1, 0, -1, -2, \ldots$

(c) $-2, 4, -8, 16, -32, 64, \ldots$

(d) $\frac{1}{2}, \frac{1}{4}, \frac{1}{8}, \frac{1}{16}, \frac{1}{32}, \frac{1}{64}, \ldots$

Solution.

(a) Let $a_i = 5i$ for $i = 1, 2, \ldots$

(b) Let $a_i = 5 - i$ for $i = 1, 2, \ldots$
Then $a_1 = 5 - 1 = 4$, $a_2 = 5 - 2 = 3$, $a_3 = 5 - 3 = 2$, and so on.

(c) 2^i would yield $2, 4, 8, 16, \ldots$ To get alternating signs, take $a_i = (-2)^i$, for $i = 1, 2, 3, \ldots$

(d) Let $a_i = \left(\frac{1}{2}\right)^i$, or $\frac{1}{2^i}$, for $i = 1, 2, \ldots$ □

EXERCISES

In Exercises 1–20:

(a) Determine the first 8 terms of the indicated sequence.
(b) Is the sequence i. increasing, ii. decreasing, iii. neither?
(c) Graph the sequence, if so indicated.

1. $a_i = i$ (Graph.)

2. $a_i = i + 2$ (Graph.)

3. $a_i = -i$ (Graph.)

4. $a_i = 10 - i$ (Graph.)

5. $a_i = 2i$ (Graph.)

6. $a_i = -2i$ (Graph.)

7. $a_i = 3i$

8. $a_i = \frac{-i}{3}$

9. $a_i = 5$, for all i

10. $a_i = -1$, for all i (Graph.)

11. $a_i = \begin{cases} 5, & \text{if } i \text{ is odd,} \\ 3, & \text{if } i \text{ is even.} \end{cases}$

12. $a_i = \begin{cases} 1, & \text{if } i \text{ is odd,} \\ 0, & \text{if } i \text{ is even.} \end{cases}$ (Graph.)

13. $a_i = \frac{1}{i}$

14. $a_i = 1 - \frac{1}{i}$

15. $a_i = 2^i$

16. $a_i = \left(\frac{-1}{2}\right)^i$

17. $a_i = 6 + 3i$ **18.** $a_i = 5 - 2i$ **19.** $a_i = 1 + \frac{i}{2}$ **20.** $a_i = -1 + \frac{i}{2}$

In Exercises 21–32 for the indicated sequences, determine the specified terms.

21. $a_i = 5i$

(a) a_4 (b) a_{10} (c) a_{20} (d) a_{100}

22. $a_i = -10i$

(a) a_1 (b) a_{12} (c) a_{100} (d) a_{200}

23. $a_i = 3 + 4i$

(a) a_1 (b) a_5 (c) a_{10} (d) a_{20}

24. $a_i = 1 - 3i$

(a) a_1 (b) a_4 (c) a_7 (d) a_{10}

25. $a_i = 2 + \frac{5i}{2}$

(a) a_1 (b) a_4 (c) a_8 (d) a_9

26. $a_i = 3 - \frac{i}{3}$

(a) a_1 (b) a_3 (c) a_9 (d) a_{27}

27. $a_i = 1 + (-1)^i$

(a) a_1 (b) a_2 (c) a_{25} (d) a_{26}

28. $a_i = -4 + (-1)^i$

(a) a_1 (b) a_2 (c) a_{99} (d) a_{100}

29. $a_i = \frac{12}{i}$

(a) a_2 (b) a_6 (c) a_{12} (d) a_{48}

30. $a_i = \frac{-144}{i}$

(a) a_2 (b) a_9 (c) a_{16} (d) a_{72}

31. $a_i = \frac{i}{i+1}$

(a) a_1 (b) a_2 (c) a_9 (d) a_{10}

32. $a_i = \frac{3i - 1}{5 + i}$

(a) a_1 (b) a_3 (c) a_5 (d) a_{15}

In each of the Exercises 33–44, describe the ith term, a_i, of the suggested sequence.

33. 2, 3, 4, 5, 6, 7, . . .

34. −1, −2, −3, −4, −5, −6, . . .

35. 5, 6, 7, 8, 9, 10, . . .

36. −3, −2, −1, 0, 1, 2, . . .

37. 3, 6, 9, 12, 15, 18, . . .

38. −7, −14, −21, −28, −35, −42, . . .

39. −4, 8, −12, 16, −20, 24, . . .

40. 5, 7, 9, 11, 13, 15, . . .

41. 2, 5, 8, 11, 14, 17, . . .

42. 3, 9, 27, 81, 243, 729, . . .

43. $2, \frac{3}{2}, \frac{4}{3}, \frac{5}{4}, \frac{6}{5}, \frac{7}{6}, \ldots$

44. 1, 4, 9, 16, 25, 36, . . .

13.2 Sigma Notation

There is a compact way of indicating addition.

$\sum_{i=1}^{n} a_i$ ***Definition.*** Let $a_1, a_2, a_3, \ldots, a_n$ be any n real numbers. Define

$$\sum_{i=1}^{n} a_i \quad \text{to stand for} \quad a_1 + a_2 + a_3 + \cdots + a_n.$$

Read

$$\sum_{i=1}^{n} a_i \quad \text{as} \quad \text{summation } a_i, \quad i = 1 \text{ to } n$$

In this notation, the symbol Σ [a stylized Greek letter Σ (sigma)] stands for "summation". Moreoever,

$$\sum_{i=1}^{n} a_i$$

indicates that the terms a_i are to be added, beginning with a_1 and ending with a_n. The variable subscript i, which ranges over the integers 1, 2, 3, . . . , n, is called the **index of summation.**

EXAMPLE 1 Let $a_1 = 4$, $a_2 = 3$, $a_3 = -2$, $a_4 = 1$, $a_5 = 5$, $a_6 = 3$. Then

$$\sum_{i=1}^{6} a_i = a_1 + a_2 + a_3 + a_4 + a_5 + a_6$$
$$= 4 + 3 + (-2) + 1 + 5 + 3$$
$$= 14$$

□

$\sum_{i=0}^{n} a_i$

Sometimes, the index of summation ranges over the integers 0, 1, 2, 3, . . . , n, instead of 1, 2, 3, . . . , n.

Definition. Let $a_0, a_1, a_2, a_3, \ldots, a_n$ be any $(n + 1)$ real numbers. Define

$$\sum_{i=0}^{n} a_i \qquad \text{to stand for} \qquad a_0 + a_1 + a_2 + a_3 + \cdots + a_n$$

EXAMPLE 2 Let $a_0 = 4$, $a_1 = -3$, $a_2 = 5$, $a_3 = -1$, $a_4 = 2$, $a_5 = 6$, $a_6 = 0$, $a_7 = 12$, $a_8 = -10$. Then

$$\sum_{i=0}^{8} a_i = a_0 + a_1 + a_2 + a_3 + a_4 + a_5 + a_6 + a_7 + a_8$$
$$= 4 + (-3) + 5 + (-1) + 2 + 6 + 0 + 12 + (-10)$$
$$= 15$$

□

$\sum_{i=k}^{n} a_i$

This "sigma notation" (or Σ-notation) can be generalized.

Definition. Let

$$a_1, a_2, a_3, \ldots, a_i, \ldots$$

be a sequence. For any integers k and n, where $1 \leq k \leq n$, define

$$\sum_{i=k}^{n} a_i \qquad \text{to stand for} \qquad a_k + a_{k+1} + a_{k+2} + \cdots + a_n.$$

Here the index of summation i ranges over the integers k, $k + 1$, $k + 2$, . . . , n. Note that if $k = 1$, the original definition applies.

EXAMPLE 3 Consider the sequence given by $a_i = 3i$.

(a) $\sum_{i=1}^{10} a_i = a_1 + a_2 + a_3 + \cdots + a_{10} = 3 + 6 + 9 + \cdots + 30$

Here $k = 1$ and $n = 10$. There are 10 terms to be added.

(b) $\sum_{i=2}^{15} a_i = a_2 + a_3 + a_4 + \cdots + a_{15} = 6 + 9 + 12 + \cdots + 45$

Here $k = 2$ and $n = 15$. The index of summation ranges over the integers 2, 3, 4, . . . , 15. There are 14 terms to be added.

(c) $\sum_{i=6}^{33} a_i = a_6 + a_7 + a_8 + \cdots + a_{33} = 18 + 21 + 24 + \cdots + 99$

Here $k = 6$ and $n = 33$. How many terms are to be added?

(d) To express the single term a_k in sigma notation, write:

$$\sum_{i=k}^{k} a_i$$

For example, for the above sequence,

$$\sum_{i=12}^{12} a_i = a_{12} = 36$$

□

In the next section you will learn a systematic way of determining the sums of Example 3.

rules for sigma notation

For any sequence, $a_1, a_2, a_3, \ldots, a_i, \ldots$ and for any integers k and n, where $1 \leq k \leq n$,

$$\sum_{i=1}^{n} a_i = \sum_{i=1}^{k} a_i + \sum_{i=k+1}^{n} a_i$$

(The sum from a_1 to a_n equals the sum from a_1 to a_k plus the sum from a_{k+1} to a_n.) In fact,

$$\sum_{i=1}^{n} a_i = a_1 + a_2 + a_3 + \cdots + a_k + a_{k+1} + a_{k+2} + \cdots + a_n$$

$$= (a_1 + a_2 + a_3 + \cdots + a_k) + (a_{k+1} + a_{k+2} + \cdots + a_n)$$

$$= \sum_{i=1}^{k} a_i + \sum_{i=k+1}^{n} a_i$$

△

Thus also,

$$\sum_{i=k+1}^{n} a_i = \sum_{i=1}^{n} a_i - \sum_{i=1}^{k} a_i$$

The notation $\sum_{i=1}^{n} a_i$ is often modified. Suppose, for example, that $a_i = 3i$, for $i = 1, 2, \ldots$. Then you can write

$$\sum_{i=1}^{n} 3i \quad \text{for} \quad \sum_{i=1}^{n} a_i$$

EXAMPLE 4 Consider the sequence of Example 3, given by $a_i = 3i$.

As you will see in the next section,

$$\sum_{i=1}^{99} a_i = \sum_{i=1}^{99} 3i = 14\,850 \qquad \text{and} \qquad \sum_{i=1}^{9} a_i = \sum_{i=1}^{9} 3i = 135$$

Therefore

$$\sum_{i=10}^{99} a_i = \sum_{i=1}^{99} a_i - \sum_{i=1}^{9} a_i = 14\,850 - 135 = 14\,715$$

In other words,

$$30 + 33 + 36 + \cdots + 297 = 14\,715$$

□

EXAMPLE 5 Write in Σ-notation:

$$3 + 5 + 7 + 9 + \cdots + 21$$

Solution. Find the (suggested) sequence whose first few terms are

$$3, 5, 7, 9, \ldots, 21.$$

This sequence is given by $a_i = 2i + 1$. Thus

$$a_1 = 2 \cdot 1 + 1 = 3 \qquad a_2 = 2 \cdot 2 + 1 = 5 \qquad a_3 = 2 \cdot 3 + 1 = 7$$

$$\ldots$$

$$a_{10} = 2 \cdot 10 + 1 = 21$$

and

$$\sum_{i=1}^{10} a_i = \sum_{i=1}^{10} (2i + 1) = 3 + 5 + 7 + \cdots + 21$$

□

You can *add* or *subtract* in Σ-notation. You can also *multiply* the terms *by a constant*.

Let $a_1, a_2, \ldots, a_n, b_1, b_2, \ldots, b_n$ be any numbers. Then

$$\sum_{i=1}^{n} (a_i + b_i) = \sum_{i=1}^{n} a_i + \sum_{i=1}^{n} b_i$$

In fact, by the Associative and Commutative Laws,

$$\begin{aligned}\sum_{i=1}^{n} (a_i + b_i) &= (a_1 + b_1) + (a_2 + b_2) + (a_3 + b_3) \\ &\quad + \cdots + (a_n + b_n) \\ &= (a_1 + a_2 + \cdots + a_n) + (b_1 + b_2 + \cdots + b_n) \\ &= \sum_{i=1}^{n} a_i + \sum_{i=1}^{n} b_i\dagger\end{aligned}$$

△

Also, for any number c,

$$\sum_{i=1}^{n} ca_i = c \sum_{i=1}^{n} a_i$$

† The proof of this and the following rule actually require Mathematical Induction. See Section 13.6 .

For, by the Distributive Laws,

$$\sum_{i=1}^{n} ca_i = ca_1 + ca_2 + ca_3 + \cdots + ca_n$$
$$= c(a_1 + a_2 + a_3 + \cdots + a_n)$$
$$= c \cdot \sum_{i=1}^{n} a_i$$

△

EXAMPLE 6 Consider the sequences given by

$$a_i = 4i \quad \text{and} \quad b_i = i + 3.$$

For each i, $4i + (i + 3) = 5i + 3$. Thus

$$\sum_{i=1}^{50} (5i + 3) = \sum_{i=1}^{50} [4i + (i + 3)] = \sum_{i=1}^{50} 4i + \sum_{i=1}^{50} (i + 3)$$
$$= 4\sum_{i=1}^{50} i + \sum_{i=1}^{50} (i + 3)$$

□

modifying the index of summation In calculus and in linear algebra it is often convenient to modify the index of summation. Observe that

$$\sum_{i=1}^{3} (i + 1) = 2 + 3 + 4 = \sum_{i=2}^{4} i.$$

This illustrates that if $a_i = i + 1$, you could let $a_{i-1} = i$ and then

$$\sum_{i=1}^{3} a_i = \sum_{i=1+1}^{3+1} a_{i-1}$$

In general, for any sequence whose general term is a_i and for any integer m,

$$\sum_{i=1}^{n} a_i = \sum_{i=1+m}^{n+m} a_{i-m} \quad \text{and} \quad \sum_{i=1}^{n} a_{i+m} = \sum_{i=1+m}^{n+m} a_i$$

EXAMPLE 7 $\frac{1}{6} + \frac{1}{7} + \frac{1}{8} + \frac{1}{9}$ can be expressed both as $\sum_{i=1}^{4} \frac{1}{i + 5}$ and as $\sum_{i=6}^{9} \frac{1}{i}$.

Here if $m = 5$, $n = 4$, and $a_i = \frac{1}{i + 5}$, then

$$\sum_{i=1}^{4} a_i = \sum_{i=1+5}^{4+5} a_{i-5} = \sum_{i=6}^{9} a_{i-5},$$

that is,

$$\sum_{i=1}^{4} \frac{1}{i + 5} = \sum_{i=1+5}^{4+5} \frac{1}{(i + 5) - 5} = \sum_{i=6}^{9} \frac{1}{i}$$

□

EXERCISES

In Exercises 1–10 express each in Σ-notation.

1. $a_1 + a_2 + a_3 + a_4$

2. $a_1 + a_2 + a_3 + \cdots + a_{30}$

3. $b_0 + b_1 + b_2 + b_3 + \cdots + b_{11}$

4. $c_1 + c_2 + c_3 + \cdots + c_{2000}$ **5.** $a_1 + a_2$ **6.** b_1 **7.** $a_5 + a_6 + a_7 + \cdots + a_{18}$

8. $b_{19} + b_{20} + b_{21}$ **9.** $c_{101} + c_{102} + c_{103} + \cdots + c_{200}$ **10.** a_{93}

In Exercises 11–20 find the numerical value of each sum.

11. $\sum_{i=1}^{6} i$ **12.** $\sum_{i=1}^{5} 2i$ **13.** $\sum_{i=0}^{6} (i + 2)$ **14.** $\sum_{i=1}^{4} 5i$ **15.** $\sum_{i=1}^{5} (3i + 2)$ **16.** $\sum_{i=1}^{6} i^2$

17. $\sum_{i=1}^{3} \frac{i}{i+1}$ **18.** $\sum_{i=0}^{4} 2^i$ **19.** $\sum_{i=4}^{10} (i + 3)$ **20.** $\sum_{i=7}^{10} (5i - 4)$

In Exercises 21–32 write in Σ-notation.

21. $3 + 4 + 5 + \cdots + 12$ **22.** $-1 + 0 + 1 + 2 + \cdots + 17$

23. $4 + 8 + 12 + 16 + \cdots + 64$ **24.** $-2 - 4 - 6 - 8 - \cdots - 34$

25. $1 + 4 + 9 + 16 + \cdots + 100$ **26.** $1^3 + 2^3 + 3^3 + 4^3 + \cdots + 20^3$

27. $\frac{1}{3} + \frac{1}{4} + \frac{1}{5} + \frac{1}{6} + \cdots + \frac{1}{92}$ **28.** $\frac{1}{2} + \frac{2}{3} + \frac{3}{4} + \frac{4}{5} + \cdots + \frac{29}{30}$

29. $4 + 6 + 8 + 10 + \cdots + 42$ **30.** $3 + 8 + 13 + 18 + \cdots + 103$

31. $\frac{1}{2} + \frac{1}{4} + \frac{1}{8} + \frac{1}{16} + \cdots + \frac{1}{256}$ **32.** $\frac{3}{2} + \frac{5}{4} + \frac{7}{8} + \frac{9}{16} + \cdots + \frac{17}{256}$

In Exercises 33–40 determine the indicated number.

33. If $7\sum_{i=1}^{60} a_i = \sum_{i=1}^{60} ca_i$, find c. **34.** If $\sum_{i=1}^{20} a_i + \sum_{i=21}^{40} a_i = \sum_{i=1}^{n} a_i$, find n.

35. If $-\sum_{i=1}^{28} a_i = \sum_{i=1}^{28} ca_i$, find c. **36.** If $\sum_{i=1}^{40} a_i = 200$ and $a_{41} = -7$, find $\sum_{i=1}^{41} a_i$.

37. If $\sum_{i=1}^{54} a_i = \sum_{i=1}^{k} a_i + \sum_{i=26}^{54} a_i$, find k. **38.** If $\sum_{i=3}^{10} b_i = \sum_{i=1}^{k} b_i - b_1 - b_2$, find k.

39. If $\sum_{i=1}^{64} 5i = \sum_{i=1}^{64} ci + 2\sum_{i=1}^{64} i$, find c. **40.** If $\sum_{i=1}^{17} a_i = \sum_{i=1}^{18} a_i$, find a_{18}.

In Exercises 41–50 determine m (and n).

41. $\sum_{i=1}^{0} a_i = \sum_{i=3}^{10} a_{i-m}$ **42.** $\sum_{i=1}^{12} a_{i+4} = \sum_{i=1+m}^{12+m} a_i$ **43.** $\sum_{i=1}^{100} a_i = \sum_{i=m}^{n} a_{i-100}$ **44.** $\sum_{i=1}^{40} a_{i+20} = \sum_{i=m}^{n} a_i$

45. $\sum_{i=1}^{5} \frac{1}{i+10} = \sum_{i=m}^{n} \frac{1}{i}$ **46.** $\sum_{i=1}^{8} \frac{1}{i} = \sum_{i=5}^{12} \frac{1}{i-m}$ **47.** $\sum_{i=7}^{10} 3i = \sum_{i=1}^{4} 3(i + m)$

48. $\sum_{i=7}^{10} 3i = \sum_{i=1}^{4} (3i + m)$ **49.** $\sum_{i=0}^{6} 2^{i+2} = \sum_{i=m}^{n} 2^i$ **50.** $\sum_{i=2}^{9} 3^{i-m} = \sum_{i=1}^{8} 3^i$

In Exercises 51–54 express the sum using + symbols.

51. $\sum_{i=5}^{10} (i - 4)$ **52.** $\sum_{i=12}^{15} \frac{i}{i+1}$ **53.** $\sum_{i=3}^{7} 3^i$ **54.** $\sum_{i=4}^{8} (2i + 1)$

In Exercises 55–60, replace a_i by an appropriate expression.

55. $\frac{1}{3} + \frac{1}{4} + \frac{1}{5} = \sum_{i=3}^{5} a_i$ **56.** $\frac{1}{3} + \frac{1}{4} + \frac{1}{5} = \sum_{i=1}^{3} a_i$ **57.** $7 + 8 + 9 + 10 = \sum_{i=7}^{10} a_i$

58. $7 + 8 + 9 + 10 = \sum_{i=1}^{4} a_i$ **59.** $\frac{1}{4} + \frac{1}{8} + \frac{1}{16} + \frac{1}{32} = \sum_{i=2}^{5} a_i$ **60.** $\frac{1}{4} + \frac{1}{8} + \frac{1}{16} + \frac{1}{32} = \sum_{i=1}^{4} a_i$

13.3 Arithmetic Progressions

In some of the sequences

$$a_1, a_2, a_3, \ldots, a_i, \ldots$$

that you have seen, each new term (other than the first) is obtained from the preceding one by adding a constant—that is, $a_{i+1} - a_i$ is *the same* for all i. For example, each of the sequences

3, 6, 9, 12, 15, . . .
(differences 3, 3, 3, 3)
3, 5, 7, 9, 11, . . .
(differences 2, 2, 2, 2)
1, 6, 11, 16, 21, 26 . . .
(differences 5, 5, 5, 5, 5)

has this property.

Definition. An **arithmetic progression** is a sequence

$$a_1, a_2, a_3, \ldots, a_i, \ldots$$

in which the difference $a_{i+1} - a_i$ is the same for $i = 1, 2, 3, \ldots$. In an arithmetic progression, $a_{i+1} - a_i$ is called the **common difference,** and is usually denoted by d.

$$d = a_{i+1} - a_i \quad \text{or} \quad a_{i+1} = a_i + d$$

Thus you add the common difference to a_i to obtain the next term, a_{i+1}. Here are some further examples of *arithmetic progressions.*

EXAMPLE 1 (a) $6, 12, 18, 24, \ldots, 6i, \ldots$
The first term is 6 and the common difference is 6.

(b) $7, 11, 15, 19, \ldots, 4i + 3, \ldots$
The first term is 7 and the common difference is 4.

(c) $1, -1, -3, -5, \ldots, -2i + 3, \ldots$
The first term is 1 and the common difference is -2.

(d) $5, 5, 5, 5, \ldots, 5, \ldots$
The first term is 5 and the common difference is 0. □

EXAMPLE 2 A sequence $a_1, a_2, a_3, \ldots, a_i, \ldots$ whose first five terms are $1, 3, 4, 8, 9$ is not an arithmetic progression because

$$a_2 - a_1 = 3 - 1 = 2, \quad \text{whereas} \quad a_3 - a_2 = 4 - 3 = 1.$$

Thus $a_{i+1} - a_i$ is not the same for all i. □

In an arithmetic progression with first term a_1 and common difference d, the terms can be listed as

$$a_1,\ a_1 + d,\ a_1 + 2d,\ a_1 + 3d,\ \ldots,\ a_1 + (i - 1)d,\ \ldots$$

For, suppose the arithmetic progression is given by

$$a_1, a_2, a_3, \ldots, a_i, \ldots$$

Then

$$a_2 - a_1 = d. \quad \text{Thus } a_2 = a_1 + d$$
$$a_3 - a_2 = d. \quad \text{Thus } a_3 = a_2 + d = a_1 + 2d$$
$$a_4 - a_3 = d. \quad \text{Thus } a_4 = a_3 + d = a_1 + 3d$$

In general,

$$a_i = a_1 + (i - 1)d$$

△

EXAMPLE 3 Determine the 17th term of the arithmetic progression with first term 4 and common difference 3.

Solution.

$$a_1 = 4, \quad d = 3$$
$$a_{17} = a_1 + (17 - 1)d = 4 + 16 \cdot 3 = 52$$

□

increasing sequences

An arithmetic progression

$$a_1, \; a_1 + d, \; a_1 + 2d, \; \ldots, \; a_1 + (i - 1)d$$

is an increasing sequence if $d > 0$; *it is a decreasing sequence if* $d < 0$.

sum of an arithmetic progression

The formula that you next derive will enable you to find the sum of the first n terms of any arithmetic progression—even if n is very large.

Consider the sum of the first four terms of the arithmetic progression

$$5, 9, 13, 17, \ldots$$

(Obviously, you could find the sum quickly by adding the four numbers. But the following is the way the formula you will use is obtained.) Denote the sum by S and write it in two ways.

$$S = 5 + 9 + 13 + 17$$
$$S = 17 + 13 + 9 + 5$$
$$\text{Add:} \quad 2S = 22 + 22 + 22 + 22$$

or $2S = 22 \cdot 4 = 88.$ Thus

$$S = \frac{1}{2} \cdot 88 = 44$$

In general, suppose you want the sum of the first n terms of an arithmetic progression $a_1, a_2, a_3, \ldots$. Then you want to find $a_1 + a_2 + a_3 + \cdots + a_{n-2} + a_{n-1} + a_n$. Again denote the sum by S. Because the terms form an arithmetic progression, $a_2 = a_1 + d$ and $a_3 = a_1 + 2d$, where d is the common difference. Moreover, $a_{n-1} = a_n - d$ and $a_{n-2} = a_n - 2d$.

Thus write S in two ways:

$$S = a_1 + (a_1 + d) + (a_1 + 2d) + \cdots + (a_n - 2d) + (a_n - d) + a_n$$
$$S = a_n + (a_n - d) + (a_n - 2d) + \cdots + (a_1 + 2d) + (a_1 + d) + a_1$$

Add: $$2S = \underbrace{(a_1 + a_n) + (a_1 + a_n) + (a_1 + a_n) + \cdots + (a_1 + a_n) + (a_1 + a_n) + (a_1 + a_n)}_{n \text{ times}}$$

$$2S = n(a_1 + a_n)$$
$$S = \frac{1}{2}n(a_1 + a_n) \qquad \triangle$$

Because the sum of the first n terms can also be written as $\sum_{i=1}^{n} a_i$, the preceeding formula for S can be written as

$$\sum_{i=1}^{n} a_i = \frac{n}{2}(a_1 + a_n).\dagger \tag{13.1}$$

In words, *the sum of the first n terms of an arithmetic progression is half the number of terms times the sum of the first and last terms.*

If you rewrite **(13.1)** as $\sum_{i=1}^{n} a_i = n\left[\frac{1}{2}(a_1 + a_n)\right]$ and note that $\frac{1}{2}(a_1 + a_n)$ is the average of the first and last terms, you could say that *the sum is the number of terms times the average of the first and last terms.*

EXAMPLE 4 Express the sum of the first 99 terms of the arithmetic progression (of Examples 3 and 4 of Section 13.2):

$$3, 6, 9, 12, 15, \ldots, 3i, \ldots$$

Solution. Here $a_1 = 3$, $n = 99$, and $a_{99} = 3 \cdot 99 = 297$. From **(13.1)**,

$$\sum_{i=1}^{99} a_i = \frac{1}{2} \cdot 99(3 + 297) = \frac{1}{\cancel{2}} \cdot 99 \cdot \overset{150}{\cancel{300}} = 14\,850 \qquad \square$$

EXAMPLE 5 (a) Find the sum of the first 50 terms of the arithmetic progression with first term -3 and 50th term 340.

(b) Find the common difference for this progression.

Solution.

(a) You are given the first and last terms. Use Formula **(13.1)** [with $n = 50$]:

$$\sum_{i=1}^{50} a_i = \frac{50}{2}(a_1 + a_{50})$$

† A rigorous proof of this rule actually requires Mathematical Induction. See Section 13.6, and in particular, Exercise 22 on page 516.

Thus

$$\sum_{i=1}^{50} a_i = 25(-3 + 340) = 25 \cdot 337 = 8425$$

(b) $$a_n = a_1 + (n - 1)d$$

Therefore $\dfrac{a_n - a_1}{n - 1} = d$ and

$$d = \frac{340 - (-3)}{49} = \frac{343}{49} = 7$$

□

EXAMPLE 6 Consider the arithmetic progression given by $a_i = 3 + 5(i - 1)$. Determine $\sum_{i=101}^{200} a_i$.

Solution. Interpret **(13.1)** to mean that the sum is half the number of terms times the sum of the first and last terms. Here the first term, a_{101}, is $3 + 5(101 - 1)$, or 503. The last term, a_{200}, is $3 + 5(200 - 1)$, or 998. The number of terms is 100. (To see this, note that $\sum_{i=1}^{200} a_i$ has 200 terms and $\sum_{i=1}^{100} a_i$ has 100 terms. When you calculate $\sum_{i=101}^{200} a_i$, you can think of this as taking 200 terms, and then discarding the first 100 terms. Thus you start with the 101st term, and 100 terms remain.) Therefore the sum is $\frac{1}{2} \cdot 100(503 + 998)$, or 75 050.

□

EXERCISES

In each of the Exercises 1–14 a sequence is indicated by means of its first few terms. (Assume a regular pattern, as suggested.) Determine whether the sequence is an arithmetic progression. If so, find the common difference.

1. $4, 8, 12, 16, \ldots$

2. $3, 6, 12, 24, 48, \ldots$

3. $6, 8, 10, 12, 14, \ldots$

4. $-3, -1, 1, 3, 5, 7, \ldots$

5. $9, 6, 3, 0, -3, -6, -9, \ldots$

6. $100, 200, 300, 400, \ldots$

7. $1, 5, 11, 15, 21, 25, \ldots$

8. $2, 4, 8, 16, 32, 64, \ldots$

9. $1, 0, 1, 0, 1, 0, \ldots$

10. $13, 17, 21, 25, 29, 33, \ldots$

11. $1, -4, -9, -14, -19, \ldots$

12. $1, 1.5, 2, 2.5, 3, 3.5, \ldots$

13. $\frac{-1}{3}, \frac{-2}{3}, -1, \frac{-4}{3}, \frac{-5}{3}, -2, \frac{-7}{3}, \ldots$

14. $2, 2 + \pi, 2 + 2\pi, 2 + 3\pi, 2 + 4\pi, \ldots$

15. Determine the 10th term of the arithmetic progression: $4, 9, 14, 19, 24, \ldots$

16. Determine the 17th term of the arithmetic progression: $-3, 1, 5, 9, 13, \ldots$

17. Determine the 80th term of the arithmetic progression: $2, -1, -4, -7, -10, \ldots$

18. Determine the 25th and 50th terms of the arithmetic progression: $-15, -7, 1, 9, 17, 25, \ldots$

19. Determine the 30th term of the arithmetic progression with first term 2 and common difference 3.

20. Determine the 40th term of the arithmetic progression with first term 9 and common difference -2.

21. Determine the 20th term of the arithmetic progression with first term 5 and 10th term 95.

22. Determine the 18th term of the arithmetic progression with first term -4 and 20th term 53.

23. Determine the 33rd term of the arithmetic progression with 2nd term 9 and 4th term 13.

24. Determine the 9th term of the arithmetic progression with 16th term 50 and 20th term 38.

In Exercises 25–28 determine the sum of the first 20 terms of the indicated arithmetic progressions.

25. $7, 9, 11, 13, 15, 17, \ldots$

26. $4, 1, -2, -5, -8, -11, \ldots$

27. $8, 8.5, 9, 9.5, 10, 10.5, \ldots$

28. $3, -2, -7, -12, -17, -22, \ldots$

In Exercises 29–32 determine the sum of the first 40 terms of the indicated arithmetic progressions.

29. $5, 14, 23, 32, 41, 50, \ldots$

30. $2 + 6(i-1), i = 1, 2, 3, \ldots$

31. $-3 + 2i, i = 1, 2, 3, \ldots$

32. $a_1 = 4, d = 5$

In Exercises 33–36 find the sum of the first 100 terms of the arithmetic progression: $a_1, a_2, a_3, \ldots, a_i, \ldots$

33. $a_1 = 4, a_{100} = 301$

34. $a_1 = 9, a_{100} = 207$

35. $a_1 = 50, a_{100} = -148$

36. $a_1 = 3, a_4 = 21$

In Exercises 37–40 evaluate (a) $\sum_{i=4}^{12} a_i$, (b) $\sum_{i=10}^{20} a_i$ for the arithmetic progression: $a_1, a_2, \ldots, a_i, \ldots$

37. $a_i = 15 + 3(i-1), i = 1, 2, 3, \ldots$

38. $a_1 = 9, a_2 = 1$

39. $a_1 = 15, a_{10} = 60$

40. $a_1 = -7, a_8 = 21$

41. Find the sum of the first 50 even positive integers.

42. Find the sum of all odd integers between 5 and 95 (inclusive).

43. A 20-year-old man earns \$9000 per year. Each year he receives an increment of \$1200 per year. How much does he earn at age 30?

13.4 Geometric Progressions

You can also determine the sum of the first n terms of a sequence

$$a_1, a_2, a_3, \ldots a_i, \ldots$$

for which the *ratio* (or quotient)

$$\frac{a_{i+1}}{a_i}$$

is *the same* for all i. For example, each of the suggested sequences

$$2, 4, 8, 16, 32, \ldots \qquad \frac{a_{i+1}}{a_i} = 2$$

$$\frac{1}{2}, \frac{1}{4}, \frac{1}{8}, \frac{1}{16}, \frac{1}{32}, \ldots \qquad \frac{a_{i+1}}{a_i} = \frac{1}{2}$$

$$-3, 1, \frac{-1}{3}, \frac{1}{9}, \frac{-1}{27}, \frac{1}{81}, \ldots \qquad \frac{a_{i+1}}{a_i} = \frac{-1}{3}$$

has this property. In this kind of sequence, each new term (other than the first) is obtained by multiplying the preceding term by a constant.

Definition. A **geometric progression** is a sequence of nonzero terms

$$a_1, a_2, a_3, \ldots, a_i, \ldots$$

in which the ratio $\dfrac{a_{i+1}}{a_i}$ is the same for $i = 1, 2, 3, \ldots$. In a geometric progression, $\dfrac{a_{i+1}}{a_i}$ is called the **common ratio,** and is usually denoted by r.

$$r = \frac{a_{i+1}}{a_i} \neq 0$$

EXAMPLE 1 Each of the indicated sequences is a geometric progression:

(a) $1, 5, 25, 125, \ldots, 5^{i-1}, \ldots$
Note that for $i = 1$, $5^{i-1} = 5^0 = 1$. The first term is 1 and the common ratio is 5. To obtain a_{i+1}, multiply a_i by 5.

(b) $2, 6, 18, 54, \ldots, 2(3^{i-1}), \ldots$
The first term is 2 and the common ratio is 3.

(c) $2, -1, \frac{1}{2}, \frac{-1}{4}, \ldots, 2\left(\frac{-1}{2}\right)^{i-1}, \ldots$
The first term is 2 and the common ratio is $\frac{-1}{2}$. □

EXAMPLE 2 The sequence indicated by

$$2, 4, 6, 8, 10, \ldots$$

is not a geometric progression because

$$\frac{a_2}{a_1} = \frac{4}{2} = 2, \qquad \text{whereas} \qquad \frac{a_3}{a_2} = \frac{6}{4} = \frac{3}{2}.$$

(Note that the sequence is an arithmetic progression with common difference 2.) □

In a geometric progression with first term a_1 and common ratio r, the terms can be listed as

$$a_1, a_1r, a_1r^2, a_1r^3, \ldots, a_1r^{i-1}, \ldots$$

For, suppose the geometric progression is given by

$$a_1, a_2, a_3, \ldots, a_i, \ldots$$

Then

$$\frac{a_2}{a_1} = r. \qquad \text{Thus } a_2 = a_1 r$$

$$\frac{a_3}{a_2} = r. \qquad \text{Thus } a_3 = a_2 r = a_1 r^2$$

$$\frac{a_4}{a_3} = r. \qquad \text{Thus } a_4 = a_3 r = a_1 r^3$$

In general,

$$a_i = a_1 r^{i-1}$$

△

EXAMPLE 3 Determine the 7th term of the geometric progression with first term 5 and common ratio 2.

Solution.

$$a_1 = 5, r = 2$$

$$a_7 = 5(2^{7-1}) = 5 \cdot 64 = 320$$

□

increasing sequences

A geometric progression

$$a_1, a_1 r, a_1 r^2, a_1 r^3, \ldots, a_1 r^{i-1}, \ldots$$

is an increasing sequence if $r > 1$; *it is a decreasing sequence if* $0 < r < 1$; *it is neither increasing nor decreasing if* $r < 0$ *or if* $r = 1$.

EXAMPLE 4

(a) The geometric progression

$$2, 4, 8, 16, 32, \ldots$$

is an increasing sequence. Here $r = 2$

(b) The geometric progression

$$2, 1, \frac{1}{2}, \frac{1}{4}, \frac{1}{8}, \ldots$$

is a decreasing sequence. Here $r = \frac{1}{2}$

(c) The geometric progression

$$2, -4, 8, -16, 32, \ldots$$

is neither increasing nor decreasing. Here $r = -2$

(d) The geometric progression

$$2, 2, 2, 2, 2, \ldots$$

is neither increasing nor decreasing. Here $r = 1$. Notice that this is also an arithmetic progression with common difference 0.

□

sum of a geometric progression

Next, you derive the formula for the sum of the first n terms of a geometric progression.

To find the sum of the first four terms of the geometric progression

$$1, 3, 9, 27, \ldots,$$

it would be simplest just to add the four numbers. However, the following approach will show the method used to develop the general formula.

Let S represent the sum. Then

$$\begin{aligned} S &= 1 + 3 + 9 + 27 \\ 3S &= \quad\; 3 + 9 + 27 + 81 \\ \text{Subtract: } -2S &= 1 \qquad\qquad\quad -81 \\ -2S &= -80 \\ S &= 40 \end{aligned}$$

Multiply both sides by the common ratio, 3.

In general, suppose you want to add the first n terms of a geometric progression $a_1, a_1r, a_1r^2, \ldots$ (where $r \neq 1$). The nth term is a_1r^{n-1}. Denote the sum by S:

$$\begin{aligned} S &= a_1 + a_1r + a_1r^2 + \cdots + a_1r^{n-1} \\ rS &= \quad\; a_1r + a_1r^2 + \cdots + a_1r^{n-1} + a_1r^n \\ \text{Subtract: } S - rS &= a_1 - a_1r^n \\ S(1 - r) &= a_1(1 - r^n) \\ S &= \frac{a_1(1 - r^n)}{1 - r} \end{aligned}$$

Now isolate the common factor on each side.

△

Thus

$$S = \sum_{i=1}^{n} a_i = \frac{a_1(1 - r^n)}{1 - r}, \quad r \neq 1†$$

If $r = 1$, then $a_1 = a_2 = a_3 = \cdots = a_n$ and

$$\sum_{i=1}^{n} a_i = na_1$$

EXAMPLE 5 Determine the sum of the first 6 terms of the sequence

$$\frac{1}{2}, \frac{1}{4}, \frac{1}{8}, \ldots, \frac{1}{2^i}, \ldots$$

Solution. This is a geometric sequence with $a_1 = \frac{1}{2}$ and $r = \frac{1}{2}$. Therefore

$$\sum_{i=1}^{6} a_i = \frac{a_1(1 - r^6)}{1 - r} = \frac{\frac{1}{2}\left[1 - \left(\frac{1}{2}\right)^6\right]}{1 - \frac{1}{2}} = \frac{\frac{1}{2}\left(1 - \frac{1}{64}\right)}{\frac{1}{2}} = \frac{63}{64}$$

□

EXAMPLE 6 Determine the sum of the first 5 terms of the geometric sequence with first term 2 and fourth term 54.

† A rigorous proof of this rule actually requires Mathematical Induction. See Section 13.6, and in particular, Exercise 23 on page 516.

Solution.

$$a_1 = 2$$
$$a_4 = a_1 r^3$$
$$54 = 2r^3$$
$$27 = r^3$$
$$3 = r$$

$$\sum_{i=1}^{5} a_i = \frac{a_1(1 - r^5)}{1 - r} = \frac{2(1 - 3^5)}{1 - 3} = \frac{\cancel{2}(-242)}{-\cancel{2}_{1}} = 242 \qquad \square$$

EXERCISES

In each of the Exercises 1–14 a sequence is indicated by means of its first few terms. (Assume a regular pattern, as suggested.) Determine whether the sequence is a geometric progression. If so, find the common ratio. Also, indicate which sequences are arithmetic progressions.

1. 5, 10, 20, 40, 80, . . .

2. 1, 4, 16, 64, 256, . . .

3. 1, 4, 7, 10, 13, . . .

4. $-2, -4, -6, -8, -10, \ldots$

5. $4, 2, 1, \frac{1}{2}, \frac{1}{4}, \ldots$

6. $9, -3, 1, \frac{-1}{3}, \frac{1}{9}, \ldots$

7. $1, -1, 1, -1, 1, -1, \ldots$

8. $7, -7, 7, -7, 7, -7, \ldots$

9. $15, -30, 60, -120, 240, \ldots$

10. 1, .1, .01, .001, .0001, . . .

11. $25, 5, 1, \frac{1}{5}, \frac{1}{25}, \ldots$

12. $\frac{2}{3}, \frac{4}{3}, \frac{8}{3}, \frac{16}{3}, \frac{32}{3}, \ldots$

13. .2, .4, .8, .16, .32, .64, . . .

14. $6, 4, \frac{8}{3}, \frac{16}{9}, \frac{32}{27}, \ldots$

15. Determine the 4th term of the geometric progression with first term 1 and common ratio 5.

16. Determine the 6th term of the geometric progression with first term 5 and common ratio 2.

17. Determine the 7th term of the geometric progression with first term 12 and common ratio $\frac{1}{2}$.

18. Determine the 10th term of the geometric progression with first term $\frac{1}{16}$ and common ratio -2.

19. Determine the 5th term of the geometric progression with 2nd term 15 and common ratio:

(a) 3 **(b)** -3 **(c)** $\frac{1}{3}$ **(d)** $\frac{-1}{3}$

20. Determine the 7th term of the geometric progression with 3rd term 24 and common ratio 4.

21. Determine the 10th term of the geometric progression with 8th term 60 006 and common ratio 4.

22. Determine the 101st term of the geometric progression with 18th term -1 and common ratio -1.

23. Determine the first term of the geometric progression with 4th and 5th terms 108 and 324, respectively.

24. Determine the 2nd term of the geometric progression with 6th term $\frac{3}{16}$ and common ratio $\frac{1}{2}$.

In Exercises 25–42 evaluate the indicated sum.

25. $\sum_{i=1}^{7} 2^i$ **26.** $\sum_{i=1}^{5} 5^i$ **27.** $\sum_{i=1}^{8} (3\cdot 2^i)$ **28.** $\sum_{i=1}^{6} 3^i(-2)$ **29.** $\sum_{i=1}^{6} 4^{i-1}$

30. $\sum_{i=1}^{7} \frac{3}{2^{i-1}}$ **31.** $\sum_{i=1}^{6} (11\cdot 2^{i-1})$ **32.** $\sum_{i=1}^{5} \left(\frac{-5}{2}\right)^{i-1}$ **33.** $\sum_{i=1}^{9} 2^{i-3}$ **34.** $\sum_{i=1}^{9} \left(\frac{1}{2}\right)^{i-4}$

35. $\sum_{i=1}^{63} 9(-1)^i$ **36.** $\sum_{i=1}^{64} 9(-1)^i$ **37.** $\sum_{i=1}^{5} (-3)^i$ **38.** $\sum_{i=1}^{5} \left(\frac{-1}{3}\right)^i$ **39.** $\sum_{i=1}^{6} (.1)^i$

40. $\sum_{i=1}^{6} (.2)^i$ **41.** $\sum_{i=1}^{8} 2^{i-1}$ **42.** $\sum_{i=3}^{8} 2^{i-1}$

In Exercises 43 and 44, use the log table, when necessary.

43. Suppose you are to work for 20 days. You are paid 1¢ for the first day. Each day afterward, your wages are doubled. How much do you receive for the entire job?

44. Suppose you are to work for 10 days. You are paid 1¢ for the first day. Each day afterward, your wages are tripled. How much do you receive for the entire job?

45. The population of tsetse flies in a swamp region is decreasing at the constant annual rate of 50%. If the number of tsetse flies present in 1976 is 64 000 000, how many will there be in 1984?

46. If you deposit \$1000 in a bank at an annual interest rate of 5% and do not withdraw, how much money do you have in the bank 3 years later?

47. For $r \neq 0$ or 1, and for each positive integer n, express $\frac{1-r^n}{1-r}$ as a sum of powers of r.

13.5 Binomial Expansion

In this section you will learn a systematic way to find the powers of a binomial, $a + b$. First you must consider some preliminary notions.

$n!$ ***Definition.*** Let n be a positive integer. Define:

$$n! = n(n-1)(n-2)\ldots 3\cdot 2\cdot 1$$

The symbol $n!$ is read: "n factorial"

EXAMPLE 1 (a) $5! = 5\cdot 4\cdot 3\cdot 2\cdot 1 = 120$

(b) $6! = 6\cdot 5\cdot 4\cdot 3\cdot 2\cdot 1 = 720$
Observe that $6! = 6(5!)$.

(c) $100! = 100\cdot 99\cdot 98\ldots 3\cdot 2\cdot 1$

(d) $1! = 1$ □

The following definition is introduced to simplify later notation.

Definition. $0! = 1$

The observation in Example 1(b) can be generalized. For $n = 0, 1, 2, \ldots,$

$$(n + 1)! = (n + 1)n!†$$

Thus, to obtain $(n + 1)!$ from $n!$, simply multiply $n!$ by $n + 1$.

For integers i and n such that $0 < i < n$,

$$\frac{n!}{i!} = \frac{n(n-1)(n-2)\ldots(i+1)\overset{1}{\cancel{i(i-1)(i-2)\ldots 3\cdot 2\cdot 1}}}{\underset{1}{\cancel{i(i-1)(i-2)\ldots 3\cdot 2\cdot 1}}}$$

$$= n(n-1)(n-2)\ldots(i+1)$$

EXAMPLE 2 (a) Evaluate $\frac{12!}{9!}$. (b) Evaluate $\frac{101!}{99!}$.

(c) Express $10 \cdot 9 \cdot 8 \cdot 7$ in terms of factorials. (d) Evaluate $\frac{20!}{18!2!}$.

Solution.

(a) $$\frac{12!}{9!} = \frac{12 \cdot 11 \cdot 10 \cdot \overset{1}{\cancel{9 \cdot 8 \cdot 7 \cdot 6 \cdot 5 \cdot 4 \cdot 3 \cdot 2 \cdot 1}}}{\underset{1}{\cancel{9 \cdot 8 \cdot 7 \cdot 6 \cdot 5 \cdot 4 \cdot 3 \cdot 2 \cdot 1}}} = 1320$$

(b) $$\frac{101!}{99!} = 101 \cdot 100 = 10\,100$$

(c) To express $10 \cdot 9 \cdot 8 \cdot 7$, let $n = 10$ and $i + 1 = 7$.

$$\frac{10!}{6!} = 10 \cdot 9 \cdot 8 \cdot 7$$

(d) $$\frac{20!}{18!2!} = \frac{20!}{18!} \cdot \frac{1}{2!} = \overset{10}{\cancel{20}} \cdot 19 \cdot \frac{1}{\cancel{2} \cdot 1} = 190$$ □

$\binom{n}{i}$ As you will see, quotients such as in Example 2(d) are important. A special symbol is given to them. (Note that $20 = 18 + 2$.)

Definition. Let i and n be integers such that $0 \leqslant i \leqslant n$. Define

$$\binom{n}{i} = \frac{n!}{i!(n-i)!}.$$

Observe that

$$\binom{n}{i} = \frac{n!}{i!(n-i)!} \quad \text{and}$$

$$\binom{n}{n-i} = \frac{n!}{(n-i)!\underbrace{[n-(n-i)]}_{i}!} = \frac{n!}{(n-i)!i!}$$

Thus

$$\binom{n}{i} = \binom{n}{n-i}$$

† See the footnote on page 511 as well as Example 5 on page 513.

EXAMPLE 3 (a) $\binom{7}{4} = \frac{7!}{4!3!} = \frac{7 \cdot \cancel{6} \cdot 5 \cdot \cancel{4!}}{\cancel{4!} \cdot \cancel{3} \cdot \cancel{2} \cdot 1} = 35$

(b) $\binom{7}{3} = \binom{7}{7-3} = \binom{7}{4} = 35$ □

EXAMPLE 4 (a) $\binom{10}{2} = \frac{10!}{2!8!} = \frac{\overset{5}{\cancel{10}} \cdot 9 \cdot \cancel{8!}}{\cancel{2} \cdot \cancel{8!}} = 45$

(b) $\binom{8}{4} = \frac{8!}{4!4!} = \frac{\overset{2}{\cancel{8}} \cdot 7 \cdot \cancel{6} \cdot 5 \cdot \cancel{4!}}{\cancel{4} \cdot \cancel{3} \cdot \cancel{2} \cdot 1 \cdot \cancel{4!}} = 70$ □

The following computations play a key role in expanding powers of $a + b$.

EXAMPLE 5 (a) $\binom{n}{0} = \binom{n}{n} = \frac{\overset{1}{\cancel{n!}}}{0!\cancel{n!}} = \frac{1}{1} = 1$

(b) $\binom{n}{1} = \binom{n}{n-1} = \frac{n!}{1!(n-1)!} = \frac{n \cdot \cancel{(n-1)!}}{\underset{1}{\cancel{(n-1)!}}} = n$

Thus

$$\binom{2}{1} = 2; \quad \binom{3}{1} = \binom{3}{2} = 3; \quad \binom{4}{1} = \binom{4}{3} = 4;$$

$$\binom{5}{1} = \binom{5}{4} = 5$$

(c) $\binom{4}{2} = \frac{\overset{4 \cdot 3 \cdot \cancel{2!}}{\cancel{4!}}}{\cancel{2!}2!} = \frac{\overset{2}{\cancel{4}} \cdot 3}{\cancel{2} \cdot 1} = 6$

(d) $\binom{5}{2} = \binom{5}{3} = \frac{\overset{5 \cdot 4}{\cancel{5!}}}{2!\cancel{3!}} = \frac{5 \cdot \overset{2}{\cancel{4}}}{\cancel{2} \cdot 1} = 10$ □

$(a + b)^n$ Next, consider the following powers of $a + b$, beginning with $(a + b)^0$. In each case, the first result is obtained by ordinary multiplication. Then the coefficients are rewritten in the form $\binom{n}{i}$. Note the pattern.

Number of Terms

$(a + b)^0 = 1$

$= \binom{0}{0} a^0 b^0$ 1

$(a + b)^1 = a + b$ 2

$= \binom{1}{0} a^1 b^0 + \binom{1}{1} a^0 b^1$

$$(a + b)^2 = a^2 + 2ab + b^2 \qquad 3$$
$$= \binom{2}{0} a^2b^0 + \binom{2}{1} a^1b^1 + \binom{2}{2} a^0b^2$$

$$(a + b)^3 = a^3 + 3a^2b + 3ab^2 + b^3 \qquad 4$$
$$= \binom{3}{0} a^3b^0 + \binom{3}{1} a^2b^1 + \binom{3}{2} a^1b^2 + \binom{3}{3} a^0b^3$$

$$(a + b)^4 = a^4 + 4a^3b + 6a^2b^2 + 4ab^3 + b^4 \qquad 5$$
$$= \binom{4}{0} a^4b^0 + \binom{4}{1} a^3b^1 + \binom{4}{2} a^2b^2 + \binom{4}{3} a^1b^3 + \binom{4}{4} a^0b^4$$

$$(a + b)^5 = a^5 + 5a^4b + 10a^3b^2 + 10a^2b^3 + 5ab^4 + b^5 \qquad 6$$
$$= \binom{5}{0} a^5b^0 + \binom{5}{1} a^4b^1 + \binom{5}{2} a^3b^2 + \binom{5}{3} a^2b^3 + \binom{5}{4} a^1b^4 + \binom{5}{5} a^0b^5$$

In each expansion of $(a + b)^n$:

1. There are $(n + 1)$ terms.
2. The exponents of a decrease by 1 and the exponents of b increase by 1 in each subsequent term.
3. The sum of the exponents of a and b in each term is n.
4. The coefficient of $a^{n-i}b^i$ is $\binom{n}{i}$.

The same pattern continues for larger n. Thus, for $n = 0, 1, 2, \ldots,$

$$(a + b)^n = \binom{n}{0} a^nb^0 + \binom{n}{1} a^{n-1}b^1 + \binom{n}{2} a^{n-2}b^2 + \cdots + \binom{n}{n-2} a^2b^{n-2} + \binom{n}{n-1} a^1b^{n-1} + \binom{n}{n} a^0b^n$$

This formula is called the **binomial expansion.** The coefficients $\binom{n}{i}$ are therefore called the **binomial coefficients.** In Σ-notation this becomes

$$(a + b)^n = \sum_{i=0}^{n} \binom{n}{i} a^{n-i}b^i$$

The binomial coefficients can be determined from the **Pascal triangle** of Figure 13.4 .

First write down the 1's along two sides of the triangle. Every other number in the triangle is the sum of the two adjacent numbers just above it. The number in the nth row and ith diagonal is the binomial coefficient $\binom{n}{i}$. This triangle can be enlarged.

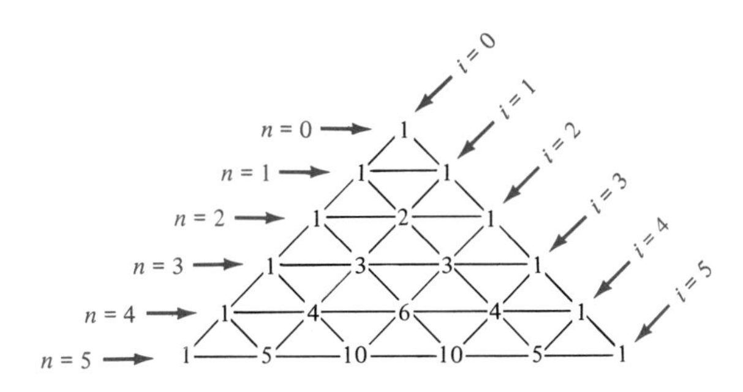

FIGURE 13.4 The Pascal triangle. The binomial coefficient $\binom{4}{2}$, or 6, lies in the row of $n = 4$ and the diagonal of $i = 2$. Note that 6 is the sum of 3 and 3, the two adjacent numbers just above 6.

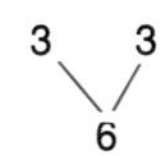

EXAMPLE 6 Determine the next row of the Pascal triangle.

Solution. The row of $n = 6$ is obtained by adding each pair of adjacent numbers of the row of $n = 5$. The end numbers (along the sides of the triangle) are again 1.

$$\begin{array}{lccccccc} n = 5 & & 1 & 5 & 10 & 10 & 5 & 1 \\ n = 6 & 1 & 6 & 15 & 20 & 15 & 6 & 1 \end{array}$$

□

EXAMPLE 7 Determine: $(x + 2)^6$

Solution. Let $a = x$ and $b = 2$ in the binomial expansion. Refer to Example 6 for the binomial coefficients $\binom{6}{i}$. Thus

$$(x + 2)^6 = \sum_{i=0}^{6} \binom{6}{i} x^{6-i}2^i$$

$$= \underbrace{x^6}_{i=0} + \underbrace{6\cdot x^5\cdot 2}_{i=1} + \underbrace{15\cdot x^4\cdot 2^2}_{i=2} + \underbrace{20\cdot x^3\cdot 2^3}_{i=3} + \underbrace{15\cdot x^2\cdot 2^4}_{i=4}$$

$$+ \underbrace{6\cdot x\cdot 2^5}_{i=5} + \underbrace{2^6}_{i=6}$$

$$= x^6 + 12x^5 + 60x^4 + 160x^3 + 240x^2 + 192x + 64$$

□

EXAMPLE 8 (a) Express $(y - 1)^{10}$ in Σ-notation.
(b) Determine the sum of the first 4 terms of this expansion.

Solution.

(a) Let $a = y$ and $b = -1$ in the binomial expansion.

$$(y-1)^{10} = \sum_{i=0}^{10} \binom{10}{i} y^{10-i}(-1)^i = \sum_{i=0}^{10} (-1)^i \binom{10}{i} y^{10-i}$$

The signs of the terms alternate, beginning with + for $(-1)^0$. $+ - + - \cdot \cdot \cdot - +$

(b) The sum of the first 4 terms is given by:

$$y^{10} - 10y^9 + \frac{10 \cdot 9}{2}y^8 - \frac{10 \cdot 9 \cdot 8}{3 \cdot 2}y^7$$
$$= y^{10} - 10y^9 + 45y^8 - 120y^7$$

□

EXAMPLE 9 Use the binomial expansion to determine $(.99)^5$ to 4 decimal places.

Solution. Let $a = 1$, $b = -.01$ in the binomial expansion.

$$(.99)^5 = (1 - .01)^5 = \sum_{i=0}^{5} \binom{5}{i} 1^{5-i}(-.01)^i = \sum_{i=0}^{5} \binom{5}{i} (-.01)^i$$

The first few terms are:

$$1 - 5(.01) + 10(.01)^2 - 10(.01)^3$$
$$= 1 - .05 + .001 - .000\,01 = .95\,099$$

Larger powers of .01 are extremely small and do not affect the first 4 decimal places. To 4 decimal places,

$$(.99)^5 = .9510$$

□

EXERCISES

In Exercises 1–8 evaluate each expression.

1. $4!$ **2.** $7!$ **3.** $8!$ **4.** $10!$ **5.** $\frac{9!}{5!}$ **6.** $\frac{12!}{5!7!}$ **7.** $\frac{10!}{6!4!}$ **8.** $\frac{(3!)^2}{(2!)^3}$

In Exercises 9–14 express each product as a quotient of the form $\frac{n!}{m!}$.

9. $5 \cdot 4$ **10.** $7 \cdot 6 \cdot 5$ **11.** $12 \cdot 11 \cdot 10 \cdot 9$ **12.** $63 \cdot 62 \cdot 61 \cdot 60 \cdot 59 \cdot 58 \cdot 57$

13. $90 \cdot 91 \cdot 92 \cdot 93 \cdot 94 \cdot 95 \cdot 96 \cdot 97 \cdot 98 \cdot 99$ **14.** $201 \cdot 202 \cdot 203 \ldots 299 \cdot 300$

In Exercises 15–30 evaluate each binomial coefficient.

15. $\binom{3}{2}$ **16.** $\binom{4}{4}$ **17.** $\binom{4}{2}$ **18.** $\binom{5}{1}$ **19.** $\binom{6}{0}$ **20.** $\binom{6}{3}$ **21.** $\binom{7}{7}$ **22.** $\binom{8}{3}$

23. $\binom{8}{6}$ **24.** $\binom{10}{3}$ **25.** $\binom{12}{5}$ **26.** $\binom{20}{10}$

27. $\binom{396}{0}$ **28.** $\binom{396}{1}$ **29.** $\binom{396}{395}$ **30.** $\binom{396}{396}$

In Exercises 31–34 suppose n is an integer and $n > 2$. Simplify each expression.

31. $\frac{n!}{(n-2)!}$ **32.** $\frac{(n+1)!}{(n-1)!}$ **33.** $\frac{(n+3)!}{(n+3)(n+2)}$ **34.** $\frac{n!}{(n-2)!(n-1)}$

35. Determine the row of $n = 7$ of the Pascal triangle.

36. Determine the row of $n = 8$ of the Pascal triangle.

37. Determine the row of $n = 9$ of the Pascal triangle.

In Exercises 38–50 use the binomial expansion to expand each power.

38. $(x + y)^3$ **39.** $(x - a)^4$ **40.** $(x - a)^5$ **41.** $(x + 1)^4$ **42.** $(x + 1)^6$ **43.** $(x - 1)^5$

44. $(x - 1)^6$ **45.** $(x + 3)^4$ **46.** $(x - 2)^6$ **47.** $(x + 2y)^4$

48. $\left(x + \frac{1}{2}\right)^4$ **49.** $(a^2 - b^2)^4$ **50.** $(xy^2z^3 - a^5b^4)^3$

In Exercises 51–60:

(a) Express each power in Σ-notation. **(b)** Determine the sum of the first 4 terms of this expansion.

51. $(x + y)^8$ **52.** $(x - y)^7$ **53.** $(x + 1)^{12}$ **54.** $(a - 1)^9$ **55.** $(x - 2)^9$

56. $(xy - 1)^{12}$ **57.** $(x^2y + 1)^8$ **58.** $(x^3y + az^2)^{10}$ **59.** $\left(x + \frac{1}{2}\right)^8$ **60.** $\left(x - \frac{a}{2}\right)^8$

In Exercises 61–68 use the binomial expansion to determine each power to 4 decimal places.

61. $(1.01)^4$ **62.** $(.99)^4$ **63.** $(1.1)^6$ **64.** $(1.1)^{10}$ **65.** $(1.01)^7$ **66.** $(.99)^8$ **67.** $(2.1)^6$

68. $(2.01)^7$

*13.6 Mathematical Induction

Often a mathematical formula, $s(n)$, is true for *every* positive integer n. For example, it will be proved in Example 3 that for *every* positive integer n,

$$1 + 2 + 3 + \cdots + n = \frac{n(n+1)}{2}. \tag{13.2}$$

Thus for each such n, if $s(n)$ is Statement (**13.2**), then

$$s(1) \text{ asserts: } \quad 1 = \frac{1(1+1)}{2}$$

$$s(2) \text{ asserts: } \quad 1 + 2 = \frac{2(2+1)}{2}$$

$$s(3) \text{ asserts: } \quad 1 + 2 + 3 = \frac{3(3+1)}{2}$$

Similarly,

$$s(20) \text{ asserts: } 1 + 2 + 3 + \ldots + 20 = \frac{20(20+1)}{2}$$

It is not difficult to establish that each individual statement, $s(1)$, $s(2)$, $s(3)$, or even with some patience, $s(20)$, is true. But there are

* Optional topic

infinitely many statements $s(n)$ of the form (**13.2**). And the claim is that *every one of them is true*.

There are also statements $s(n)$ that are true for all numbers n up to, say, 40, and then false for $n = 41$. [See Example 6.]

In order to verify that a statement is true for *every one* of the *infinitely many* positive integers, 1, 2, 3, . . . , the following *Principle of Mathematical Induction* is frequently utilized.

Principle of Mathematical Induction.

Let $s(n)$ be a statement for each positive integer n. Suppose

(A) $s(1)$ is true;

(B) whenever $s(k)$ is true, then $s(k + 1)$ is also true.

Then $s(n)$ is true for every positive integer n. [See Figure 13.5 .]

Be careful of (**B**). It does *not* say that $s(k)$ is true. What it *does say* is that *if* $s(k)$ is true, *then* $s(k + 1)$ must also be true. The *assumption* that $s(k)$ is true is known as the **Inductive Hypothesis.**

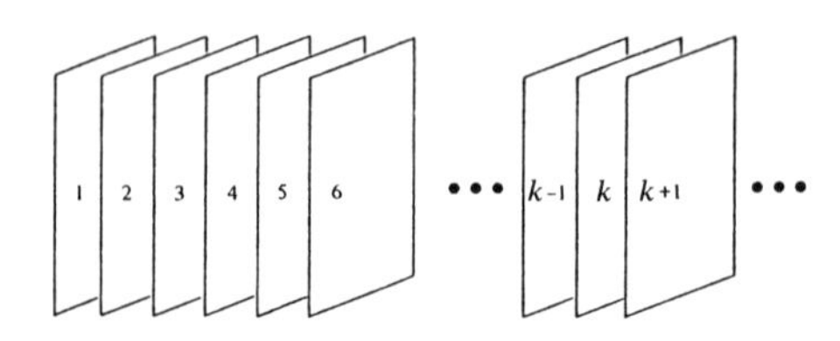

FIGURE 13.5 Imagine an "infinite deck of cards" on which the positive integers are listed. Knock over the first card. As each card falls, the next card also falls. If you neglect the time factor, every card must fall.

If either Property (**A**) or (**B**) fails, then $s(n)$ is not true for *every* positive integer—that is, $s(n)$ is false for at least one positive integer.

Before utilizing the Principle of Mathematical Induction, here is a very obvious example in which Property (**B**) *fails*.

EXAMPLE 1 For each positive integer n, let

$s(n)$ be the statement: $n \neq 5$

Then

$s(1)$ is true because $1 \neq 5$ $s(2)$ is true because $2 \neq 5$

$s(3)$ is true because $3 \neq 5$ $s(4)$ is true because $4 \neq 5$.

However,

$s(5)$ is *false*, because $5 = 5$

Thus $s(4)$ is *true*, but $s(4 + 1)$ is *false*. Therefore Property (**B**) fails, despite the fact that $s(6)$, $s(7)$, $s(8)$ . . . are all true. □

As a first example of the use of Mathematical Induction, consider a rule of exponents first stated on p. 189 and subsequently used throughout the book.

EXAMPLE 2 Use the Principle of Mathematical Induction to show that for all real numbers a and b and for every positive integer n,

$$(ab)^n = a^n b^n \tag{13.3}$$

Solution. For each such n, let $s(n)$ be Statement **(13.3)**. You must show that both Properties **(A)** and **(B)** hold true.

$$s(1) \text{ asserts:} \qquad (ab)^1 = a^1 b^1$$

Each side of this equation simplifies to ab, so that $s(1)$ is true.

Suppose that $s(k)$ is true, that is, suppose

$$(ab)^k = a^k b^k \tag{13.4}$$

Then to establish $s(k + 1)$, use the Inductive Hypothesis, **(13.4)**. Thus

$$\begin{aligned} (ab)^{k+1} &= (ab)^k \cdot (ab) && \text{by the definition of exponentiation}^\dagger \\ &= (a^k b^k) \cdot (ab) && \text{by (13.4)} \\ &= (a^k \cdot a) \cdot (b^k \cdot b) && \text{by Associativity and Commutativity} \\ &= a^{k+1} b^{k+1} && \text{by the definition of exponentiation} \end{aligned}$$

Both Properties **(A)** and **(B)** have now been established. Therefore $s(n)$ is true for every positive integer n. □

Often the use of Mathematical Induction requires some computational dexterity, as in the next example.

EXAMPLE 3 Use the Principle of Mathematical Induction to show that for every positive integer n, Statement **(13.2)** is true:

$$1 + 2 + 3 + \cdots + n = \frac{n \cdot (n + 1)}{2}$$

Solution. For each such n, let $s(n)$ be Statement **(13.2)**. You must show that both Properties **(A)** and **(B)** hold true.

$$s(1): \qquad 1 = \frac{1(1 + 1)}{2}$$

This is obviously true.

† Actuality, the definition of exponentiation is an example of a "Definition by Induction" (or a "recursive definition"):

$$x^1 = x; \qquad x^{n+1} = x^n \cdot x$$

Another example of a Definition by Induction is

$$1! = 1; \qquad (n + 1)! = (n + 1) \cdot n!$$

The relationship between the Principle of Mathematical Induction and its application to Definition by Induction is beyond the level of this book. For a discussion, see M. M. Zuckerman, *Sets and Transfinite Numbers* (New York: Macmillan, 1974), pp. 79–80.

Suppose that $s(k)$ is true, that is, suppose

$$1 + 2 + 3 + \cdots + k = \frac{k(k+1)}{2} \tag{13.5}$$

The statement $s(k+1)$ that must be tested is:

$$1 + 2 + 3 + \cdots + k + (k+1) = \frac{(k+1)[(k+1)+1]}{2}$$

that is,

$$(1 + 2 + 3 + \cdots + k) + (k+1) = \frac{(k+1)(k+2)}{2} \tag{13.6}$$

Use the Inductive Hypothesis, **(13.5)**, to establish the truth of **(13.6)**.

$$(1 + 2 + 3 + \cdots + k) + (k+1)$$

$$= \frac{k(k+1)}{2} + (k+1) \quad \text{by (13.5). Next, combine as a single fraction.}$$

$$= \frac{k(k+1) + 2(k+1)}{2} \quad \text{Factor the numerator.}$$

$$= \frac{(k+1)(k+2)}{2}$$

Thus, whenever $s(k)$ is true, then $s(k+1)$ is also true. Both Properties **(A)** and **(B)** hold true; consequently $s(n)$ is true for every positive integer n.

Observe that

$$1, 2, 3, \ldots, n, n+1, \ldots$$

is an arithmetic progression. Thus **(13.2)** can also be established by applying Formula **(13.1)** on p. 496 for the sum of the first n terms of an arithmetic progression. The proof of Formula **(13.1)** is Exercise 22. □

Now that you've seen two proofs by Induction, let us review the procedure. Suppose you establish Properties **(A)** and **(B)**. By **(A)**, $s(1)$ is true. Use **(B)** with $k = 1$ and $k + 1 = 2$ to obtain $s(2)$. Use **(B)** again with $k = 2$ and $k + 1 = 3$ to obtain $s(3)$. Similarly, $s(4), s(5), s(6), \ldots$ are all established.

EXAMPLE 4 Prove that for every positive integer n,

$$1^2 + 2^2 + 3^2 + \cdots + n^2 = \frac{n(n+1)(2n+1)}{6} \tag{13.7}$$

Solution. For each such n, let $s(n)$ be Statement **(13.7)**.

$$s(1): \quad 1^2 = \frac{1(1+1)(2\cdot 1+1)}{6}$$

This is true because

$$1^2 = 1 \quad \text{and} \quad \frac{1(1+1)(2\cdot 1+1)}{6} = \frac{2\cdot 3}{6} = 1.$$

Suppose $s(k)$ is true—that is, suppose

$$1^2 + 2^2 + 3^2 + \cdots + k^2 = \frac{k(k+1)(2k+1)}{6} \tag{13.8}$$

Does it follow that $s(k + 1)$ is also true—that is, letting $n = k + 1$ in **(13.7)**, does

$$1^2 + 2^2 + 3^2 + \cdots + k^2 + (k + 1)^2 = \frac{(k + 1)[(k + 1) + 1][2(k + 1) + 1]}{6}\ ?$$

Observe that the right-hand side is $\dfrac{(k + 1)(k + 2)(2k + 3)}{6}$.

Thus consider the equation:

$$1^2 + 2^2 + \cdots + k^2 + (k + 1)^2 = \frac{(k + 1)(k + 2)(2k + 3)}{6} \tag{13.9}$$

Use the Inductive Hypothesis, **(13.8)**, to establish **(13.9)**.

$$\begin{aligned}(1^2 + 2^2 + \cdots + k^2) + (k + 1)^2 &= \frac{k(k + 1)(2k + 1)}{6} + (k + 1)^2 && \text{Combine as a single fraction.}\\ &= \frac{k(k + 1)(2k + 1) + 6(k + 1)^2}{6} && \text{Now factor the numerator.}\\ &= \frac{(k + 1)[k(2k + 1) + 6(k + 1)]}{6}\\ &= \frac{(k + 1)(2k^2 + k + 6k + 6)}{6}\\ &= \frac{(k + 1)(2k^2 + 7k + 6)}{6} && \text{Factor the numerator.}\\ &= \frac{(k + 1)(k + 2)(2k + 3)}{6}\end{aligned}$$

□

EXAMPLE 5 Prove that for every positive integer n,

$$(n + 1)! = (n + 1) \cdot n! \tag{13.10}$$

Solution. Let $s(n)$ be Statement **(13.10)**:

$$s(1): \quad (1 + 1)! \stackrel{?}{=} (1 + 1) \cdot 1!$$

This is true because

$$2! \stackrel{\checkmark}{=} 2 \cdot 1$$

Now suppose $s(k)$ is true, that is, suppose

$$(k + 1)! = (k + 1) \cdot k!$$

For $n = k + 1$,

$$\begin{aligned}[(k + 1) + 1]! &= (k + 2)!\\ &= (k + 2)(k + 1)k(k - 1) \ldots 3 \cdot 2 \cdot 1\\ &= (k + 2)[(k + 1)k(k - 1) \ldots 3 \cdot 2 \cdot 1]\\ &= (k + 2)(k + 1)!\\ &= [(k + 1) + 1](k + 1)!\end{aligned}$$

Thus, $s(k + 1)$ is true, and by Induction, $s(n)$ is true for every positive integer n. □

Sometimes a formula may hold for the first m positive integers, but fail for $m + 1$. An example of this phenomenon is the so-called "prime-construction formula".

Recall that a prime p is an integer greater than 1 that is not divisible by any positive integer other than 1 and p. Thus 2, 3, 5, 7, and 11 are the smallest primes.

EXAMPLE 6 Let $s(n)$ be the statement:

$$n^2 - n + 41 \text{ is a prime.}$$

Observe that

$$1^2 - 1 + 41 = 41 \qquad 2^2 - 2 + 41 = 43 \qquad 3^2 - 3 + 41 = 47$$

$$4^2 - 4 + 41 = 53 \qquad 5^2 - 5 + 41 = 61 \qquad 6^2 - 6 + 41 = 71$$

The numbers 41, 43, 47, 53, 61, and 71 are all primes. (Check this.) In fact, $s(n)$ is true for $n = 1, 2, 3, \ldots, 39, 40$. However,

$$41^2 - 41 + 41 = 41^2$$

Clearly, 41^2 is not a prime because 41 divides 41^2. Thus, $s(41)$ is false. □

In Exercise 15, you will show that for every positive integer m, there are statements $s(n)$ that are true for $n = 1, 2, \ldots, m$, but false for all $n > m$.

EXERCISES

1. Prove by Mathematical Induction that for every positive integer n,

$$1\cdot 2 + 2\cdot 3 + 3\cdot 4 + \cdots + n(n+1) = \frac{n(n+1)(n+2)}{3}.$$

2. Determine:

(a) $1\cdot 2 + 2\cdot 3 + 3\cdot 4 + \cdots + 9\cdot 10$ **(b)** $1\cdot 2 + 2\cdot 3 + 3\cdot 4 + \cdots + 17\cdot 18$

(c) $1\cdot 2 + 2\cdot 3 + 3\cdot 4 + \cdots + 100\cdot 101$ **(d)** $1\cdot 2 + 2\cdot 3 + 3\cdot 4 + \cdots + 1000\cdot 1001$

(e) $10\cdot 11 + 11\cdot 12 + 12\cdot 13 + \cdots + 17\cdot 18$

3. Prove by Mathematical Induction that for every positive integer n,

$$1 + 4 + 7 + \cdots + (3n - 2) = \frac{n(3n-1)}{2}.$$

4. Determine:

(a) $1 + 4 + 7 + \cdots + 25$ **(b)** $2 + 3 + 5 + 6 + 8 + 9 + \cdots + 23 + 24$

(c) $1 + 4 + 7 + \cdots + 100$ **(d)** $2 + 3 + 5 + 6 + 8 + 9 + \cdots + 98 + 99$

5. Prove by Mathematical Induction that for every positive integer n, $n < 2^n$.

6. Which of the Properties **(A)**, **(B)** on page 510 hold true for the following:

(a) $n^2 \leqslant n$ **(b)** $n^2 \geqslant 10n$ **(c)** $2^n \leqslant n + 1$ **(d)** $n + 1 = n$ **(e)** $n \geqslant 1$

7. Prove by Mathematical Induction that for every positive integer n, $2^{n+3} < (n+3)!$

8. Determine:

(a) $1 + 2 + 3 + \cdots + 10$

(b) $1 + 2 + 3 + \cdots + 49$

(c) $1 + 2 + 3 + \cdots + 100$

(d) $1 + 2 + 3 + \cdots + 1000$

(e) $101 + 102 + 103 + \cdots + 1000$

9. Determine:

(a) $1^2 + 2^2 + 3^2 + \cdots + 8^2$

(b) $1^2 + 2^2 + 3^2 + \cdots + 20^2$

(c) $1^2 + 2^2 + 3^2 + \cdots + 100^2$

(d) $1^2 + 2^2 + 3^2 + \cdots + 1000^2$

(e) $21^2 + 22^2 + 23^2 + \cdots + 100^2$

10. Prove by Mathematical Induction that for every positive integer n,

$$1^3 + 2^3 + 3^3 + \cdots + n^3 = \left[\frac{n(n+1)}{2}\right]^2$$

11. Determine:

(a) $1^3 + 2^3 + 3^3 + 4^3 + 5^3 + 6^3$

(b) $1^3 + 2^3 + 3^3 + \cdots + 15^3$

(c) $1^3 + 2^3 + 3^3 + \cdots + 100^3$

(d) $1^3 + 2^3 + 3^3 + \cdots + 1000^3$

(e) $16^3 + 17^3 + 18^3 + \cdots + 1000^3$

12. Prove by Mathematical Induction that for every positive integer n,

$$\frac{1}{1\cdot 2} + \frac{1}{2\cdot 3} + \frac{1}{3\cdot 4} + \cdots + \frac{1}{n(n+1)} = \frac{n}{n+1}$$

13. Determine:

(a) $\frac{1}{1\cdot 2} + \frac{1}{2\cdot 3} + \frac{1}{3\cdot 4} + \frac{1}{4\cdot 5} + \frac{1}{5\cdot 6} + \frac{1}{6\cdot 7}$

(b) $\frac{1}{1\cdot 2} + \frac{1}{2\cdot 3} + \frac{1}{3\cdot 4} + \cdots + \frac{1}{40\cdot 41}$

(c) $\frac{1}{10\cdot 11} + \frac{1}{11\cdot 12} + \frac{1}{12\cdot 13} + \cdots + \frac{1}{99\cdot 100}$

(d) $\frac{1}{200\cdot 201} + \frac{1}{201\cdot 202} + \frac{1}{202\cdot 203} + \cdots + \frac{1}{999\cdot 1000}$

14. Prove by Mathematical Induction that for every positive integer n,

$$1 + 3 + 5 + \cdots + (2n - 1) = n^2$$

15. Prove that for every positive integer m, the statement,

$$(n-1)(n-2)\cdots(n-m) = 0$$

is true for $n = 1, 2, 3, \ldots, m$, but is false for all $n > m$.

In Exercises 16–23, prove by Mathematical Induction that the following statements are true for every positive integer n.

16. For all positive real numbers $a_1, a_2, \ldots, a_n, \ldots$,

$$\log(a_1 a_2 \ldots a_n) = \log a_1 + \log a_2 + \cdots + \log a_n$$

17. For every complex number z, if $\bar{z}$ is the complex conjugate of z, then

$$\overline{z^n} = (\bar{z})^n.$$

18. De Moivre's Theorem: For every real number θ and for every positive real number r,

$$[r(\cos\theta + i\sin\theta)]^n = r^n(\cos n\theta + i\sin n\theta)$$

19. For every positive real number h,

$$(1 + h)^n \geqslant 1 + nh$$

20. Let $a_1, a_2, \ldots, a_n, \ldots; b_1, b_2, \ldots, b_n, \ldots$ be real numbers. Then

$$\sum_{i=1}^{n}(a_i + b_i) = \sum_{i=1}^{n} a_i + \sum_{i=1}^{n} b_i$$

21. Let $c, a_1, a_2, \ldots$ be real numbers. Then

$$\sum_{i=1}^{n} ca_i = c\sum_{i=1}^{n} a_i$$

22. For an arithmetic progression $a_1, a_2, \ldots, a_i, \ldots,$

$$\sum_{i=1}^{n} a_i = \frac{n}{2}(a_1 + a_n)$$

23. For a geometric progression, $a_1, a_2, \ldots, a_i, \ldots$ with common ratio r ($\neq 1$),

$$\sum_{i=1}^{n} a_i = \frac{a_1(1 - r^n)}{1 - r}$$

24. **(a)** Use the Principle of Mathematical Induction to establish the following statement, known as the **Well-Ordering Principle:** Every nonempty set of positive integers contains a smallest integer.†

(b) Find a counterexample if "positive" is deleted.

Review Exercises for Chapter 13

1. **(a)** Determine the first 8 terms of the indicated sequence.
(b) Is the sequence increasing, decreasing, or neither of these?
(c) Graph the sequence.

i. $a_i = i + 3$ **ii.** $a_i = 1 - i$ **iii.** $a_i = \frac{2}{i}$

2. Consider the sequence given by $a_i = 4i + 1$, $i = 1, 2, \ldots$. Determine:
(a) a_2 **(b)** a_4 **(c)** a_{10} **(d)** a_{20}

3. Describe the ith term of the suggested sequence whose first six terms are 5, 7, 9, 11, 13, 15.

4. Express $a_1 + a_2 + a_3 + a_4 + a_5$ in Σ-notation.

5. Express $b_7 + b_8 + b_9 + b_{10}$ in Σ-notation.

6. Determine the numerical value of $\sum_{i=1}^{4} 3i$.

7. Write in Σ-notation: $3 + 6 + 9 + 12 + 15 + 18$.

8. Determine m (and n).

(a) $\sum_{i=1}^{10} a_i + \sum_{i=11}^{20} a_i = \sum_{i=1}^{m} a_i$ **(b)** $\sum_{i=1}^{6} \frac{1}{i} = \sum_{i=4}^{9} \frac{1}{i - m}$ **(c)** $\sum_{i=1}^{5} 3^{i+2} = \sum_{i=m}^{n} 3^i$

9. Determine the 12th term of the arithmetic progression indicated by: 40, 45, 50, 55, 60, 65, . . .

10. Determine the sum of the first 20 terms of the arithmetic progression indicated by:

6, 10, 14, 18, 22, 26, . . .

11. Find the sum of the first 30 odd positive integers.

12. Determine the 5th term of the geometric progression with first term 3 and common ratio $\frac{1}{3}$.

13. Determine the first term of the geometric progression with 4th term 2000 and 5th term 10 000.

14. Evaluate: **(a)** $\sum_{i=1}^{6} 3^i$ **(b)** $\sum_{i=1}^{7} \frac{5}{2^{i-1}}$

15. Evaluate: **(a)** $\frac{10!}{5!3!}$ **(b)** $\binom{5}{3}$ **(c)** $\binom{20}{4}$

16. Use the binomial expansion to expand each power:

(a) $(x + a)^3$ **(b)** $(x - 1)^4$ **(c)** $(2xy^2 + 1)^5$

17. **(a)** Express $(x - 2)^{10}$ in Σ-notation.
(b) Determine the sum of the first 4 terms of this expansion.

† See the footnote on p. 83.

18. Use the binomial expansion to determine $(1.1)^5$ to 4 decimal places.

19. Prove each statement by Mathematical Induction:

(a) For every positive integer n, $n^2 + n$ is even.

(b) For every positive integer n, $\frac{1}{1\cdot 4} + \frac{1}{4\cdot 7} + \frac{1}{7\cdot 10} + \cdots + \frac{1}{(3n-2)(3n+1)} = \frac{n}{3n+1}$

Appendix: Tables

Table 1 Powers and Roots

n	n^2	$\sqrt{n}$	n^3	$\sqrt[3]{n}$	n	n^2	$\sqrt{n}$	n^3	$\sqrt[3]{n}$
1	1	1.000	1	1.000	**51**	2,601	7.141	132,651	3.708
2	4	1.414	8	1.260	**52**	2,704	7.211	140,608	3.733
3	9	1.732	27	1.442	**53**	2,809	7.280	148,877	3.756
4	16	2.000	64	1.587	**54**	2,916	7.348	157,464	3.780
5	25	2.236	125	1.710	**55**	3,025	7.416	166,375	3.803
6	36	2.449	216	1.817	**56**	3,136	7.483	175,616	3.826
7	49	2.646	343	1.913	**57**	3,249	7.550	185,193	3.849
8	64	2.828	512	2.000	**58**	3,364	7.616	195,112	3.871
9	81	3.000	729	2.080	**59**	3,481	7.681	205,379	3.893
10	100	3.162	1,000	2.154	**60**	3,600	7.746	216,000	3.915
11	121	3.317	1,331	2.224	**61**	3,721	7.810	226,981	3.936
12	144	3.464	1,728	2.289	**62**	3,844	7.874	238,328	3.958
13	169	3.606	2,197	2.351	**63**	3,969	7.937	250,047	3.979
14	196	3.742	2,744	2.410	**64**	4,096	8.000	262,144	4.000
15	225	3.873	3,375	2.466	**65**	4,225	8.062	274,625	4.021
16	256	4.000	4,096	2.520	**66**	4,356	8.124	287,496	4.041
17	289	4.123	4,913	2.571	**67**	4,489	8.185	300,763	4.062
18	324	4.243	5,832	2.621	**68**	4,624	8.246	314,432	4.082
19	361	4.359	6,859	2.668	**69**	4,761	8.307	328,509	4.102
20	400	4.472	8,000	2.714	**70**	4,900	8.367	343,000	4.121
21	441	4.583	9,261	2.759	**71**	5,041	8.426	357,911	4.141
22	484	4.690	10,648	2.802	**72**	5,184	8.485	373,248	4.160
23	529	4.796	12,167	2.844	**73**	5,329	8.544	389,017	4.179
24	576	4.899	13,824	2.884	**74**	5,476	8.602	405,224	4.198
25	625	5.000	15,625	2.924	**75**	5,625	8.660	421,875	4.217
26	676	5.099	17,576	2.962	**76**	5,776	8.718	438,976	4.236
27	729	5.196	19,683	3.000	**77**	5,929	8.775	456,533	4.254
28	784	5.292	21,952	3.037	**78**	6,084	8.832	474,552	4.273
29	841	5.385	24,389	3.072	**79**	6,241	8.888	493,039	4.291
30	900	5.477	27,000	3.107	**80**	6,400	8.944	512,000	4.309
31	961	5.568	29,791	3.141	**81**	6,561	9.000	531,441	4.327
32	1,024	5.657	32,768	3.175	**82**	6,724	9.055	551,368	4.344
33	1,089	5.745	35,937	3.208	**83**	6,889	9.110	571,787	4.362
34	1,156	5.831	39,304	3.240	**84**	7,056	9.165	592,704	4.380
35	1,225	5.916	42,875	3.271	**85**	7,225	9.220	614,125	4.397
36	1,296	6.000	46,656	3.302	**86**	7,396	9.274	636,056	4.414
37	1,369	6.083	50,653	3.332	**87**	7,569	9.327	658,503	4.431
38	1,444	6.164	54,872	3.362	**88**	7,744	9.381	681,472	4.448
39	1,521	6.245	59,319	3.391	**89**	7,921	9.434	704,969	4.465
40	1,600	6.325	64,000	3.420	**90**	8,100	9.487	729,000	4.481
41	1,681	6.403	68,921	3.448	**91**	8,281	9.539	753,571	4.498
42	1,764	6.481	74,088	3.476	**92**	8,464	9.592	778,688	4.514
43	1,849	6.557	79,507	3.503	**93**	8,649	9.644	804,357	4.531
44	1,936	6.633	85,184	3.530	**94**	8,836	9.695	830,584	4.547
45	2,025	6.708	91,125	3.557	**95**	9,025	9.747	857,375	4.563
46	2,116	6.782	97,336	3.583	**96**	9,216	9.798	884,736	4.579
47	2,209	6.856	103,823	3.609	**97**	9,409	9.849	912,673	4.595
48	2,304	6.928	110,592	3.634	**98**	9,604	9.899	941,192	4.610
49	2,401	7.000	117,649	3.659	**99**	9,801	9.950	970,299	4.626
50	2,500	7.071	125,000	3.684	**100**	10,000	10.000	1,000,000	4.642

Table 2 Common Logarithms

x	0	1	2	3	4	5	6	7	8	9
1.0	.0000	.0043	.0086	.0128	.0170	.0212	.0253	.0294	.0334	.0374
1.1	.0414	.0453	.0492	.0531	.0569	.0607	.0645	.0682	.0719	.0755
1.2	.0792	.0828	.0864	.0899	.0934	.0969	.1004	.1038	.1072	.1106
1.3	.1139	.1173	.1206	.1239	.1271	.1303	.1335	.1367	.1399	.1430
1.4	.1461	.1492	.1523	.1553	.1584	.1614	.1644	.1673	.1703	.1732
1.5	.1761	.1790	.1818	.1847	.1875	.1903	.1931	.1959	.1987	.2014
1.6	.2041	.2068	.2095	.2122	.2148	.2175	.2201	.2227	.2253	.2279
1.7	.2304	.2330	.2355	.2380	.2405	.2430	.2455	.2480	.2504	.2529
1.8	.2553	.2577	.2601	.2625	.2648	.2672	.2695	.2718	.2742	.2765
1.9	.2788	.2810	.2833	.2856	.2878	.2900	.2923	.2945	.2967	.2989
2.0	.3010	.3032	.3054	.3075	.3096	.3118	.3139	.3160	.3181	.3201
2.1	.3222	.3243	.3263	.3284	.3304	.3324	.3345	.3365	.3385	.3404
2.2	.3424	.3444	.3464	.3483	.3502	.3522	.3541	.3560	.3579	.3598
2.3	.3617	.3636	.3655	.3674	.3692	.3711	.3729	.3747	.3766	.3784
2.4	.3802	.3820	.3838	.3856	.3874	.3892	.3909	.3927	.3945	.3962
2.5	.3979	.3997	.4014	.4031	.4048	.4065	.4082	.4099	.4116	.4133
2.6	.4150	.4166	.4183	.4200	.4216	.4232	.4249	.4265	.4281	.4298
2.7	.4314	.4330	.4346	.4362	.4378	.4393	.4409	.4425	.4440	.4456
2.8	.4472	.4487	.4502	.4518	.4533	.4548	.4564	.4579	.4594	.4609
2.9	.4624	.4639	.4654	.4669	.4683	.4698	.4713	.4728	.4742	.4757
3.0	.4771	.4786	.4800	.4814	.4829	.4843	.4857	.4871	.4886	.4900
3.1	.4914	.4928	.4942	.4955	.4969	.4983	.4997	.5011	.5024	.5038
3.2	.5051	.5065	.5079	.5092	.5105	.5119	.5132	.5145	.5159	.5172
3.3	.5185	.5198	.5211	.5224	.5237	.5250	.5263	.5276	.5289	.5302
3.4	.5315	.5328	.5340	.5353	.5366	.5378	.5391	.5403	.5416	.5428
3.5	.5441	.5453	.5465	.5478	.5490	.5502	.5514	.5527	.5539	.5551
3.6	.5563	.5575	.5587	.5599	.5611	.5623	.5635	.5647	.5658	.5670
3.7	.5682	.5694	.5705	.5717	.5729	.5740	.5752	.5763	.5775	.5786
3.8	.5798	.5809	.5821	.5832	.5843	.5855	.5866	.5877	.5888	.5899
3.9	.5911	.5922	.5933	.5944	.5955	.5966	.5977	.5988	.5999	.6010
4.0	.6021	.6031	.6042	.6053	.6064	.6075	.6085	.6096	.6107	.6117
4.1	.6128	.6138	.6149	.6160	.6170	.6180	.6191	.6201	.6212	.6222
4.2	.6232	.6243	.6253	.6263	.6274	.6284	.6294	.6304	.6314	.6325
4.3	.6335	.6345	.6355	.6365	.6375	.6385	.6395	.6405	.6415	.6425
4.4	.6435	.6444	.6454	.6464	.6474	.6484	.6493	.6503	.6513	.6522
4.5	.6532	.6542	.6551	.6561	.6571	.6580	.6590	.6599	.6609	.6618
4.6	.6628	.6637	.6646	.6656	.6665	.6675	.6684	.6693	.6702	.6712
4.7	.6721	.6730	.6739	.6749	.6758	.6767	.6776	.6785	.6794	.6803
4.8	.6812	.6821	.6830	.6839	.6848	.6857	.6866	.6875	.6884	.6893
4.9	.6902	.6911	.6920	.6928	.6937	.6946	.6955	.6964	.6972	.6981
5.0	.6990	.6998	.7007	.7016	.7024	.7033	.7042	.7050	.7059	.7067
5.1	.7076	.7084	.7093	.7101	.7110	.7118	.7126	.7135	.7143	.7152
5.2	.7160	.7168	.7177	.7185	.7193	.7202	.7210	.7218	.7226	.7235
5.3	.7243	.7251	.7259	.7267	.7275	.7284	.7292	.7300	.7308	.7316
5.4	.7324	.7332	.7340	.7348	.7356	.7364	.7372	.7380	.7388	.7396
x	**0**	**1**	**2**	**3**	**4**	**5**	**6**	**7**	**8**	**9**

Table 2 Common Logarithms (*Continued*)

x	0	1	2	3	4	5	6	7	8	9
5.5	.7404	.7412	.7419	.7427	.7435	.7443	.7451	.7459	.7466	.7474
5.6	.7482	.7490	.7497	.7505	.7513	.7520	.7528	.7536	.7543	.7551
5.7	.7559	.7566	.7574	.7582	.7589	.7597	.7604	.7612	.7619	.7627
5.8	.7634	.7642	.7649	.7657	.7664	.7672	.7679	.7686	.7694	.7701
5.9	.7709	.7716	.7723	.7731	.7738	.7745	.7752	.7760	.7767	.7774
6.0	.7782	.7789	.7796	.7803	.7810	.7818	.7825	.7832	.7839	.7846
6.1	.7853	.7860	.7868	.7875	.7882	.7889	.7896	.7903	.7910	.7917
6.2	.7924	.7931	.7938	.7945	.7952	.7959	.7966	.7973	.7980	.7987
6.3	.7993	.8000	.8007	.8014	.8021	.8028	.8035	.8041	.8048	.8055
6.4	.8062	.8069	.8075	.8082	.8089	.8096	.8102	.8109	.8116	.8122
6.5	.8129	.8136	.8142	.8149	.8156	.8162	.8169	.8176	.8182	.8189
6.6	.8195	.8202	.8209	.8215	.8222	.8228	.8235	.8241	.8248	.8254
6.7	.8261	.8267	.8274	.8280	.8287	.8293	.8299	.8306	.8312	.8319
6.8	.8325	.8331	.8338	.8344	.8351	.8357	.8363	.8370	.8376	.8382
6.9	.8388	.8395	.8401	.8407	.8414	.8420	.8426	.8432	.8439	.8445
7.0	.8451	.8457	.8463	.8470	.8476	.8482	.8488	.8494	.8500	.8506
7.1	.8513	.8519	.8525	.8531	.8537	.8543	.8549	.8555	.8561	.8567
7.2	.8573	.8579	.8585	.8591	.8597	.8603	.8609	.8615	.8621	.8627
7.3	.8633	.8639	.8645	.8651	.8657	.8663	.8669	.8675	.8681	.8686
7.4	.8692	.8698	.8704	.8710	.8716	.8722	.8727	.8733	.8739	.8745
7.5	.8751	.8756	.8762	.8768	.8774	.8779	.8785	.8791	.8797	.8802
7.6	.8808	.8814	.8820	.8825	.8831	.8837	.8842	.8848	.8854	.8859
7.7	.8865	.8871	.8876	.8882	.8887	.8893	.8899	.8904	.8910	.8915
7.8	.8921	.8927	.8932	.8938	.8943	.8949	.8954	.8960	.8965	.8971
7.9	.8976	.8982	.8987	.8993	.8998	.9004	.9009	.9015	.9020	.9025
8.0	.9031	.9036	.9042	.9047	.9053	.9058	.9063	.9069	.9074	.9079
8.1	.9085	.9090	.9096	.9101	.9106	.9112	.9117	.9122	.9128	.9133
8.2	.9138	.9143	.9149	.9154	.9159	.9165	.9170	.9175	.9180	.9186
8.3	.9191	.9196	.9201	.9206	.9212	.9217	.9222	.9227	.9232	.9238
8.4	.9243	.9248	.9253	.9258	.9263	.9269	.9274	.9279	.9284	.9289
8.5	.9294	.9299	.9304	.9309	.9315	.9320	.9325	.9330	.9335	.9340
8.6	.9345	.9350	.9355	.9360	.9365	.9370	.9375	.9380	.9385	.9390
8.7	.9395	.9400	.9405	.9410	.9415	.9420	.9425	.9430	.9435	.9440
8.8	.9445	.9450	.9455	.9460	.9465	.9469	.9474	.9479	.9484	.9489
8.9	.9494	.9499	.9504	.9509	.9513	.9518	.9523	.9528	.9533	.9538
9.0	.9542	.9547	.9552	.9557	.9562	.9566	.9571	.9576	.9581	.9586
9.1	.9590	.9595	.9600	.9605	.9609	.9614	.9619	.9624	.9628	.9633
9.2	.9638	.9643	.9647	.9652	.9657	.9661	.9666	.9671	.9675	.9680
9.3	.9685	.9689	.9694	.9699	.9703	.9708	.9713	.9717	.9722	.9727
9.4	9731	9736	.9741	.9745	.9750	.9754	.9759	.9763	.9768	.9773
9.5	.9777	.9782	.9786	.9791	.9795	.9800	.9805	.9809	.9814	.9818
9.6	.9823	.9827	.9832	.9836	.9841	.9845	.9850	.9854	.9859	.9863
9.7	.9868	.9872	.9877	.9881	.9886	.9890	.9894	.9899	.9903	.9908
9.8	.9912	.9917	.9921	.9926	.9930	.9934	.9939	.9943	.9948	.9952
9.9	.9956	.9961	.9965	.9969	.9974	.9978	.9983	.9987	.9991	.9996
x	0	1	2	3	4	5	6	7	8	9

Table 3 Trigonometric Functions

Degrees	Radians	Sine	Tangent	Cotangent	Cosine		
0°00′	.0000	.0000	.0000	undefined	1.0000	1.5708	**90°00′**
10′	.0029	.0029	.0029	343.77	1.0000	1.5679	50′
20′	.0058	.0058	.0058	171.89	1.0000	1.5650	40′
30′	.0087	.0087	.0087	114.59	1.0000	1.5621	30′
40′	.0116	.0116	.0116	85.940	.9999	1.5592	20′
50′	.0145	.0145	.0145	68.750	.9999	1.5563	10′
1°00′	.0175	.0175	.0175	57.290	.9998	1.5533	**89°00′**
10′	.0204	.0204	.0204	49.104	.9998	1.5504	50′
20′	.0233	.0233	.0233	42.964	.9997	1.5475	40′
30′	.0262	.0262	.0262	38.188	.9997	1.5446	30′
40′	.0291	.0291	.0291	34.368	.9996	1.5417	20′
50′	.0320	.0320	.0320	31.242	.9995	1.5388	10′
2°00′	.0349	.0349	.0349	28.636	.9994	1.5359	**88°00′**
10′	.0378	.0378	.0378	26.432	.9993	1.5330	50′
20′	.0407	.0407	.0407	24.542	.9992	1.5301	40′
30′	.0436	.0436	.0437	22.904	.9990	1.5272	30′
40′	.0465	.0465	.0466	21.470	.9989	1.5243	20′
50′	.0495	.0494	.0495	20.206	.9988	1.5213	10′
3°00′	.0524	.0523	.0524	19.081	.9986	1.5184	**87°00′**
10′	.0553	.0552	.0553	18.075	.9985	1.5155	50′
20′	.0582	.0581	.0582	17.169	.9983	1.5126	40′
30′	.0611	.0610	.0612	16.350	.9981	1.5097	30′
40′	.0640	.0640	.0641	15.605	.9980	1.5068	20′
50′	.0669	.0669	.0670	14.924	.9978	1.5039	10′
4°00′	.0698	.0698	.0699	14.301	.9976	1.5010	**86°00′**
10′	.0727	.0727	.0729	13.727	.9974	1.4981	50′
20′	.0756	.0756	.0758	13.197	.9971	1.4952	40′
30′	.0785	.0785	.0787	12.706	.9969	1.4923	30′
40′	.0814	.0814	.0816	12.251	.9967	1.4893	20′
50′	.0844	.0843	.0846	11.826	.9964	1.4864	10′
5°00′	.0873	.0872	.0875	11.430	.9962	1.4835	**85°00′**
10′	.0902	.0901	.0904	11.059	.9959	1.4806	50′
20′	.0931	.0929	.0934	10.712	.9957	1.4777	40′
30′	.0960	.0958	.0963	10.385	.9954	1.4748	30′
40′	.0989	.0987	.0992	10.078	.9951	1.4719	20′
50′	.1018	.1016	.1022	9.7882	.9948	1.4690	10′
6°00′	.1047	.1045	.1051	9.5144	.9945	1.4661	**84°00′**
10′	.1076	.1074	.1080	9.2553	.9942	1.4632	50′
20′	.1105	.1103	.1110	9.0098	.9939	1.4603	40′
30′	.1134	.1132	.1139	8.7769	.9936	1.4573	30′
40′	.1164	.1161	.1169	8.5555	.9932	1.4544	20′
50′	.1193	.1190	.1198	8.3450	.9929	1.4515	10′
7°00′	.1222	.1291	.1228	8.1443	.9925	1.4486	**83°00′**
10′	.1251	.1248	.1257	7.9530	.9922	1.4457	50′
20′	.1280	.1276	.1287	7.7704	.9918	1.4428	40′
30′	.1309	.1305	.1317	7.5958	.9914	1.4399	30′
40′	.1338	.1334	.1346	7.4287	.9911	1.4370	20′
50′	.1367	.1363	.1376	7.2687	.9907	1.4341	10′
8°00′	.1396	.1392	.1405	7.1154	.9903	1.4312	**82°00′**
10′	.1425	.1421	.1435	6.9682	.9899	1.4283	50′
20′	.1454	.1449	.1465	6.8269	.9894	1.4254	40′
30′	.1484	.1478	.1495	6.6912	.9890	1.4224	30′
40′	.1513	.1507	.1524	6.5606	.9886	1.4195	20′
50′	.1542	.1536	.1554	6.4348	.9881	1.4166	10′
9°00′	.1571	.1564	.1584	6.3138	.9877	1.4137	**81°00′**
		Cosine	**Cotangent**	**Tangent**	**Sine**	**Radians**	**Degrees**

Table 3 Trigonometric Functions (*Continued*)

Degrees	Radians	Sine	Tangent	Cotangent	Cosine		
9°00′	.1571	.1564	.1584	6.3138	.9877	1.4137	**81°00′**
10′	.1600	.1593	.1614	6.1970	.9872	1.4108	50′
20′	.1629	.1622	.1644	6.0844	.9868	1.4079	40′
30′	.1658	.1650	.1673	5.9758	.9863	1.4050	30′
40′	.1687	.1679	.1703	5.8708	.9858	1.4021	20′
50′	.1716	.1708	.1733	5.7694	.9853	1.3992	10′
10°00′	.1745	.1736	.1763	5.6713	.9848	1.3963	**80°00′**
10′	.1774	.1765	.1793	5.5764	.9843	1.3934	50′
20′	.1804	.1794	.1823	5.4845	.9838	1.3904	40′
30′	.1833	.1822	.1853	5.3955	.9833	1.3875	30′
40′	.1862	.1851	.1883	5.3093	.9827	1.3846	20′
50′	.1891	.1880	.1914	5.2257	.9822	1.3817	10′
11°00′	.1920	.1908	.1944	5.1446	.9816	1.3788	**79°00′**
10′	.1949	.1937	.1974	5.0658	.9811	1.3759	50′
20′	.1978	.1965	.2004	4.9894	.9805	1.3730	40′
30′	.2007	.1994	.2035	4.9152	.9799	1.3701	30′
40′	.2036	.2022	.2065	4.8430	.9793	1.3672	20′
50′	.2065	.2051	.2095	4.7729	.9787	1.3643	10′
12°00′	.2094	.2079	.2126	4.7046	.9781	1.3614	**78°00′**
10′	.2123	.2108	.2156	4.6382	.9775	1.3584	50′
20′	.2153	.2136	.2186	4.5736	.9769	1.3555	40′
30′	.2182	.2164	.2217	4.5107	.9763	1.3526	30′
40′	.2211	.2193	.2247	4.4494	.9757	1.3497	20′
50′	.2240	.2221	.2278	4.3897	.9750	1.3468	10′
13°00′	.2269	.2250	.2309	4.3315	.9744	1.3439	**77°00′**
10′	.2298	.2278	.2339	4.2747	.9737	1.3410	50′
20′	.2327	.2306	.2370	4.2193	.9730	1.3381	40′
30′	.2356	.2334	.2401	4.1653	.9724	1.3352	30′
40′	.2385	.2363	.2432	4.1126	.9717	1.3323	20′
50′	.2414	.2391	.2462	4.0611	.9710	1.3294	10′
14°00′	.2443	.2419	.2493	4.0108	.9703	1.3265	**76°00′**
10′	.2473	.2447	.2524	3.9617	.9696	1.3235	50′
20′	.2502	.2476	.2555	3.9136	.9689	1.3206	40′
30′	.2531	.2504	.2586	3.8667	.9681	1.3177	30′
40′	.2560	.2532	.2617	3.8208	.9674	1.3148	20′
50′	.2589	.2560	.2648	3.7760	.9667	1.3119	10′
15°00′	.2618	.2588	.2679	3.7321	.9659	1.3090	**75°00′**
10′	.2647	.2616	.2711	3.6891	.9652	1.3061	50′
20′	.2676	.2644	.2742	3.6470	.9644	1.3032	40′
30′	.2705	.2672	.2773	3.6059	.9636	1.3003	30′
40′	.2734	.2700	.2805	3.5656	.9628	1.2974	20′
50′	.2763	.2728	.2836	3.5261	.9621	1.2945	10′
16°00′	.2793	.2756	.2867	3.4874	.9613	1.2915	**74°00′**
10′	.2822	.2784	.2899	3.4495	.9605	1.2886	50′
20′	.2851	.2812	.2931	3.4124	.9596	1.2857	40′
30′	.2880	.2840	.2962	3.3759	.9588	1.2828	30′
40′	.2909	.2868	.2994	3.3402	.9580	1.2799	20′
50′	.2938	.2896	.3026	3.3052	.9572	1.2770	10′
17°00′	.2967	.2924	.3057	3.2709	.9563	1.2741	**73°00′**
10′	.2996	.2952	.3089	3.2371	.9555	1.2712	50′
20′	.3025	.2979	.3121	3.2041	.9546	1.2683	40′
30′	.3054	.3007	.3153	3.1716	.9537	1.2654	30′
40′	.3083	.3035	.3185	3.1397	.9528	1.2625	20′
50′	.3113	.3062	.3217	3.1084	.9520	1.2595	10′
18°00′	.3142	.3090	.3249	3.0777	.9511	1.2566	**72°00′**
		Cosine	**Cotangent**	**Tangent**	**Sine**	**Radians**	**Degrees**

Table 3 Trigonometric Functions (*Continued*)

Degrees	Radians	Sine	Tangent	Cotangent	Cosine		
18°00′	.3142	.3090	.3249	3.0777	.9511	1.2566	**72°00′**
10′	.3171	.3118	.3281	3.0475	.9502	1.2537	50′
20′	.3200	.3145	.3314	3.0178	.9492	1.2508	40′
30′	.3229	.3173	.3346	2.9887	.9483	1.2479	30′
40′	.3258	.3201	.3378	2.9600	.9474	1.2450	20′
50′	.3287	.3228	.3411	2.9319	.9465	1.2421	10′
19°00′	.3316	.3256	.3443	2.9042	.9455	1.2392	**71°00′**
10′	.3345	.3283	.3476	2.8770	.9446	1.2363	50′
20′	.3374	.3311	.3508	2.8502	.9436	1.2334	40′
30′	.3403	.3338	.3541	2.8239	.9426	1.2305	30′
40′	.3432	.3365	.3574	2.7980	.9417	1.2275	20′
50′	.3462	.3393	.3607	2.7725	.9407	1.2246	10′
20°00′	.3491	.3420	.3640	2.7475	.9397	1.2217	**70°00′**
10′	.3520	.3448	.3673	2.7228	.9387	1.2188	50′
20′	.3549	.3475	.3706	2.6985	.9377	1.2159	40′
30′	.3578	.3502	.3739	2.6746	.9367	1.2130	30′
40′	.3607	.3529	.3772	2.6511	.9356	1.2101	20′
50′	.3636	.3557	.3805	2.6279	.9346	1.2072	10′
21°00′	.3665	.3584	.3839	2.6051	.9336	1.2043	**69°00′**
10′	.3694	.3611	.3872	2.5826	.9325	1.2014	50′
20′	.3723	.3638	.3906	2.5605	.9315	1.1985	40′
30′	.3752	.3665	.3939	2.5386	.9304	1.1956	30′
40′	.3782	.3692	.3973	2.5172	.9293	1.1926	20′
50′	.3811	.3719	.4006	2.4960	.9283	1.1869	10′
22°00′	.3840	.3746	.4040	2.4751	.9272	1.1868	**68°00′**
10′	.3869	.3773	.4074	2.4545	.9261	1.1839	50′
20′	.3898	.3800	.4108	2.4342	.9250	1.1810	40′
30′	.3927	.3827	.4142	2.4142	.9239	1.1781	30′
40′	.3956	.3854	.4176	2.3945	.9228	1.1752	20′
50′	.3985	.3881	.4210	2.3750	.9216	1.1723	10′
23°00′	.4014	.3907	.4245	2.3559	.9205	1.1694	**67°00′**
10′	.4043	.3934	.4279	2.3369	.9194	1.1665	50′
20′	.4072	.3961	.4314	2.3183	.9182	1.1636	40′
30′	.4102	.3987	.4348	2.2998	.9171	1.1606	30′
40′	.4131	.4014	.4383	2.2817	.9159	1.1577	20′
50′	.4160	.4041	.4417	2.2637	.9147	1.1548	10′
24°00′	.4189	.4067	.4452	2.2460	.9135	1.1519	**66°00′**
10′	.4128	.4094	.4487	2.2286	.9124	1.1490	50′
20′	.4247	.4120	.4522	2.2113	.9112	1.1461	40′
30′	.4276	.4147	.4557	2.1943	.9100	1.1432	30′
40′	.4305	.4173	.4592	2.1775	.9088	1.1403	20′
50′	.4334	.4200	.4628	2.1609	.9075	1.1374	10′
25°00′	.4363	.4226	.4663	2.1445	.9063	1.1345	**65°00′**
10′	.4392	.4253	.4699	2.1283	.9051	1.1316	50′
20′	.4422	.4279	.4734	2.1123	.9038	1.1286	40′
30′	.4451	.4305	.4770	2.0965	.9026	1.1257	30′
40′	.4480	.4331	.4806	2.0809	.9013	1.1228	20′
50′	.4509	.4358	.4841	2.0655	.9001	1.1199	10′
26°00′	.4538	.4384	.4877	2.0503	.8988	1.1170	**64°00′**
10′	.4567	.4410	.4913	2.0353	.8975	1.1141	50′
20′	.4596	.4436	.4950	2.0204	.8962	1.1112	40′
30′	.4625	.4462	.4986	2.0057	.8949	1.1083	30′
40′	.4654	.4488	.5022	1.9912	.8936	1.1054	20′
50′	.4683	.4514	.5059	1.9768	.8923	1.1025	10′
27°00′	.4712	.4540	.5095	1.9626	.8910	1.0996	**63°00′**
		Cosine	**Cotangent**	**Tangent**	**Sine**	**Radians**	**Degrees**

Table 3 Trigonometric Functions (*Continued*)

Degrees	Radians	Sine	Tangent	Cotangent	Cosine		
27°00′	.4712	.4540	.5095	1.9626	.8910	1.0996	**63°00′**
10′	.4741	.4566	.5132	1.9486	.8897	1.0966	50′
20′	.4771	.4592	.5169	1.9347	.8884	1.0937	40′
30′	.4800	.4617	.5206	1.9210	.8870	1.0908	30′
40′	.4829	.4643	.5243	1.9074	.8857	1.0879	20′
50′	.4858	.4669	.5280	1.8940	.8843	1.0850	10′
28°00′	.4887	.4695	.5317	1.8807	.8829	1.0821	**62°00′**
10′	.4916	.4720	.5354	1.8676	.8816	1.0792	50′
20′	.4945	.4746	.5392	1.8546	.8802	1.0763	40′
30′	.4974	.4772	.5430	1.8418	.8788	1.0734	30′
40′	.5003	.4797	.5467	1.8291	.8774	1.0705	20′
50′	.5032	.4823	.5505	1.8165	.8760	1.0676	10′
29°00′	.5061	.4848	.5543	1.8040	.8746	1.0647	**61°00′**
10′	.5091	.4874	.5581	1.7917	.8732	1.0617	50′
20′	.5120	.4899	.5619	1.7796	.8718	1.0588	40′
30′	.5149	.4924	.5658	1.7675	.8704	1.0559	30′
40′	.5178	.4950	.5696	1.7556	.8689	1.0530	20′
50′	.5207	.4975	.5735	1.7437	.8675	1.0501	10′
30°00′	.5236	.5000	.5774	1.7321	.8660	1.0472	**60°00′**
10′	.5265	.5025	.5812	1.7205	.8646	1.0443	50′
20′	.5294	.5050	.5851	1.7090	.8631	1.0414	40′
30′	.5323	.5075	.5890	1.6977	.8616	1.0385	30′
40′	.5352	.5100	.5930	1.6864	.8601	1.0356	20′
50′	.5381	.5125	.5969	1.6753	.8587	1.0327	10′
31°00′	.5411	.5150	.6009	1.6643	.8572	1.0297	**59°00′**
10′	.5440	.5175	.6048	1.6534	.8557	1.0268	50′
20′	.5469	.5200	.6088	1.6426	.8542	1.0239	40′
30′	.5498	.5225	.6128	1.6319	.8526	1.0210	30′
40′	.5527	.5250	.6168	1.6212	.8511	1.0181	20′
50′	.5556	.5275	.6208	1.6107	.8496	1.0152	10′
32°00′	.5585	.5299	.6249	1.6003	.8480	1.0123	**58°00′**
10′	.5614	.5324	.6289	1.5900	.8465	1.0094	50′
20′	.5643	.5348	.6330	1.5789	.8450	1.0065	40′
30′	.5672	.5373	.6371	1.5697	.8434	1.0036	30′
40′	.5701	.5398	.6412	1.5597	.8418	1.0007	20′
50′	.5730	.5422	.6453	1.5497	.8403	.9977	10′
33°00′	.5760	.5446	.6494	1.5399	.8387	.9948	**57°00′**
10′	.5789	.5471	.6536	1.5301	.8371	.9919	50′
20′	.5818	.5495	.6577	1.5204	.8355	.9890	40′
30′	.5847	.5519	.6619	1.5108	.8339	.9861	30′
40′	.5876	.5544	.6661	1.5013	.8323	.9832	20′
50′	.5905	.5568	.6703	1.4919	.8307	.9803	10′
34°00′	.5934	.5592	.6745	1.4826	.8290	.9774	**56°00′**
10′	.5963	.5616	.6787	1.4733	.8274	.9745	50′
20′	.5992	.5640	.6830	1.4641	.8258	.9716	40′
30′	.6021	.5664	.6873	1.4550	.8241	.9687	30′
40′	.6050	.5688	.6916	1.4460	.8225	.9657	20′
50′	.6080	.5712	.6959	1.4370	.8208	.9628	10′
35°00′	.6109	.5736	.7002	1.4281	.8192	.9599	**55°00′**
10′	.6138	.5760	.7046	1.4193	.8175	.9570	50′
20′	.6167	.5783	.7089	1.4106	.8158	.9541	40′
30′	.6196	.5807	.7133	1.4019	.8141	.9512	30′
40′	.6225	.5831	.7177	1.3934	.8124	.9483	20′
50′	.6254	.5854	.7221	1.3848	.8107	.9454	10′
36°00′	.6283	.5878	.7265	1.3764	.8090	.9425	**54°00′**
		Cosine	**Cotangent**	**Tangent**	**Sine**	**Radians**	**Degrees**

Table 3 Trigonometric Functions (*Continued*)

Degrees	Radians	Sine	Tangent	Cotangent	Cosine		
36°00′	.6283	.5878	.7265	1.3764	.8090	.9425	**54°00′**
10′	.6312	.5901	.7310	1.3680	.8073	.9396	50′
20′	.6341	.5925	.7355	1.3597	.8056	.9367	40′
30′	.6370	.5948	.7400	1.3514	.8039	.9338	30′
40′	.6400	.5972	.7445	1.3432	.8021	.9308	20′
50′	.6429	.5995	.7490	1.3351	.8004	.9279	10′
37°00′	.6458	.6018	.7536	1.3270	.7986	.9250	**53°00′**
10′	.6487	.6041	.7581	1.3190	.7969	.9221	50′
20′	.6516	.6065	.7627	1.3111	.7951	.9192	40′
30′	.6545	.6088	.7673	1.3032	.7934	.9163	30′
40′	.6574	.6111	.7720	1.2954	.7916	.9134	20′
50′	.6603	.6134	.7766	1.2876	.7898	.9105	10′
38°00′	.6632	.6157	.7813	1.2799	.7880	.9076	**52°00′**
10′	.6661	.6180	.7860	1.2723	.7862	.9047	50′
20′	.6690	.6202	.7907	1.2647	.7844	.9018	40′
30′	.6720	.6225	.7954	1.2572	.7826	.8988	30′
40′	.6749	.6248	.8002	1.2497	.7808	.8959	20′
50′	.6778	.6271	.8050	1.2423	.7790	.8930	10′
39°00′	.6807	.6293	.8098	1.2349	.7771	.8901	**51°00′**
10′	.6836	.6316	.8146	1.2276	.7753	.8872	50′
20′	.6865	.6338	.8195	1.2203	.7735	.8843	40′
30′	.6894	.6361	.8243	1.2131	.7716	.8814	30′
40′	.6923	.6383	.8292	1.2059	.7698	.8785	20′
50′	.6952	.6406	.8342	1.1988	.7679	.8756	10′
40°00′	.6981	.6428	.8391	1.1918	.7660	.8727	**50°00′**
10′	.7010	.6450	.8441	1.1847	.7642	.8698	50′
20′	.7039	.6472	.8491	1.1778	.7623	.8668	40′
30′	.7069	.6494	.8541	1.1708	.7604	.8639	30′
40′	.7098	.6517	.8591	1.1640	.7585	.8610	20′
50′	.7127	.6539	.8642	1.1571	.7566	.8581	10′
41°00′	.7156	.6561	.8693	1.1504	.7547	.8552	**49°00′**
10′	.7185	.6583	.8744	1.1436	.7528	.8523	50′
20′	.7214	.6604	.8796	1.1369	.7509	.8494	40′
30′	.7243	.6626	.8847	1.1303	.7490	.8465	30′
40′	.7272	.6648	.8899	1.1237	.7470	.8436	20′
50′	.7301	.6670	.8952	1.1171	.7451	.8407	10′
42°00′	.7330	.6691	.9004	1.1106	.7431	.8378	**48°00′**
10′	.7359	.6713	.9057	1.1041	.7412	.8348	50′
20′	.7389	.6734	.9110	1.0977	.7392	.8319	40′
30′	.7418	.6756	.9163	1.0913	.7373	.8290	30′
40′	.7447	.6777	.9217	1.0850	.7353	.8261	20′
50′	.7476	.6799	.9271	1.0786	.7333	.8232	10′
43°00′	.7505	.6820	.9325	1.0724	.7314	.8203	**47°00′**
10′	.7534	.6841	.9380	1.0661	.7294	.8174	50′
20′	.7563	.6862	.9435	1.0599	.7274	.8145	40′
30′	.7592	.6884	.9490	1.0538	.7254	.8116	30′
40′	.7621	.6905	.9545	1.0477	.7234	.8087	20′
50′	.7650	.6926	.9601	1.0416	.7214	.8058	10′
44°00′	.7679	.6947	.9657	1.0355	.7193	.8029	**46°00′**
10′	.7709	.6967	.9713	1.0295	.7173	.7999	50′
20′	.7738	.6988	.9770	1.0235	.7153	.7970	40′
30′	.7767	.7009	.9827	1.0176	.7133	.7941	30′
40′	.7796	.7030	.9884	1.0117	.7112	.7912	20′
50′	.7825	.7050	.9942	1.0058	.7092	.7883	10′
45°00′	.7854	.7071	1.0000	1.0000	.7071	.7854	**45°00′**
		Cosine	**Cotangent**	**Tangent**	**Sine**	**Radians**	**Degrees**

Table 4 Common Logarithms of Trigonometric Functions

Degrees	log sin	log cos	log tan	log cot	
0°00′	undefined	.0000	undefined	undefined	**90°00′**
10	.4637 − 3	.0000	.4637 − 3	2.5363	50
20	.7648 − 3	.0000	.7648 − 3	2.2352	40
30	.9408 − 3	.0000	.9409 − 3	2.0591	30
40	.0658 − 2	.0000	.0658 − 2	1.9342	20
50	.1627 − 2	.0000	.1627 − 2	1.8373	10
1°00′	.2419 − 2	.9999 − 1	.2419 − 2	1.7581	**89°00′**
10	.3088 − 2	.9999 − 1	.3089 − 2	1.6911	50
20	.3668 − 2	.9999 − 1	.3669 − 2	1.6331	40
30	.4179 − 2	.9999 − 1	.4181 − 2	1.5819	30
40	.4637 − 2	.9998 − 1	.4638 − 2	1.5362	20
50	.5050 − 2	.9998 − 1	.5053 − 2	1.4947	10
2°00′	.5428 − 2	.9997 − 1	.5431 − 2	1.4569	**88°00′**
10	.5776 − 2	.9997 − 1	.5779 − 2	1.4221	50
20	.6097 − 2	.9996 − 1	.6101 − 2	1.3899	40
30	.6397 − 2	.9996 − 1	.6401 − 2	1.3599	30
40	.6677 − 2	.9995 − 1	.6682 − 2	1.3318	20
50	.6940 − 2	.9995 − 1	.6945 − 2	1.3055	10
3°00′	.7188 − 2	.9994 − 1	.7194 − 2	1.2806	**87°00′**
10	.7423 − 2	.9993 − 1	.7429 − 2	1.2571	50
20	.7645 − 2	.9993 − 1	.7652 − 2	1.2348	40
30	.7857 − 2	.9992 − 1	.7865 − 2	1.2135	30
40	.8059 − 2	.9991 − 1	.8067 − 2	1.1933	20
50	.8251 − 2	.9990 − 1	.8261 − 2	1.1739	10
4°00′	.8436 − 2	.9989 − 1	.8446 − 2	1.1554	**86°00′**
10	.8613 − 2	.9989 − 1	.8624 − 2	1.1376	50
20	.8783 − 2	.9988 − 1	.8795 − 2	1.1205	40
30	.8946 − 2	.9987 − 1	.8960 − 2	1.1040	30
40	.9104 − 2	.9986 − 1	.9118 − 2	1.0882	20
50	.9256 − 2	.9985 − 1	.9272 − 2	1.0728	10
5°00′	.9403 − 2	.9983 − 1	.9420 − 2	1.0580	**85°00′**
10	.9545 − 2	.9982 − 1	.9563 − 2	1.0437	50
20	.9682 − 2	.9981 − 1	.9701 − 2	1.0299	40
30	.9816 − 2	.9980 − 1	.9836 − 2	1.0164	30
40	.9945 − 2	.9979 − 1	.9966 − 2	1.0034	20
50	.0070 − 1	.9977 − 1	.0093 − 1	.9907	10
6°00′	.1092 − 1	.9976 − 1	.0216 − 1	.9784	**84°00′**
10	.0311 − 1	.9975 − 1	.0336 − 1	.9664	50
20	.0426 − 1	.9973 − 1	.0453 − 1	.9547	40
30	.0539 − 1	.9972 − 1	.0567 − 1	.9433	30
40	.0648 1	.9971 − 1	.0678 − 1	.9322	20
50	.0755 − 1	.9969 − 1	.0786 − 1	.9214	10
7°00′	.0859 − 1	.9968 − 1	.0891 − 1	.9109	**83°00′**
10	.0961 − 1	.9966 − 1	.0995 − 1	.9005	50
20	.1060 − 1	.9964 − 1	.1096 − 1	.8904	40
30	.1157 − 1	.9963 − 1	.1194 − 1	.8806	30
40	.1252 − 1	.9961 − 1	.1291 − 1	.8709	20
50	.1345 − 1	.9959 − 1	.1385 − 1	.8615	10
8°00′	.1436 − 1	.9958 − 1	.1478 − 1	.8522	**82°00′**
10	.1525 − 1	.9956 − 1	.1569 − 1	.8431	50
20	.1612 − 1	.9954 − 1	.1658 − 1	.8342	40
30	.1697 − 1	.9952 − 1	.1745 − 1	.8255	30
40	.1781 − 1	.9950 − 1	.1831 − 1	.8169	20
50	.1863 − 1	.9948 − 1	.1915 − 1	.8085	10
9°00′	.1943 − 1	.9946 − 1	.1977 − 1	.8003	**81°00′**
	log cos	**log sin**	**log cot**	**log tan**	**Degrees**

Table 4 Common Logarithms of Trigonometric Functions (*Continued*)

Degrees	log sin	log cos	log tan	log cot	
9°00′	.1943 − 1	.9946 − 1	.1997 − 1	.8003	**81°00′**
10	.2022 − 1	.9944 − 1	.2078 − 1	.7922	50
20	.2100 − 1	.9942 − 1	.2158 − 1	.7842	40
30	.2176 − 1	.9940 − 1	.2236 − 1	.7764	30
40	.2251 − 1	.9938 − 1	.2313 − 1	.7687	20
50	.2324 − 1	.9936 − 1	.2389 − 1	.7611	10
10°00′	.2397 − 1	.9934 − 1	.2463 − 1	.7537	**80°00′**
10	.2468 − 1	.9931 − 1	.2536 − 1	.7464	50
20	.2538 − 1	.9929 − 1	.2609 − 1	.7391	40
30	.2606 − 1	.9927 − 1	.2680 − 1	.7320	30
40	.2674 − 1	.9924 − 1	.2750 − 1	.7250	20
50	.2740 − 1	.9922 − 1	.2819 − 1	.7181	10
11°00′	.2806 − 1	.9919 − 1	.2887 − 1	.7113	**79°00′**
10	.2870 − 1	.9917 − 1	.2953 − 1	.7047	50
20	.2934 − 1	.9914 − 1	.3020 − 1	.6980	40
30	.2997 − 1	.9912 − 1	.3085 − 1	.6915	30
40	.3058 − 1	.9909 − 1	.3149 − 1	.6851	20
50	.3119 − 1	.9907 − 1	.3212 − 1	.6788	10
12°00′	.3179 − 1	.9904 − 1	.3275 − 1	.6725	**78°00′**
10	.3238 − 1	.9901 − 1	.3336 − 1	.6664	50
20	.3296 − 1	.9899 − 1	.3397 − 1	.6603	40
30	.3353 − 1	.9896 − 1	.3458 − 1	.6542	30
40	.3410 − 1	.9893 − 1	.3517 − 1	.6483	20
50	.3466 − 1	.9890 − 1	.3576 − 1	.6424	10
13°00′	.3521 − 1	.9887 − 1	.3634 − 1	.6366	**77°00′**
10	.3575 − 1	.9884 − 1	.3691 − 1	.6309	50
20	.3629 − 1	.9881 − 1	.3748 − 1	.6252	40
30	.3682 − 1	.9878 − 1	.3804 − 1	.6196	30
40	.3734 − 1	.9875 − 1	.3859 − 1	.6141	20
50	.3786 − 1	.9872 − 1	.3914 − 1	.6086	10
14°00′	.3837 − 1	.9869 − 1	.3968 − 1	.6032	**76°00′**
10	.3887 − 1	.9866 − 1	.4021 − 1	.5979	50
20	.3937 − 1	.9863 − 1	.4074 − 1	.5926	40
30	.3986 − 1	.9859 − 1	.4127 − 1	.5873	30
40	.4035 − 1	.9856 − 1	.4178 − 1	.5822	20
50	.4083 − 1	.9853 − 1	.4230 − 1	.5770	10
15°00′	.4130 − 1	.9849 − 1	.4281 − 1	.5719	**75°00′**
10	.4177 − 1	.9846 − 1	.4331 − 1	.5669	50
20	.4223 − 1	.9843 − 1	.4381 − 1	.5619	40
30	.4269 − 1	.9839 − 1	.4430 − 1	.5570	30
40	.4314 − 1	.9836 − 1	.4479 − 1	.5521	20
50	.4359 − 1	.9832 − 1	.4527 − 1	.5473	10
16°00′	.4403 − 1	.9828 − 1	.4575 − 1	.5425	**74°00′**
10	.4447 − 1	.9825 − 1	.4622 − 1	.5378	50
20	.4491 − 1	.9821 − 1	.4669 − 1	.5331	40
30	.4533 − 1	.9817 − 1	.4716 − 1	.5284	30
40	.4576 − 1	.9814 − 1	.4762 − 1	.5238	20
50	.4618 − 1	.9810 − 1	.4808 − 1	.5192	10
17°00′	.4659 − 1	.9806 − 1	.4853 − 1	.5147	**73°00′**
10	.4700 − 1	.9802 − 1	.4898 − 1	.5102	50
20	.4741 − 1	.9798 − 1	.4943 − 1	.5057	40
30	.4781 − 1	.9794 − 1	.4987 − 1	.5013	30
40	.4821 − 1	.9790 − 1	.5031 − 1	.4969	20
50	.4861 − 1	.9786 − 1	.5075 − 1	.4925	10
18°00′	.4900 − 1	.9782 − 1	.5118 − 1	.4882	**72°00′**
	log cos	**log sin**	**log cot**	**log tan**	**Degrees**

Table 4 Common Logarithms of Trigonometric Functions (*Continued*)

Degrees	log sin	log cos	log tan	log cot	
18°00′	.4900 − 1	.9782 − 1	.5118 − 1	.4882	**72°00′**
10	.4939 − 1	.9778 − 1	.5161 − 1	.4839	50
20	.4977 − 1	.9774 − 1	.5203 − 1	.4797	40
30	.5015 − 1	.9770 − 1	.5245 − 1	.4755	30
40	.5052 − 1	.9765 − 1	.5287 − 1	.4713	20
50	.5090 − 1	.9761 − 1	.5329 − 1	.4671	10
19°00′	.5126 − 1	.9757 − 1	.5370 − 1	.4630	**71°00′**
10	.5163 − 1	.9752 − 1	.5411 − 1	.4589	50
20	.5199 − 1	.9748 − 1	.5451 − 1	.4549	40
30	.5235 − 1	.9743 − 1	.5491 − 1	.4509	30
40	.5270 − 1	.9739 − 1	.5531 − 1	.4469	20
50	.5306 − 1	.9734 − 1	.5571 − 1	.4429	10
20°00′	.5341 − 1	.9730 − 1	.5611 − 1	.4389	**70°00′**
10	.5375 − 1	.9725 − 1	.5650 − 1	.4350	50
20	.5409 − 1	.9721 − 1	.5689 − 1	.4311	40
30	.5443 − 1	.9716 − 1	.5727 − 1	.4273	30
40	.5477 − 1	.9711 − 1	.5766 − 1	.4234	20
50	.5510 − 1	.9706 − 1	.5804 − 1	.4196	10
21°00′	.5543 − 1	.9702 − 1	.5842 − 1	.4158	**69°00′**
10	.5576 − 1	.9697 − 1	.5879 − 1	.4121	50
20	.5609 − 1	.9692 − 1	.5917 − 1	.4083	40
30	.5641 − 1	.9687 − 1	.5954 − 1	.4046	30
40	.5673 − 1	.9682 − 1	.5991 − 1	.4009	20
50	.5704 − 1	.9677 − 1	.6028 − 1	.3972	10
22°00′	.5736 − 1	.9672 − 1	.6064 − 1	.3936	**68°00′**
10	.5767 − 1	.9667 − 1	.6100 − 1	.3900	50
20	.5798 − 1	.9661 − 1	.6136 − 1	.3864	40
30	.5828 − 1	.9656 − 1	.6172 − 1	.3828	30
40	.5859 − 1	.9651 − 1	.6208 − 1	.3792	20
50	.5889 − 1	.9646 − 1	.6243 − 1	.3757	10
23°00′	.5919 − 1	.9640 − 1	.6279 − 1	.3721	**67°00′**
10	.5948 − 1	.9635 − 1	.6314 − 1	.3686	50
20	.5978 − 1	.9629 − 1	.6348 − 1	.3652	40
30	.6007 − 1	.9624 − 1	.6383 − 1	.3617	30
40	.6036 − 1	.9618 − 1	.6417 − 1	.3583	20
50	.6065 − 1	.9613 − 1	.6452 − 1	.3548	10
24°00′	.6093 − 1	.9607 − 1	.6486 − 1	.3514	**66°00′**
10	.6121 − 1	.9602 − 1	.6520 − 1	.3480	50
20	.6149 − 1	.9596 − 1	.6553 − 1	.3447	40
30	.6177 − 1	.9590 − 1	.6587 − 1	.3413	30
40	.6205 − 1	.9584 − 1	.6620 − 1	.3380	20
50	.6232 − 1	.9579 − 1	.6654 − 1	.3346	10
25°00′	.6259 − 1	.9573 − 1	.6687 − 1	.3313	**65°00′**
10	.6286 − 1	.9567 − 1	.6720 − 1	.3280	50
20	.6313 − 1	.9561 − 1	.6752 − 1	.3248	40
30	.6340 − 1	.9555 − 1	.6785 − 1	.3215	30
40	.6366 − 1	.9549 − 1	.6817 − 1	.3183	20
50	.6392 − 1	.9543 − 1	.6850 − 1	.3150	10
26°00′	.6418 − 1	.9537 − 1	.6882 − 1	.3118	**64°00′**
10	.6444 − 1	.9530 − 1	.6914 − 1	.3086	50
20	.6470 − 1	.9524 − 1	.6946 − 1	.3054	40
30	.6495 − 1	.9518 − 1	.6977 − 1	.3023	30
40	.6521 − 1	.9512 − 1	.7009 − 1	.2991	20
50	.6546 − 1	.9505 − 1	.7040 − 1	.2960	10
27°00′	.6570 − 1	.9499 − 1	.7072 − 1	.2928	**63°00′**
	log cos	**log sin**	**log cot**	**log tan**	**Degrees**

Table 4 Common Logarithms of Trigonometric Functions (*Continued*)

Degrees	log sin	log cos	log tan	log cot	
27°00′	.6570 − 1	.9499 − 1	.7072 − 1	.2928	**63°00′**
10	.6595 − 1	.9492 − 1	.7103 − 1	.2897	50
20	.6620 − 1	.9486 − 1	.7134 − 1	.2866	40
30	.6644 − 1	.9479 − 1	.7165 − 1	.2835	30
40	.6668 − 1	.9473 − 1	.7196 − 1	.2804	20
50	.6692 − 1	.9466 − 1	.7226 − 1	.2774	10
28°00′	.6716 − 1	.9459 − 1	.7257 − 1	.2743	**62°00′**
10	.6740 − 1	.9453 − 1	.7287 − 1	.2713	50
20	.6763 − 1	.9446 − 1	.7317 − 1	.2683	40
30	.6787 − 1	.9439 − 1	.7348 − 1	.2652	30
40	.6810 − 1	.9432 − 1	.7378 − 1	.2622	20
50	.6833 − 1	.9425 − 1	.7408 − 1	.2592	10
29°00′	.6856 − 1	.9418 − 1	.7438 − 1	.2562	**61°00′**
10	.6878 − 1	.9411 − 1	.7467 − 1	.2533	50
20	.6901 − 1	.9404 − 1	.7497 − 1	.2503	40
30	.6923 − 1	.9397 − 1	.7526 − 1	.2474	30
40	.6946 − 1	.9390 − 1	.7556 − 1	.2444	20
50	.6968 − 1	.9383 − 1	.7585 − 1	.2415	10
30°00′	.6990 − 1	.9375 − 1	.7614 − 1	.2386	**60°00′**
10	.7012 − 1	.9368 − 1	.7644 − 1	.2356	50
20	.7033 − 1	.9361 − 1	.7673 − 1	.2327	40
30	.7055 − 1	.9353 − 1	.7701 − 1	.2299	30
40	.7076 − 1	.9346 − 1	.7730 − 1	.2270	20
50	.7097 − 1	.9338 − 1	.7759 − 1	.2241	10
31°00′	.7118 − 1	.9331 − 1	.7788 − 1	.2212	**59°00′**
10	.7139 − 1	.9323 − 1	.7816 − 1	.2184	50
20	.7160 − 1	.9315 − 1	.7845 − 1	.2155	40
30	.7181 − 1	.9308 − 1	.7873 − 1	.2127	30
40	.7201 − 1	.9300 − 1	.7902 − 1	.2098	20
50	.7222 − 1	.9292 − 1	.7930 − 1	.2070	10
32°00′	.7242 − 1	.9284 − 1	.7958 − 1	.2042	**58°00′**
10	.7262 − 1	.9276 − 1	.7986 − 1	.2014	50
20	.7282 − 1	.9268 − 1	.8014 − 1	.1986	40
30	.7302 − 1	.9260 − 1	.8042 − 1	.1958	30
40	.7322 − 1	.9252 − 1	.8070 − 1	.1930	20
50	.7342 − 1	.9244 − 1	.8097 − 1	.1903	10
33°00′	.7361 − 1	.9236 − 1	.8125 − 1	.1875	**57°00′**
10	.7380 − 1	.9228 − 1	.8153 − 1	.1847	50
20	.7400 − 1	.9219 − 1	.8180 − 1	.1820	40
30	.7419 − 1	.9211 − 1	.8208 − 1	.1792	30
40	.7438 − 1	.9203 − 1	.8235 − 1	.1765	20
50	.7457 − 1	.9194 − 1	.8263 − 1	.1737	10
34°00′	.7476 − 1	.9186 − 1	.8290 − 1	.1710	**56°00′**
10	.7494 − 1	.9177 − 1	.8317 − 1	.1683	50
20	.7513 − 1	.9169 − 1	.8344 − 1	.1656	40
30	.7531 − 1	.9160 − 1	.8371 − 1	.1629	30
40	.7550 − 1	.9151 − 1	.8398 − 1	.1602	20
50	.7568 − 1	.9142 − 1	.8425 − 1	.1575	10
35°00′	.7586 − 1	.9134 − 1	.8452 − 1	.1548	**55°00′**
10	.7604 − 1	.9125 − 1	.8479 − 1	.1521	50
20	.7622 − 1	.9116 − 1	.8506 − 1	.1494	40
30	.7640 − 1	.9107 − 1	.8533 − 1	.1467	30
40	.7657 − 1	.9098 − 1	.8559 − 1	.1441	20
50	.7675 − 1	.9089 − 1	.8586 − 1	.1414	10
36°00′	.7692 − 1	.9080 − 1	.8613 − 1	.1387	**54°00′**
	log cos	**log sin**	**log cot**	**log tan**	**Degrees**

Table 4 Common Logarithms of Trigonometric Functions (*Continued*)

Degrees	log sin	log cos	log tan	log cot	
36°00′	.7692 − 1	.9080 − 1	.8613 − 1	.1387	**54°00′**
10	.7710 − 1	.9070 − 1	.8639 − 1	.1361	50
20	.7727 − 1	.9061 − 1	.8666 − 1	.1334	40
30	.7744 − 1	.9052 − 1	.8692 − 1	.1308	30
40	.7761 − 1	.9042 − 1	.8718 − 1	.1282	20
50	.7778 − 1	.9033 − 1	.8745 − 1	.1255	10
37°00′	.7795 − 1	.9023 − 1	.8771 − 1	.1229	**53°00′**
10	.7811 − 1	.9014 − 1	.8797 − 1	.1203	50
20	.7828 − 1	.9004 − 1	.8824 − 1	.1176	40
30	.7844 − 1	.8995 − 1	.8850 − 1	.1150	30
40	.7861 − 1	.8985 − 1	.8876 − 1	.1124	20
50	.7877 − 1	.8975 − 1	.8902 − 1	.1098	10
38°00′	.7893 − 1	.8965 − 1	.8928 − 1	.1072	**52°00′**
10	.7910 − 1	.8955 − 1	.8954 − 1	.1046	50
20	.7926 − 1	.8945 − 1	.8980 − 1	.1020	40
30	.7941 − 1	.8935 − 1	.9006 − 1	.0994	30
40	.7957 − 1	.8925 − 1	.9032 − 1	.0968	20
50	.7973 − 1	.8915 − 1	.9058 − 1	.0942	10
39°00′	.7989 − 1	.8905 − 1	.9084 − 1	.0916	**51°00′**
10	.8004 − 1	.8895 − 1	.9110 − 1	.0890	50
20	.8020 − 1	.8884 − 1	.9135 − 1	.0865	40
30	.8035 − 1	.8874 − 1	.9161 − 1	.0839	30
40	.8050 − 1	.8864 − 1	.9187 − 1	.0813	20
50	.8066 − 1	.8853 − 1	.9212 − 1	.0788	10
40°00′	.8081 − 1	.8843 − 1	.9238 − 1	.0762	**50°00′**
10	.8096 − 1	.8832 − 1	.9264 − 1	.0736	50
20	.8111 − 1	.8821 − 1	.9289 − 1	.0711	40
30	.8125 − 1	.8810 − 1	.9315 − 1	.0685	30
40	.8140 − 1	.8800 − 1	.9341 − 1	.0659	20
50	.8155 − 1	.8789 − 1	.9366 − 1	.0634	10
41°00′	.8169 − 1	.8778 − 1	.9392 − 1	.0608	**49°00′**
10	.8184 − 1	.8767 − 1	.9417 − 1	.0583	50
20	.8198 − 1	.8756 − 1	.9443 − 1	.0557	40
30	.8213 − 1	.8745 − 1	.9468 − 1	.0532	30
40	.8227 − 1	.8733 − 1	.9494 − 1	.0506	20
50	.8241 − 1	.8722 − 1	.9519 − 1	.0481	10
42°00′	.8255 − 1	.8711 − 1	.9544 − 1	.0456	**48°00′**
10	.8269 − 1	.8699 − 1	.9570 − 1	.0430	50
20	.8283 − 1	.8688 − 1	.9595 − 1	.0405	40
30	.8297 − 1	.8676 − 1	.9621 − 1	.0379	30
40	.8311 − 1	.8665 − 1	.9646 − 1	.0354	20
50	.8324 − 1	.8653 − 1	.9671 − 1	.0329	10
43°00′	.8338 − 1	.8641 − 1	.9697 − 1	.0303	**47°00′**
10	.8351 − 1	.8629 − 1	.9722 − 1	.0278	50
20	.8365 − 1	.8618 − 1	.9747 − 1	.0253	40
30	.8378 − 1	.8606 − 1	.9772 − 1	.0228	30
40	.8391 − 1	.8594 − 1	.9798 − 1	.0202	20
50	.8405 − 1	.8582 − 1	.9823 − 1	.0177	10
44°00′	.8418 − 1	.8569 − 1	.9848 − 1	.0152	**46°00′**
10	.8431 − 1	.8557 − 1	.9874 − 1	.0126	50
20	.8444 − 1	.8545 − 1	.9899 − 1	.0101	40
30	.8457 − 1	.8532 − 1	.9924 − 1	.0076	30
40	.8469 − 1	.8520 − 1	.9949 − 1	.0051	20
50	.8482 − 1	.8507 − 1	.9975 − 1	.0025	10
45°00′	.8495 − 1	.8495 − 1	.0000	.0000	**45°00′**
	log cos	**log sin**	**log cot**	**log tan**	**Degrees**

Chapter 1

Section 1.1, p. 6

1. (a) $x + 5$ (b) $x - 8$ (c) $3x$ (d) $\frac{1}{4}x$ or $\frac{x}{4}$ (e) $x + 3$ (f) $x - 10$ **3.** term **5.** term
7. term **9.** not a term **11.** term **13.** -19 **15.** -7 **17.** π **19.** $\frac{2}{3}$ **21.** $10xy$ **23.** $-xy$
25. polynomial **27.** polynomial **29.** not a polynomial **31.** polynomial **33.** 17 **35.** -6
37. (a) 1991 (b) 11 (c) 1 099 901 (d) 1.001 **39.** -3 **41.** 1 **43.** -133 **45.** (a) -1 (b) 0
47. 121 square feet **49.** 670 square centimeters **51.** 1200π cubic inches **53.** 240 kilometers
55. (a) 64 feet (b) 48 feet (c) 0 feet

Section 1.2, p. 11

1. similar **3.** similar **5.** not similar **9.** $14, \frac{1}{2}|x^2, -x^2|x, \frac{1}{2}x$ **11.** $-r^2st, 2r^2ts|rs^2t, 3s^2rt|4rst^2$
13. $9a$ **15.** $14b$ **17.** $2b + 1$ **19.** $5r - 5s$ **21.** $6a + 4b$ **23.** $3w + x + 2z$ **25.** $-2a$
27. $-2a - 5b$ **29.** $abc - ab + ac - 4$ **31.** $x + 2y$ **33.** $x + 2z$ **35.** $3b$ **37.** $2m - r + 4$
39. $7a + 3c$ **41.** $4a - 4b + 3$ **43.** $5x + 5y - z$ **45.** x^3 **47.** $-3x - 3z$ **49.** 40

Section 1.3, p. 15

1. x^3 **3.** b^{12} **5.** z^9 **7.** y^4z^3 **9.** $42a^3b^4$ **11.** $2x - 2y$ **13.** $x^2 + 3x$ **15.** $-5x^3 - 5x^3y$
17. $3r^3s^2t^2 - 18r^2s^2t^3 + 6r^5s^2t^{12}$ **19.** $m^2 + 5m - 14$ **21.** $x^2 + 12x + 36$ **23.** $x^2 - 81$ **25.** $u^8 - 1$
27. $x^3 + 6x^2 + 4x - 5$ **29.** $x^5 + x^3y - 2x^2y - 2y^2$ **31.** $-10a^2 + 36ab - 18b^2$
33. $x^2 + 2xy + y^2 + xz + yz$ **35.** $2a^2 + 7ab + 3b^2 + 5ac + 15bc$ **37.** $2a - 3b - 6c$ **39.** $3x^2 - 6xy + 12x$
41. $x^2 - x + y^2 + y - 4$ **43.** $-2a^2 + 4ab + b^2$ **45.** $2x^2$ **47.** $x^3 - 16x^2 + 79x - 129$
49. $a^3 + 6a^2 + 11a + 6$ **51.** $10x^4 - 32x^3 + 18x^2 - 32x + 4$ **53.** $x^6 + 3x^5 - x^4 + 6x^3 + 4x^2 - 3x + 2$
55. $a^3 + 13a^2 + 16a + 6$ **57.** $2x^2 + 7x + 5$

Section 1.4, p. 21

1. 2^3 **3.** $2^3 \cdot 5$ **5.** $2^2 \cdot 3 \cdot 7$ **7.** $2^2 \cdot 5^2$ **9.** 5^3 **11.** $2^5 \cdot 5$ **13.** 8 **15.** 10 **17.** 12 **19.** 2
21. 25 **23.** 30 **25.** $5(a - 3)$ **27.** $x(3x + 5)$ **29.** $2x(x + 4)$ **31.** $12z^7(9z^3 - 8)$ **33.** $4a^2b^2(2b + 3a)$

35. $4(2x^2 + x + 3)$ **37.** $7b^2(b^3 + 2b^2 - 1)$ **39.** $xz(7xy^3 + 10yz^3 - 5z^2)$ **41.** $(x + 1)(x - 1)$
43. $(1 + z)(1 - z)$ **45.** $(x + 3z)(x - 3z)$ **47.** $(12r + 11t)(12r - 11t)$ **49.** $(x^2 + 2y^3)(x^2 - 2y^3)$
51. $(11a^4 + 10b^3)(11a^4 - 10b^3)$ **53.** $3(a + 2x)(a - 2x)$ **55.** $x(x + 1)(x - 1)$ **57.** $3a(x + y)(x - y)$
59. $(4y^6 + 1)(2y^3 + 1)(2y^3 - 1)$ **61.** $(2x + 3 + y)(2x + 3 - y)$ **63.** $4ab$

Section 1.5, p. 27

1. $(x + 2)(x + 1)$ **3.** $(x + 2)^2$ **5.** $(m + 2)(m - 1)$ **7.** $(s + 4)(s - 3)$ **9.** $(a - 5)^2$ **11.** $(2x + 1)(x + 1)$
13. $(2m - 1)(m + 3)$ **15.** $(3x - 1)^2$ **17.** $(x + y)^2$ **19.** $(2u + v)^2$ **21.** $(c + d)(u + v)$
23. $(y - a)(y + b)$ **25.** $2(2a - 3b)(x + y)$ **27.** $(a + b)(a + 5)$ **29.** $(a - 1)(a^2 + a + 1)$
31. $(x - 2z)(x^2 + 2xz + 4z^2)$ **33.** $(y + z)(y^2 - yz + z^2)(y - z)(y^2 + yz + z^2)$
35. $(100 - z^3)(10\,000 + 100z^3 + z^6)$ **37.** $5(a + 2)^2$ **39.** DOES NOT FACTOR. **41.** $5(3x + 1)^2$
43. $-2(y + 7)(y + 5)$ **45.** DOES NOT FACTOR. **47.** $(x + y)(x^2 - xy + y^2)(a + b)(a - b)$
49. $4y^2(2y - 5)(y + 5)$ **51.** $(a + b)(x^2 + xy + y^2)$ **53.** $20(x^2y^2 + 2)(x^4y^4 - 2x^2y^2 + 4)$
55. $(a + b)(a - b)(x + 1)^2$ **57.** $d^2(abc - 1)(a^2b^2c^2 + abc + 1)(a^6b^6c^6 + a^3b^3c^3 + 1)$ **59.** DOES NOT FACTOR.

Section 1.6, p. 33

1. $\frac{1}{2}$ **3.** $\frac{-\pi}{4}$ **5.** $\frac{-5}{8}$ **7.** 8 **9.** $\frac{1}{25}$ **11.** $\frac{3}{5}$ **13.** 3 **15.** -7 **17.** $\frac{5}{8}$ **19.** (a) -1 (b) $\frac{-1}{3}$
(c) $\frac{-1}{3}$ **21.** 9 meters per second per second **23.** (a) 2000 (b) 1000 **25.** (a) $\frac{29}{2}$ or 14.5
(b) \$181.25 **27.** a^3c **29.** $\frac{-2}{3bc}$ **31.** $\frac{1}{2a}$ **33.** $a + b$ **35.** $\frac{x + y}{(x - y)(m + n)}$ **37.** $ab(b + 1)$
39. $\frac{1}{y(y^2z^4 + 1)}$ **41.** $-(3a^3 + 2b)$ **43.** $\frac{1}{5n + p - 1}$ **45.** $\frac{a - b}{a + b}$ **47.** $\frac{2(m - n)}{m + n}$ **49.** $\frac{a + 5}{a + 2}$ **51.** -1
53. $\frac{(a + 2)(a - 2)}{3}$ **55.** $\frac{r + s}{y - x}$ **57.** $\frac{-27}{(2x - 5)^2}$ **59.** $\frac{-x^2 + 6x + 4}{(x^2 + x + 1)^2}$ **61.** $\frac{2(x^2 - x - 4)}{(2x - 1)^3}$

Section 1.7, p. 39

1. $\frac{3}{10}$ **3.** $\frac{-7}{50}$ **5.** $\frac{56}{3}$ **7.** $\frac{160}{117}$ **9.** $\frac{xa}{yb}$ **11.** $\frac{b^5c^3d^3}{3a^6}$ **13.** $\frac{4(x + a)}{c(x - a)}$ **15.** $\frac{(x + a)^3}{3(x - a)}$
17. $-a^2(x - a)(x + 1)$ **19.** $\frac{(x - 4)(a + 1)}{5(1 - a)(x + 1)}$ **21.** $\frac{(2x - 3)(x - 1)}{7a^2x}$ **23.** $\frac{-(a - b)^2}{3ab(a + b)}$ **25.** $\frac{15}{2}$ **27.** $\frac{32}{27}$
29. $\frac{-5}{6}$ **31.** $\frac{14}{5}$ **33.** $\frac{5a^2c}{3b}$ **35.** $\frac{a^2cy}{27bx}$ **37.** $\frac{(x - a)(x + a)}{4}$ **39.** $\frac{(x + 1)^3}{(x - 3)^2}$ **41.** $\frac{b(a - 2)}{a^2(a + 2)}$
43. $\frac{9(y - 3)}{4}$ **45.** $\frac{4(1 - y)(a - 1)}{a(a^2 + a + 1)}$ **47.** $\frac{(x + 1)^2}{x^2}$ **49.** $\frac{(x + 1)^2(x + 3)}{x(x - 1)}$ **51.** $(x + 5)^2(x - 5)$
53. $\left(a + \frac{1}{4}\right)\left(a - \frac{1}{4}\right)$ **55.** $\left(\frac{2y}{3} + \frac{1}{5}\right)\left(\frac{2y}{3} - \frac{1}{5}\right)$ **57.** $\frac{-14(x + 4)}{(x - 3)^3}$ **59.** $\frac{2x(x + 1)(x - 1)(-x^6 + 3x^4 + 3x^2 - 1)}{(x^4 + 1)^3}$

Section 1.8, p. 45

1. 20 **3.** 36 **5.** 20 **7.** 2000 **9.** xy^3 **11.** $x^2y^2z^2$ **13.** $a(x + 3)(x - 3)$ **15.** $(u + a)^3(u - a)$
17. (a) 6 (b) $\frac{1}{2} = \frac{3}{6}, \frac{1}{3} = \frac{2}{6}$ **19.** (a) 24 (b) $\frac{5}{8} = \frac{15}{24}, \frac{-1}{12} = \frac{-2}{24}$ **21.** (a) 12 (b) $\frac{1}{4} = \frac{3}{12}, \frac{5}{6} = \frac{10}{12}$

23. (a) 300 (b) $\frac{8}{3 \cdot 5^2} = \frac{32}{300}, \frac{-7}{2 \cdot 3 \cdot 5} = \frac{-70}{300}, \frac{-3}{2^2 \cdot 5^2} = \frac{-9}{300}$ **25.** (a) x^2 (b) $\frac{b}{x} = \frac{bx}{x^2}$ **27.** (a) $x^2(x-a)^2$ (b) $\frac{1}{x^2(x-a)} = \frac{x-a}{x^2(x-a)^2}, \frac{-1}{x(x-a)^2} = \frac{-x}{x^2(x-a)^2}$ **29.** (a) $(x+3)^2(3-x)$ (b) $\frac{1}{9-x^2} = \frac{x+3}{(x+3)^2(3-x)}, \frac{x}{x+3} = \frac{x(x+3)(3-x)}{(x+3)^2(3-x)}, \frac{x^2}{x^2+6x+9} = \frac{x^2(3-x)}{(x+3)^2(3-x)}$ **31.** 1 **33.** $\frac{7\pi}{12}$ **35.** $\frac{17}{48}$ **37.** 1 **39.** $\frac{221}{375}$ **41.** $\frac{3}{m}$ **43.** $\frac{(5+x)^2}{x^4}$ **45.** $\frac{bc+ac-ab}{abc}$ **47.** $\frac{3}{x-a}$ **49.** $\frac{x+3}{(x+4)(x+1)}$ **51.** $\frac{-8}{(x+5)(x-5)}$ **53.** $5x(x^4-x+1)$ **55.** $\frac{5-x}{x^2(x+4)}$ **57.** $\left(x+\frac{1}{2}\right)^2$ or $\frac{(2x+1)^2}{4}$ **59.** $\left(x-\frac{1}{3}\right)^2$ or $\frac{(3x-1)^2}{9}$

Section 1.9, p. 48

1. $\frac{5}{16}$ **3.** 6 **5.** 2 **7.** $\frac{2}{5}$ **9.** $\frac{16}{63}$ **11.** $\frac{3}{4}$ **13.** $\frac{15}{8}$ **15.** $\frac{13}{34}$ **17.** $\frac{-4}{33}$ **19.** $\frac{ac}{b}$ **21.** $\frac{ad}{bc}$ **23.** $\frac{x}{az}$ **25.** $\frac{abcxy}{z}$ **27.** $\frac{6ab^2c^3}{49x^3y^3}$ **29.** $\frac{y+x}{y(x-1)}$ **31.** $\frac{1}{a^2-a+1}$ **33.** -2 **35.** -6 **37.** $5y^2$ **39.** $\frac{2x-y}{2x}$ **41.** $\frac{2-x}{x+2}$ **43.** $\frac{n-m}{n+m}$ **45.** $-xy$ **47.** $\frac{2x-y-xy+2y^2}{(x-2y)^3}$ **49.** $\frac{-1}{2(2+h)}$

Review Exercises, p. 49

1. (a) $6x^2y$ (b) $24xyz$ (c) $\frac{1}{4}xy$ (d) $-4xyz$ **2.** (a), (b), and (d) **3.** (a) 4 (b) -2 (c) 13 (d) -11 **4.** 10 000 square feet **5.** $6t+4x+3y-z$ **6.** $9x+4y-5$ **7.** $x^3y^2+x^3y-3x^4$ **8.** $2m^2+mn-n^2$ **9.** $-3x^3-2x^2+2x-2$ **10.** $x^4-2x^3+x^2$ **11.** $2^2 \cdot 3^2 \cdot 5^2$ **12.** 25 **13.** (a) $5x(x-5)$ (b) DOES NOT FACTOR. (c) $(y-5)(y+4)$ (d) $s^2(t+4)(t-4)$ (e) $(a+2)(a^2-2a+4)$ (f) $(2x+3)(2x+1)$ (g) $(x+2a)(x+a)$ (h) $(a-b)(x-y)$ (i) DOES NOT FACTOR. **14.** -12 **15.** (a) $4xz$ (b) $\frac{4x(x+y)}{3y(x-y)}$ (c) $2x-3y$ (d) $\frac{x+1}{x-6}$ (e) $\frac{-(x+a)}{a^2+ax+x^2}$ **16.** (a) $\frac{(a-x)(x+2)}{x+a}$ (b) $\frac{(a+b)(1-b)}{3(a-b)}$ (c) $\frac{(x^2+y^2)(x+y)}{3}$ **17.** (a) $\frac{3}{x}$ (b) $\frac{1+y^2-z}{x^2yz^2}$ (c) $\frac{1-2x}{(x+3)(x-3)}$ **18.** $\frac{1}{(x+1)^2(x-1)}$ **19.** (a) $\frac{4xb}{3az}$ (b) $x+6$

Chapter 2

Section 2.1, p. 57 (*The checks are given below.*)

1. no, **ii.** yes **3.** yes, **i.** false **5.** no, **ii.** no **7.** no, **ii.** yes **9.** root **11.** not a root **13.** not a root **15.** root **17.** root **19.** not a root **21.** not a root **23.** 5 **25.** 2 **27.** 10 **29.** $\frac{1}{9}$ **31.** 3 **33.** 5 **35.** 9 **37.** 5 **39.** $\frac{13}{40}$ **41.** -4 **43.** 4 **45.** 22 **47.** -2 **49.** 0 **51.** -5 **53.** -2

55. No. 0 is a root of the first equation, but not of the second. Restrictions must be placed on the Addition Principle, as stated on p. 54. **57.** No. 0 is a root of the second equation, but not of the first.

Checks.

37. $5-[2-3(5)] \stackrel{?}{=} 18$
$5-[2-15] \stackrel{?}{=} 18$
$5+13 \stackrel{?}{=} 18$
$18 \stackrel{\checkmark}{=} 18$

39. $10\left(\frac{13}{40}-\frac{1}{4}\right) \stackrel{?}{=} \frac{3}{4}$
$10\frac{(13-10)}{40} \stackrel{?}{=} \frac{3}{4}$
$\frac{3}{4} \stackrel{\checkmark}{=} \frac{3}{4}$

43. $4-[2-(4-3)] \stackrel{?}{=} 7-4$
$4-[2-1] \stackrel{?}{=} 3$
$3 \stackrel{\checkmark}{=} 3$

47. $(-2)^2+5(-2) \stackrel{?}{=} (-2)^2-10$
$4-10 \stackrel{\checkmark}{=} 4-10$

Section 2.2, p. 59 (*The checks are given below.*)

1. -18 **3.** $\frac{-4}{3}$ **5.** 20 **7.** $\frac{19}{12}$ **9.** 8 **11.** 20 **13.** 12 **15.** $\frac{3}{2}$ **17.** 7 **19.** -6 **21.** $\frac{-1}{2}$

23. 3 **25.** 11 **27.** 5 **29.** -3 **31.** 16 **33.** 4 **35.** $\frac{1}{4}$ **37.** $\frac{45}{11}$ **39.** 3 **41.** 4 **43.** 6

45. $\frac{8}{7}$ **47.** $\frac{10}{7}$ **49.** $\frac{x}{x-2}=5+\frac{2}{x-2}$

Multiply both sides by $x-2$. (See the check.)

$x=5(x-2)+2$
$x=5x-10+2$
$8=4x$
$2=x$

Check.

$\frac{2}{2-2} \stackrel{?}{=} 5+\frac{2}{2-2}$
$\frac{2}{0} \stackrel{?}{=} 5+\frac{2}{0}$

But $\frac{2}{0}$ is *undefined*. Thus 2 is *not* a root of the given equation. (In fact, $\frac{x}{x-2}$ and $\frac{2}{x-2}$ are not defined when $x=2$.) When you multiplied the given equation by $x-2$, you effectively multiplied both sides by 0. Note that upon multiplication of both sides by 0, the *false* statement $1=2$ becomes the *true* statement $0=0$.

Checks.

1. $\frac{-18}{9} \stackrel{?}{=} -2$
$-2 \stackrel{\checkmark}{=} -2$

13. $\frac{5}{6}(12)+\frac{3(12)}{4} \stackrel{?}{=} \frac{5(12)}{3}-1$
$10+9 \stackrel{?}{=} 20-1$
$19 \stackrel{\checkmark}{=} 19$

17. $\frac{35}{7} \stackrel{?}{=} 5$
$5 \stackrel{\checkmark}{=} 5$

21. $\frac{\frac{-1}{2}}{\frac{-1}{2}+1} \stackrel{?}{=} -1$
$\frac{\frac{-1}{2}}{\frac{1}{2}} \stackrel{?}{=} -1$
$\frac{-1}{2}\cdot\frac{2}{1} \stackrel{?}{=} -1$
$-1 \stackrel{\checkmark}{=} -1$

29. $\frac{-3+7}{-3+2} \stackrel{?}{=} -4$
$\frac{4}{-1} \stackrel{?}{=} -4$
$-4 \stackrel{\checkmark}{=} -4$

33. $\frac{2}{4}+\frac{6}{4} \stackrel{?}{=} 2$
$\frac{1}{2}+\frac{3}{2} \stackrel{?}{=} 2$
$\frac{4}{2} \stackrel{?}{=} 2$
$2 \stackrel{\checkmark}{=} 2$

Section 2.3, p. 67

1. $\frac{1}{2}$ hour **3.** $2\frac{1}{4}$ hours **5.** 50 kilometers **7.** 75 miles per hour **9.** 70 miles per hour **11.** 750 feet per minute **13.** \$33.60 **15.** \$3570 **17.** \$687.66 **19.** \$1600 **21.** 7% **23.** 16 gallons **25.** 36 liters **27.** $33\frac{1}{3}$ ounces **29.** $15\frac{1}{3}\%$ **31.** 54 pounds **33.** 30 pounds at \$2.75 per pound; 20 pounds at \$3.25 per pound **35.** \$3.80 per pound **37.** 6 hours **39.** 6 days **41.** $\frac{36}{11}$ hours **43.** 30 minutes **45.** 90 minutes

Section 2.4, p. 72 (*The checks are given below.*)

1. $\frac{c}{5}$ **3.** $b - c - 13$ **5.** $\frac{9}{3-a}$ **7.** $\frac{3+b}{1-2a}$ **9.** $\frac{9b+a}{3}$ **11.** $\frac{4b+6c}{33}$ **13.** $\frac{4}{a+b}$ **15.** $\frac{-a(bc+1)}{bc}$ **17.** $\frac{9}{4a^2b}$ **19.** $4 - \frac{a}{b(7-a)}$ **21.** $\frac{a-4-b}{2}$ **23.** $\frac{3}{y_2+y_3-9}$ **25.** (a) $1 + t - 2x$ (b) $\frac{1+t-y}{2}$ **27.** (a) $\frac{cz-by}{a}$ (b) $\frac{cz-ax}{b}$ **29.** (a) $\frac{1+3xz}{5a}$ (b) $\frac{5ay-1}{3x}$ **31.** (a) $\frac{9-y^2-yz}{y+z}$ (b) $\frac{9-xy-y^2}{x+y}$ **33.** (a) $\frac{-bt+b^2+1+a^2}{a}$ (b) $\frac{a^2-ax+1+b^2}{b}$ **35.** (a) $\frac{Kz-3b(x-1)}{5z}$ (b) $\frac{3b(x-1)}{K-5t}$ **37.** $\frac{c}{2\pi}$ **39.** $\frac{2A}{h}$ **41.** $\frac{V}{wh}$ **43.** $\frac{3V}{\pi r^2}$ **45.** (a) $\frac{c(2gh-v^2)}{2g}$ (b) $\frac{35}{16}$

Checks.

1. $5\left(\frac{c}{5}\right) \stackrel{?}{=} c$

$c \stackrel{\checkmark}{=} c$

5. $a\left(\frac{9}{3-a}\right) + 7 \stackrel{?}{=} 3\left(\frac{9}{3-a}\right) - 2$

$a\left(\frac{9}{3-a}\right) + 9 \stackrel{?}{=} 3\left(\frac{9}{3-a}\right)$

$\frac{9a + 9(3-a)}{3-a} \stackrel{?}{=} \frac{27}{3-a}$

$9a + 27 - 9a \stackrel{?}{=} 27$

$27 \stackrel{\checkmark}{=} 27$

15. $\dfrac{\frac{-a(bc+1)}{bc}}{\frac{-a(bc+1)}{bc} + a} \stackrel{?}{=} bc + 1$

$\dfrac{\frac{-a(bc+1)}{bc}}{\frac{-abc - a + abc}{bc}} \stackrel{?}{=} bc + 1$

$\frac{-a(bc+1)}{bc} \cdot \frac{bc}{-a} \stackrel{?}{=} bc + 1$

$bc + 1 \stackrel{\checkmark}{=} bc + 1$

Section 2.5, p. 79

1.

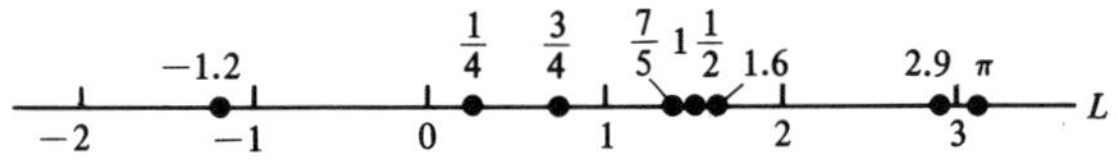

FIGURE 2A

3. < **5.** > **7.** < **9.** < **11.** > **13.** > **15.** < **17.** left **19.** right **21.** right **23.** right **25.** right **27.** $-8 < -6 < 0 < \frac{1}{2} < 2 < 8$ **29.** $-2.71 < -2.701 < -2.7 < -2.693 < -2.69 < -2.689$ **31.** $\geq$ **33.** $\leq$ **35.** $\leq$ **37.** $\leq$ **39.** (a) **41.** (g) **43.** (f) **45.** (d) **47.** true **49.** true **51.** true **53.** true **55.** false **59.** $(-\infty, -6)$ **61.** $[8, \infty)$

Section 2.6, p. 82

1. $<$ **3.** $<$ **5.** $>$ **7.** $>$ **9.** $<$ **11.** $>$ **13.** $>$ **15.** $>$ **17.** $>$ **19.** $<$ **21.** $<$
23. $x + 2 < 4 + 2$ **25.** $2a \leq 2(4)$ **27.** $-2x \geq -2(5)$ because -2 is *negative* and multiplication by a negative number reverses the sense of the inequality. **29.** $2x < 3$. Therefore, $x < \frac{3}{2} < 2$. Thus $x < 2$
31. $0 < x + y < 90$

Section 2.7, p. 87

1. $(-\infty, 4)$ **3.** $(-\infty, 13)$ **5.** $(-\infty, 4]$ **7.** $(1, \infty)$ **9.** $(-\infty, 3)$ **11.** $(8, \infty)$ **13.** $(-1, \infty)$
15. $(-\infty, -1)$ **17.** $\left[\frac{-1}{2}, \infty\right)$ **19.** $\left(\frac{3}{4}, \infty\right)$ **21.** $(-\infty, -3]$ **23.** $\left[\frac{-9}{4}, \infty\right)$ **25.** $\left(-\infty, \frac{-11}{4}\right)$
27. $\left(-\infty, \frac{-1}{2}\right]$ **29.** $\left[\frac{-14}{27}, \infty\right)$ **31.** $\left(-\infty, \frac{1}{2}\right)$ **33.** $\left[\frac{1}{4}, \infty\right)$ **35.** $\left[\frac{-2}{5}, \infty\right)$ **37.** $(0, 6)$
39. $[-9, -3]$ **41.** $(3, 4)$ **43.** $(-3, 3)$ **45.** $\left(-1, \frac{7}{2}\right)$ **47.** $\left(\frac{-8}{5}, 2\right)$ **49.** $[-2, 1]$ **51.** $(-4, 0)$
53. $\left(\frac{-1}{5}, 1\right)$ **55.** $\left(\frac{-1}{2}, \frac{1}{2}\right)$ **57.** $[120, 165]$ **59.** $[32, 41]$

Section 2.8, p. 92 *(The checks are given below.)*

1. 15 **3.** 0 **5.** 8, -8 **7.** $-3, 7$ **9.** $-8, -2$ **11.** true **13.** false; let $a = b = 1$ **15.** 7, -7
17. no root **19.** 2, -2 **21.** 7, -5 **23.** 2, -6 **25.** 5, -4 **27.** 4, 16 **29.** $\frac{-1}{5}$
31. $\frac{1}{2}$ $\left(\frac{-1}{6}\right.$ is not a root.$\left.\right)$ **33.** $\frac{4}{3}$ (6 is not a root.) **35.** $\frac{4}{3}$ $\left(\frac{-6}{5}\right.$ is not a root.$\left.\right)$
37. no root $\left(\frac{-1}{2}\right.$ is not a root.$\left.\right)$ **39.** The roots are the nonnegative numbers. **41.** $1, \frac{-1}{3}$ **43.** $\frac{9}{7}$ and $\frac{-9}{5}$ **45.** $\frac{-1}{2}$
47. 2 and $\frac{-2}{3}$ **49.** -7 and $\frac{5}{7}$ **51.** $\frac{1}{2}$ and $\frac{-7}{2}$ **53.** $\frac{-1}{2}$

Checks.

21. $|7 - 1| \stackrel{?}{=} 6$, $6 \stackrel{\checkmark}{=} 6$ | $|-5 - 1| \stackrel{?}{=} 6$, $|-6| \stackrel{?}{=} 6$, $6 \stackrel{\checkmark}{=} 6$

25. $|2(5) - 1| \stackrel{?}{=} 9$, $9 \stackrel{\checkmark}{=} 9$ | $|2(-4) - 1| \stackrel{?}{=} 9$, $|-9| \stackrel{?}{=} 9$, $9 \stackrel{\checkmark}{=} 9$

31. $\left|2\left(\frac{1}{2}\right) + 1\right| \stackrel{?}{=} 4\left(\frac{1}{2}\right)$, $2 \stackrel{\checkmark}{=} 2$ | $\left|2\left(\frac{-1}{6}\right) + 1\right| \stackrel{?}{=} 4\left(\frac{-1}{6}\right)$, $\left|\frac{2}{3}\right| \stackrel{?}{=} \frac{-2}{3}$, $\frac{2}{3} \stackrel{x}{=} \frac{-2}{3}$

33. $\left|\frac{4}{3} + 1\right| \stackrel{?}{=} 5 - 2\left(\frac{4}{3}\right)$, $\frac{7}{3} \stackrel{\checkmark}{=} \frac{7}{3}$ | $|6 + 1| \stackrel{?}{=} 5 - 2(6)$, $7 \stackrel{x}{=} -7$

35. $\left|\frac{4}{3} + 5\right| \stackrel{?}{=} 4\left(\frac{4}{3}\right) + 1$, $\frac{19}{3} \stackrel{\checkmark}{=} \frac{19}{3}$ | $\left|\left(\frac{-6}{5}\right) + 5\right| \stackrel{?}{=} 4\left(\frac{-6}{5}\right) + 1$, $\frac{19}{5} \stackrel{x}{=} \frac{-19}{5}$

37. $\left|\frac{-1}{2} + 1\right| \stackrel{?}{=} \frac{-1}{2}$, $\frac{1}{2} \stackrel{x}{=} \frac{-1}{2}$

Section 2.9, p. 97

1. $(-10, 10)$ **3.** $[-4, 4]$ **5.** $(-6, 8)$ **7.** $(-11, 5)$ **9.** $\left(\frac{-7}{2}, \frac{9}{2}\right)$ **11.** $(-2, 3)$ **13.** $\left(\frac{-7}{5}, 1\right)$

15. no solution **17.** $\left(\frac{-7}{3}, 3\right)$ **19.** $(-\infty, -5) \cup (5, \infty)$ **21.** $(-\infty, -40] \cup [40, \infty)$

23. $(-\infty, -1] \cup [1, \infty)$ **25.** $(-\infty, -5) \cup (7, \infty)$ **27.** $\left(-\infty, \frac{-5}{3}\right) \cup (1, \infty)$ **29.** $(-\infty, -3) \cup \left(\frac{9}{2}, \infty\right)$

31. $(-\infty, 2) \cup \left(\frac{10}{3}, \infty\right)$ **33.** all real numbers **35.** all real numbers **37.** 5 **39.** 9 **41.** 5 **43.** 4

45. -6 **47.** 5 **49.** $\frac{1}{2}$ **51.** 36 **53.** false; let $a = -1$ **55.** true **57.** true

59.
$$|x - 2| < \frac{1}{10}$$
$$5|x - 2| < 5 \cdot \frac{1}{10}$$
$$|5(x - 2)| < \frac{5}{10}$$
$$|5x - 10| < \frac{1}{2}$$

Review Exercises, p. 98

1. (a) root (b) not a root (c) root (d) root **2.** (a) 1 (b) $\frac{12}{5}$ (c) 3 (d) 5 (e) 3 (f) 4

Checks.

(a) $2(1) + 1 \stackrel{?}{=} 5 - 2(1)$

$3 \stackrel{\checkmark}{=} 3$

(d) $\frac{5 + 3}{8} \stackrel{?}{=} \frac{5 - 2}{3}$

$1 \stackrel{\checkmark}{=} 1$

3. an hour and a half **4.** \$1500 at 5%, \$3000 at 8%

5. 24 pounds **6.** $\frac{24}{7}$ hours **7.** (a) $\frac{3b - a}{5}$ (b) $\frac{8x}{2xz - 1}$ (c) i. $\frac{-1}{2t + A}$ ii. $\frac{-(As + 1)}{zs}$

8. $\frac{1}{6} < \frac{1}{5} < \frac{1}{3} < \frac{2}{5} < \frac{1}{2} < \frac{2}{3}$ **9.** (a) -5 (b) 5 (c) 0 (d) -5 **10.** (a) 12 (b) 4 (c) 0 (d) $\frac{3}{4}$

11. (a) false (b) true (c) false (d) true **12.** (a) $[5, 7)$ (b) $(-2, 3)$ (c) $[-1, \infty)$ (d) $(-\infty, 3)$

13. (a) $>$ (b) $<$ (c) $>$ **14.** (a) $[1, \infty)$ (b) $(2, \infty)$ (c) $[\ 1, 3]$ **15.** (a) 2 and 10
(b) -10 and 2

16. (a) 0, 14 (b) $\frac{1}{2}$ (c) 1

Check for $\frac{1}{2}$:

$$\left|2 - \frac{1}{2}\right| \stackrel{?}{=} 3 \cdot \frac{1}{2}$$
$$\frac{3}{2} \stackrel{\checkmark}{=} \frac{3}{2}$$

Note that -1 is not a root because
$$|2 - x| = x - 2$$
only if $x \geq 2$. In fact,
$$|2 - (-1)| \stackrel{?}{=} 3(-1)$$
$$3 \stackrel{x}{=} -3$$

Check: $|1 + 1| \stackrel{?}{=} |1 - 3|$
$$2 \stackrel{\checkmark}{=} 2$$

17. (a) $(-3, 5)$ (b) $(-\infty, -6] \cup [0, \infty)$ (c) -2 **18.** (a) 3 (b) 3 **19.** $\frac{-1}{2}$

Chapter 3

Section 3.1, p. 105

1. function **3.** (a) $\{-6, -5, -4, -3, -2, -1\}$ (b) $\{0, 1\}$ **5.** (a) 1, 2, 3, and 4 (b) 0 (c) 0
7. (a) 2 (b) 4 (c) 6 (d) 8 **9.** (a) -10 (b) -5 (c) 0 (d) -20 **11.** (a) -2 (b) 1
(c) 4 (d) $\frac{2}{3}$ **13.** (a) 1 (b) 1 (c) 1 (d) 13 **15.** (a) 1 (b) $\frac{1}{4}$ (c) 4 (d) 10
17. (a) 0 (b) $\frac{-1}{3}$ (c) $\frac{2}{3}$ (d) $\frac{40}{9}$ **19.** (a) 4 (b) 0 (c) 10 (d) 0
21. the set of all real numbers other than $\frac{1}{2}$ **23.** the set of all nonzero real numbers
25. the set of all real numbers other than 2 and -2 **27.** (a) (0, 4) (b) (0, 4) **29.** (a) $(-2, 2)$ (b) $(-1, 3)$
31. (a) $[1, \infty)$ (b) $[3, \infty)$ **33.** (a) [4, 10] (b) [0, 6] **35.** (a) $V = f(r) = \frac{4\pi}{3}r^3$ (b) $\frac{4\pi}{3}$ (c) $\frac{500\pi}{3}$
37. (a) $A = f(h) = \frac{h^2}{3}$ (b) $A = g(b) = \frac{3b^2}{4}$ (c) 108 (d) 243 **39.** (a) $f(h) = h + 5$ (b) $g(H) = H - 5$
41. (a) $f(t) = 4t + 2$ (b) 210 **43.** (a) 192 feet (b) 256 feet (c) 192 feet (d) 0 feet (e) [0, 8]
45. $f(x) = (x - 6)\left(\frac{240}{x} - 4\right)$ or $\frac{4(x - 6)(60 - x)}{x}$ [square inches] **47.** $A = f(l) = l(100 - l)$ [square meters]
49. (a) $V = f(x) = x(50 - 2x)^2$ [cubic centimeters] (b) (0, 25)

Section 3.2, p. 118

1. $P = (1, 1), Q = (-1, -1), R = (3, -2), S = \left(5, \frac{1}{2}\right), T = (-4, -3), U = \left(0, \frac{-9}{2}\right), V = \left(\frac{5}{2}, -5\right)$
2.–8. See Figure 3A. **9.** IV **11.** III **13.** I

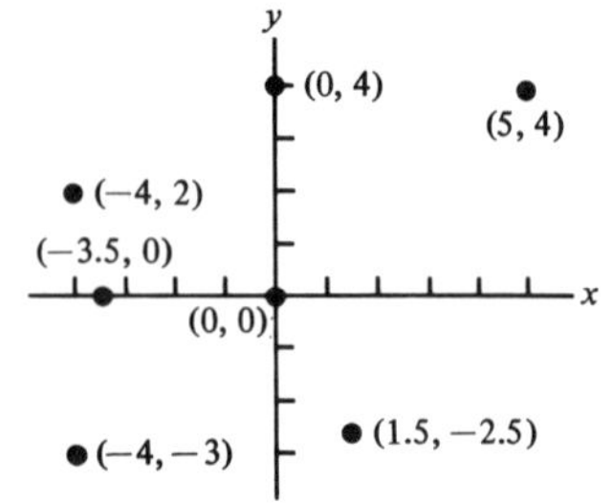

FIGURE 3A

15.

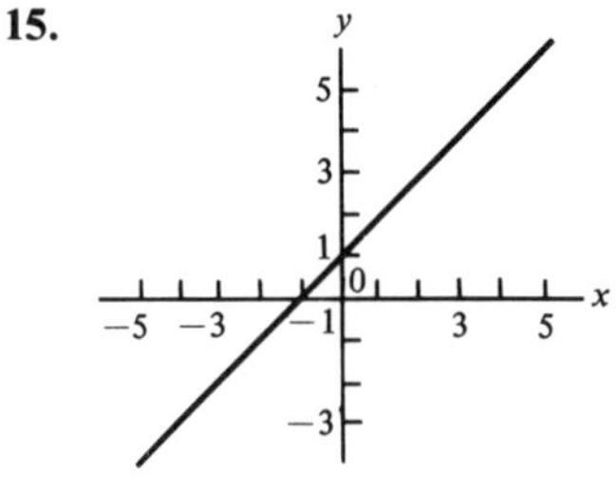

FIGURE 3B $y = x + 1$

17.

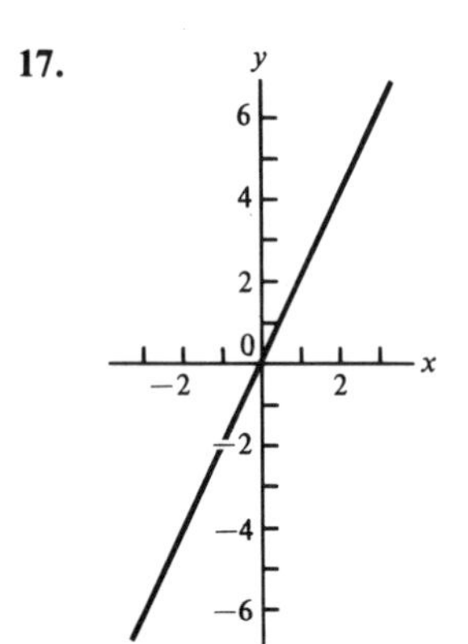

FIGURE 3C $f(x) = 2x$

19.

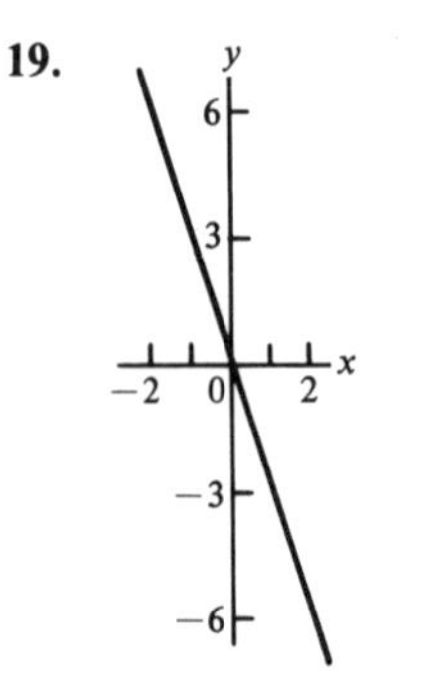

FIGURE 3D $h(x) = -3x$

21.

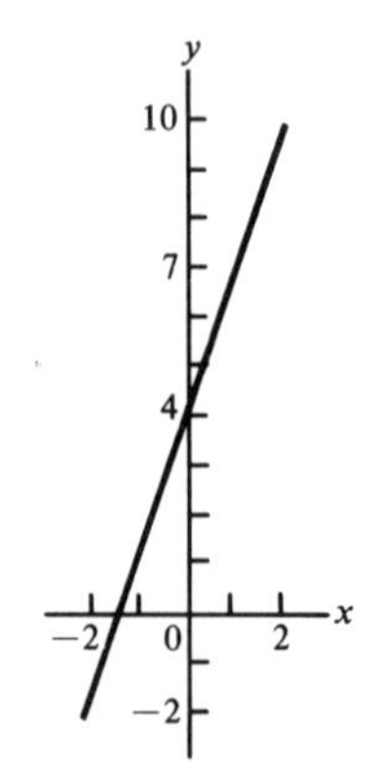

FIGURE 3E $y = 3x + 4$

23.

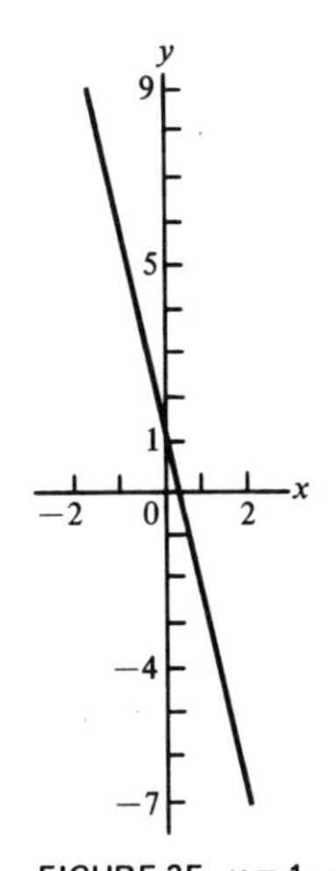

FIGURE 3F $y = 1 - 4x$

25.

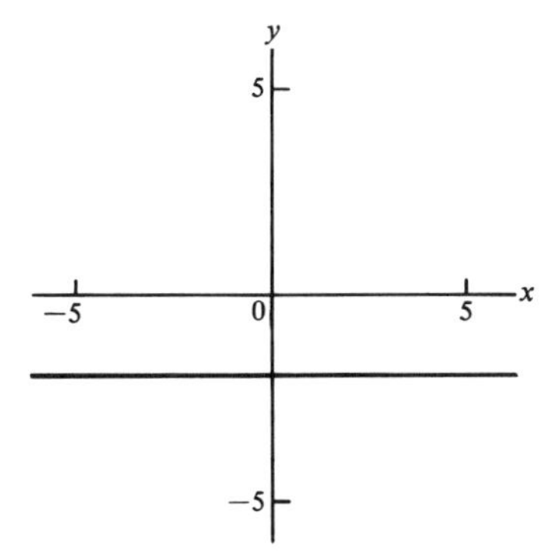

FIGURE 3G $y \equiv -2$

27.

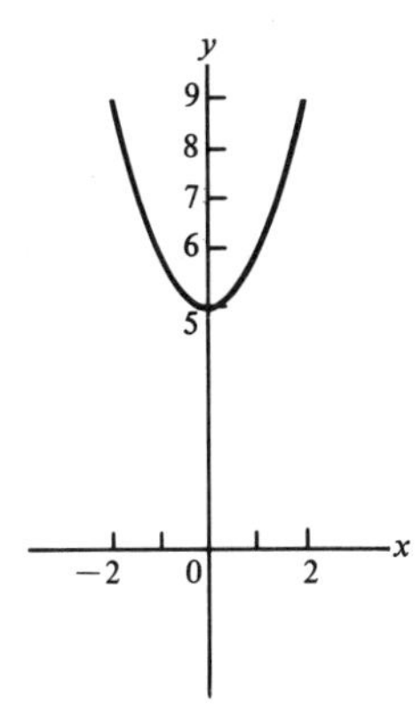

FIGURE 3H $F(x) = x^2 + 5$

29.

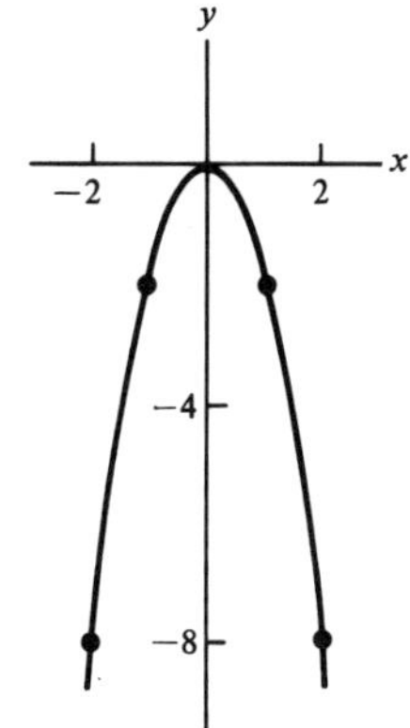

FIGURE 3I $H(x) = -2x^2$

31.

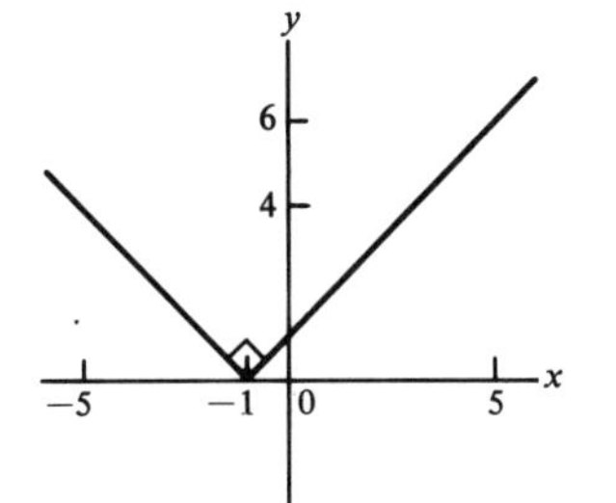

FIGURE 3J $y = |x + 1|$

33.

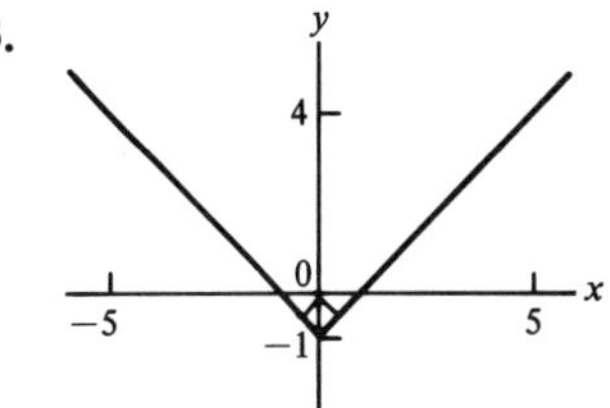

FIGURE 3K $F(x) = |x| - 1$

35.

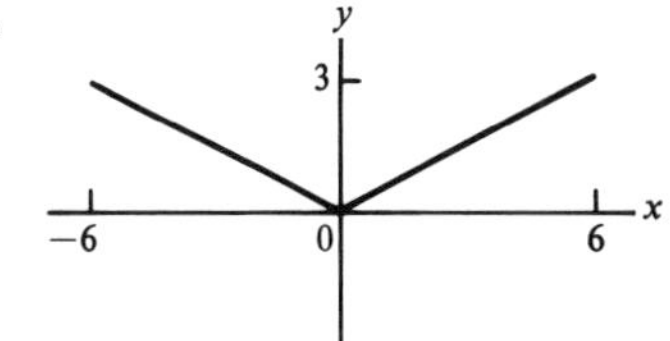

FIGURE 3L $h(x) = \left|\frac{x}{2}\right|$

37.

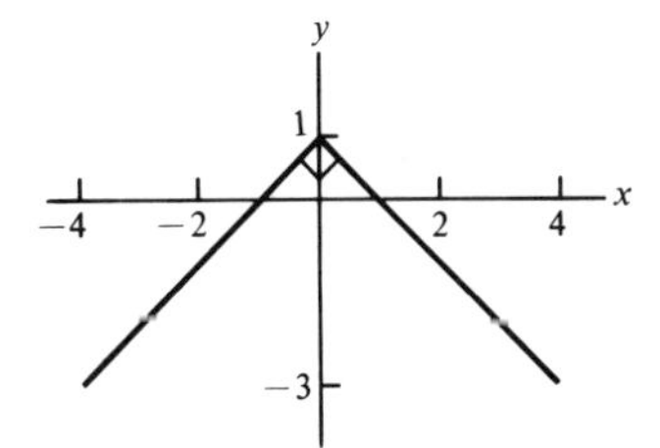

FIGURE 3M $y = \begin{cases} x + 1, \text{ if } x < 0 \\ 1 - x, \text{ if } x \geqslant 0 \end{cases}$

39.

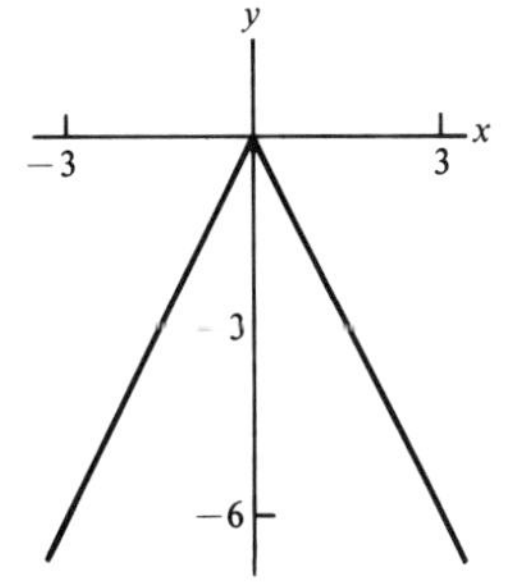

FIGURE 3N $y = \begin{cases} 2x, \text{ if } x \leqslant 0 \\ -2x, \text{ if } x > 0 \end{cases}$

41.

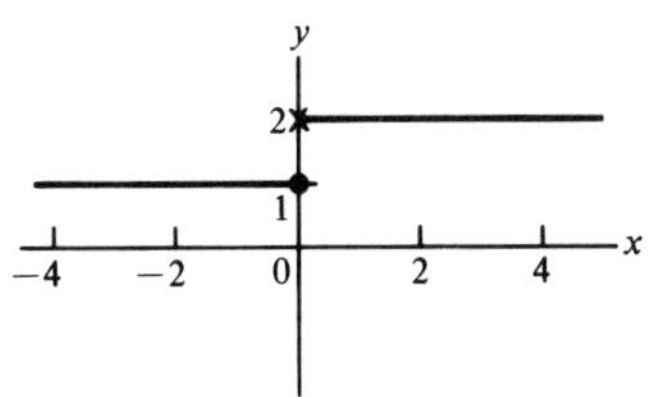

FIGURE 3O $G(x) = \begin{cases} 1, \text{ if } x \leqslant 0 \\ 2, \text{ if } x > 0 \end{cases}$

43.

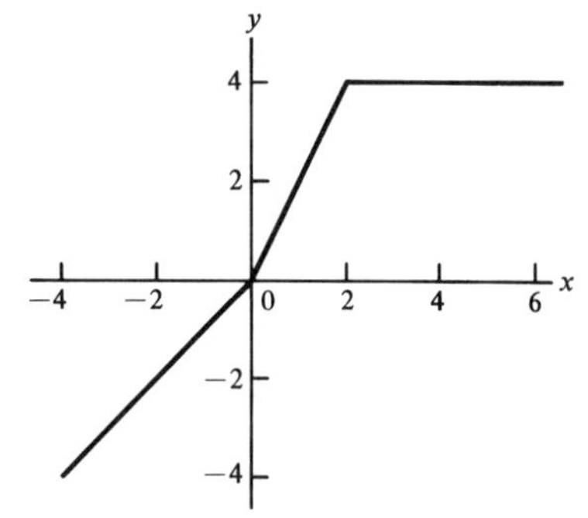

FIGURE 3P $y = \begin{cases} x, & \text{if } x < 0 \\ 2x, & \text{if } 0 \leqslant x \leqslant 2 \\ 4, & \text{if } x > 2 \end{cases}$

45.

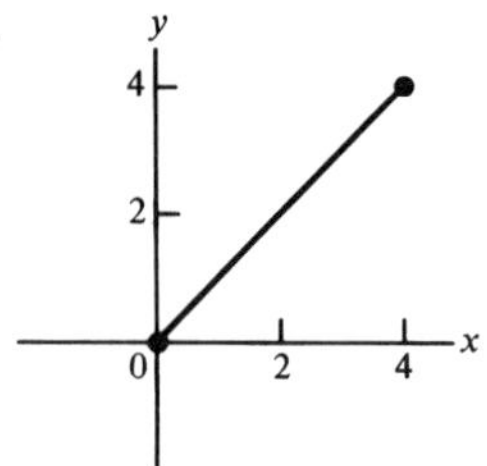

FIGURE 3Q (a) $y = x$, $0 \leqslant x \leqslant 4$
(b), (c) domain = range = $[0, 4]$

47.

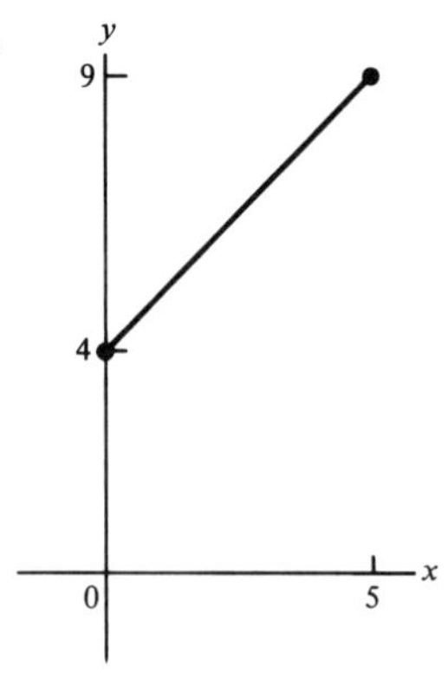

FIGURE 3R (a) $y = x + 4$, $0 \leqslant x \leqslant 5$
(b) domain = $[0, 5]$ (c) range = $[4, 9]$

49.

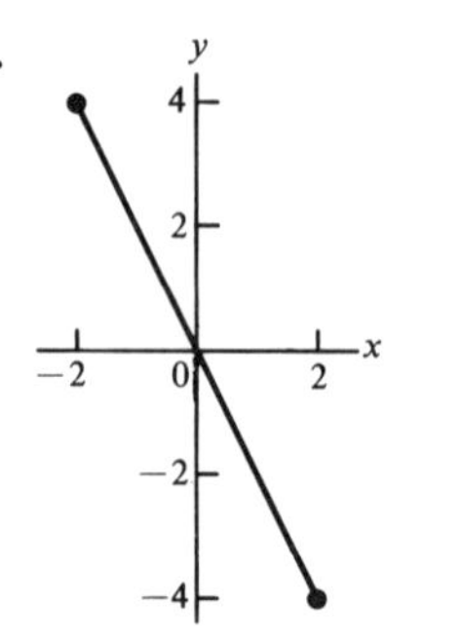

FIGURE 3S (a) $y = -2x$, $-2 \leqslant x \leqslant 2$
(b) domain = $[-2, 2]$ (c) range = $[-4, 4]$

51.

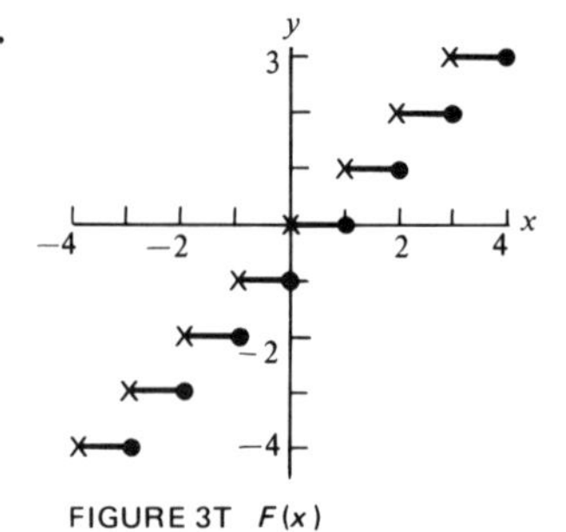

FIGURE 3T $F(x)$
(a) $F(0) = -1$
(b) $F(1) = 0$
(c) $F(-1) = -2$
(d) $F\left(\frac{1}{2}\right) = 0$
(e) $F\left(-\frac{1}{2}\right) = -1$

53.

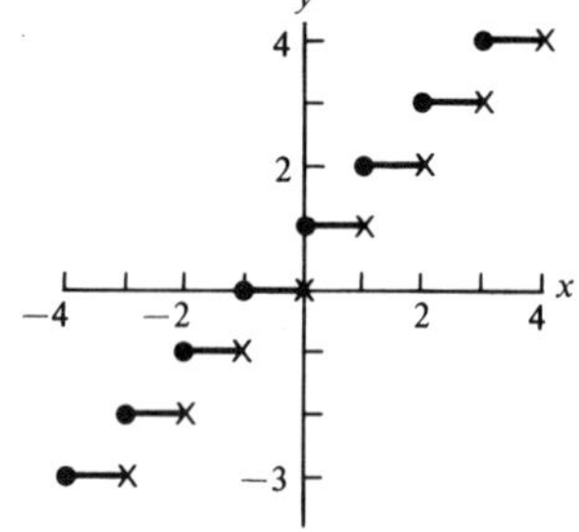

FIGURE 3U $H(x)$
(a) $H(0) = 1$
(b) $H(1) = 2$
(c) $H(-1) = 0$
(d) $H\left(\frac{1}{2}\right) = 1$
(e) $H\left(-\frac{1}{2}\right) = 0$

55.

FIGURE 3V

Section 3.3, p. 126

1. (a) 50 (b) -50 (c) 5π (d) $5x + 10$ (e) $5x - 15$ **3.** (a) $10x + 10$ (b) $10x + 10h$ (c) 10 (d) $10t$ (e) $10t^2$ **5.** (a) $t^2 + 4t + 4$ (b) $t^2 + 2th + h^2$ (c) $2t + h$ (d) t^2 (e) t^4 **7.** (a) -1 (b) 4 (c) x (d) $\dfrac{1}{x+h}$ (e) $\dfrac{-1}{(x+h)x}$ **9.** (a) 1 (b) $\dfrac{2}{3}$ (c) 2 (d) $\dfrac{x}{x+1}$ (e) $\dfrac{-1}{(x+h+1)(x+1)}$ **11.** (a) 126 (b) $x^2 + 5x$ (c) $x^4 + 3x^2 - 4$ (d) $\dfrac{1 + 3x - 4x^2}{x^2}$ (e) $x^2 - 3x - 4$ (f) $2x + h + 3$ **13.** no **15.** yes **17.** yes **19.** (a) 4 (b) 7 (c) 12 (d) $[0, 2]$ **21.** function **23.** function **25.** not a function **27.** function **29.** function

31. not a function **33.** (a) $-3, 3$ (b) 3 (c) yes (d) no **35.** (a) none (b) 3 (c) yes (d) no **37.** (a) $-4, -2, 0, 2, 4$ (b) 0 (c) yes (d) no **39.** (a) $[-5, 0]$ (b) 0 (c) no (d) no **41.** (a) 0 (b) 0 (c) no (d) yes **43.** (a) 2 (b) -4 (c) no (d) no
45. (a) $-3, 3$ (b) -9 (c) yes (d) no **47.** (a) $-1, 1$ (b) -1 (c) yes (d) no
49. For all x, $f(-x) = (-x)^4 = x^4 = f(x)$ **51.** For all x, $h(-x) = \dfrac{(-x)^2 + 1}{(-x)^4 - 2} = \dfrac{x^2 + 1}{x^4 - 2} = h(x)$
53. For all x, $g(-x) = (-x)^5 - 2(-x)^3 + 4(-x) = -x^5 + 2x^3 - 4x = -g(x)$ **55.** (a) 5 (b) 3 (c) 0 (d) $-\pi$ **57.** 22 meters per second **59.** \$4000.05 per tractor

Section 3.4, p. 135

1. (a) $2x + 3$ (b) 7 (c) 1 (d) 1 (e) -1 (f) -1 **3.** (a) $3x + 3$ (b) 9 (c) $x - 3$ (d) -3 (e) $3 - x$ (f) 4 **5.** (a) $\dfrac{4x^3 + 4x + 1}{x^2 + 1}$ (b) $\dfrac{41}{5}$ (c) $\dfrac{4x^3 + 4x - 1}{x^2 + 1}$ (d) -1
(e) $\dfrac{1 - 4x^3 - 4x}{x^2 + 1}$ (f) $\dfrac{9}{2}$ **7.** (a) $2x^2 + x + 1$ (b) 11 (c) $3 - x$ (d) 3 (e) $x - 3$ (f) -4
9. (a) $6x$ (b) 24 (c) $-3x - 6$ (d) -9 (e) $3x^2 + 6x$ (f) 0 **11.** (a) 8 (b) 8 (c) $3x - 12$ (d) -9 (e) $16 - 4x$ (f) 16 **13.** (a) $4x + 2$ (b) 18 (c) $3 - 9x$ (d) -6 (e) $6x^2 + x - 1$ (f) -1 **15.** (a) $2x^2 + 4$ (b) 36 (c) $3x^2 - 3$ (d) 0 (e) $-x^4 - x^2 + 2$ (f) 2
17. (a) $\dfrac{x - 4}{x + 2}$, $x \neq -2$ (b) $-\dfrac{1}{2}$ (c) $\dfrac{x + 2}{x - 4}$, $x \neq 4$ (d) 0 (e) all real numbers other than -2
(f) all real numbers other than 4 **19.** (a) $\dfrac{x^2 + 1}{x^2 - 1}$, $x \neq -1, 1$ (b) $\dfrac{5}{3}$ (c) $\dfrac{x^2 - 1}{x^2 + 1}$ (d) $\dfrac{3}{5}$
(e) $x \neq -1, 1$ (f) all real numbers **21.** (b) **23.** (a) **25.** (a) **27.** (c) **29.** (b) **31.** (c)
33. (a) $x + 2$ (b) 3 (c) $x + 2$ (d) 3 (e) all real numbers (f) all real numbers
35. (a) $-8x - 4$ (b) -12 (c) $-8x + 2$ (d) -6 (e) all real numbers (f) all real numbers
37. (a) $|x + 2|$ (b) 3 (c) $|x| + 2$ (d) 3 (e) all real numbers (f) all real numbers
39. (a) $\dfrac{1}{x^2}$, $x \neq 0$ (b) 1 (c) $\dfrac{1}{x^2}$, $x \neq 0$ (d) 1 (e) all nonzero (real) numbers
(f) all nonzero numbers **41.** (a) 0 (b) 0 (c) 4 (d) 4 (e) all real numbers
(f) all real numbers **43.** (a) $\dfrac{-1}{x^2}$, $x \neq 0$ (b) $-\dfrac{1}{4}$ (c) $\dfrac{-1}{(x + h)^2}$, $x \neq -h$ (d) no **45.** $2x - 1$
47. (a) 12 (b) 14 (c) 16 (d) 30 (e) $f(-2) = 8$ (f) not enough information
49. Let P = perimeter. $f(P) = \dfrac{25}{2}(P - 50)$ **51.** $f(t) = \dfrac{\pi t^4}{(t + 1)^2}$ [square centimeters]

Section 3.5, p. 141

1. one-to-one **3.** not one-to-one **5.** one-to-one **7.** one-to-one **9.** (a) 1 (b) 2 (c) 3 (d) 4
11. $f(0) = f(1) = 4$ **13.** $f(1) = f(-1) = -2$ **15.** $f(1) = f(-1) = 1$ **17.** $f(1) = f(2) = 1$ **19.** $y = \dfrac{x}{2}$
21. $y = x - 8$ **23.** $y = 9 - x$ **25.** $y = \dfrac{x - 7}{3}$ **27.** $\dfrac{7 - x}{2}$ **29.** $y = 4x - 2$ **31.** (a) $f^{-1}(x) = x - 2$, $2 \leq x \leq 3$ (b) $[2, 3]$ (c) $[0, 1]$ **33.** (a) $f^{-1}(x) = \dfrac{x - 1}{3}$, $-2 \leq x \leq 7$ (b) $[-2, 7]$ (c) $[-1, 2]$

Review Exercises, p. 143

1. (a) 9 (b) 25 (c) 6 (d) 4 **2.** (a) 0 (b) 3 (c) $\frac{-3}{5}$ (d) $\frac{-9}{11}$ **3.** $x \neq 4$
4. (a) $f(s) = s^3$ (b) 125 cubic inches (c) .008 cubic inch **5.** (a) Let t be the number of hours of computer time. $f(t) = 300 + 200t$ (b) \$1300 (c) 11 hours **6.** $P = (3, 2)$, $Q = (2, 3)$, $R = (-1, -3)$, $S = (4.5, -3)$, $T = (4.5, 0)$, $U = (0, -.5)$, $V = (4, -2)$ **7.** See Figure 3W. **8.** (a) I (b) II (c) III (d) y-axis (e) IV (f) I **9.** (a) **10.** See Figure 3X. **11.** (a) -2 (b) $3x + 4$ (c) $6x + 1$ (d) $3x + 3h + 1$ (e) 3 **12.** yes **13.** (b), (c), (e)
14. **i.** (a) 6 (b) 6 (c) 0 (d) -2 **ii.** (a) 19 (b) 1 (c) -8 (d) 4 **iii.** (a) -1 (b) -3 (c) 0 (d) 0
15. (a) $\frac{1 + 5x}{x}$, $x \neq 0$ (b) 4 (c) $\frac{1}{x + 5}$, $x \neq -5$ (d) $\frac{1}{4}$ (e) $x \neq 0$ (f) $x \neq -5$ **16.** (a)
17. $y = \frac{x + 4}{5}$ **18.** $f(-1) = f(1) = 0$

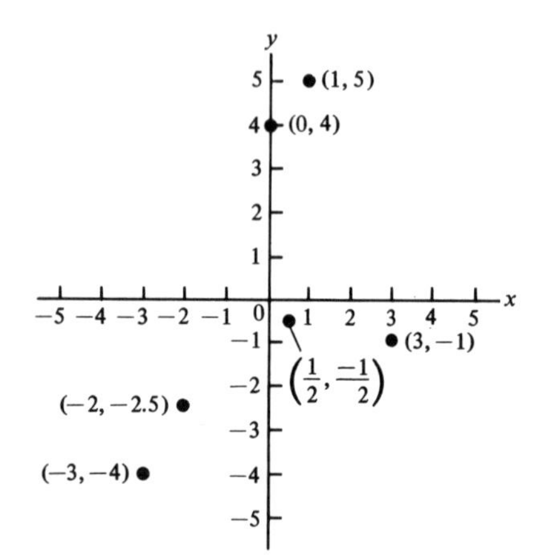

FIGURE 3W

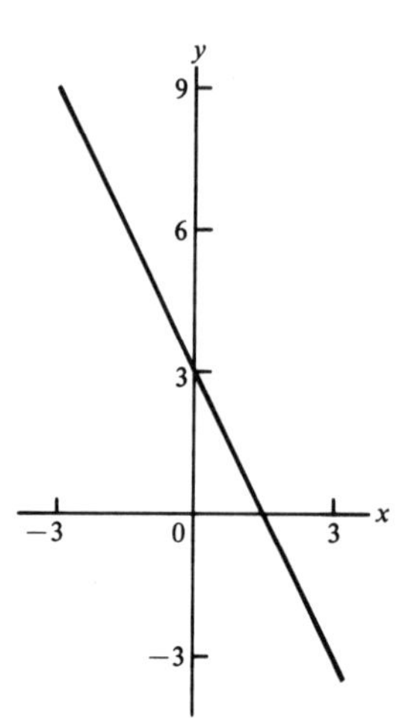

FIGURE 3X (a) $y = 3 - 2x$

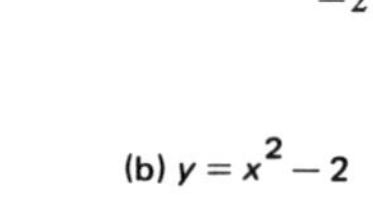

(b) $y = x^2 - 2$

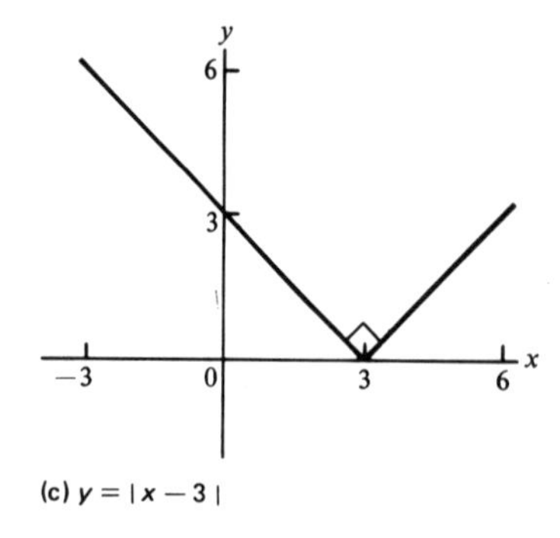

(c) $y = |x - 3|$

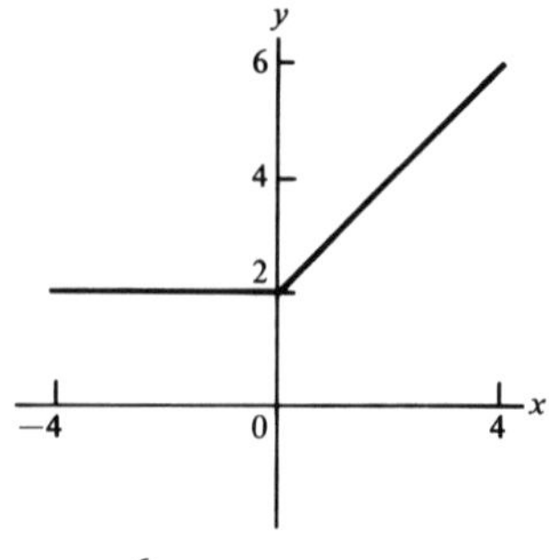

(d) $y = \begin{cases} 2, & \text{if } x \leq 0 \\ x + 2, & \text{if } x > 0 \end{cases}$

Chapter 4

Section 4.1, p. 149

1. 10 **3.** 2 **5.** 1 **7.** $\frac{42}{5}$ **9.** 4, -4 **11.** \$1.05 **13.** \$18 **15.** \$1.25 **17.** \$31.40 **19.** 45

21. 21 meters, 6 meters **23.** 24 **25.** $a = 3, c = 16.5$ **27.** $a = 9, c = 15$ **29.** $\frac{8}{3}$ **31.** $A = \frac{5l^2}{7}$

33. $l = \frac{10t}{3}$ **35.** $V = \pi r^2(8 - 1.6r)$

Section 4.2, p. 156

1. 1 **3.** $\frac{-1}{6}$ **5.** -7 **7.** $\frac{3}{5}$ **9.** $\frac{-1}{12}$ **11.** $\frac{1}{2}$ **13.** 13 **15.** -15 **17.** $y - 4 = 3x$

19. $y - 3 = 2(x + 1)$ **21.** $y - 2 = \frac{1}{2}(x - 1)$ **23.** $y + 1 = -3(x + 1)$ **25.** $y = \frac{3}{4}$

27. (a) $y - 3 = 2(x - 3)$ (b) See Figure 4A. **29.** (a) $y - 5 = 4(x - 1)$ (b) See Figure 4B.

31. (a) $y - 2 = \frac{-1}{2}(x - 4)$ (b) See Figure 4C. **33.** (a) neither horizontal nor vertical (b) neither horizontal nor vertical (c) horizontal (d) vertical (e) horizontal (f) vertical

35. $y = 5$ **37.** $x = 1$ **39.** $x_0 = \frac{4}{m}$

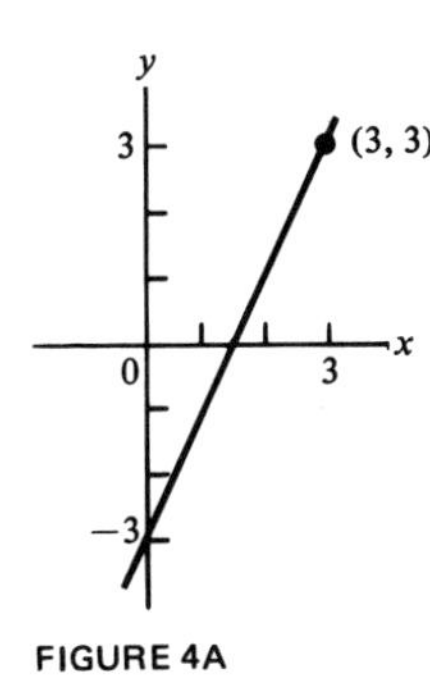

FIGURE 4A

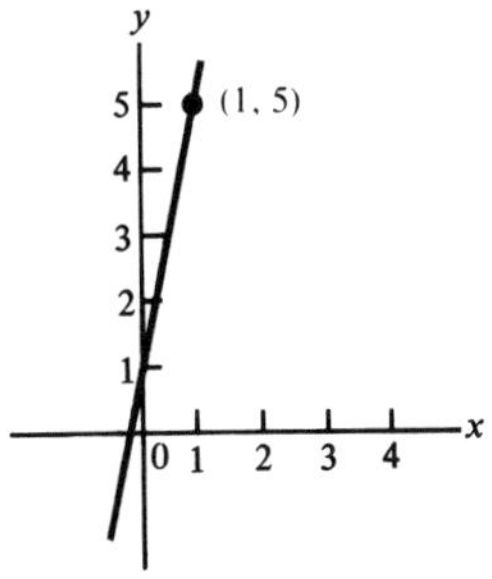

FIGURE 4B

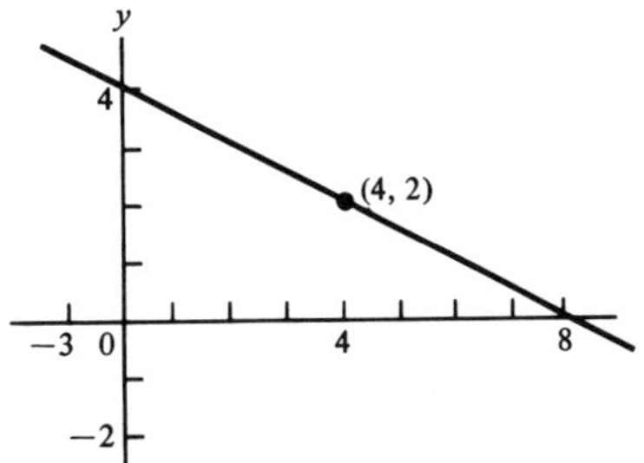

FIGURE 4C

Section 4.3, p. 159

1. (a) -7 (b) 14 **3.** (a) $\frac{5}{2}$ (b) -5 **5.** (a) $\frac{-5}{2}$ (b) $\frac{-5}{3}$ **7.** $y = 2x + 3$ **9.** $y = 3x + 1$

11. $y = \frac{-3}{5}x + \frac{29}{5}$ **13.** $x - y + 10 = 0$ **15.** $8x - y = 0$ **17.** $y - 5 = 0$ **19.** (a) 5 (b) See Figure 4D. **21.** (a) 2 (b) See Figure 4E. **23.** (a) $\frac{-1}{5}$ (b) See Figure 4F.

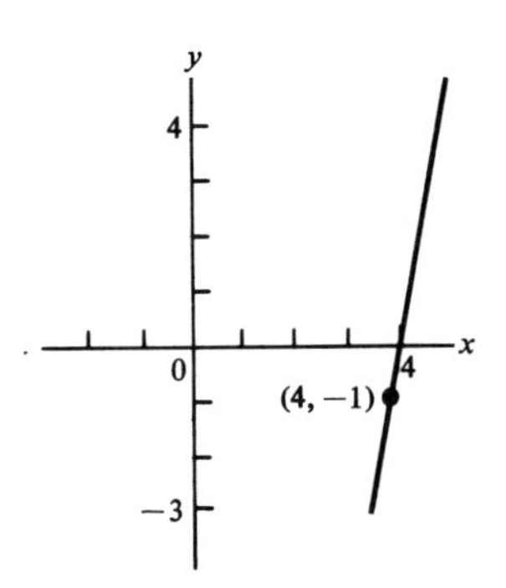

FIGURE 4D

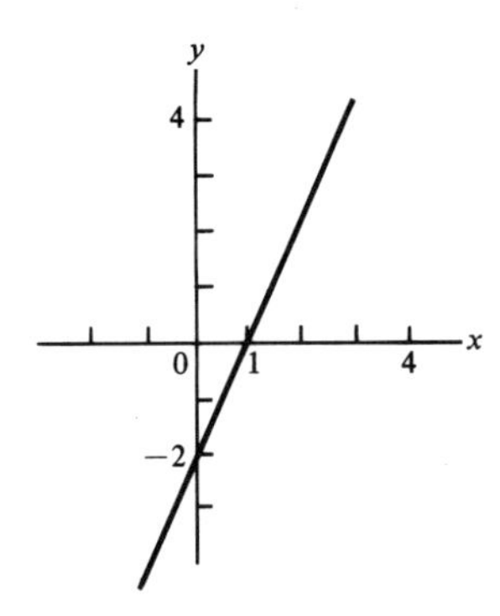

FIGURE 4E

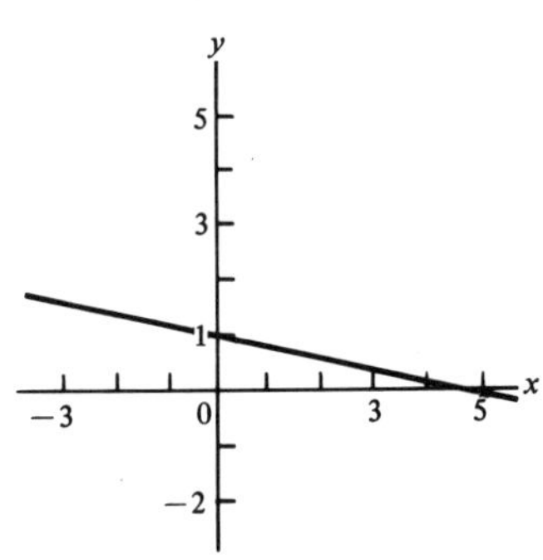

FIGURE 4F

25. $y + 1 = \frac{3}{5}(x + 1)$ **27.** $y = -4x + 5$ **29.** $y = -2$ **31.** $y = 5$ **33.** linear **35.** linear

37. nonlinear **39.** linear **41.** 10 **43.** 5 **45.** $f^{-1}(x) = \frac{1}{m}x - \frac{b}{m}$ **47.** $F = \frac{9}{5}C + 32;\ C = \frac{5}{9}F - \frac{160}{9}$

49. $\frac{-b}{m}$

Section 4.4, p. 164

1. (a) **3.** (b) **5.** (c) **7.** (a) **9.** (a) **11.** (c) **13.** (c) **15.** (c) **17.** (b) **19.** (b)

21. (c) **23.** (a) **25.** (a) $(y - 1) = 2(x - 1)$ (b) $(y - 1) = \frac{-1}{2}(x - 1)$ **27.** (a) $y + 3 = -2x$

(b) $y + 3 = \frac{x}{2}$ **29.** (a) $y - 5 = -(x - 2)$ (b) $y - 5 = x - 2$ **31.** (a) $y = 8$ (b) $x = 2$

33. (a) $x = 0$, (b) $y = 0$ **35.** $L_1 \| L_4$; $L_1 \perp L_5$, $L_4 \perp L_5$, $L_2 \perp L_3$

37. $L_1 \| L_3$, $L_1 \| L_5$, $L_3 \| L_5$, $L_2 \| L_4$; $L_1 \perp L_2$, $L_1 \perp L_4$, $L_3 \perp L_2$, $L_3 \perp L_4$, $L_5 \perp L_2$, $L_5 \perp L_4$

39. Not necessarily! For, L_1 can be the same line as L_3. However, if these lines are distinct, then they are parallel.

41. $L_1 \perp L_3$

Section 4.5, p. 168

1. (a) **3.** (a), (d) **5.** (a), (d) **7.** (a), (b), (c) **9.** (a), (b), (c), (d) **11.** (a), (c) **13.** (a), (d)
15. (a), (b), (c), (d) **17.** (a), (b), (d) **19.** (b), (d)

21.

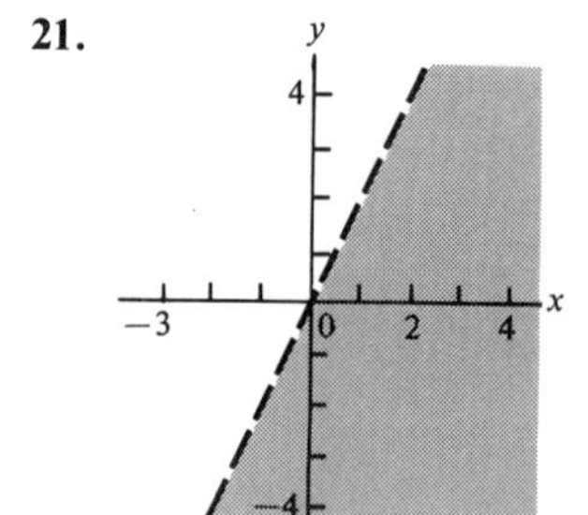

FIGURE 4G $y < 2x$

23.

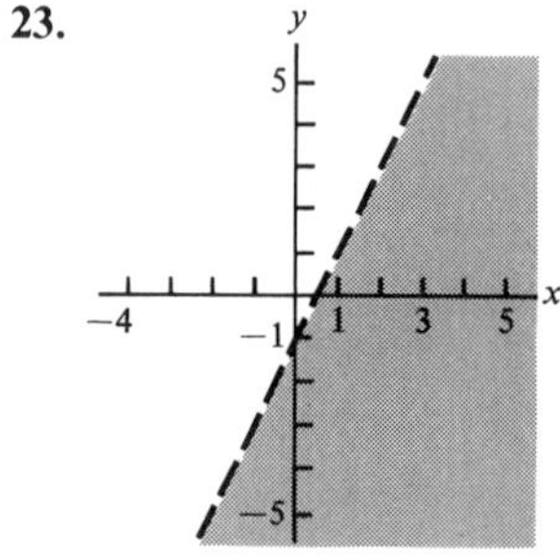

FIGURE 4H $y < 2x - 1$

25.

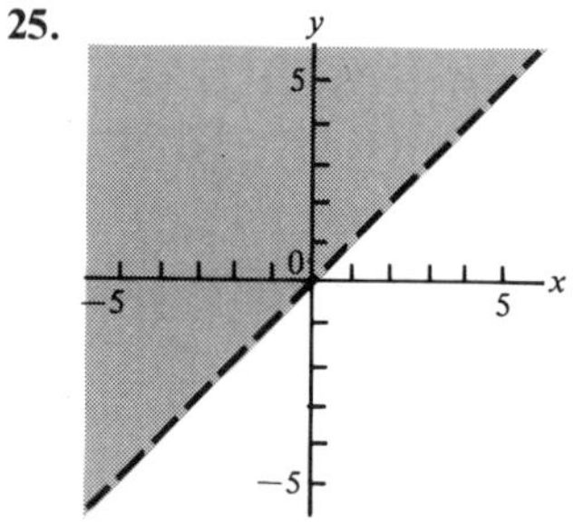

FIGURE 4I $y > x$

27.

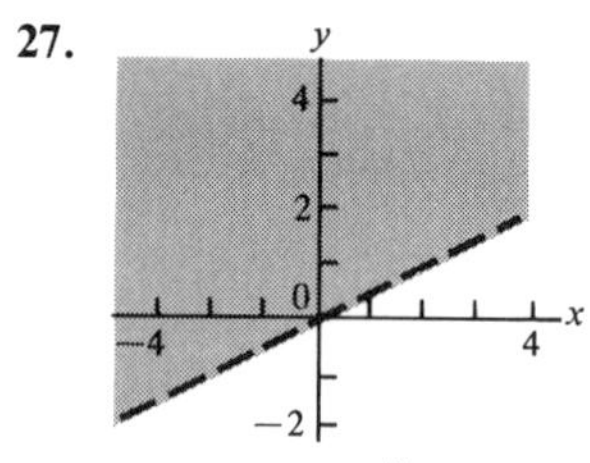

FIGURE 4J $y > \frac{x}{2}$

29.

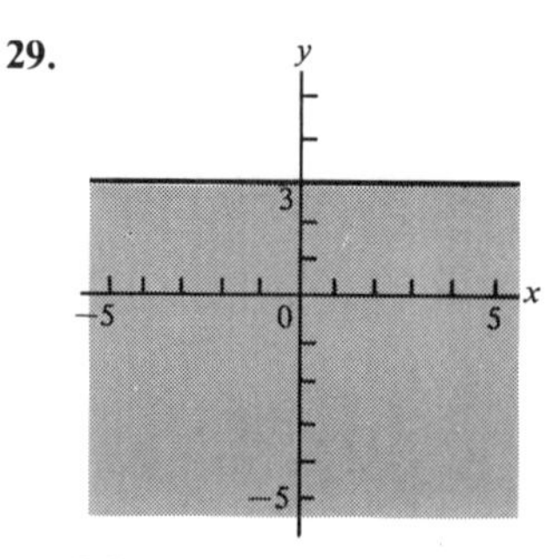

FIGURE 4K $y \leqslant 3$

31.

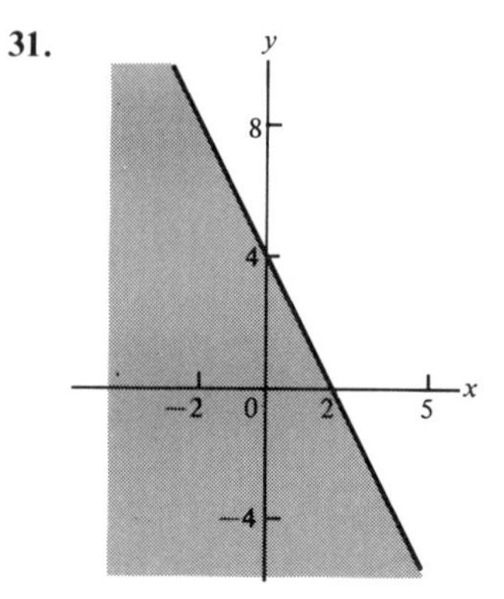

FIGURE 4L $y \leqslant 4 - 2x$

33.

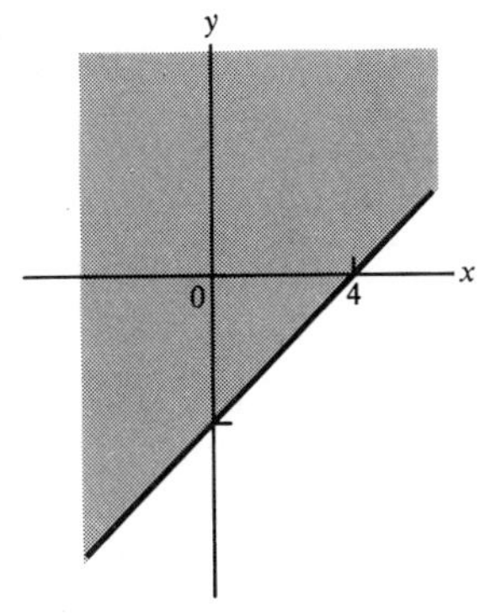

FIGURE 4M $y \geqslant x-4$

35.

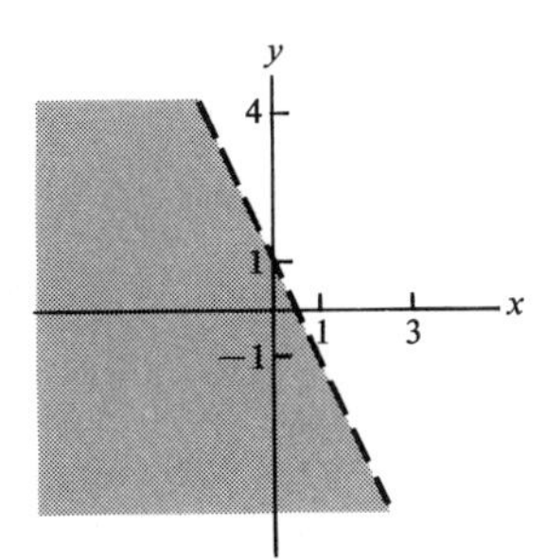

FIGURE 4N $2x+y<1$

Section 4.6, p. 174

1. 3 **3.** $\frac{1}{3}$ **5.** 12 **7.** $\frac{-1}{2}$ **9.** 1 **11.** 6 **13.** 48 **15.** $\frac{1}{8}$ **17.** 2 **19.** (a) $y=3x$ (b) See Figure 4O. **21.** 0 **23.** 60 000 **25.** 8 **27.** 9 feet **29.** 8 **31.** $\frac{1}{6}$ **33.** 2 **35.** 3 **37.** 20 **39.** -50 **41.** 1 **43.** $\frac{1}{4}$ **45.** (a) $y=\frac{6}{x}$ (b) See Figure 4P. (c) hyperbola **47.** no **49.** directly **51.** directly **53.** 400 **55.** 5 ohms

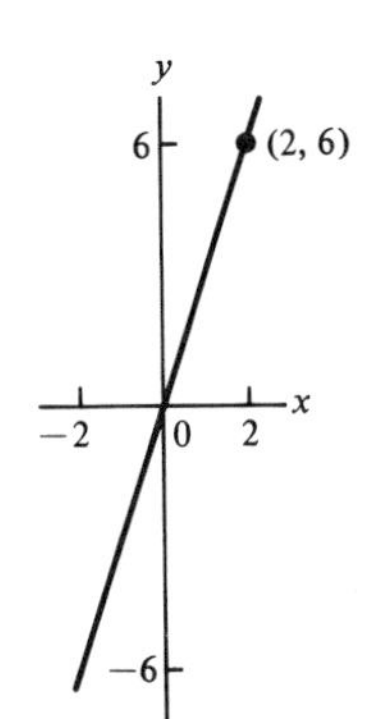

FIGURE 4O

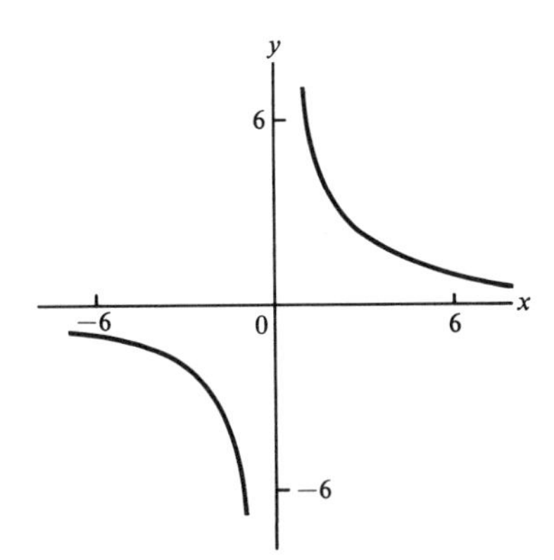

FIGURE 4P

Section 4.7, p. 179

1. 4 **3.** $\frac{1}{2}$ **5.** 36 **7.** 4 **9.** 144 **11.** 90 **13.** 250 **15.** 20 **17.** -6 **19.** 3 **21.** $\frac{1}{2}$ **23.** 6 **25.** 90 **27.** 32 **29.** 8 **31.** 30 centimeters **33.** 27 **35.** 10 000 pounds **37.** 540 000 kilometers **39.** $\frac{4}{3}$

Review Exercises, p. 181

1. (a) 1 (b) −7 (c) 6 **2.** 62 cents **3.** $a = 12, b = \frac{13}{3}$ **4.** $\frac{-3}{4}$ **5.** (a) $y + 1 = 2(x - 3)$ (b) See Figure 4Q. **6.** (a) horizontal (b) vertical (c) neither **7.** (a) −1 (b) 2 **8.** $y = -x + 7$ **9.** $y = 2(x - 2)$ or $y + 4 = 2x$ **10.** (a) $y - 5 = 2(x - 2)$ (b) $y - 5 = -\frac{1}{2}(x - 2)$ **11.** (a) $L_2 \parallel L_4$ (b) $L_1 \perp L_2, L_1 \perp L_4, L_3 \perp L_5$ **12.** (a), (b), (d) **13.** See Figure 4R. **14.** −24 **15.** 45 **16.** −1 **17.** 2 or −2 **18.** 20 **19.** 60 cents

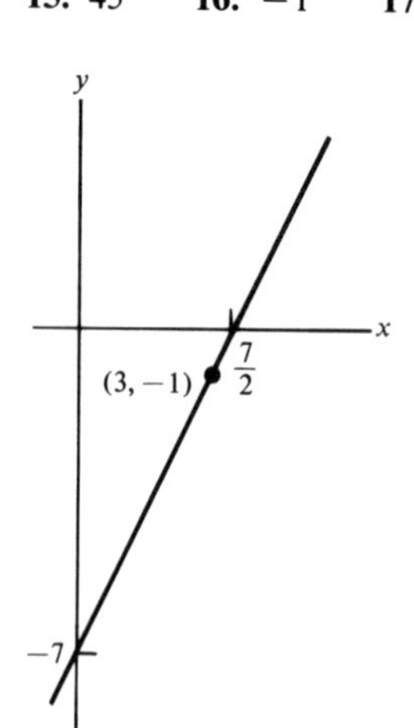

FIGURE 4Q $y + 1 = 2(x - 3)$

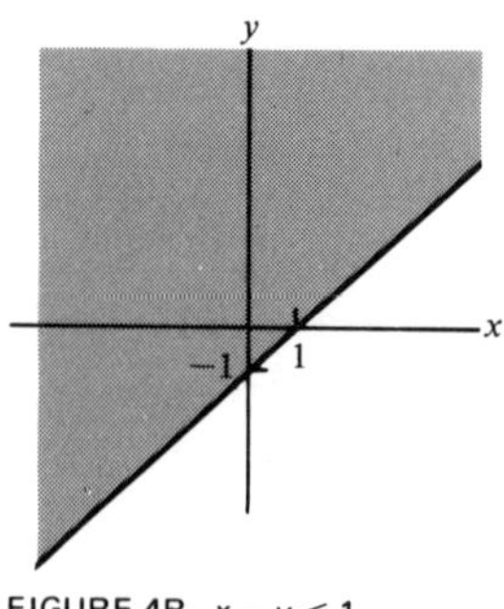

FIGURE 4R $x - y \leqslant 1$

Chapter 5

Section 5.1, p. 186

1. $\frac{1}{10}$ **3.** 1 **5.** $\frac{1}{32}$ **7.** 1 **9.** $\frac{1}{144}$ **11.** $\frac{1}{10\,000}$ **13.** $\frac{-1}{5}$ **15.** $\frac{-1}{8}$ **17.** 2π **19.** $\frac{5}{2}$ **21.** 1 **23.** $\frac{1}{12}$ **25.** 64 **27.** −1 **29.** −1 **31.** 0 **33.** −2 **35.** −1 **37.** −5 **41.** 1 **43.** $a^{-2}b^{-1}$ **45.** $(x + y)^{-1}$ **47.** x^{10} **49.** (a) 1 (b) $\frac{1}{2}$ (c) 2 (d) $\frac{-1}{10}$ (e) $x \neq 0$ (f) odd (g) See Figure 5A. **51.** (a) 4 (b) 4 (c) 4 (d) 4 (e) $x \neq 0$ (f) even (g) See Figure 5B.

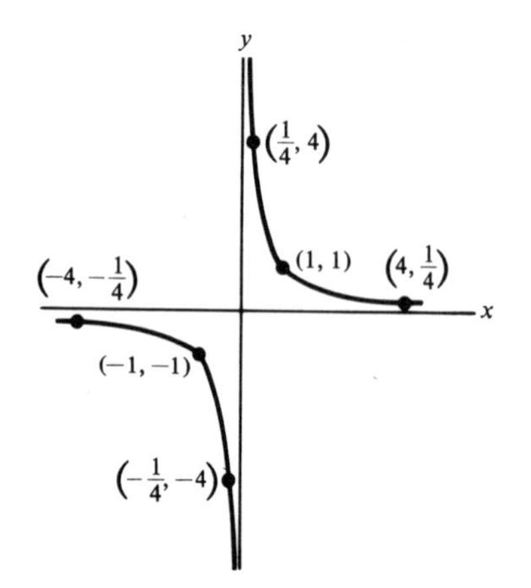

FIGURE 5A $f(x) = x^{-1}$

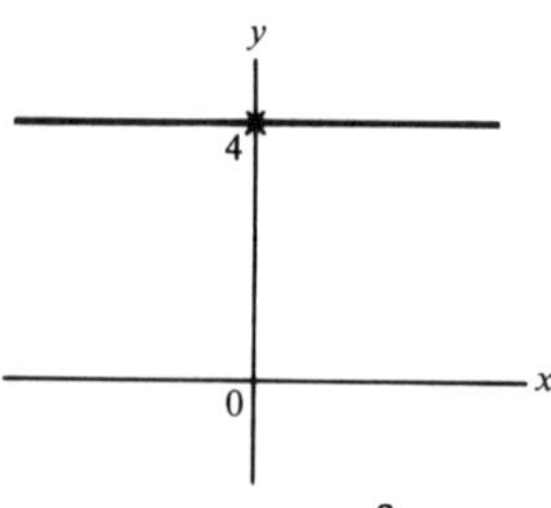

FIGURE 5B $F(x) = 4x^0$

Section 5.2, p. 190

1. a^3 **3.** a^9 **5.** $\frac{1}{a}$ **7.** $\frac{1}{a^{10}}$ **9.** a^{10} **11.** a^4x^2 **13.** $\frac{a^3}{100}$ **15.** b^2 **17.** $4x^2y^2$ **19.** $2a\pi^2$
21. $\frac{2a^7}{b^6}$ **23.** a^{18} **25.** $\frac{1}{x+y}$ **27.** $\frac{10}{3}$ **29.** $\frac{\pi^2+1}{\pi}$ **31.** $\frac{y^2-2y+3}{y^3}$ **33.** 0 **35.** $\frac{5}{3x^2y^2}$
37. $b-a$ **39.** $\frac{(b^4+a^3)(b^4-a^3)}{a^3b^2}$ **41.** $\frac{a^5+b^2}{a(a^2b^3-1)}$ **43.** (a) 64 (b) 64

Section 5.3, p. 194

1. 1.49×10 **3.** 5.084×10^3 **5.** 6.95×10^{-1} **7.** 9×10^{-1} **9.** 2.48×10^5 **11.** $8.200\,01 \times 10^{-4}$
13. 2.79×10^{-1} **15.** 6×10^{-1} **17.** 4×10^8 **19.** 83.4 **21.** 444 000 000 **23.** 5.15 **25.** .8181
27. .062 **29.** 11.1 **31.** 389.624 **33.** 10^{10} **35.** 6×10^{12} **37.** 1 **39.** $.5 \times 10^2$ **41.** (b), (c), (d)
43. 1.5×10^{15} miles

Section 5.4, p. 200

1. 6 **3.** 8 **5.** 12 **7.** 5 **9.** 2 **11.** 16 **13.** 1 **15.** 4 **17.** $\frac{1}{4}$ **19.** $\frac{1}{10}$ **21.** $\frac{1}{2}$ **23.** $\frac{1}{3}$

25. $\frac{1}{27}$ **27.** $\frac{1}{1000}$ **29.** $\frac{1}{6}$ **31.** 13 centimeters **33.** 7 centimeters
35. $(5 \text{ centimeters})^2 + (6 \text{ centimeters})^2 \neq (8 \text{ centimeters})^2$ **37.** 60 inches **39.** 12 centimeters
41. Let A be the area and r the length of the radius of the circle. Then $r = \left(\frac{A}{\pi}\right)^{1/2}$ **43.** 1 **45.** 1
47. (a) all real numbers (b) $[0, \infty]$ **49.** (a) all real numbers (b) all real numbers
51. (a) all real numbers (b) $[-4, \infty)$ **53.** (a) all real numbers (b) $[0, \infty)$ **55.** (a) $x \neq 4$ (b) $(0, \infty)$
57. $A = l(400 - l^2)^{1/2}$ **59.** $f(t) = (49t^2 + 10\,000)^{1/2}$

Section 5.5, p. 203

1. x **3.** $x^{9/4}$ **5.** $y^{1/2}$ **7.** a^4 **9.** a **11.** $x^{3/4}$ **13.** $y^{1/4}$ **15.** $x^{5/6}$ **17.** $x^{1/2}$ **19.** $x^{5/12}$
21. $b^{1/2}$ **23.** $a^{3/2}$ **25.** $y^{5/12}$ **27.** y **29.** $\frac{c^{1/2}}{a^{1/3}b^{2/3}}$ **31.** $x^{10/3}y^{4/3}$ **33.** $x^5z^{1/6}$ **35.** $3 \cdot 3^{1/2}$
37. $4 \cdot 2^{1/2} \cdot a^2$ **39.** $\frac{1}{5 \cdot 3^{1/2}ab^{5/2}}$ **41.** $4x^{1/2}$ **43.** $\frac{8}{27}$ **45.** 25 **47.** 3 **49.** 25 **51.** $x^{-1/2}(x+2)^2$
53. $a^{-5/3}(a+1)(a-1)$ **55.** $(x^2+4)^{-5/6}(x^2+3)$ **57.** $x^{-1/2}$ **59.** $(y+4)^{-2}(y+5)(y+3)$
61. $5 \cdot 3^{1/2}$ centimeters

Section 5.6, p. 208

1. 5 **3.** 2 **5.** 1 **7.** $(\sqrt[3]{b})^2$ **9.** $\sqrt{a}\sqrt[3]{b}$ **11.** $\sqrt{a+b}$ **13.** $x^{1/2}y^{1/4}$ **15.** $x^{1/3}y$ **17.** $(xyz)^{2/3}$
19. $2 + x^{1/2}$ **21.** $5a$ **23.** $\frac{2a^3b}{c^4}$ **25.** .12 **27.** $\frac{1}{125a^6}$ **29.** $2ab^2\sqrt[4]{2a^3}$ **31.** $4\sqrt[4]{5}$ **33.** $\sqrt{13}$
35. $-\frac{\sqrt{3}}{10}$ **37.** $3\sqrt{2}$ **39.** $6\sqrt{7}$ **41.** 0 **43.** $\frac{ab\sqrt{c}(5b-2)}{2}$ **45.** $3\sqrt{3} + 3 - \sqrt{6}$ **47.** $x + 4\sqrt{xy} + 4y$
49. $x + x^{1/2}$ **51.** $x^{1/2} - 2x^{1/4}y^{1/4} + y^{1/2}$ **53.** $4 + 5\sqrt{3}$ **55.** $\frac{\sqrt{b}(\sqrt{2}-a)}{a^3}$ **57.** $\frac{4b-a}{2}$

59. (a) $\sqrt{2x-1}$; $x \geq \frac{1}{2}$ (b) $2\sqrt{x}-1$; $x \geq 0$ (c) 3 (d) $2\sqrt{5}-1$ **61.** (a) 1 (b) $\sqrt{2}$ (c) $\sqrt{3}$ (d) 64

Section 5.7, p. 212

1. $\frac{2\sqrt{3}}{3}$ **3.** $\frac{7(5^{1/2})}{5}$ **5.** $\frac{3\sqrt{2}}{4}$ **7.** $\frac{\sqrt{10}}{6}$ **9.** $\frac{\sqrt{a}}{a}$ **11.** $\frac{x\sqrt{ab}}{ab}$ **13.** $\frac{-5\sqrt{x}}{2x}$ **15.** $\frac{x^{1/2}(1+2x)}{x}$
17. $\frac{\sqrt{y}-y\sqrt{2}}{2y}$ **19.** $\frac{9^{1/3}}{3}$ **21.** $\frac{3\sqrt[3]{x}}{x}$ **23.** $\frac{x^{1/2}(x-a)^{2/3}}{x-a}$ **25.** $-1+\sqrt{2}$ **27.** $\frac{5(3+\sqrt{2})}{7}$
29. $\frac{4(1-\sqrt{a})}{1-a}$ **31.** $\frac{\sqrt{x}(1-\sqrt{x})}{1-x}$ **33.** $\frac{-3(\sqrt{x}-1)}{x-1}$ **35.** $\frac{-5\sqrt{x-y}}{x-y}$ **37.** $\frac{3(a+\sqrt{b+1})}{a^2-b-1}$
39. $\frac{\sqrt{x}(\sqrt{x}-\sqrt{y})}{x-y}$ **41.** $\frac{6(3-2\sqrt{x})}{9-4x}$ **43.** $\frac{2(5-7\sqrt{2})}{73}$ **45.** $\frac{2(7\sqrt{2}-5\sqrt{3})}{23}$ **47.** $\frac{(a+b)(a\sqrt{x}+b\sqrt{y})}{a^2x-b^2y}$
49. $\frac{\sqrt{x+y}}{x+y}$ **51.** $\frac{\sqrt{a}(\sqrt{x}+\sqrt{y})}{x-y}$ **53.** $\frac{1}{\sqrt{x+h+1}+\sqrt{x+1}}$ **55.** $\frac{9}{\sqrt{1+9(x+h)}+\sqrt{1+9x}}$
57. $\frac{1}{(x+h)^{2/3}+(x+h)^{1/3}x^{1/3}+x^{2/3}}$ **59.** $\frac{1}{\sqrt{x}+\sqrt{x}}=\frac{1}{2\sqrt{x}}=\frac{1}{2\sqrt{x}}\cdot\frac{\sqrt{x}}{\sqrt{x}}=\frac{\sqrt{x}}{2x}$

Section 5.8, p. 217

1. 2 **3.** 3 **5.** 5 **7.** $3\sqrt{2}$ **9.** 5 **11.** $2\sqrt{13}$ **13.** $.2\sqrt{2}$ **15.** (4, 5) **17.** $\left(0, \frac{5}{12}\right)$ **19.** (3, 3)
21. (0, 0) **23.** (3, −11) **25.** the points on the y-axis **27.** (a) $|x_1 - x_0|$ (b) $|y_1 - y_0|$
29. Let $A = (2, 1)$, $B = (6, 4)$, $C = (9, 0)$, $D = (5, -3)$. Then $\text{dist}(A, B) = \text{dist}(B, C) = \text{dist}(C, D) = \text{dist}(D, A) = 5$. Also, $\overline{AB} \perp \overline{BC}$ because slope $\overline{AB} = \frac{3}{4}$ and slope $\overline{BC} = \frac{-4}{3}$; therefore $\angle ABC$ is a right angle. Thus, A, B, C, and D are the vertices of a square.

Review Exercises, p. 218

1. (a) $\frac{1}{16}$ (b) $\frac{5}{3}$ (c) $\frac{7}{16}$ **2.** (a) −3 (b) −2 (c) −4 **3.** (a) $\frac{1}{a^6}$ (b) a (c) $\frac{64}{b^2}$
4. (a) 6.0494×10^3 (b) 1.018×10^{-1} (c) 6.39×10^{-4} **5.** (a) 719 (b) .638 (c) .000 121
6. (a) 8 (b) $\frac{1}{1000}$ (c) $8a^{1/8}b^{1/2}c$ (d) $8c^3$ **7.** (a) $\sqrt{x\sqrt[3]{y}}$ (b) $(\sqrt[4]{x+y})^3$ **8.** (a) $(xyz)^{3/5}$
(b) $\frac{(x+y)^{1/3}}{x^{1/2}}$ **9.** (a) $8ab^2$ (b) $3a^2bc^4$ (c) $5-a-2c$ **10.** (a) $\frac{5\sqrt{3}}{3}$ (b) $x^{5/4}+x^{3/4}-x$
(c) $8\sqrt{2}$ **11.** (a) $\sqrt{2}+\sqrt{5}$ (b) $3-5x+\sqrt{2xy^2}$ **12.** (a) $\frac{5\sqrt{3x}}{3x}$ (b) $\frac{1-\sqrt{3}}{2}$ (c) $\frac{x(\sqrt{x}+2\sqrt{y})}{x-4y}$
13. (a) $x \geq 2$ (b) $f(x) \geq 1$ **14.** $10\sqrt{3}$ in. **15.** (a) $\sqrt{53}$ (b) $\left(2, \frac{3}{2}\right)$

Chapter 6

Section 6.1, p. 223 *(The checks are given below.)*

1. 0, 2 **3.** 1, 2 **5.** $\frac{2}{3}$, 2 **7.** A, B **9.** $\frac{B}{A}, \frac{-B}{A}$ **11.** 2, 3, 4 **13.** 0, 3 **15.** $0, \frac{9}{4}$ **17.** 1, 7

19. 2, 3 **21.** 7, −3 **23.** 4 **25.** $\frac{1}{2}, -3$ **27.** $1, \frac{-1}{5}$ **29.** $\frac{1}{2}, \frac{-1}{4}$ **31.** 1 **33.** 1, 2 **35.** 4, −1

37. $-\frac{1}{2}$ (Why *not* 3?) **39.** 0, 3, −5 **41.** $x^2 + x - 12 = 0$ **43.** $3x^2 - 5x + 2 = 0$ **45.** $x^2 - 6x + 9 = 0$

47. 4, −1 **49.** −9

Checks.

17. *for* 1:

$$1^2 - 8(1) + 7 \stackrel{?}{=} 0$$
$$0 \stackrel{\checkmark}{=} 0$$

for 7:

$$7^2 - 8(7) + 7 \stackrel{?}{=} 0$$
$$49 - 56 + 7 \stackrel{?}{=} 0$$
$$0 \stackrel{\checkmark}{=} 0$$

19. *for* 2:

$$2^2 - 5(2) + 6 \stackrel{?}{=} 0$$
$$4 - 10 + 6 \stackrel{?}{=} 0$$
$$0 \stackrel{\checkmark}{=} 0$$

for 3:

$$3^2 - 5(3) + 6 \stackrel{?}{=} 0$$
$$9 - 15 + 6 \stackrel{?}{=} 0$$
$$0 \stackrel{\checkmark}{=} 0$$

Section 6.2, p. 226 (*The checks are given below.*)

1. ± 1 **3.** $\pm 2\sqrt{2}$ **5.** ± 2 **7.** $\pm 2\sqrt{7}$ **9.** $\pm 4B$ **11.** $\pm 9AB^2C^3$ **13.** 2, −8 **15.** 2, 3

17. 2, −6 **19.** $\pm A$ **21.** $\frac{\pm B}{A}$ **23.** $\frac{\pm 7AB}{2}$ **25.** $C \pm \frac{4}{AB}$ **27.** 0, −2 **29.** $1 \pm 2\sqrt{3}$

31. $B \pm AC$ **33.** ± 4 **35.** $0, \pm 1$ **37.** $\pm 1, \pm 3$ **39.** $(2 \pm \sqrt{15}, 0)$ **41.** ± 3 **43.** $-\frac{5}{3}$

Checks.

13. *for* 2:

$$(2 + 3)^2 \stackrel{?}{=} 25$$
$$5^2 \stackrel{?}{=} 25$$
$$25 \stackrel{\checkmark}{=} 25$$

for −8:

$$(-8 + 3)^2 \stackrel{?}{=} 25$$
$$(-5)^2 \stackrel{?}{=} 25$$
$$25 \stackrel{\checkmark}{=} 25$$

15. *for* 2:

$$(2 \cdot 2 - 5)^2 \stackrel{?}{=} 1$$
$$(-1)^2 \stackrel{?}{=} 1$$
$$1 \stackrel{\checkmark}{=} 1$$

for 3:

$$(2 \cdot 3 - 5)^2 \stackrel{?}{=} 1$$
$$1^2 \stackrel{?}{=} 1$$
$$1 \stackrel{\checkmark}{=} 1$$

Section 6.3, p. 230 (*The checks are given below.*)

1. $3i$ **3.** $2\sqrt{2}i$ **5.** $\frac{1}{2}i$ **7.** $\frac{2}{3}i$ **9.** (a) 1 (b) 1 (c) $1 - i$ **11.** (a) −2 (b) 4 (c) $-2 - 4i$

13. (a) 0 (b) 2 (c) $-2i$ **15.** (a) $\frac{1}{2}$ (b) $\sqrt{2}$ (c) $\frac{1}{2} - \sqrt{2}i$ **17.** (a) $-\sqrt{2}$ (b) $\sqrt{3}$

(c) $-\sqrt{2} - \sqrt{3}i$ **19.** $\pm 4i$ **21.** $\pm 2\sqrt{5}i$ **23.** $\pm Ai$ **25.** $2 \pm 2i$ **27.** $-5 \pm \sqrt{3}i$ **29.** $\frac{-2 \pm \sqrt{7}i}{3}$

31. $\frac{-3 \pm 2\sqrt{6}i}{4}$ **33.** $\frac{-B \pm Ci}{A}$ **35.** $3 \pm \sqrt{2}i$ **37.** $\pm 6i$ **39.** $\pm 2, \pm 2i$ **41.** $0, \pm i$

Checks.

19. *for* $4i$:

$$(4i)^2 \stackrel{?}{=} -16$$
$$16(-1) \stackrel{?}{=} -16$$
$$-16 \stackrel{\checkmark}{=} -16$$

for $-4i$:

$$(-4i)^2 \stackrel{?}{=} -16$$
$$16(-1) \stackrel{?}{=} -16$$
$$-16 \stackrel{\checkmark}{=} -16$$

25. *for* $2 + 2i$:

$$(2 + 2i - 2)^2 \stackrel{?}{=} -4$$
$$(2i)^2 \stackrel{?}{=} -4$$
$$4(-1) \stackrel{?}{=} -4$$
$$-4 \stackrel{\checkmark}{=} -4$$

for $2 - 2i$:

$$(2 - 2i - 2)^2 \stackrel{?}{=} -4$$
$$(-2i)^2 \stackrel{?}{=} -4$$
$$4(-1) \stackrel{?}{=} -4$$
$$-4 \stackrel{\checkmark}{=} -4$$

Section 6.4, p. 234 (*The checks are given below.*)

1. $1 \pm \sqrt{2}$ **3.** $-2 \pm \sqrt{6}$ **5.** $2 \pm \sqrt{5}$ **7.** $4 \pm \sqrt{3}$ **9.** $1 \pm \sqrt{5}$ **11.** $\frac{-1 \pm \sqrt{5}}{2}$ **13.** $\frac{-5 \pm \sqrt{17}}{2}$

15. $\frac{5 \pm \sqrt{5}}{2}$ **17.** $\frac{-1 \pm \sqrt{17}}{4}$ **19.** $\frac{-1 \pm \sqrt{19}}{3}$ **21.** $-2 \pm \sqrt{2}$ **23.** $-1 \pm i$ **25.** $\frac{3 \pm \sqrt{3}i}{2}$

27. $-1 \pm \frac{\sqrt{2}}{2}$ **29.** $\frac{3 \pm \sqrt{7}}{2}$ **31.** $-1 \pm \frac{\sqrt{5}}{2}$ **33.** $\frac{-1 \pm \sqrt{2}i}{2}$ **35.** $-1 \pm \sqrt{2}$

Checks.

7. *for* $4+\sqrt{3}$:

$$(4+\sqrt{3})^2 - 8(4+\sqrt{3}) + 13 \stackrel{?}{=} 0$$
$$16 + 8\sqrt{3} + 3 - 32 - 8\sqrt{3} + 13 \stackrel{?}{=} 0$$
$$0 \stackrel{\checkmark}{=} 0$$

for $4-\sqrt{3}$:

$$(4-\sqrt{3})^2 - 8(4-\sqrt{3}) + 13 \stackrel{?}{=} 0$$
$$16 - 8\sqrt{3} + 3 - 32 + 8\sqrt{3} + 13 \stackrel{?}{=} 0$$
$$0 \stackrel{\checkmark}{=} 0$$

9. *for* $1+\sqrt{5}$:

$$(1+\sqrt{5})^2 - 2(1+\sqrt{5}) - 4 \stackrel{?}{=} 0$$
$$1 + 2\sqrt{5} + 5 - 2 - 2\sqrt{5} - 4 \stackrel{?}{=} 0$$
$$0 \stackrel{\checkmark}{=} 0$$

for $1-\sqrt{5}$:

$$(1-\sqrt{5})^2 - 2(1-\sqrt{5}) - 4 \stackrel{?}{=} 0$$
$$1 - 2\sqrt{5} + 5 - 2 + 2\sqrt{5} - 4 \stackrel{?}{=} 0$$
$$0 \stackrel{\checkmark}{=} 0$$

Section 6.5, p. 236 (*The checks are given below.*)

1. (a) 53 (b) 2 distinct real roots **3.** (a) 5 (b) 2 distinct real roots **5.** (a) 0 (b) exactly 1 real root **7.** (a) 17 (b) 2 distinct real roots **9.** (a) 0 (b) exactly 1 real root

11. (a) 112 (b) 2 distinct real roots **13.** $\frac{-1 \pm \sqrt{7}i}{2}$ **15.** $1 \pm \sqrt{3}$ **17.** $\frac{5 \pm \sqrt{65}}{2}$ **19.** $\frac{\pm\sqrt{2}i}{2}$

21. $\frac{-9 \pm \sqrt{97}}{4}$ **23.** $\frac{-5 \pm \sqrt{10}}{5}$ **25.** -7 **27.** $\frac{1}{2}$ **29.** -6 **31.** $-1, -2$ **33.** $\frac{B}{2}, -B$

35. ± 8 **37.** ± 2 **39.** ± 4 **41.** $(2x-3)^2 = 0$

$$2x - 3 = 0$$
$$2x = 3$$
$$x = \frac{3}{2}$$

43. $\frac{-1 \pm \sqrt{7}}{3}$

Checks.

15. *for* $1+\sqrt{3}$:

$$(1+\sqrt{3})^2 - 2(1+\sqrt{3}) - 2 \stackrel{?}{=} 0$$
$$1 + 2\sqrt{3} + 3 - 2 - 2\sqrt{3} - 2 \stackrel{?}{=} 0$$
$$0 \stackrel{\checkmark}{=} 0$$

for $1-\sqrt{3}$:

$$(1-\sqrt{3})^2 - 2(1-\sqrt{3}) - 2 \stackrel{?}{=} 0$$
$$1 - 2\sqrt{3} + 3 - 2 + 2\sqrt{3} - 2 \stackrel{?}{=} 0$$
$$0 \stackrel{\checkmark}{=} 0$$

27. *for* $\frac{1}{2}$:

$$(\tfrac{1}{2})^2 + \tfrac{1}{4} \stackrel{?}{=} \tfrac{1}{2}$$
$$\tfrac{1}{4} + \tfrac{1}{4} \stackrel{?}{=} \tfrac{1}{2}$$
$$\tfrac{2}{4} \stackrel{?}{=} \tfrac{1}{2}$$
$$\tfrac{1}{2} \stackrel{\checkmark}{=} \tfrac{1}{2}$$

Section 6.6, p. 241

1. 2, 10 **3.** 5, -5 **5.** -1 **7.** $2 \pm \sqrt{3}$ **9.** $-8, -7$ **11.** 10 centimeters by 5 centimeters **13.** 50 meters by 20 meters, 40 meters by 25 meters **15.** 20 feet by 8 feet **17.** 60 miles per hour **19.** 5 seconds **21.** 6 hours **23.** 15 seconds **25.** 6 or 10 **27.** 50 meters

Section 6.7, p. 246

1. 2

Check: $\sqrt{2+2} \stackrel{?}{=} 2$

$$\sqrt{4} \stackrel{?}{=} 2$$
$$2 \stackrel{\checkmark}{=} 2$$

3. 7

Check: $\sqrt{2(7)+2} \stackrel{?}{=} 4$

$$\sqrt{16} \stackrel{?}{=} 4$$
$$4 \stackrel{\checkmark}{=} 4$$

5. 17

Check: $\sqrt{17+8} - 5 \stackrel{?}{=} 0$

$$\sqrt{25} - 5 \stackrel{?}{=} 0$$
$$5 - 5 \stackrel{\checkmark}{=} 0$$

7. 3

Check: $[13(3) - 3]^{1/2} - 6 \stackrel{?}{=} 0$

$$36^{1/2} = 6$$
$$6 \stackrel{\checkmark}{=} 6$$

9. 5

Check: $\sqrt{2(5)+2} \stackrel{?}{=} 2\sqrt{3}$

$$\sqrt{12} \stackrel{?}{=} 2\sqrt{3}$$
$$2\sqrt{3} \stackrel{\checkmark}{=} 2\sqrt{3}$$

11. 5

Check: $5 + \sqrt{5-1} \stackrel{?}{=} 7$

$$\sqrt{4} \stackrel{?}{=} 2$$
$$2 \stackrel{\checkmark}{=} 2$$

(10 is not a root.)

$$10 + \sqrt{10-1} \stackrel{?}{=} 7$$
$$\sqrt{9} \stackrel{?}{=} -3$$
$$3 \stackrel{x}{=} -3$$

13. 5

Check: $\sqrt{2(5)-1} - 8 \stackrel{?}{=} -5$

$$\sqrt{9} \stackrel{?}{=} 3$$
$$3 \stackrel{\checkmark}{=} 3$$

(13 is not a root.)

$$\sqrt{2(13)-1} - 8 \stackrel{?}{=} -13$$
$$\sqrt{25} \stackrel{?}{=} -5$$
$$5 \stackrel{x}{=} -5$$

15. -4

Check: $\sqrt{1-2(-4)}+(-4)+1 \overset{?}{=} 0$; $\sqrt{9} \overset{?}{=} 3$; $3 \overset{\checkmark}{=} 3$

(0 is not a root.) $\sqrt{1-2(0)}+0+1 \overset{?}{=} 0$; $\sqrt{1} \overset{?}{=} -1$; $1 \overset{x}{=} -1$

17. 2

Check: $\sqrt{2(2)}-\sqrt{2-1} \overset{?}{=} 1$; $2-1 \overset{\checkmark}{=} 1$

19. 0

Check: $\sqrt{4(0)+1}+3=\sqrt{16+0}$; $1+3 \overset{\checkmark}{=} 4$

(20 is not a root.) $\sqrt{4(20)+1}+3 \overset{?}{=} \sqrt{16+20}$; $\sqrt{81}+3 \overset{?}{=} \sqrt{36}$; $9+3 \overset{x}{=} 6$

21. 16

Check: $\sqrt{1+3(16)}-3 \overset{?}{=} \sqrt{16}$; $\sqrt{49}-3 \overset{?}{=} 4$; $7-3 \overset{\checkmark}{=} 4$

(1 is not a root.) $\sqrt{1+3(1)}-3 \overset{?}{=} \sqrt{1}$; $2-3 \overset{x}{=} 1$

23. 4

Check: $\sqrt{4}\sqrt{4-3} \overset{?}{=} 2$; $2\cdot 1 \overset{\checkmark}{=} 2$

(-1 is not a root.) $\sqrt{-1}\sqrt{-1-3} \overset{?}{=} 2$; $i\cdot 2i \overset{?}{=} 2$; $-2 \overset{x}{=} 2$

25. 13

Check: $\sqrt{13+3}\sqrt{13-9} \overset{?}{=} 8$; $4\cdot 2 \overset{\checkmark}{=} 8$

(-7 is not a root.) $\sqrt{-7+3}\sqrt{-7-9} \overset{?}{=} 8$; $2i\cdot 4i \overset{?}{=} 8$; $-8 \overset{x}{=} 8$

27. $\frac{1}{3}$

Check: $\sqrt{3\left(\frac{1}{3}\right)}\sqrt{9\left(\frac{1}{3}\right)+1} \overset{?}{=} 2$; $\sqrt{1}\sqrt{4} \overset{?}{=} 2$; $1\cdot 2 \overset{\checkmark}{=} 2$

$\left(\frac{-4}{9}\ \textit{is not a root.}\right)$ $\sqrt{3\left(\frac{-4}{9}\right)}\sqrt{9\left(\frac{-4}{9}\right)+1} \overset{?}{=} 2$; $\sqrt{\frac{-4}{3}}\sqrt{-3} \overset{?}{=} 2$; $\frac{2\sqrt{3}i}{3}\cdot\sqrt{3}i \overset{?}{=} 2$; $-2 \overset{x}{=} 2$

29. 10

Check: $5\sqrt{3(10)+6} \overset{?}{=} 3\sqrt{9(10)+10}$; $5\sqrt{36} \overset{?}{=} 3\sqrt{100}$; $5\cdot 6 \overset{\checkmark}{=} 3\cdot 10$

31. 6

Check: $(6+2)^{1/3} \overset{?}{=} 2$; $8^{1/3} \overset{?}{=} 2$; $2 \overset{\checkmark}{=} 2$

33. 4, 20

Check for 4: $\sqrt{48(4)+64}-4 \overset{?}{=} 12$; $\sqrt{256}-4 \overset{?}{=} 12$; $16-4 \overset{\checkmark}{=} 12$

for 20: $\sqrt{48(20)+64}-20 \overset{?}{=} 12$; $\sqrt{1024}-20 \overset{?}{=} 12$; $32-20 \overset{\checkmark}{=} 12$

Section 6.8, p. 251

1. (a) upward (b) $(-2, 0)$ (c) $x=-2$ (d) See Figure 6A. (e) -2 **3.** (a) upward (b) $(0, 2)$ (c) $x=0$ (the y-axis) (d) See Figure 6B. (e) no root **5.** (a) upward (b) $(2, 3)$ (c) $x=2$ (d) See Figure 6C. (e) no root **7.** (a) upward (b) $(0, 0)$ (c) $x=0$ (d) See Figure 6D. (e) 0

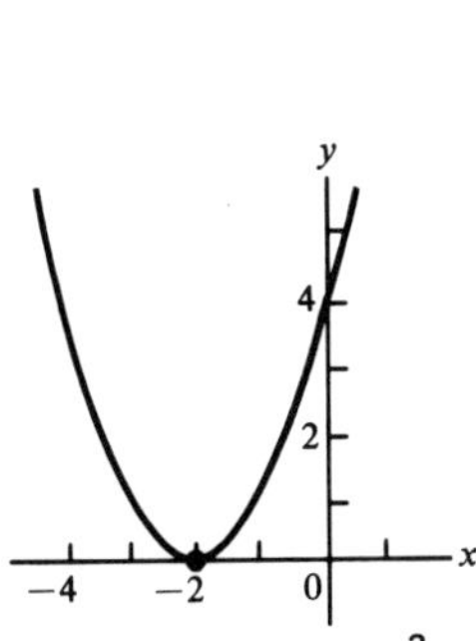

FIGURE 6A $y=(x+2)^2$

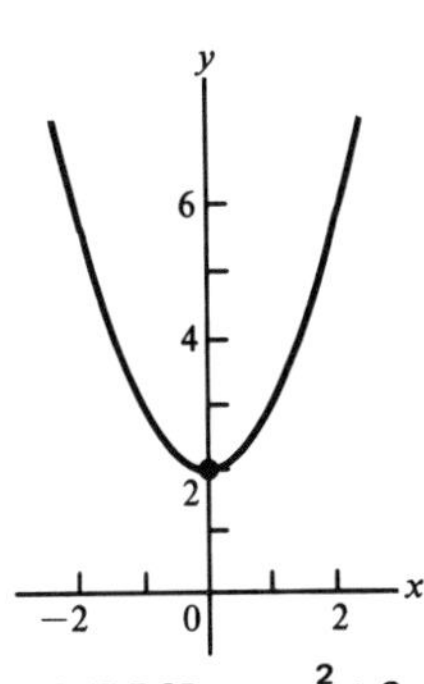

FIGURE 6B $y=x^2+2$

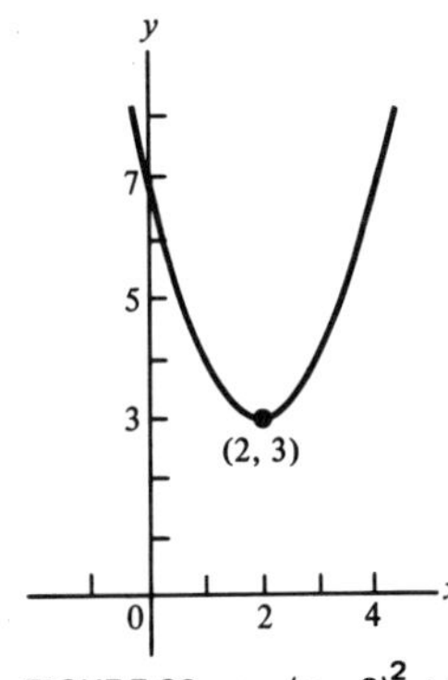

FIGURE 6C $y=(x-2)^2+3$

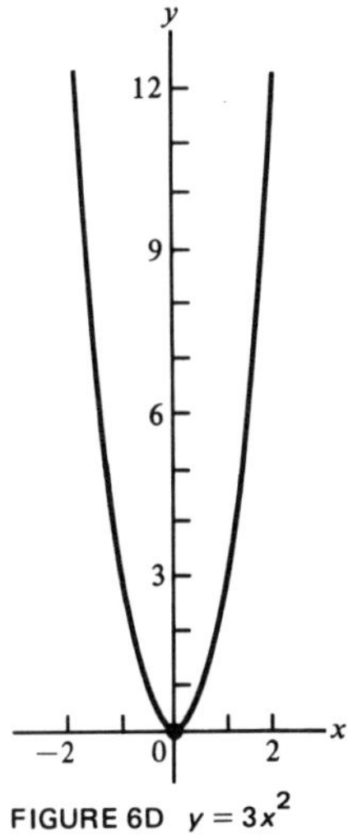

FIGURE 6D $y=3x^2$

9. (a) upward (b) (0, 0) (c) $x = 0$ (d) See Figure 6E. (e) 0 **11.** (a) upward (b) $(0, -4)$ (c) $x = 0$ (d) See Figure 6F. (e) One root lies between -2 and -1; the other lies between 1 and 2. **13.** (a) upward (b) (3, 0) (c) $x = 3$ (d) See Figure 6G. (e) 3 **15.** (a) upward (b) $(0, -4)$ (c) $x = 0$ (d) See Figure 6H. (e) One root lies between -2 and -1; the other root lies between 1 and 2. **17.** (a) upward (b) $(-1, 4)$ (c) $x = -1$ (d) See Figure 6I. (e) no root **19.** (a) upward (b) (1, 0) (c) $x = 1$ (d) See Figure 6J. (e) 1 **21.** (a) upward (b) $\left(-\frac{1}{2}, \frac{3}{4}\right)$ (c) $x = -\frac{1}{2}$ (d) See Figure 6K. (e) no root **23.** (a) upward (b) $(-2, 0)$ (c) $x = -2$ (d) See Figure 6L. (e) -2 **25.** (a) downward (b) (0, 5) (c) $x = 0$ (d) See Figure 6M. (e) -1 and 1 **27.** (a) $y = x^2$ (b) $y = (x - 4)^2$ (c) $y = x^2 - 4$ (d) $y = -(x - 4)^2$ **29.** 60 meters by 120 meters **31.** 5625 square feet

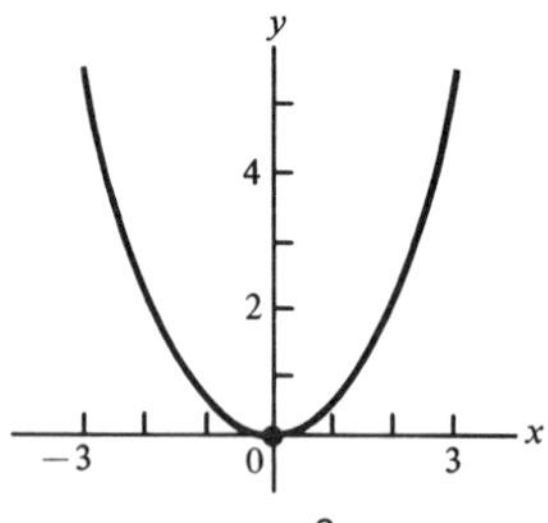

FIGURE 6E $y = \frac{x^2}{2}$

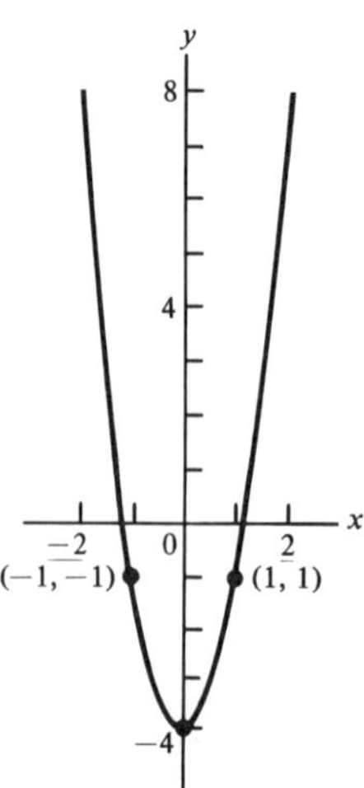

FIGURE 6F $y = 3x^2 - 4$

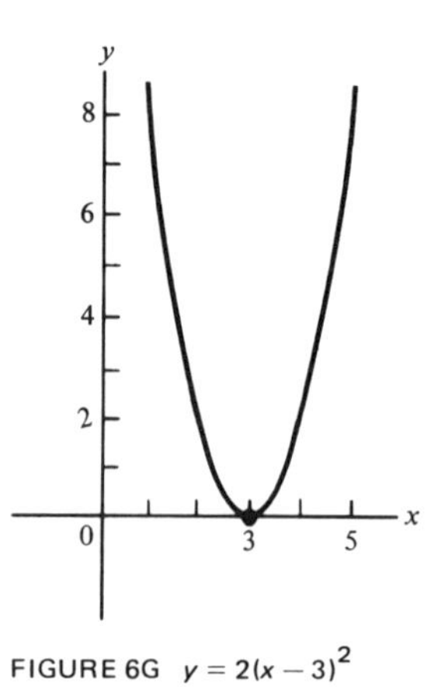

FIGURE 6G $y = 2(x - 3)^2$

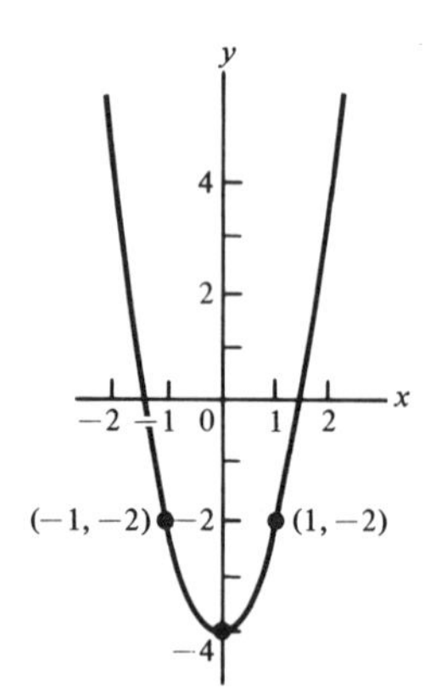

FIGURE 6H $y = 2(x^2 - 2)$

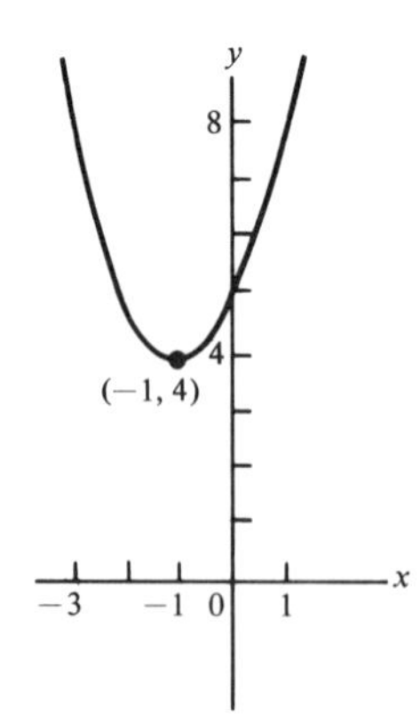

FIGURE 6I $y = x^2 + 2x + 5$
$= (x + 1)^2 + 4$

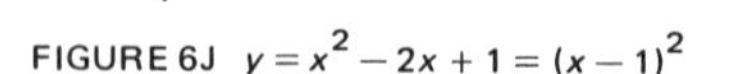
FIGURE 6J $y = x^2 - 2x + 1 = (x - 1)^2$

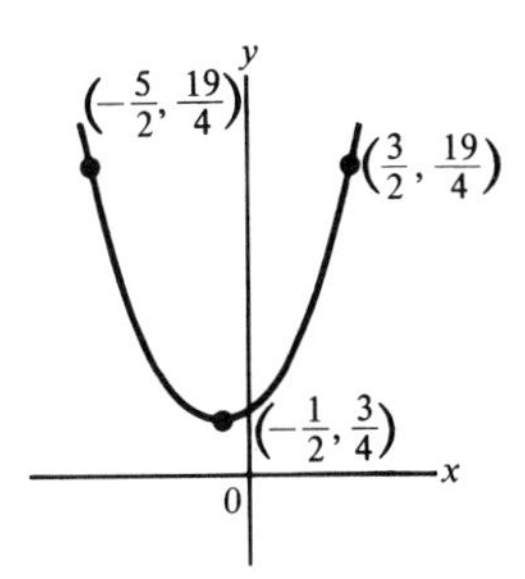

FIGURE 6K $y = x^2 + x + 1 = \left(x + \frac{1}{2}\right)^2 + \frac{3}{4}$

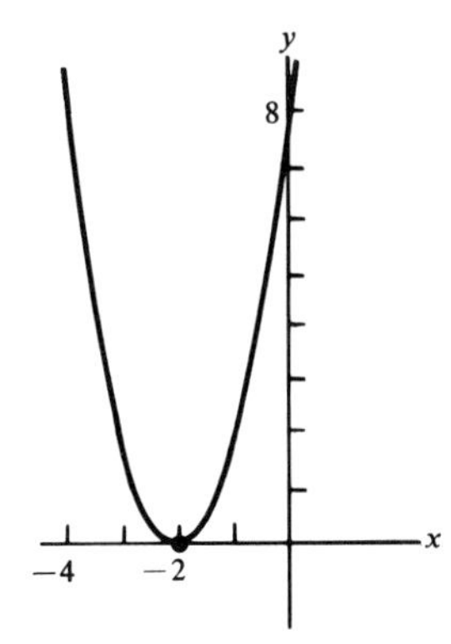

FIGURE 6L $y = 2x^2 + 8x + 8 = 2(x + 2)^2$

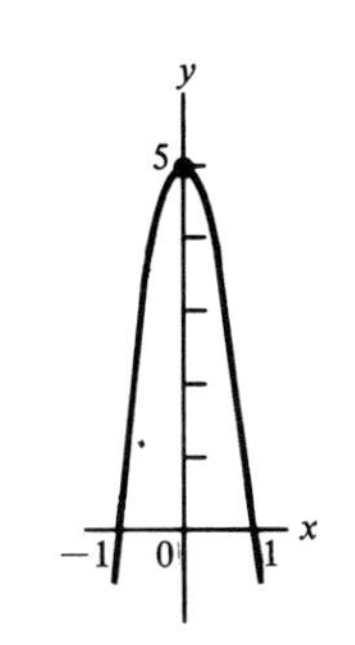

FIGURE 6M $y = 5 - 5x^2 = 5(1 - x^2)$

Section 6.9, p. 256

1. $(-7, 3)$ **3.** $(0, 5)$ **5.** $[0, 4]$ **7.** no solution **9.** all x **11.** $(-\infty, -3) \cup (-1, \infty)$ **13.** $(2, 5)$ **15.** $[-2, 6]$ **17.** $\left(-1, \frac{-1}{2}\right)$ **19.** $\left(-\infty, \frac{1}{2}\right) \cup \left(\frac{3}{2}, \infty\right)$ **21.** $(-\infty, 0) \cup (3, \infty)$ **23.** $(-\infty, -4) \cup (0, \infty)$ **25.** $(-\infty, -1) \cup (0, \infty)$ **27.** $(-\infty, -4) \cup [-3, \infty)$ **29.** $(-2, \infty)$ **31.** $(-\infty, -10) \cup (-5, \infty)$ **33.** $(-\infty, -2)$ **35.** $(-\infty, -2) \cup (1, \infty)$ **37.** $(-\infty, -5) \cup (-4, \infty)$ **39.** all x **41.** $(0, 2)$ **43.** $(-5, -2)$ **45.** $(-1, 2)$

Section 6.10, p. 258

1. (a) $(0, 0)$ (b) 3 **3.** (a) $(0, 0)$ (b) $2\sqrt{2}$ **5.** (a) $(2, 4)$ (b) 7 **7.** (a) $(-8, 7)$ (b) 12 **9.** (a) $\left(\frac{1}{4}, \frac{1}{2}\right)$ (b) $\frac{1}{2}$ **11.** (a) $(-2.7, 1.7)$ (b) .1 **13.** $x^2 + y^2 = 16$ **15.** $(x - 2)^2 + (y - 3)^2 = 25$ **17.** $\left(x - \frac{1}{4}\right)^2 + \left(y - \frac{1}{2}\right)^2 = \frac{1}{4}$ **19.** See Figure 6N. **21.** See Figure 6O. **23.** $(x + 5)^2 + y^2 = 25$; center $(-5, 0)$, radius 5 **25.** $(x + 1)^2 + (y + 3)^2 = 10$; center $(-1, -3)$, radius $\sqrt{10}$ **27.** $(x + 3)^2 + (y + 9)^2 = 90$; center $(-3, -9)$, radius $3\sqrt{10}$ **29.** $(x + 3)^2 + \left(y + \frac{5}{2}\right)^2 = \frac{61}{4}$; center $\left(-3, \frac{-5}{2}\right)$, radius $\frac{\sqrt{61}}{2}$ **31.** $(x + 2)^2 + \left(y - \frac{3}{2}\right)^2 = \frac{29}{4}$; center $\left(-2, \frac{3}{2}\right)$, radius $\frac{\sqrt{29}}{2}$ **33.** $(x + 2)^2 + (y + 10)^2 = 104$; center $(-2, -10)$, radius $2\sqrt{26}$

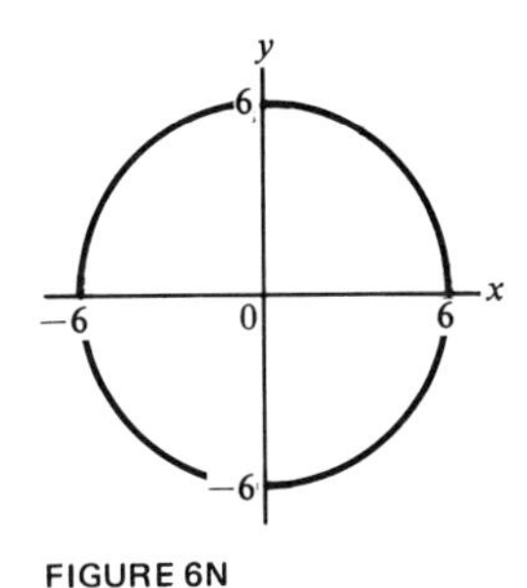

FIGURE 6N

FIGURE 6O

Section 6.11, p. 265

1. (5, 0), (−5, 0), (0, 4), (0, −4). [See Figure 6P.] **3.** (3, 0), (−3, 0), (0, 2), (0, −2)
5. (13, 0), (−13, 0), (0, 12), (0, −12). [See Figure 6Q.] **7.** (10, 0), (−10, 0), (0, 6), (0, −6). [See Figure 6R.]
9. (1, 0), (−1, 0), (0, 2), (0, −2) **11.** $(\sqrt{13}, 0), (-\sqrt{13}, 0), (0, 2), (0, -2)$ **13.** (2, 0), (−2, 0), (0, 3), (0, −3)
15. (a) (3, 0), (−3, 0) (b) $y = \frac{4}{3}x, y = \frac{-4}{3}x$ (c) [See Figure 6S.] **17.** (a) (1, 0), (−1, 0)
(b) $y = 2x, y = -2x$ **19.** (a) (0, 4), (0, −4) (b) $y = \frac{4}{3}x, y = \frac{-4}{3}x.$ (c) [See Figure 6T.]
21. (a) (0, 3), (0, −3) (b) $y = 3x, y = -3x$ **23.** (a) (3, 0), (−3, 0) (b) $y = \frac{2}{3}x, y = \frac{-2}{3}x$
(c) [See Figure 6U.] **25.** (a) (2, 0), (−2, 0) (b) $y = \frac{1}{2}x, y = \frac{-1}{2}x$ **27.** (a) (5, 0), (−5, 0)
(b) $y = \frac{2\sqrt{6}}{5}x, y = \frac{-2\sqrt{6}}{5}x$ **29.** i. **31.** v. **33.** iii. **35.** ii. **37.** ii. **39.** iv.
41. $\frac{x^2}{16} + \frac{y^2}{36} = 1$ **43.** $\frac{x^2}{9} - \frac{4y^2}{9} = 1$

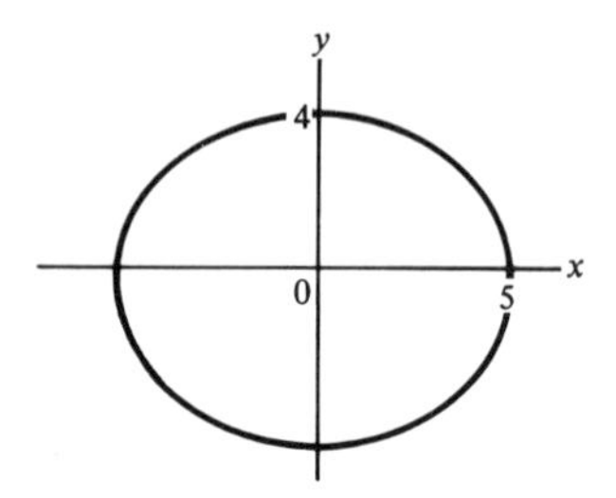

FIGURE 6P $\frac{x^2}{25} + \frac{y^2}{16} = 1$

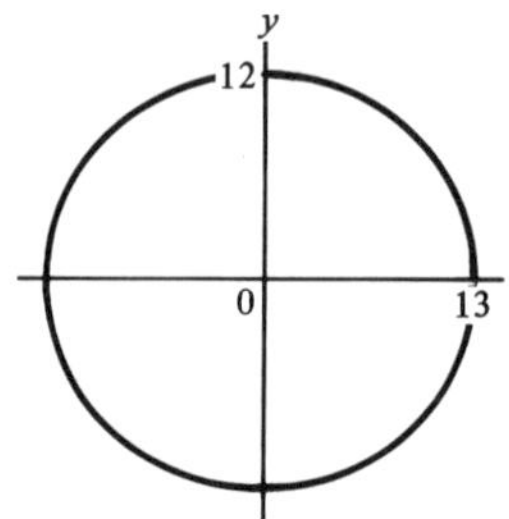

FIGURE 6Q $\frac{x^2}{169} + \frac{y^2}{144} = 1$

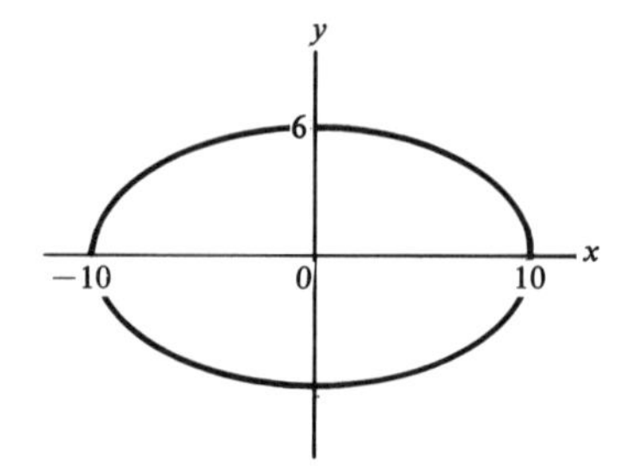

FIGURE 6R $\frac{x^2}{100} + \frac{y^2}{36} = 1$

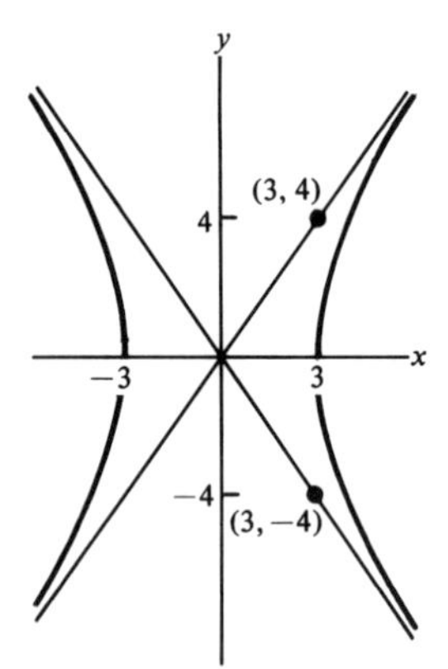

FIGURE 6S $\frac{x^2}{9} - \frac{y^2}{16} = 1$

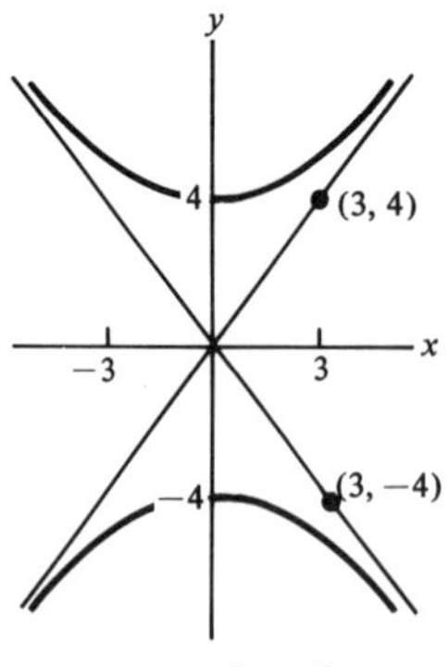

FIGURE 6T $\frac{y^2}{16} - \frac{x^2}{9} = 1$

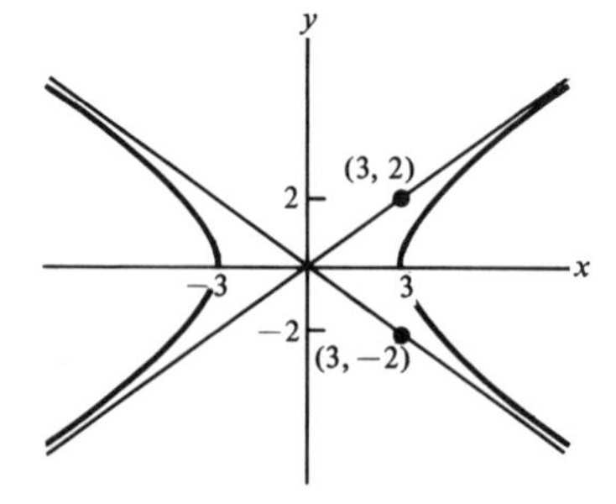

FIGURE 6U $4x^2 - 9y^2 = 36$

Review Exercises, p. 266

1. (a) 2, −3 (b) −1, −4 (c) 3, −2 **2.** $x^2 + x - 20 = 0$ **3.** (a) $6i$ (b) $\frac{\sqrt{2}}{3}i$ **4. i.** (a) 3
(b) −4 (c) $3 + 4i$ **ii.** (a) −8 (b) 0 (c) −8 **iii.** (a) 2 (b) $\frac{1}{2}$ (c) $\frac{4 - i}{2}$ **5.** (a) $\pm 5A$

(b) $\frac{5}{2}, \frac{-7}{2}$ (c) $\pm 4\sqrt{3}i$ **6.** (a) $-3 \pm 2\sqrt{5}$ (b) $\frac{5 \pm \sqrt{21}}{2}$ (c) $\frac{-4 \pm \sqrt{10}}{2}$ **7.** **i.** (a) 5
(b) 2 distinct real roots **ii.** (a) -11 (b) 2 complex conjugate roots **iii.** (a) 0 (b) exactly 1 real root
8. (a) $-5 \pm 2\sqrt{6}$ (b) $\frac{9 \pm 3\sqrt{5}}{2}$ (c) $\frac{5 \pm \sqrt{5}}{2}$ (d) $\frac{5 \pm \sqrt{15}i}{4}$ **9.** $\frac{9 \pm \sqrt{77}}{2}$ **10.** 9 inches by 7 inches

11. (a) 6
Check: $\sqrt{6+3} - 3 \stackrel{?}{=} 0$
$3 - 3 \stackrel{?}{=} 0$
$0 \stackrel{\checkmark}{=} 0$

(b) (0 is not a root.)
$\sqrt{0+1} - \sqrt{0} + 1 \stackrel{?}{=} 0$
$1 - 0 + 1 \stackrel{?}{=} 0$
$2 \stackrel{x}{=} 0$

(c) 1
Check: $\sqrt{1}\sqrt{1+3} \stackrel{?}{=} 2$
$1 \cdot 2 \stackrel{?}{=} 2$
$2 \stackrel{\checkmark}{=} 2$

(-4 is not a root.)
$\sqrt{-4}\sqrt{-4+3} \stackrel{?}{=} 2$
$2i \cdot i \stackrel{?}{=} 2$
$-2 \stackrel{x}{=} 2$

12. **i.** (a) upward (b) $(-2, 0)$ (c) $x = -2$ (d) See Figure 6V. (e) -2 **ii.** (a) downward (b) $(1, 2)$
(c) $x = 1$ (d) See Figure 6W. (e) One root lies between -1 and 0; the other root lies between 2 and 3.
iii. (a) upward (b) $(3, -9)$ (c) $x = 3$ (d) See Figure 6X. (e) 0, 6 **13.** (a) $(-\infty, -3) \cup (-2, \infty)$
(b) $\left(-\frac{1}{4}, \frac{1}{4}\right)$ (c) $(-\infty, -1) \cup (2, \infty)$ **14.** **i.** (a) $(-3, 5)$ (b) $2\sqrt{2}$ **ii.** (a) $\left(-3, \frac{5}{2}\right)$ (b) $\frac{\sqrt{61}}{2}$
15. (a) $(10, 0), (-10, 0), (0, 6), (0, -6)$ (b) See Figure 6Y. **16.** (a) $(5, 0), (-5, 0)$ (b) $y = \frac{3}{5}x, y = \frac{-3}{5}x$
(c) See Figure 6Z. **17.** (a) iii. (b) i. (c) iv. (d) v. (e) vi.

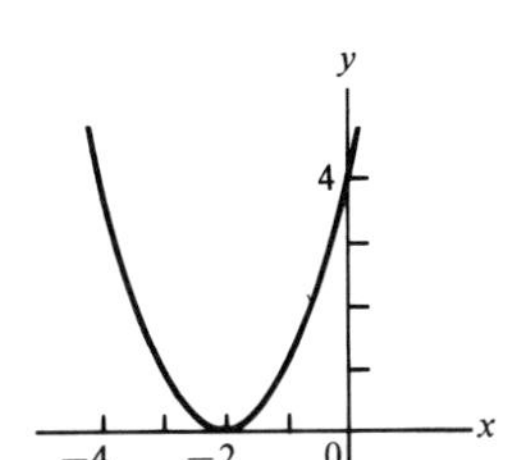

FIGURE 6V $y = (x+2)^2$

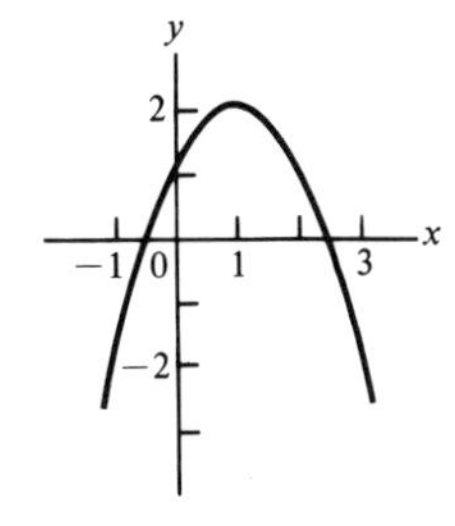

FIGURE 6W $y = 2 - (x-1)^2$

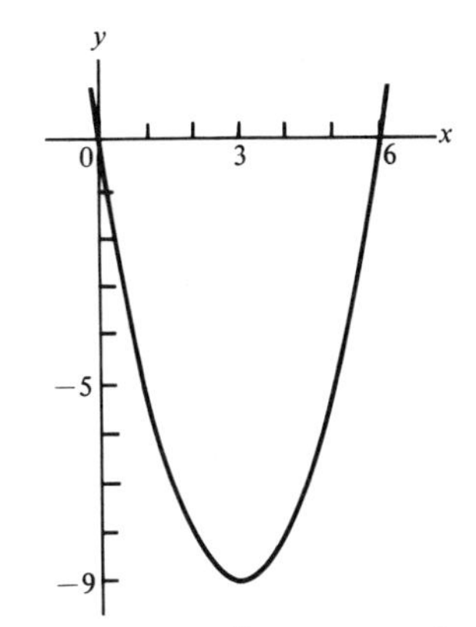

FIGURE 6X $y = x^2 - 6x = (x-3)^2 - 9$

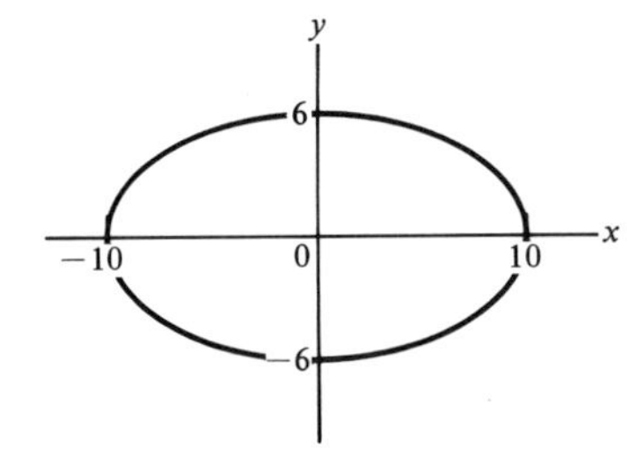

FIGURE 6Y $\frac{x^2}{100} + \frac{y^2}{36} = 1$

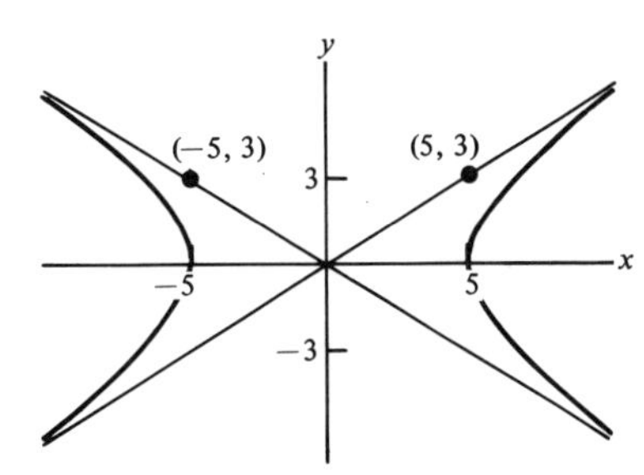

FIGURE 6Z $\frac{x^2}{25} - \frac{y^2}{9} = 1$

Chapter 7

Section 7.1, p. 271

1. $\log_2 16 = 4$ **3.** $\log_{10} 1\,000\,000 = 6$ **5.** $\log_{10} \frac{1}{10} = -1$ **7.** $\log_4 2 = \frac{1}{2}$ **9.** $\log_{16} 8 = \frac{3}{4}$

11. $\log_9 \frac{1}{3} = \frac{-1}{2}$ **13.** $\log_{1/3} 9 = -2$ **15.** $\log_{4/7} \frac{7}{4} = -1$ **17.** $\log_{.1} .0001 = 4$ **19.** $\log_{.01} .1 = \frac{1}{2}$

21. $2^4 = 16$ **23.** $8^{1/3} = 2$ **25.** $7^{-1} = \frac{1}{7}$ **27.** $49^{1/2} = 7$ **29.** $10^{-2} = .01$ **31.** $81^{3/4} = 27$ **33.** 2

35. 4 **37.** 3 **39.** $\frac{3}{2}$ **41.** -6 **43.** 0 **45.** 3 **47.** -3 **49.** 25 **51.** 1 **53.** $\sqrt{3}$ **55.** 0

57. 2 **59.** .8 **61.** .2 **63.** .1

Section 7.2, p. 275

3. (a) $6 = 2 + 4$ (b) $\log_b n_1 n_2 = \log_b n_1 + \log_b n_2$. Here $b = 2$, $n_1 = 4$, $n_2 = 16$ **5.** (a) $3 = 5 - 2$
(b) $\log_b \frac{n_1}{n_2} = \log_b n_1 - \log_b n_2$. Here $b = 10$, $n_1 = 100\,000$, $n_2 = 100$ **7.** (a) $6 = 2 \cdot 3$
(b) $\log_b n^r = r \log_b n$. Here $b = 2$, $n = 8$, $r = 2$ **9.** (a) $4 = 2 \cdot 2$
(b) $\log_b n^r = r \log_b n$. Here $b = 3$, $n = 9$, $r = 2$ **13.** $\log_b x - \log_b y$ **15.** $\log_{10} a + \log_{10} b - \log_{10} c$

17. $3 \log_{10} x$ **19.** $\log_5 x + 4 \log_5 y$ **21.** $2 \log_3 x + \frac{1}{2} \log_3 y - \log_3 z$ **23.** $\frac{1}{2}(\log_b 5 + \log_b x - \log_b 3)$

25. $4 \log_{10} a + 2 \log_{10} b - 3 \log_{10} c$ **27.** $\frac{1}{2} \log_b x - \frac{1}{3} \log_b y + \frac{3}{4} \log_b z$

29. $-5\left(\frac{1}{2} \log_b x + 2 \log_b y - 3 \log z\right)$ **31.** .78 **33.** 1.40 **35.** $-.18$ **37.** 1.08 **39.** .28 **41.** $-.48$

43. $\frac{\log_{10} 5}{\log_{10} 2} = 2.32$, whereas $\log_{10} 5 - \log_{10} 2 = .40$ **45.** $\log_b \frac{x}{y}$ **47.** $\log_b \frac{10x}{a}$ **49.** $\log_b \frac{10x}{a}$

51. $\log_b m^3 n^{1/2}$ **53.** $\log_b\left(\frac{rt^{1/3}}{s^2}\right)^{1/5}$ **55.** $\log_{10}(ab^3)^{1/5}\left(\frac{c}{d}\right)^{1/3}$ **57.** $5 - \log_{10} 6$

Section 7.3, p. 281

1. .0128 **3.** .6345 **5.** .4771 **7.** .9731 **9.** .7803 **11.** .8621 **13.** .0374 **15.** 1 **17.** 4
19. -2 **21.** -4 **23.** 6 **25.** 8 **27.** 6 **29.** 3 **31.** 2.9405 **33.** 1.2279 **35.** 8.4082
37. $.9708 - 1$ **39.** $.1335 - 2$ **41.** .8055 **43.** $.9201 - 4$ **45.** 1.7924 **47.** $.6920 - 3$ **49.** $.9222 - 3$

Section 7.4, p. 284

1. 1.97 **3.** 7.44 **5.** 2.72 **7.** 3.65 **9.** 36 500 **11.** .003 65 **13.** 6.09 **15.** 6 090 000
17. .0609 **19.** 10.9 **21.** 956 **23.** .007 06 **25.** .0956 **27.** 1.07 **29.** 3.65 **31.** 446 000
33. 644 000 000

Section 7.5, p. 289

1. (a) 3 (b) 9 (c) 27 (d) $\frac{1}{3}$ **3.** (a) 100 (b) $\frac{1}{100}$ (c) 10 (d) $\frac{1}{10}$ **5.** (a) $\frac{1}{4}$ (b) $\frac{1}{16}$

(c) 4 (d) 16 **7.** (a) 0 (b) 1 (c) 2 (d) 5 **9.** (a) 2 (b) $\frac{1}{2}$ (c) 3 (d) $\frac{3}{2}$ **11.** (a) $\frac{1}{2}$
(b) -1 (c) $-\frac{1}{2}$ (d) $\frac{3}{2}$ **13.** (a) 8^x (b) 8 (c) 64 (d) $3(2^x)$ (e) 6 (f) 12 **15.** (a) $2^{(x^2)}$
(b) 16 (c) 512 (d) 4^x (e) 16 (f) 64 **17.** 6^x **19.** See Figure 7A. **21.** (a) increasing
(b) all real numbers (c) all positive (real) numbers **23.** (a) decreasing (b) all real numbers
(c) all positive numbers **25.** (a) decreasing (b) all real numbers (c) all negative numbers
27. (a) decreasing (b) all real numbers (c) all positive numbers **29.** (a) neither (b) all real numbers
(c) all numbers $y \geq 1$ **31.** (a) 2000 (b) 16 000 (c) 3

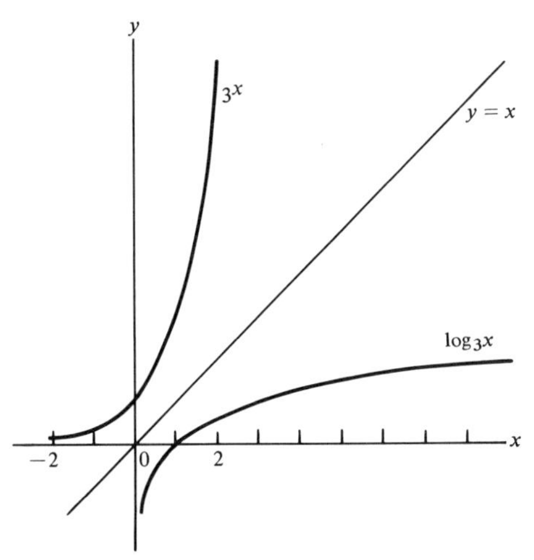

FIGURE 7A

Section 7.6, p. 293

1. .6133 **3.** .9098 **5.** 1.6167 **7.** .8473 − 1 **9.** 5.0153 **11.** .8864 − 3 **13.** .8452
15. .8452 − 1 **17.** .5550 **19.** .5550 − 1 **21.** 3.9338 **23.** .9300 − 3 **25.** 7.5963 **27.** .9225 − 5
29. 9.9913 **31.** 4.001 **33.** 9.004 **35.** 95.42 **37.** 97 180 **39.** .099 63 **41.** 6.653 **43.** 66 530
45. 6.668 **47.** .066 68 **49.** 1 064 000 **51.** .2472 **53.** 1.549 **55.** .9998 **57.** 4 641 000
59. .083 64

Section 7.7, p. 298 (*Interpolation was used to obtain these answers.*)

1. 55.01 **3.** 911.2 **5.** 486.8 **7.** 45 390 **9.** 47 480 000 **11.** 4.748 **13.** .005 398 **15.** .000 969 4
17. 20 440 000 **19.** 166.2 **21.** 2.801×10^{-11} **23.** 1.951 **25.** .7488 **27.** .8294 **29.** 18.71
31. .013 40 **33.** $2655 **35.** 1.325×10^{12}

Section 7.8, p. 302

1. 2.21 **3.** 1.16 **5.** .53 **7.** .51 **9.** 1.16 **11.** −.67 **13.** 10 000 **15.** −5 **17.** .025
19. 2000 **21.** $\frac{10}{3}$ **23.** 3.86 hours **25.** 8.31 years **27.** 20 **29.** 391 000 000 **31.** 1.59 **33.** 1.83
35. 2.47 **37.** 1.72 **41.** $\frac{1}{2}\log_2 29$ **43.** $\frac{\log_2 93}{\log_2 10}$ **45.** $\frac{\log_{12} 9.02}{\log_{12} 7}$ **47.** x **49.** $\frac{\log 2}{\log 1.5}$
51. $\log\left(\frac{5+\sqrt{29}}{2}\right)$ **53.** $\log(4 \pm \sqrt{15})$ **55.** $\frac{\log(2 \pm \sqrt{3})}{\log 2}$ **57.** $\sqrt{11} - 1$

Review Exercises, p. 304

1. (a) $\log_{10} 1000 = 3$ (b) $\log_{1/2} \frac{1}{4} = 2$ (c) $\log_{16} 4 = \frac{1}{2}$ **2.** (a) 2 (b) 4 (c) 2 **3.** (a) 2 (b) 5 (c) $\frac{1}{2}$ **4.** $2 \log_b 2 + \log_b 3 + \log_b 5$ **5.** $\log_b \frac{xy^3}{z}$ **6.** .9253 **7.** (a) 2 (b) -4 **8.** (a) 2.7882 (b) .8000 − 2 **9.** (a) 4590 (b) .394 **10.** (a) 8 (b) 64 (c) $\frac{1}{64}$ (d) 2 **11.** (a) 1 (b) $\frac{1}{2}$ (c) $-\frac{1}{2}$ (d) $\frac{3}{2}$ **12.** (a) .9141 (b) .8641 − 2 **13.** (a) 5.005 (b) 95.44 **14.** (a) 58.36 (b) .622 (c) .0141 **15.** 200 000 **16.** 6.06 **17.** 7.75 **18.** $\frac{\log_3 7}{\log_3 2}$

Chapter 8

Section 8.1, p. 310

1. 35°12′ **3.** 100°45′ **5.** 42.5° **7.** 75.25° **9.** 60° **11.** 37°48′ **13.** (a) $\frac{3}{5}$ (b) $\frac{4}{5}$ (c) $\frac{3}{4}$ (d) $\frac{4}{5}$ (e) $\frac{3}{5}$ (f) $\frac{4}{3}$ **15.** (a) $\frac{\sqrt{2}}{2}$ (b) $\frac{\sqrt{2}}{2}$ (c) 1 (d) $\frac{\sqrt{2}}{2}$ (e) $\frac{\sqrt{2}}{2}$ (f) 1 **17.** (a) $\frac{\sqrt{5}}{5}$ (b) $\frac{2\sqrt{5}}{5}$ (c) $\frac{1}{2}$ (d) $\frac{2\sqrt{5}}{5}$ (e) $\frac{\sqrt{5}}{5}$ (f) 2 **19.** (a) $\frac{\sqrt{10}}{5}$ (b) $\frac{\sqrt{15}}{5}$ (c) $\sqrt{\frac{2}{3}}$ or $\frac{\sqrt{6}}{3}$ (d) $\frac{\sqrt{15}}{5}$ (e) $\frac{\sqrt{10}}{5}$ (f) $\sqrt{\frac{3}{2}}$ or $\frac{\sqrt{6}}{2}$ **21.** $\cos A = \frac{3}{5}$, $\tan A = \frac{4}{3}$, $\csc A = \frac{5}{4}$, $\sec A = \frac{5}{3}$, $\cot A = \frac{3}{4}$
23. $\sin A = \frac{4}{5}$, $\cos A = \frac{3}{5}$, $\csc A = \frac{5}{4}$, $\sec A = \frac{5}{3}$, $\cot A = \frac{3}{4}$ **25.** $\cos A = \frac{\sqrt{15}}{4}$, $\tan A = \frac{\sqrt{15}}{15}$, $\csc A = 4$, $\sec A = \frac{4\sqrt{15}}{15}$, $\cot A = \sqrt{15}$ **27.** $\sin A = \frac{1}{5}$, $\cos A = \frac{2\sqrt{6}}{5}$, $\tan A = \frac{\sqrt{6}}{12}$, $\sec A = \frac{5\sqrt{6}}{12}$, $\cot A = 2\sqrt{6}$
29. $\sin A = \frac{\sqrt{91}}{10}$, $\cos A = \frac{3}{10}$, $\tan A = \frac{\sqrt{91}}{3}$, $\csc A = \frac{10\sqrt{91}}{91}$, $\cot A = \frac{3\sqrt{91}}{91}$ **31.** $b = 5, c = 5\sqrt{2}$
33. $a = b = 5\sqrt{2}$ **35.** $a = 9, b = 9\sqrt{3}$ **37.** $\frac{1}{2}$ **39.** $\frac{1}{2}$ **41.** $\sqrt{3}$ **43.** $\frac{2\sqrt{3}}{3}$ **45.** 1
47. (a) 14.1 feet (b) 14.1 feet **49.** 433 feet **51.** 63.6 feet **53.** $(2850\sqrt{3} + 1350)$, or approximately 6286.2, square meters
55. (a) $10 \sin A$ (b) $10(1 + \cos A)$ (c) $\frac{1000\pi}{3} \sin^2 A(1 + \cos A)$

Section 8.2, p. 321

1. .4226 **3.** .1584 **5.** .1736 **7.** .2447 **9.** .9827 **11.** .0233 **13.** 11° **15.** 34° **17.** 68°
19. 8°10′ **21.** 11°50′ **23.** 83°10′ **25.** $\angle A = 35°$, $\angle B = 55°$, $c = 12.2$ **27.** $\angle A = 66°30'$, $\angle B = 23°30'$, $a = 36.7$ **29.** $\angle B = 47°40'$, $a = 16.8$, $b = 18.5$ **31.** $\angle A = 83°10'$, $a = 150.2$, $c = 151.3$ **33.** 51°20′ **35.** (a) 19.4 feet (b) 14.1 feet **37.** 1.0 mile **39.** 22.3 inches **41.** 51°20′
43. 255.0 feet **45.** 15.8 feet or 15 feet, 10 inches **47.** .3352 **49.** .8426 **51.** .0858 **53.** .0178
55. 24°35′ **57.** 41°12′ **59.** 77°25′ **61.** $\angle A = 47°39'$, $\angle B = 42°21'$, $c = 26.0$ **63.** $\angle A = 17°54'$, $a = 15.2$, $b = 47.1$

Section 8.3, p. 333

1. yes **3.** no **5.** yes **7.** yes **9.** $\frac{\pi}{4}$ **11.** $\frac{4\pi}{5}$ **13.** $\frac{3\pi}{2}$ **15.** 5π **17.** $-\frac{5\pi}{6}$ **19.** 60°

21. 45° **23.** 315° **25.** 126° **27.** -1080° **29.** π **31.** $\frac{\pi}{6}$ **33.** $\frac{7\pi}{4}$ **35.** $\frac{\pi}{4}$ **37.** $\sin\frac{2\pi}{3}=\frac{\sqrt{3}}{2}$, $\cos\frac{2\pi}{3}=-\frac{1}{2}$, $\tan\frac{2\pi}{3}=-\sqrt{3}$, $\csc\frac{2\pi}{3}=\frac{2\sqrt{3}}{3}$, $\sec\frac{2\pi}{3}=-2$, $\cot\frac{2\pi}{3}=-\frac{\sqrt{3}}{3}$ **39.** $\sin\frac{3\pi}{2}=-1$, $\cos\frac{3\pi}{2}=0$, $\tan\frac{3\pi}{2}$ is undefined, $\csc\frac{3\pi}{2}=-1$, $\sec\frac{3\pi}{2}$ is undefined, $\cot\frac{3\pi}{2}=0$ **41.** $\sin\left(-\frac{3\pi}{2}\right)=1$, $\cos\left(-\frac{3\pi}{2}\right)=0$, $\tan\left(-\frac{3\pi}{2}\right)$ is undefined, $\csc\left(-\frac{3\pi}{2}\right)=1$, $\sec\left(-\frac{3\pi}{2}\right)$ is undefined, $\cot\left(-\frac{3\pi}{2}\right)=0$

43. $\sin\frac{7\pi}{4}=-\frac{\sqrt{2}}{2}$, $\cos\frac{7\pi}{4}=\frac{\sqrt{2}}{2}$, $\tan\frac{7\pi}{4}=-1$, $\csc\frac{7\pi}{4}=-\sqrt{2}$, $\sec\frac{7\pi}{4}=\sqrt{2}$, $\cot\frac{7\pi}{4}=-1$

45. $\sin(-\pi)=0$, $\cos(-\pi)=-1$, $\tan(-\pi)=0$, $\csc(-\pi)$ is undefined, $\sec(-\pi)=-1$, $\cot(-\pi)$ is undefined.

47. $\sin\frac{9\pi}{4}=\frac{\sqrt{2}}{2}$, $\cos\frac{9\pi}{4}=\frac{\sqrt{2}}{2}$, $\tan\frac{9\pi}{4}=1$, $\csc\frac{9\pi}{4}=\sqrt{2}$, $\sec\frac{9\pi}{4}=\sqrt{2}$, $\cot\frac{9\pi}{4}=1$ **49.** (a) .2588 (b) .9659 (c) .2679 (d) 3.7321 **51.** (a) .9239 (b) $-.3827$ (c) -2.4142 (d) $-.4142$

53. (a) $-.9659$ (b) $-.2588$ (c) 3.7321 (d) .2679 **55.** (a) $-\frac{\pi}{2}$ (b) -6π **57.** $\tan\theta=$ length $\overline{PR}$

Section 8.4, p. 345

1.

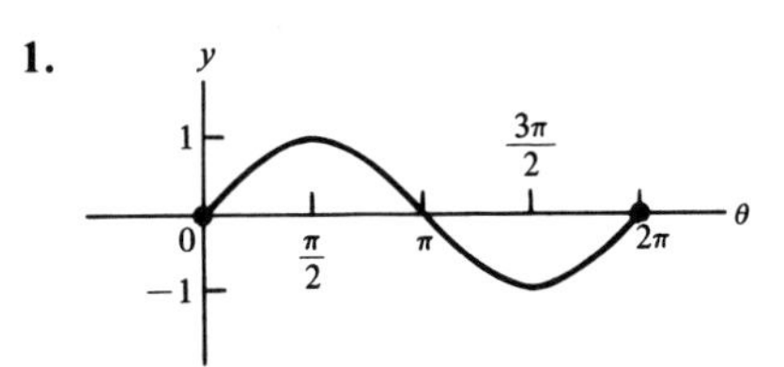

FIGURE 8A $\sin\theta,\ 0 \leqslant \theta \leqslant 2\pi$

3.

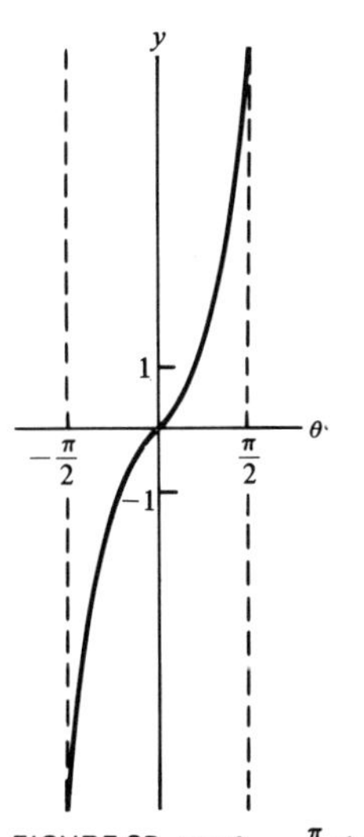

FIGURE 8B $\tan\theta,\ -\frac{\pi}{2} < \theta < \frac{\pi}{2}$

5.

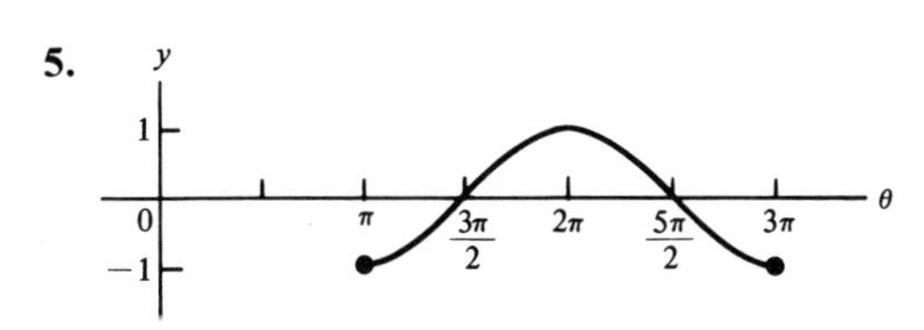

FIGURE 8C $\cos\theta,\ \pi \leqslant \theta \leqslant 3\pi$

7.

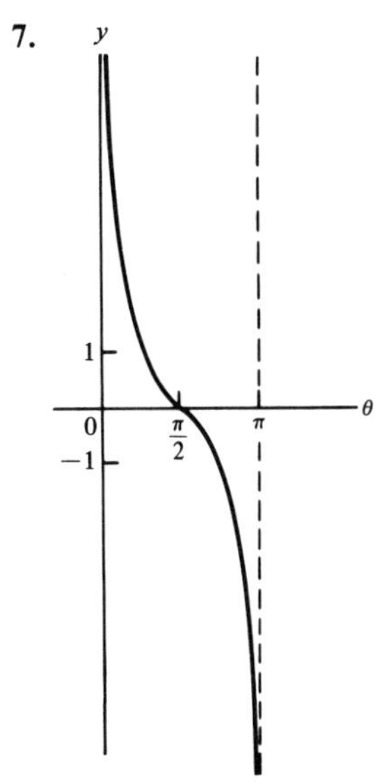

FIGURE 8D $\cot\theta,\ 0 < \theta < \pi$

9.

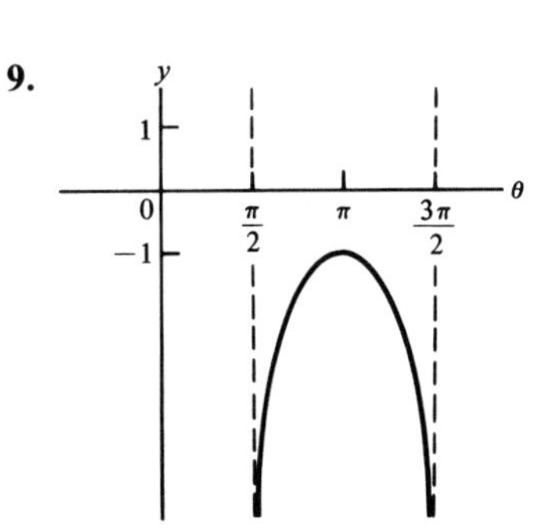

FIGURE 8E $\sec\theta,\ \frac{\pi}{2} < \theta < \frac{3\pi}{2}$

11. (See Figure 8.59(a), page 341.)

13. (See Figure 8.61(b), page 344.)

15. sine, cosine **17.** tangent, secant **19.** sine, cosine **21.** sine, tangent **23.** cotangent, cosecant **25.** cosine, cotangent **27.** tangent, cotangent **29.** sine, cosine, tangent, cotangent **31.** all six functions **33.** tangent, cotangent **35.** sine, cosecant **37.** sine, tangent, cosecant, cotangent **39.** sine and cosecant cosine and secant, tangent and cotangent **41.** $f =$ cosine, $g =$ sine; $f =$ secant, $g =$ cosecant

Section 8.5, p. 353

1. (a) $\frac{2\pi}{3}$ (b) 1 **3.** (a) 2π (b) 2 **5.** (a) $\frac{\pi}{2}$ (b) $\frac{1}{2}$ **7.** (a) 20π (b) 6 **9.** (a) 1 (b) 1

11.

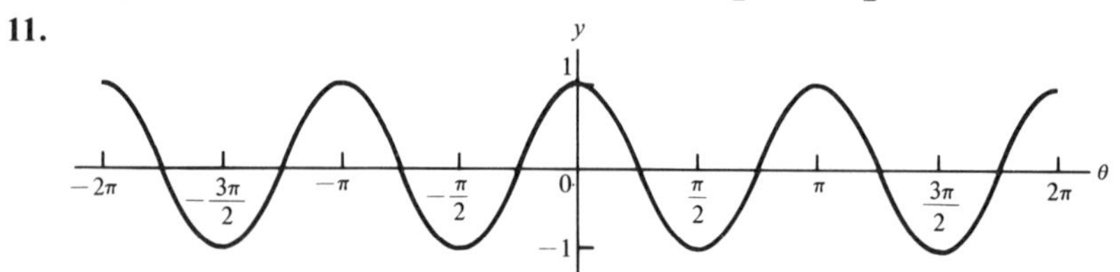

FIGURE 8H $\cos 2\theta$

13.

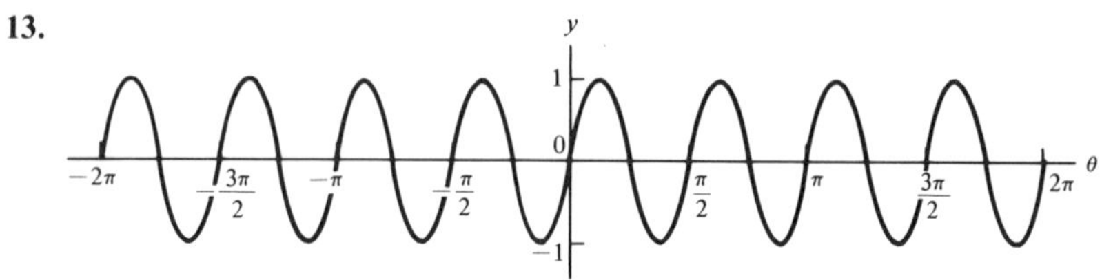

FIGURE 8I $\sin 4\theta$

15.

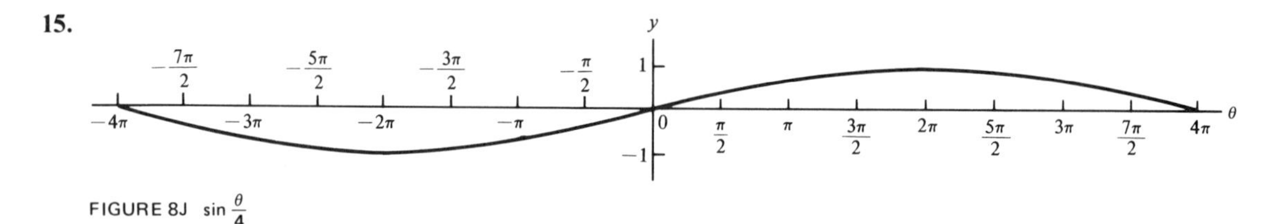

FIGURE 8J $\sin \frac{\theta}{4}$

17.

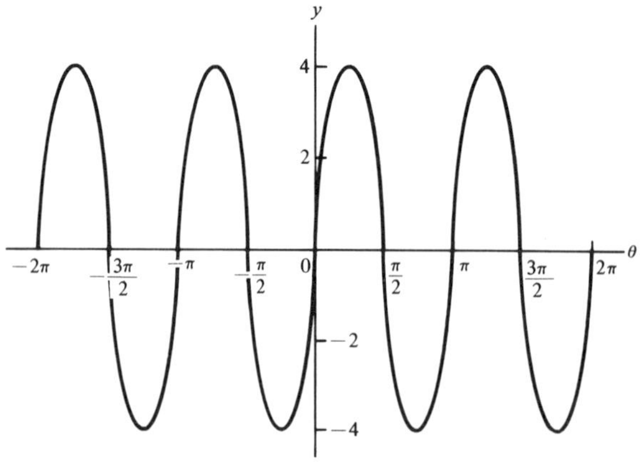

FIGURE 8K $4 \sin 2\theta$

19.

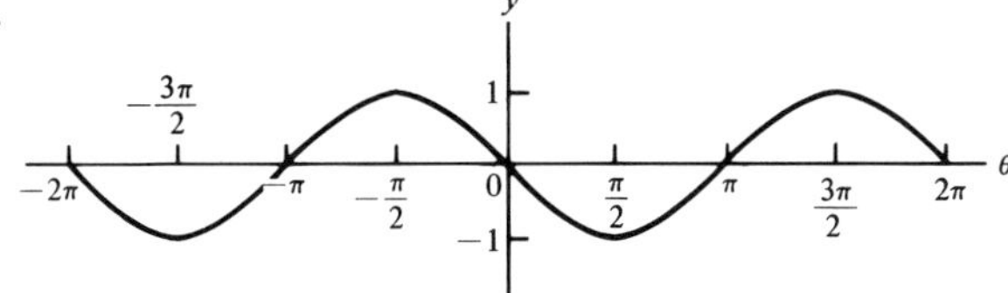

FIGURE 8L $-\sin\theta$

21.

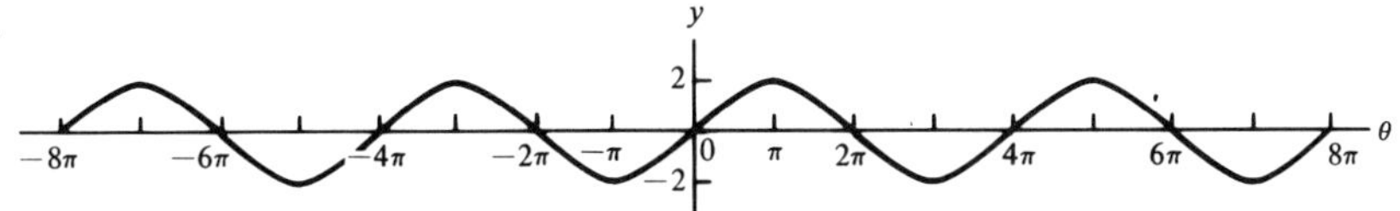

FIGURE 8M $2\sin\frac{\theta}{2}$

23.

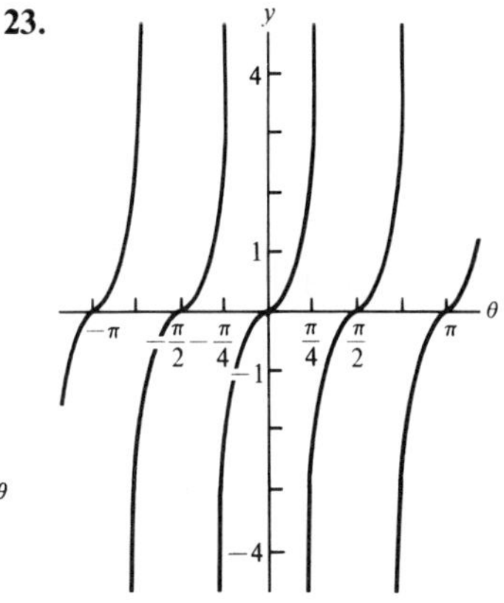

FIGURE 8N $\tan 2\theta$

25. (a) $\frac{\pi}{3}$ (b) right **27.** (a) $-\pi$ (b) left **29.** (a) -1 (b) left **31.** (a) 2π (b) $\frac{\pi}{2}$ (c) 1 (d) See Figure 8O. **33.** (a) 2π (b) $\frac{\pi}{6}$ (c) 1 (d) See Figure 8P. **35.** (a) 2π (b) $\frac{\pi}{3}$ (c) 1 (d) See Figure 8Q. **37.** (a) 2π (b) $\frac{\pi}{4}$ (c) 2 (d) See Figure 8R. **39.** (a) π (b) $\frac{\pi}{8}$ (c) 1 (d) See Figure 8S. **41.** (a) 2π (b) $\frac{\pi}{2}$ (c) 3 (d) See Figure 8T. **43.** (a) π (b) $-\frac{\pi}{2}$ (c) 1 (d) See Figure 8U. **45.** (a) π (b) $\frac{\pi}{2}$ (c) undefined (d) See Figure 8V.

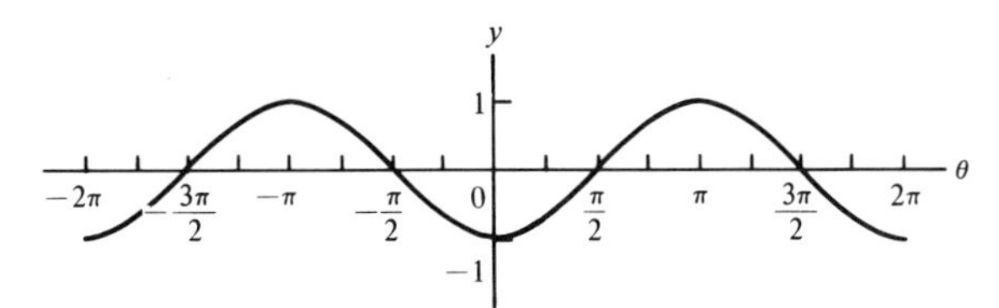

FIGURE 8O $\sin\left(\theta - \frac{\pi}{2}\right)$

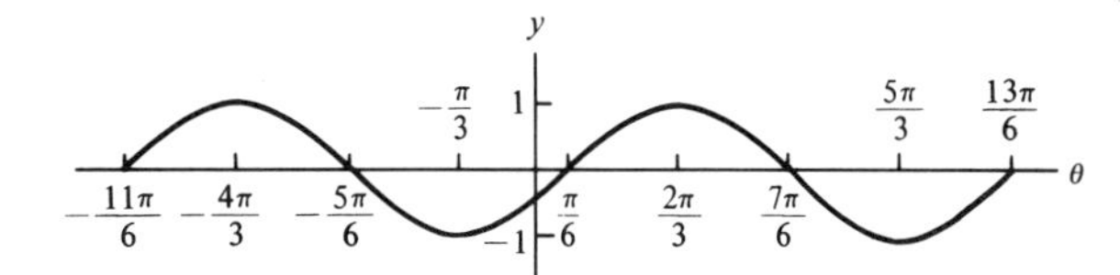

FIGURE 8P $\sin\left(\theta - \frac{\pi}{6}\right)$

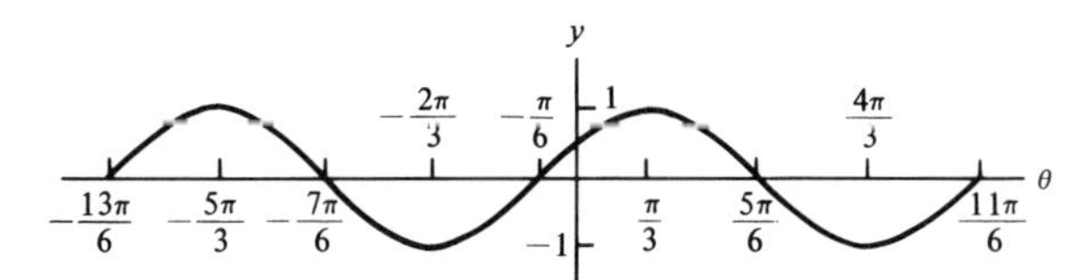

FIGURE 8Q $\cos\left(\theta - \frac{\pi}{3}\right)$

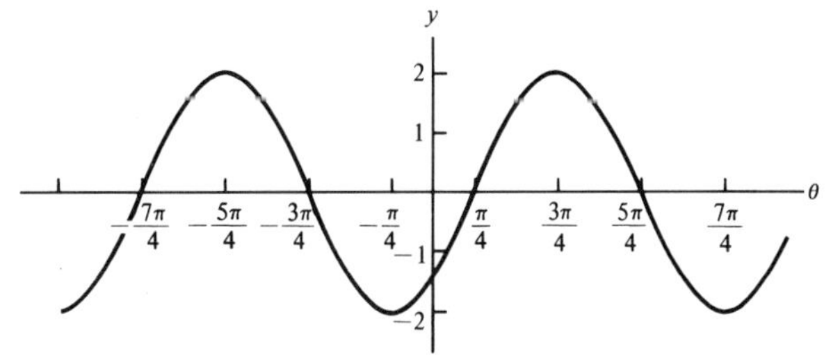

FIGURE 8R $2\sin\left(\theta - \frac{\pi}{4}\right)$

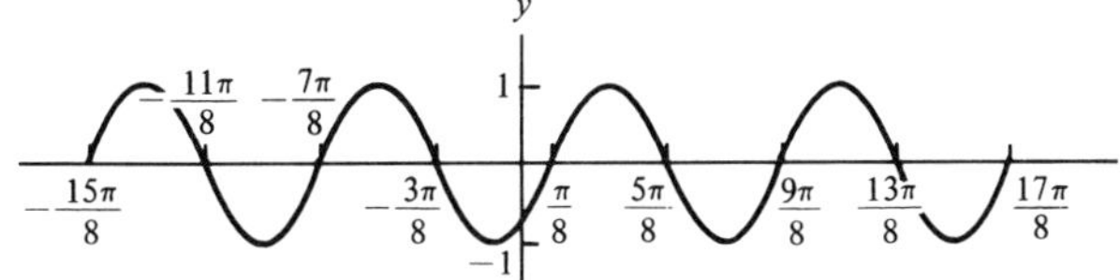

FIGURE 8S $\sin\left(2\theta - \frac{\pi}{4}\right)$

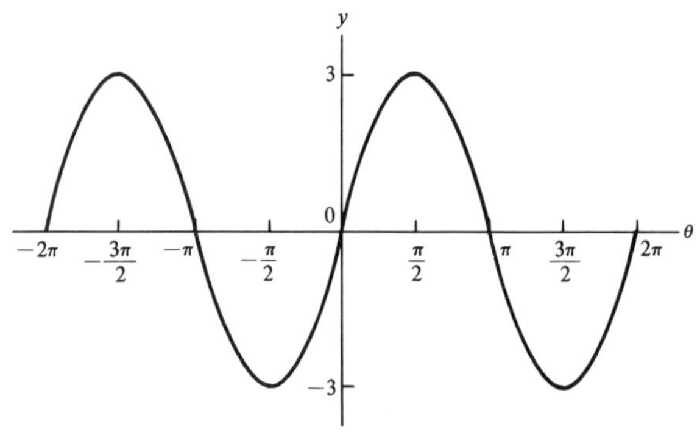

FIGURE 8T $3\cos\left(\theta - \frac{\pi}{2}\right)$

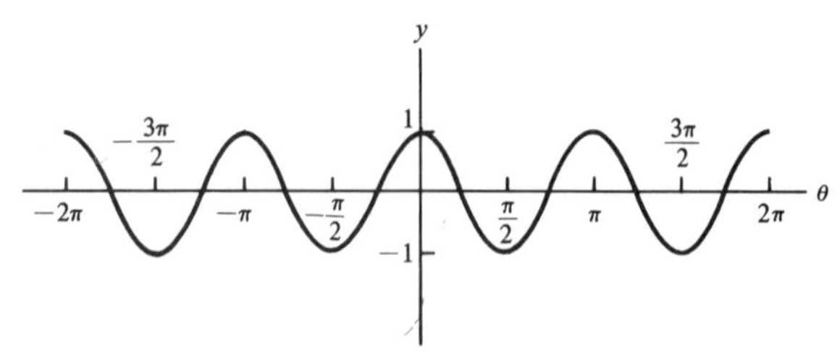

FIGURE 8U $-\cos(2\theta + \pi)$

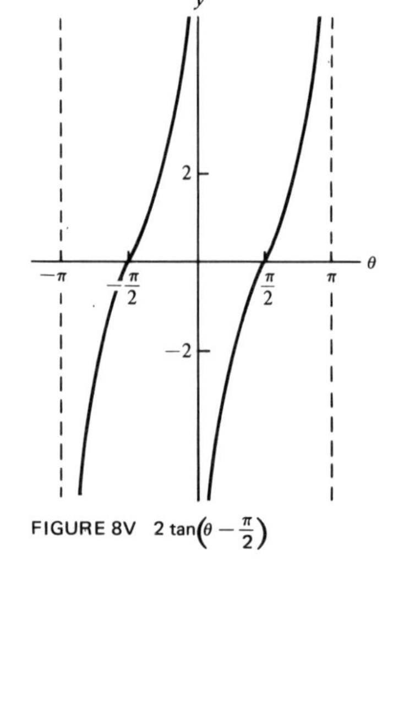

FIGURE 8V $2\tan\left(\theta - \frac{\pi}{2}\right)$

47.

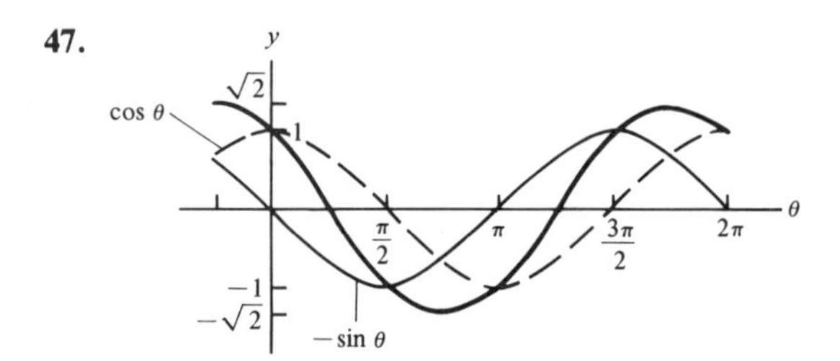

FIGURE 8W $\cos\theta - \sin\theta$ (heavy curve)

49.

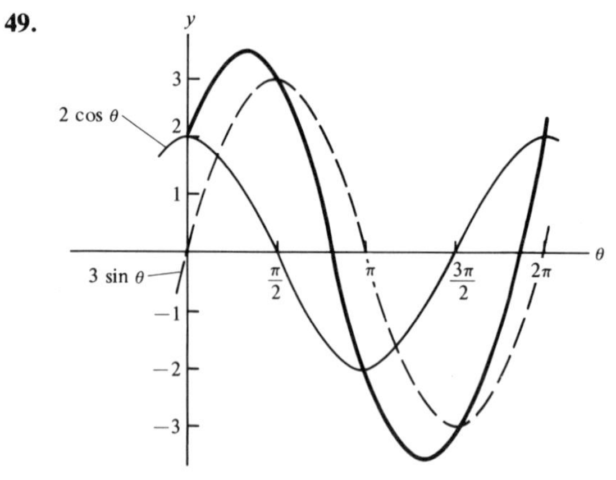

FIGURE 8X $2\cos\theta + 3\sin\theta$ (heavy curve)

53. $\left[-\frac{\pi}{6}, \frac{\pi}{6}\right]$ **55.** $\left[-\frac{\pi}{3}, \frac{\pi}{3}\right]$

Section 8.6, p. 359

1. $\angle C = 106^\circ$, $a = 8.2$, $b = 13.9$ **3.** $\angle B = 41^\circ$, $a = 19.8$, $c = 52.8$ **5.** no triangle **7.** $\angle B = 90^\circ$, $\angle C = 60^\circ$, $c = 33.1$ **9.** no triangle **11.** $\angle A = 6^\circ 40'$, $\angle C = 165^\circ 20'$, $c = 32.7$ **13.** $\angle B = 39^\circ 10'$, $\angle A = 93^\circ 40'$, $a = 23.4$ **15.** no triangle **17.** $\angle A = 109^\circ 50'$, $a = 38.7$, $b = 27.7$
19. 2 triangles: $\angle B = 53^\circ 50'$, $\angle C = 98^\circ 40'$, $c = 25.7$ and $\angle B' = 126^\circ 10'$, $\angle C' = 26^\circ 20'$, $c' = 11.5$
21. $\angle A = 32^\circ$, $\angle C = 13^\circ$, $c = 5.1$ **23.** no triangle **25.** 2.8 miles from A and 2.4 miles from B
27. 17.5 centimeters and 16 centimeters **29.** (a) 10.9 meters (b) 13.9 meters

Section 8.7, p. 362

1. 14 **3.** 72° **5.** 37 **7.** 49° **9.** 56° **11.** 54 **13.** 48° **15.** 123
17. $\angle A = 44°$, $\angle B = 57°$, $\angle C = 78°$ (using the Law of Cosines in each case. Due to errors in approximating, $\angle A + \angle B + \angle C = 189°$) **19.** $c = 8$, $\angle A = 121°$, $\angle B = 39°$ (See Figure 8Z.) **21.** $a = 22$, $\angle B = 35°$, $\angle C = 127°$ **23.** 12 inches, 32 inches **25.** two measure 77°; two measure 103°
27. 23 meters

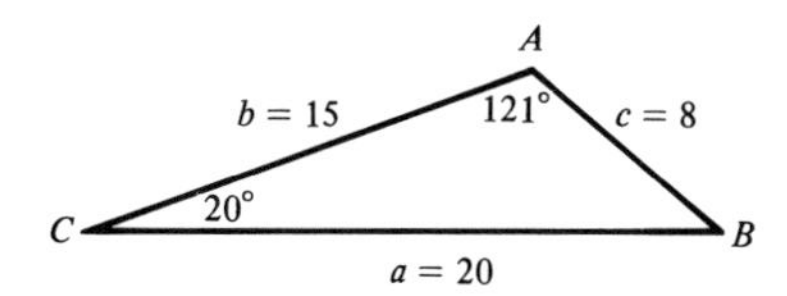

FIGURE 8Z

Review Exercises, p. 363

1. (a) $\frac{2\sqrt{13}}{13}$, (b) $\frac{3\sqrt{13}}{13}$, (c) $\frac{2}{3}$, (d) $\frac{3\sqrt{13}}{13}$, (e) $\frac{2\sqrt{13}}{13}$, (f) $\frac{3}{2}$ **2.** $\cos A = \frac{4}{5}$, $\tan A = \frac{3}{4}$, $\csc A = \frac{5}{3}$, $\sec A = \frac{5}{4}$, $\cot A = \frac{4}{3}$ **3.** $a = \frac{5\sqrt{3}}{3}$, $c = \frac{10\sqrt{3}}{3}$ **4.** (a) $\frac{\sqrt{3}}{2}$, (b) $\frac{\sqrt{2}}{2}$, (c) $\sqrt{3}$, (d) 2 **5.** $42\sqrt{3}$ (or 72.7) meters **6.** (a) .4540 (b) .7112 (c) 3.7760 **7.** (a) 33°10′ (b) 50°30′ (c) 14°40′ **8.** $b = 13.2$, $\angle A = 48°40'$, $\angle B = 41°20'$ **9.** (a) .8459 (b) .9500 (c) 68°30′
10. (a) $\frac{5\pi}{12}$ (b) $\frac{5\pi}{4}$ (c) $-\frac{2\pi}{3}$ **11.** (a) 120° (b) 54° (c) −270°
12. (a) $\sin\frac{\pi}{4} = \frac{\sqrt{2}}{2}$, $\cos\frac{\pi}{4} = \frac{\sqrt{2}}{2}$, $\tan\frac{\pi}{4} = 1$, $\csc\frac{\pi}{4} = \sqrt{2}$, $\sec\frac{\pi}{4} = \sqrt{2}$, $\cot\frac{\pi}{4} = 1$ (b) $\sin\frac{7\pi}{6} = -\frac{1}{2}$, $\cos\frac{7\pi}{6} = -\frac{\sqrt{3}}{2}$, $\tan\frac{7\pi}{6} = \frac{\sqrt{3}}{3}$, $\csc\frac{7\pi}{6} = -2$, $\sec\frac{7\pi}{6} = -\frac{2\sqrt{3}}{3}$, $\cot\frac{7\pi}{6} = \sqrt{3}$ (c) $\sin\left(-\frac{\pi}{2}\right) = -1$, $\cos\left(-\frac{\pi}{2}\right) = 0$, $\tan\left(-\frac{\pi}{2}\right)$ is undefined, $\csc\left(-\frac{\pi}{2}\right) = -1$, $\sec\left(-\frac{\pi}{2}\right)$ is undefined, $\cot\left(-\frac{\pi}{2}\right) = 0$
13. See Figure 8AA. **14.** See Figure 8BB. **15.** (a) 4π (b) 3 (c) See Figure 8CC. **16.** (a) $-\frac{\pi}{2}$ (b) left (c) 2π (d) See Figure 8DD. **17.** $\angle A = 70°20'$, $a = 19.2$, $c = 18.8$
18. two triangles: $\angle A = 54°50'$, $\angle C = 88°40'$, $c = 14.8$ and $\angle A' = 125°10'$, $\angle C' = 18°20'$, $c' = 4.7$
19. $\angle A = 43°40'$, $\angle B = 50°10'$, $\angle C = 86°10'$ **20.** 31 centimeters, 62 centimeters

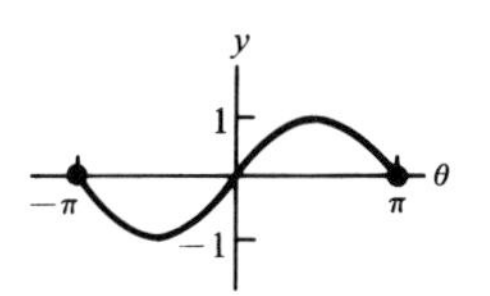

FIGURE 8AA $\sin\theta$, $-\pi \leq \theta \leq \pi$

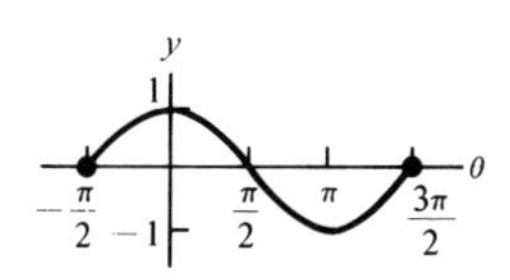

FIGURE 8BB $\cos\theta$, $-\frac{\pi}{2} \leq \theta \leq \frac{3\pi}{2}$

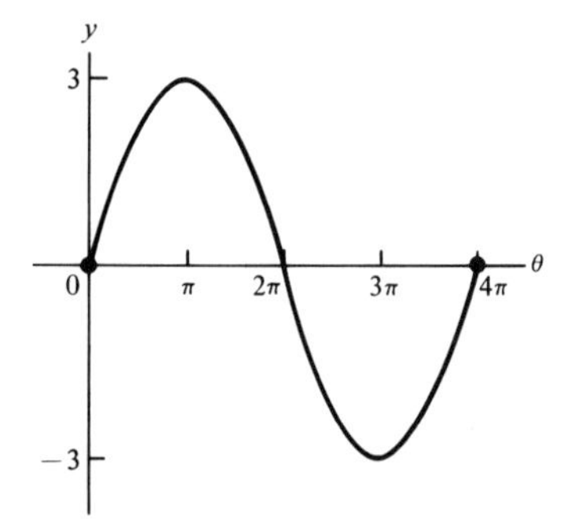

FIGURE 8CC $3\sin\frac{\theta}{2}$, $0 \leq \theta \leq 4\pi$

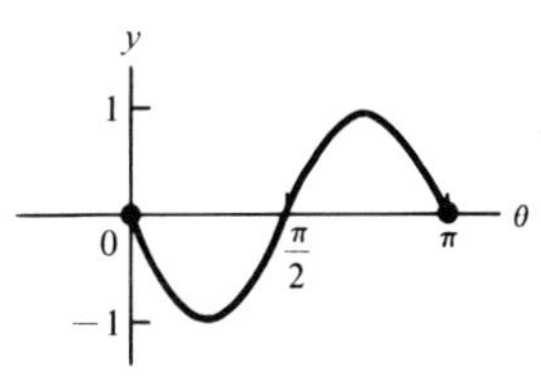

FIGURE 8DD $\cos\left(\theta + \frac{\pi}{2}\right)$, $0 \leq \theta \leq 2\pi$

Chapter 9

Section 9.1, p. 369

1. $\cos^2\theta$ **3.** $\cot^2 A$ **5.** 1 **7.** $-\cos^3 A$ **9.** $2\sec\theta$ **11.** $\cos\theta\cdot\sec\theta = \cos\theta\cdot\dfrac{1}{\cos\theta} = 1$

13. $\cot^2 x - \csc^2 x + 1 = (\cot^2 x + 1) - \csc^2 x = \csc^2 x - \csc^2 x = 0$

15. $\dfrac{\cos^2 A}{\sin A} = \dfrac{\cos A}{\sin A}\cdot\cos A = \cot A\cdot\dfrac{1}{\sec A} = \dfrac{\cot A}{\sec A}$

17. $\tan y + \cot y = \dfrac{\sin y}{\cos y} + \dfrac{\cos y}{\sin y} = \dfrac{\sin^2 y + \cos^2 y}{\cos y\sin y} = \dfrac{1}{\cos y\sin y}$

19. $(\sin A - \cos A)^2 = \sin^2 A - 2\sin A\cos A + \cos^2 A = (\sin^2 A + \cos^2 A) - 2\sin A\cos A = 1 - 2\sin A\cos A$

21. $(\sin^2\theta - \cos^2\theta)(\tan^2\theta + 1) = (\sin^2\theta - \cos^2\theta)\sec^2\theta = \sin^2\theta\cdot\sec^2\theta - \cos^2\theta\sec^2\theta = \dfrac{\sin^2\theta}{\cos^2\theta} - 1 = \tan^2\theta - 1$

23. Left side $= \dfrac{\sin A + \cos A}{2}\cdot\dfrac{\sin A - \cos A}{\sin A - \cos A} = \dfrac{\sin^2 A - \cos^2 A}{2(\sin A - \cos A)}$

Right side $= \dfrac{\sin^2 A - \frac{1}{2}}{\sin A - \cos A}\cdot\dfrac{2}{2} = \dfrac{2\sin^2 A - 1}{2(\sin A - \cos A)} = \dfrac{\sin^2 A + (\sin^2 A - 1)}{2(\sin A - \cos A)} = \dfrac{\sin^2 A - \cos^2 A}{2(\sin A - \cos A)}$

25. Left side $= \dfrac{\sin A}{1 + \sin A}\cdot\dfrac{1 + \csc A}{1 + \csc A} = \dfrac{\sin A + \sin A\csc A}{(1 + \sin A)(1 + \csc A)} = \dfrac{\sin A + 1}{(1 + \sin A)(1 + \csc A)}$

Right side $= \dfrac{1}{1 + \csc A}\cdot\dfrac{1 + \sin A}{1 + \sin A} = \dfrac{\sin A + 1}{(1 + \sin A)(1 + \csc A)}$

27. $\csc C - \sin C = \dfrac{1}{\sin C} - \sin C = \dfrac{1 - \sin^2 C}{\sin C} = \dfrac{\cos^2 C}{\sin C} = \dfrac{\cos C}{\sin C}\cdot\cos C = \cot C\cdot\cos C$

29. $\dfrac{-1}{\cot A + \csc A} = \dfrac{-1}{\csc A + \cot A}\cdot\dfrac{\csc A - \cot A}{\csc A - \cot A} = \dfrac{\cot A - \csc A}{\csc^2 A - \cot^2 A} = \dfrac{\cot A - \csc A}{1} = \cot A - \csc A$

31. Right side $= \dfrac{1(1 + \cos\theta) + 1(1 - \cos\theta)}{(1 - \cos\theta)(1 + \cos\theta)} = \dfrac{2}{1 - \cos^2\theta} = \dfrac{2}{\sin^2\theta} = 2\csc^2\theta$

33. Left side $= \dfrac{1 + \cot\theta}{1 + \tan\theta}\cdot\dfrac{1 - \tan\theta}{1 - \tan\theta} = \dfrac{1 - \tan\theta + \cot\theta - \cot\theta\tan\theta}{1 - \tan^2\theta} = \dfrac{1 - \tan\theta + \cot\theta - 1}{1 - \tan^2\theta} = \dfrac{\cot\theta - \tan\theta}{1 - \tan^2\theta}$

Right side $= \dfrac{\cot\theta - 1}{1 - \tan\theta}\cdot\dfrac{1 + \tan\theta}{1 + \tan\theta} = \dfrac{\cot\theta + \cot\theta\tan\theta - 1 - \tan\theta}{1 - \tan^2\theta} = \dfrac{\cot\theta + 1 - 1 - \tan\theta}{1 - \tan^2\theta} = \dfrac{\cot\theta - \tan\theta}{1 - \tan^2\theta}$

35. $\cos^4 B - \sin^4 B = (\cos^2 B + \sin^2 B)(\cos^2 B - \sin^2 B) = 1(\cos^2 B - \sin^2 B) = \cos^2 B - \sin^2 B$

37. Left side $= \dfrac{\cos^2 x + 1}{\cos x + \dfrac{1}{\cos x}} = \dfrac{\cos^2 x + 1}{\dfrac{\cos^2 x + 1}{\cos x}} = \cos x$

Right side $= \dfrac{\sin^2 x}{\dfrac{1}{\cos x} - \cos x} = \dfrac{\sin^2 x}{\dfrac{1 - \cos^2 x}{\cos x}} = \dfrac{\sin^2 x}{\dfrac{\sin^2 x}{\cos x}} = \cos x$

39. $\cos\theta = \begin{cases} \sqrt{1 - \sin^2\theta} & \text{for } 0 \le \theta \le \dfrac{\pi}{2} \text{ and for } \dfrac{3\pi}{2} \le \theta < 2\pi \\ -\sqrt{1 - \sin^2\theta} & \text{for } \dfrac{\pi}{2} < \theta < \dfrac{3\pi}{2} \end{cases}$

41. (a) and (b) $\sqrt{1 + \cot^2\theta}$ (c) and (d) $-\sqrt{1 + \cot^2\theta}$

43. $\dfrac{1 - \cos\theta}{\theta} = \dfrac{1 - \cos\theta}{\theta}\cdot\dfrac{1 + \cos\theta}{1 + \cos\theta} = \dfrac{1 - \cos^2\theta}{\theta(1 + \cos\theta)} = \dfrac{\sin^2\theta}{\theta(1 + \cos\theta)} = \dfrac{\sin\theta}{\theta}\cdot\dfrac{\sin\theta}{1 + \cos\theta}$

45. $\sqrt{x^2 - a^2} = \sqrt{a^2\sec^2\theta - a^2} = \sqrt{a^2(\sec^2\theta - 1)} = \sqrt{a^2\tan^2\theta} = a\tan\theta$

Section 9.2, p. 375

1. (a) $\frac{\sqrt{2}(1-\sqrt{3})}{4}$ (b) $\frac{\sqrt{2}(1+\sqrt{3})}{4}$ (c) $-\frac{1}{2}(1+\sqrt{3})^2$ **3.** (a) $\frac{(1+\sqrt{3})\sqrt{2}}{4}$ (b) $\frac{(1-\sqrt{3})\sqrt{2}}{4}$ (c) $-\frac{(3-\sqrt{3})^2}{6}$ **5.** (a) $\frac{\sqrt{2}}{2}$ (b) $-\frac{\sqrt{2}}{2}$ (c) -1 **7.** (a) $-\frac{(1+\sqrt{3})\sqrt{2}}{4}$ (b) $\frac{(1-\sqrt{3})\sqrt{2}}{4}$ (c) $\frac{(1+\sqrt{3})^2}{2}$ **9.** (a) $\frac{(1+\sqrt{3})\sqrt{2}}{4}$ (b) $\frac{(1-\sqrt{3})\sqrt{2}}{4}$ (c) $-\frac{(3-\sqrt{3})^2}{6}$ **11.** (a) -1 (b) $\frac{7}{25}$ (c) 0 (d) $\frac{24}{25}$ (e) 0 (f) $\frac{24}{7}$ (g) (negative) x-axis (h) I **13.** (a) $\frac{119}{169}$ (b) -1 (c) $\frac{120}{169}$ (d) 0 (e) $\frac{120}{119}$ (f) 0 (g) I (h) (negative) x-axis **15.** (a) $\frac{4}{5}$ (b) 0 (c) $-\frac{3}{5}$ (d) 1 (e) $-\frac{3}{4}$ (f) undefined (g) IV (h) (positive) y-axis

17. $\cos(\theta+\pi) = \cos\theta\cos\pi - \sin\theta\sin\pi = \cos\theta(-1) - \sin\theta\cdot 0 = -\cos\theta$

19. $\sin(\pi-\theta) = \sin\pi\cos\theta - \cos\pi\sin\theta = 0\cdot\cos\theta - (-1)\sin\theta = \sin\theta$

21. $\tan(\theta+\pi) = \frac{\tan\theta + \tan\pi}{1 - \tan\theta\tan\pi} = \frac{\tan\theta + 0}{1 - \tan\theta\cdot 0} = \frac{\tan\theta}{1} = \tan\theta$ **23.** 0 **25.** $\frac{\sqrt{3}}{2}$ **27.** 1

29. $\tan(A-B) = \frac{\sin(A-B)}{\cos(A-B)} = \frac{\sin A\cos B - \cos A\sin B}{\cos A\cos B + \sin A\sin B} = \frac{\frac{\sin A\cos B}{\cos A\cos B} - \frac{\cos A\sin B}{\cos A\cos B}}{\frac{\cos A\cos B}{\cos A\cos B} + \frac{\sin A\sin B}{\cos A\cos B}} = \frac{\tan A - \tan B}{1 + \tan A\tan B}$, for $A, B, A-B \neq \pm\frac{\pi}{2}, \pm\frac{3\pi}{2}, \pm\frac{5\pi}{2}, \ldots$

31. $\tan\left(\theta+\frac{\pi}{2}\right) = \frac{\sin\left(\theta+\frac{\pi}{2}\right)}{\cos\left(\theta+\frac{\pi}{2}\right)} = \frac{\sin\theta\cos\frac{\pi}{2} + \cos\theta\sin\frac{\pi}{2}}{\cos\theta\cos\frac{\pi}{2} - \sin\theta\sin\frac{\pi}{2}} = \frac{\cos\theta}{-\sin\theta} = -\cot\theta,\quad \theta \neq 0, \pm\pi, \pm 2\pi, \pm 3\pi, \ldots$

33. $\cos\left(\theta-\frac{\pi}{4}\right) = \cos\theta\cos\frac{\pi}{4} + \sin\theta\sin\frac{\pi}{4} = \frac{\sqrt{2}}{2}\cos\theta + \frac{\sqrt{2}}{2}\sin\theta = \frac{\sqrt{2}}{2}(\cos\theta + \sin\theta)$

35. $\frac{f(x+h)-f(x)}{h} = \frac{\sin(x+h) - \sin x}{h} = \frac{\sin x\cos h + \cos x\sin h - \sin x}{h} = \frac{\sin x\cos h - \sin x}{h} + \frac{\cos x\sin h}{h}$
$= \sin x\left(\frac{\cos h - 1}{h}\right) + \cos x\left(\frac{\sin h}{h}\right)$

Section 9.3, p. 382

1. $\cos 80^\circ$ **3.** $\frac{1}{2}\tan 34^\circ$ **5.** $\frac{\sqrt{2}}{4}$ **7.** $-\frac{\sqrt{2}}{2}$ **9.** $\cos 21^\circ$ **11.** $\tan 2A$ **13.** $\frac{1}{2}\sqrt{2+\sqrt{3}}$ **15.** $\frac{1+\cos 10\theta}{2}$ **17.** (a) $-\frac{7}{25}$ (b) $\frac{24}{25}$ (c) $-\frac{24}{7}$ (d) $\frac{2\sqrt{5}}{5}$ (e) $\frac{\sqrt{5}}{5}$ (f) $\frac{1}{2}$ **19.** (a) $-\frac{15}{17}$ (b) $\frac{8}{17}$ (c) $-\frac{8}{15}$ (d) $\sqrt{\frac{17+\sqrt{17}}{34}}$ (e) $\sqrt{\frac{17-\sqrt{17}}{34}}$ (f) $\frac{\sqrt{17}-1}{4}$ **21.** (a) $-\frac{7}{25}$ (b) $-\frac{24}{25}$ (c) $\frac{24}{7}$ (d) $-\frac{2\sqrt{5}}{5}$ (e) $\frac{\sqrt{5}}{5}$ (f) $-\frac{1}{2}$ **23.** $\frac{1}{2}\left(\frac{1}{2} + \cos 40^\circ\right)$ **25.** $4\left(\sin\frac{4\pi}{5} + \sin\frac{2\pi}{5}\right)$ **27.** $2\cos 44^\circ\cos 4^\circ$ **29.** $-2\sin 19^\circ\sin 7^\circ$ **31.** $-\frac{\sqrt{3}}{2}$ **33.** $\frac{\sqrt{6}}{2}$ **35.** $\frac{1}{2}\left(\frac{\sqrt{2}}{2} + \sin 13^\circ\right)$

37. $\dfrac{1+\sin 76^\circ}{2}$ **39.** $-\dfrac{1}{4}$ **41.** $\sec 2A = \dfrac{1}{\cos 2A} = \dfrac{1}{\cos^2 A - \sin^2 A} = \dfrac{\dfrac{1}{\cos^2 A}}{1 - \dfrac{\sin^2 A}{\cos^2 A}} = \dfrac{\sec^2 A}{1 - \tan^2 A}$

43. $2\sin^2 2A = 1 - \cos 2(2A) = 1 - \cos 4A$ **45.** $\dfrac{\tan A}{2} = \dfrac{1}{2}\tan 2\left(\dfrac{A}{2}\right) = \dfrac{1}{2}\cdot\dfrac{2\tan\dfrac{A}{2}}{1-\tan^2\dfrac{A}{2}} = \dfrac{\tan\dfrac{A}{2}}{1-\tan^2\dfrac{A}{2}}$

47. $\sin 3A = \sin(A+2A) = \sin A\cos 2A + \cos A \sin 2A = \sin A(1 - 2\sin^2 A) + \cos A \cdot 2\sin A\cos A$
$= \sin A - 2\sin^3 A + 2\sin A(1-\sin^2 A) = \sin A - 2\sin^3 A + 2\sin A - 2\sin^3 A = 3\sin A - 4\sin^3 A$

49. Right side $= \dfrac{2\sin A\sin B}{2\cos A\cos B} = \tan A\tan B$

51. $8\sin^4 A = 8\left(\dfrac{1-\cos 2A}{2}\right)^2 = 2(1 - 2\cos 2A + \cos^2 2A)$
$= 2\left(1 - 2\cos 2A + \dfrac{1+\cos 4A}{2}\right)$
$= 2\left(\dfrac{2 - 4\cos 2A + 1 + \cos 4A}{2}\right)$
$= 3 - 4\cos 2A + \cos 4A$
$= \cos 4A - 4\cos 2A + 3$

53. (a) $1 \geq \cos A$ for all A. Therefore, $1 - \cos A \geq \cos A - \cos A = 0$

(b) Let n be any integer. Suppose $2n\pi < A < (2n+1)\pi$. Then $n\pi < A/2 < n\pi + \pi/2$. Thus, if A is in Quadrant I or II or on the positive y-axis, then $\sin A > 0$. It follows that $A/2$ is in Quadrant I or III, and therefore $\tan A/2 > 0$.
Suppose $(2n+1)\pi < A < (2n+2)\pi$. Then $n\pi + \pi/2 < A/2 < (n+1)\pi$. Thus, if A is in Quadrant III or IV or on the negative y-axis, then $\sin A < 0$. It follows that $A/2$ is in Quadrant II or IV, and therefore, $\tan A/2 < 0$.

(c) $\tan\dfrac{A}{2} = \begin{cases} \sqrt{\tan^2\dfrac{A}{2}}, & \text{if } \tan\dfrac{A}{2} > 0 \\ -\sqrt{\tan^2\dfrac{A}{2}}, & \text{if } \tan\dfrac{A}{2} < 0 \end{cases}$

Because $1 - \cos A \geq 0$ for all A, $\dfrac{1-\cos A}{\sin A} = \begin{cases} \sqrt{\tan^2\dfrac{A}{2}}, & \text{if } \sin A > 0 \\ -\sqrt{\tan^2\dfrac{A}{2}}, & \text{if } \sin A < 0 \end{cases}$

By part (b), $\tan\dfrac{A}{2} = \dfrac{1-\cos A}{\sin A}$, $A \neq 0, \pm\pi, \pm 2\pi, \ldots$

Section 9.4, p. 388

1. $0, \pi$ **3.** $\dfrac{\pi}{3}, \dfrac{4\pi}{3}$ **5.** $\dfrac{\pi}{6}, \dfrac{\pi}{3}, \dfrac{7\pi}{6}, \dfrac{4\pi}{3}$ **7.** $\dfrac{\pi}{4}, \dfrac{3\pi}{4}$ **9.** $\dfrac{\pi}{9}, \dfrac{5\pi}{9}, \dfrac{7\pi}{9}, \dfrac{11\pi}{9}, \dfrac{13\pi}{9}, \dfrac{17\pi}{9}$ **11.** $0, \pi$ **13.** $\dfrac{\pi}{6}, \dfrac{\pi}{2}, \dfrac{5\pi}{6}$

15. $\dfrac{\pi}{3}, \dfrac{2\pi}{3}, \dfrac{4\pi}{3}, \dfrac{5\pi}{3}$ **17.** $\dfrac{\pi}{3}, \dfrac{2\pi}{3}, \dfrac{4\pi}{3}, \dfrac{5\pi}{3}$ **19.** $\dfrac{\pi}{6}, \dfrac{\pi}{4}, \dfrac{3\pi}{4}, \dfrac{11\pi}{6}$ **21.** $0, \dfrac{\pi}{3}, \pi, \dfrac{5\pi}{3}$ **23.** $\dfrac{\pi}{6}, \dfrac{5\pi}{6}$

25. $\dfrac{3\pi}{2} + 2n\pi$, n an integer **27.** $\dfrac{\pi}{4} + n\pi$ or $\dfrac{3\pi}{4} + n\pi$, n an integer **29.** $\dfrac{\pi}{3} + 2n\pi$ or $\dfrac{5\pi}{3} + 2n\pi$, n an integer

31. $2n\pi$, n an integer **33.** $2n\pi$ or $\dfrac{\pi}{3} + 2n\pi$ or $\dfrac{5\pi}{3} + 2n\pi$, n an integer **35.** $n\pi$, n an integer

37. $\dfrac{2\pi}{3} + 2n\pi$ or $\dfrac{4\pi}{3} + 2n\pi$, n an integer **39.** $\dfrac{\pi}{2} + n\pi$ or $\dfrac{3\pi}{4} + n\pi$, n an integer **41.** $\dfrac{49\pi}{180}, \dfrac{131\pi}{180}$

43. $\dfrac{5\pi}{18}, \dfrac{23\pi}{18}$ **45.** $\dfrac{19\pi}{45}, \dfrac{16\pi}{15}, \dfrac{71\pi}{45}, \dfrac{29\pi}{15}$ **47.** $\dfrac{14\pi}{45}, \dfrac{59\pi}{45}$ **49.** no root **51.** $\left(\dfrac{\pi}{6}, \dfrac{5\pi}{6}\right)$ **53.** $\left[\dfrac{\pi}{4}, \dfrac{\pi}{2}\right) \cup \left[\dfrac{5\pi}{4}, \dfrac{3\pi}{2}\right)$

Section 9.5, p. 395

1. $\frac{\pi}{3}$ **3.** $\frac{\pi}{4}$ **5.** $\frac{\pi}{6}$ **7.** 0 **9.** 0 **11.** $-\frac{\pi}{3}$ **13.** $\frac{4\pi}{45}$ **15.** $\frac{59\pi}{180}$ **17.** $-\frac{3\pi}{10}$ **19.** $\frac{1}{3}$ **21.** $\frac{4}{5}$
23. $\frac{\sqrt{7}}{4}$ **25.** -1 **27.** $\frac{\pi}{6}$ **29.** $\frac{\pi}{4}$ **31.** $-\frac{7}{25}$ **33.** $\frac{1}{2}$ **35.** $\frac{\pi}{4}$ **37.** 0 **39.** $\frac{\pi}{2}$ **41.** $\frac{1}{x}, x \neq 0$

43.

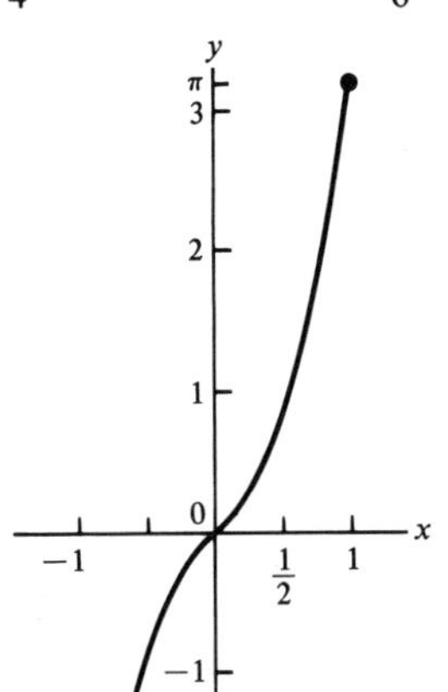

FIGURE 9A $y = 2 \text{ arc sin } x, \quad -1 \leqslant x \leqslant 1$

45.

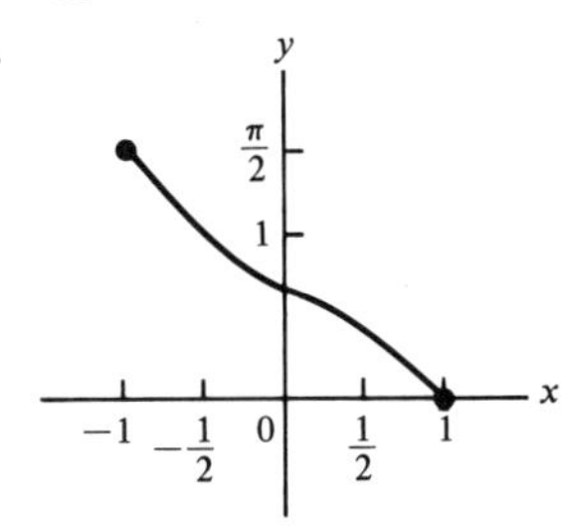

FIGURE 9B $y = \frac{1}{2} \text{ arc cos } x, \quad -1 \leqslant x \leqslant 1$

47.

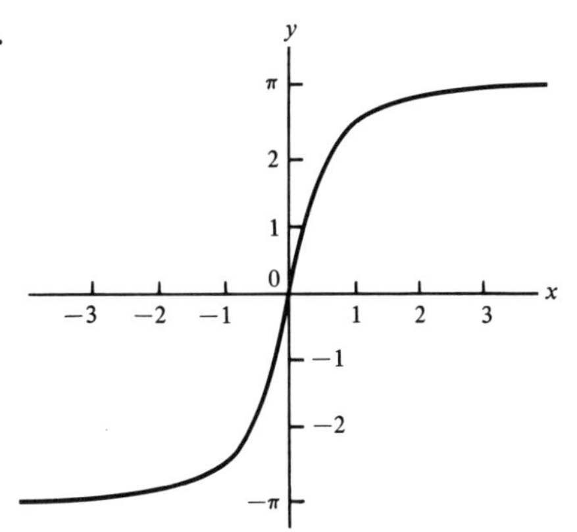

FIGURE 9C $y = \text{arc tan } 4x$

49.

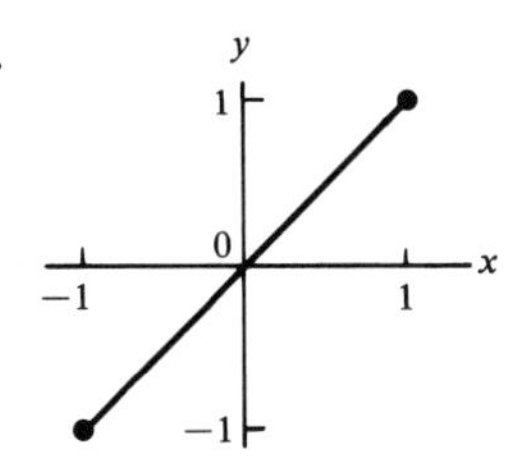

FIGURE 9D $y = \sin (\text{arc sin } x) = x, \quad -1 \leqslant x \leqslant 1$

51. $\frac{\sqrt{2}}{4}$ **53.** (a) $y = \text{arc cos } x$ (b) $y = \cos x$ (c) $y = \text{arc sin } x$ (d) $y = \text{arc tan } x$

Section 9.6, p. 402

1. I **3.** IV **5.** IV **7.** (positive) x-axis **9.** III **11.** I **13.** (c) **15.** (b), (c) **17.** (b), (d)
19. $\left(2, \frac{7\pi}{6}\right)$ **21.** $\left(3, \frac{2\pi}{3}\right)$ **23.** $\left(2, \frac{\pi}{2}\right)$ **25.** $\left(1, \frac{7\pi}{4}\right)$ **27.** $(2\sqrt{3}, 2)$ **29.** (2, 2) **31.** (6, 0) **33.** (0, 6)
35. $(-1, 0)$ **37.** $(-10, 0)$ **39.** $\left(\sqrt{2}, \frac{\pi}{4}\right)$ **41.** $\left(2, \frac{\pi}{6}\right)$ **43.** $\left(5\sqrt{2}, \frac{7\pi}{4}\right)$ **45.** $(\sqrt{5}, \text{arc tan } 2)$ **47.** (8, 0)
49. $(8, \pi)$ or $(-8, 0)$ **51.** $\theta = \frac{\pi}{4}$ **53.** $\theta = 0$ **55.** $r = 1$ **57.** points other than the origin with $y \geq 0$

59.

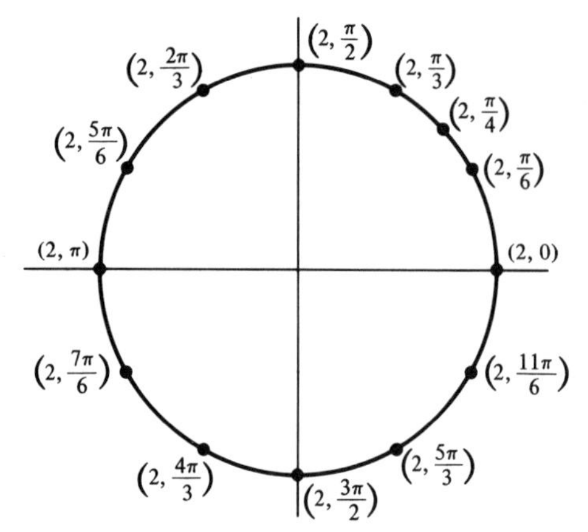

FIGURE 9E $r = 2$

61. **63.**

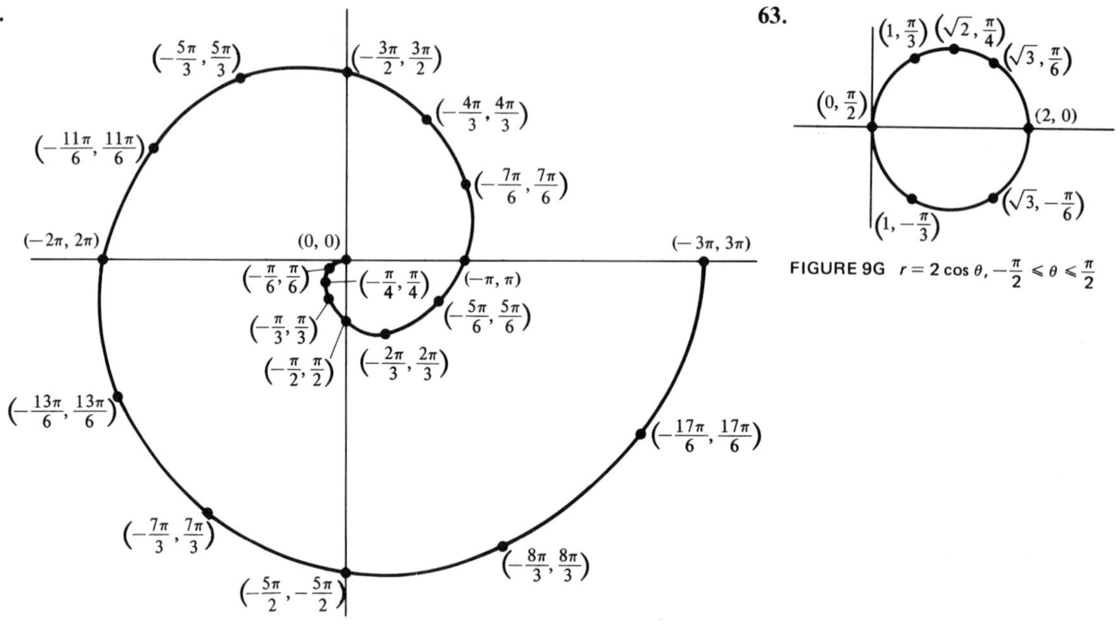

FIGURE 9F $r = -\theta,\ 0 \leqslant \theta \leqslant 3\pi$

FIGURE 9G $r = 2\cos\theta, -\frac{\pi}{2} \leqslant \theta \leqslant \frac{\pi}{2}$

Review Exercises, p. 404

1. (a) 1 (b) 1

2. (a) $\dfrac{\cos^2\theta}{1+\sin\theta} = \dfrac{1-\sin^2\theta}{1+\sin\theta} = \dfrac{(1+\sin\theta)(1-\sin\theta)}{1+\sin\theta} = 1-\sin\theta$

(b) $\cos^2 x - \sin^2 x = (\cos^2 x - \sin^2 x)\dfrac{\cos^2 x}{\cos^2 x} = \dfrac{\cos^2 x - \sin^2 x}{\cos^2 x}\cdot\dfrac{1}{\sec^2 x} = \dfrac{1-\tan^2 x}{1+\tan^2 x}$

(c) $\dfrac{1+\tan A}{\sec A + \csc A} = \dfrac{1+\dfrac{\sin A}{\cos A}}{\dfrac{1}{\cos A}+\dfrac{1}{\sin A}} = \dfrac{\dfrac{\cos A + \sin A}{\cos A}}{\dfrac{\sin A + \cos A}{\cos A \sin A}} = \dfrac{\cos A + \sin A}{\cos A}\cdot\dfrac{\cos A \sin A}{\cos A + \sin A} = \sin A$

(d) $\cot^2 B \sec^2 B - 1 = \cot^2 B(1+\tan^2 B) - 1 = \cot^2 B + \cot^2 B \tan^2 B - 1 = \cot^2 B + 1 - 1 = \cot^2 B$

3. (a) $\frac{\sqrt{2}(1+\sqrt{3})}{4}$ (b) $2+\sqrt{2}$ **4.** (a) -1 (b) $-\frac{24}{25}$ (c) 0 (d) (negative) x-axis **5.** 1 **6.** $\frac{\sqrt{3}}{2}$

7. $-2 \sin 23°30' \sin 10°30'$ **8.** $\sin 4\theta = \sin 2(2\theta) = 2 \sin 2\theta \cos 2\theta = 2 \sin 2\theta(\cos^2 \theta - \sin^2 \theta) =$
$2 \sin 2\theta \cos^2 \theta - 2 \sin 2\theta \sin^2 \theta = 2 \sin 2\theta \cos^2 \theta - 2 \sin 2\theta(1 - \cos^2 \theta) =$
$2 \sin 2\theta \cos^2 \theta - 2 \sin 2\theta + 2 \sin 2\theta \cos^2 \theta = 4 \sin 2\theta \cos^2 \theta - 2 \sin 2\theta$ **9.** (a) $\frac{7\pi}{6}, \frac{11\pi}{6}$ (b) $\frac{\pi}{6}, \frac{5\pi}{6}, \frac{7\pi}{6}, \frac{11\pi}{6}$

(c) $\frac{\pi}{6}, \frac{5\pi}{6}, \frac{3\pi}{2}$ **10.** (a) $\frac{\pi}{12} + \frac{n\pi}{2}$, n an integer (b) $(2n+1)\pi$ or $\frac{\pi}{2} + n\pi$, n an integer

(c) $\frac{3\pi}{2} + 2n\pi$, n an integer **11.** (a) $\frac{\pi}{4}$ (b) π (c) $-\frac{\pi}{4}$ **12.** (a) $-\frac{\pi}{10}$ (b) $\frac{5\pi}{12}$

13. (a) $\frac{2}{5}$ (b) $\frac{\sqrt{3}}{2}$ (c) $-\frac{\pi}{6}$

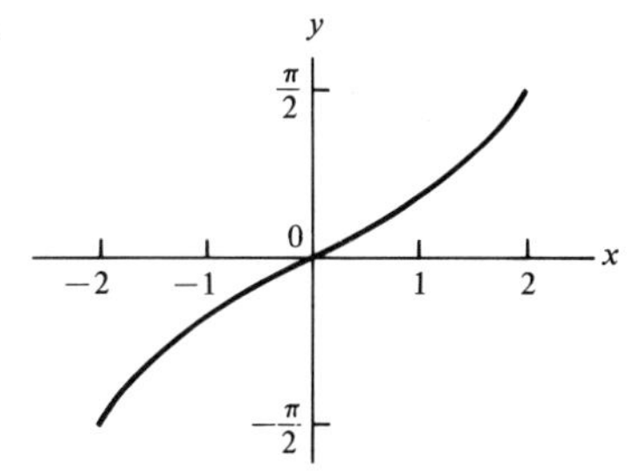

FIGURE 9H $y = \arcsin \frac{x}{2}, -2 \leqslant x \leqslant 2$

15. (a) $(3, 2\pi)$ or $(-3, 0)$ (b) $\left(1, \frac{5\pi}{4}\right)$ (c) $\left(1, \frac{\pi}{2}\right)$ (d) $(0, \theta)$ for any θ **16.** (a) $\left(\frac{3}{2}, \frac{3\sqrt{3}}{2}\right)$

(b) $\left(-\frac{5\sqrt{2}}{2}, \frac{5\sqrt{2}}{2}\right)$ (c) $(1, 0)$ (d) $(0, -3)$ **17.** (a) $\left(4\sqrt{2}, \frac{\pi}{4}\right)$ (b) $(2, 0)$ (c) $\left(2, \frac{2\pi}{3}\right)$

(d) $\left(5, \frac{3\pi}{2}\right)$

Chapter 10

Section 10.1, p. 410

1. $8i$ **3.** $7 - 12i$ **5.** $\frac{1}{2} - \frac{5}{4}i$ **7.** $-1 + 6i$ **9.** $\frac{1}{4} - \frac{1}{4}i$ **11.** 20 **13.** $6 + 21i$ **15.** $5 + i$

17. $63 - 8i$ **19.** -25 **21.** $-11 - 2i$ **23.** $18 - 26i$ **25.** $\frac{3}{2} - \frac{1}{2}i$ **27.** $2 - \frac{3}{2}i$ **29.** $\frac{4}{5} + \frac{3}{5}i$

31. $\frac{6}{169} - \frac{5}{338}i$ **33.** $-\frac{3}{5} + \frac{1}{5}i$ **35.** -1 **37.** i **39.** $-1 - i$ **41.** $-i$ **43.** i **45.** no

47. (a) $(z + \sqrt{6}i)(z - \sqrt{6}i)$ (b) $\pm\sqrt{6}i$

49. (a) $\overline{z_1 + z_2} = \overline{(1+i) + (4-2i)} = \overline{5 - i} = 5 + i$
$\overline{z_1} + \overline{z_2} = (1 - i) + (4 + 2i) = 5 + i$

(b) $\overline{z_1 - z_2} = \overline{(1+i) - (4-2i)} = \overline{-3 + 3i} = -3 - 3i$
$\overline{z_1} - \overline{z_2} = (1 - i) - (4 + 2i) = -3 - 3i$

(c) $\overline{z_1 \cdot z_2} = \overline{(1+i)(4-2i)} = \overline{6 + 2i} = 6 - 2i$
$\overline{z_1} \cdot \overline{z_2} = (1 - i)(4 + 2i) = 6 - 2i$

(d) $\overline{\left(\frac{z_1}{z_2}\right)} = \overline{\left(\frac{1+i}{4-2i}\right)} = \overline{\left(\frac{1}{10} + \frac{3}{10}i\right)} = \frac{1}{10} - \frac{3}{10}i$
$\frac{\overline{z_1}}{\overline{z_2}} = \frac{1-i}{4+2i} = \frac{(1-i)(4-2i)}{20} = \frac{1}{10} - \frac{3}{10}i$

(e) $\overline{z_2^{\,2}} = \overline{(4-2i)^2} = 12 + 16i$
$\overline{z_2}^{\,2} = (4 + 2i)^2 = 12 + 16i$

51. (a) x (b) y

Section 10.2, p. 413

1. $P = 1 + 3i, Q = 4 - i, R = 5, S = i, T = -3i, U = -3 - 2i, V = \frac{3}{2}, W = -2 + 4i$

3.–12.

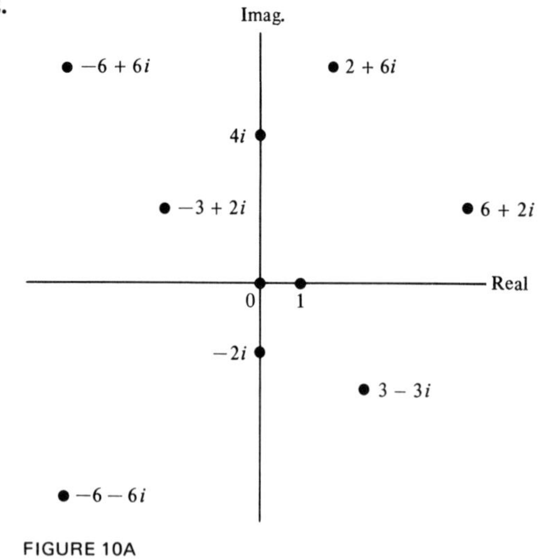

FIGURE 10A

13.

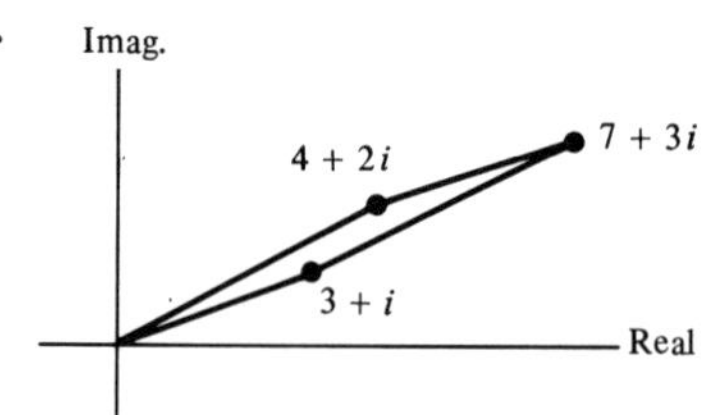

FIGURE 10B $(4 + 2i) + (3 + i) = 7 + 3i$

15.

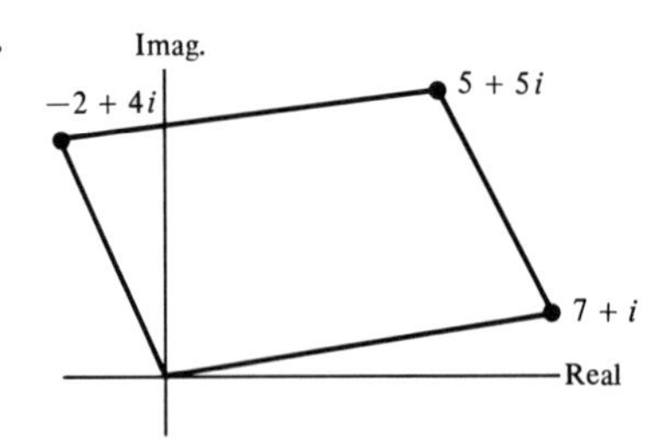

FIGURE 10C $(7 + i) + (-2 + 4i) = 5 + 5i$

17.

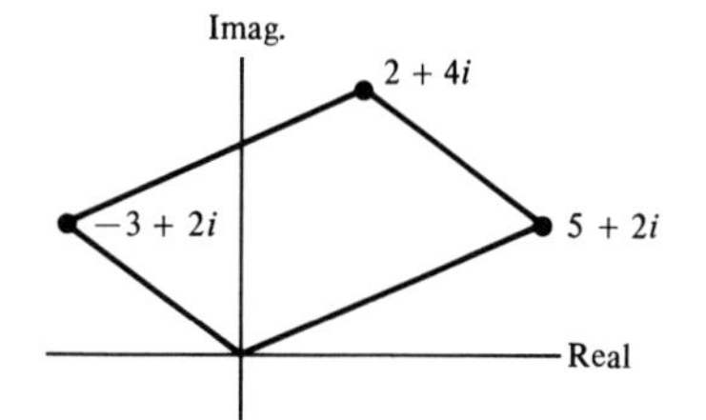

FIGURE 10D $(-3 + 2i) + (5 + 2i) = 2 + 4i$

19.

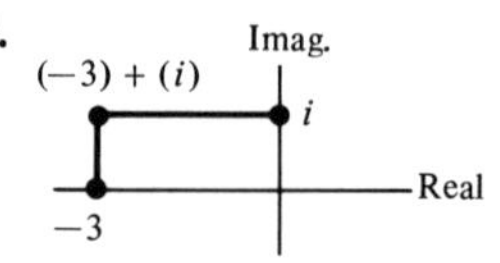

FIGURE 10E $(-3) + (i) = -3 + i$

21.

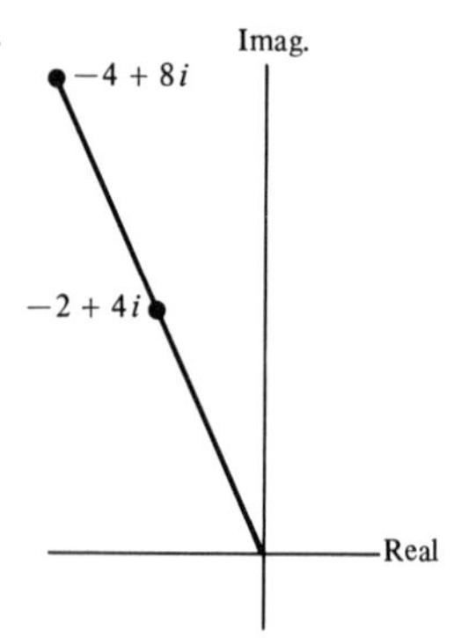

FIGURE 10F $(-2 + 4i) + (-2 + 4i) = -4 + 8i$

23.

Imag. 2 + 4i 0 Real −2 + 4i

FIGURE 10G $(2 + 4i) - (2 + 4i) = 0$

Section 10.3, p. 418

1. (a) 6 (b) $\frac{\pi}{4}$ (c) See Figure 10H. **3.** (a) 2 (b) $\frac{\pi}{5}$ (c) See Figure 10H. **5.** (a) 1 (b) $-\frac{3\pi}{5}$

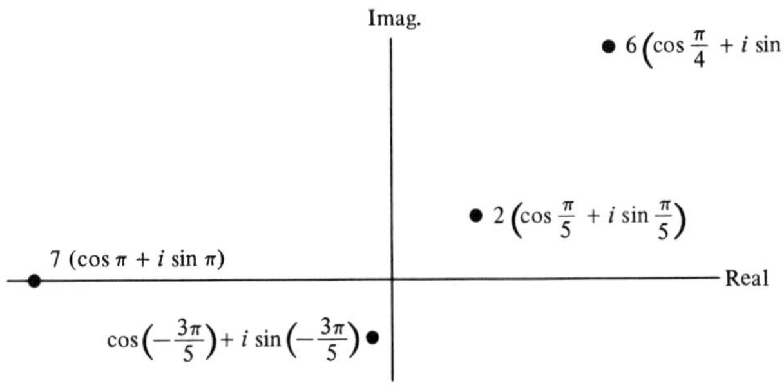

Imag.

$6\left(\cos\frac{\pi}{4} + i\sin\frac{\pi}{4}\right)$

$2\left(\cos\frac{\pi}{5} + i\sin\frac{\pi}{5}\right)$

$7(\cos\pi + i\sin\pi)$

Real

$\cos\left(-\frac{3\pi}{5}\right) + i\sin\left(-\frac{3\pi}{5}\right)$

FIGURE 10H

(c) See Figure 10H. **7.** (a) 7 (b) π (c) See Figure 10H. **9.** (a) $\sqrt{2}\left(\cos\frac{\pi}{4} + i\sin\frac{\pi}{4}\right)$ (b) $\sqrt{2}$ (c) $\frac{\pi}{4}$ **11.** (a) $2\sqrt{2}\left(\cos\frac{3\pi}{4} + i\sin\frac{3\pi}{4}\right)$ (b) $2\sqrt{2}$ (c) $\frac{3\pi}{4}$ **13.** (a) $10\left[\cos\left(-\frac{\pi}{3}\right) + i\sin\left(-\frac{\pi}{3}\right)\right]$ (b) 10 (c) $-\frac{\pi}{3}$ **15.** (a) $8(\cos\pi + i\sin\pi)$ (b) 8 (c) π **17.** (a) $\frac{1}{2}\left(\cos\frac{\pi}{2} + i\sin\frac{\pi}{2}\right)$ (b) $\frac{1}{2}$ (c) $\frac{\pi}{2}$ **19.** (a) $\frac{3\sqrt{2}}{2} + \frac{3\sqrt{2}}{2}i$ (b) $\frac{3\sqrt{2}}{2}$ (c) $\frac{3\sqrt{2}}{2}$ **21.** (a) $\frac{9\sqrt{3}}{2} + \frac{9}{2}i$ (b) $\frac{9\sqrt{3}}{2}$ (c) $\frac{9}{2}$
23. (a) $-6\sqrt{3} + 6i$ (b) $-6\sqrt{3}$ (c) 6 **25.** (a) -8 (b) -8 (c) 0 **27.** (a) 1 (b) 1 (c) 0

29. (a) $4\left(\cos\frac{\pi}{6} - i\sin\frac{\pi}{6}\right)$ (b) $4\left(-\cos\frac{\pi}{6} - i\sin\frac{\pi}{6}\right)$ (c) $\frac{1}{4}\left(\cos\frac{\pi}{6} - i\sin\frac{\pi}{6}\right)$
31. (a) $12\left(\cos\frac{3\pi}{4} - i\sin\frac{3\pi}{4}\right)$ (b) $12\left(-\cos\frac{3\pi}{4} - i\sin\frac{3\pi}{4}\right)$ (c) $\frac{1}{12}\left(\cos\frac{3\pi}{4} - i\sin\frac{3\pi}{4}\right)$
33. (a) $10\left(\cos\frac{\pi}{2} - i\sin\frac{\pi}{2}\right)$ (b) $10\left(-\cos\frac{\pi}{2} - i\sin\frac{\pi}{2}\right)$ (c) $\frac{1}{10}\left(\cos\frac{\pi}{2} - i\sin\frac{\pi}{2}\right)$
35. (a) $\pi\left(\cos\frac{7\pi}{8} - i\sin\frac{7\pi}{8}\right)$ (b) $\pi\left(-\cos\frac{7\pi}{8} - i\sin\frac{7\pi}{8}\right)$ (c) $\frac{1}{\pi}\left(\cos\frac{7\pi}{8} - i\sin\frac{7\pi}{8}\right)$
37. $10\left(\cos\frac{5\pi}{9} + i\sin\frac{5\pi}{9}\right)$ **39.** $14\left(\cos\frac{8\pi}{15} + i\sin\frac{8\pi}{15}\right)$ **41.** $3\left[\cos\left(-\frac{\pi}{5}\right) + i\sin\left(-\frac{\pi}{5}\right)\right]$
43. $20(\cos\pi + i\sin\pi) = -20$ **45.** $6i$ **47.** $\frac{5\sqrt{2}}{2} - \frac{5\sqrt{2}}{2}i$

49. (a) $(1 - \sqrt{3}) + (1 + \sqrt{3})i$
(b) $1 + i = \sqrt{2}\left(\cos\frac{\pi}{4} + i\sin\frac{\pi}{4}\right)$, $1 + \sqrt{3}i = 2\left(\cos\frac{\pi}{3} + i\sin\frac{\pi}{3}\right)$;
$\sqrt{2}\left(\cos\frac{\pi}{4} + i\sin\frac{\pi}{4}\right)2\left(\cos\frac{\pi}{3} + i\sin\frac{\pi}{3}\right) = 2\sqrt{2}\left(\cos\frac{7\pi}{12} + i\sin\frac{7\pi}{12}\right)$
(c) From part (b), $x = 2\sqrt{2}\cos\left(\frac{\pi}{4} + \frac{\pi}{3}\right) = 2\sqrt{2}\left(\cos\frac{\pi}{4}\cos\frac{\pi}{3} - \sin\frac{\pi}{4}\sin\frac{\pi}{3}\right) =$
$2\sqrt{2}\left(\frac{\sqrt{2} - \sqrt{6}}{4}\right) = \frac{4 - 4\sqrt{3}}{4} = 1 - \sqrt{3}$; $y = 2\sqrt{2}\sin\left(\frac{\pi}{4} + \frac{\pi}{3}\right) = 2\sqrt{2}\left(\frac{\sqrt{2} + \sqrt{6}}{4}\right) = 1 + \sqrt{3}$
51. (a) $15 - 15\sqrt{3}i$ (b) $5 = 5(\cos 0 + i\sin 0)$, $3 - 3\sqrt{3}i = 6\left[\cos\left(-\frac{\pi}{3}\right) + i\sin\left(-\frac{\pi}{3}\right)\right]$;
$5(\cos 0 + i\sin 0)\cdot 6\left[\cos\left(-\frac{\pi}{3}\right) + i\sin\left(-\frac{\pi}{3}\right)\right] = 30\left[\cos\left(-\frac{\pi}{3}\right) + i\sin\left(-\frac{\pi}{3}\right)\right]$
(c) From part (b), $x = 30\cdot\frac{1}{2} = 15$; $y = 30\left(-\frac{\sqrt{3}}{2}\right) = -15\sqrt{3}$ **53.** (a) $3i$

(b) $6+6\sqrt{3}i = 12\left(\cos\frac{\pi}{3} + i\sin\frac{\pi}{3}\right)$, $2\sqrt{3}-2i = 4\left[\cos\left(-\frac{\pi}{6}\right) + i\sin\left(-\frac{\pi}{6}\right)\right]$;

$$\frac{12\left(\cos\frac{\pi}{3} + i\sin\frac{\pi}{3}\right)}{4\left[\cos\left(-\frac{\pi}{6}\right) + i\sin\left(-\frac{\pi}{6}\right)\right]} = 3\left(\cos\frac{\pi}{2} + i\sin\frac{\pi}{2}\right)$$

(c) $x = 3\cos\frac{\pi}{2} = 0$; $y = 3\sin\frac{\pi}{2} = 3$

55. counterclockwise, 90

Section 10.4, p. 424

1. (a) $-54+54i$ (b) $(3+3i)^3 = \left[3\sqrt{2}\left(\cos\frac{\pi}{4} + i\sin\frac{\pi}{4}\right)\right]^3 = 54\sqrt{2}\left(\cos\frac{3\pi}{4} + i\sin\frac{3\pi}{4}\right)$

3. $9\left(\cos\frac{2\pi}{3} + i\sin\frac{2\pi}{3}\right)$ **5.** $10000\left[\cos\left(-\frac{\pi}{2}\right) + i\sin\left(-\frac{\pi}{2}\right)\right]$ **7.** $2\sqrt{2}\left(\cos\frac{3\pi}{4} + i\sin\frac{3\pi}{4}\right)$

9. $256\left(\cos\frac{8\pi}{3} + i\sin\frac{8\pi}{3}\right)$ **11.** $\cos 3\pi + i\sin 3\pi$ **13.** $2i$ **15.** $8i$ **17.** 1

19. $\cos\frac{\pi}{12} + i\sin\frac{\pi}{12}$, $\cos\frac{13\pi}{12} + i\sin\frac{13\pi}{12}$ [See Figure 10I.] **21.** $\cos\frac{\pi}{4} + i\sin\frac{\pi}{4}$, $\cos\frac{7\pi}{12} + i\sin\frac{7\pi}{12}$, $\cos\frac{11\pi}{12} + i\sin\frac{11\pi}{12}$, $\cos\frac{5\pi}{4} + i\sin\frac{5\pi}{4}$, $\cos\frac{19\pi}{12} + i\sin\frac{19\pi}{12}$, $\cos\frac{23\pi}{12} + i\sin\frac{23\pi}{12}$ [See Figure 10J.]

23. $\cos\frac{\pi}{20} + i\sin\frac{\pi}{20}$, $\cos\frac{\pi}{4} + i\sin\frac{\pi}{4}$, $\cos\frac{9\pi}{20} + i\sin\frac{9\pi}{20}$, $\cos\frac{13\pi}{20} + i\sin\frac{13\pi}{20}$, $\cos\frac{17\pi}{20} + i\sin\frac{17\pi}{20}$, $\cos\frac{21\pi}{20} + i\sin\frac{21\pi}{20}$, $\cos\frac{5\pi}{4} + i\sin\frac{5\pi}{4}$, $\cos\frac{29\pi}{20} + i\sin\frac{29\pi}{20}$, $\cos\frac{33\pi}{20} + i\sin\frac{33\pi}{20}$, $\cos\frac{37\pi}{20} + i\sin\frac{37\pi}{20}$

25. $\frac{\sqrt{2}}{2} + \frac{\sqrt{2}}{2}i$, $-\frac{\sqrt{2}}{2} - \frac{\sqrt{2}}{2}i$ **27.** $2+2\sqrt{3}i$, $-2-2\sqrt{3}i$ **29.** 2, $\sqrt{2}+\sqrt{2}i$, $2i$, $-\sqrt{2}+\sqrt{2}i$, -2, $-\sqrt{2}-\sqrt{2}i$, $-2i$, $\sqrt{2}-\sqrt{2}i$ **31.** $\cos\frac{2k\pi}{5} + i\sin\frac{2k\pi}{5}$ for $k = 0, 1, 2, 3, 4$

33. $\cos\frac{k\pi}{5} + i\sin\frac{k\pi}{5}$ for $k = 0, 1, 2, \ldots, 9$ **35.** $-\frac{1}{2} + \frac{\sqrt{3}}{2}i$, $-\frac{1}{2} - \frac{\sqrt{3}}{2}i$, 1 **37.** 1, $\frac{\sqrt{2}}{2} + \frac{\sqrt{2}}{2}i$, i, $-\frac{\sqrt{2}}{2} + \frac{\sqrt{2}}{2}i$, -1, $-\frac{\sqrt{2}}{2} - \frac{\sqrt{2}}{2}i$, $-i$, $\frac{\sqrt{2}}{2} - \frac{\sqrt{2}}{2}i$ **39.** $2i$, $-2i$ **41.** $\frac{3\sqrt{2}}{2} + \frac{3\sqrt{2}}{2}i$, $-\frac{3\sqrt{2}}{2} - \frac{3\sqrt{2}}{2}i$ **43.** $\sqrt{3}+i$, $2i$, $-\sqrt{3}+i$, $-\sqrt{3}-i$, $-2i$, $\sqrt{3}-i$

45. $1+i$, $\frac{1-\sqrt{3}}{2} + \frac{1+\sqrt{3}}{2}i$, $\frac{-\sqrt{3}-1}{2} + \frac{\sqrt{3}-1}{2}i$, $-1-i$, $\frac{\sqrt{3}-1}{2} + \frac{-\sqrt{3}-1}{2}i$, $\frac{1+\sqrt{3}}{2} + \frac{1-\sqrt{3}}{2}i$

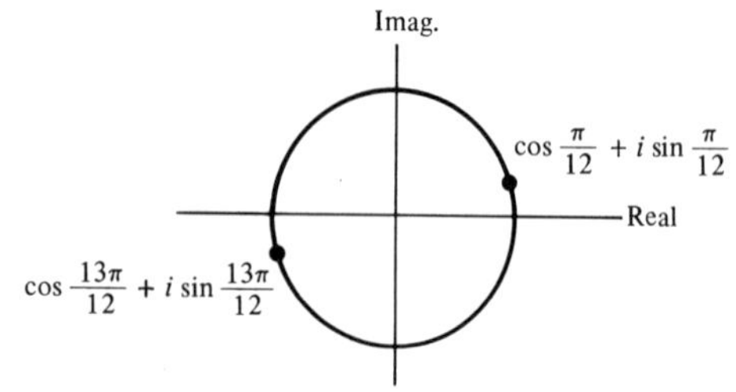

FIGURE 10I

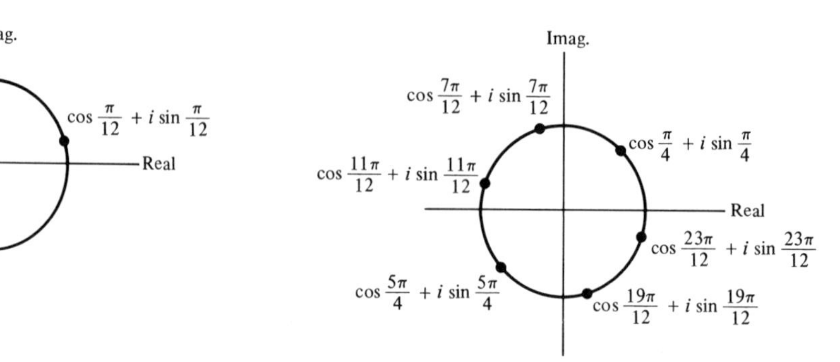

FIGURE 10J

47. By De Moivre's Theorem, $(\cos\theta + i\sin\theta)^2 = \cos 2\theta + i\sin 2\theta$. But also,

$$(\cos\theta + i\sin\theta)^2 = (\cos\theta + i\sin\theta)(\cos\theta + i\sin\theta) = \cos^2\theta - \sin^2\theta + 2\cos\theta\sin\theta\, i.$$

Equate real parts and imaginary parts: $\cos 2\theta = \cos^2\theta - \sin^2\theta$, $\sin 2\theta = 2\sin\theta\cos\theta$

49. (a) 0 (b) 0 (c) For an even integer n, the sum of the nth roots of unity equals 0.

Review Exercises, p. 426

1. (a) $2 - i$ (b) $11 - 7i$ (c) $\frac{1}{10} + \frac{13}{10}i$ (d) $3 - 4i$ **2.** See Figure 10K. **3.** See Figure 10L.

4. (a) 3 (b) $\frac{3\pi}{4}$ (c) See Figure 10M. **5.** (a) $4\sqrt{2}\left(\cos\frac{7\pi}{4} + i\sin\frac{7\pi}{4}\right)$ (b) $4\sqrt{2}$ (c) $\frac{7\pi}{4}$

6. (a) $-1 - \sqrt{3}i$ (b) -1 (c) $-\sqrt{3}$ **7.** (a) $3\left(\cos\frac{\pi}{4} - i\sin\frac{\pi}{4}\right)$ (b) $3\left(-\cos\frac{\pi}{4} - i\sin\frac{\pi}{4}\right)$

(c) $\frac{1}{3}\left(\cos\frac{\pi}{4} - i\sin\frac{\pi}{4}\right)$ **8.** (a) $\frac{3}{2}\left[\cos\left(-\frac{\pi}{6}\right) + i\sin\left(-\frac{\pi}{6}\right)\right]$ (b) $\frac{3\sqrt{3}}{4} - \frac{3}{4}i$ **9.** (a) $-8i$

(b) $8[\cos(-\pi) + i\sin(-\pi)]$ or $-8i$ **10.** $16\left(\cos\frac{4\pi}{3} + i\sin\frac{4\pi}{3}\right)$ **11.** $-2 - 2i$

12. $2\left[\cos\left(\frac{\pi}{6} + k\frac{\pi}{6}\right) + i\sin\left(\frac{\pi}{6} + k\frac{\pi}{6}\right)\right]$, $k = 0, 1, 2, 3, 4, 5$ **13.** $\cos\frac{k\pi}{4} + i\sin\frac{k\pi}{4}$, $k = 0, 1, 2, \ldots, 7$

14. $\frac{3\sqrt{3}}{2} + \frac{3}{2}i$, $-\frac{3\sqrt{3}}{2} + \frac{3}{2}i$, $-3i$

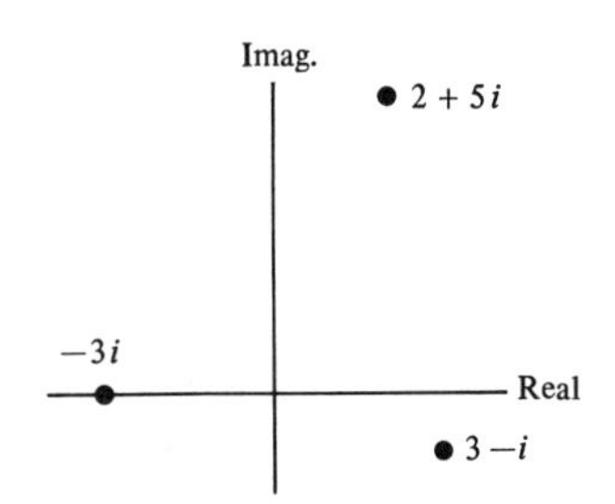

FIGURE 10K

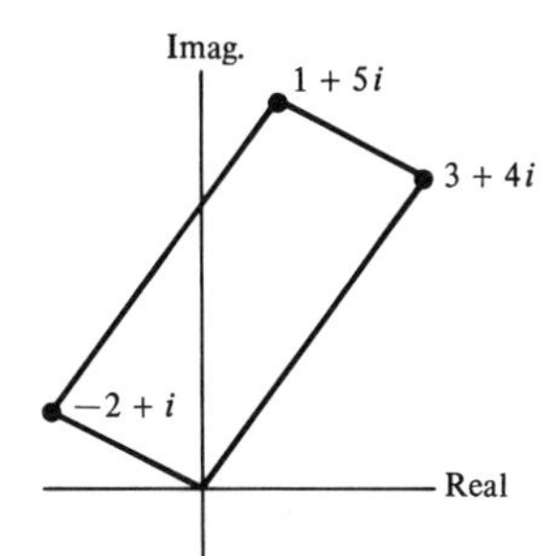

FIGURE 10L $(3 + 4i) + (-2 + i) = 1 + 5i$

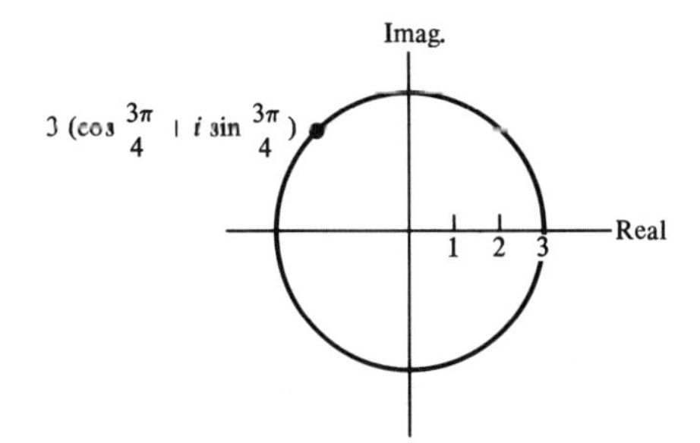

FIGURE 10M

Chapter 11

Section 11.1, p. 432 (*The checks are given below.*)

1. $z + 10$ **3.** $a^2 + a + 2$ **5.** $x^2 - 3x + 5$ **7.** $2x + 4$ **9.** $t^2 - 2t - 9$ **11.** $a + 5$ **13.** $m + 1$

15. $x^2 - x + 2$ **17.** $x + 2 + \frac{1}{x+2}$ **19.** $x - 9 + \frac{7}{x}$ **21.** $x + 5 + \frac{-4}{x+4}$ **23.** $c + 2$

25. $2x^2 - 2x + 3 + \frac{-2x-1}{x^2+x+1}$ **27.** $y^2 - 2 + \frac{2y-8}{y^2-2}$ **29.** $1 + \frac{1}{z+1}$ **31.** $1 + \frac{-1}{10t+1}$ **33.** $t - 3$

35. $z^2 + 2z - 1 + \frac{-10z+12}{2z^2-z+4}$ **37.** $a^3 + a^2 + 5a + 4 + \frac{a^2-5a-2}{a^3-a^2+1}$ **39.** $x^2 + 5x + 6$

Checks.

3.
$$\begin{array}{rl} a^2 + a + 2 & \leftarrow \text{quotient} \\ a + 4 & \leftarrow \text{divisor} \\ \hline a^3 + a^2 + 2a & \\ 4a^2 + 4a + 8 & \\ \hline a^3 + 5a^2 + 6a + 8 & \leftarrow \text{dividend} \end{array}$$

13.
$$\begin{array}{rl} m + 1 & \leftarrow \text{quotient} \\ 2m^2 - 5 & \leftarrow \text{divisor} \\ \hline 2m^3 + 2m^2 & \\ -5m - 5 & \\ \hline 2m^3 + 2m^2 - 5m - 5 & \leftarrow \text{dividend} \end{array}$$

17.
$$\begin{array}{rl} x + 2 & \leftarrow \text{quotient} \\ x + 2 & \leftarrow \text{divisor} \\ \hline x^2 + 4x + 4 & \\ + \qquad 1 & \leftarrow \text{remainder} \\ \hline x^2 + 4x + 5 & \leftarrow \text{dividend} \end{array}$$

25.
$$\begin{array}{rl} 2x^2 - 2x + 3 & \leftarrow \text{quotient} \\ x^2 + x + 1 & \leftarrow \text{divisor} \\ \hline 2x^4 - 2x^3 + 3x^2 & \\ 2x^3 - 2x^2 + 3x & \\ 2x^2 - 2x + 3 & \\ \hline 2x^4 + 3x^2 + x + 3 & \\ + \qquad -2x - 1 & \leftarrow \text{remainder} \\ \hline 2x^4 + 3x^2 - x + 2 & \leftarrow \text{dividend} \end{array}$$

27.
$$\begin{array}{rl} y^2 - 2 & \leftarrow \text{quotient} \\ y^2 - 2 & \leftarrow \text{divisor} \\ \hline y^4 - 4y^2 + 4 & \\ + \qquad 2y - 8 & \leftarrow \text{remainder} \\ \hline y^4 - 4y^2 + 2y - 4 & \leftarrow \text{dividend} \end{array}$$

Section 11.2, p. 438

1. $\underline{-1}\rfloor\ 1\ 2\ 4$ **3.** $\underline{0}\rfloor\ 1\ 5\ 1\ 1\ -1$ **5.** $\underline{-4}\rfloor\ 1\ -3\ 0\ 4$ **7.** $2x + 2$, remainder 1
9. $x^3 + x$ (no remainder) **11.** $x^5 + x^2 + 2$, remainder 3 **13.** $x + 3$ **15.** $y - 3$ **17.** $x^2 + x + 1$
19. $t^2 + 1$ **21.** $x^3 - x^2 + 2x - 3$ **23.** $a^2 + 2a + 3$ **25.** $y^7 + y^6 + y^5 + y^4 + y^3 + y^2 + y + 1$
27. $m^3 - m^2 + 2m - 2 + \frac{3}{m+1}$ **29.** $a^2 + 2a + 1$ **31.** $z^3 + 3z^2 + 3 + \frac{9}{z-3}$ **33.** (a) 8 (b) no
35. (a) 4 (b) no **37.** (a) 0 (b) yes **39.** (a) 0 (b) yes **41.** (a) -128 (b) no **43.** (a) 0 (b) yes **45.** (a) 5 (b)
$$\begin{array}{r|rrrr} \underline{3} & 1 & -3 & 1 & 2 \\ & & 3 & 0 & 3 \\ \hline & 1 & 0 & 1 & \lfloor 5 \end{array}$$
47. (a) 28 (b)
$$\begin{array}{r|rrrrr} \underline{-2} & 1 & 0 & 2 & -1 & 2 \\ & & -2 & 4 & -12 & 26 \\ \hline & 1 & -2 & 6 & -13 & \lfloor 28 \end{array}$$
49. $k = 8$
51. $k = -2$ **53.** (a) $1^4 - 1 = 0$ (b) $1^6 - 1 = 0$ (c) $1^{10} - 1 = 0$

Section 11.3, p. 446

1. simple: 0, 2, -7 **3.** simple: $1 - i$; multiplicity 2: 4 **5.** simple: $\frac{2}{3}$; multiplicity 2: $\frac{1}{2}$
7. simple: $\pm 2i$; multiplicity 3: 0 **9.** $x^3 - 7x^2 + 14x - 8$
11. $x^4 - (6 + 8i)x^3 - (7 - 48i)x^2 + (96 - 72i)x - 144$ **13.** $x(x + 2)(x - 1 - i)(x - 1 + i)$
15. $(x - 4)^3(x - 4 + 3i)^2$ **17.** (a) $\pm 1, -2$ (b) none **19.** (a) $0, \pm 2, -1$ (b) none **21.** (a) ± 3 (b) $\pm i$ **23.** (a) 2 (b) $-1 \pm 2i$ **25.** (a) $\frac{1}{2}, \pm 1$ (b) none **27.** (a) -1 (multiplicity 2), $-\frac{1}{2}, \frac{1}{3}$ (b) none **29.** (a) $\frac{2}{5}, 1$ (b) $\frac{-1 \pm \sqrt{3}i}{2}$ **31.** (a) 1 (multiplicity 2), -2 (b) none
33. (a) $(x - 2)(x - i)(x + i) = x^3 - 2x^2 + x - 2$ (b) $-2x^3 + 4x^2 - 2x + 4$
35. $P(x) = x(x - 1 - 3i)(x - 1 + 3i) = x^3 - 2x^2 + 10x$ **37.** 10 **39.** $(x - 5)(x^2 + 1)$ **41.** $x(x - 1)(x^2 + 5)$
43. $(x - 2)(x + 4)(x^2 + x + 1)$
45. Nonreal complex roots of polynomials with real coefficients occur in conjugate pairs.
47. $(x - i)^2(x + i) = x^3 - ix^2 + x - i$
49. By the Rational Zeros Theorem, the only possible *rational* zeros of the polynomial $x^2 - 2$ are $\pm 1, \pm 2$. Substitution shows that these are not roots. Thus, the roots $\pm\sqrt{2}$ are irrational. **51.** 2

53. From Equation (11.1),

$$a_{n-1}c^{n-1}d + \cdots + a_1cd + a_0d^n = -a_nc^n$$
$$d(a_{n-1}c^{n-1} + \cdots + a_1cd^{n-2} + a_0d^{n-1}) = -a_nc^n$$

Consider the unique prime factorization of d: $d = \pm p_1^{l_1}p_2^{l_2}\cdots p_s^{l_s}$ Each prime power $p_i^{l_i}$ is a factor of d, and hence of $-a_nc^n$. Because c and d have no factors in common, each prime power $p_i^{l_i}$ must divide a_n. Therefore d, the product of these prime powers, is a factor of a_n.

Section 11.4, p. 454

1. origin **3.** no symmetry **5.** origin **7.** (a) $x = 0, 5$ (b) $x > 5$ (c) $0 < x < 5$, $x < 0$ **9.** (a) $x = -4, 0, 1$ (b) $x > 1$, $-4 < x < 0$ (c) $0 < x < 1$, $x < -4$ **11.** (a) $x = -2, 2$ (b) $x > 2$, $-2 < x < 2$ (c) $x < -2$ **13.** (a) $x = 0, 1, 2, 3$ (b) $x > 3$, $1 < x < 2$, $x < 0$ (c) $2 < x < 3$, $0 < x < 1$ **15.** (a) $y = 0$ when $x = -2, 0, 2$ (b) $x > 2$, $x < -2$ (c) $0 < x < 2$, $-2 < x < 0$ **17.** (a) $x = \frac{5}{2}$ (b) $x > \frac{5}{2}$ (c) $x < \frac{5}{2}$ **19.** (a) $y \to \infty$ (b) $y \to -\infty$

21. (a) $y \to -\infty$ (b) $y \to \infty$ **23.** (a) $y \to \infty$ (b) $y \to -\infty$ **25.** (a) $y \to \infty$ (b) $y \to -\infty$ **27.** (a) none (b) $y = 0$ when $x = 0, 3$; $y > 0$ when $x > 3$; $y < 0$ when $0 < x < 3$ and when $x < 0$ (c) when $x \to \infty$, then $y \to \infty$; when $x \to -\infty$, then $y \to -\infty$ (d) See Figure 11A. **29.** (a) none (b) $y = 0$ when $x = 0, 2, 4$; $y > 0$ when $x > 4$ and when $0 < x < 2$; $y < 0$ when $2 < x < 4$ and when $x < 0$ (c) when $x \to \infty$, then $y \to \infty$; when $x \to -\infty$, then $y \to -\infty$ (d) See Figure 11B. **31.** (a) none (b) $y = 0$ when $x = 0, 1, 4$; $y > 0$ when $x > 4$, when $0 < x < 1$, and when $x < 0$; $y < 0$ when $1 < x < 4$ (c) when $x \to \infty$, then $y \to \infty$; when $x \to -\infty$, then $y \to \infty$ (d) See Figure 11C. **33.** (a) none

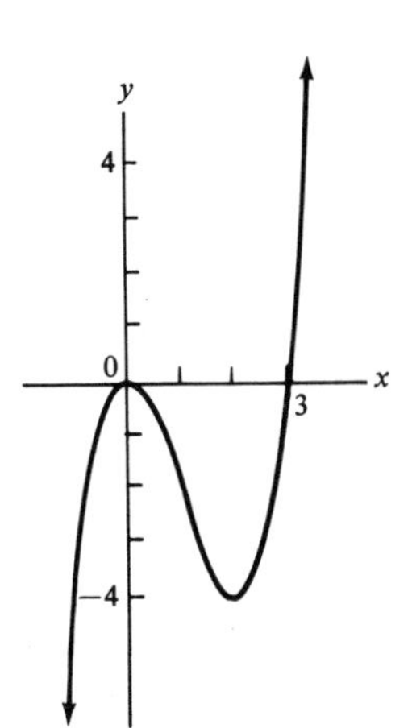

FIGURE 11A $y = x^2(x-3)$

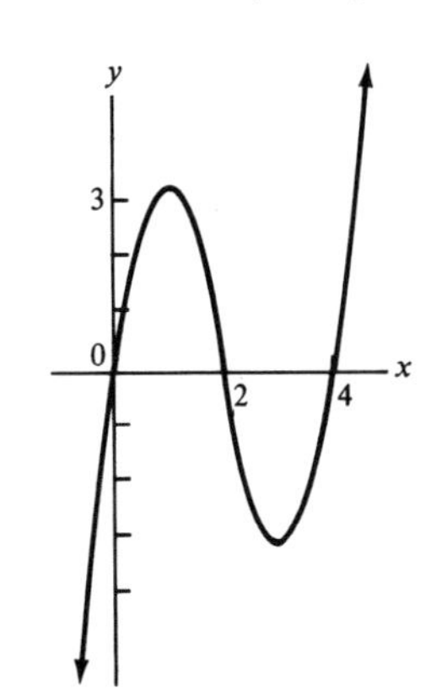

FIGURE 11B $y = x(x-2)(x-4)$

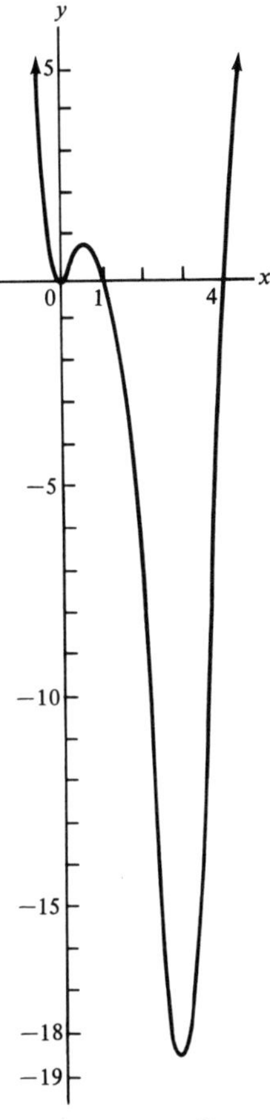

FIGURE 11C $y = x^2(x-1)(x-4)$

(b) $y = 0$ when $x = -2, -1, 2$; $y > 0$ when $x > 2$ and when $x < -2$; $y < 0$ when $-1 < x < 2$ and when $-2 < x < -1$ (c) when $x \to \infty$, then $y \to \infty$; when $x \to -\infty$, then $y \to \infty$ (d) See Figure 11D.

35. (a) origin symmetry (b) $y = 0$ when $x = -2, -1, 0, 1, 2$; $y > 0$ when $x > 2$, when $0 < x < 1$, and when $-2 < x < -1$; $y < 0$ when $1 < x < 2$, when $-1 < x < 0$, and when $x < -2$.
(c) when $x \to \infty$, then $y \to \infty$; when $x \to -\infty$, then $y \to -\infty$ (d) See Figure 11E.

37. (a) y-axis symmetry (b) $y = 0$ when $x = -\sqrt{2}, 0, \sqrt{2}$; $y > 0$ when $x > \sqrt{2}$ and when $x < -\sqrt{2}$; $y < 0$ when $0 < x < \sqrt{2}$ and when $-\sqrt{2} < x < 0$ (c) when $x \to \infty$, then $y \to \infty$; when $x \to -\infty$, then $y \to \infty$ (d) See Figure 11F. **39.** (a) none (b) $y = 0$ when $x = -1, 0, \frac{4}{3}$; $y > 0$ when $x > \frac{4}{3}$; $y < 0$ when $0 < x < \frac{4}{3}$, when $-1 < x < 0$, and when $x < -1$ (c) when $x \to \infty$, then $y \to \infty$; when $x \to -\infty$, then $y \to -\infty$ (d) See Figure 11G. **41.** (a) origin symmetry (b) $y = 0$ when $x = 0$; $y > 0$ when $x > 0$; $y < 0$ when $x < 0$ (c) when $x \to \infty$, then $y \to \infty$; when $x \to -\infty$, then $y \to -\infty$ (d) See Figure 11H. **43.** (a) none (b) $y = 0$ when $x = 0, 2$; $y > 0$ when $x > 2$, when $0 < x < 2$, and when $x < 0$ (y is never negative.) (c) when $x \to \infty$, then $y \to \infty$; when $x \to -\infty$, then $y \to \infty$ (d) See Figure 11I. **45.** (a) none (b) $y = 0$ when $x = 1, 3$; $y > 0$ when $x > 3$; $y < 0$ when $1 < x < 3$ and when $x < 1$ (c) when $x \to \infty$, then $y \to \infty$; when $x \to -\infty$, then $y \to -\infty$ (d) See Figure 11J. **47.** (a) none (b) $y = 0$ when $x = -2, 5$; $y > 0$ when $x > 5$ and when $x < -2$; $y < 0$ when $-2 < x < 5$ (c) when $x \to \infty$, then $y \to \infty$; when $x \to -\infty$, then $y \to \infty$ (d) See Figure 11K. **49.** (a) none (b) $y = 0$ when $x = \frac{1}{2}$; $y > 0$ when $x > \frac{1}{2}$, $y < 0$ when $x < \frac{1}{2}$ (c) when $x \to \infty$, then $y \to \infty$; when $x \to -\infty$, then $y \to -\infty$ (d) See Figure 11L. **51.** $y = x^4 + 1$

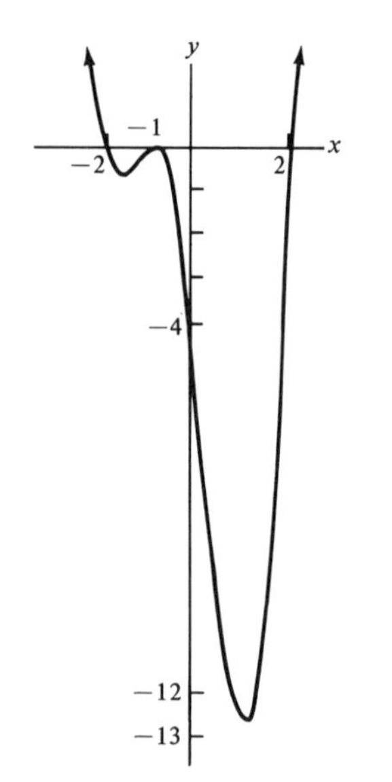

FIGURE 11D $y = (x + 1)^2 (x - 2)(x + 2)$

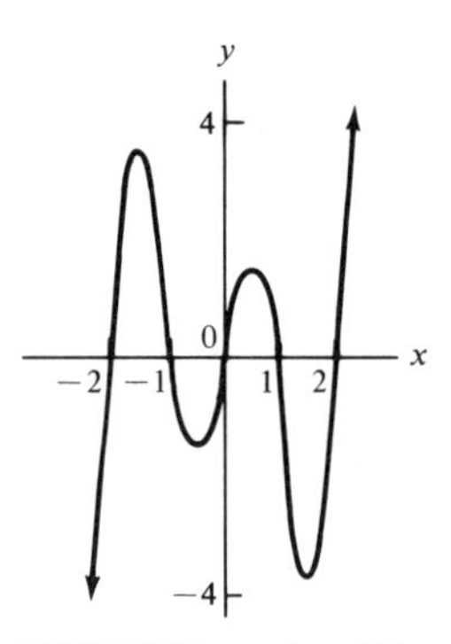

FIGURE 11E $y = (x + 2)(x + 1)\,x\,(x - 1)(x - 2)$

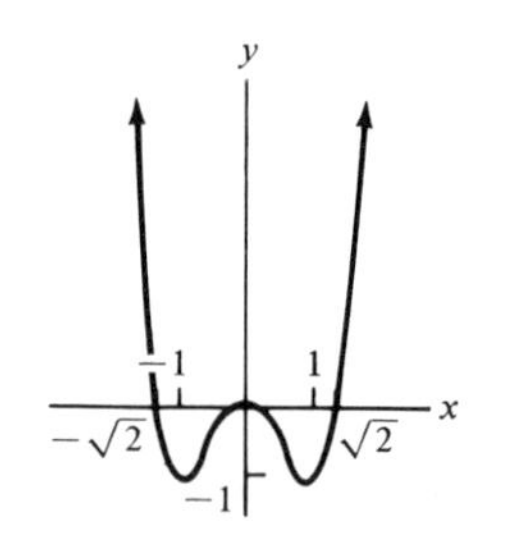

FIGURE 11F $y = x^2 (x^2 - 2)$

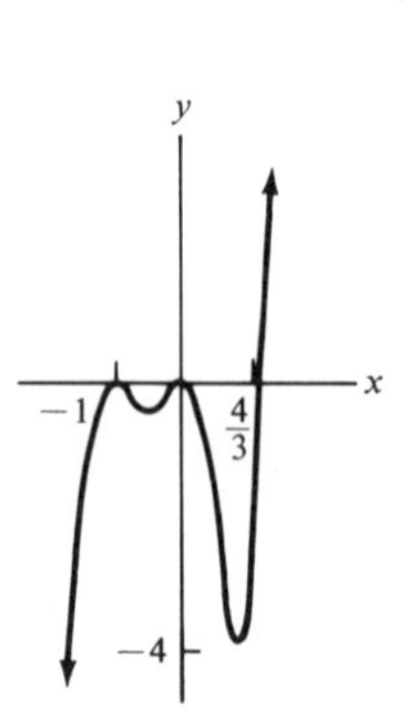

FIGURE 11G $y = x^2(3x - 4)(x + 1)^2$

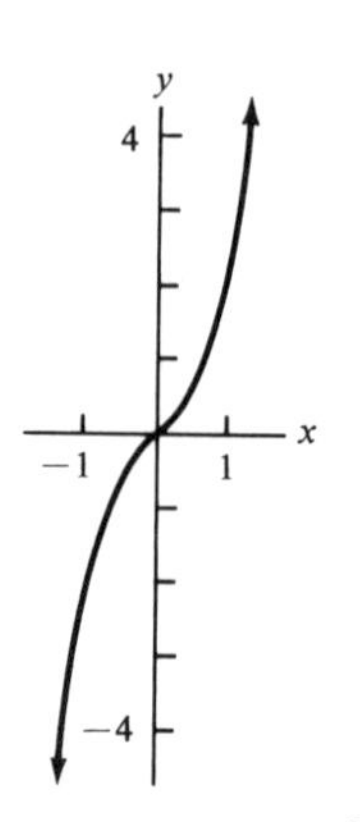

FIGURE 11H $y = x^3 + x$

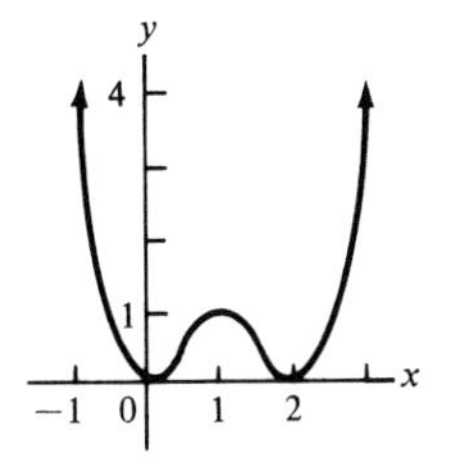

FIGURE 11I $y = x^4 - 4x^3 + 4x^2$

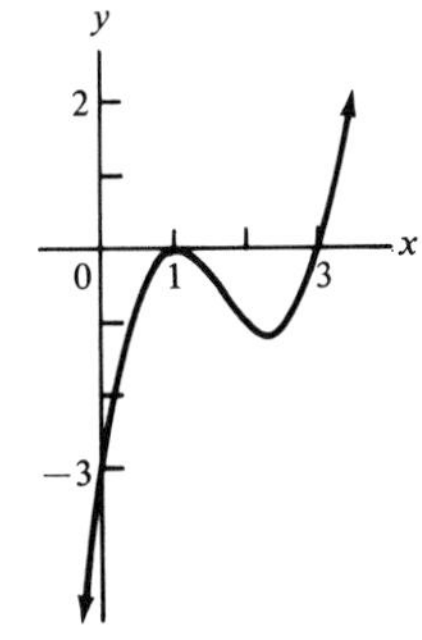

FIGURE 11J $y = x^3 - 5x^2 + 7x - 3$

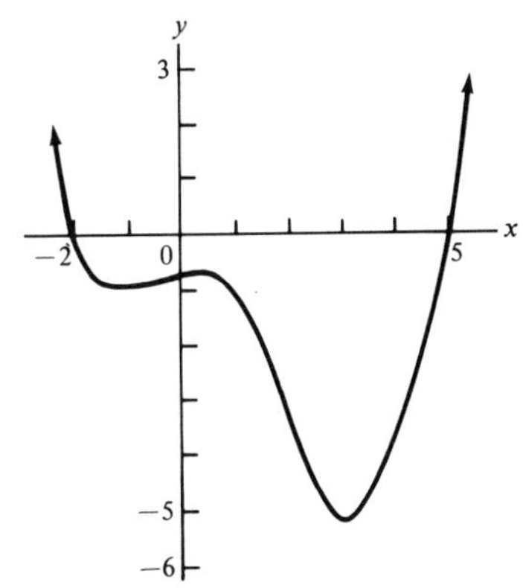

FIGURE 11K $20y = x^4 - 3x^3 - 9x^2 - 3x - 10$

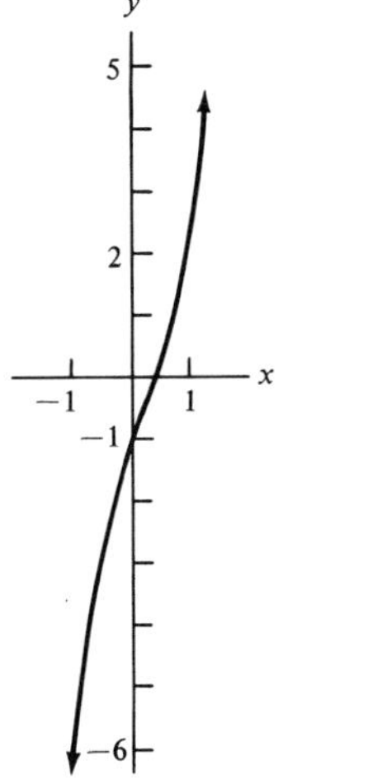

FIGURE 11L $y = 2x^3 - x^2 + 2x - 1$

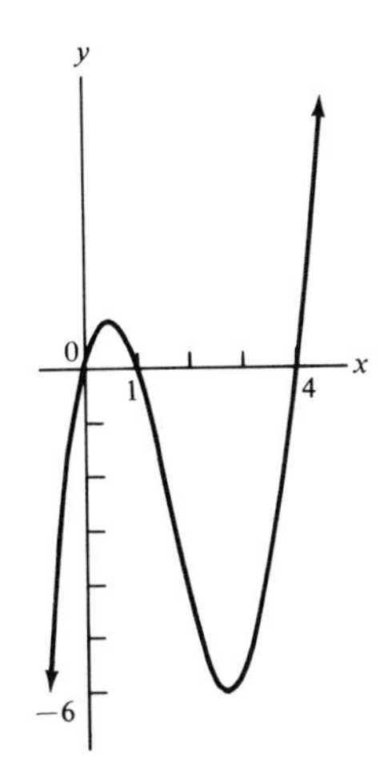

FIGURE 11M $y = x(x-1)(x-4)$

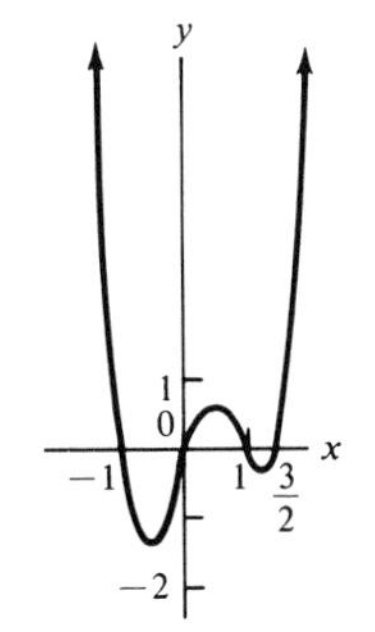

FIGURE 11N $y = (2x-3)(x-1)x(x+1)$

Review Exercises, p. 455

1. (a) $y^2 + 3$ (b) $m + 5 + \dfrac{-10}{m^2 + 2}$ (c) $c^2 + 1 + \dfrac{c+1}{c^3 + 4}$ **2.** (a) $x + 7$ (b) $t + 4 + \dfrac{6}{t - 9}$ (c) $x^2 + 4$ (d) $z^2 + 9z + 5 + \dfrac{5}{z - 1}$ **3. i.** (a) 6 (b) no **ii.** (a) 8 (b) no **iii.** (a) -2 (b) no **4.** 0

5. (a) -3 (simple); 1 (multiplicity 2); i (multiplicity 3) (b) $\pm 3^{1/4}$ (simple), $\pm 3^{1/4} i$ (simple); 0 (multiplicity 2) (c) $\dfrac{-1 \pm \sqrt{3}i}{2}$ (simple), ± 4 (multiplicity 2) **6.** $x^4 + 2x^3 + x^2 + 12x + 20$ **7. i.** (a) 2 (b) $\pm i$ **ii.** (a) $\dfrac{-1}{3}, \dfrac{1}{2}, 2$ (b) none **8.** $x^4 - 6x^3 + 15x^2 - 18x + 10$ **9.** $(x-3)(x+1)(x^2+4)$ **10.** (c)

11. i. origin **ii.** y-axis **iii.** neither **12. i.** (a) $x = -3, 1, 4$ (b) $x > 4$, $-3 < x < 1$ (c) $1 < x < 4$, $x < -3$ **ii.** (a) $x = -2, 5$ (b) $x > 5$, $-2 < x < 5$ (c) $x < -2$ **iii.** (a) $x = -4, 0, 4$ (b) $x > 4$, $-4 < x < 0$ (c) $0 < x < 4$, $x < -4$ **13. i.** (a) $y \to \infty$ (b) $y \to -\infty$ **ii.** (a) $y \to \infty$ (b) $y \to \infty$ **iii.** (a) $y \to -\infty$ (b) $y \to \infty$ **14. i.** (a) no symmetry (b) $y = 0$ when $x = 0, 1, 4$; $y > 0$ when $x > 4$ and when $0 < x < 1$; $y < 0$ when $1 < x < 4$ and when $x < 0$ (c) when $x \to \infty$, then $y \to \infty$; when $x \to -\infty$, then $y \to -\infty$ (d) See Figure 11M. **ii.** (a) no symmetry (b) $y = 0$ when $x = -1, 0, 1, \frac{3}{2}$; $y > 0$ when $x > \frac{3}{2}$, when $0 < x < 1$, and when $x < -1$; $y < 0$ when $1 < x < \frac{3}{2}$ and when $-1 < x < 0$ (c) when $x \to \infty$, then $y \to \infty$; when $x \to -\infty$, then $y \to \infty$ (d) See Figure 11N.

iii. (a) y-axis symmetry (b) $y = 0$ when $x = -1, 0, 1$; $y > 0$ when $x > 1$ and when $x < -1$; $y < 0$ when $0 < x < 1$ and when $-1 < x < 0$ (c) when $x \to \infty$, then $y \to \infty$; when $x \to -\infty$, then $y \to \infty$ (d) See Figure 11O. **iv.** (a) origin symmetry (b) $y = 0$ when $x = -1, 0, 1$; $y > 0$ when $0 < x < 1$ and when $x < -1$; $y < 0$ when $x > 1$ and when $-1 < x < 0$ (c) when $x \to \infty$, then $y \to -\infty$; when $x \to -\infty$, then $y \to \infty$ (d) See Figure 11P.

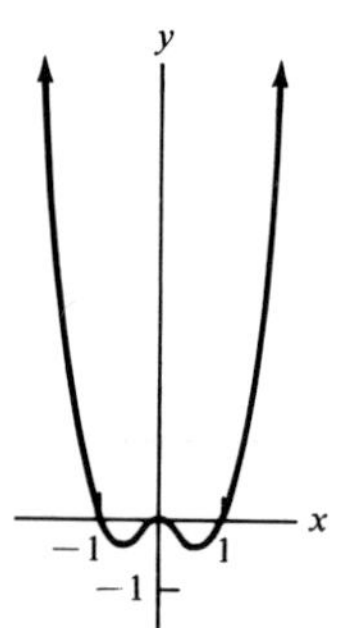

FIGURE 11O $y = x^2(x^2 - 1)$

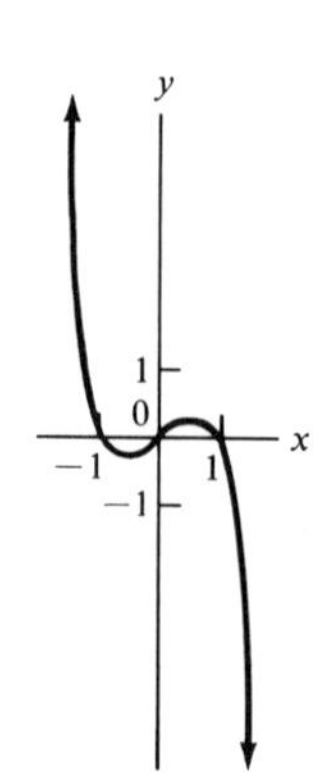

FIGURE 11P $y = x^3 - x^5$

Chapter 12

Section 12.1, p. 463

1.

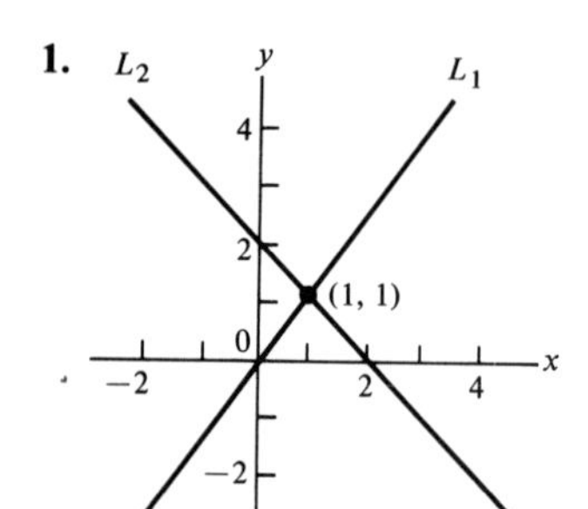

FIGURE 12A

3.

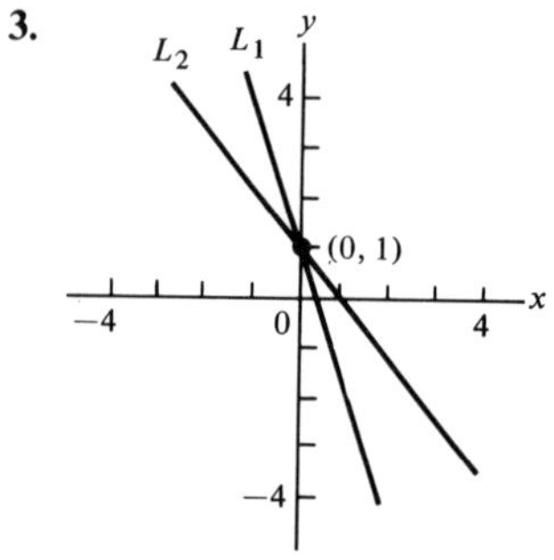

FIGURE 12B

5.

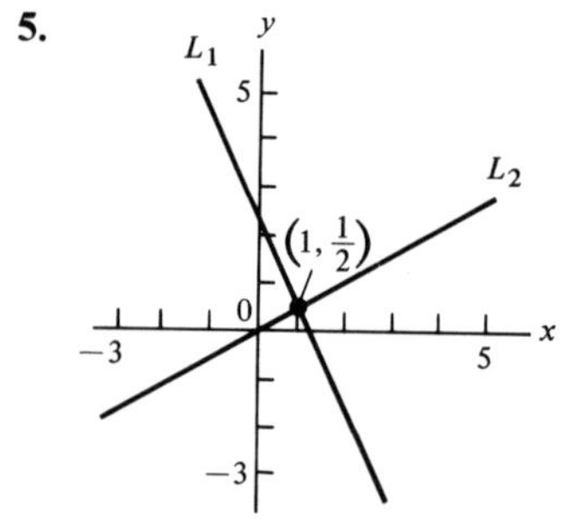

FIGURE 12C

7.

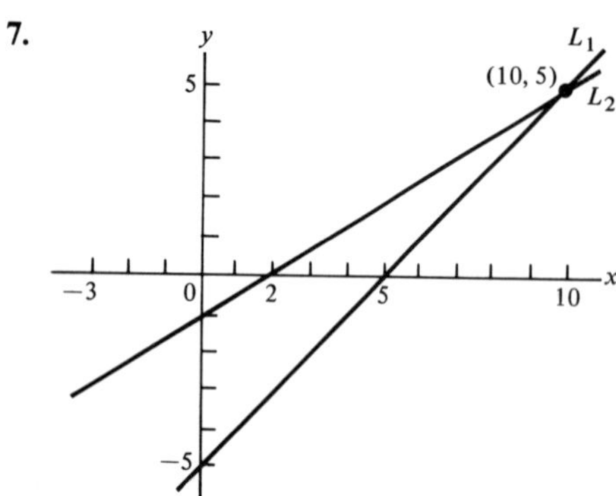

FIGURE 12D

9.

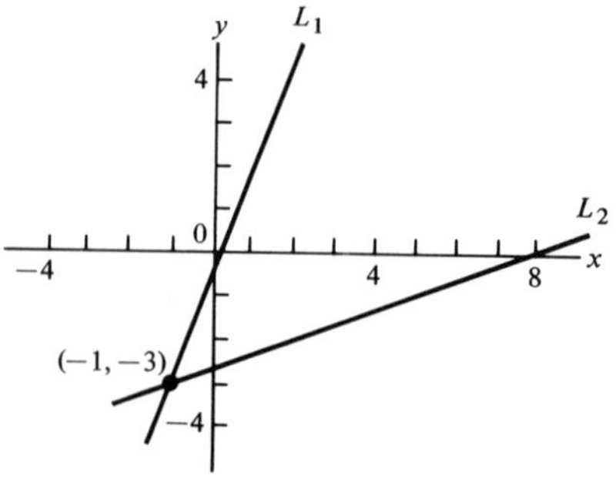

FIGURE 12E

11. (1, 2) **13.** (5, 2) **15.** (6, 0) **17.** $\left(\frac{1}{2}, \frac{1}{4}\right)$ **19.** (2, 5) **21.** $\left(\frac{\pi}{3}, \frac{\pi}{6}\right)$ **23.** (9, −3) **25.** (4, 3)

27. (0, 0) **29.** (6, −2) **31.** (6, −3) **33.** (3, −4) **35.** (5, −5) **37.** (11, 12) **39.** 17, 13
41. Ted is 12; his brother is 6. **43.** The dictionary is $1\frac{1}{2}$ inches thick. The encyclopedia is 2 inches thick.
45. 8 pounds **47.** $a = 4, b = -3$ **49.** $a = 1, b = 1$ **51.** $a = 1, b = 1$ **53.** $x = \frac{\pi}{6}, y = 0$

Section 12.2, p. 468

1. (1, 1, 4) **3.** (1, −1, 3) **5.** (6, −1, −2) **7.** (8, 2, 3) **9.** (1, 4, 8) **11.** (4, 7, 2) **13.** (20, 1, 2)
15. (1, 9, −2) **17.** (−2, −1, −4) **19.** $\left(2, \frac{1}{2}, -1\right)$ **21.** 2, 4, and 6 **23.** 6 nickels, 3 dimes, and 8 quarters
25. $\left(x - \frac{7}{4}\right)^2 + \left(y - \frac{9}{4}\right)^2 = \frac{65}{8}$ **27.** $A = 1, B = 1$ **29.** $A = 2, B = -2$ **31.** $A = 2, B = 1, C = 1$
33. $A = 2, B = C = 0$ **35.** $A = 1, B = C = -1$ **37.** $A = 0, B = C = 1$

Section 12.3, p. 472

1. (a) (3, 4), (3, −4) (b) $x^2 + y^2 = 25$ (circle), $x = 3$ (line) (c) See Figure 12F. **3.** (a) (0, 0), $\left(\frac{-1}{8}, \frac{-1}{8}\right)$
(b) $y = -8x^2$ (parabola), $y = x$ (line) **5.** (a) (0, 7), (7, 0) (b) $x^2 + y^2 = 49$ (circle), $y = 7 - x$ (line)
(c) See Figure 12G. **7.** (a) (0, 0), $\left(\frac{-1}{2}, \frac{1}{2}\right)$ (b) $x = -2y^2$ (parabola), $y = -x$ (line)
9. (a) (0, 10), (6, −8) (b) $x^2 + y^2 = 100$ (circle), $y = 10 - 3x$ (line) **11.** (a) (0, 0), (1, −1)
(b) $y = -x^2$ (parabola), $x + y = 0$ (line) (c) See Figure 12H. **13.** (a) (−1, 1), (−1, −1)
(b) $x^2 + y^2 = 2$ (circle), $x = -y^2$ (parabola)
15. (a) No intersection. Apply the quadratic formula to the equation $5x^2 - 8x + 12 = 0$, obtained by substituting $2x$ for y in the equation $(x + 2)^2 + (y - 3)^2 = 1$. The roots are not real. (b) $(x + 2)^2 + (y - 3)^2 = 1$ (circle), $y = 2x$ (line) **17.** (a) (2, 0) (b) $x^2 + y^2 = 4$ (circle), $(x - 4)^2 + y^2 = 4$ (circle) (c) See Figure 12I.
19. (a) $\left(1, \frac{\sqrt{3}}{2}\right), \left(1, \frac{-\sqrt{3}}{2}\right)$ (b) $\frac{x^2}{4} + y^2 = 1$ (ellipse), $x = 1$ (line) **21.** (a) (−1, 0)

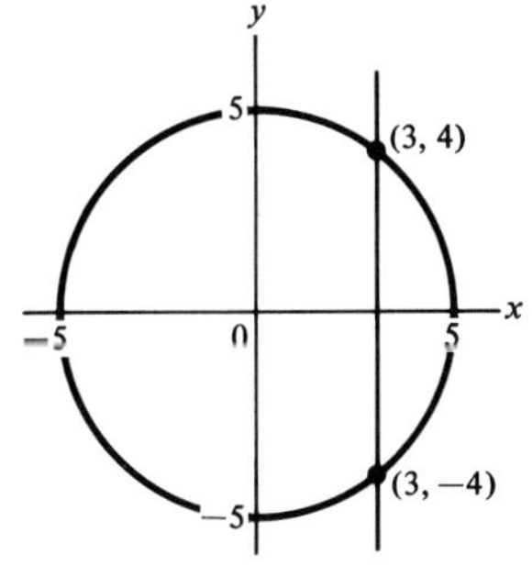

FIGURE 12F

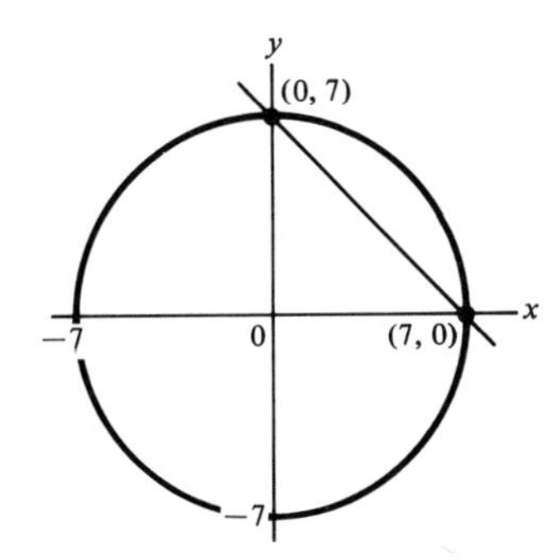

FIGURE 12G

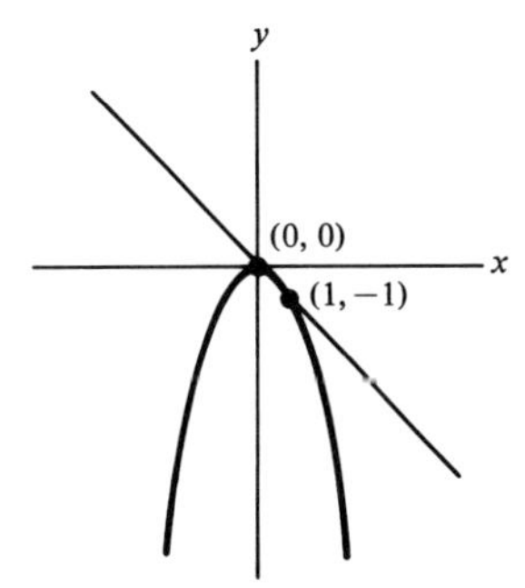

FIGURE 12H

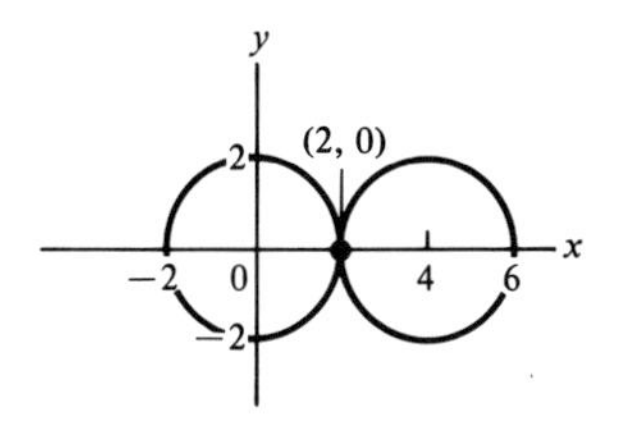

FIGURE 12I

(b) $x^2 - y^2 = 1$ (hyperbola), $x = -1$ (line) (c) See Figure 12J. **23.** (a) $\left(\frac{12}{5}, \frac{-12}{5}\right), \left(\frac{-12}{5}, \frac{12}{5}\right)$
(b) $\frac{x^2}{16} + \frac{y^2}{9} = 1$ (ellipse), $y = -x$ (line) **25.** (a) no intersection (b) $x^2 + y^2 = 4$ (circle),
$\frac{x^2}{9} + \frac{y^2}{16} = 1$ (ellipse) (c) See Figure 12K. **27.** (a) $(8, 0), (-8, 0)$ (b) $\frac{x^2}{64} - \frac{y^2}{36} = 1$ (hyperbola),
$\frac{x^2}{64} + \frac{y^2}{36} = 1$ (ellipse) **29.** (a) $(-4, 0)$ (b) $\frac{x^2}{16} + \frac{y^2}{16} = 1$ (circle); $(x - 2)^2 + y^2 = 36$ (circle)
(c) See Figure 12L.

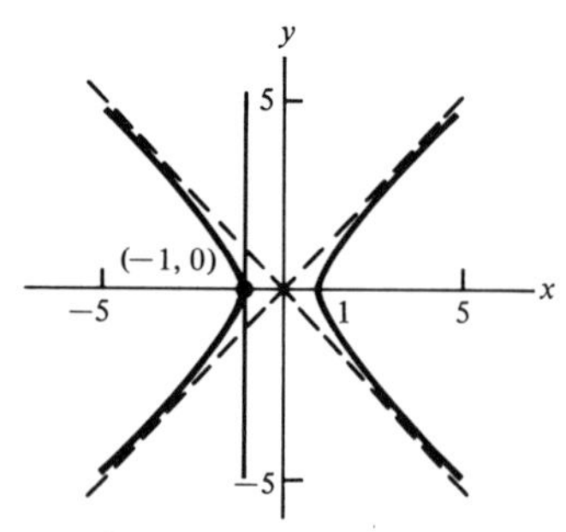

FIGURE 12J

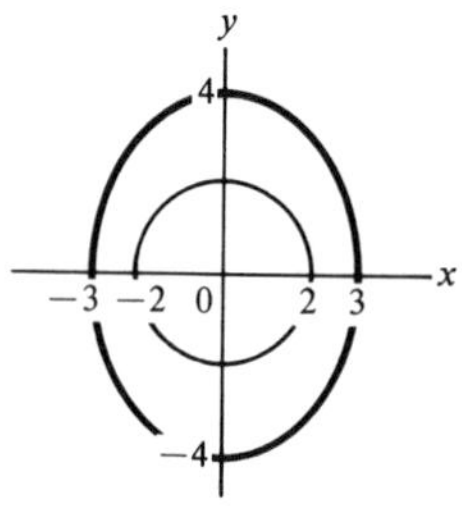

FIGURE 12K

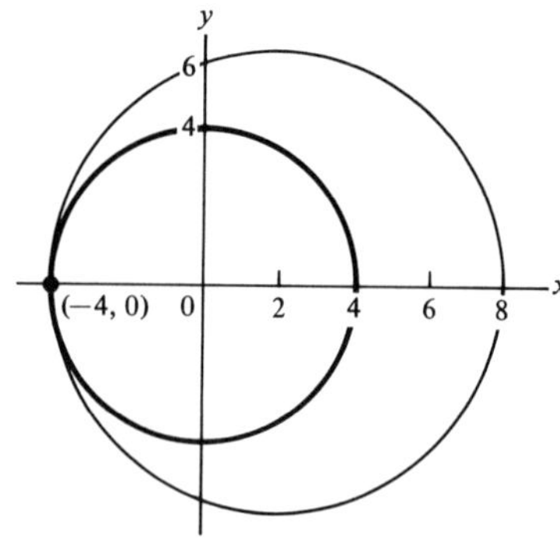

FIGURE 12L

Section 12.4, p. 476

1. -1 **3.** 6 **5.** 0 **7.** -2 **9.** 80 **11.** $\frac{-2}{5}$ **13.** 1 **15.** 2 **17.** 10 **19.** -1 **21.** 9
23. 0, 3 **25.** $\frac{\pi}{2}$ **27.** 0 **29.** $\begin{vmatrix} a & c \\ b & d \end{vmatrix} = -\begin{vmatrix} c & a \\ d & b \end{vmatrix}$ **31.** (a) Cramer's Rule applies. (b) (3, 2)
33. (a) Cramer's Rule applies. (b) $\left(\frac{1}{3}, \frac{1}{3}\right)$ **35.** (a) Cramer's Rule does not apply.
37. (a) Cramer's Rule applies. (b) $\left(-7, \frac{4}{3}\right)$ **39.** (a) Cramer's Rule applies. (b) (0, 1)
41. (a) Cramer's Rule applies. (b) $\left(\frac{10}{3}, \frac{-5}{2}\right)$ **43.** (a) Cramer's Rule applies. (b) $(-1, 2)$

Section 12.5, p. 482

1. -1 **3.** 2 **5.** 0 **7.** 0 **9.** -60 **11.** $-.08$ **13.** 2 **15.** 5 **17.** 0
19. $\begin{vmatrix} ka_1 & b_1 & c_1 \\ ka_2 & b_2 & c_2 \\ ka_3 & b_3 & c_3 \end{vmatrix} = k\begin{vmatrix} a_1 & b_1 & c_1 \\ a_2 & b_2 & c_2 \\ a_3 & b_3 & c_3 \end{vmatrix}$ **21.** (a) Cramer's Rule applies. (b) (1, 1, 1)
23. (a) Cramer's Rule applies. (b) $(-1, 2, 0)$ **25.** (a) Cramer's Rule applies. (b) $(6, 0, -6)$
27. (a) Cramer's Rule does not apply. **29.** (a) Cramer's Rule does not apply. **31.** (a) Cramer's Rule applies.
(b) $(-1, -1, -1)$

Review Exercises, p. 483

1. See Figure 12M. **2.** (a) $(1, -1)$ (b) $(-2, 2)$ (c) (0, 4) **3.** $a = 2, b = -3$ **4.** (a) $(1, 2, -1)$
(b) (1, 5, 2) (c) $(-1, 2, 7)$ **5.** 7, 11, 13 **6.** (a) $A = 1, \quad B = 2$ (b) $A = \frac{1}{2}, \quad B = -\frac{2}{3}, \quad C = \frac{1}{6}$

(c) $A = -\frac{1}{4}$, $B = \frac{1}{4}$, $C = 0$, $D = \frac{1}{2}$ **7. i.** (a) $(3\sqrt{2}, 3\sqrt{2})$, $(-3\sqrt{2}, -3\sqrt{2})$ (b) $x^2 + y^2 = 36$ (circle) $y = x$ (line) (c) See Figure 12N. **ii.** (a) $(0, 0)$, $\left(\frac{1}{2}, 1\right)$ (b) $y = 4x^2$ (parabola), $y = 2x$ (line) **iii.** (a) $(0, 2)$, $(0, -2)$ (b) $\frac{x^2}{9} + \frac{y^2}{4} = 1$ (ellipse), $x^2 + y^2 = 4$ (circle) (c) See Figure 12O.

8. (a) 2 (b) 14 (c) -11 **9.** 2 **10. i.** (a) Cramer's Rule applies. (b) $(0, 4)$ **ii.** (a) Cramer's Rule does not apply. **iii.** (a) Cramer's Rule applies. (b) $(1, 1)$ **11.** (a) 2 (b) 2 **12.** 5 **13. i.** (a) Cramer's Rule applies. (b) $(2, 0, -1)$ **ii.** (a) Cramer's Rule does not apply.

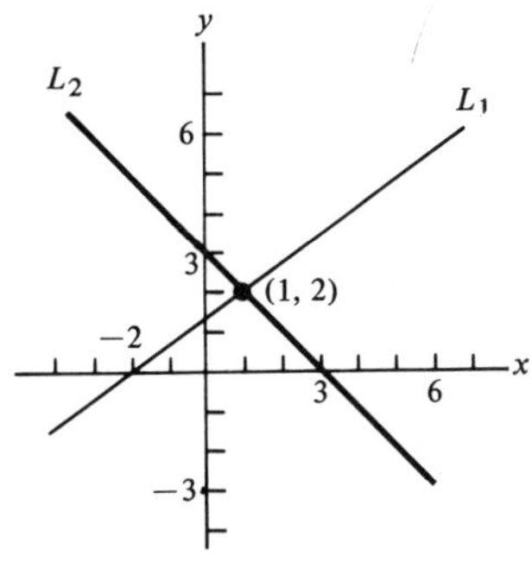

FIGURE 12M

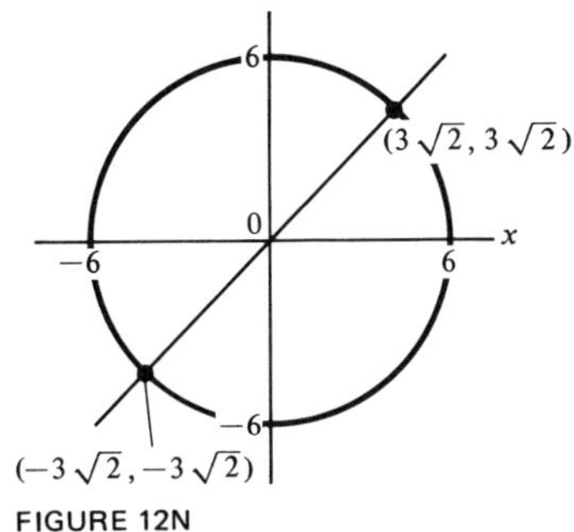

FIGURE 12N

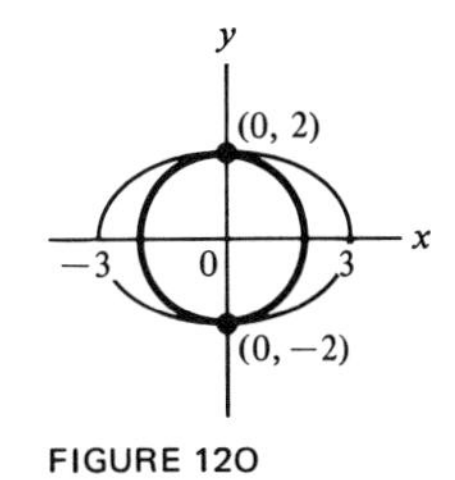

FIGURE 12O

Chapter 13

Section 13.1, p. 487

1. (a) 1, 2, 3, 4, 5, 6, 7, 8 (b) i (c) See Figure 13A. **3.** (a) $-1, -2, -3, -4, -5, -6, -7, -8$ (b) ii (c) See Figure 13B. **5.** (a) 2, 4, 6, 8, 10, 12, 14, 16 (b) i (c) See Figure 13C. **7.** (a) 3, 6, 9, 12, 15, 18, 21, 24 (b) i **9.** (a) 5, 5, 5, 5, 5, 5, 5, 5 (b) iii **11.** (a) 5, 3, 5, 3, 5, 3, 5, 3 (b) iii **13.** (a) $1, \frac{1}{2}, \frac{1}{3}, \frac{1}{4}, \frac{1}{5}, \frac{1}{6}, \frac{1}{7}, \frac{1}{8}$ (b) ii **15.** (a) 2, 4, 8, 16, 32, 64, 128, 256 (b) i

17. (a) 9, 12, 15, 18, 21, 24, 27, 30 (b) i **19.** (a) $\frac{3}{2}, 2, \frac{5}{2}, 3, \frac{7}{2}, 4, \frac{9}{2}, 5$ (b) i **21.** (a) 20 (b) 50 (c) 100 (d) 500 **23.** (a) 7 (b) 23 (c) 43 (d) 83 **25.** (a) $\frac{9}{2}$ (b) 12 (c) 22 (d) $\frac{49}{2}$

27. (a) 0 (b) 2 (c) 0 (d) 2 **29.** (a) 6 (b) 2 (c) 1 (d) $\frac{1}{4}$ **31.** (a) $\frac{1}{2}$ (b) $\frac{2}{3}$ (c) $\frac{9}{10}$ (d) $\frac{10}{11}$ **33.** $a_i = i + 1$ **35.** $a_i = i + 4$ **37.** $a_i = 3i$ **39.** $a_i = (-1)^i \cdot 4i$ **41.** $a_i = -1 + 3i$

43. $a_i = \frac{i+1}{i}$

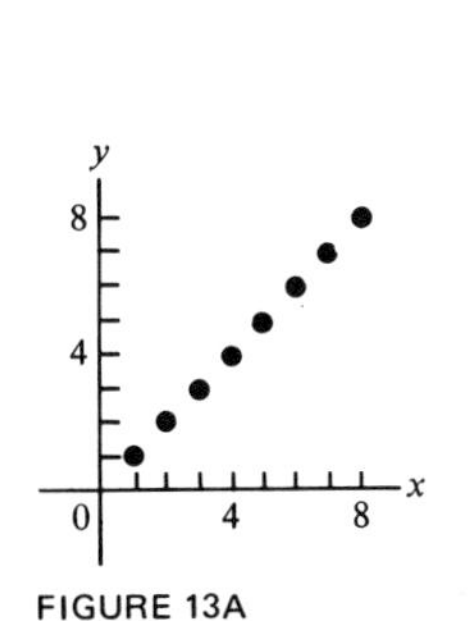

FIGURE 13A

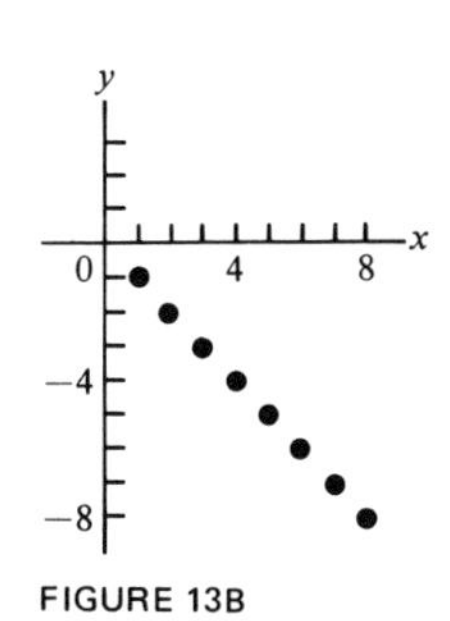

FIGURE 13B

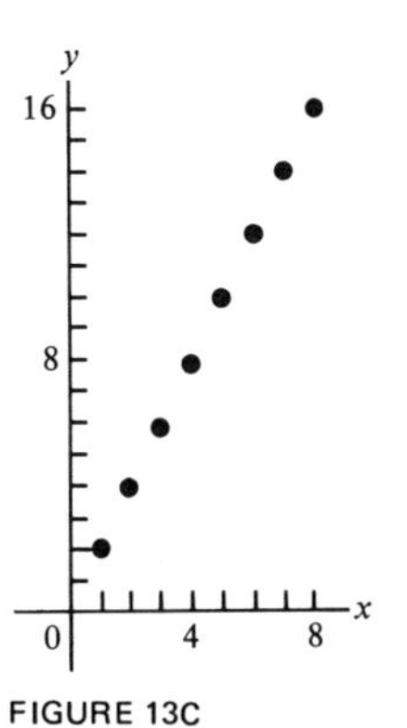

FIGURE 13C

Section 13.2, p. 492

1. $\sum_{i=1}^{4} a_i$ **3.** $\sum_{i=0}^{11} b_i$ **5.** $\sum_{i=1}^{2} a_i$ **7.** $\sum_{i=5}^{18} a_i$ **9.** $\sum_{i=101}^{200} c_i$ **11.** 21 **13.** 35 **15.** 55 **17.** $\frac{23}{12}$

19. 70 **21.** $\sum_{i=3}^{12} i$ **23.** $\sum_{i=1}^{16} 4i$ **25.** $\sum_{i=1}^{10} i^2$ **27.** $\sum_{i=3}^{92} \frac{1}{i}$ **29.** $\sum_{i=2}^{21} 2i$ **31.** $\sum_{i=1}^{8} \frac{1}{2^i}$ **33.** 7 **35.** -1

37. 25 **39.** 3 **41.** 2 **43.** $m = 101, n = 200$ **45.** $m = 11, n = 15$ **47.** $m = 6$ **49.** $m = 2, n = 8$

51. $(5-4)+(6-4)+(7-4)+(8-4)+(9-4)+(10-4)$ **53.** $3^3+3^4+3^5+3^6+3^7$ **55.** $\frac{1}{i}$ **57.** i

59. $\frac{1}{2^i}$

Section 13.3, p. 497

1. arithmetic progression with common difference 4 **3.** arithmetic progression with common difference 2
5. arithmetic progression with common difference -3 **7.** not an arithmetic progression
9. not an arithmetic progression **11.** arithmetic progression with common difference -5

13. arithmetic progression with common difference $\frac{-1}{3}$ **15.** 49 **17.** -235 **19.** 89 **21.** 195 **23.** 71

25. 520 **27.** 255 **29.** 7220 **31.** 1520 **33.** 15 250 **35.** -4900 **37.** (a) 324 (b) 627
39. (a) 450 (b) 935 **41.** 2550 **43.** $21 000 per year

Section 13.4, p. 502

1. geometric progression with common ratio 2 **3.** not a geometric progression; arithmetic progression

5. geometric progression with common ratio $\frac{1}{2}$ **7.** geometric progression with common ratio -1

9. geometric progression with common ratio -2 **11.** geometric progression with common ratio $\frac{1}{5}$

13. not a geometric progression **15.** 125 **17.** $\frac{3}{16}$ **19.** (a) 405 (b) -405 (c) $\frac{5}{9}$ (d) $\frac{-5}{9}$

21. 960 096 **23.** 4 **25.** 254 **27.** 1530 **29.** 1365 **31.** 693 **33.** $\frac{511}{4}$ **35.** -9 **37.** -183

39. .111 111 **41.** 255 **43.** $10 485.75 **45.** 250 000 **47.** $\sum_{i=0}^{n-1} r^i$

Section 13.5, p. 508

1. 24 **3.** 40 320 **5.** 3024 **7.** 210 **9.** $\frac{5!}{3!}$ **11.** $\frac{12!}{8!}$ **13.** $\frac{99!}{89!}$ **15.** 3 **17.** 6 **19.** 1

21. 1 **23.** 28 **25.** 792 **27.** 1 **29.** 396 **31.** $n(n-1)$ **33.** $(n+1)!$ **35.** 1 7 21 35 35 21 7 1
37. 1 9 36 84 126 126 84 36 9 1 **39.** $x^4 - 4x^3a + 6x^2a^2 - 4xa^3 + a^4$ **41.** $x^4 + 4x^3 + 6x^2 + 4x + 1$
43. $x^5 - 5x^4 + 10x^3 - 10x^2 + 5x - 1$ **45.** $x^4 + 12x^3 + 54x^2 + 108x + 81$

47. $x^4 + 8x^3y + 24x^2y^2 + 32xy^3 + 16y^4$ **49.** $a^8 - 4a^6b^2 + 6a^4b^4 - 4a^2b^6 + b^8$ **51.** (a) $\sum_{i=0}^{8} \binom{8}{i} x^{8-i}y^i$

(b) $x^8 + 8x^7y + 28x^6y^2 + 56x^5y^3$ **53.** (a) $\sum_{i=0}^{12} \binom{12}{i} x^{12-i}$ (b) $x^{12} + 12x^{11} + 66x^{10} + 220x^9$

55. (a) $\sum_{i=0}^{9} (-1)^i \binom{9}{i} x^{9-i}2^i$ (b) $x^9 - 18x^8 + 144x^7 - 672x^6$ **57.** (a) $\sum_{i=0}^{8} \binom{8}{i} x^{16-2i}y^{8-i}$

(b) $x^{16}y^8 + 8x^{14}y^7 + 28x^{12}y^6 + 56x^{10}y^5$ **59.** (a) $\sum_{i=0}^{8} \binom{8}{i} \frac{x^{8-i}}{2^i}$ (b) $x^8 + 4x^7 + 7x^6 + 7x^5$

61. 1.0406 **63.** 1.7716 **65.** 1.0721 **67.** 85.7661

Section 13.6, p. 514

1. Let $s(n)$ be the statement, $1 \cdot 2 + 2 \cdot 3 + 3 \cdot 4 + \cdots + n(n+1) = \frac{n(n+1)(n+2)}{3}$. $s(1)$: $1 \cdot 2 = \frac{1(1+1)(1+2)}{3}$.
This is true because both sides equal 2. Suppose $s(k)$ is true, that is, suppose
$1 \cdot 2 + 2 \cdot 3 + 3 \cdot 4 + \cdots + k(k+1) = \frac{k(k+1)(k+2)}{3}$. Then

$$\begin{aligned}1 \cdot 2 + 2 \cdot 3 + 3 \cdot 4 + \cdots + k(k+1) + (k+1)(k+2) &= \frac{k(k+1)(k+2)}{3} + (k+1)(k+2) \\ &= \frac{k(k+1)(k+2) + 3(k+1)(k+2)}{3} \\ &= \frac{(k+1)(k+2)[k+3]}{3}\end{aligned}$$

Thus, $s(k+1)$ is true, and by Induction, $s(n)$ is true for every positive integer n.

3. Let $s(n)$ be the statement, $1 + 4 + 7 + \cdots + (3n-2) = \frac{n(3n-1)}{2}$. $s(1)$: $1 = \frac{1(3 \cdot 1 - 1)}{2}$. This is true because the right side equals 1. Suppose $s(k)$ is true, that is, suppose $1 + 4 + 7 + \cdots + (3k-2) = \frac{k(3k-1)}{2}$.

Then
$$\begin{aligned}1 + 4 + 7 + \cdots + (3k-2) + [3(k+1) - 2] &= \frac{k(3k-1)}{2} + [3(k+1) - 2] \\ &= \frac{k(3k-1)}{2} + [3k + 3 - 2] \\ &= \frac{k(3k-1)}{2} + (3k+1) \\ &= \frac{k(3k-1) + 2(3k+1)}{2} \\ &= \frac{3k^2 - k + 6k + 2}{2} \\ &= \frac{3k^2 + 5k + 2}{2} \\ &= \frac{(k+1)(3k+2)}{2} \\ &= \frac{(k+1)[3(k+1) - 1]}{2}\end{aligned}$$

Thus, $s(k+1)$ is true, and by Induction, $s(n)$ is true for every positive integer n.

5. Let $s(n)$ be the statement, $n < 2^n$. $s(1)$: $1 < 2^1$. This is true. Suppose $s(k)$ is true, that is, suppose $k < 2^k$. Then $k + 1 \le k + k = 2k < 2 \cdot 2^k = 2^{k+1}$. Thus, $s(k+1)$ is true, and by Induction, $s(n)$ is true for every positive integer n.

7. Let $s(n)$ be the statement, $2^{n+3} < (n+3)!$ $s(1)$: $2^{1+3} < (1+3)!$ Note that $2^{1+3} = 2^4 = 16$ and $(1+3)! = 4! = 24$. Thus, $s(1)$ is true. Suppose $s(k)$ is true, that is, suppose $2^{k+3} < (k+3)!$ Then $2^{(k+1)+3} = 2^{k+4} = 2 \cdot 2^{k+3} < 2 \cdot (k+3)! < (k+4) \cdot (k+3)! = (k+4)! = [(k+1)+3]!$
Thus, $s(k+1)$ is true, and by Induction, $s(n)$ is true for every positive integer n.

9. (a) 204 (b) 2870 (c) 338 350 (d) 333 833 500 (e) 335 480 **11.** (a) 441 (b) 14 400 (c) 25 502 500 (d) 250 500 250 000 (e) 250 500 235 600 **13.** (a) $\frac{6}{7}$ (b) $\frac{40}{41}$ (c) $\frac{9}{100}$ (d) $\frac{1}{250}$

15. Let m be any positive integer. If $n = 1, 2, 3, \ldots, m$, then one of the factors of $(n-1)(n-2)\cdots(n-m)$ is $n-n$, which equals 0. Thus, the product is 0. If $n > m$, then each factor is positive; hence the product is positive.

17. Let z be a complex number and let $s(n)$ be the statement, $\overline{z^n} = \bar{z}^n$. $s(1)$: $\overline{z^1} = \bar{z}^1$. This is true because both sides equal $\bar{z}$. Suppose $s(k)$ is true, that is, suppose $\overline{z^k} = \bar{z}^k$. Then $\overline{z^{k+1}} = \overline{z \cdot z^k} = \bar{z} \cdot \overline{z^k} = \bar{z} \cdot \bar{z}^k = \bar{z}^{k+1}$. Thus $s(k+1)$ is true, and by Induction, $s(n)$ is true for every positive integer n.

19. Let $s(n)$ be the statement, for every positive real number h, $(1+h)^n \geq 1 + nh$. $s(1)$: For every positive real number h, $(1+h)^1 \geq 1 + 1 \cdot h$. This is true because for every h, $1 + h = 1 + 1 \cdot h$. Suppose $s(k)$ is true, that is, suppose that for every positive real number h, $(1+h)^k \geq 1 + kh$. Then for every positive h, $(1+h)^{k+1} = (1+h)^k(1+h) \geq (1+kh)(1+h) = 1 + kh + h + kh^2 = [1 + (k+1)h] + kh^2 \geq 1 + (k+1)h$, because $kh^2 \geq 0$. Thus $s(k+1)$ is true, and by Induction, $s(n)$ is true for every positive integer n.

21. Let $c, a_1, a_2, \ldots$ be real numbers. Let $s(n)$ be the statement, $\sum_{i=1}^{n} ca_i = c\sum_{i=1}^{n} a_i$. $s(1)$: $\sum_{i=1}^{1} ca_i = c\sum_{i=1}^{1} a_i$. This is true because both sides equal ca_1. Suppose $s(k)$ is true, that is, suppose $\sum_{i=1}^{k} ca_i = c\sum_{i=1}^{k} a_i$.

Then $\sum_{i=1}^{k+1} ca_i = \sum_{i=1}^{k} ca_i + ca_{k+1} = c\sum_{i=1}^{k} a_i + ca_{k+1} = c\left(\sum_{i=1}^{k} a_i + a_{k+1}\right) = c\sum_{i=1}^{k+1} a_i$. Thus, $s(k+1)$ is true, and by Induction, $s(n)$ is true for every positive integer n.

23. Let $a_1, a_2, \ldots$ be real numbers and let $r(\neq 1)$ be a real number. Let $s(n)$ be the statement, $\sum_{i=1}^{n} a_i = \dfrac{a_1(1-r^n)}{1-r}$.

s_1: $\sum_{i=1}^{1} a_i = \dfrac{a_1(1-r^1)}{1-r}$. This is true because both sides equal a_1. Suppose $s(k)$ is true, that is, suppose $\sum_{i=1}^{k} a_i = \dfrac{a_1(1-r^k)}{1-r}$. Then

$$\begin{aligned}\sum_{i=1}^{k+1} a_i &= \sum_{i=1}^{k} a_i + a_{k+1} \\ &= \frac{a_1(1-r^k)}{1-r} + a_{k+1} \\ &= \frac{a_1(1-r^k)}{1-r} + a_1r^k \\ &= \frac{a_1(1-r^k) + a_1r^k(1-r)}{1-r} \\ &= \frac{a_1 - a_1r^k + a_1r^k - a_1r^{k+1}}{1-r} \\ &= \frac{a_1 - a_1r^{k+1}}{1-r} \\ &= \frac{a_1(1-r^{k+1})}{1-r}\end{aligned}$$

Thus, $s(k+1)$ is true, and by Induction, $s(n)$ is true for every positive integer n.

Review Exercises, p. 516

1. **i.** (a) 4, 5, 6, 7, 8, 9, 10, 11 (b) increasing (c) See Figure 13D. **ii.** (a) 0, −1, −2, −3, −4, −5, −6, −7 (b) decreasing (c) See Figure 13E. **iii.** (a) $2, 1, \frac{2}{3}, \frac{1}{2}, \frac{2}{5}, \frac{1}{3}, \frac{2}{7}, \frac{1}{4}$ (b) decreasing (c) See Figure 13F.

2. (a) 9 (b) 17 (c) 41 (d) 81 **3.** $a_i = 3 + 2i$ **4.** $\sum_{i=1}^{5} a_i$ **5.** $\sum_{i=7}^{10} b_i$ or $\sum_{i=1}^{4} b_{i+6}$ **6.** 30

7. $\sum_{i=1}^{6} 3i$ **8.** (a) 20 (b) 3 (c) $m = 3, n = 7$ **9.** 95 **10.** 880 **11.** 900 **12.** $\frac{1}{27}$ **13.** 16

14. (a) 1092 (b) $\frac{635}{64}$ **15.** (a) 5040 (b) 10 (c) 4845 **16.** (a) $x^3 + 3x^2a + 3xa^2 + a^3$ (b) $x^4 - 4x^3 + 6x^2 - 4x + 1$ (c) $32x^5y^{10} + 80x^4y^8 + 80x^3y^6 + 40x^2y^4 + 10xy^2 + 1$

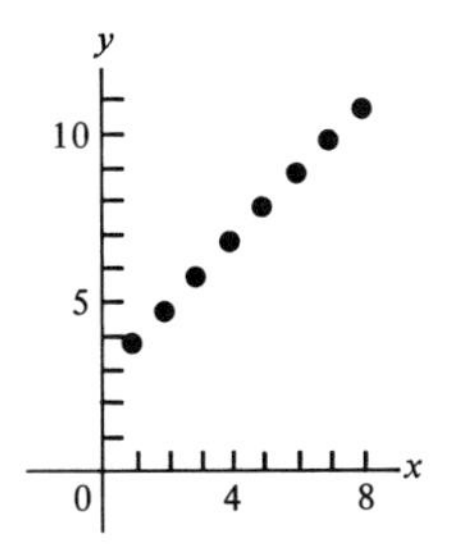

FIGURE 13D

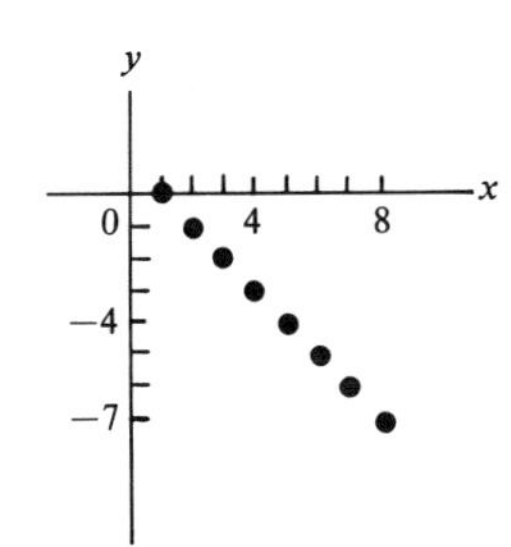

FIGURE 13E

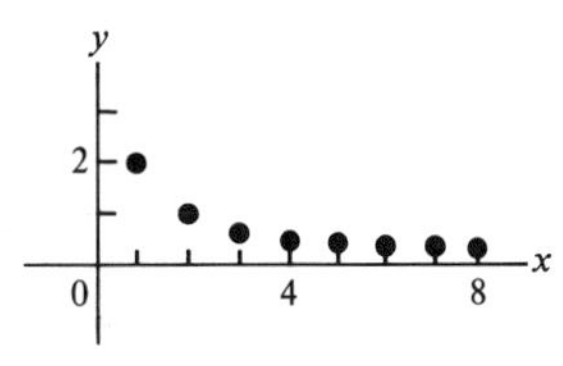

FIGURE 13F

17. (a) $\sum_{i=0}^{10} (-1)^i \binom{10}{i} x^{10-i} 2^i$ (b) $x^{10} - 20x^9 + 180x^8 - 960x^7$ **18.** 1.6105

19. (a) Let $s(n)$ be the statement, $n^2 + n$ is even. $s(1)$: $1^2 + 1$ is even. This is true because $1^2 + 1 = 2$, and 2 is even. Suppose $s(k)$ is true, that is, suppose $k^2 + k$ is even. Then $k^2 + k = 2l$ for some integer l, and

$$\begin{aligned}(k+1)^2 + (k+1) &= (k^2 + 2k + 1) + (k + 1)\\ &= (k^2 + k) + (2k + 2)\\ &= 2l + 2(k+1)\\ &= 2[l + k + 1]\end{aligned}$$

Thus $(k+1)^2 + (k+1)$ is even, and $s(k+1)$ is true. Therefore, by Induction, $s(n)$ is true for every positive integer n.

Note: Part (a) can also be proved without using Induction. You can use the fact that the product of even integers is even. Also, the sum of odd integers as well as the sum of even integers is even.

(b) Let $s(n)$ be the statement, $\frac{1}{1\cdot 4} + \frac{1}{4\cdot 7} + \frac{1}{7\cdot 10} + \cdots + \frac{1}{(3n-2)(3n+1)} = \frac{n}{3n+1}$. $s(1)$: $\frac{1}{1\cdot 4} = \frac{1}{3\cdot 1 + 1}$. This is true. Suppose $s(k)$ is true, that is, suppose $\frac{1}{1\cdot 4} + \frac{1}{4\cdot 7} + \frac{1}{7\cdot 10} + \cdots + \frac{1}{(3k-2)(3k+1)} = \frac{k}{3k+1}$.

Then

$$\begin{aligned}&\frac{1}{1\cdot 4} + \frac{1}{4\cdot 7} + \frac{1}{7\cdot 10} + \cdots + \frac{1}{(3k-2)(3k+1)} + \frac{1}{[3(k+1)-2][3(k+1)+1]}\\ &= \frac{k}{3k+1} + \frac{1}{[3(k+1)-2][3(k+1)+1]}\\ &= \frac{k}{3k+1} + \frac{1}{(3k+1)(3k+4)}\\ &= \frac{k(3k+4)+1}{(3k+1)(3k+4)}\\ &= \frac{3k^2+4k+1}{(3k+1)(3k+4)}\\ &= \frac{(3k+1)(k+1)}{(3k+1)(3k+4)}\\ &= \frac{k+1}{3k+4}\\ &= \frac{k+1}{3(k+1)+1}\end{aligned}$$

Thus $s(k+1)$ is true, and by Induction, $s(n)$ is true for every positive integer n.

Index

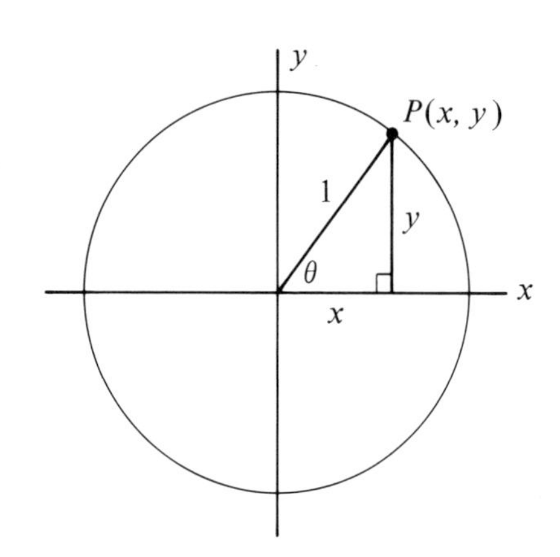

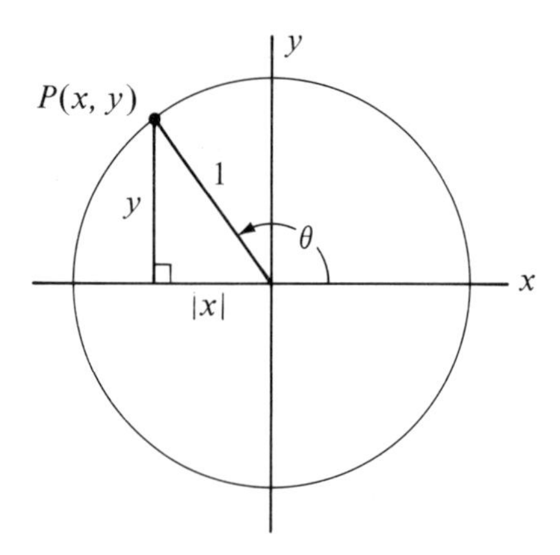

UNIT CIRCLE:

$$x^2 + y^2 = 1$$

$$\cos\theta = x, \quad \sin\theta = y$$

$\tan\theta = \dfrac{\sin\theta}{\cos\theta}$, if $\theta \neq \pm\dfrac{\pi}{2}, \pm\dfrac{3\pi}{2}, \ldots$

$\sec\theta = \dfrac{1}{\cos\theta}$, if $\theta \neq \pm\dfrac{\pi}{2}, \pm\dfrac{3\pi}{2}, \ldots$

$\csc\theta = \dfrac{1}{\sin\theta}$, if $\theta \neq 0, \pm\pi, \pm 2\pi, \ldots$

$\cot\theta = \dfrac{\cos\theta}{\sin\theta}$, if $\theta \neq 0, \pm\pi, \pm 2\pi, \ldots$

$\cos^2\theta + \sin^2\theta = 1$

$1 + \tan^2\theta = \sec^2\theta$, if $\theta \neq \pm\dfrac{\pi}{2}, \pm\dfrac{3\pi}{2}, \ldots$

$\cot^2\theta + 1 = \csc^2\theta$, if $\theta \neq 0, \pm\pi, \pm 2\pi, \ldots$

$\cos(-\theta) = \cos\theta$

$\sin(-\theta) = -\sin\theta$

$\tan(-\theta) = -\tan\theta$, if $\theta \neq \pm\dfrac{\pi}{2}, \pm\dfrac{3\pi}{2}, \ldots$

$\cos(A - B) = \cos A \cos B + \sin A \sin B$

$\cos(A + B) = \cos A \cos B - \sin A \sin B$

$\sin(A + B) = \sin A \cos B + \cos A \sin B$

$\sin(A - B) = \sin A \cos B - \cos A \sin B$

$\tan(A + B) = \dfrac{\tan A + \tan B}{1 - \tan A \tan B}, \quad \left.\begin{matrix} A \\ B \\ A + B \end{matrix}\right\} \neq \pm\dfrac{\pi}{2}, \pm\dfrac{3\pi}{2}, \ldots$

$\tan(A - B) = \dfrac{\tan A - \tan B}{1 + \tan A \tan B}, \quad \left.\begin{matrix} A \\ B \\ A - B \end{matrix}\right\} \neq \pm\dfrac{\pi}{2}, \pm\dfrac{3\pi}{2}, \ldots$

$\cos\theta = \sin\left(\dfrac{\pi}{2} - \theta\right)$

$\sin\theta = \cos\left(\dfrac{\pi}{2} - \theta\right)$

$\cos(\theta + 2\pi) = \cos\theta$

$\sin(\theta + 2\pi) = \sin\theta$

$\cos(\theta + \pi) = \cos(\pi - \theta) = -\cos\theta$